AF709327

Tidal Current Tables 2005

Pacific Coast of North America and Asia

INDEX OF TIDE TABLE COVERAGE

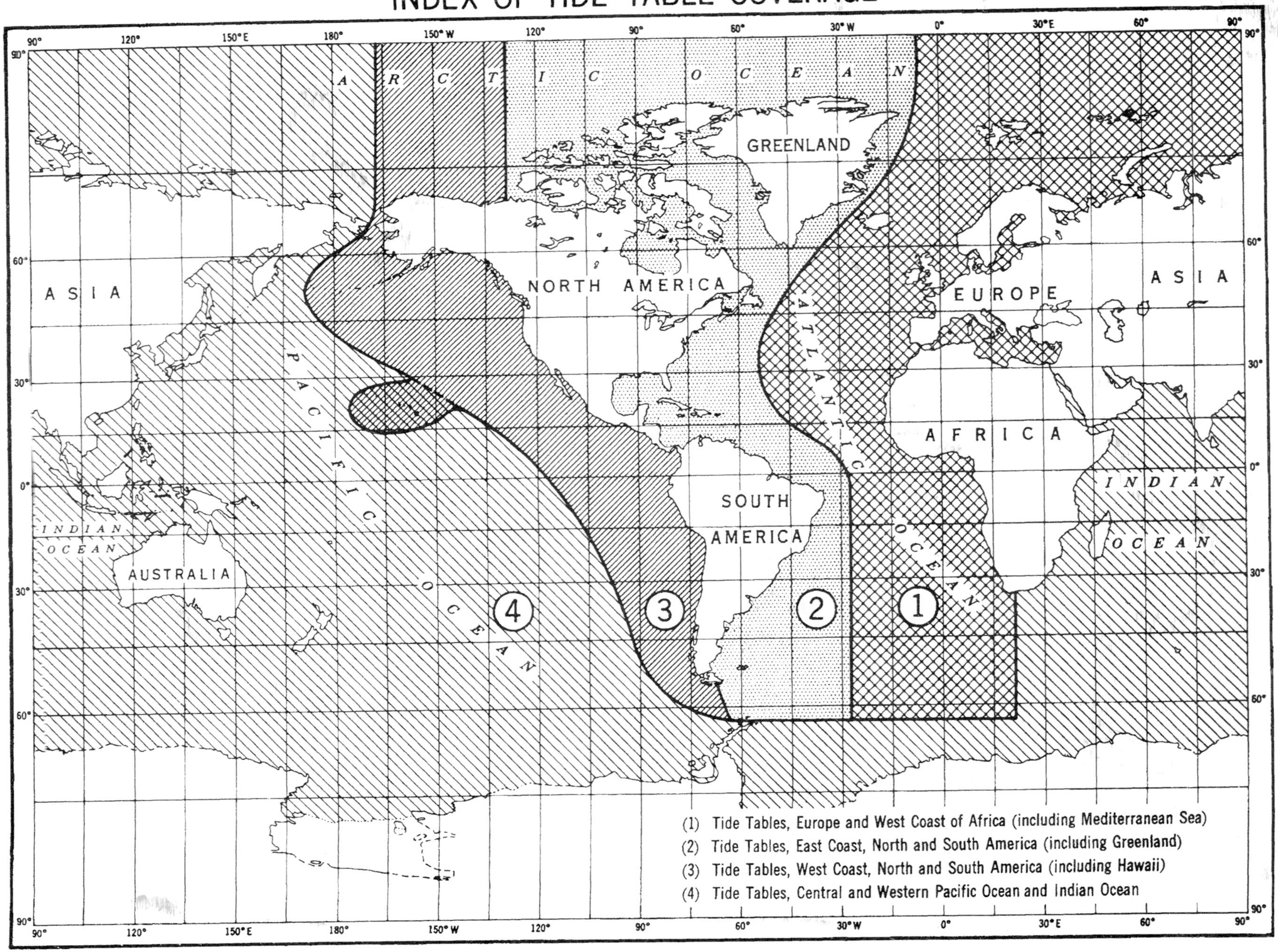

Tidal Current Tables 2005

Pacific Coast of North America and Asia

Issued 2004

Formerly published by the
National Ocean Service, NOS,
a division of the
National Oceanic and Atmospheric Administration, NOAA.

Accepted by the U.S. Coast Guard
as fulfilling federal commercial fishing vessel regulations
(Title 46 CFR, Part 28.225, Navigational Information).

International Marine / McGraw-Hill

Camden, Maine • New York • Chicago • San Francisco • Lisbon • London • Madrid •
Mexico City • Milan • New Delhi • San Juan • Seoul • Singapore • Sydney • Toronto

Other Publications

Other publications containing NOS data available from International Marine:

Tide Tables 2005: East Coast of North and South America, Including Greenland

Tide Tables 2005: West Coast of North and South America, Including the Hawaiian Islands and the Alaskan Supplement

Tide Tables 2005: Central and Western Pacific Ocean and Indian Ocean

Tide Tables 2005: Europe and West Coast of Africa, Including the Mediterranean Sea

Tidal Current Tables 2005: Atlantic Coast of North America

The McGraw-Hill Companies

Visit us at: www.internationalmarine.com

1 2 3 4 5 6 7 8 9 0 QPD QPD 0 1 0 9 8 7 6 5 4

Copyright © 2005 by International Marine

All rights reserved. The name "International Marine" and the International Marine and logo are trademarks of The McGraw-Hill Companies. Printed in the United States of America.

Questions regarding the content of this book should be addressed to:

International Marine
P.O. Box 220
Camden, ME 04843
207-236-4837

Questions regarding the ordering of this book and other International Marine titles should be addressed to:

The McGraw-Hill Companies
Customer Service Department
P.O. Box 547
Blacklick, OH 43004
Retail customers: 1-800-262-4729
Bookstores: 1-800-722-4726

This book is printed on acid-free paper.

SOURCES OF ADDITIONAL INFORMATION

THE NATIONAL OCEAN SERVICE IS NO LONGER PRINTING AND DISTRIBUTING THE TIDE AND TIDAL CURRENT TABLES

Tide and Tidal current data continue to be updated, generated and published by the NOAA/National Ocean Service; however, the printing and distribution in book-form is now done by private companies working from information provided by NOS.

NOS now offers two new vehicles for obtaining predictions. First, the complete set of Tables as camera-ready page-images will be available on CD-ROM. The CD-ROM vehicle is primarily intended for use by private printers who wish to print in book-form the full set of Tables for distribution to resellers and the general public. Second, for domestic tide reference stations, limited predictions are available on the NOS, Center for Operational Oceanographic Products and Services (CO-OPS), web site,(http://co-ops. nos.noaa.gov).

In addition to predictions, the web site provides updated information on the status of the Tables as they are finalized each year. Notices concerning the most recent Table updates and publication cut-off dates are included.

For the names of companies printing and distributing the Tables, please call or write to:

National Ocean Service
Products and Services Division, N/OPS3
1305 East-West Highway
Silver Spring, MD 20910
301-713-2815, fax 301-713-4500

PUBLICATIONS:

United States Coast Pilots, Distance Tables, and Nautical Charts may be ordered from:

FAA Distribution Division, AVN-530
National Aeronautical Charting Office
Greenbelt, MD 20770-1479
(301) 436-8301
(800) 638-8972 toll free, U.S. Only

A list of authorized sales agents is published in the Nautical Chart Catalogs or may be obtained on request from the National Ocean Service. The publications may also be purchased across-the-counter at the NOAA, Distribution Branch office listed above.

TECHNICAL ASSISTANCE:

Technical questions relating to tide and current predictions, as well as requests for special predictions, should be addressed to:

National Ocean Service
Products and Services Division, N/OPS3
1305 East-West Highway
Silver Spring, MD 20910
(301) 713-2815

SOURCES OF ADDITIONAL INFORMATION

Technical questions relating to actual tide observations, tidal datums, and other information necessary for engineering projects should be addressed to:

National Ocean Service
Products and Services Division, N/OPS3
1305 East-West Highway
Silver Spring, MD 20910
(301) 713-2877

Technical questions relating to other publications and nautical charts should be addressed to:

National Ocean Service
Customer Affairs Branch
1315 East-West Highway.
Silver Spring, MD 20910
(301) 713-2729

WEBSITES

Center for Operational Oceanographic Products and Services
(PORTS * Predictions * Observations * Bench Marks * Tides Online * Great Lakes Online)

www.co-ops.nos.noaa.gov

Coastal Services Center - www.csc.noaa.gov

Marine Chart Division - www.chartmaker.ncd.noaa.gov

Marine Predictions Center - www.mpc.ncep.noaa.gov

National Centers for Environmental Predictions - www.ncep.noaa.gov

National Climatic Data Center - www.ncdc.noaa.gov

National Data Buoy Center - www.ndbc.noaa.gov

National Geodetic Survey - www.ngs.noaa.gov

National Geophysical Data Center - www.ngdc.noaa.gov

National Ocean Service - www.nos.noaa.gov

National Oceanic and Atmospheric Administration - www.noaa.gov

National Oceanographic Data Center - www.nodc.noaa.gov

National Weather Service - www.nws.noaa.gov

U.S. Coast Guard - www.uscg.mil

U.S. Geological Survey - www.usgs.gov

U.S. Naval Observatory - www.usno.navy.mil

U.S. Naval Oceanographic Office - www.navo.navy.mil

CORRECTIONS:

Corrections to this publication, after the date of printing, may appear in the Notice to Mariners. They may also appear in the Local Notice to Mariners, published weekly, by the various United States Coast Guard Districts.

Contents

PREFACE

Current tables for the use of mariners have been published by the National Ocean Service (formerly the Coast and Geodetic Survey) since 1890. Tables for the Pacific coast first appeared in 1898 as a part of the tide tables and consisted of brief directions for obtaining the times of the currents for a few locations from the times of high and low waters. Daily predictions of slack water for two stations were given for the year 1899, and by 1923 the tables had so expanded that they were then issued as a separate publication entitled *Current Tables: Pacific Coast.* A companion volume, *Current Tables: Atlantic Coast,* was also issued that year. In 1926 the predictions for the Pacific coast were extended to include the times and velocities of maximum current.

Using information supplied by the National Ocean Service, International Marine is now publishing, unaltered and unabridged, the following tide tables and tidal current tables:

Tide Tables 2005: East Coast of North and South America, Including Greenland

Tide Tables 2005: West Coast of North and South America, Including the Hawaiian Islands and the Alaskan Supplement

Tide Tables 2005: Central and Western Pacific Ocean and Indian Ocean

Tide Tables 2005: Europe and West Coast of Africa, Including the Mediterranean Sea

Tidal Current Tables 2005: Atlantic Coast of North America

Tidal Current Tables 2005: Pacific Coast of North America and Asia

In the preparation of these tables, all available observations were used. In some cases, however, the observations were insufficient for obtaining final results, and as further information becomes available it will be included in subsequent editions. All persons using these tables are invited to send information or suggestions for increasing their usefulness to the National Ocean Service, Products and Services Division, 1305 East-West Highway, N/OPS3, Silver Spring, Maryland 20910, U.S.A. The data for lightship stations are based on observations obtained through the cooperation of the U.S. Coast Guard. In accordance with cooperative arrangements full predictions for Race Rocks, Seymour Narrows, Burrard Inlet, and Active Pass were furnished by the Canadian Hydrographic Service. The bureau of Coast and Geodetic Survey, Philippines, supplies the predictions for Iloilo, San Juanico and San Bernardino Straits, and Cebu Harbor. The Japanese Hydrographic Office furnished the predictions for Tokyo Wan entrance, Akashi Kaikyo, Naruto, Kurushima Kaikyo, Kanmon Kaikyo, and Tomogashima Suido. The Hydrographic Department, England furnished Basilan Strait.

Daily predicted times of slack water and predicted times and velocities of maximum current (flood and ebb) are presented in Table 1 for a number of reference stations. Similar predictions for many other locations may be obtained by applying the correction factors listed in Table 2 to the predictions of the appropriate reference station. The speed of a current at times between slack water and maximum current may be approximated by the use of Table 3. The duration of weak current near the time of slack water may be computed by the use of Table 4.

Note: Data from the Philippines were not available at the time of publication. Therefore, this volume does not include tables for Cebu Harbor, Iloilo Strait, San Bernardino Strait, or San Juanico Strait (pages 146–163). See Important Notices, page xii.

INTRODUCTION

As the gravitational tugs of the sun and moon regularly raise and lower the height of the sea, creating the tides, there is a corresponding horizontal flow of water—the tidal currents. This flow has an effect on any boat that is in it. Whether "riding the current," "bucking the current," or being "swept aside by the current," if current exists, the boater should know something about it. The importance of knowing depends on the speed of the craft. A 20-knot powerboat will hardly notice the 2-knot current that may be all-important to the sailor on a day of light wind.

Tidal currents are largely a function of geography. In some places the currents are weak enough to have little practical effect. In other areas, they are strong enough to be notorious—San Francisco Bay and the Golden Gate, for example, or near The Race in Long Island Sound. There are also nontidal currents, such as the Gulf Stream, which have a substantial effect on offshore passages.

Current is a flow of water characterized by its speed and direction. The speed of a river current is speed with respect to the riverbank; the speed of an ocean current is measured with respect to the coast or the seabed. Speed is often, but not necessarily, expressed in knots. The direction of a current is the *direction to which it flows:* a current flowing from north to south is a *southerly* current. (This is opposite to the convention for wind, according to which a wind that blows from north to south is called a *northerly* wind.) The direction of a current can be expressed by the angle it makes with true north. Thus, a current at 45° is one that flows *toward* the northeast. In standard piloting texts the nautical term *drift* is used for the speed of the current, and *set* for its direction, but we will just use speed and direction here.

How can the boater tell by looking that currents exist? That is not so easy as it might appear. Current is the motion of water *with respect to something*—the seabed, for example, or things attached to it, or the shore itself. But usually the seabed isn't visible, nor can you detect the motion of water with respect to a distant shore. The main visual evidence of currents is their effect on objects *attached* to the seabed—for example, the "bow wave" in front of a buoy or fixed navaid, the buoy's angle from the vertical, or the direction in which anchored boats lie on those windless days when they are held by current alone. If these indicators are not available, there is no way of telling, by simply looking at the water, whether current is flowing or not.

Even if there is a fixed navaid the current may be too weak to generate a noticeable flow pattern at its base. Still, the navaid can be helpful by providing a fixed point of reference. If you were to throw a float in the water you could detect a current, note its direction, and even calculate its speed by watching the float move past the fixed mark.

Now I am *not* suggesting that boaters regularly stop to do hydrographic experiments on currents. But I personally have done simple experiments from time to time, and they are quite instructive in two ways.

First, to detect current with any reliability using a float, you must make sure that the float is carried along by the water and not by the wind. It does *not* do to throw a piece of wood overboard, or to observe a bit of floating orange peel. You can't look at seaweed or plastic debris, or whatever, moving past a fixture in the water, and conclude that current is flowing. To detect current, you need a float that extends well into the water, with little surface area exposed. A plastic bottle, almost filled with water so that it barely floats, is satisfactory, or you can deploy a fender weighted by a piece of chain.

The second point is that you can't tell the direction of current from the direction of small waves or ripples on the water's surface. As you see waves moving in a certain direction, it is very natural to assume there is a current pushing them along. Early on I made the assumption myself, and I've often heard others do the same, but there is no truth in it.

The direction of travel of ripples or small waves on the surface of the water *does not indicate* the direction of currents. Such waves are wind-generated and they do not constitute a net horizontal motion of water. They are simply periodic up and down variations of the form of the water's surface. You realize sharply the distinction between current flow and wave direction when you see a current float bucking the waves, moving in just the opposite direction to their travel.

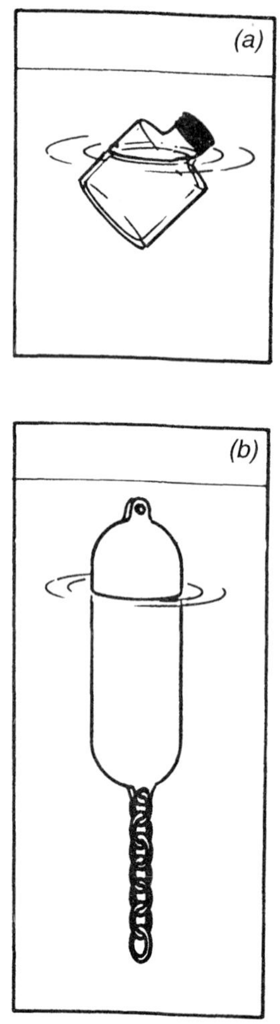

Figure 1. (Lisa Scott)

The fact that a buoy leans tells you with reasonable reliability that current exists, and the direction of lean is, of course, the direction of the flow. But it is not easy to correlate the amount of lean with the speed of the current. Every buoy is a law unto itself. In an article entitled "Guessing the Current" in the British magazine *Yachting Monthly* (October 1983), a series of different buoys were photographed in the same current conditions; the degree of lean among them varied considerably. The editor's heading for the article described the conclusions well: "Buoys in fast tidal currents should provide a ready means of estimating the rate—it is not so simple as it seems."

A current may happen to be directed *along*, or may be in the *exact opposite* direction to a course line. In the first case it speeds up the boat, and in the second it slows it down. Usually speeding or slowing doesn't have any dangerous consequences, although it may spoil schedules. The real problem with current comes when it acts at an *angle* to the course and partly tends to push the boat aside—when it is a *deflecting current.*

A typical piloting problem generated by a deflecting current is illustrated in Figure 2. The skipper has somehow fixed his initial position, I, and wants to sail from it to a destination D. He finds from the chart that the line between initial point and destination—the course line—is 47° magnetic. At the tiller, he gets onto that course and begins his journey. But in the presence of current he will not arrive at the destination, no matter how scrupulously he steers at 47°, since the boat is constantly being pushed sideways. In the presence of current, a boat *points* in one direction, but *moves* in another. This crabbing motion has an angle A, *the deflection angle,* between the direction the boat points (its *heading*) and the direction it moves (its *track*). (Note that angle A is not a universal angle; its size depends not only on the boat's heading, but also on the boat's speed. A faster boat will be less deflected—have a smaller angle A—than a slower one.)

There are other reasons besides tidal current that can make a boat move as in Figure 2, with an angle between heading and track. For example, a sailboat on the wind has *leeway;* as it travels, it is pushed aside by the current of ocean air acting on its sails and hull much as the tidal currents of ocean water push aside its hull. In a sailboat, the deflection angle A may then be the result of the combined effect of current and leeway. If there is no current, the angle would be due to leeway only.

There are two unhappy effects of a deflecting current on an attempted compass course. As we have seen, the first is that the skipper doesn't arrive at his destination; he passes right by it and leaves it off somewhere to port or starboard. The second is that the boat moves not on the original course line but on the line of the *course made good.* Presumably the original course line was checked out on the chart as being a safe one. But if the skipper is not aware of current, he isn't where he thinks he is, and the course made good may be full of dangers and trouble spots. That is the real problem.

It is not only on a compass course that a deflecting current can spell trouble. Of course, if the pilot is sailing visually with the final destination always in sight, and points toward it, he avoids the first problem of a compass course—he doesn't pass by his destination. But the second problem remains. The boat may be swept aside onto dangers. Figure 3 depicts the curved path of a boat that *always* steers directly toward its destination but that comes to grief on a portion of its unintentionally curved path. Although it seems a simple and obvious point that with current present you may not get to the destination no matter how carefully you steer toward it, this is often overlooked in practice.

You can avoid this kind of mishap by knowing something about the tidal current. Does it exist as a deflecting current for the projected course? If so, to what side does it push the boat? That may be all you need to know. If the current pushes the boat toward a side with no dangers, you can afford to be unconcerned; simply relax and recognize that the trip may take longer than planned.

On the other hand, if the current pushes the boat to a side of possible danger, you'll want to know its effects more precisely. Is it large enough to push the boat into danger? If you are sailing by the compass, you'll want to know the accurate deflection angle A so that you can compensate for it in your steering and—with some luck—not miss the destination.

—Excerpted from *The Practical Pilot,* by Leonard Eyges (International Marine, 1989)

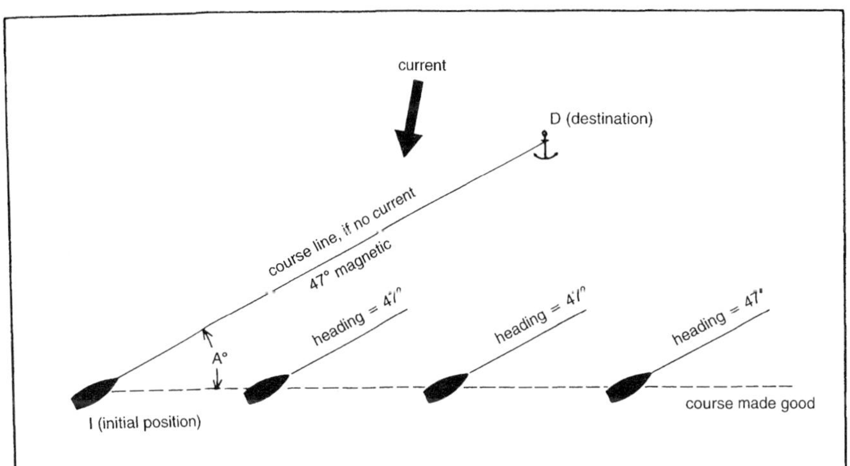

Figure 2. From the chart the course line from I to D is found to be 47° magnetic. Although the skipper keeps to this heading, the boat, being "set" by current, misses destination D. (Lisa Scott)

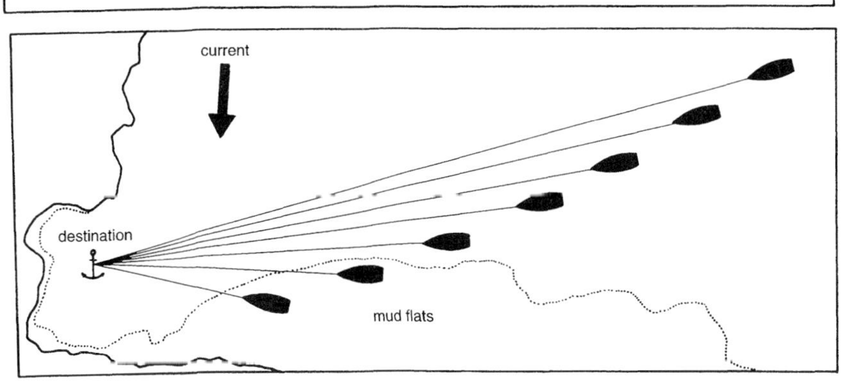

Figure 3. This skipper carefully points his boat toward the destination at all times, but never gets there; he runs aground on the mud flat instead. (Lisa Scott)

IMPORTANT NOTICES

The 2005 daily tidal current predictions for the Philippine locations, ILOILO STRAIT, CEBU, HARBOR, SAN JUANICO STRAIT, and SAN BERNARDINO STRAIT do not appear in the publication. Daily tidal current predictions for those locations are normally supplied to the National Ocean Service by the Bureau of Coast and Geodetic Survey, Republic of the Philippines in accordance with cooperative arrangements for the exchange of tidal predictions. Their 2004 predictions were not forwarded in time to appear in this publication. Those predictions usually appear on pages 142 through 159. Pages 142 through 159 have been omitted from this publication.

Daylight-saving time is not used in this publication. All daily tidal current predictions and predictions compiled by the use of Table 2 data are based on the standard time meridian indicated by each location. Predicted times may be converted to daylight-saving time, where necessary, by adding 1 hour to these data. In converting times from the Astronomical Data page on the inside back cover, it should be remembered that daylight saving time is based on a meridian 15° east of the normal standard meridian for a particular place.

NOS, in partnership with other agencies and institutions, has established a series of Physical Oceanographic Real Time Systems (PORTS®) in selected areas. These PORTS® sites provide constantly updated information on tide and tidal current conditions, water temperature, and weather conditions. This information is updated every six minutes. PORTS® sites are currently in operation at several major harbors with future sites to be added. The information is accessible through a computer data connection or by a voice response system at the following numbers:

PORTS® SITES	VOICE ACCESS	INTERNET ACCESS
NARRAGANSETT BAY	866-75-PORTS (866-757-6787)	http://co-ops.nos.noaa.gov
NEW HAVEN		"
NEW YORK/NEW JERSEY	866-21-PORTS (866-217-6787)	"
DELAWARE RIVER & BAY	DISCONTINUED	"
CHESAPEAKE BAY	866-CH-PORTS (866-247-6787)	"
TAMPA BAY	866-TB-PORTS (866-827-6787)	"
HOUSTON/GALVESTON	866-HG-PORTS (866-447-6787)	"
SAN FRANCISCO BAY	866-SB-PORTS (866-727-6787)	"
LOS ANGELES/LONG BEACH		"
TACOMA		"
PORT OF ANCHORAGE	866-AK-PORTS (866-257-6787)	"
SOO LOCKS		"

PUBLISHED CAUTIONARY NOTICES

Published in Local Notice to Mariners and United States Coast Pilot Notices

CHANGES TO 2004 AND FUTURE EDITIONS OF THE NOS TIDE TABLES

The National Ocean Service's, Center for Operational Oceanographic Products and Services (CO-OPS) is continuing to work on updating tidal data for the 1983-2001 Tidal Epoch. The updated information will begin to appear in the 2004 edition of the published Tide Tables and is expected to be completed for the 2005 Tide Tables. In conjunction with the 1983-2001 Tidal Epoch update, CO-OPS has started a comprehensive review of the secondary stations listed in the published Tide Tables. As a result of this review, there will be numerous changes to the stations listed in the "Table 2 - Tidal Differences and Other Constants" pages of the published Tide Tables and in the CO-OPS web products. These changes will include the addition of new stations, removal of obsolete stations, and updating information for other existing stations. These changes will begin to appear in the 2004 edition of the published Tide Tables and are expected to continue for several years.

IMPORTANT NOTICES

Tables in which U.S. stations will be affected by the 1983-2001 Epoch and Table 2 station review include:

- Tide Tables - East Coast of North and South America, Including Greenland
- Tide Tables - West Coast of North and South America, Including the Hawaii Islands
- Tide Tables - Central and Western Pacific Ocean and Indian Ocean

Issued October 1, 2003

CHANGES TO 2004 EDITIONS OF THE NOS TIDAL CURRENT TABLES

Four new reference stations have been added to, and two reference stations updated from, the National Ocean Services' Tidal Current Tables for 2004. Table 2 "time" and "velocity" correction factors at secondary stations which are affected by these changes have been updated based on the new reference station data.

Tidal Current Tables 2004 – Pacific Coast of North America and Asia

1. Wrangell Narrows, Alaska (updated)
2. Sergius Narrows, Alaska (updated)
3. Snow Passage, Alaska (new)

Tidal Current Tables 2004 – Atlantic Coast of North America

1. Brandywine Shoal Light, Delaware Bay (new)
2. Reedy Point, Delaware Bay (new)
3. Philadelphia (Penns Landing), Delaware River (new)

Issued October 1, 2003

CHANGES TO THE 2003 EDITION OF THE NOS TIDE AND TIDAL CURRENT TABLES

A number of new reference stations have been added to, and three reference stations deleted from, the National Ocean Services' Tide and Tidal Current Tables for 2003. Table 2 "time" and "height" correction factors effected by these changes have been updated based on the new reference stations.

Tide Tables 2003 – East Coast of North and South America Including Greenland

- Bar Harbor, Maine
- Montauk, Fort Pond Bay, New York
- Duck Pier, North Carolina
- Port Canaveral, Florida
- Naples, Florida
- Charlotte Amalie, St. Thomas Island
- Montauk Point, New York - replaced by Montauk, Fort Pond Bay
- Oregon Inlet Marina, North Carolina
- Canaveral Harbor Entrance - deleted, replaced by Port Canaveral
- Cedar Key, Florida

Tide Tables 2003 – West Coast of North and South America Including the Hawaiian Islands

- La Libertad, Ecuador
- Port San Luis Wharf, California
- Arena Cove, California
- Isla San Cristobal, Galapagos Islands
- Monterey, Monterey Bay, California
- Charleston, Coos Bay, Oregon

Tide Tables 2003 – Central and Western Pacific Ocean and Indian Ocean

- Chuuk, Moen Island, Truk Atoll
- Suva, Suva Harbor
- Dublon Island, Truk Islands - deleted, replaced by Chuuk

Tidal Current Tables 2003 – Atlantic Coast of North America

- Quonset Point, Narragansett Bay, Rhode Island

Issued October 1, 2002

IMPORTANT NOTICES

UPDATED TIDE PREDICTIONS FOR ALASKA DO TO ANOMALOUS SEA LEVEL TRENDS

The Center for Operational Oceanographic Products and Services (CO-OPS) of the National Ocean Service (NOS) has updated the tidal datums and predictions for locations in Cook Inlet, Alaska, because of anomalous sea level trends compared with other geographic regions of the United States. The new levels are based on the last five years of sea level data rather than the 1960-1978 National Tidal Datum Epoch (NTDE). The ranges are still based on the 1960-1978 NTDE and applied to the five year (1994-1998) Mean Tide Level (MTL) to compute the other datums. This procedure was necessary to ensure that these tidal datums accurately represent the existing stand of sea level.

The tide predictions for Anchorage, Kodiak, Niskiski, and Seldovia, Alaska, and their associated secondary stations have been updated in the 2001 edition of the NOS Tide Tables - West Coast of North and South America including the Hawaiian Islands.

Issued October 2, 2000

REVISED TIDAL CURRENT PREDICTIONS FOR SAN FRANCISCO BAY

In order to provide the most accurate tidal current predictions for San Francisco Bay possible, the Center for Operational Oceanographic Products and Services (CO-OPS) of the National Ocean Service (NOS) has added four (4) new tidal current reference stations to the 2001 edition of the NOS Tidal Current Tables - Pacific Coast of North America and Asia. The historic tidal current reference station for the bay, "San Francisco (Golden Gate)," is located outside of the bay itself along the approach channel. This location has been renamed "San Francisco (Outside)" and is still used as the reference station for ocean-side coastal stations and the southern most part of the bay itself. Carquinez Strait (West End Bridge) remains a reference station for upper San Pablo Bay.

The four new Tidal Current reference stations for San Francisco Bay are:

1. Golden Gate Bridge - West central area.
2. Oakland, Yerba Buena Island - Oakland Harbor area.
3. Richmond - East central area.
4. Benica Bridge - Suisun Bay.

Most of the secondary locations inside San Francisco Bay are now referred to one of these four new reference stations within the bay. These changes parallel the NOS Physical Oceanographic Real Time System (PORTS) that provides real time tide, tidal current, and meteorological information for San Francisco Bay to local mariners.

Issued October 2, 2000

TIDAL CURRENT PREDICTIONS INSIDE U.S. ESTUARIES

At present there are four U.S. estuaries with operational Physical Oceanographic Real Time Systems (PORTS) installed. PORTS systems are presently being installed in several additional estuaries. Over the next ten years there are projected to be twenty or more additional systems installed. In the past, the tidal current reference station has always been located at the entrance to each estuary. All tidal current secondary stations both inside and outside (along the coast) have been referred to the reference station at the entrance to the estuary. This will no longer be the case in estuaries with an operational PORTS system.

Estuaries with an operational PORTS system will have at least two reference stations. One will be the historic station at the entrance to the estuary. All secondary stations along the coast will continue to be referred to this station. The second tidal current reference station will be the primary PORTS station within the estuary. All secondary locations within the estuary itself will be referred to this location. Depending on the circulation dynamics of the estuary, daily tidal current predictions may be provided for one or more additional stations within the estuary.

IMPORTANT NOTICES

Tidal current predictions for Galveston Bay, TX, have been converted to this new format in the year 2000 Atlantic Coast Current Table (ACCT). Tidal current predictions for the year 2001 for San Francisco Bay, CA; Tampa Bay, FL; and New York Harbor, NY are scheduled for conversion to this new format for the 2001 Tidal Current Tables. This format will be extended to each estuary as each new PORTS™ system is installed and becomes fully operational.

(Issued October 1, 1999)

MAJOR QUALITY ASSURANCE STUDY OF NOS TIDAL CURRENT PREDICTIONS

A major statistical study of subordinate tidal current stations listed in Table 2 of both the Atlantic Coast Current Table (ACCT) and the Pacific Coast Current Table (PCCT) has been completed by the Center for Operational Oceanographic Products and Services (CO-OPS) of the National Ocean Service (NOS). The reasons for the study were:

- All major estuaries are constantly undergoing construction, dredging, and other activities that critically effect the times, speeds, and patterns of the tidal currents within the estuary.
- Older observations were very often of very short duration due to the type of instrumentation available at the time and the costs involved.
- Instrumentation for obtaining continuous current observations have only been available for about ten years and are extremely expensive. These Acoustic Doppler type meters are employed in and are what makes the Physical Oceanographic Real Time Systems (PORTS) possible. Operational PORTS systems providing real time tide, current, and meteorological data to mariners are presently available in only four major estuaries.
- The extremely high cost of obtaining new tidal current observations as well as verifying the accuracy of presently published data preclude the updating of subordinate stations in the published Tidal Current Tables.

Therefore the two criteria the study concentrated on were:

- The age of the data upon which the subordinate station time and speed differences are based.
- The duration of the observations upon which the subordinate station time and speed differences are based.

Just because the original data is older than an arbitrary date or the duration of observations is less than an arbitrary length does not, in and of itself, mean that the published correction factors are in error. However, it does greatly increase both the possibility and probability of error. The NOS does not have the funding, personnel, and other resources to verify or update tidal current subordinate stations. Since one of the prime responsibilities of the NOS is Marine Safety and Navigation, this study was ordered.

The results of this study are being evaluated. A major policy decision should be forth coming sometime after January 1, 2000. ***This decision could result in the removal of fifty percent (50%) or more of the tidal current subordinate stations now listed in NOS Tidal Current Tables! This reduction in the number of tidal current subordinate stations could occur as early as the year 2001 Tidal Current Tables.***

(Issued October 1, 1999)

CHANGES IN TIDE PREDICTIONS FOR SUISUN BAY, CA

In order to provide more accurate tide predictions for subordinate stations in upper San Francisco Bay, the National Ocean Services (NOS) has replaced San Francisco (Golden Gate) with Port Chicago as the reference station for all stations above Carquinez Strait and within Suisun Bay. This change becomes effective beginning with the year 2000 edition of West Coast Tide Table (WCTT). Subordinate tide stations for the rest of San Francisco Bay will continue to be referred to San Francisco (Golden Gate).

(Issued October 1, 1999)

IMPORTANT NOTICES

CHIGNIK, ALASKA

The US Army Corps of Engineers (USACOE) is planning the construction of a Small Boat Harbor in Chignik, AK. The construction will include dredging and the construction of a breakwater. Official published Tide and Tidal Current predictions will be degraded once the project begins. Tidal Currents will be effected the most. From the beginning of the project until a resurvey of the area can be completed, Tide and Tidal Current predictions should be used with caution. Tidal Current predictions should be used only with extreme caution. Therefore, until such time as a resurvey of the area is conducted, the National Oceanic and Atmospheric Administration, National Ocean Service will be unable to provide the accurate Tide and Tidal Current predictions necessary for marine safety and navigation in this area.

(Issued May 30, 1997)

KUSKOKWIM BAY AND RIVER, ALASKA

The National Ocean Service's (NOS) official published time and height corrections for this area (Table 2 in the Tide Tables West Coast of North and South America) in recent years have been based on the daily predictions for Nushagak Bay, AK, the nearest NOS reference station. These published values, however, do not provide the most accurate corrections. The shape of the tide curves varies considerably along the Alaskan coast. The previously published corrections based on Matarani, Peru, provide more accurate results for this area because the shape of the tide curves closely match. The corrections based on Matarani are:

Location Name	**Position**		**Differences**			
	Lat.	Long.	Time		Height	
	N	W	High	Low	High	Low
	° ′	° ′	h. m.	h. m.	ft.	ft.
Goodnews Bay entrance	59 03	161 49	+0 59	+0 51	*2.83	*2.00
Carter Spit	59 19	161 57	+1 19	+1 24	*3.63	*2.33
Eek Channel, off Quinhagak	59 45	162 15	+2 39	+3 05	*4.25	*1.67
Warehouse Creek entrance	59 56	162 05	+3 05	+3 50	*4.38	*1.67
Kuskokwak Creek entrance	60 02	162 10	+3 53	+4 40	*4.21	*1.67
Popokamute	60 04	162 25	+4 12	+5 05	*3.67	*1.67
Apokak Creek entrance	60 08	162 10	+4 13	+5 10	*4.13	*1.67
Bethel	60 48	161 45	+8 51	+11 11	+0.3	+0.1

(Issued May 30, 1997)

NEPTUNE BEACH, WASHINGTON

Pudget Sound Pilots report that observed tidal currents in the vicinity of Neptune Beach, WA deviate significantly from official published predictions. Reliable sources report that the observed velocities are close to double the predicted values and that the times are up to 1 hour earlier than predicted. Extreme caution should be exercised in this vicinity by all vessels especially tankers passing through the area approaching oil refineries. Funding for a resurvey of the area and/or the installation of a real-time monitoring system is not presently available. Therefore, until such time as real-time system is installed or a resurvey of the area conducted, the National Oceanic and Atmospheric Administration, National Ocean Service will be unable to provide the accurate Tidal Current predictions necessary for marine safety and navigation in this area.

(Issued May 30, 1997)

GRAYS HARBOR, WASHINGTON

Tidal Currents in Grays Harbor have been significantly altered by dredging and construction activities. Tidal predictions for the Tidal Reference Station at Aberdeen have been updated to reflect these changes. Tidal Current predictions for this area should be considered questionable and potentially dangerous to rely upon. Funding for a real-time system to monitor the Tidal Currents or a resurvey of this area is not available at this time. Therefore, until such time as a real-time system is installed or a resurvey of the area conducted, the National Oceanic and Atmospheric Administration, National Ocean Service will be unable to provide accurate Tidal Current predictions necessary for marine safety and navigation in this area.

(Issued June 5, 1996)

HUMBOLDT BAY, CALIFORNIA

The US Coast Guard (USCG) and the Harbor Safety Committee (HSC) for Humboldt Bay have indicated that projects by the US Army Corps of Engineers (COE) since 1985 have degraded the accuracy of the Tide and Tidal Current predictions for Humboldt Bay. The primary tide station for Humboldt Bay is located on North Spit on the ocean side and thus does not reflect the changes that have occurred within the bay itself. Tidal Current predictions within Humboldt Bay should be used with extreme caution. The last update of secondary tide stations within the bay occurred in 1985. A hydrographic survey of Humboldt Bay is scheduled to begin in the last quarter of 1996. Some of the secondary tide locations within the bay will be updated in conjunction with this survey. Funding for a real-time system to monitor the Tidal Currents or a new Tidal Current survey is not presently available. Therefore, until such time as a real-time system is installed or a resurvey of the area conducted, the National Oceanic and Atmospheric Administration, National Ocean Service will be unable to provide accurate Tidal Current predictions necessary for marine safety and navigation in this area.

(Issued June 5, 1996)

SAN DIEGO, CALIFORNIA

The US Army Corps of Engineers (COE) is planning a dredging project for the US Navy in the area of the North Island Naval Base in San Diego Harbor. This project calls for both deepening and widening the channel to accommodate larger naval vessels. Such actions in the past in other areas have resulted in dramatic changes in the observed Tidal Currents of those areas. Once dredging operations commence, the Tidal Current predictions for this region should be considered questionable and potentially dangerous to rely upon. Tidal predictions will also be affected but to a lesser degree. Funding for a real-time system to monitor the Tidal Currents during the project and a resurvey of the area after COE operations are complete are presently not available. Therefore, once COE operations begin and until such time as a real-time system is installed or a resurvey of the area conducted, the National Oceanic and Atmospheric Administration, National Ocean Service will be unable to provide accurate Tidal Current predictions necessary for marine safety and navigation in this area.

(Issued June 5, 1996)

MOSS LANDING, CALIFORNIA

The U.S. Coast Guard has reported investigating an incident at the Pacific Gas and Electric Co. terminal located off Moss Landing. The terminal is located at the head of Monterey Canyon. The incident was apparently caused by strong currents moving landward through Monterey Canyon and resulted in a "hydraulic jump" where the canyon narrows and finally ends. USCG reports indicate that a tanker was "set on" the oil terminal buoy at the facility. The chain that connects the buoy and the transfer hose became fouled around the propeller of the tanker. Extreme caution should be exercised while in this area particularly while engaged in transfer operations.

(Issued October 21, 1991)

SAN FRANCISCO BAY, CALIFORNIA

Effective October 21, 1991, the Tidal Current Charts - San Francisco Bay will be officially withdrawn from distribution. A recent evaluation shows that the San Francisco Bay charts, last revised in 1969, are inadequate when compared to the more recent information as published in the Tidal Current Tables - Pacific Coast of North America and Asia. Due to resource constraints, NOS plans no further revision of its tidal current chart series.

(Issued October 17, 1991)

LIST OF REFERENCE STATIONS

Station Names	Page	Updated	Data Series
Active Pass, British Columbia	64		
Admiralty Inlet, Washington	44	1948	123 days (12/11/1908 - 3/19/1943)
Akashi Kaikyo, Japan	122		
Akutan Pass, Aleutian Islands	102	1985	24 days beginning 7/31/1950
Basilan Strait, Philippines	142		
Benecia Bridge, Suisan Bay, California	28	2001	8 months (½/1996 - 9/3/1996)
Burrard Inlet (First Narrows), British Columbia	68		
Carquinez Strait, California	24	1989	224 days (4/3/1980 - 11/12/1980)
Cebu Harbor, Philippines**	150		
Changjiang Entrance, China	134		
Deception Pass, Washington	52	1933	29 days (9/9/1925 - 10/27/1925)
Golden Gate Bridge, California	12	2001	7 months (11/1/1997 - 5/31/1998)
Grays Harbor Entrance, Washington	32	1952	29 days beginning 3/25/1950
Iloilo Strait, Philippines**	146		
Isanotski Strait (False Pass Cannery), Alaska	92	1985	Form 444 (8/18/1925)
Kanmon Kaikyo, Japan	130		
Kurushima Kaikyo, Japan	126		
Kvichak Bay (off Naknek River Entrance), Alaska	106	1985	14 days 9/16/1946
Naruto, Japan	118		
North Inian Pass, Alaska	88	1985	104 days (1901)
Oakland, Yerba Buena Island,	16	2001	1 year (1999)
Race Rocks, British Columbia	40		
Richmond (Long Wharf), California	20	2001	1 year (1999)
Rosario Strait, Washington	56	1967	29 days beginning 3/10/1965
San Bernardino Strait, Philippines**	159*,160		
San Diego Bay Entrance, California	4	1936	29 days beginning 8/24/1934
San Francisco Bay Entrance, California	8	1990	7 days beginning 10/19/1923
San Juan Channel (south entrance), Washington	60	1966	29 days beginning 5/21/1964
San Juanico Strait, Philippines**	154		
Sergius Narrows, Alaska	84	2004	1 month (4/2/2002 - 5/7/2002)
Seymour Narrows, British Columbia	72		
Snow Passage, Alaska	76	2004	1 month (4/5/2001 - 5/15/2001)
Strait of Juan de Fuca Entrance	36	1945	Inferred from Admiralty Inlet station
The Narrows, Puget Sound, Washington	48	1948	28 days beginning 1/19/1944
Tokyo Wan Entrance, Japan	110		
Tomogashima Suido, Japan	114		
Unimak Pass, Aleutian Islands	97*,98	1985	33 days beginning 6/14/1950
Wrangell Narrows, Alaska	80	2004	1 month (5/17/2002 - 6/19/2002)
Wusong Kou, China	138		

* Explanation precedes the predictions.

** Daily predictions for this station were omitted.

*** New reference station.

TABLE 1.— DAILY CURRENT PREDICTIONS

EXPLANATION OF TABLE

This table gives the predicted times of slack water and the predicted times and speeds of maximum current-flood and ebb-for each day of the year at a number of stations on the Pacific coast of North America. The times are given in hours and minutes and the speeds in knots.

Time.— The kind of time used for the predictions at each reference station is indicated by the time meridian at the bottom of each page. **Daylight-saving time is not used in this publication**. If daylight-saving time is required, add one (1) hour to the predicted time.

Slack water and maximum current.— The columns headed "Slack" contain the predicted times at which there is no current; or, in other words, the times at which the current has stopped setting in a given direction and is about to begin to set in the opposite direction. Offshore, where the current is rotary, slack water denotes the time of minimum current. Beginning with the slack water before flood the current increases in speed until the strength or maximum speed of the flood current is reached; it then decreases until the following slack water or slack before ebb. The ebb current now begins, increases to a maximum speed, and then decreases to the next slack. The predicted times and speeds of maximum current are given in the columns headed "Maximum." Flood speeds are marked with an "F," the ebb speeds with an "E." An entry in the "Slack" column will be slack, flood begins if the maximum current which follows it is marked "F." Otherwise the entry will be slack, ebb begins.

Direction of set.— The terms flood and ebb do not in all cases clearly indicate the direction of the current, the approximate direction toward which the currents flow are given at the top of each page to distinguish the two streams.

Number of slacks and strengths.—There are usually four slacks and four maximums each day. If one is missing in a given day, it will occur soon after midnight as the first slack or maximum of the following day. At some stations where the diurnal inequality is large, there may be on certain days a continuous flood or ebb current with varying speed throughout half the day giving only two slacks and two maximums on that particular day.

Current and tide. – It is important to note that the predicted slacks and strengths given in this table refer to the horizontal motion of the water and not to the vertical rise and fall of the tide. The relation of current to tide is not constant, but varies from place to place, and the time of slack water does not generally coincide with the time of high or low water, nor does the time of maximum speed of the current usually coincide with the time of most rapid change in the vertical height of the tide. At stations located on a tidal river or bay the time of slack water may differ from 1 to 3 hours from the time of high or low water. The times of high and low waters are given in theTide Tables published by the National Ocean Service.

Variations from predictions.— In using this table, bear in mind that actual times of slack or maximum occasionally differ from the predicted times by as much as half an hour and in rare instances the difference may be as much as an hour. Comparisons of predicted with observed times of slack water indicate that more than 90 percent of the slack waters occurred within half an hour of the predicted times. To make sure, therefore, of getting the full advantage of a favorable current or slack water, the navigator should reach the entrance or strait at least half an hour before the predicted time of the desired condition of current. Currents are frequently disturbed by wind or variations in river discharge. On days when the current is affected by such disturbing influences the times and speeds will differ from those given in the table, but local knowledge will enable one to make proper allowance for these effects.

Typical current curves.—The variations in the tidal current from day to day and from place to place are illustrated on the opposite page by the current curves for representative ports. Flood current is represented by the solid line curve above the zero speed (slack water) line and the ebb current by the broken line curve below the slack water line. The outstanding feature of the currents in this region is the diurnal inequality, i.e., the differences in speed of two consecutive flood or two consecutive ebb maximums. This inequality varies directly with the Moon's declination; consequently it tends to disappear when the Moon is near the Equator. By reference to the curves it will be noted that at certain places the inequality is chiefly in the flood currents. At Seymour Narrows the two floods of a day sometimes differ by 5 knots. At other places the inequality is chiefly in the ebb currents, while at still other places there is a marked inequality in both flood and ebb currents. The effect of the inequality at some places is such that there are times when the current may be erratic (marked by an asterisk) or one flood or ebb current of the day may be quite weak. Therefore, in using the predictions of the current it is essential to carefully note the speeds as well as the times. A detailed explanation of the predictions for Unimak Pass is given on the page immediately preceding the predictions.

TYPICAL CURRENT CURVES FOR REFERENCE STATIONS
(Flood: Solid Line, Ebb: Broken Line)

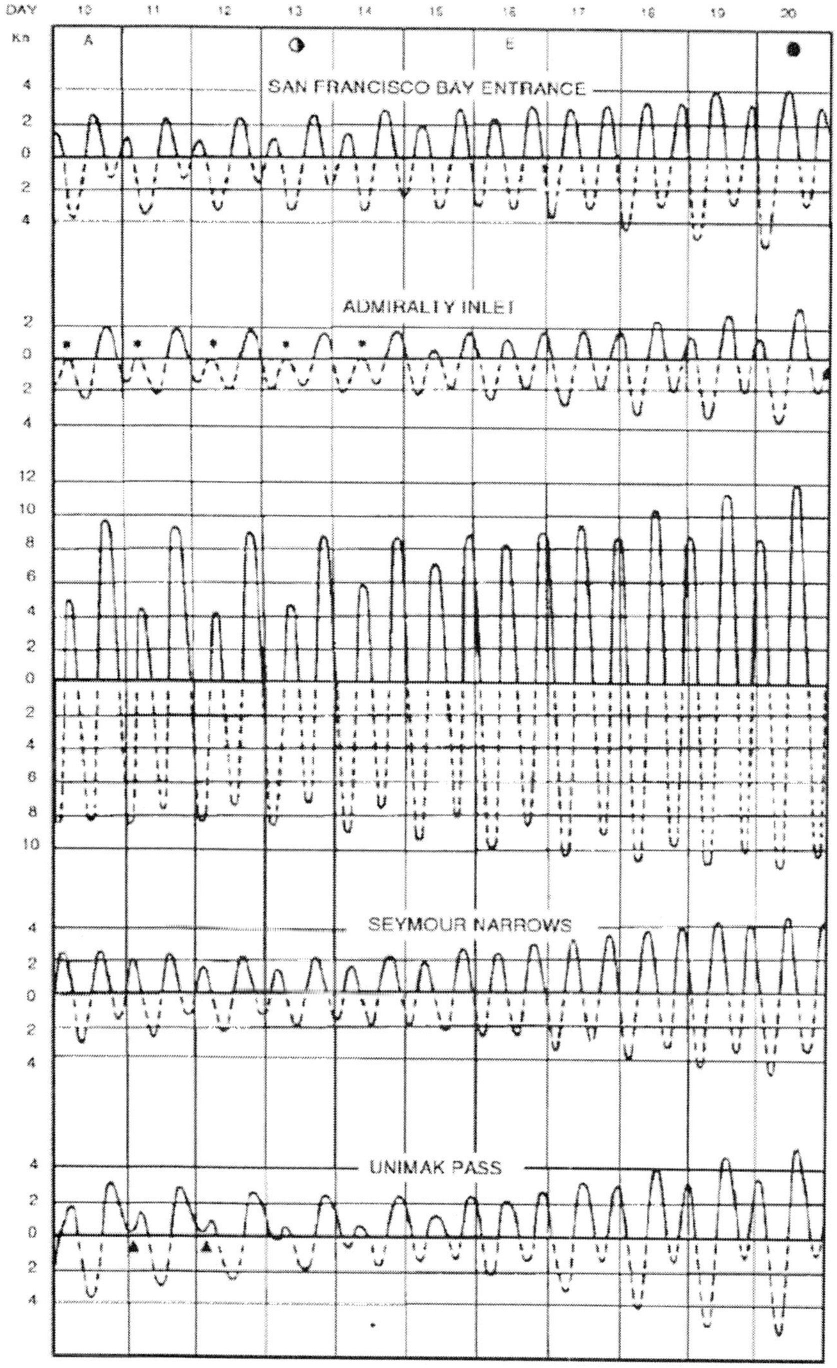

• Current weak and variable
▲ Minimum flood. *See explanation on page 93*
A discussion of these curves is given on the preceding page.

Lunar data: A – Moon in apogee
◑ – last quarter
E – Moon on Equator
● – new moon

San Diego Bay Entrance (off Ballast Point), Calif., 2005

F–Flood, Dir. 355° True E–Ebb, Dir. 175° True

January

Day	Slack h m	Maximum h m	knots
1 Sa	0147	0409	0.6E
	0635	0917	0.6F
	1206	1612	1.3E
	1945	2255	0.9F
2 Su	0234	0513	0.7E
	0805	1031	0.4F
	1309	1710	1.1E
	2028	2350	0.9F
3 M ◑	0322	0623	0.9E
	0954	1222	0.4F
	1449	1819	0.9E
	2118		
4 Tu		0051	0.9F
	0411	0730	1.2E
	1126	1409	0.5F
	1642	1933	0.8E
	2213		
5 W		0151	1.0F
	0458	0831	1.6E
	1228	1522	0.8F
	1812	2041	0.7E
	2310		
6 Th		0246	1.2F
	0546	0924	2.0E
	1319	1618	1.2F
	1921	2141	0.8E
7 F	0005	0336	1.4F
	0633	1013	2.4E
	1404	1706	1.5F
	2016	2233	0.9E
8 Sa	0056	0424	1.6F
	0719	1100	2.7E
	1449	1751	1.8F
	2104	2322	1.1E
9 Su	0145	0510	1.8F
	0806	1146	3.0E
	1532	1834	1.9F
	2149		
10 M ●		0008	1.2E
	0233	0555	1.9F
	0852	1231	3.1E
	1616	1918	2.0F
	2233		
11 Tu		0054	1.2E
	0321	0641	1.9F
	0938	1315	3.0E
	1659	2001	1.9F
	2317		
12 W		0141	1.2E
	0411	0727	1.8F
	1024	1400	2.8E
	1742	2045	1.8F
13 Th	0002	0230	1.2E
	0506	0816	1.5F
	1112	1446	2.4E
	1824	2131	1.6F
14 F	0049	0322	1.2E
	0608	0910	1.2F
	1202	1534	2.0E
	1908	2220	1.4F
15 Sa	0139	0420	1.1E
	0724	1017	0.8F
	1300	1627	1.5E
	1953	2316	1.2F
16 Su ◐	0233	0528	1.1E
	0904	1147	0.6F
	1417	1729	1.0E
	2042		
17 M		0020	1.1F
	0330	0644	1.2E
	1058	1334	0.5F
	1600	1846	0.7E
	2140		
18 Tu		0130	1.0F
	0428	0802	1.3E
	1222	1503	0.7F
	1748	2010	0.5E
	2248		
19 W		0235	1.0F
	0523	0906	1.6E
	1319	1606	0.9F
	1909	2123	0.5E
	2352		
20 Th		0329	1.0F
	0612	0957	1.8E
	1402	1654	1.2F
	2003	2217	0.6E
21 F	0044	0414	1.1F
	0656	1038	2.0E
	1438	1733	1.3F
	2043	2257	0.7E
22 Sa	0125	0452	1.2F
	0734	1113	2.2E
	1509	1807	1.4F
	2115	2331	0.8E
23 Su	0200	0524	1.3F
	0809	1146	2.3E
	1538	1837	1.5F
	2144		
24 M		0002	0.9E
	0231	0553	1.4F
	0841	1216	2.3E
	1604	1904	1.5F
	2212		
25 Tu ○		0032	1.0E
	0300	0619	1.4F
	0911	1245	2.3E
	1629	1929	1.5F
	2238		
26 W		0101	1.0E
	0330	0644	1.4F
	0940	1314	2.3E
	1653	1952	1.4F
	2304		
27 Th		0131	1.0E
	0401	0710	1.3F
	1008	1344	2.2E
	1718	2014	1.3F
	2329		
28 F		0203	1.0E
	0437	0738	1.2F
	1037	1414	2.0E
	1742	2037	1.3F
	2355		
29 Sa		0238	1.0E
	0519	0811	1.1F
	1110	1447	1.7E
	1808	2102	1.2F
30 Su	0024	0319	1.0E
	0611	0852	0.9F
	1148	1525	1.3E
	1837	2134	1.1F
31 M	0100	0410	1.0E
	0720	0948	0.6F
	1240	1613	1.0E
	1911	2215	0.9F

February

Day	Slack h m	Maximum h m	knots
1 Tu ◑	0147	0516	1.1E
	0857	1118	0.4F
	1412	1721	0.6E
	1956	2316	0.8F
2 W	0250	0638	1.2E
	1055	1349	0.4F
	1639	1855	0.4E
	2104		
3 Th		0050	0.8F
	0405	0800	1.5E
	1218	1522	0.8F
	1827	2028	0.4E
	2234		
4 F		0223	1.0F
	0517	0908	1.9E
	1313	1618	1.2F
	1930	2138	0.6E
	2354		
5 Sa		0330	1.2F
	0618	1004	2.4E
	1358	1703	1.6F
	2015	2231	0.9E
6 Su	0056	0424	1.6F
	0713	1053	2.7E
	1440	1743	1.9F
	2054	2317	1.2E
7 M	0149	0511	1.8F
	0802	1137	3.0E
	1519	1822	2.1F
	2131		
8 Tu ●		0000	1.4E
	0237	0555	2.0F
	0849	1219	3.0E
	1557	1859	2.1F
	2207		
9 W		0042	1.6E
	0323	0638	2.0F
	0933	1259	2.9E
	1633	1935	2.1F
	2243		
10 Th		0123	1.7E
	0410	0720	1.9F
	1017	1339	2.7E
	1708	2010	2.0F
	2318		
11 F		0204	1.7E
	0459	0803	1.7F
	1100	1418	2.3E
	1741	2045	1.7F
	2355		
12 Sa		0248	1.6E
	0552	0849	1.3F
	1145	1458	1.8E
	1813	2121	1.4F
13 Su	0033	0335	1.5E
	0654	0943	0.9F
	1236	1542	1.2E
	1845	2200	1.1F
14 M	0116	0430	1.3E
	0818	1100	0.6F
	1345	1635	0.7E
	1919	2250	0.8F
15 Tu ◐	0210	0542	1.1E
	1022	1307	0.4F
	1545	1753	0.3E
	2008		
16 W		0021	0.6F
	0321	0715	1.1E
	1207	1456	0.6F
		1952	*
17 Th		0208	0.6F
	0440	0842	1.3E
	1305	1559	0.9F
	1920	2126	0.3E
	2349		
18 F		0318	0.8F
	0547	0941	1.6E
	1345	1641	1.1F
	1957	2215	0.6E
19 Sa	0048	0405	1.0F
	0639	1022	1.8E
	1417	1715	1.3F
	2025	2248	0.8E
20 Su	0127	0443	1.2F
	0720	1056	2.1E
	1444	1744	1.5F
	2050	2317	1.0E
21 M	0159	0514	1.4F
	0756	1126	2.2E
	1509	1810	1.6F
	2114	2344	1.1E
22 Tu	0227	0541	1.5F
	0827	1154	2.3E
	1532	1833	1.6F
	2136		
23 W ○		0010	1.3E
	0254	0606	1.5F
	0857	1221	2.3E
	1553	1854	1.6F
	2158		
24 Th		0037	1.4E
	0323	0631	1.6F
	0926	1248	2.3E
	1614	1913	1.6F
	2218		
25 F		0104	1.5E
	0353	0656	1.5F
	0955	1316	2.1E
	1635	1931	1.6F
	2238		
26 Sa		0133	1.5E
	0427	0724	1.5F
	1026	1344	1.9E
	1656	1950	1.5F
	2259		
27 Su		0205	1.6E
	0507	0757	1.3F
	1100	1415	1.6E
	1718	2012	1.4F
	2324		
28 M		0242	1.5E
	0555	0836	1.1F
	1141	1451	1.2E
	1743	2040	1.2F
	2355		

March

Day	Slack h m	Maximum h m	knots
1 Tu		0328	1.5E
	0657	0929	0.8F
	1237	1536	0.7E
	1811	2116	1.0F
2 W	0037	0431	1.4E
	0828	1059	0.5F
	1425	1647	0.3E
	1850	2212	0.8F
3 Th ◑	0144	0601	1.3E
	1034	1353	0.5F
		1844	*
4 F		0012	0.6F
	0325	0739	1.5E
	1203	1518	0.9F
	1841	2031	0.3E
	2235		
5 Sa		0221	0.8F
	0459	0856	1.9E
	1258	1608	1.3F
	1924	2137	0.7E
6 Su	0005	0330	1.2F
	0610	0953	2.3E
	1341	1648	1.7F
	1958	2225	1.1E
7 M	0104	0422	1.6F
	0706	1040	2.6E
	1419	1724	2.0F
	2030	2306	1.5E
8 Tu	0153	0507	1.9F
	0756	1121	2.8E
	1455	1758	2.1F
	2102	2345	1.8E
9 W	0238	0549	2.0F
	0841	1200	2.8E
	1527	1830	2.2F
	2133		
10 Th ●		0022	2.0E
	0321	0629	2.1F
	0924	1237	2.6E
	1558	1901	2.1F
	2203		
11 F		0059	2.1E
	0404	0708	1.9F
	1006	1313	2.3E
	1627	1931	1.9F
	2233		
12 Sa		0136	2.1E
	0448	0748	1.7F
	1048	1349	1.9E
	1653	1958	1.7F
	2303		
13 Su		0214	2.0E
	0535	0829	1.3F
	1131	1424	1.4E
	1718	2025	1.4F
	2333		
14 M		0254	1.7E
	0628	0916	0.9F
	1220	1502	0.9E
	1741	2051	1.1F
15 Tu	0006	0341	1.5E
	0738	1022	0.6F
	1329	1550	0.4E
	1804	2120	0.7F
16 W	0048	0444	1.2E
	0929	1238	0.4F
		1709	*
		2214	0.4F
17 Th ◐	0157	0618	1.0E
	1129	1436	0.6F
		1941	*
18 F		0138	0.3F
	0346	0803	1.1E
	1230	1534	0.9F
	1904	2119	0.3E
	2355		
19 Sa		0258	0.6F
	0513	0909	1.4E
	1308	1611	1.1F
	1926	2156	0.6E
20 Su	0043	0345	0.8F
	0611	0951	1.7E
	1338	1642	1.4F
	1949	2224	0.9E
21 M	0117	0421	1.1F
	0655	1025	1.9E
	1404	1708	1.5F
	2011	2250	1.2E
22 Tu	0145	0451	1.3F
	0732	1054	2.1E
	1427	1732	1.7F
	2032	2315	1.4E
23 W	0213	0519	1.5F
	0805	1123	2.1E
	1448	1753	1.7F
	2053	2341	1.6E
24 Th	0241	0546	1.6F
	0838	1151	2.1E
	1509	1812	1.7F
	2112		
25 F ○		0008	1.8E
	0310	0613	1.7F
	0910	1219	2.1E
	1529	1830	1.7F
	2131		
26 Sa		0035	1.9E
	0342	0641	1.6F
	0943	1247	1.9E
	1550	1849	1.7F
	2151		
27 Su		0105	2.0E
	0418	0712	1.6F
	1019	1318	1.6E
	1611	1910	1.6F
	2214		
28 M		0138	2.0E
	0459	0748	1.4F
	1100	1351	1.3E
	1634	1934	1.4F
	2241		
29 Tu		0217	2.0E
	0548	0832	1.1F
	1150	1431	0.9E
	1700	2005	1.2F
	2314		
30 W		0305	1.8E
	0651	0932	0.8F
	1304	1523	0.5E
	1730	2044	1.0F
	2359		
31 Th		0410	1.6E
	0821	1123	0.6F
		1648	*
		2147	0.7F

Time meridian 120° W. 0000 is midnight. 1200 is noon. Times are not adjusted for Daylight Saving Time.
* Current weak and variable.

San Diego Bay Entrance (off Ballast Point), Calif., 2005

F–Flood, Dir. 355° True   E–Ebb, Dir. 175° True

April

Day	Slack h m	Maximum h m	knots
1 F	0111	0542	1.5E
	1013	1344	0.8F
		1853	*
2 Sa		0020	0.5F
	0304	0721	1.6E
	1134	1455	1.1F
	1822	2029	0.5E
	2258		
3 Su		0218	0.8F
	0445	0836	1.9E
	1228	1542	1.5F
	1856	2126	0.9E
4 M	0014	0323	1.2F
	0558	0932	2.2E
	1311	1620	1.8F
	1928	2209	1.4E
5 Tu	0107	0413	1.6F
	0656	1019	2.3E
	1348	1655	2.0F
	1958	2248	1.8E
6 W	0152	0457	1.8F
	0746	1059	2.4E
	1420	1727	2.1F
	2027	2325	2.1E
7 Th	0235	0539	1.9F
	0832	1137	2.3E
	1450	1757	2.1F
	2056		
8 F		0000	2.3E
	0316	0618	1.9F
	0916	1213	2.1E
	1518	1825	2.0F
	2123		
9 Sa		0034	2.4E
	0357	0657	1.8F
	0958	1247	1.7E
	1543	1851	1.8F
	2151		
10 Su		0109	2.3E
	0438	0735	1.6F
	1040	1321	1.4E
	1606	1915	1.5F
	2218		
11 M		0144	2.2E
	0521	0815	1.3F
	1124	1356	1.0E
	1628	1938	1.3F
	2245		
12 Tu		0222	1.9E
	0609	0900	0.9F
	1216	1433	0.6E
	1649	2001	1.0F
	2313		
13 W		0305	1.6E
	0709	1003	0.6F
	1331	1522	0.3E
	1710	2027	0.7F
	2347		
14 Th		0402	1.3E
	0835	1200	0.5F
		1642	*
		2106	0.4F
15 F	0040	0523	1.1E
	1019	1348	0.6F
		1858	*
16 Sa		0037	*
		0700	1.1E
	1127	1446	0.9F
	1816	2034	0.3E
	2331		
17 Su		0216	0.4F
	0417	0812	1.3E
	1210	1525	1.1F
	1839	2114	0.7E
18 M	0019	0308	0.7F
	0526	0901	1.5E
	1242	1555	1.3F
	1901	2144	1.0E
19 Tu	0053	0347	1.0F
	0617	0940	1.7E
	1309	1621	1.5F
	1923	2213	1.3E
20 W	0123	0421	1.2F
	0659	1013	1.8E
	1334	1645	1.6F
	1944	2240	1.6E
21 Th	0153	0452	1.4F
	0739	1045	1.8E
	1357	1706	1.7F
	2005	2308	1.9E
22 F	0223	0523	1.5F
	0817	1117	1.8E
	1420	1727	1.7F
	2026	2338	2.1E
23 Sa	0256	0555	1.6F
	0856	1149	1.7E
	1443	1748	1.7F
	2048		
24 Su		0009	2.3E
	0332	0629	1.7F
	0936	1222	1.5E
	1507	1812	1.7F
	2113		
25 M		0042	2.4E
	0411	0706	1.6F
	1020	1258	1.3E
	1533	1839	1.6F
	2142		
26 Tu		0120	2.4E
	0456	0749	1.4F
	1110	1338	1.0E
	1602	1911	1.4F
	2215		
27 W		0204	2.3E
	0549	0842	1.2F
	1213	1426	0.7E
	1635	1949	1.2F
	2256		
28 Th		0257	2.1E
	0654	0954	1.0F
	1340	1530	0.4E
	1718	2039	0.9F
	2349		
29 F		0405	1.8E
	0815	1137	0.9F
		1701	*
		2203	0.6F
30 Sa	0106	0528	1.7E
	0942	1312	1.1F
	1651	1846	0.4E
	2104		

May

Day	Slack h m	Maximum h m	knots
1 Su		0024	0.6F
	0251	0655	1.7E
	1053	1417	1.3F
	1739	2006	0.7E
	2302		
2 M		0202	0.8F
	0426	0807	1.8E
	1147	1505	1.6F
	1817	2102	1.2E
3 Tu	0011	0307	1.1F
	0541	0904	1.9E
	1231	1546	1.8F
	1850	2147	1.6E
4 W	0103	0400	1.4F
	0643	0952	1.9E
	1308	1621	1.9F
	1921	2226	2.0E
5 Th	0149	0447	1.6F
	0736	1034	1.8E
	1341	1654	1.9F
	1951	2303	2.3E
6 F	0232	0530	1.7F
	0825	1113	1.7E
	1410	1724	1.9F
	2020	2338	2.4E
7 Sa	0313	0611	1.7F
	0911	1150	1.4E
	1436	1752	1.7F
	2048		
8 Su		0013	2.5E
	0352	0650	1.6F
	0955	1225	1.2E
	1501	1818	1.6F
	2116		
9 M		0047	2.4E
	0432	0729	1.4F
	1039	1300	0.9E
	1525	1842	1.4F
	2143		
10 Tu		0122	2.2E
	0512	0809	1.2F
	1125	1336	0.7E
	1549	1907	1.2F
	2212		
11 W		0159	2.0E
	0557	0854	1.0F
	1219	1417	0.5E
	1615	1933	0.9F
	2242		
12 Th		0242	1.7E
	0647	0952	0.8F
	1329	1507	0.3E
	1647	2006	0.7F
	2317		
13 F		0333	1.5E
	0749	1111	0.7F
		1618	*
		2055	0.4F
14 Sa	0003	0437	1.3E
	0858	1236	0.7F
		1751	*
		2247	*
15 Su		0551	1.2E
	1001	1338	0.9F
	1706	1917	0.4E
	2221		
16 M		0102	0.3F
	0300	0700	1.2E
	1051	1422	1.0F
	1737	2013	0.7E
	2334		
17 Tu		0214	0.5F
	0423	0758	1.3E
	1130	1456	1.2F
	1804	2054	1.0E
18 W	0019	0304	0.7F
	0529	0845	1.3E
	1203	1526	1.4F
	1828	2129	1.4E
19 Th	0056	0347	1.0F
	0624	0928	1.4E
	1233	1553	1.5F
	1853	2202	1.8E
20 F	0131	0426	1.2F
	0714	1007	1.4E
	1302	1619	1.6F
	1918	2236	2.1E
21 Sa	0207	0504	1.4F
	0802	1045	1.4E
	1331	1645	1.6F
	1945	2311	2.4E
22 Su	0244	0542	1.6F
	0848	1124	1.3E
	1401	1714	1.7F
	2014	2348	2.6E
23 M	0325	0623	1.6F
	0936	1204	1.2E
	1432	1747	1.7F
	2048		
24 Tu		0027	2.7E
	0409	0707	1.6F
	1027	1247	1.0E
	1507	1823	1.6F
	2125		
25 W		0111	2.6E
	0457	0756	1.5F
	1123	1334	0.8E
	1546	1904	1.5F
	2207		
26 Th		0159	2.5E
	0551	0853	1.4F
	1226	1429	0.6E
	1633	1952	1.3F
	2255		
27 F		0254	2.3E
	0651	1000	1.3F
	1339	1535	0.5E
	1736	2054	1.0F
	2353		
28 Sa		0356	2.1E
	0756	1115	1.2F
	1453	1653	0.5E
	1907	2219	0.7F
29 Su	0105	0507	1.8E
	0902	1228	1.3F
	1557	1817	0.7E
	2105		
30 M		0006	0.7F
	0233	0620	1.7E
	1004	1300	1.4F
	1648	1931	1.0E
	2248		
31 Tu		0138	0.8F
	0403	0730	1.6E
	1058	1423	1.5F
	1731	2031	1.4E

June

Day	Slack h m	Maximum h m	knots
1 W	0002	0249	1.0F
	0523	0831	1.5E
	1144	1508	1.6F
	1809	2121	1.8E
2 Th	0059	0349	1.2F
	0632	0925	1.3E
	1225	1548	1.7F
	1844	2204	2.1E
3 F	0147	0440	1.4F
	0732	1012	1.2E
	1301	1624	1.6F
	1918	2244	2.3E
4 Sa	0231	0526	1.5F
	0826	1055	1.1E
	1333	1657	1.6F
	1950	2321	2.4E
5 Su	0312	0609	1.5F
	0914	1134	0.9E
	1402	1728	1.5F
	2021	2357	2.4E
6 M	0351	0648	1.4F
	0959	1211	0.8E
	1431	1757	1.4F
	2052		
7 Tu		0032	2.3E
	0428	0727	1.3F
	1041	1247	0.7E
	1459	1825	1.3F
	2123		
8 W		0108	2.2E
	0505	0805	1.2F
	1123	1324	0.6E
	1530	1854	1.1F
	2155		
9 Th		0144	2.1E
	0543	0844	1.1F
	1208	1405	0.5E
	1604	1925	1.0F
	2227		
10 F		0223	1.9E
	0622	0928	1.0F
	1258	1451	0.4E
	1646	2000	0.8F
	2302		
11 Sa		0306	1.7E
	0705	1017	0.9F
	1354	1545	0.3E
	1742	2045	0.6F
	2341		
12 Su		0354	1.5E
	0749	1111	0.9F
	1449	1649	0.4E
	1902	2148	0.4F
13 M	0032	0448	1.3E
	0836	1207	0.9F
	1539	1757	0.5E
	2047	2320	0.3F
14 Tu	0142	0548	1.2E
	0922	1258	1.0F
	1620	1901	0.7E
	2227		
15 W		0059	0.4F
	0311	0649	1.1E
	1007	1342	1.0F
	1655	1956	1.1E
	2337		
16 Th		0216	0.5F
	0439	0748	1.0E
	1049	1422	1.2F
	1728	2043	1.4E
17 F	0028	0315	0.8F
	0554	0843	1.0E
	1130	1459	1.3F
	1801	2126	1.8E
18 Sa	0112	0405	1.1F
	0658	0934	1.0E
	1210	1536	1.4F
	1836	2208	2.2E
19 Su	0154	0452	1.3F
	0755	1021	1.0E
	1250	1614	1.5F
	1913	2251	2.5E
20 M	0237	0536	1.5F
	0847	1108	1.0E
	1331	1654	1.6F
	1953	2334	2.7E
21 Tu	0320	0621	1.6F
	0937	1154	1.0E
	1413	1735	1.7F
	2035		
22 W		0018	2.8E
	0406	0707	1.7F
	1027	1241	1.0E
	1458	1820	1.7F
	2120		
23 Th		0104	2.8E
	0453	0755	1.7F
	1117	1330	0.9E
	1547	1907	1.6F
	2207		
24 F		0153	2.7E
	0541	0845	1.7F
	1210	1423	0.9E
	1642	1959	1.4F
	2257		
25 Sa		0243	2.5E
	0631	0939	1.6F
	1305	1522	0.9E
	1747	2058	1.2F
	2352		
26 Su		0337	2.2E
	0723	1036	1.5F
	1401	1626	0.9E
	1908	2210	0.9F
27 M	0055	0436	1.9E
	0815	1137	1.4F
	1458	1737	1.0E
	2046	2338	0.7F
28 Tu	0210	0541	1.5E
	0909	1238	1.4F
	1552	1849	1.2E
	2228		
29 W		0112	0.7F
	0339	0650	1.2E
	1003	1336	1.3F
	1642	1957	1.5E
	2351		
30 Th		0234	0.8F
	0511	0759	1.0E
	1056	1431	1.3F
	1729	2050	1.7F

Time meridian 120° W. 0000 is midnight. 1200 is noon. Times are not adjusted for Daylight Saving Time.
* Current weak and variable.

San Diego Bay Entrance (off Ballast Point), Calif., 2005

F–Flood, Dir. 355° True E–Ebb, Dir. 175° True

July

Day	Slack h m	Maximum h m	knots
1 F	0055	0342	1.0F
	0632	0903	0.8E
	1146	1520	1.3F
	1812	2146	2.0E
2 Sa	0146	0438	1.2F
	0739	0959	0.8E
	1231	1603	1.3F
	1852	2230	2.1E
3 Su	0230	0525	1.3F
	0833	1046	0.7E
	1311	1643	1.3F
	1930	2310	2.3E
4 M	0309	0606	1.4F
	0917	1127	0.7E
	1348	1719	1.3F
	2007	2346	2.3E
5 Tu	0345	0643	1.4F
	0955	1203	0.7E
	1422	1751	1.3F
	2041		
6 W ●		0021	2.3E
	0418	0717	1.4F
	1029	1237	0.7E
	1454	1821	1.3F
	2114		
7 Th		0054	2.2E
	0448	0749	1.3F
	1102	1311	0.7E
	1527	1849	1.2F
	2145		
8 F		0127	2.2E
	0518	0819	1.2F
	1135	1346	0.7E
	1603	1918	1.1F
	2216		
9 Sa		0201	2.0E
	0547	0849	1.2F
	1209	1424	0.7E
	1642	1949	1.0F
	2247		
10 Su		0235	1.9E
	0617	0920	1.1F
	1245	1505	0.7E
	1728	2025	0.9F
	2320		
11 M		0312	1.6E
	0648	0952	1.0F
	1323	1552	0.7E
	1826	2109	0.7F
	2359		
12 Tu		0354	1.4E
	0722	1029	1.0F
	1403	1646	0.7E
	1940	2208	0.5F
13 W	0050	0443	1.1E
	0759	1114	0.9F
	1447	1748	0.9E
	2114	2336	0.4F
14 Th ◐	0210	0544	0.9E
	0843	1208	0.9F
	1535	1855	1.1E
	2251		
15 F		0128	0.4F
	0400	0655	0.7E
	0934	1310	0.9F
	1625	1959	1.4E
16 Sa	0004	0255	0.7F
	0541	0808	0.6E
	1033	1412	1.0F
	1716	2057	1.8E
17 Su	0058	0357	1.0F
	0657	0913	0.7E
	1132	1508	1.2F
	1806	2149	2.2E
18 M	0145	0446	1.3F
	0755	1009	0.8E
	1228	1559	1.4F
	1855	2238	2.5E
19 Tu	0229	0531	1.6F
	0844	1059	0.9E
	1320	1647	1.6F
	1943	2324	2.8E
20 W	0312	0614	1.8F
	0928	1146	1.1E
	1410	1734	1.8F
	2030		
21 Th ○		0009	3.0E
	0354	0656	1.9F
	1010	1232	1.2E
	1459	1820	1.9F
	2117		
22 F		0053	3.0E
	0436	0737	1.9F
	1051	1318	1.3E
	1550	1906	1.8F
	2203		
23 Sa		0138	2.8E
	0517	0819	1.9F
	1134	1405	1.3E
	1644	1955	1.7F
	2251		
24 Su		0223	2.6E
	0558	0903	1.8F
	1218	1455	1.3E
	1743	2047	1.4F
	2342		
25 M		0310	2.1E
	0639	0948	1.6F
	1304	1550	1.3E
	1852	2148	1.1F
26 Tu	0038	0400	1.7E
	0722	1039	1.4F
	1354	1652	1.3E
	2018	2307	0.8F
27 W ◑	0149	0459	1.2E
	0808	1137	1.1F
	1449	1803	1.3E
	2204		
28 Th		0049	0.6F
	0324	0610	0.8E
	0903	1247	1.0F
	1549	1921	1.4E
	2341		
29 F		0227	0.7F
	0515	0735	0.5E
	1010	1400	1.0F
	1649	2034	1.6E
30 Sa	0050	0340	0.9F
	0646	0857	0.5E
	1123	1503	1.0F
	1746	2133	1.8E
31 Su	0140	0435	1.2F
	0747	0959	0.6E
	1224	1555	1.1F
	1835	2220	2.0E

August

Day	Slack h m	Maximum h m	knots
1 M	0221	0517	1.3F
	0830	1044	0.7E
	1312	1638	1.2F
	1919	2259	2.1E
2 Tu	0256	0553	1.4F
	0903	1120	0.8E
	1350	1714	1.3F
	1957	2334	2.2E
3 W	0326	0625	1.5F
	0932	1151	0.9E
	1423	1745	1.4F
	2031		
4 Th ●		0005	2.3E
	0353	0653	1.5F
	0959	1221	1.0E
	1453	1812	1.4F
	2102		
5 F		0034	2.3E
	0418	0718	1.5F
	1024	1250	1.0E
	1524	1838	1.4F
	2132		
6 Sa		0103	2.2E
	0441	0741	1.4F
	1049	1320	1.1E
	1555	1904	1.3F
	2200		
7 Su		0131	2.1E
	0504	0802	1.4F
	1113	1350	1.1E
	1630	1931	1.2F
	2229		
8 M		0201	1.9E
	0526	0822	1.3F
	1137	1424	1.1E
	1709	2001	1.1F
	2300		
9 Tu		0232	1.6E
	0550	0845	1.2F
	1204	1501	1.1E
	1756	2038	0.9F
	2335		
10 W		0307	1.3E
	0616	0912	1.1F
	1234	1546	1.1E
	1857	2127	0.7F
11 Th	0022	0349	1.0E
	0646	0946	0.9F
	1314	1645	1.1E
	2020	2241	0.4F
12 F ◐	0138	0449	0.6E
	0725	1037	0.8F
	1410	1801	1.2E
	2212		
13 Sa		0101	0.4F
	0357	0616	0.4E
	0823	1200	0.7F
	1524	1924	1.4E
	2347		
14 Su		0253	0.7F
	0557	0753	0.3E
	0953	1344	0.8F
	1642	2037	1.7E
15 M	0046	0353	1.0F
	0704	0908	0.5E
	1121	1500	1.1F
	1748	2136	2.2E
16 Tu	0133	0438	1.4F
	0749	1004	0.8E
	1228	1556	1.4F
	1845	2226	2.5E
17 W	0214	0518	1.7F
	0828	1051	1.1E
	1323	1645	1.7F
	1936	2311	2.8E
18 Th	0253	0556	2.0F
	0904	1135	1.4E
	1412	1730	1.9F
	2024	2354	3.0E
19 F ○	0330	0632	2.1F
	0939	1216	1.6E
	1500	1814	2.0F
	2110		
20 Sa		0035	2.9E
	0406	0708	2.1F
	1015	1258	1.8E
	1547	1857	2.0F
	2155		
21 Su		0115	2.7E
	0442	0744	2.0F
	1050	1340	1.8E
	1637	1942	1.8F
	2240		
22 M		0156	2.3E
	0516	0819	1.8F
	1127	1424	1.8E
	1730	2030	1.5F
	2328		
23 Tu		0238	1.8E
	0549	0856	1.6F
	1205	1511	1.7E
	1831	2124	1.1F
24 W	0022	0323	1.3E
	0623	0936	1.2F
	1248	1605	1.5E
	1949	2238	0.7F
25 Th	0133	0417	0.8E
	0700	1025	0.9F
	1340	1714	1.3E
	2136		
26 F ◑		0032	0.5F
	0325	0533	0.4E
	0751	1149	0.6F
	1449	1842	1.3E
	2326		
27 Sa		0223	0.7F
		0727	*
		1339	0.6F
	1611	2013	1.4E
28 Su	0035	0333	0.9F
	0658	0906	0.3E
	1132	1455	0.8F
	1723	2118	1.6E
29 M	0121	0420	1.2F
	0737	0958	0.6E
	1234	1548	1.0F
	1820	2204	1.8E
30 Tu	0157	0456	1.4F
	0806	1033	0.8E
	1315	1627	1.2F
	1904	2240	2.0E
31 W	0226	0526	1.5F
	0832	1102	1.0E
	1348	1700	1.4F
	1942	2311	2.2E

September

Day	Slack h m	Maximum h m	knots
1 Th	0252	0553	1.6F
	0855	1129	1.2E
	1417	1729	1.5F
	2015	2339	2.2E
2 F	0315	0616	1.6F
	0918	1155	1.3E
	1445	1755	1.5F
	2045		
3 Sa ●		0006	2.2E
	0336	0637	1.6F
	0939	1222	1.4E
	1513	1820	1.5F
	2114		
4 Su		0032	2.1E
	0356	0655	1.6F
	0959	1248	1.5E
	1542	1844	1.5F
	2143		
5 M		0059	2.0E
	0415	0712	1.5F
	1018	1316	1.6E
	1614	1911	1.4F
	2213		
6 Tu		0127	1.8E
	0435	0730	1.4F
	1037	1346	1.6E
	1651	1940	1.3F
	2245		
7 W		0156	1.5E
	0455	0749	1.3F
	1059	1420	1.6E
	1735	2016	1.0F
	2323		
8 Th		0229	1.1E
	0518	0814	1.2F
	1125	1502	1.5E
	1831	2103	0.8F
9 F	0015	0310	0.7E
	0543	0846	1.0F
	1202	1559	1.3E
	1952	2220	0.5F
10 Sa	0150	0414	0.3E
	0616	0933	0.7F
	1259	1722	1.3E
	2150		
11 Su ◐		0112	0.5F
		0605	*
		1111	0.5F
	1437	1901	1.4E
	2328		
12 M		0249	0.8F
	0615	0757	0.3E
	0953	1342	0.7F
	1621	2022	1.7E
13 Tu	0027	0339	1.2F
	0656	0907	0.6E
	1134	1459	1.0F
	1737	2122	2.1E
14 W	0110	0419	1.6F
	0729	0956	1.0E
	1236	1553	1.5F
	1837	2210	2.5E
15 Th	0149	0455	1.9F
	0801	1038	1.5E
	1326	1639	1.8F
	1928	2253	2.7E
16 F	0224	0529	2.1F
	0832	1117	1.8E
	1412	1722	2.0F
	2015	2333	2.7E
17 Sa ○	0258	0601	2.2F
	0904	1156	2.1E
	1457	1804	2.1F
	2100		
18 Su		0012	2.6E
	0330	0633	2.2F
	0935	1234	2.2E
	1541	1845	2.0F
	2145		
19 M		0050	2.3E
	0400	0704	2.0F
	1006	1312	2.3E
	1627	1928	1.8F
	2230		
20 Tu		0127	1.9E
	0429	0734	1.8F
	1038	1352	2.2E
	1716	2012	1.5F
	2317		
21 W		0206	1.4E
	0456	0804	1.5F
	1110	1434	1.9E
	1811	2103	1.1F
22 Th	0011	0247	0.9E
	0523	0834	1.1F
	1145	1523	1.6E
	1921	2214	0.7F
23 F	0129	0339	0.4E
	0550	0909	0.7F
	1228	1626	1.3E
	2103		
24 Sa ◑		0017	0.5F
		0503	*
		1018	0.4F
	1337	1758	1.2E
	2255		
25 Su		0209	0.7F
		0732	*
		1318	0.3F
	1524	1940	1.2E
26 M	0002	0310	1.0F
	0642	0903	0.4E
	1143	1439	0.6F
	1653	2048	1.4E
27 Tu	0044	0350	1.2F
	0706	0940	0.7E
	1232	1528	0.9F
	1754	2133	1.7E
28 W	0117	0421	1.4F
	0729	1008	1.0E
	1306	1606	1.1F
	1839	2207	1.9E
29 Th	0143	0449	1.6F
	0751	1034	1.2E
	1335	1638	1.3F
	1917	2237	2.0E
30 F	0207	0513	1.7F
	0813	1100	1.5E
	1402	1706	1.5F
	1951	2305	2.1E

Time meridian 120° W. 0000 is midnight. 1200 is noon. Times are not adjusted for Daylight Saving Time.

* Current weak and variable.

San Diego Bay Entrance (off Ballast Point), Calif., 2005

F–Flood, Dir. 355° True E–Ebb, Dir. 175° True

October

Day	Slack h m	Maximum h m	knots
1 Sa	0228	0534	1.7F
	0833	1125	1.7E
	1430	1733	1.6F
	2023	2333	2.0E
2 Su	0248	0553	1.7F
	0852	1151	1.8E
	1458	1759	1.6F
	2055		
3 M ●		0000	1.9E
	0308	0610	1.7F
	0911	1217	2.0E
	1528	1826	1.6F
	2128		
4 Tu		0028	1.7E
	0327	0627	1.6F
	0929	1246	2.0E
	1601	1855	1.5F
	2202		
5 W		0057	1.5E
	0347	0646	1.5F
	0949	1316	2.0E
	1639	1927	1.3F
	2240		
6 Th		0128	1.2E
	0408	0708	1.4F
	1013	1352	1.9E
	1724	2007	1.1F
	2327		
7 F		0205	0.8E
	0430	0735	1.2F
	1042	1436	1.8E
	1822	2101	0.8F
8 Sa	0035	0252	0.5E
	0456	0810	1.0F
	1122	1536	1.6E
	1943	2236	0.6F
9 Su		0409	*
		0903	0.7F
	1223	1701	1.4E
	2130		
10 M ◐		0108	0.7F
		0612	*
		1109	0.4F
	1411	1841	1.5E
	2258		
11 Tu		0225	1.0F
	0555	0755	0.4E
	1018	1339	0.6F
	1604	2001	1.8E
	2354		
12 W		0312	1.4F
	0627	0855	0.9E
	1143	1451	1.0F
	1724	2100	2.1E
13 Th	0038	0350	1.7F
	0658	0940	1.3E
	1239	1544	1.5F
	1825	2148	2.3E
14 F	0115	0425	2.0F
	0728	1020	1.8E
	1326	1630	1.8F
	1918	2231	2.3E
15 Sa	0149	0458	2.1F
	0758	1058	2.2E
	1410	1713	1.9F
	2007	2310	2.3E
16 Su	0221	0529	2.1F
	0828	1135	2.4E
	1453	1755	2.0F
	2053	2348	2.1E
17 M ○	0250	0559	2.1F
	0857	1211	2.5E
	1536	1836	1.9F
	2138		
18 Tu		0025	1.8E
	0317	0627	1.9F
	0927	1247	2.5E
	1619	1917	1.7F
	2223		
19 W		0101	1.4E
	0343	0655	1.6F
	0956	1325	2.4E
	1705	2000	1.4F
	2312		
20 Th		0139	1.0E
	0408	0721	1.4F
	1026	1404	2.1E
	1755	2049	1.1F
21 F	0009	0220	0.6E
	0431	0747	1.0F
	1057	1450	1.8E
	1857	2157	0.7F
22 Sa	0131	0312	0.3E
	0454	0816	0.7F
	1133	1547	1.4E
	2020	2346	0.6F
23 Su		0435	*
		0901	0.4F
	1228	1708	1.2E
	2157		
24 M ◑		0127	0.7F
		0652	*
		1225	*
		1842	1.1E
	2305		
25 Tu		0226	0.9F
	0559	0823	0.4E
	1125	1400	0.4F
	1600	1955	1.3E
	2349		
26 W		0306	1.2F
	0021	0902	0.7E
	1212	1454	0.7F
	1710	2045	1.5E
27 Th	0022	0337	1.4F
	0644	0932	1.1E
	1245	1535	0.9F
	1802	2123	1.6E
28 F	0050	0404	1.5F
	0706	0959	1.4E
	1315	1609	1.2F
	1846	2157	1.7E
29 Sa	0115	0428	1.6F
	0727	1026	1.7E
	1344	1641	1.3F
	1926	2229	1.7E
30 Su	0138	0449	1.7F
	0748	1054	1.9E
	1413	1711	1.5F
	2003	2300	1.7E
31 M	0200	0510	1.7F
	0808	1121	2.1E
	1444	1742	1.6F
	2041	2331	1.6E

November

Day	Slack h m	Maximum h m	knots
1 Tu ●	0222	0529	1.7F
	0829	1151	2.3E
	1517	1813	1.6F
	2120		
2 W		0002	1.4E
	0244	0551	1.6F
	0852	1222	2.4E
	1553	1847	1.5F
	2201		
3 Th		0036	1.2E
	0308	0615	1.6F
	0918	1257	2.4E
	1634	1926	1.4F
	2247		
4 F		0113	0.9E
	0334	0643	1.4F
	0948	1338	2.3E
	1723	2013	1.2F
	2344		
5 Sa		0157	0.6E
	0403	0718	1.3F
	1025	1426	2.1E
	1822	2116	1.0F
6 Su	0103	0254	0.4E
	0441	0802	1.0F
	1112	1528	1.9E
	1936	2250	0.9F
7 M		0416	*
		0909	0.7F
	1218	1646	1.7E
	2100		
8 Tu ◐		0032	1.0F
	0420	0601	0.3E
	0758	1120	0.5F
	1358	1813	1.6E
	2214		
9 W		0143	1.2F
	0510	0730	0.6E
	1020	1321	0.7F
	1543	1929	1.7E
	2312		
10 Th		0234	1.5F
	0547	0831	1.1E
	1141	1436	1.0F
	1706	2031	1.8E
	2357		
11 F		0316	1.7F
	0021	0918	1.0E
	1237	1533	1.3F
	1813	2122	1.8E
12 Sa	0037	0353	1.9F
	0654	1000	2.0E
	1325	1622	1.6F
	1911	2208	1.8E
13 Su	0112	0427	2.0F
	0725	1039	2.4E
	1410	1707	1.7F
	2003	2250	1.7E
14 M	0144	0500	1.9F
	0756	1116	2.6E
	1452	1750	1.8F
	2052	2329	1.5E
15 Tu ○	0213	0530	1.9F
	0827	1153	2.6E
	1534	1832	1.7F
	2138		
16 W		0006	1.2E
	0241	0559	1.7F
	0857	1229	2.6E
	1615	1913	1.5F
	2225		
17 Th		0044	1.0E
	0307	0627	1.5F
	0927	1306	2.4E
	1658	1956	1.3F
	2313		
18 F		0122	0.7E
	0334	0655	1.3F
	0958	1345	2.2E
	1743	2042	1.1F
19 Sa	0007	0203	0.5E
	0401	0723	1.0F
	1030	1427	1.9E
	1833	2138	0.9F
20 Su	0114	0253	0.3E
	0434	0756	0.7F
	1105	1517	1.6E
	1932	2252	0.7F
21 M		0400	*
		0843	0.5F
	1149	1617	1.3E
	2038		
22 Tu		0013	0.8F
		0530	*
		1025	*
		1728	1.2E
	2140		
23 W ◑		0118	0.9F
	0448	0700	0.4E
		1246	*
		1839	1.2E
	2231		
24 Th		0205	1.0F
	0521	0801	0.7E
	1133	1402	0.4F
	1605	1940	1.2E
	2312		
25 F		0242	1.2F
	0549	0843	1.0E
	1217	1456	0.7F
	1715	2030	1.2E
	2346		
26 Sa		0313	1.3F
	0815	0918	1.4E
	1253	1539	0.9F
	1813	2113	1.3E
27 Su	0017	0340	1.4F
	0640	0951	1.7E
	1326	1618	1.1F
	1903	2153	1.3E
28 M	0046	0406	1.5F
	0704	1023	2.0E
	1359	1654	1.3F
	1950	2231	1.3E
29 Tu	0114	0432	1.6F
	0730	1056	2.3E
	1433	1731	1.5F
	2035	2308	1.2E
30 W	0143	0459	1.6F
	0758	1131	2.5E
	1510	1808	1.6F
	2119	2346	1.1E

December

Day	Slack h m	Maximum h m	knots
1 Th ●	0212	0528	1.6F
	0829	1208	2.6E
	1550	1848	1.6F
	2206		
2 F		0025	1.0E
	0245	0601	1.6F
	0903	1248	2.6E
	1634	1931	1.5F
	2256		
3 Sa		0109	0.8E
	0321	0638	1.5F
	0942	1333	2.6E
	1723	2021	1.4F
	2352		
4 Su		0158	0.7E
	0403	0722	1.4F
	1026	1423	2.4E
	1816	2119	1.3F
5 M	0056	0256	0.6E
	0457	0814	1.1F
	1117	1519	2.1E
	1916	2228	1.2F
6 Tu	0207	0406	0.5E
	0613	0925	0.8F
	1220	1624	1.9E
	2019	2342	1.2F
7 W	0315	0528	0.6E
	0803	1105	0.6F
	1342	1736	1.6E
	2122		
8 Th ◐		0050	1.3F
	0412	0650	0.9E
	1003	1253	0.7F
	1517	1850	1.5E
	2220		
9 F		0149	1.4F
	0459	0758	1.3E
	1131	1418	0.9F
	1648	1958	1.4E
	2311		
10 Sa		0238	1.5F
	0541	0854	1.7E
	1235	1524	1.1F
	1806	2058	1.3E
	2357		
11 Su		0323	1.6F
	0620	0942	2.1E
	1327	1619	1.4F
	1912	2151	1.2E
12 M	0038	0403	1.7F
	0657	1025	2.3E
	1413	1708	1.5F
	2009	2237	1.1E
13 Tu	0115	0440	1.6F
	0732	1104	2.5E
	1455	1752	1.6F
	2059	2319	1.0E
14 W	0148	0514	1.6F
	0807	1142	2.6E
	1535	1834	1.6F
	2145	2358	0.9E
15 Th ○	0220	0547	1.5F
	0840	1219	2.5E
	1614	1913	1.5F
	2227		
16 F		0035	0.8E
	0251	0618	1.4F
	0913	1255	2.4E
	1651	1951	1.3F
	2307		
17 Sa		0112	0.7E
	0323	0648	1.3F
	0946	1331	2.2E
	1728	2028	1.2F
	2348		
18 Su		0150	0.6E
	0357	0718	1.1F
	1018	1408	2.0E
	1805	2107	1.1F
19 M	0032	0232	0.5E
	0436	0751	0.9F
	1051	1448	1.8E
	1843	2150	1.0F
20 Tu	0121	0321	0.4E
	0526	0830	0.7F
	1127	1531	1.6E
	1923	2239	0.9F
21 W	0215	0419	0.4E
	0634	0921	0.5F
	1208	1621	1.3E
	2007	2334	0.9F
22 Th	0309	0527	0.5E
	0814	1041	0.3F
	1307	1718	1.1E
	2053		
23 F ◑		0031	0.9F
	0356	0637	0.7E
		1235	*
		1823	0.9E
	2141		
24 Sa		0123	0.9F
	0438	0739	0.9E
	1137	1407	0.4F
	1619	1928	0.8E
	2228		
25 Su		0209	1.0F
	0514	0831	1.3E
	1228	1511	0.6F
	1744	2029	0.8E
	2314		
26 M		0249	1.1F
	0549	0916	1.6E
	1309	1602	0.9F
	1851	2122	0.8E
	2357		
27 Tu		0327	1.3F
	0624	0957	2.0E
	1348	1645	1.2F
	1946	2210	0.9E
28 W	0037	0404	1.4F
	0700	1038	2.3E
	1426	1726	1.4F
	2035	2254	0.9E
29 Th	0117	0441	1.5F
	0738	1118	2.6E
	1505	1806	1.6F
	2120	2337	1.0E
30 F ●	0158	0520	1.7F
	0818	1200	2.8E
	1546	1847	1.7F
	2204		
31 Sa		0020	1.0E
	0239	0601	1.7F
	0900	1242	2.9E
	1628	1929	1.7F
	2248		

Time meridian 120° W. 0000 is midnight. 1200 is noon. Times are not adjusted for Daylight Saving Time.
* Current weak and variable.

San Francisco Bay Entrance (Outside), Calif., 2005

F–Flood, Dir. 065° True   E–Ebb, Dir. 245° True

January

Day	Slack h m	Maximum h m	knots
1 Sa		0210	2.9F
	0540	0803	2.2E
	1135	1410	1.6F
	1645	2007	3.2E
	2340		
2 Su		0251	2.7F
	0616	0853	2.6E
	1240	1511	1.6F
	1756	2100	2.8E
3 M ◑	0022	0336	2.6F
	0653	0944	3.0E
	1342	1616	1.8F
	1915	2156	2.4E
4 Tu	0109	0423	2.5F
	0731	1036	3.5E
	1440	1723	2.2F
	2034	2256	2.1E
5 W	0201	0513	2.5F
	0812	1129	4.0E
	1535	1828	2.7F
	2147	2357	1.9E
6 Th	0258	0605	2.5F
	0856	1223	4.6E
	1627	1929	3.3F
	2254		
7 F		0058	1.9E
	0355	0658	2.6F
	0944	1316	5.1E
	1719	2025	3.8F
	2353		
8 Sa		0156	1.9E
	0450	0750	2.7F
	1034	1408	5.6E
	1810	2118	4.2F
9 Su	0048	0251	2.1E
	0544	0842	2.9F
	1126	1501	5.9E
	1901	2208	4.4F
10 M ●	0139	0343	2.2E
	0637	0935	3.0F
	1219	1552	6.0E
	1951	2257	4.5F
11 Tu	0226	0434	2.4E
	0731	1028	3.1F
	1314	1643	5.9E
	2040	2345	4.5F
12 W	0312	0525	2.5E
	0829	1122	3.1F
	1410	1733	5.5E
	2128		
13 Th		0033	4.3F
	0356	0616	2.7E
	0931	1219	2.9F
	1509	1824	4.9E
	2216		
14 F		0120	4.0F
	0440	0708	2.9E
	1036	1321	2.7F
	1612	1917	4.1E
	2304		
15 Sa		0207	3.6F
	0523	0802	3.1E
	1145	1427	2.5F
	1721	2011	3.3E
	2354		
16 Su ◐		0257	3.1F
	0607	0858	3.3E
	1255	1541	2.4F
	1836	2110	2.5E
17 M	0046	0349	2.7F
	0652	0955	3.4E
	1401	1659	2.4F
	1953	2215	1.9E
18 Tu	0142	0443	2.4F
	0738	1055	3.6E
	1504	1811	2.6F
	2108	2328	1.5E
19 W	0241	0539	2.2F
	0826	1153	3.8E
	1600	1912	2.9F
	2216		
20 Th		0046	1.4E
	0339	0633	2.1F
	0915	1246	4.0E
	1651	2006	3.1F
	2313		
21 F		0148	1.4E
	0431	0723	2.1F
	1002	1333	4.1E
	1738	2052	3.3F
22 Sa	0002	0230	1.4E
	0517	0807	2.2F
	1047	1416	4.3E
	1821	2134	3.4F
23 Su	0045	0302	1.5E
	0558	0847	2.3F
	1129	1454	4.5E
	1900	2210	3.4F
24 M	0123	0332	1.7E
	0635	0924	2.3F
	1209	1532	4.6E
	1937	2243	3.4F
25 Tu ○	0158	0405	1.8E
	0711	1001	2.4F
	1247	1609	4.6E
	2011	2312	3.4F
26 W	0232	0440	2.0E
	0747	1039	2.4F
	1325	1646	4.5E
	2043	2341	3.3F
27 Th	0304	0517	2.2E
	0826	1118	2.3F
	1403	1724	4.3E
	2113		
28 F		0012	3.2F
	0335	0556	2.4E
	0909	1201	2.2F
	1445	1804	3.9E
	2143		
29 Sa		0045	3.0F
	0406	0638	2.7E
	0958	1247	2.1F
	1532	1847	3.5E
	2213		
30 Su		0121	2.9F
	0437	0723	2.9E
	1052	1339	2.0F
	1630	1934	3.0E
	2247		
31 M		0201	2.6F
	0511	0810	3.1E
	1153	1438	2.0F
	1740	2027	2.4E
	2328		

February

Day	Slack h m	Maximum h m	knots
1 Tu ◑		0247	2.4F
	0549	0902	3.4E
	1259	1545	2.1F
	1902	2126	1.9E
2 W	0019	0339	2.1F
	0634	0959	3.7E
	1406	1659	2.3F
	2025	2230	1.6E
3 Th	0124	0438	2.0F
	0727	1059	4.1E
	1511	1812	2.8F
	2140	2337	1.5E
4 F	0236	0540	2.1F
	0825	1159	4.6E
	1610	1918	3.3F
	2244		
5 Sa		0044	1.6E
	0343	0641	2.4F
	0925	1259	5.1E
	1705	2015	3.8F
	2339		
6 Su		0144	1.9E
	0443	0739	2.7F
	1024	1355	5.5E
	1757	2105	4.2F
7 M	0027	0238	2.3E
	0537	0833	3.1F
	1121	1448	5.8E
	1845	2151	4.5F
8 Tu ●	0111	0327	2.6E
	0630	0926	3.4F
	1216	1538	5.8E
	1931	2234	4.5F
9 W	0152	0414	3.0E
	0722	1018	3.6F
	1310	1626	5.6E
	2016	2316	4.4F
10 Th	0231	0500	3.3E
	0815	1109	3.5F
	1405	1713	5.1E
	2058	2356	4.1F
11 F	0309	0545	3.6E
	0911	1202	3.4F
	1501	1800	4.4E
	2141		
12 Sa		0038	3.7F
	0348	0631	3.7E
	1009	1258	3.1F
	1601	1848	3.5E
	2224		
13 Su		0120	3.2F
	0427	0719	3.7E
	1111	1400	2.7F
	1707	1939	2.7E
	2311		
14 M		0207	2.6F
	0510	0811	3.6E
	1217	1510	2.5F
	1819	2036	1.9E
15 Tu ◐	0006	0259	2.1F
	0557	0907	3.4E
	1325	1631	2.4F
	1937	2141	1.4E
16 W	0110	0400	1.8F
	0650	1009	3.3E
	1432	1749	2.5F
	2053	2306	1.1E
17 Th	0221	0508	1.6F
	0749	1117	3.4E
	1534	1853	2.7F
	2200		
18 F		0054	1.1E
	0324	0614	1.7F
	0847	1223	3.6E
	1628	1946	3.0F
	2253		
19 Sa		0148	1.3E
	0417	0709	1.9F
	0941	1316	3.8E
	1716	2031	3.2F
	2337		
20 Su		0220	1.5E
	0502	0754	2.2F
	1030	1359	4.1E
	1757	2109	3.3F
21 M	0013	0243	1.8E
	0541	0832	2.4F
	1115	1436	4.3E
	1834	2141	3.4F
22 Tu	0047	0308	2.1E
	0618	0908	2.6F
	1156	1510	4.5E
	1907	2208	3.5F
23 W ○	0117	0338	2.4E
	0653	0944	2.8F
	1235	1545	4.5E
	1938	2233	3.4F
24 Th	0146	0410	2.8E
	0729	1020	2.9F
	1314	1620	4.3E
	2006	2300	3.4F
25 F	0213	0445	3.1E
	0805	1058	2.9F
	1354	1657	4.0E
	2032	2329	3.2F
26 Sa	0239	0522	3.3E
	0844	1139	2.8F
	1437	1737	3.6E
	2059		
27 Su		0000	3.0F
	0305	0602	3.5E
	0927	1223	2.7F
	1526	1820	3.1E
	2128		
28 M		0036	2.7F
	0334	0645	3.7E
	1017	1314	2.6F
	1625	1907	2.6E
	2203		

March

Day	Slack h m	Maximum h m	knots
1 Tu		0116	2.4F
	0407	0733	3.7E
	1115	1413	2.4F
	1736	2001	2.0E
	2246		
2 W		0205	2.0F
	0449	0827	3.7E
	1225	1523	2.4F
	1858	2103	1.5E
	2345		
3 Th ◑		0305	1.7F
	0544	0928	3.8E
	1340	1643	2.5F
	2020	2212	1.3E
4 F	0109	0414	1.7F
	0654	1034	4.0E
	1451	1803	2.9F
	2132	2327	1.3E
5 Sa	0236	0526	1.9F
	0808	1142	4.4E
	1554	1908	3.4F
	2229		
6 Su		0037	1.7E
	0345	0634	2.3F
	0918	1246	4.8E
	1649	2000	3.9F
	2317		
7 M		0136	2.2E
	0442	0734	2.9F
	1021	1343	5.2E
	1738	2045	4.2F
	2358		
8 Tu		0225	2.8E
	0534	0829	3.4F
	1119	1434	5.3E
	1823	2126	4.4F
9 W	0036	0309	3.4E
	0624	0919	3.8F
	1214	1522	5.3E
	1905	2204	4.4F
10 Th ●	0111	0351	3.9E
	0712	1008	3.9F
	1307	1606	4.9E
	1945	2240	4.2F
11 F	0146	0431	4.2E
	0800	1056	3.9F
	1400	1650	4.4E
	2024	2317	3.8F
12 Sa	0220	0512	4.4E
	0849	1144	3.7F
	1454	1734	3.7E
	2103	2354	3.3F
13 Su	0254	0554	4.3E
	0940	1235	3.4F
	1550	1819	2.9E
	2143		
14 M		0034	2.7F
	0330	0638	4.1E
	1035	1330	3.0F
	1651	1908	2.2E
	2228		
15 Tu		0118	2.1F
	0410	0726	3.8E
	1136	1434	2.6F
	1800	2002	1.5E
	2325		
16 W		0210	1.6F
	0457	0820	3.4E
	1244	1555	2.3F
	1915	2106	1.0E
17 Th ◐	0040	0315	1.3F
	0557	0923	3.1E
	1355	1719	2.3F
	2030	2233	0.9E
18 F	0200	0433	1.2F
	0708	1034	3.0E
	1501	1826	2.5F
	2133		
19 Sa		0042	1.1E
	0307	0550	1.4F
	0817	1148	3.2E
	1557	1917	2.8F
	2221		
20 Su		0126	1.4E
	0400	0649	1.8F
	0917	1247	3.5E
	1644	1959	3.1F
	2300		
21 M		0151	1.8E
	0443	0734	2.2F
	1010	1331	3.8E
	1723	2033	3.3F
	2333		
22 Tu		0211	2.2E
	0522	0813	2.5F
	1057	1408	4.0E
	1758	2101	3.4F
23 W	0002	0237	2.7E
	0559	0850	2.9F
	1141	1443	4.1E
	1829	2126	3.4F
24 Th	0030	0307	3.2E
	0634	0926	3.1F
	1223	1518	4.0E
	1858	2151	3.4F
25 F ○	0056	0339	3.6E
	0709	1002	3.3F
	1305	1554	3.9E
	1926	2218	3.3F
26 Sa	0121	0414	4.0E
	0744	1040	3.4F
	1348	1633	3.6E
	1953	2248	3.1F
27 Su	0145	0450	4.3E
	0822	1121	3.4F
	1435	1713	3.2E
	2021	2322	2.9F
28 M	0212	0530	4.4E
	0904	1205	3.3F
	1527	1757	2.7E
	2052	2359	2.5F
29 Tu	0241	0614	4.4E
	0953	1256	3.1F
	1627	1846	2.2E
	2130		
30 W		0042	2.1F
	0318	0703	4.3E
	1052	1356	2.9F
	1738	1942	1.7E
	2219		
31 Th		0135	1.8F
	0406	0800	4.1E
	1203	1509	2.7F
	1856	2047	1.3E
	2332		

Time meridian 120° W. 0000 is midnight. 1200 is noon. Times are not adjusted for Daylight Saving Time.

San Francisco Bay Entrance (Outside), Calif., 2005

F–Flood, Dir. 065° True   E–Ebb, Dir. 245° True

April

Day	Slack (h m)	Maximum (h m)	knots
1 F		0241	1.5F
	0511	0904	4.0E
	1319	1632	2.8F
	2011	2201	1.2E
2 Sa	0116	0400	1.5F
	0635	1015	4.0E
	1431	1749	3.1F
	2113	2321	1.5E
3 Su	0241	0520	1.8F
	0759	1126	4.1E
	1533	1849	3.5F
	2203		
4 M		0032	2.1E
	0346	0632	2.4F
	0914	1232	4.4E
	1626	1937	3.9F
	2245		
5 Tu		0124	2.8E
	0440	0732	3.0F
	1019	1329	4.5E
	1713	2018	4.1F
	2322		
6 W		0208	3.5E
	0530	0825	3.6F
	1117	1419	4.5E
	1756	2056	4.1F
	2356		
7 Th		0247	4.1E
	0616	0913	4.0F
	1212	1504	4.3E
	1836	2131	4.0F
8 F	0029	0325	4.6E
	0701	0959	4.1F
	1304	1547	3.9E
	1914	2205	3.7F
9 Sa	0102	0403	4.9E
	0744	1043	4.1F
	1356	1629	3.4E
	1951	2240	3.3F
10 Su	0134	0441	4.9E
	0828	1128	3.9F
	1446	1710	2.9E
	2028	2316	2.9F
11 M	0206	0521	4.7E
	0914	1213	3.6F
	1539	1753	2.3E
	2106	2355	2.3F
12 Tu	0239	0603	4.4E
	1002	1303	3.1F
	1635	1839	1.8E
	2148		
13 W		0037	1.8F
	0316	0649	3.9E
	1058	1401	2.7F
	1738	1930	1.3E
	2242		
14 Th		0128	1.4F
	0401	0741	3.5E
	1201	1512	2.3F
	1846	2030	1.0E
15 F	0001	0230	1.1F
	0501	0840	3.1E
	1310	1634	2.3F
	1953	2143	0.8E

Day	Slack (h m)	Maximum (h m)	knots
16 Sa	0128	0347	1.0F
	0618	0946	2.9E
	1416	1742	2.4F
	2051	2330	1.1E
17 Su	0239	0509	1.2F
	0736	1055	2.9E
	1512	1833	2.6F
	2135		
18 M		0030	1.5E
	0334	0615	1.6F
	0844	1157	3.1E
	1558	1913	2.9F
	2212		
19 Tu		0059	2.1E
	0420	0706	2.0F
	0943	1248	3.3E
	1638	1945	3.0F
	2244		
20 W		0128	2.7E
	0500	0749	2.5F
	1036	1331	3.4E
	1713	2012	3.2F
	2313		
21 Th		0159	3.3E
	0537	0828	3.0F
	1124	1411	3.4E
	1746	2040	3.2F
	2340		
22 F		0232	3.9E
	0613	0906	3.4F
	1211	1450	3.4E
	1817	2109	3.2F
23 Sa	0006	0307	4.4E
	0648	0944	3.7F
	1257	1530	3.3E
	1847	2140	3.2F
24 Su	0033	0344	4.8E
	0725	1024	3.8F
	1345	1611	3.0E
	1918	2214	3.0F
25 M	0101	0423	5.1E
	0805	1107	3.9F
	1435	1655	2.7E
	1951	2251	2.7F
26 Tu	0132	0505	5.1E
	0849	1154	3.8F
	1529	1741	2.3E
	2028	2333	2.4F
27 W	0208	0551	5.0E
	0940	1246	3.6F
	1628	1832	1.9E
	2112		
28 Th		0020	2.1F
	0252	0643	4.8E
	1039	1347	3.3F
	1734	1929	1.5E
	2212		
29 F		0118	1.8F
	0346	0740	4.4E
	1146	1456	3.1F
	1843	2035	1.4E
	2340		
30 Sa		0228	1.5F
	0458	0845	4.1E
	1256	1612	3.1F
	1946	2149	1.5E

May

Day	Slack (h m)	Maximum (h m)	knots
1 Su	0117	0349	1.6F
	0625	0954	3.9E
	1403	1721	3.3F
	2040	2308	1.9E
2 M	0236	0514	1.9F
	0752	1106	3.7E
	1503	1817	3.5F
	2125		
3 Tu		0014	2.6E
	0340	0628	2.5F
	0908	1213	3.7E
	1555	1905	3.6F
	2205		
4 W		0104	3.4E
	0433	0728	3.1F
	1016	1312	3.6E
	1643	1946	3.7F
	2242		
5 Th		0146	4.1E
	0522	0821	3.6F
	1117	1403	3.4E
	1726	2024	3.6F
	2317		
6 F		0224	4.6E
	0606	0908	4.0F
	1212	1449	3.2E
	1807	2100	3.4F
	2351		
7 Sa		0301	5.0E
	0649	0952	4.1F
	1303	1531	2.9E
	1845	2135	3.2F
8 Su	0024	0338	5.1E
	0730	1034	4.1F
	1352	1611	2.6E
	1923	2210	2.9F
9 M	0057	0415	5.1E
	0811	1115	3.9F
	1440	1650	2.3E
	2000	2246	2.5F
10 Tu	0130	0454	4.9E
	0854	1157	3.6F
	1528	1731	1.9E
	2037	2325	2.1F
11 W	0204	0536	4.5E
	0938	1242	3.2F
	1618	1815	1.6E
	2117		
12 Th		0007	1.8F
	0241	0620	4.1E
	1027	1332	2.9F
	1712	1903	1.3E
	2205		
13 F		0055	1.4F
	0324	0708	3.7E
	1121	1428	2.6F
	1809	1956	1.1E
	2314		
14 Sa		0151	1.2F
	0417	0802	3.4E
	1219	1531	2.4F
	1904	2056	1.1E
15 Su	0038	0258	1.0F
	0525	0900	3.0E
	1317	1632	2.4F
	1955	2201	1.3E

Day	Slack (h m)	Maximum (h m)	knots
16 M	0156	0412	1.1F
	0645	1000	2.8E
	1410	1724	2.5F
	2037	2303	1.8E
17 Tu	0258	0526	1.4F
	0802	1100	2.7E
	1458	1807	2.6F
	2115	2355	2.4E
18 W	0348	0627	1.8F
	0911	1157	2.7E
	1542	1844	2.7F
	2148		
19 Th		0038	3.1E
	0431	0719	2.4F
	1012	1249	2.7E
	1622	1919	2.8F
	2219		
20 F		0118	3.7E
	0511	0804	2.9F
	1108	1337	2.7E
	1700	1954	2.9F
	2250		
21 Sa		0158	4.4E
	0549	0847	3.4F
	1201	1423	2.7E
	1737	2030	2.9F
	2321		
22 Su		0238	4.9E
	0628	0929	3.9F
	1251	1508	2.6E
	1814	2108	2.9F
	2354		
23 M		0319	5.3E
	0709	1012	4.1F
	1342	1554	2.5E
	1852	2148	2.9F
24 Tu	0029	0402	5.6E
	0753	1058	4.2F
	1433	1640	2.4E
	1931	2231	2.7F
25 W	0109	0448	5.7E
	0840	1146	4.2F
	1526	1729	2.1E
	2016	2317	2.6F
26 Th	0153	0536	5.5E
	0932	1239	4.0F
	1621	1821	1.9E
	2108		
27 F		0009	2.3F
	0244	0628	5.2E
	1027	1335	3.8F
	1717	1917	1.8E
	2215		
28 Sa		0108	2.1F
	0343	0724	4.7E
	1126	1436	3.6F
	1814	2019	1.9E
	2338		
29 Su		0217	1.9F
	0455	0825	4.2E
	1227	1539	3.4F
	1907	2127	2.1E
30 M	0104	0336	1.8F
	0617	0930	3.6E
	1327	1640	3.3F
	1955	2236	2.6E
31 Tu	0220	0501	2.1F
	0742	1038	3.1E
	1425	1736	3.2F
	2040	2341	3.2E

June

Day	Slack (h m)	Maximum (h m)	knots
1 W	0325	0617	2.5F
	0902	1148	2.8E
	1519	1826	3.2F
	2122		
2 Th		0035	3.8E
	0420	0721	3.1F
	1013	1254	2.5E
	1610	1912	3.1F
	2202		
3 F		0121	4.3E
	0510	0816	3.5F
	1116	1351	2.4E
	1658	1954	2.9F
	2241		
4 Sa		0202	4.7E
	0555	0904	3.8F
	1212	1439	2.2E
	1743	2034	2.8F
	2318		
5 Su		0241	4.9E
	0638	0947	4.0F
	1301	1521	2.1E
	1825	2112	2.7F
	2355		
6 M		0319	5.0E
	0719	1028	3.9F
	1347	1559	2.0E
	1904	2149	2.5F
7 Tu	0032	0357	5.0E
	0759	1107	3.8F
	1431	1636	1.9E
	1941	2226	2.3F
8 W	0108	0435	4.8E
	0839	1145	3.6F
	1514	1714	1.7E
	2017	2305	2.1F
9 Th	0144	0515	4.6E
	0920	1224	3.4F
	1557	1754	1.6E
	2055	2345	1.9F
10 F	0221	0557	4.4E
	1001	1304	3.1F
	1640	1837	1.6E
	2139		
11 Sa		0029	1.7F
	0302	0641	4.0E
	1043	1347	2.9F
	1724	1924	1.6E
	2235		
12 Su		0119	1.5F
	0348	0727	3.6E
	1126	1431	2.7F
	1808	2014	1.7E
	2344		
13 M		0216	1.3F
	0445	0817	3.2E
	1210	1518	2.5F
	1850	2107	1.9E
14 Tu	0059	0320	1.2F
	0555	0911	2.8E
	1256	1605	2.4F
	1930	2202	2.3E
15 W	0207	0430	1.4F
	0716	1008	2.4E
	1341	1652	2.4F
	2008	2256	2.8E

Day	Slack (h m)	Maximum (h m)	knots
16 Th	0305	0540	1.7F
	0836	1107	2.1E
	1434	1740	2.4F
	2046	2347	3.3E
17 F	0356	0643	2.3F
	0948	1207	2.0E
	1524	1826	2.4F
	2124		
18 Sa		0037	4.0E
	0442	0738	2.8F
	1053	1304	2.0E
	1614	1912	2.5F
	2202		
19 Su		0124	4.6E
	0526	0829	3.4F
	1151	1359	2.0E
	1702	1958	2.6F
	2243		
20 M		0212	5.1E
	0611	0916	3.9F
	1244	1450	2.1E
	1748	2044	2.7F
	2326		
21 Tu		0259	5.6E
	0657	1003	4.3F
	1335	1539	2.2E
	1833	2130	2.9F
22 W	0011	0346	5.9E
	0743	1050	4.5F
	1424	1628	2.3E
	1920	2218	2.9F
23 Th	0059	0435	6.0E
	0831	1138	4.5F
	1512	1717	2.3E
	2010	2308	2.9F
24 F	0150	0524	5.9E
	0920	1226	4.4F
	1559	1808	2.4E
	2107		
25 Sa		0001	2.8F
	0244	0615	5.5E
	1010	1316	4.2F
	1646	1900	2.5E
	2212		
26 Su		0100	2.6F
	0344	0707	4.8E
	1100	1406	3.8F
	1732	1955	2.6E
	2324		
27 M		0204	2.4F
	0452	0803	4.1E
	1152	1458	3.5F
	1818	2054	2.9E
28 Tu	0040	0318	2.2F
	0609	0902	3.2E
	1246	1552	3.1F
	1904	2155	3.2E
29 W	0154	0440	2.3F
	0731	1007	2.5E
	1343	1648	2.8F
	1951	2258	3.5E
30 Th	0301	0601	2.5F
	0853	1119	1.9E
	1442	1745	2.6F
	2038	2359	3.9E

Time meridian 120° W. 0000 is midnight. 1200 is noon. Times are not adjusted for Daylight Saving Time.

San Francisco Bay Entrance (Outside), Calif., 2005

F–Flood, Dir. 065° True E–Ebb, Dir. 245° True

July

Day	Slack (h m)	Maximum (h m)	knots
1 F	0401	0710	2.9F
	1009	1238	1.7E
	1540	1839	2.4F
	2124		
2 Sa		0055	4.2E
	0454	0807	3.3F
	1114	1349	1.6E
	1636	1930	2.3F
	2210		
3 Su		0143	4.4E
	0542	0857	3.6F
	1208	1443	1.6E
	1726	2016	2.3F
	2255		
4 M		0226	4.6E
	0626	0941	3.7F
	1255	1522	1.7E
	1810	2058	2.3F
	2337		
5 Tu		0306	4.7E
	0708	1021	3.8F
	1337	1553	1.7E
	1850	2136	2.4F
6 W ●	0017	0343	4.8E
	0747	1057	3.7F
	1415	1623	1.8E
	1927	2213	2.4F
7 Th	0056	0420	4.8E
	0824	1130	3.6F
	1451	1656	1.9E
	2002	2250	2.3F
8 F	0133	0457	4.7E
	0859	1202	3.5F
	1527	1732	2.0E
	2039	2328	2.3F
9 Sa	0210	0535	4.5E
	0933	1233	3.3F
	1601	1810	2.1E
	2119		
10 Su		0008	2.1F
	0248	0615	4.2E
	1005	1306	3.1F
	1635	1851	2.2E
	2205		
11 M		0052	1.9F
	0331	0656	3.8E
	1037	1341	2.9F
	1708	1935	2.4E
	2300		
12 Tu		0142	1.8F
	0421	0741	3.3E
	1109	1420	2.7F
	1742	2022	2.6E
13 W	0002	0238	1.6F
	0523	0830	2.7E
	1145	1502	2.5F
	1818	2112	2.9E
14 Th ◐	0109	0342	1.6F
	0640	0924	2.2E
	1228	1550	2.2F
	1857	2205	3.2E
15 F	0214	0453	1.8F
	0806	1025	1.8E
	1322	1643	2.1F
	1941	2301	3.5E
16 Sa	0316	0606	2.2F
	0927	1130	1.5E
	1427	1739	2.0F
	2030	2358	4.0E
17 Su	0412	0714	2.8F
	1038	1235	1.5E
	1534	1837	2.1F
	2122		
18 M		0055	4.6E
	0505	0813	3.4F
	1139	1337	1.6E
	1635	1933	2.4F
	2216		
19 Tu		0150	5.1E
	0555	0904	3.9F
	1231	1434	1.9E
	1730	2026	2.7F
	2309		
20 W		0243	5.6E
	0644	0952	4.3F
	1319	1526	2.2E
	1821	2118	3.0F
21 Th ○	0002	0333	6.0E
	0731	1037	4.6F
	1403	1614	2.5E
	1912	2208	3.3F
22 F	0055	0422	6.1E
	0816	1120	4.7F
	1445	1701	2.8E
	2004	2300	3.4F
23 Sa	0149	0510	5.8E
	0901	1203	4.5F
	1525	1748	3.1E
	2100	2352	3.4F
24 Su	0244	0559	5.3E
	0945	1246	4.3F
	1605	1836	3.4E
	2159		
25 M		0048	3.2F
	0343	0648	4.6E
	1030	1330	3.8F
	1645	1925	3.5E
	2302		
26 Tu		0148	2.9F
	0447	0739	3.7E
	1115	1416	3.3F
	1727	2017	3.6E
27 W ◑	0010	0256	2.6F
	0559	0834	2.7E
	1205	1505	2.8F
	1812	2113	3.6E
28 Th	0121	0414	2.4F
	0718	0935	1.9E
	1302	1601	2.3F
	1901	2214	3.6E
29 F	0231	0538	2.5F
	0841	1048	1.4E
	1408	1704	2.0F
	1954	2320	3.6E
30 Sa	0337	0652	2.7F
	0958	1230	1.1E
	1518	1811	1.9F
	2050		
31 Su		0027	3.8E
	0435	0753	3.0F
	1104	1356	1.2E
	1620	1913	1.9F
	2145		

August

Day	Slack (h m)	Maximum (h m)	knots
1 M		0127	4.0E
	0526	0843	3.3F
	1155	1446	1.4E
	1712	2004	2.1F
	2236		
2 Tu		0215	4.2E
	0611	0927	3.5F
	1237	1520	1.6E
	1756	2047	2.3F
	2323		
3 W		0254	4.4E
	0652	1004	3.6F
	1313	1542	1.8E
	1835	2124	2.4F
4 Th ●	0005	0329	4.5E
	0728	1037	3.6F
	1347	1605	2.0E
	1911	2200	2.6F
5 F	0045	0403	4.6E
	0801	1105	3.6F
	1417	1634	2.2E
	1946	2235	2.6F
6 Sa	0123	0437	4.6E
	0832	1130	3.5F
	1447	1706	2.5E
	2022	2311	2.6F
7 Su	0200	0512	4.4E
	0900	1157	3.4F
	1515	1741	2.8E
	2059	2349	2.6F
8 M	0239	0549	4.1E
	0927	1225	3.2F
	1542	1818	3.0E
	2139		
9 Tu		0030	2.5F
	0321	0628	3.6E
	0953	1257	3.0F
	1609	1858	3.2E
	2224		
10 W		0115	2.3F
	0409	0711	3.1E
	1020	1332	2.7F
	1637	1942	3.3E
	2316		
11 Th		0207	2.1F
	0508	0758	2.5E
	1053	1413	2.4F
	1711	2030	3.4E
12 F ◐	0017	0307	2.0F
	0621	0852	1.9E
	1134	1501	2.0F
	1751	2124	3.5E
13 Sa	0127	0417	2.1F
	0746	0953	1.5E
	1229	1558	1.8F
	1844	2223	3.7E
14 Su	0238	0536	2.3F
	0910	1101	1.3E
	1346	1704	1.8F
	1947	2327	4.0E
15 M	0345	0652	2.8F
	1021	1212	1.3E
	1511	1812	2.0F
	2054		
16 Tu		0031	4.5E
	0444	0755	3.4F
	1119	1320	1.6E
	1621	1915	2.3F
	2159		
17 W		0131	5.0E
	0537	0846	3.9F
	1208	1418	2.0E
	1719	2014	2.8F
	2300		
18 Th		0227	5.5E
	0625	0932	4.3F
	1250	1508	2.6E
	1812	2108	3.3F
	2357		
19 F ○		0318	5.7E
	0711	1014	4.5F
	1330	1555	3.1E
	1903	2159	3.7F
20 Sa	0052	0406	5.7E
	0754	1054	4.5F
	1407	1639	3.6E
	1955	2250	3.9F
21 Su	0147	0453	5.3E
	0835	1133	4.4F
	1443	1723	4.0E
	2047	2341	3.8F
22 M	0242	0539	4.7E
	0916	1212	4.0F
	1519	1807	4.2E
	2141		
23 Tu		0034	3.7F
	0340	0626	3.9E
	0958	1252	3.5F
	1557	1853	4.2E
	2238		
24 W		0130	3.3F
	0442	0715	3.1E
	1041	1336	2.9F
	1636	1941	4.1E
	2339		
25 Th		0233	2.9F
	0549	0808	2.2E
	1130	1424	2.4F
	1721	2033	3.8E
26 F ◑	0046	0346	2.6F
	0703	0907	1.5E
	1230	1521	1.9F
	1813	2132	3.5E
27 Sa	0157	0511	2.5F
	0822	1020	1.1E
	1344	1629	1.5F
	1914	2240	3.4E
28 Su	0306	0627	2.6F
	0937	1229	1.0E
	1500	1746	1.5F
	2019	2355	3.4E
29 M	0408	0728	2.8F
	1038	1343	1.2E
	1603	1854	1.7F
	2122		
30 Tu		0104	3.6E
	0500	0818	3.1F
	1125	1426	1.5E
	1654	1947	2.0F
	2217		
31 W		0154	3.8E
	0545	0859	3.3F
	1202	1454	1.8E
	1737	2030	2.3F
	2306		

September

Day	Slack (h m)	Maximum (h m)	knots
1 Th		0232	4.1E
	0624	0934	3.4F
	1235	1512	2.1E
	1816	2107	2.6F
	2350		
2 F		0305	4.2E
	0658	1002	3.4F
	1305	1535	2.5E
	1852	2142	2.8F
3 Sa ●	0031	0338	4.2E
	0729	1027	3.4F
	1333	1603	2.8E
	1927	2217	2.9F
4 Su	0111	0412	4.1E
	0757	1050	3.4F
	1359	1634	3.2E
	2002	2252	3.0F
5 M	0151	0447	3.9E
	0823	1116	3.2F
	1424	1709	3.5E
	2037	2330	3.0F
6 Tu	0232	0524	3.6E
	0848	1145	3.0F
	1448	1745	3.8E
	2115		
7 W		0010	2.9F
	0317	0603	3.2E
	0914	1217	2.8F
	1513	1824	3.9E
	2156		
8 Th		0054	2.8F
	0407	0647	2.7E
	0943	1253	2.5F
	1542	1908	3.9E
	2245		
9 F		0145	2.6F
	0506	0735	2.2E
	1018	1336	2.1F
	1618	1957	3.9E
	2344		
10 Sa		0244	2.5F
	0618	0830	1.7E
	1103	1427	1.8F
	1704	2053	3.8E
11 Su ◐	0055	0355	2.4F
	0737	0933	1.3E
	1208	1531	1.6F
	1806	2156	3.9E
12 M	0210	0515	2.6F
	0854	1043	1.2E
	1340	1643	1.6F
	1921	2303	4.1E
13 Tu	0319	0630	3.0F
	0958	1156	1.4E
	1507	1756	1.9F
	2039		
14 W		0010	4.4E
	0419	0730	3.4F
	1049	1302	1.9E
	1614	1904	2.5F
	2149		
15 Th		0112	4.7E
	0511	0819	3.9F
	1133	1358	2.6E
	1711	2004	3.0F
	2253		
16 F		0208	5.0E
	0559	0902	4.1F
	1211	1445	3.3E
	1803	2058	3.6F
	2353		
17 Sa ○		0259	5.0E
	0643	0942	4.2F
	1248	1529	3.9E
	1853	2149	4.0F
18 Su	0049	0347	4.8E
	0725	1020	4.2F
	1323	1612	4.4E
	1941	2239	4.2F
19 M	0144	0433	4.4E
	0805	1057	3.9F
	1357	1654	4.7E
	2030	2329	4.1F
20 Tu	0240	0519	3.8E
	0845	1136	3.5F
	1432	1736	4.8E
	2120		
21 W		0019	3.9F
	0336	0605	3.1E
	0926	1216	3.0F
	1509	1820	4.6E
	2213		
22 Th		0112	3.5F
	0435	0653	2.4E
	1011	1259	2.4F
	1549	1907	4.3E
	2309		
23 F		0211	3.1F
	0538	0744	1.8E
	1103	1349	1.9F
	1634	1958	3.9E
24 Sa ◑	0012	0320	2.7F
	0646	0842	1.3E
	1208	1447	1.5F
	1729	2056	3.5E
25 Su	0121	0439	2.5F
	0757	0954	1.0E
	1325	1558	1.3F
	1836	2201	3.2E
26 M	0228	0552	2.5F
	0902	1152	1.0E
	1439	1717	1.3F
	1947	2313	3.2E
27 Tu	0330	0651	2.7F
	0956	1304	1.3E
	1539	1826	1.6F
	2053		
28 W		0021	3.3E
	0421	0738	2.9F
	1039	1341	1.7E
	1629	1920	2.0F
	2151		
29 Th		0113	3.5E
	0505	0817	3.1F
	1114	1404	2.1E
	1712	2004	2.3F
	2243		
30 F		0154	3.6E
	0543	0848	3.1F
	1145	1428	2.6E
	1751	2043	2.7F
	2331		

Time meridian 120° W. 0000 is midnight. 1200 is noon. Times are not adjusted for Daylight Saving Time.

San Francisco Bay Entrance (Outside), Calif., 2005

F–Flood, Dir. 065° True E–Ebb, Dir. 245° True

October

Day	Slack h m	Maximum h m	knots
1 Sa		0231	3.6E
	0617	0915	3.2F
	1214	1455	3.1E
	1828	2120	2.9F
2 Su	0016	0307	3.6E
	0647	0940	3.2F
	1241	1527	3.5E
	1904	2156	3.1F
3 M	0100	0343	3.5E
	0716	1006	3.1F
	1306	1600	3.9E
●	1939	2233	3.3F
4 Tu	0143	0420	3.3E
	0744	1035	2.9F
	1331	1636	4.2E
	2015	2312	3.4F
5 W	0228	0500	3.0E
	0811	1108	2.7F
	1356	1714	4.4E
	2053	2353	3.3F
6 Th	0316	0542	2.6E
	0841	1143	2.5F
	1425	1756	4.4E
	2136		
7 F		0040	3.2F
	0410	0628	2.2E
	0915	1223	2.2F
	1459	1842	4.4E
	2227		
8 Sa		0132	3.0F
	0511	0719	1.8E
	0958	1311	1.9F
	1542	1933	4.3E
	2327		
9 Su		0233	2.9F
	0619	0817	1.5E
	1056	1408	1.7F
	1637	2031	4.1E
10 M	0036	0343	2.8F
	0728	0921	1.4E
	1219	1517	1.6F
◐	1748	2135	4.0E
11 Tu	0146	0457	3.0F
	0832	1031	1.5E
	1351	1632	1.7F
	1911	2243	4.0E
12 W	0250	0603	3.2F
	0925	1141	1.9E
	1506	1747	2.1F
	2031	2349	4.1E
13 Th	0348	0657	3.5F
	1010	1242	2.6E
	1607	1855	2.7F
	2143		
14 F		0051	4.2E
	0439	0744	3.7F
	1050	1333	3.3E
	1702	1955	3.2F
	2249		
15 Sa		0147	4.2E
	0526	0826	3.8F
	1127	1419	4.0E
	1752	2049	3.8F
	2349		

Day	Slack h m	Maximum h m	knots
16 Su		0239	4.0E
	0610	0905	3.8F
	1203	1501	4.6E
	1840	2140	4.1F
17 M	0047	0327	3.7E
	0652	0943	3.6F
	1238	1543	5.0E
○	1927	2228	4.2F
18 Tu	0142	0413	3.3E
	0733	1022	3.3F
	1313	1624	5.1E
	2013	2316	4.2F
19 W	0236	0458	2.8E
	0815	1101	2.9F
	1349	1706	5.0E
	2100		
20 Th		0004	3.9F
	0330	0544	2.4E
	0858	1143	2.5F
	1427	1750	4.7E
	2150		
21 F		0055	3.6F
	0426	0631	1.9E
	0945	1228	2.0F
	1508	1837	4.3E
	2243		
22 Sa		0151	3.2F
	0524	0722	1.5E
	1040	1318	1.6F
	1555	1927	3.9E
	2341		
23 Su		0253	2.8F
	0624	0819	1.2E
	1146	1417	1.3F
	1651	2023	3.5E
24 M	0043	0401	2.6F
	0724	0924	1.1E
	1300	1524	1.2F
◑	1758	2123	3.2E
25 Tu	0144	0506	2.6F
	0819	1038	1.3E
	1409	1638	1.3F
	1909	2226	3.1E
26 W	0240	0600	2.7F
	0905	1145	1.6E
	1509	1746	1.5F
	2018	2326	3.0E
27 Th	0330	0644	2.8F
	0944	1229	2.1E
	1559	1843	1.9F
	2120		
28 F		0020	3.0E
	0413	0720	2.8F
	1019	1305	2.6E
	1643	1931	2.3F
	2217		
29 Sa		0107	3.0E
	0451	0751	2.9F
	1050	1339	3.1E
	1723	2014	2.7F
	2309		
30 Su		0151	3.0E
	0520	0820	2.9F
	1118	1414	3.7E
	1801	2055	3.0F
	2359		
31 M		0232	2.9E
	0600	0851	2.9F
	1146	1449	4.2E
	1838	2134	3.3F

November

Day	Slack h m	Maximum h m	knots
1 Tu	0048	0313	2.8E
	0632	0923	2.8F
	1214	1527	4.5E
●	1915	2214	3.5F
2 W	0136	0355	2.6E
	0704	0957	2.7F
	1243	1606	4.8E
	1954	2256	3.6F
3 Th	0225	0439	2.4E
	0737	1035	2.5F
	1314	1648	4.9E
	2037	2341	3.6F
4 F	0316	0525	2.1E
	0814	1116	2.3F
	1350	1733	4.9E
	2124		
5 Sa		0030	3.6F
	0412	0614	1.9E
	0858	1203	2.1F
	1433	1822	4.8E
	2217		
6 Su		0125	3.4F
	0510	0708	1.7E
	0953	1256	1.9F
	1525	1916	4.6E
	2316		
7 M		0225	3.3F
	0611	0807	1.6E
	1106	1359	1.7F
	1628	2015	4.4E
8 Tu	0018	0329	3.3F
	0708	0911	1.7E
	1231	1509	1.7F
◐	1744	2118	4.1E
9 W	0120	0432	3.3F
	0800	1018	2.1E
	1351	1625	1.9F
	1907	2222	3.9E
10 Th	0219	0529	3.4F
	0846	1121	2.6E
	1458	1740	2.3F
	2026	2327	3.6E
11 F	0313	0620	3.5F
	0928	1217	3.3E
	1557	1848	2.8F
	2139		
12 Sa		0029	3.4E
	0404	0706	3.5F
	1007	1307	4.0E
	1650	1947	3.4F
	2246		
13 Su		0126	3.2E
	0451	0749	3.4F
	1044	1352	4.6E
	1740	2041	3.8F
	2347		
14 M		0219	2.9E
	0536	0830	3.2F
	1121	1435	5.0E
	1827	2131	4.0F
15 Tu	0044	0308	2.0E
	0621	0910	3.0F
	1158	1517	5.2E
○	1912	2219	4.1F

Day	Slack h m	Maximum h m	knots
16 W	0138	0354	2.4E
	0704	0951	2.8F
	1236	1558	5.1E
	1957	2305	4.0F
17 Th	0230	0439	2.1E
	0748	1032	2.5F
	1314	1641	5.0E
	2043	2351	3.8F
18 F	0321	0524	1.8E
	0832	1115	2.1F
	1354	1725	4.7E
	2130		
19 Sa		0039	3.5F
	0411	0610	1.6E
	0920	1201	1.8F
	1437	1811	4.3E
	2219		
20 Su		0129	3.2F
	0503	0658	1.4E
	1014	1251	1.6F
	1524	1859	4.0E
	2310		
21 M		0222	2.9F
	0554	0751	1.3E
	1117	1346	1.4F
	1618	1950	3.6E
22 Tu	0003	0316	2.8F
	0644	0846	1.4E
	1225	1447	1.3F
	1720	2044	3.3E
23 W	0054	0407	2.7F
	0729	0943	1.6E
	1332	1553	1.3F
◑	1829	2140	3.0E
24 Th	0143	0454	2.7F
	0810	1038	2.0E
	1431	1700	1.5F
	1939	2235	2.8E
25 F	0229	0536	2.6F
	0846	1127	2.5E
	1524	1801	1.8F
	2047	2330	2.6E
26 Sa	0312	0614	2.6F
	0920	1211	3.1E
	1610	1855	2.2F
	2150		
27 Su		0022	2.4E
	0353	0651	2.6F
	0953	1253	3.6E
	1652	1944	2.7F
	2248		
28 M		0111	2.3E
	0432	0728	2.6F
	1024	1334	4.1E
	1733	2029	3.1F
	2343		
29 Tu		0159	2.3E
	0511	0806	2.6F
	1056	1415	4.6E
	1813	2112	3.4F
30 W	0035	0246	2.2E
	0549	0845	2.6F
	1130	1457	5.0E
	1854	2156	3.7F

December

Day	Slack h m	Maximum h m	knots
1 Th	0126	0332	2.1E
	0628	0926	2.5F
	1206	1541	5.2E
●	1938	2242	3.8F
2 F	0218	0420	2.0E
	0710	1010	2.5F
	1247	1627	5.3E
	2024	2330	3.9F
3 Sa	0310	0509	1.9E
	0755	1057	2.4F
	1331	1716	5.3E
	2113		
4 Su		0021	3.9F
	0402	0601	1.9E
	0849	1148	2.3F
	1422	1807	5.2E
	2205		
5 M		0114	3.8F
	0455	0655	1.9E
	0953	1246	2.1F
	1520	1901	4.9E
	2259		
6 Tu		0209	3.7F
	0546	0753	2.0E
	1108	1350	2.0F
	1627	1958	4.4E
	2355		
7 W		0304	3.6F
	0634	0854	2.3E
	1225	1501	2.0F
	1743	2058	3.9E
8 Th	0050	0359	3.5F
	0720	0955	2.8E
	1339	1616	2.2F
◐	1903	2200	3.4E
9 F	0144	0452	3.3F
	0803	1054	3.3E
	1445	1732	2.5F
	2023	2305	2.9E
10 Sa	0238	0543	3.2F
	0845	1150	3.9E
	1544	1840	3.0F
	2137		
11 Su		0009	2.5E
	0330	0631	3.1F
	0927	1241	4.4E
	1637	1941	3.4F
	2245		
12 M		0110	2.3E
	0421	0717	2.9F
	1008	1328	4.8E
	1727	2035	3.7F
	2346		
13 Tu		0206	2.1E
	0510	0802	2.7F
	1049	1413	5.0E
	1814	2125	3.9F
14 W	0041	0256	1.9E
	0557	0846	2.6F
	1130	1456	5.0E
	1900	2211	3.9F
15 Th	0131	0341	1.8E
	0642	0929	2.4F
	1211	1539	5.0E
○	1944	2255	3.8F

Day	Slack h m	Maximum h m	knots
16 F	0218	0423	1.7E
	0726	1011	2.3F
	1252	1621	4.8E
	2027	2337	3.6F
17 Sa	0303	0504	1.6E
	0809	1053	2.1F
	1333	1703	4.6E
	2110		
18 Su		0019	3.4F
	0347	0546	1.6E
	0854	1137	1.9F
	1415	1746	4.4E
	2153		
19 M		0100	3.2F
	0430	0630	1.6E
	0943	1223	1.8F
	1459	1830	4.1E
	2235		
20 Tu		0141	3.0F
	0512	0716	1.7E
	1039	1313	1.6F
	1548	1916	3.7E
	2317		
21 W		0223	2.9F
	0553	0804	1.8E
	1141	1408	1.5F
	1644	2005	3.3E
	2359		
22 Th		0305	2.7F
	0632	0853	2.1E
	1245	1509	1.4F
	1750	2055	2.9E
23 F	0041	0347	2.6F
	0709	0944	2.4E
	1347	1613	1.5F
◑	1902	2149	2.5E
24 Sa	0124	0431	2.5F
	0745	1034	2.9E
	1443	1718	1.8F
	2015	2244	2.1E
25 Su	0209	0515	2.4F
	0821	1123	3.4E
	1533	1819	2.2F
	2125	2341	1.9E
26 M	0255	0600	2.4F
	0858	1212	3.9E
	1620	1914	2.6F
	2229		
27 Tu		0037	1.8E
	0343	0645	2.4F
	0937	1259	4.4E
	1705	2005	3.1F
	2327		
28 W		0131	1.8E
	0430	0731	2.4F
	1017	1347	4.8E
	1751	2054	3.5F
29 Th	0021	0222	1.9E
	0517	0817	2.5F
	1100	1434	5.2E
	1836	2141	3.8F
30 F	0113	0313	1.9E
	0003	0004	2.6F
	1145	1522	5.5E
●	1922	2228	4.1F
31 Sa	0202	0402	2.0E
	0651	0953	2.7F
	1233	1611	5.7E
	2010	2315	4.2F

Time meridian 120° W. 0000 is midnight. 1200 is noon. Times are not adjusted for Daylight Saving Time.

Golden Gate Bridge, Calif., 2005

F–Flood, Dir. 055° True   E–Ebb, Dir. 237° True

January

Day	Slack h m	Maximum h m	knots
1 Sa		0142	2.8F
	0531	0821	1.0E
	1038	1343	1.4F
	1609	1912	1.5E
	2238		
2 Su		0225	2.7F
	0604	0907	1.1E
	1147	1445	1.3F
	1719	2037	1.2E
	2316		
3 M ◑		0312	2.6F
	0639	0952	1.3E
	1308	1555	1.3F
	1845	2148	1.0E
4 Tu	0000	0403	2.5F
	0717	1036	1.6E
	1423	1709	1.6F
	2012	2251	0.8E
5 W	0051	0455	2.6F
	0757	1120	1.9E
	1524	1816	2.0F
	2127	2351	0.8E
6 Th	0145	0548	2.7F
	0839	1203	2.3E
	1616	1912	2.5F
	2231		
7 F		0046	0.8E
	0242	0640	2.9F
	0923	1246	2.7E
	1705	2002	3.0F
	2327		
8 Sa		0139	0.9E
	0340	0731	3.0F
	1009	1329	3.0E
	1752	2050	3.4F
9 Su	0018	0232	1.0E
	0439	0821	3.1F
	1057	1415	3.2E
	1838	2138	3.8F
10 M ●	0107	0325	1.1E
	0539	0912	3.1F
	1146	1506	3.3E
	1923	2224	4.0F
11 Tu	0154	0418	1.2E
	0639	1003	3.1F
	1236	1559	3.2E
	2006	2310	4.1F
12 W	0239	0510	1.3E
	0739	1055	3.0F
	1327	1653	3.1E
	2049	2355	4.0F
13 Th	0323	0600	1.4E
	0840	1147	2.7F
	1421	1747	2.7E
	2130		
14 F		0040	3.8F
	0407	0651	1.5E
	0944	1242	2.4F
	1519	1845	2.3E
	2210		
15 Sa		0127	3.5F
	0451	0744	1.5E
	1054	1343	2.0F
	1626	1949	1.8E
	2249		
16 Su ◐		0216	3.1F
	0539	0839	1.5E
	1214	1456	1.7F
	1746	2053	1.3E
	2331		
17 M		0308	2.8F
	0628	0934	1.5E
	1337	1620	1.6F
	1924	2155	1.0E
18 Tu	0018	0404	2.5F
	0719	1031	1.6E
	1450	1740	1.8F
	2059	2257	0.8E
19 W	0115	0501	2.4F
	0808	1129	1.8E
	1550	1843	2.2F
	2213	2357	0.7E
20 Th	0217	0556	2.3F
	0854	1222	1.9E
	1640	1934	2.5F
	2310		
21 F		0053	0.7E
	0316	0645	2.4F
	0936	1308	2.1E
	1723	2019	2.8F
	2358		
22 Sa		0143	0.8E
	0410	0730	2.4F
	1016	1351	2.3E
	1801	2100	3.0F
23 Su	0041	0231	0.9E
	0458	0812	2.4F
	1056	1431	2.5E
	1836	2138	3.2F
24 M	0120	0318	1.0E
	0542	0853	2.5F
	1134	1511	2.6E
	1908	2213	3.3F
25 Tu ○	0155	0404	1.1E
	0625	0935	2.5F
	1213	1550	2.6E
	1938	2245	3.3F
26 W	0227	0447	1.2E
	0706	1016	2.4F
	1252	1627	2.5E
	2005	2315	3.3F
27 Th	0254	0527	1.2E
	0747	1058	2.4F
	1332	1659	2.4E
	2030	2347	3.3F
28 F	0319	0605	1.2E
	0828	1140	2.2F
	1413	1725	2.2E
	2056		
29 Sa		0021	3.2F
	0345	0639	1.2E
	0911	1225	2.0F
	1459	1753	1.9E
	2123		
30 Su		0057	3.0F
	0414	0708	1.3E
	0958	1315	1.8F
	1551	1832	1.5E
	2155		
31 M		0139	2.8F
	0447	0729	1.4E
	1054	1413	1.6F
	1657	1931	1.1E
	2230		

February

Day	Slack h m	Maximum h m	knots
1 Tu ◑		0226	2.6F
	0528	0813	1.5E
	1208	1523	1.5F
	1822	2105	0.8E
	2313		
2 W		0319	2.4F
	0617	0916	1.6E
	1340	1641	1.7F
	1958	2223	0.6E
3 Th	0007	0417	2.4F
	0710	1027	1.9E
	1456	1755	2.1F
	2123	2331	0.6E
4 F	0112	0519	2.5F
	0806	1136	2.2E
	1556	1855	2.7F
	2228		
5 Sa		0032	0.7E
	0224	0619	2.6F
	0900	1236	2.6E
	1648	1946	3.2F
	2319		
6 Su		0127	0.9E
	0337	0715	2.9F
	0954	1329	2.9E
	1736	2033	3.6F
7 M	0003	0220	1.2E
	0446	0809	3.1F
	1047	1421	3.1E
	1821	2118	4.0F
8 Tu ●	0044	0310	1.4E
	0548	0901	3.2F
	1140	1512	3.2E
	1904	2202	4.1F
9 W	0123	0359	1.6E
	0646	0953	3.2F
	1232	1602	3.1E
	1945	2245	4.2F
10 Th	0202	0446	1.8E
	0741	1044	3.1F
	1325	1651	2.8E
	2024	2327	4.0F
11 F	0240	0530	1.8E
	0835	1135	2.9F
	1419	1740	2.5E
	2100		
12 Sa		0008	3.7F
	0319	0613	1.8E
	0931	1227	2.5F
	1516	1830	2.0E
	2135		
13 Su		0051	3.3F
	0400	0659	1.7E
	1030	1325	2.1F
	1620	1926	1.5E
	2208		
14 M		0135	2.9F
	0443	0750	1.6E
	1141	1433	1.8F
	1738	2027	1.0E
	2243		
15 Tu ◐		0225	2.4F
	0532	0848	1.5E
	1302	1557	1.6F
	1921	2129	0.7E
	2325		
16 W		0320	2.1F
	0627	0950	1.4E
	1419	1720	1.8F
	2057	2233	0.5E
17 Th	0024	0421	1.9F
	0726	1054	1.5E
	1522	1824	2.1F
	2205	2335	0.5E
18 F	0146	0524	1.8F
	0821	1156	1.7E
	1612	1914	2.5F
	2255		
19 Sa		0033	0.7E
	0303	0621	2.0F
	0911	1247	1.9E
	1654	1956	2.8F
	2335		
20 Su		0123	0.9E
	0402	0709	2.1F
	0956	1331	2.2E
	1730	2032	3.0F
21 M	0010	0209	1.1E
	0450	0752	2.3F
	1038	1411	2.4E
	1802	2105	3.2F
22 Tu	0040	0252	1.2E
	0533	0834	2.5F
	1118	1450	2.5E
	1832	2136	3.3F
23 W ○	0106	0333	1.4E
	0613	0915	2.6F
	1159	1528	2.5E
	1859	2205	3.4F
24 Th	0129	0411	1.5E
	0650	0956	2.6F
	1239	1604	2.4E
	1924	2235	3.4F
25 F	0151	0444	1.6E
	0726	1037	2.6F
	1319	1636	2.3E
	1949	2307	3.3F
26 Sa	0215	0509	1.7E
	0802	1119	2.6F
	1402	1702	2.0E
	2015	2341	3.2F
27 Su	0241	0517	1.7E
	0841	1203	2.4F
	1448	1728	1.7E
	2044		
28 M		0018	3.1F
	0312	0536	1.9E
	0924	1252	2.2F
	1541	1806	1.4E
	2116		

March

Day	Slack h m	Maximum h m	knots
1 Tu		0100	2.8F
	0348	0613	1.9E
	1016	1349	2.0F
	1647	1901	1.0E
	2153		
2 W		0147	2.5F
	0433	0703	1.9E
	1125	1457	1.9F
	1814	2044	0.6E
	2239		
3 Th ◑		0244	2.3F
	0527	0808	1.9E
	1300	1616	2.0F
	1956	2207	0.5E
	2340		
4 F		0348	2.1F
	0632	0944	1.9E
	1426	1733	2.4F
	2118	2317	0.6E
5 Sa	0102	0457	2.2F
	0740	1120	2.2E
	1530	1834	2.9F
	2212		
6 Su		0019	0.9E
	0235	0604	2.4F
	0845	1228	2.5E
	1624	1924	3.4F
	2255		
7 M		0114	1.2E
	0354	0705	2.8F
	0946	1323	2.8E
	1712	2010	3.8F
	2332		
8 Tu		0203	1.6E
	0458	0800	3.1F
	1044	1414	2.9E
	1756	2053	4.0F
9 W	0008	0249	1.9E
	0554	0853	3.3F
	1138	1503	2.9E
	1837	2135	4.1F
10 Th ●	0044	0333	2.1E
	0646	0944	3.3F
	1232	1551	2.7E
	1916	2216	4.0F
11 F	0119	0415	2.2E
	0736	1034	3.3F
	1325	1638	2.4E
	1953	2256	3.8F
12 Sa	0155	0455	2.2E
	0824	1123	3.0F
	1418	1725	2.1E
	2027	2335	3.4F
13 Su	0232	0532	2.1E
	0913	1213	2.7F
	1515	1813	1.6E
	2058		
14 M		0015	3.0F
	0310	0608	1.9E
	1004	1307	2.3F
	1618	1906	1.2E
	2129		
15 Tu		0057	2.5F
	0350	0649	1.7E
	1104	1411	1.9F
	1738	2005	0.8E
	2203		
16 W		0144	2.1F
	0436	0750	1.4E
	1217	1531	1.7F
	1922	2108	0.5E
	2248		
17 Th ◐		0238	1.7F
	0532	0907	1.3E
	1334	1652	1.8F
	2046	2211	0.5E
	2356		
18 F		0342	1.5F
	0638	1018	1.3E
	1439	1756	2.1F
	2144	2313	0.6E
19 Sa	0140	0452	1.4F
	0746	1123	1.5E
	1530	1844	2.5F
	2226		
20 Su		0010	0.8E
	0259	0557	1.6F
	0846	1218	1.7E
	1611	1923	2.8F
	2259		
21 M		0059	1.1E
	0355	0649	1.9F
	0936	1303	2.0E
	1647	1955	3.0F
	2326		
22 Tu		0142	1.3E
	0441	0733	2.2F
	1021	1344	2.1E
	1718	2024	3.1F
	2349		
23 W		0220	1.6E
	0521	0814	2.5F
	1104	1423	2.2E
	1746	2053	3.2F
24 Th	0009	0256	1.8E
	0557	0855	2.7F
	1146	1501	2.2E
	1813	2122	3.3F
25 F ○	0030	0328	1.9E
	0631	0936	2.8F
	1227	1537	2.1E
	1839	2155	3.3F
26 Sa	0053	0352	2.0E
	0705	1018	2.9F
	1310	1610	1.9E
	1906	2229	3.3F
27 Su	0119	0401	2.2E
	0740	1100	2.9F
	1354	1640	1.7E
	1935	2306	3.2F
28 M	0149	0422	2.4E
	0818	1145	2.8F
	1443	1711	1.5E
	2007	2345	3.0F
29 Tu	0223	0455	2.5E
	0902	1234	2.6F
	1539	1752	1.1E
	2043		
30 W		0028	2.7F
	0303	0537	2.4E
	0955	1331	2.4F
	1648	1859	0.8E
	2125		
31 Th		0118	2.4F
	0351	0628	2.3E
	1102	1438	2.3F
	1815	2040	0.6E
	2219		

Time meridian 120° W. 0000 is midnight. 1200 is noon. Times are not adjusted for Daylight Saving Time.

Golden Gate Bridge, Calif., 2005

F–Flood, Dir. 055° True   E–Ebb, Dir. 237° True

April

Day	Slack h m	Maximum h m	knots
1		0217	2.1F
F	0450	0735	2.0E
	1228	1553	2.4F
◑	1950	2155	0.6E
	2335		
2		0327	1.9F
Sa	0602	0930	1.9E
	1351	1706	2.7F
	2056	2302	0.8E
3	0120	0441	2.0F
Su	0722	1105	2.1E
	1457	1806	3.1F
	2141		
4		0003	1.2E
M	0255	0554	2.3F
	0836	1212	2.3E
	1552	1856	3.5F
	2219		
5		0055	1.6E
Tu	0404	0657	2.7F
	0943	1308	2.4E
	1640	1941	3.8F
	2254		
6		0141	2.0E
W	0501	0753	3.1F
	1043	1358	2.5E
	1724	2023	3.9F
	2329		
7		0224	2.3E
Th	0552	0844	3.3F
	1139	1446	2.4E
	1804	2104	3.9F
8	0003	0304	2.5E
F	0640	0935	3.4F
	1233	1533	2.2E
●	1842	2144	3.7F
9	0037	0343	2.5E
Sa	0727	1023	3.3F
	1326	1620	1.9E
	1918	2223	3.4F
10	0112	0419	2.4E
Su	0812	1111	3.1F
	1420	1707	1.6E
	1951	2302	3.1F
11	0148	0452	2.3E
M	0856	1159	2.8F
	1518	1756	1.3E
	2022	2342	2.7F
12	0224	0522	2.1E
Tu	0941	1250	2.5F
	1623	1848	0.9E
	2055		
13		0023	2.3F
W	0303	0553	1.8E
	1030	1349	2.2F
	1743	1946	0.7E
	2134		
14		0108	1.8F
Th	0346	0638	1.6E
	1127	1458	2.0F
	1911	2048	0.5E
	2226		
15		0201	1.4F
F	0439	0822	1.3E
	1233	1610	2.0F
	2020	2148	0.5E
	2346		

Day	Slack h m	Maximum h m	knots
16		0305	1.2F
Sa	0546	0939	1.3E
	1337	1713	2.2F
◐	2108	2248	0.7E
17	0133	0417	1.2F
Su	0704	1044	1.4E
	1430	1801	2.5F
	2144	2342	1.0E
18	0248	0529	1.4F
M	0815	1140	1.5E
	1513	1838	2.7F
	2212		
19		0029	1.3E
Tu	0341	0626	1.7F
	0914	1229	1.7E
	1550	1909	2.9F
	2233		
20		0109	1.6E
W	0425	0712	2.1F
	1004	1312	1.8E
	1622	1938	3.0F
	2252		
21		0145	1.9E
Th	0503	0754	2.4F
	1050	1353	1.8E
	1652	2008	3.1F
	2312		
22		0217	2.1E
F	0539	0835	2.7F
	1134	1431	1.8E
	1720	2041	3.2F
	2336		
23		0242	2.3E
Sa	0613	0917	2.9F
	1218	1509	1.7E
	1750	2116	3.2F
24	0003	0256	2.5E
Su	0648	1000	3.1F
	1303	1546	1.5E
○	1823	2155	3.2F
25	0033	0315	2.7E
M	0725	1045	3.1F
	1351	1623	1.4E
	1857	2235	3.1F
26	0108	0348	2.9E
Tu	0806	1131	3.1F
	1443	1705	1.1E
	1935	2318	2.9F
27	0147	0428	2.9E
W	0853	1221	3.0F
	1542	1759	0.9E
	2018		
28		0004	2.6F
Th	0231	0514	2.8E
	0946	1316	2.9F
	1650	1914	0.7E
	2110		
29		0056	2.3F
F	0322	0608	2.5E
	1047	1419	2.8F
	1808	2030	0.7E
	2216		
30		0158	2.0F
Sa	0424	0710	2.2E
	1158	1526	2.9F
◑	1923	2139	0.8E
	2348		

May

Day	Slack h m	Maximum h m	knots
1		0310	1.8F
Su	0541	0922	1.9E
	1311	1632	3.0F
	2018	2242	1.1E
2	0139	0428	1.8F
M	0708	1044	1.9E
	1416	1732	3.3F
	2101	2340	1.5E
3	0300	0544	2.2F
Tu	0830	1150	2.0E
	1512	1823	3.5F
	2139		
4		0031	2.0E
W	0403	0649	2.6F
	0942	1247	2.0E
	1602	1909	3.6F
	2215		
5		0116	2.3E
Th	0457	0745	3.0F
	1045	1338	1.9E
	1647	1951	3.6F
	2249		
6		0157	2.6E
F	0546	0836	3.2F
	1142	1426	1.8E
	1728	2032	3.5F
	2324		
7		0236	2.7E
Sa	0632	0925	3.3F
	1237	1513	1.6E
	1806	2112	3.3F
	2359		
8		0313	2.7E
Su	0716	1013	3.3F
	1330	1601	1.4E
●	1842	2152	3.1F
9	0035	0349	2.6E
M	0759	1100	3.2F
	1425	1650	1.2E
	1916	2232	2.8F
10	0110	0422	2.4E
Tu	0841	1146	3.0F
	1523	1739	1.0E
	1951	2312	2.4F
11	0147	0453	2.2E
W	0920	1232	2.7F
	1626	1831	0.8E
	2029	2353	2.1F
12	0225	0525	2.0E
Th	1000	1322	2.5F
	1733	1926	0.7E
	2115		
13		0038	1.7F
F	0307	0608	1.8E
	1041	1416	2.4F
	1840	2024	0.6E
	2213		
14		0129	1.4F
Sa	0355	0734	1.5E
	1126	1514	2.3F
	1937	2120	0.7E
	2332		
15		0229	1.1F
Su	0457	0858	1.4E
	1217	1608	2.3F
	2019	2215	0.8E

Day	Slack h m	Maximum h m	knots
16	0108	0339	1.0F
M	0617	1001	1.3E
	1308	1657	2.4F
◐	2051	2306	1.1E
17	0223	0453	1.2F
Tu	0739	1058	1.3E
	1355	1738	2.6F
	2114	2352	1.4E
18	0318	0558	1.6F
W	0848	1151	1.3E
	1437	1814	2.7F
	2134		
19		0032	1.7E
Th	0403	0649	2.0F
	0945	1238	1.4E
	1515	1849	2.9F
	2155		
20		0107	2.0E
F	0443	0734	2.4F
	1036	1321	1.4E
	1550	1925	3.0F
	2220		
21		0136	2.3E
Sa	0521	0817	2.7F
	1124	1402	1.3E
	1627	2003	3.1F
	2249		
22		0157	2.6E
Su	0558	0900	3.0F
	1211	1444	1.3E
	1704	2043	3.1F
	2322		
23		0217	2.8E
M	0637	0945	3.2F
	1259	1527	1.2E
○	1744	2125	3.1F
	2359		
24		0248	3.1E
Tu	0718	1031	3.4F
	1350	1614	1.1E
	1827	2210	3.0F
25	0039	0327	3.2E
W	0802	1118	3.5F
	1444	1706	1.0E
	1913	2257	2.9F
26	0120	0413	3.2E
Th	0848	1208	3.5F
	1540	1803	0.9E
	2006	2346	2.6F
27	0211	0502	3.0E
F	0938	1300	3.4F
	1640	1906	0.9E
	2108		
28		0040	2.3F
Sa	0305	0559	2.6E
	1031	1356	3.3F
	1742	2012	0.9E
	2223		
29		0142	2.0F
Su	0409	0720	2.2E
	1128	1455	3.2F
	1840	2115	1.1E
	2359		
30		0254	1.8F
M	0527	0905	1.9E
	1228	1555	3.2F
◑	1932	2214	1.4E
31	0136	0414	1.8F
Tu	0657	1018	1.7E
	1330	1653	3.2F
	2017	2311	1.8E

June

Day	Slack h m	Maximum h m	knots
1	0252	0534	2.1F
W	0826	1123	1.6E
	1427	1747	3.3F
	2058		
2		0003	2.1E
Th	0354	0641	2.5F
	0943	1222	1.5E
	1519	1836	3.3F
	2136		
3		0050	2.4E
F	0448	0737	2.9F
	1049	1315	1.4E
	1607	1920	3.2F
	2214		
4		0132	2.6E
Sa	0537	0827	3.1F
	1147	1405	1.3E
	1650	2002	3.1F
	2251		
5		0212	2.6E
Su	0622	0916	3.2F
	1242	1454	1.2E
	1730	2044	3.0F
	2328		
6		0250	2.6E
M	0705	1002	3.2F
	1335	1543	1.1E
●	1809	2125	2.8F
7	0004	0329	2.6E
Tu	0745	1046	3.2F
	1427	1632	1.0E
	1848	2206	2.5F
8	0041	0406	2.5E
W	0822	1128	3.1F
	1519	1721	0.9E
	1928	2247	2.3F
9	0119	0443	2.4E
Th	0857	1209	3.0F
	1610	1810	0.8E
	2011	2328	2.0F
10	0157	0519	2.2E
F	0929	1249	2.9F
	1700	1900	0.8F
	2100		
11		0012	1.8F
Sa	0238	0559	2.0E
	1000	1331	2.7F
	1748	1952	0.8E
	2155		
12		0100	1.5F
Su	0324	0656	1.7E
	1033	1414	2.6F
	1832	2044	0.8E
	2301		
13		0156	1.2F
M	0420	0811	1.5E
	1109	1459	2.5F
	1909	2134	1.0E
14	0021	0300	1.1F
Tu	0533	0916	1.2E
	1150	1545	2.5F
◐	1939	2223	1.2E
15	0141	0411	1.2F
W	0659	1015	1.1E
	1235	1632	2.5F
	2006	2308	1.4E

Day	Slack h m	Maximum h m	knots
16	0244	0523	1.4F
Th	0819	1110	1.0E
	1322	1718	2.6F
	2033	2350	1.8E
17	0336	0623	1.8F
F	0927	1203	1.0E
	1408	1803	2.7F
	2103		
18		0027	2.1E
Sa	0422	0714	2.3F
	1025	1252	1.0E
	1454	1847	2.8F
	2136		
19		0058	2.4E
Su	0504	0800	2.7F
	1118	1338	1.0E
	1540	1931	2.9F
	2213		
20		0125	2.7E
M	0546	0845	3.1F
	1208	1425	1.0E
	1628	2016	3.0F
	2253		
21		0156	3.0E
Tu	0628	0931	3.4F
	1257	1514	1.0E
○	1718	2103	3.0F
	2336		
22		0235	3.2E
W	0711	1017	3.6F
	1345	1606	1.0E
	1811	2152	3.0F
23	0021	0321	3.3E
Th	0755	1104	3.8F
	1434	1658	1.0E
	1906	2242	2.9F
24	0109	0411	3.2E
F	0839	1150	3.9F
	1523	1751	1.1E
	2006	2333	2.7F
25	0200	0505	3.0E
Sa	0923	1238	3.8F
	1611	1846	1.2E
	2111		
26		0027	2.4F
Su	0256	0808	2.0E
	1008	1328	3.6F
	1700	1944	1.3E
	2225		
27		0128	2.1F
M	0400	0725	2.1E
	1054	1420	3.4F
	1751	2042	1.4E
	2351		
28		0238	1.8F
Tu	0517	0843	1.7E
	1144	1516	3.2F
◑	1841	2140	1.6E
29	0120	0358	1.8F
W	0649	0951	1.4E
	1238	1612	3.0F
	1931	2237	1.8E
30	0237	0521	2.0F
Th	0824	1056	1.1E
	1337	1709	2.9F
	2017	2333	2.0E

Time meridian 120° W. 0000 is midnight. 1200 is noon. Times are not adjusted for Daylight Saving Time.

Golden Gate Bridge, Calif., 2005

F–Flood, Dir. 055° True E–Ebb, Dir. 237° True

July

Day	Slack h m	Maximum h m	knots
1 F	0342	0631	2.3F
	0947	1157	1.0E
	1434	1803	2.8F
	2101		
2 Sa		0025	2.2E
	0437	0727	2.7F
	1054	1254	1.0E
	1528	1852	2.8F
	2143		
3 Su		0112	2.4E
	0525	0817	2.9F
	1151	1345	0.9E
	1618	1936	2.7F
	2223		
4 M		0154	2.5E
	0608	0903	3.1F
	1241	1435	0.9E
	1704	2019	2.7F
	2303		
5 Tu		0235	2.5E
	0648	0946	3.2F
	1328	1523	0.9E
	1748	2101	2.6F
	2341		
6 W ●		0315	2.6E
	0724	1026	3.3F
	1412	1611	1.0E
	1831	2143	2.5F
7 Th	0020	0356	2.6E
	0758	1104	3.3F
	1453	1658	1.0E
	1913	2225	2.3F
8 F	0058	0435	2.5E
	0828	1138	3.2F
	1530	1742	1.0E
	1956	2306	2.2F
9 Sa	0138	0512	2.4E
	0855	1211	3.2F
	1605	1827	1.0E
	2041	2349	2.0F
10 Su	0219	0548	2.2E
	0921	1246	3.0F
	1637	1912	1.0E
	2128		
11 M		0034	1.7F
	0304	0625	1.9E
	0949	1322	2.9F
	1707	1957	1.1E
	2220		
12 Tu		0124	1.5F
	0355	0715	1.5E
	1020	1403	2.7F
	1740	2043	1.1E
	2321		
13 W		0222	1.3F
	0459	0824	1.2E
	1055	1447	2.5F
	1815	2129	1.2E
14 Th ◐	0039	0330	1.2F
	0622	0930	0.9E
	1135	1536	2.4F
	1853	2215	1.4E
15 F	0159	0446	1.4F
	0754	1032	0.7E
	1222	1628	2.4F
	1934	2301	1.7E
16 Sa	0304	0557	1.8F
	0914	1132	0.7E
	1315	1722	2.5F
	2016	2346	2.0E
17 Su	0358	0654	2.2F
	1020	1228	0.7E
	1412	1815	2.6F
	2100		
18 M		0029	2.4E
	0446	0743	2.7F
	1113	1320	0.8E
	1511	1907	2.8F
	2146		
19 Tu		0110	2.7E
	0531	0829	3.2F
	1200	1410	0.9E
	1611	1957	2.9F
	2233		
20 W		0154	3.0E
	0615	0914	3.6F
	1244	1501	1.0E
	1712	2048	3.1F
	2322		
21 Th ○		0241	3.2E
	0657	0959	3.9F
	1326	1551	1.2E
	1811	2139	3.1F
22 F	0012	0331	3.3E
	0739	1044	4.1F
	1408	1640	1.4E
	1910	2230	3.1F
23 Sa	0103	0424	3.1E
	0821	1127	4.1F
	1449	1728	1.5E
	2009	2321	2.9F
24 Su	0156	0516	2.9E
	0901	1211	4.0F
	1531	1816	1.6E
	2110		
25 M		0015	2.6F
	0253	0612	2.5E
	0940	1256	3.7F
	1614	1906	1.7E
	2215		
26 Tu		0113	2.3F
	0356	0714	2.0E
	1020	1344	3.3F
	1659	2001	1.6E
	2330		
27 W ◑		0220	1.9F
	0510	0821	1.5E
	1101	1436	2.9F
	1749	2059	1.6E
28 Th	0055	0340	1.8F
	0644	0926	1.0E
	1147	1531	2.6F
	1843	2159	1.6E
29 F	0216	0506	1.9F
	0827	1031	0.8E
	1245	1631	2.4F
	1937	2301	1.7E
30 Sa	0324	0618	2.2F
	0950	1135	0.7E
	1354	1731	2.3F
	2030		
31 Su		0002	1.9E
	0419	0714	2.5F
	1053	1234	0.7E
	1502	1827	2.3F
	2118		

August

Day	Slack h m	Maximum h m	knots
1 M		0054	2.1E
	0506	0801	2.9F
	1142	1327	0.8E
	1601	1915	2.4F
	2202		
2 Tu		0138	2.3E
	0546	0843	3.1F
	1224	1415	0.9E
	1652	1959	2.4F
	2244		
3 W		0219	2.4E
	0623	0922	3.2F
	1302	1501	1.0E
	1738	2042	2.5F
	2324		
4 Th ●		0259	2.5E
	0655	0958	3.3F
	1336	1546	1.2E
	1820	2123	2.5F
5 F	0004	0339	2.6E
	0725	1030	3.4F
	1406	1628	1.3E
	1859	2204	2.5F
6 Sa	0043	0417	2.5E
	0752	1101	3.4F
	1434	1708	1.3E
	1938	2245	2.4F
7 Su	0123	0454	2.4E
	0817	1131	3.3F
	1459	1745	1.4E
	2017	2326	2.3F
8 M	0205	0527	2.2E
	0841	1203	3.2F
	1524	1820	1.4E
	2057		
9 Tu		0009	2.1F
	0249	0556	1.9E
	0908	1237	3.0F
	1552	1848	1.4E
	2140		
10 W		0056	1.9F
	0338	0628	1.5E
	0937	1316	2.8F
	1623	1903	1.4E
	2229		
11 Th		0150	1.6F
	0438	0718	1.1E
	1010	1400	2.5F
	1701	1935	1.4E
	2334		
12 F ◐		0255	1.5F
	0558	0847	0.7E
	1048	1450	2.3F
	1746	2033	1.5E
13 Sa	0103	0411	1.5F
	0737	1001	0.5E
	1136	1548	2.2F
	1840	2150	1.7E
14 Su	0227	0530	1.9F
	0908	1109	0.5E
	1239	1650	2.2F
	1937	2307	2.0E
15 M	0331	0632	2.4F
	1012	1210	0.6E
	1352	1751	2.4F
	2033		
16 Tu		0011	2.3E
	0423	0723	2.9F
	1059	1305	0.8E
	1507	1849	2.6F
	2128		
17 W		0105	2.7E
	0510	0808	3.4F
	1138	1355	1.1E
	1617	1944	2.9F
	2222		
18 Th		0155	3.0E
	0554	0852	3.8F
	1216	1443	1.4E
	1721	2036	3.1F
	2315		
19 F ○		0245	3.1E
	0636	0935	4.1F
	1253	1530	1.7E
	1818	2128	3.3F
20 Sa	0008	0334	3.1E
	0716	1018	4.2F
	1330	1615	1.9E
	1913	2220	3.3F
21 Su	0101	0424	2.9E
	0756	1100	4.1F
	1407	1658	2.0E
	2006	2311	3.1F
22 M	0155	0513	2.6E
	0833	1141	3.9F
	1446	1740	2.0E
	2101		
23 Tu		0002	2.9F
	0252	0603	2.2E
	0910	1224	3.5F
	1526	1823	2.0E
	2158		
24 W		0058	2.5F
	0354	0659	1.6E
	0945	1308	3.1F
	1609	1912	1.8E
	2304		
25 Th		0202	2.1F
	0509	0800	1.2E
	1022	1357	2.6F
	1657	2012	1.6E
26 F ◑	0024	0321	1.8F
	0647	0905	0.8E
	1104	1452	2.2F
	1753	2119	1.5E
27 Sa	0146	0448	1.8F
	0831	1010	0.6E
	1203	1554	1.9F
	1856	2228	1.5E
28 Su	0256	0559	2.1F
	0944	1114	0.5E
	1332	1702	1.8F
	1959	2336	1.6E
29 M	0351	0653	2.5F
	1037	1215	0.7E
	1455	1805	1.9F
	2055		
30 Tu		0032	1.8E
	0436	0737	2.8F
	1117	1307	0.9E
	1556	1857	2.1F
	2144		
31 W		0117	2.1E
	0513	0815	3.1F
	1151	1352	1.1E
	1645	1942	2.3F
	2228		

September

Day	Slack h m	Maximum h m	knots
1 Th		0157	2.3E
	0547	0849	3.2F
	1220	1434	1.3E
	1728	2023	2.4F
	2310		
2 F		0236	2.4E
	0617	0920	3.3F
	1246	1514	1.5E
	1807	2104	2.6F
	2350		
3 Sa ●		0315	2.4E
	0645	0950	3.3F
	1310	1552	1.6E
	1843	2144	2.6F
4 Su	0031	0353	2.4E
	0710	1019	3.3F
	1332	1627	1.7E
	1917	2225	2.7F
5 M	0112	0429	2.2E
	0734	1050	3.3F
	1356	1655	1.8E
	1951	2305	2.6F
6 Tu	0154	0502	2.0E
	0800	1123	3.2F
	1421	1707	1.8E
	2027	2347	2.5F
7 W	0238	0530	1.7E
	0827	1158	3.0F
	1450	1718	1.9E
	2106		
8 Th		0033	2.3F
	0328	0558	1.3E
	0857	1237	2.7F
	1523	1749	1.9E
	2153		
9 F		0125	2.0F
	0428	0643	0.9E
	0931	1321	2.4F
	1603	1832	1.9E
	2251		
10 Sa		0229	1.8F
	0548	0820	0.6E
	1012	1414	2.2F
	1653	1929	1.8E
11 Su ◐	0013	0343	1.9F
	0730	0942	0.4E
	1108	1516	2.0F
	1755	2047	1.8E
12 M	0146	0500	2.2F
	0857	1051	0.5E
	1225	1625	2.0F
	1905	2241	1.9E
13 Tu	0256	0605	2.7F
	0949	1153	0.8E
	1359	1734	2.2F
	2014	2357	2.2E
14 W	0352	0657	3.2F
	1028	1248	1.1E
	1524	1837	2.5F
	2118		
15 Th		0055	2.6E
	0441	0742	3.6F
	1104	1336	1.5E
	1631	1934	2.9F
	2217		
16 F		0146	2.8E
	0525	0825	3.9F
	1138	1421	1.9E
	1728	2027	3.2F
	2313		
17 Sa ○		0235	2.8E
	0607	0907	4.1F
	1213	1504	2.2E
	1820	2119	3.4F
18 Su	0008	0324	2.7E
	0647	0948	4.1F
	1249	1546	2.3E
	1910	2210	3.4F
19 M	0102	0413	2.5E
	0726	1030	3.9F
	1325	1625	2.4E
	1959	2300	3.3F
20 Tu	0157	0501	2.1E
	0802	1110	3.6F
	1403	1703	2.4E
	2049	2351	3.0F
21 W	0254	0550	1.7E
	0837	1152	3.2F
	1442	1739	2.2E
	2141		
22 Th		0044	2.6F
	0357	0644	1.3E
	0911	1235	2.7F
	1523	1818	1.9E
	2239		
23 F		0146	2.2F
	0514	0743	0.9E
	0948	1321	2.2F
	1609	1912	1.6E
	2348		
24 Sa ◑		0300	2.0F
	0653	0846	0.6E
	1033	1415	1.8F
	1703	2037	1.4E
25 Su	0104	0421	2.0F
	0822	0950	0.5E
	1142	1519	1.5F
	1809	2153	1.3E
26 M	0212	0530	2.2F
	0922	1053	0.6E
	1330	1631	1.4F
	1923	2300	1.4E
27 Tu	0307	0622	2.5F
	1005	1152	0.8E
	1451	1742	1.5F
	2029	2358	1.6E
28 W	0351	0704	2.8F
	1039	1242	1.1E
	1548	1838	1.8F
	2124		
29 Th		0046	1.8E
	0428	0738	3.0F
	1107	1324	1.4E
	1634	1923	2.1F
	2212		
30 F		0128	2.0E
	0500	0809	3.1F
	1130	1403	1.7E
	1714	2004	2.4F
	2256		

Time meridian 120° W. 0000 is midnight. 1200 is noon. Times are not adjusted for Daylight Saving Time.

Golden Gate Bridge, Calif., 2005

F–Flood, Dir. 055° True E–Ebb, Dir. 237° True

October

Day	Slack h m	Maximum h m	knots
1 Sa		0208	2.1E
	0530	0837	3.2F
	1151	1439	1.9E
	1750	2044	2.6F
	2338		
2 Su		0247	2.1E
	0557	0906	3.2F
	1212	1512	2.0E
	1824	2124	2.8F
3 M ●	0020	0326	2.0E
	0622	0937	3.2F
	1235	1539	2.1E
	1856	2205	2.9F
4 Tu	0102	0403	1.8E
	0649	1011	3.2F
	1300	1553	2.2E
	1929	2246	2.9F
5 W	0145	0438	1.6E
	0717	1046	3.1F
	1328	1606	2.3E
	2005	2329	2.8F
6 Th	0232	0509	1.3E
	0748	1124	2.9F
	1401	1635	2.5E
	2044		
7 F		0015	2.6F
	0324	0543	1.0E
	0822	1205	2.7F
	1438	1713	2.5E
	2131		
8 Sa		0107	2.5F
	0427	0638	0.7E
	0901	1251	2.4F
	1521	1759	2.3E
	2228		
9 Su		0208	2.3F
	0546	0812	0.5E
	0950	1346	2.0F
	1615	1857	2.1E
	2341		
10 M ◐		0317	2.3F
	0717	0927	0.5E
	1059	1452	1.8F
	1721	2018	1.9E
11 Tu	0104	0428	2.6F
	0826	1033	0.7E
	1235	1606	1.8F
	1841	2222	1.9E
12 W	0215	0532	3.0F
	0911	1133	1.1E
	1419	1720	2.1F
	2001	2337	2.1E
13 Th	0314	0625	3.4F
	0948	1227	1.5E
	1534	1828	2.5F
	2112		
14 F		0037	2.3E
	0405	0711	3.7F
	1024	1313	2.0E
	1634	1926	2.9F
	2215		
15 Sa		0129	2.4E
	0451	0754	3.9F
	1058	1356	2.3E
	1727	2019	3.3F
	2314		

Day	Slack h m	Maximum h m	knots
16 Su		0219	2.3E
	0534	0836	3.9F
	1134	1437	2.6E
	1816	2111	3.5F
17 M ○	0011	0308	2.2E
	0615	0918	3.8F
	1210	1516	2.7E
	1904	2201	3.5F
18 Tu	0106	0357	1.9E
	0653	0959	3.5F
	1246	1554	2.7E
	1951	2251	3.4F
19 W	0202	0446	1.6E
	0730	1041	3.2F
	1324	1630	2.5E
	2038	2340	3.1F
20 Th	0301	0536	1.3E
	0805	1122	2.8F
	1402	1704	2.3E
	2126		
21 F		0032	2.8F
	0406	0629	1.0E
	0842	1204	2.4F
	1442	1738	2.0E
	2215		
22 Sa		0128	2.5F
	0522	0726	0.7E
	0922	1250	1.9F
	1526	1822	1.7E
	2309		
23 Su		0232	2.2F
	0645	0827	0.6E
	1015	1342	1.5F
	1616	1953	1.4E
24 M ◑	0009	0342	2.2F
	0754	0928	0.6E
	1136	1444	1.2F
	1721	2114	1.3E
25 Tu	0110	0445	2.3F
	0845	1026	0.7E
	1320	1557	1.1F
	1840	2219	1.3E
26 W	0204	0537	2.5F
	0922	1121	1.0E
	1436	1713	1.3F
	1958	2318	1.4E
27 Th	0249	0618	2.7F
	0952	1210	1.3E
	1531	1815	1.6F
	2101		
28 F		0009	1.5E
	0328	0652	2.8F
	1015	1251	1.6E
	1616	1903	2.0F
	2154		
29 Sa		0055	1.7E
	0402	0721	3.0F
	1035	1328	1.9E
	1655	1944	2.3F
	2242		
30 Su		0137	1.7E
	0432	0751	3.0F
	1056	1402	2.1E
	1731	2025	2.6F
	2327		
31 M		0218	1.8E
	0502	0823	3.1F
	1119	1430	2.3E
	1805	2105	2.8F

November

Day	Slack h m	Maximum h m	knots
1 Tu ●	0010	0258	1.5E
	0532	0858	3.1F
	1145	1449	2.4E
	1839	2147	3.0F
2 W	0055	0337	1.4E
	0603	0935	3.1F
	1214	1503	2.6E
	1914	2229	3.1F
3 Th	0141	0417	1.2E
	0637	1014	3.0F
	1247	1530	2.8E
	1952	2313	3.1F
4 F	0230	0457	1.0E
	0713	1056	2.8F
	1324	1607	2.9E
	2034		
5 Sa		0000	3.0F
	0324	0544	0.8E
	0754	1140	2.6F
	1406	1650	2.8E
	2121		
6 Su		0051	3.0F
	0426	0646	0.7E
	0843	1229	2.3F
	1453	1739	2.7E
	2214		
7 M		0148	2.9F
	0534	0759	0.6E
	0944	1326	2.0F
	1549	1839	2.3E
	2315		
8 Tu ◐		0250	2.9F
	0644	0907	0.8E
	1104	1434	1.8F
	1659	2009	2.0E
9 W	0023	0353	3.0F
	0741	1009	1.0E
	1250	1549	1.7F
	1824	2201	1.9E
10 Th	0130	0454	3.2F
	0826	1107	1.4E
	1423	1708	2.0F
	1951	2313	1.9E
11 F	0231	0549	3.4F
	0906	1200	1.9E
	1532	1818	2.4F
	2109		
12 Sa		0015	1.9E
	0324	0638	3.6F
	0943	1248	2.3E
	1630	1918	2.9F
	2217		
13 Su		0109	1.9E
	0413	0723	3.6F
	1021	1331	2.6E
	1722	2012	3.2F
	2318		
14 M		0200	1.8E
	0458	0806	3.6F
	1058	1411	2.8E
	1810	2103	3.4F
15 Tu ○	0016	0250	1.6E
	0540	0849	3.4F
	1135	1451	2.8E
	1857	2153	3.4F

Day	Slack h m	Maximum h m	knots
16 W	0113	0341	1.4E
	0621	0931	3.2F
	1213	1530	2.7E
	1943	2242	3.4F
17 Th	0210	0431	1.2E
	0659	1013	2.9F
	1251	1608	2.6E
	2027	2329	3.2F
18 F	0308	0522	1.0E
	0738	1055	2.6F
	1330	1644	2.4E
	2109		
19 Sa		0016	3.0F
	0409	0613	0.8E
	0819	1138	2.2F
	1409	1721	2.2E
	2150		
20 Su		0105	2.8F
	0512	0707	0.7E
	0906	1222	1.8F
	1451	1805	1.9E
	2229		
21 M		0156	2.6F
	0614	0803	0.7E
	1004	1312	1.4F
	1539	1916	1.6E
	2309		
22 Tu		0250	2.5F
	0709	0858	0.7E
	1119	1410	1.2F
	1638	2033	1.4E
	2353		
23 W ◑		0343	2.4F
	0753	0952	0.9E
	1250	1519	1.0F
	1754	2137	1.3E
24 Th	0040	0432	2.4F
	0828	1043	1.1E
	1407	1634	1.1F
	1919	2235	1.2E
25 F	0127	0516	2.5F
	0855	1131	1.4E
	1505	1744	1.4F
	2034	2329	1.2E
26 Sa	0211	0554	2.6F
	0917	1214	1.7E
	1552	1838	1.8F
	2135		
27 Su		0020	1.2E
	0251	0630	2.7F
	0940	1251	2.0E
	1634	1924	2.2F
	2229		
28 M		0106	1.2E
	0329	0707	2.8F
	1005	1324	2.2E
	1712	2006	2.5F
	2318		
29 Tu		0149	1.2E
	0405	0744	2.9F
	1033	1350	2.5E
	1749	2047	2.8F
30 W	0004	0232	1.1E
	0443	0823	2.9F
	1104	1408	2.7E
	1826	2130	3.1F

December

Day	Slack h m	Maximum h m	knots
1 Th ●	0050	0315	1.0E
	0522	0905	3.0F
	1139	1432	2.9E
	1904	2214	3.2F
2 F	0138	0401	0.9E
	0604	0948	2.9F
	1218	1508	3.1E
	1944	2259	3.4F
3 Sa	0227	0448	0.9E
	0649	1034	2.8F
	1300	1550	3.1E
	2026	2345	3.5F
4 Su	0318	0539	0.9E
	0740	1122	2.6F
	1346	1638	3.1E
	2111		
5 M		0033	3.5F
	0410	0635	0.9E
	0838	1213	2.4F
	1437	1730	2.8E
	2158		
6 Tu		0124	3.4F
	0504	0737	0.9E
	0946	1311	2.1F
	1535	1832	2.4E
	2249		
7 W		0219	3.3F
	0558	0839	1.1E
	1109	1417	1.8F
	1646	2010	2.0E
	2344		
8 Th ◐		0317	3.3F
	0650	0938	1.4E
	1247	1533	1.8F
	1812	2138	1.7E
9 F	0043	0415	3.2F
	0739	1035	1.7E
	1414	1654	2.0F
	1944	2247	1.5E
10 Sa	0143	0511	3.2F
	0824	1130	2.0E
	1523	1809	2.3F
	2109	2351	1.4E
11 Su	0241	0604	3.2F
	0906	1222	2.4E
	1022	1911	2.8F
	2222		
12 M		0049	1.3E
	0334	0653	3.2F
	0947	1308	2.6E
	1714	2005	3.1F
	2326		
13 Tu		0142	1.3E
	0424	0739	3.2F
	1027	1352	2.7E
	1802	2055	3.3F
14 W	0024	0233	1.2E
	0510	0823	3.0F
	1107	1434	2.7E
	1848	2143	3.4F
15 Th ○	0118	0324	1.1E
	0554	0907	2.9F
	1147	1515	2.7E
	1930	2229	3.4F

Day	Slack h m	Maximum h m	knots
16 F	0211	0415	1.0E
	0637	0950	2.6F
	1226	1556	2.6E
	2010	2313	3.3F
17 Sa	0302	0504	0.9E
	0720	1033	2.4F
	1305	1637	2.5E
	2047	2354	3.2F
18 Su	0350	0552	0.9E
	0804	1115	2.2F
	1345	1716	2.3E
	2119		
19 M		0034	3.1F
	0436	0640	0.9E
	0852	1158	1.9F
	1427	1758	2.1E
	2149		
20 Tu		0113	2.9F
	0520	0730	0.9E
	0944	1245	1.6F
	1512	1848	1.8E
	2218		
21 W		0154	2.7F
	0602	0820	0.9E
	1045	1337	1.3F
	1606	1951	1.5E
	2250		
22 Th		0236	2.6F
	0641	0910	1.0E
	1159	1439	1.1F
	1714	2054	1.3E
	2327		
23 F ◑		0321	2.5F
	0715	0959	1.2E
	1319	1549	1.1F
	1838	2153	1.1E
24 Sa	0009	0408	2.4F
	0746	1046	1.4E
	1427	1705	1.3F
	2005	2250	0.9E
25 Su	0055	0455	2.4F
	0815	1132	1.6E
	1523	1811	1.7F
	2118	2345	0.9E
26 M	0143	0541	2.5F
	0846	1213	1.9E
	1610	1902	2.1F
	2220		
27 Tu		0036	0.8E
	0230	0626	2.6F
	0919	1248	2.2E
	1653	1947	2.5F
	2312		
28 W		0124	0.8E
	0317	0711	2.7F
	0954	1318	2.5E
	1733	2030	2.8F
29 Th	0000	0210	0.8E
	0404	0755	2.8F
	1033	1345	2.8E
	1813	2114	3.2F
30 F ●	0044	0257	0.9E
	0454	0841	2.9F
	1115	1419	3.0E
	1853	2157	3.5F
31 Sa	0128	0346	0.9E
	0546	0929	2.9F
	1159	1500	3.2E
	1933	2241	3.7F

Time meridian 120° W. 0000 is midnight. 1200 is noon. Times are not adjusted for Daylight Saving Time.

Oakland, Yerba Buena Island, Calif., 2005

F–Flood, Dir. 167° True   E–Ebb, Dir. 338° True

January

Day	Slack h m	Maximum h m	knots
1 Sa		0033	1.3F
	0418	0635	0.9E
	0930	1241	1.2F
	1534	1826	1.3E
	2124		
2 Su		0115	1.3F
	0455	0719	1.0E
	1017	1333	1.1F
	1638	1924	1.1E
	2214		
3 M		0202	1.3F
	0536	0806	1.0E
	1116	1434	1.0F
	1802	2028	0.9E
	2311		
4 Tu		0253	1.3F
	0620	0856	1.2E
	1222	1554	0.9F
	1933	2138	0.7E
5 W	0016	0350	1.2F
	0707	0949	1.3E
	1330	1741	1.0F
	2056	2250	0.7E
6 Th	0125	0449	1.2F
	0756	1042	1.5E
	1435	1906	1.3F
	2205	2359	0.7E
7 F	0235	0549	1.3F
	0848	1135	1.6E
	1535	2008	1.6F
	2303		
8 Sa		0102	0.9E
	0342	0648	1.3F
	0941	1227	1.7E
	1631	2100	1.8F
	2355		
9 Su		0159	1.0E
	0444	0746	1.4F
	1035	1319	1.8E
	1723	2146	2.0F
10 M	0042	0251	1.1E
	0540	0842	1.5F
	1129	1412	1.9E
	1813	2229	2.0F
11 Tu	0128	0339	1.2E
	0632	0934	1.6F
	1222	1504	1.9E
	1900	2310	2.0F
12 W	0212	0424	1.3E
	0721	1022	1.6F
	1314	1555	1.9E
	1946	2348	2.0F
13 Th	0255	0509	1.3E
	0809	1110	1.6F
	1406	1646	1.8E
	2031		
14 F		0026	1.9F
	0338	0554	1.3E
	0858	1200	1.5F
	1501	1739	1.6E
	2115		
15 Sa		0103	1.7F
	0422	0640	1.2E
	0951	1254	1.3F
	1602	1836	1.4E
	2201		
16 Su		0142	1.6F
	0508	0729	1.2E
	1050	1357	1.1F
	1713	1939	1.1E
	2251		
17 M		0223	1.5F
	0554	0819	1.1E
	1156	1525	1.0F
	1833	2049	0.9E
	2346		
18 Tu		0309	1.3F
	0641	0910	1.1E
	1305	1712	1.1F
	1953	2208	0.8E
19 W	0047	0401	1.2F
	0728	1002	1.1E
	1410	1823	1.3F
	2105	2323	0.8E
20 Th	0151	0456	1.2F
	0814	1052	1.1E
	1507	1920	1.5F
	2204		
21 F		0024	0.9E
	0253	0549	1.1F
	0900	1137	1.2E
	1554	2009	1.7F
	2254		
22 Sa		0117	1.0E
	0350	0639	1.1F
	0945	1218	1.2E
	1634	2052	1.7F
	2338		
23 Su		0203	1.1E
	0441	0727	1.2F
	1028	1259	1.3E
	1708	2130	1.7F
24 M	0017	0242	1.1E
	0525	0811	1.3F
	1109	1340	1.3E
	1738	2200	1.6F
25 Tu	0053	0315	1.1E
	0603	0852	1.3F
	1147	1420	1.4E
	1806	2221	1.6F
26 W	0126	0342	1.1E
	0637	0931	1.4F
	1224	1501	1.5E
	1833	2235	1.5F
27 Th	0155	0408	1.1E
	0706	1008	1.4F
	1301	1543	1.6E
	1902	2253	1.5F
28 F	0221	0436	1.1E
	0733	1045	1.4F
	1340	1625	1.5E
	1933	2317	1.5F
29 Sa	0246	0508	1.2E
	0801	1122	1.3F
	1422	1710	1.4E
	2007	2349	1.5F
30 Su	0313	0544	1.2E
	0836	1204	1.2F
	1512	1800	1.2E
	2045		
31 M		0028	1.4F
	0345	0627	1.2E
	0920	1254	1.1F
	1620	1857	0.9E
	2130		

February

Day	Slack h m	Maximum h m	knots
1 Tu		0115	1.3F
	0425	0715	1.3E
	1017	1359	0.9F
	1756	2005	0.7E
	2228		
2 W		0210	1.2F
	0515	0809	1.3E
	1128	1546	0.9F
	1935	2122	0.6E
	2347		
3 Th		0314	1.1F
	0615	0908	1.3E
	1251	1759	1.1F
	2054	2245	0.6E
4 F	0116	0424	1.1F
	0722	1010	1.4E
	1413	1903	1.5F
	2156	2358	0.8E
5 Sa	0235	0534	1.2F
	0829	1113	1.5E
	1523	1955	1.8F
	2247		
6 Su		0057	1.0E
	0341	0639	1.3F
	0932	1212	1.6E
	1622	2041	2.0F
	2332		
7 M		0148	1.2E
	0438	0738	1.5F
	1030	1308	1.8E
	1714	2124	2.1F
8 Tu	0015	0233	1.4E
	0527	0832	1.6F
	1124	1402	1.9E
	1800	2202	2.1F
9 W	0057	0315	1.4E
	0613	0921	1.7F
	1215	1452	1.9E
	1842	2237	2.1F
10 Th	0136	0354	1.5E
	0657	1007	1.7F
	1304	1541	1.9E
	1921	2308	2.0F
11 F	0215	0432	1.5E
	0741	1051	1.6F
	1354	1628	1.7E
	1959	2337	1.8F
12 Sa	0252	0511	1.4E
	0825	1137	1.5F
	1446	1718	1.5E
	2037		
13 Su		0007	1.7F
	0329	0551	1.4E
	0912	1228	1.3F
	1546	1812	1.2E
	2118		
14 M		0042	1.5F
	0407	0634	1.3E
	1004	1331	1.1F
	1659	1914	0.9E
	2208		
15 Tu		0123	1.3F
	0449	0720	1.1E
	1105	1514	1.0F
	1823	2029	0.7E
	2310		
16 W		0213	1.1F
	0538	0811	1.0E
	1216	1701	1.1F
	1942	2157	0.7E
17 Th	0024	0313	1.0F
	0635	0906	1.0E
	1327	1805	1.3F
	2049	2313	0.8E
18 F	0137	0420	1.0F
	0736	1006	0.9E
	1429	1857	1.5F
	2142		
19 Sa		0010	1.0E
	0242	0525	1.0F
	0834	1105	1.0E
	1521	1942	1.7F
	2226		
20 Su		0058	1.1E
	0337	0621	1.1F
	0927	1156	1.1E
	1605	2021	1.7F
	2306		
21 M		0139	1.2E
	0423	0710	1.3F
	1014	1241	1.2E
	1643	2055	1.7F
	2343		
22 Tu		0213	1.3E
	0502	0753	1.4F
	1056	1323	1.3E
	1715	2121	1.6F
23 W	0015	0240	1.2E
	0536	0833	1.5F
	1134	1404	1.4E
	1744	2138	1.6F
24 Th	0044	0302	1.2E
	0605	0910	1.5F
	1212	1443	1.5E
	1810	2154	1.5F
25 F	0109	0327	1.3E
	0631	0946	1.5F
	1249	1524	1.5E
	1836	2214	1.5F
26 Sa	0130	0355	1.3E
	0656	1022	1.5F
	1329	1605	1.4E
	1904	2240	1.5F
27 Su	0151	0427	1.4E
	0724	1058	1.4F
	1415	1650	1.3E
	1935	2312	1.5F
28 M	0217	0503	1.5E
	0759	1139	1.3F
	1512	1740	1.0E
	2013	2352	1.4F

March

Day	Slack h m	Maximum h m	knots
1 Tu	0250	0546	1.5E
	0843	1230	1.1F
	1630	1841	0.7E
	2102		
2 W		0042	1.2F
	0334	0636	1.5E
	0938	1349	1.0F
	1806	1954	0.5E
	2212		
3 Th		0143	1.1F
	0430	0734	1.4E
	1049	1620	1.0F
	1933	2118	0.5E
	2350		
4 F		0254	1.0F
	0543	0838	1.3E
	1224	1748	1.3F
	2039	2245	0.7E
5 Sa	0121	0411	1.0F
	0705	0948	1.3E
	1357	1843	1.6F
	2133	2350	1.0E
6 Su	0233	0526	1.2F
	0820	1058	1.4E
	1508	1930	1.9F
	2219		
7 M		0041	1.2E
	0332	0630	1.4F
	0926	1201	1.6E
	1605	2013	2.0F
	2302		
8 Tu		0126	1.4E
	0423	0727	1.6F
	1023	1257	1.7E
	1652	2052	2.1F
	2342		
9 W		0206	1.5E
	0509	0819	1.7F
	1115	1349	1.8E
	1734	2127	2.0F
10 Th	0021	0244	1.6E
	0552	0907	1.7F
	1205	1437	1.8E
	1811	2157	2.0F
11 F	0057	0319	1.6E
	0634	0952	1.7F
	1254	1523	1.7E
	1847	2222	1.8F
12 Sa	0130	0354	1.6E
	0714	1036	1.6F
	1344	1609	1.5E
	1922	2248	1.7F
13 Su	0201	0428	1.6E
	0754	1120	1.4F
	1438	1658	1.3E
	1959	2317	1.6F
14 M	0232	0504	1.5E
	0833	1210	1.3F
	1540	1752	1.0E
	2043	2354	1.4F
15 Tu	0304	0544	1.4E
	0915	1314	1.1F
	1652	1857	0.8E
	2138		
16 W		0039	1.2F
	0343	0628	1.2E
	1003	1459	1.0F
	1809	2015	0.6E
	2249		
17 Th		0134	1.0F
	0434	0719	1.0E
	1106	1633	1.1F
	1919	2140	0.7E
18 F	0008	0238	0.9F
	0543	0818	0.9E
	1222	1734	1.3F
	2017	2250	0.9E
19 Sa	0120	0348	0.9F
	0659	0924	0.8E
	1335	1822	1.5F
	2106	2343	1.1E
20 Su	0221	0457	1.0F
	0807	1031	0.9E
	1435	1903	1.6F
	2149		
21 M		0027	1.2E
	0311	0555	1.2F
	0904	1130	1.0E
	1524	1939	1.6F
	2227		
22 Tu		0103	1.3E
	0354	0645	1.3F
	0953	1218	1.2E
	1604	2008	1.6F
	2301		
23 W		0132	1.3E
	0431	0729	1.4F
	1037	1301	1.3E
	1639	2029	1.6F
	2331		
24 Th		0155	1.3E
	0504	0811	1.5F
	1118	1342	1.4E
	1709	2048	1.5F
	2356		
25 F		0219	1.4E
	0533	0851	1.5F
	1200	1423	1.4E
	1736	2109	1.5F
26 Sa	0018	0246	1.5E
	0600	0930	1.5F
	1243	1505	1.3E
	1804	2136	1.5F
27 Su	0039	0317	1.6E
	0627	1008	1.5F
	1331	1549	1.2E
	1834	2207	1.5F
28 M	0103	0352	1.7E
	0658	1049	1.4F
	1425	1637	1.0E
	1911	2244	1.4F
29 Tu	0134	0431	1.8E
	0735	1136	1.3F
	1530	1732	0.8E
	1958	2329	1.3F
30 W	0213	0516	1.7E
	0821	1241	1.2F
	1647	1837	0.6E
	2102		
31 Th		0025	1.1F
	0302	0609	1.6E
	0918	1419	1.2F
	1806	1954	0.6E
	2229		

Time meridian 120° W. 0000 is midnight. 1200 is noon. Times are not adjusted for Daylight Saving Time.

Oakland, Yerba Buena Island, Calif., 2005

F–Flood, Dir. 167° True E–Ebb, Dir. 338° True

April

Day	Slack h m	Maximum h m	knots
1 F		0130	1.0F
	0407	0710	1.4E
	1033	1610	1.3F
	1915	2117	0.7E
2 Sa	0000	0244	0.9F
	0530	0819	1.3E
	1211	1719	1.5F
	2012	2232	0.9E
3 Su	0117	0401	1.0F
	0657	0933	1.3E
	1337	1811	1.7F
	2101	2329	1.2E
4 M	0220	0515	1.2F
	0812	1045	1.4E
	1442	1856	1.9F
	2146		
5 Tu		0015	1.4E
	0314	0619	1.4F
	0916	1148	1.5E
	1534	1937	2.0F
	2227		
6 W		0056	1.5E
	0404	0715	1.6F
	1012	1243	1.6E
	1618	2013	2.0F
	2306		
7 Th		0134	1.6E
	0449	0809	1.7F
	1105	1332	1.6E
	1658	2044	1.9F
	2341		
8 F		0209	1.7E
	0532	0859	1.7F
	1157	1419	1.5E
	1734	2110	1.8F
9 Sa	0013	0242	1.7E
	0613	0947	1.6F
	1249	1506	1.3E
	1810	2136	1.7F
10 Su	0043	0315	1.7E
	0650	1032	1.5F
	1342	1553	1.2E
	1848	2204	1.5F
11 M	0111	0348	1.6E
	0724	1118	1.4F
	1438	1043	1.0E
	1930	2238	1.4F
12 Tu	0139	0423	1.6E
	0756	1206	1.3F
	1537	1739	0.8E
	2021	2319	1.2F
13 W	0211	0502	1.4E
	0829	1306	1.2F
	1640	1844	0.7E
	2123		
14 Th		0008	1.1F
	0252	0546	1.3E
	0908	1423	1.2F
	1743	1958	0.7E
	2233		
15 F		0105	0.9F
	0347	0639	1.1E
	1002	1542	1.2F
	1843	2110	0.8E
	2316		
16 Sa		0208	0.9F
	0459	0740	0.9E
	1115	1643	1.3F
	1935	2211	0.9E
17 Su	0050	0314	0.9F
	0619	0846	0.9E
	1233	1730	1.4F
	2021	2302	1.1E
18 M	0145	0420	1.0F
	0731	0954	0.9E
	1338	1808	1.4F
	2103	2342	1.2E
19 Tu	0234	0520	1.2F
	0831	1055	1.0E
	1430	1837	1.5F
	2139		
20 W		0014	1.2E
	0317	0613	1.3F
	0925	1147	1.1E
	1513	1902	1.5F
	2211		
21 Th		0040	1.3E
	0356	0703	1.4F
	1015	1234	1.2E
	1551	1927	1.5F
	2239		
22 F		0106	1.4E
	0432	0751	1.4F
	1104	1318	1.2E
	1625	1955	1.5F
	2304		
23 Sa		0135	1.5E
	0504	0839	1.4F
	1154	1404	1.1E
	1658	2026	1.4F
	2328		
24 Su		0208	1.7E
	0535	0926	1.5F
	1247	1451	1.0E
	1734	2102	1.4F
	2354		
25 M		0244	1.8E
	0606	1013	1.5F
	1342	1540	0.9E
	1815	2142	1.4F
26 Tu	0027	0324	1.9E
	0642	1103	1.5F
	1441	1033	0.8E
	1906	2227	1.3F
27 W	0106	0408	1.9E
	0725	1201	1.4F
	1542	1732	0.7E
	2008	2319	1.2F
28 Th	0153	0457	1.8E
	0815	1310	1.4F
	1645	1838	0.7E
	2121		
29 F		0018	1.1F
	0251	0553	1.6E
	0918	1425	1.5F
	1748	1951	0.8E
	2238		
30 Sa		0123	1.0F
	0402	0656	1.4E
	1035	1537	1.5F
	1845	2101	0.9E
	2353		

May

Day	Slack h m	Maximum h m	knots
1 Su		0233	1.0F
	0524	0806	1.3E
	1157	1639	1.7F
	1937	2204	1.1E
2 M	0059	0347	1.1F
	0645	0918	1.3E
	1308	1730	1.8F
	2025	2257	1.3E
3 Tu	0159	0500	1.3F
	0757	1028	1.3E
	1405	1813	1.9F
	2109	2343	1.5E
4 W	0254	0606	1.4F
	0902	1130	1.4E
	1454	1852	1.8F
	2149		
5 Th		0023	1.6E
	0345	0708	1.5F
	1002	1225	1.3E
	1537	1925	1.8F
	2225		
6 F		0059	1.6E
	0433	0807	1.6F
	1059	1315	1.2E
	1618	1954	1.7F
	2259		
7 Sa		0132	1.7E
	0516	0903	1.6F
	1155	1405	1.1E
	1658	2023	1.5F
	2330		
8 Su		0205	1.7E
	0555	0955	1.5F
	1250	1454	1.0E
	1739	2055	1.4F
	2358		
9 M		0238	1.7E
	0628	1042	1.5F
	1344	1545	0.9E
	1825	2130	1.3F
10 Tu	0027	0313	1.6E
	0657	1127	1.4F
	1436	1636	0.8E
	1915	2211	1.2F
11 W	0058	0350	1.6E
	0724	1208	1.4F
	1527	1730	0.8E
	2011	2256	1.1F
12 Th	0136	0430	1.5E
	0755	1249	1.3F
	1617	1827	0.8E
	2109	2345	1.0F
13 F	0222	0516	1.3E
	0835	1331	1.3F
	1708	1926	0.8E
	2209		
14 Sa		0040	1.0F
	0317	0609	1.2E
	0925	1418	1.3F
	1757	2022	0.9E
	2308		
15 Su		0137	1.0F
	0422	0708	1.1E
	1028	1508	1.3F
	1845	2113	0.9E
16 M	0005	0236	1.0F
	0534	0811	1.0E
	1135	1556	1.3F
	1928	2159	1.0E
17 Tu	0059	0336	1.0F
	0646	0914	1.0E
	1236	1638	1.3F
	2008	2239	1.1E
18 W	0149	0438	1.1F
	0753	1016	1.0E
	1329	1716	1.4F
	2043	2312	1.2E
19 Th	0235	0540	1.2F
	0856	1112	1.0E
	1415	1752	1.4F
	2114	2345	1.3E
20 F	0319	0640	1.2F
	0957	1205	1.0E
	1458	1828	1.4F
	2143		
21 Sa		0019	1.5E
	0359	0741	1.3F
	1057	1256	0.9E
	1541	1907	1.4F
	2212		
22 Su		0055	1.7E
	0438	0842	1.4F
	1155	1349	0.8E
	1627	1950	1.3F
	2244		
23 M		0135	1.8E
	0516	0938	1.5F
	1252	1442	0.8E
	1717	2037	1.3F
	2321		
24 Tu		0217	1.9E
	0555	1031	1.6F
	1348	1537	0.8E
	1813	2127	1.2F
25 W	0004	0303	1.9E
	0639	1122	1.7F
	1441	1632	0.8E
	1915	2219	1.2F
26 Th	0053	0352	1.9E
	0728	1214	1.7F
	1534	1730	0.8E
	2019	2313	1.2F
27 F	0149	0445	1.8E
	0824	1307	1.7F
	1626	1831	0.9E
	2123		
28 Sa		0011	1.2F
	0250	0543	1.7E
	0925	1401	1.7F
	1718	1932	1.0E
	2227		
29 Su		0113	1.2F
	0359	0646	1.5E
	1030	1456	1.7F
	1810	2031	1.1E
	2331		
30 M		0217	1.2F
	0513	0752	1.4E
	1133	1549	1.8F
	1859	2127	1.2E
31 Tu	0035	0327	1.2F
	0629	0900	1.3E
	1231	1638	1.8F
	1945	2219	1.4E

June

Day	Slack h m	Maximum h m	knots
1 W	0136	0445	1.2F
	0741	1007	1.2E
	1323	1721	1.7F
	2028	2305	1.5E
2 Th	0234	0600	1.3F
	0850	1111	1.1E
	1411	1758	1.7F
	2107	2346	1.5E
3 F	0329	0709	1.4F
	0956	1209	1.0E
	1458	1833	1.6F
	2144		
4 Sa		0022	1.6E
	0419	0812	1.5F
	1058	1304	0.9E
	1544	1907	1.4F
	2218		
5 Su		0056	1.6E
	0502	0908	1.6F
	1156	1358	0.9E
	1632	1944	1.3F
	2250		
6 M		0130	1.6E
	0540	0958	1.6F
	1249	1451	0.8E
	1723	2025	1.2F
	2323		
7 Tu		0206	1.6E
	0610	1043	1.6F
	1338	1541	0.8E
	1815	2108	1.2F
	2357		
8 W		0244	1.5E
	0637	1121	1.5F
	1423	1628	0.8E
	1906	2152	1.1F
9 Th	0035	0324	1.5E
	0704	1152	1.5F
	1504	1712	0.8E
	1956	2237	1.1F
10 F	0117	0407	1.5E
	0736	1216	1.4F
	1544	1756	0.9E
	2044	2323	1.1F
11 Sa	0203	0454	1.4E
	0816	1241	1.4F
	1625	1839	0.9E
	2131		
12 Su		0011	1.1F
	0252	0544	1.3E
	0901	1314	1.4F
	1706	1922	0.9E
	2219		
13 M		0102	1.1F
	0348	0639	1.2E
	0951	1353	1.4F
	1746	2004	0.9E
	2310		
14 Tu		0155	1.0F
	0450	0737	1.1E
	1044	1435	1.4F
	1825	2046	1.0E
15 W	0003	0252	1.0F
	0602	0837	1.0E
	1137	1520	1.4F
	1902	2128	1.1E
16 Th	0056	0356	1.0F
	0719	0938	0.9E
	1229	1607	1.4F
	1937	2211	1.2E
17 F	0149	0509	1.0F
	0836	1041	0.8E
	1320	1654	1.4F
	2012	2254	1.4E
18 Sa	0239	0630	1.1F
	0948	1142	0.7E
	1413	1742	1.3F
	2049	2337	1.6E
19 Su	0328	0746	1.3F
	1053	1241	0.7E
	1510	1832	1.3F
	2129		
20 M		0022	1.7E
	0416	0850	1.5F
	1153	1340	0.7E
	1611	1926	1.2F
	2215		
21 Tu		0108	1.8E
	0504	0943	1.7F
	1247	1437	0.8E
	1714	2022	1.2F
	2304		
22 W		0158	1.9E
	0552	1031	1.8F
	1337	1532	0.9E
	1815	2118	1.3F
	2356		
23 Th		0249	1.9E
	0643	1115	1.9F
	1424	1625	1.0E
	1914	2212	1.3F
24 F	0051	0342	1.9E
	0735	1159	1.9F
	1510	1716	1.1E
	2010	2305	1.4F
25 Sa	0148	0436	1.8E
	0827	1242	1.9F
	1556	1807	1.1E
	2106	2359	1.4F
26 Su	0247	0532	1.7E
	0918	1326	1.9F
	1642	1859	1.2E
	2203		
27 M		0056	1.3F
	0349	0632	1.6E
	1009	1409	1.8F
	1729	1951	1.2E
	2303		
28 Tu		0157	1.2F
	0458	0734	1.4E
	1100	1453	1.8F
	1816	2043	1.3E
29 W	0007	0308	1.1F
	0613	0839	1.2E
	1151	1537	1.7F
	1901	2134	1.3E
30 Th	0111	0437	1.1F
	0729	0947	1.0E
	1243	1621	1.6F
	1943	2222	1.4E

Time meridian 120° W. 0000 is midnight. 1200 is noon. Times are not adjusted for Daylight Saving Time.

Oakland, Yerba Buena Island, Calif., 2005

F–Flood, Dir. 167° True E–Ebb, Dir. 338° True

July

Day	Slack h m	Maximum h m	knots
1 F	0214	0602	1.3F
	0844	1056	0.9E
	1335	1705	1.5F
	2024	2306	1.4E
2 Sa	0312	0710	1.4F
	0952	1200	0.8E
	1430	1748	1.4F
	2104	2346	1.4E
3 Su	0403	0808	1.6F
	1054	1259	0.8E
	1525	1832	1.3F
	2143		
4 M		0024	1.4E
	0446	0900	1.6F
	1147	1354	0.8E
	1621	1918	1.2F
	2223		
5 Tu		0102	1.4E
	0523	0945	1.7F
	1234	1444	0.9E
	1715	2006	1.1F
	2302		
6 W ●		0142	1.4E
	0554	1025	1.7F
	1317	1529	0.9E
	1804	2052	1.2F
	2343		
7 Th		0223	1.4E
	0622	1057	1.6F
	1355	1607	0.9E
	1849	2135	1.2F
8 F	0023	0305	1.5E
	0651	1119	1.5F
	1430	1641	0.9E
	1930	2217	1.2F
9 Sa	0103	0349	1.5E
	0722	1135	1.5F
	1504	1713	0.9E
	2008	2258	1.3F
10 Su	0145	0433	1.5E
	0757	1155	1.5F
	1536	1746	1.0E
	2045	2341	1.2F
11 M	0229	0520	1.4E
	0834	1224	1.5F
	1607	1822	1.0E
	2124		
12 Tu		0025	1.2F
	0318	0610	1.3E
	0914	1300	1.5F
	1639	1901	1.0E
	2207		
13 W		0114	1.1F
	0416	0704	1.1E
	0957	1341	1.5F
	1714	1944	1.1E
	2258		
14 Th ◐		0210	1.0F
	0531	0804	0.9E
	1045	1426	1.4F
	1751	2030	1.2E
	2356		
15 F		0318	0.9F
	0701	0908	0.7E
	1139	1516	1.3F
	1832	2119	1.3E
16 Sa	0058	0454	0.9F
	0827	1018	0.6E
	1241	1611	1.3F
	1918	2210	1.5E
17 Su	0200	0637	1.1F
	0941	1127	0.6E
	1350	1710	1.2F
	2009	2303	1.6E
18 M	0301	0746	1.4F
	1043	1232	0.7E
	1500	1811	1.2F
	2104	2356	1.7E
19 Tu	0400	0841	1.6F
	1137	1332	0.8E
	1608	1912	1.2F
	2201		
20 W		0050	1.8E
	0456	0928	1.8F
	1225	1426	0.9E
	1709	2012	1.3F
	2258		
21 Th ○		0144	1.8E
	0549	1011	2.0F
	1310	1516	1.1E
	1805	2108	1.4F
	2354		
22 F		0238	1.9E
	0639	1051	2.0F
	1354	1603	1.2E
	1858	2201	1.5F
23 Sa	0048	0332	1.9E
	0726	1129	2.0F
	1436	1647	1.3E
	1948	2251	1.5F
24 Su	0142	0424	1.9E
	0811	1206	2.0F
	1518	1732	1.3E
	2040	2342	1.5F
25 M	0237	0518	1.7E
	0854	1242	1.9F
	1600	1818	1.3E
	2133		
26 Tu		0036	1.3F
	0337	0614	1.5E
	0938	1319	1.8F
	1643	1905	1.3E
	2231		
27 W ◑		0138	1.2F
	0445	0714	1.3E
	1024	1358	1.7F
	1727	1954	1.3E
	2335		
28 Th		0257	1.1F
	0603	0820	1.0E
	1115	1441	1.5F
	1811	2043	1.3E
29 F	0042	0439	1.1F
	0723	0933	0.8E
	1212	1529	1.4F
	1857	2134	1.3E
30 Sa	0148	0557	1.3F
	0837	1048	0.8E
	1314	1622	1.3F
	1944	2224	1.3E
31 Su	0247	0658	1.5F
	0942	1154	0.8E
	1416	1717	1.2F
	2032	2312	1.3E

August

Day	Slack h m	Maximum h m	knots
1 M	0339	0750	1.6F
	1036	1251	0.9E
	1517	1811	1.2F
	2120	2358	1.3E
2 Tu	0423	0837	1.7F
	1123	1341	1.0E
	1612	1902	1.2F
	2207		
3 W		0041	1.3E
	0501	0918	1.7F
	1204	1424	1.0E
	1701	1950	1.2F
	2251		
4 Th ●		0124	1.3E
	0534	0953	1.7F
	1243	1502	1.0E
	1743	2034	1.3F
	2332		
5 F		0206	1.4E
	0604	1020	1.6F
	1317	1532	1.0E
	1821	2115	1.3F
6 Sa	0011	0248	1.5E
	0632	1035	1.5F
	1348	1558	1.0E
	1854	2154	1.4F
7 Su	0049	0330	1.5E
	0701	1049	1.5F
	1416	1625	1.0E
	1925	2232	1.3F
8 M	0128	0412	1.5E
	0730	1111	1.5F
	1440	1655	1.1E
	1955	2310	1.3F
9 Tu	0209	0456	1.4E
	0801	1139	1.5F
	1504	1729	1.2E
	2028	2351	1.2F
10 W	0258	0544	1.2E
	0836	1215	1.5F
	1532	1808	1.2E
	2108		
11 Th		0038	1.1F
	0359	0638	1.0E
	0916	1257	1.4F
	1606	1853	1.3E
	2158		
12 F ◐		0135	1.0F
	0524	0740	0.7E
	1005	1346	1.3F
	1649	1943	1.4E
	2258		
13 Sa		0255	0.9F
	0659	0850	0.6E
	1110	1442	1.2F
	1741	2037	1.4E
14 Su	0010	0510	1.0F
	0821	1006	0.5E
	1229	1545	1.1F
	1842	2137	1.4E
15 M	0127	0632	1.3F
	0926	1120	0.7E
	1348	1653	1.2F
	1947	2238	1.5E
16 Tu	0241	0728	1.5F
	1021	1222	0.8E
	1458	1759	1.2F
	2053	2339	1.6E
17 W	0347	0817	1.8F
	1109	1315	1.0E
	1600	1902	1.4F
	2156		
18 Th		0037	1.7E
	0444	0901	1.9F
	1153	1404	1.2E
	1655	2000	1.5F
	2253		
19 F ○		0133	1.8E
	0534	0941	2.0F
	1236	1449	1.3E
	1745	2055	1.6F
	2347		
20 Sa		0227	1.9E
	0619	1017	2.0F
	1317	1531	1.4E
	1834	2146	1.6F
21 Su	0040	0318	1.9E
	0701	1051	2.0F
	1356	1612	1.4E
	1922	2235	1.6F
22 M	0132	0408	1.8E
	0741	1122	1.9F
	1435	1652	1.4E
	2011	2325	1.5F
23 Tu	0228	0459	1.6E
	0821	1154	1.8F
	1512	1733	1.4E
	2102		
24 W		0020	1.3F
	0328	0554	1.3E
	0903	1229	1.7F
	1551	1817	1.4E
	2156		
25 Th		0126	1.2F
	0438	0656	1.0E
	0950	1309	1.5F
	1632	1903	1.3E
	2256		
26 F ◑		0255	1.1F
	0557	0806	0.8E
	1047	1356	1.3F
	1718	1952	1.2E
27 Sa	0003	0431	1.2F
	0714	0925	0.7E
	1153	1450	1.2F
	1812	2046	1.1E
28 Su	0109	0539	1.4F
	0821	1039	0.8E
	1302	1551	1.1F
	1910	2144	1.1E
29 M	0211	0633	1.5F
	0917	1140	0.9E
	1407	1654	1.1F
	2008	2243	1.1E
30 Tu	0303	0721	1.6F
	1005	1231	1.1E
	1504	1752	1.2F
	2103	2336	1.1E
31 W	0349	0803	1.7F
	1047	1314	1.1E
	1553	1843	1.3F
	2152		

September

Day	Slack h m	Maximum h m	knots
1 Th		0023	1.2E
	0429	0840	1.7F
	1126	1352	1.2E
	1636	1930	1.3F
	2236		
2 F		0107	1.3E
	0504	0910	1.6F
	1201	1423	1.1E
	1714	2013	1.4F
	2317		
3 Sa ●		0148	1.4E
	0534	0930	1.5F
	1232	1448	1.1E
	1747	2053	1.4F
	2356		
4 Su		0229	1.4E
	0602	0944	1.5F
	1259	1512	1.1E
	1816	2131	1.4F
5 M	0035	0309	1.4E
	0629	1003	1.5F
	1321	1539	1.2E
	1843	2208	1.4F
6 Tu	0116	0351	1.4E
	0656	1029	1.5F
	1341	1610	1.3E
	1910	2245	1.3F
7 W	0202	0434	1.2E
	0726	1100	1.5F
	1405	1645	1.4E
	1943	2326	1.2F
8 Th	0256	0523	1.0E
	0801	1138	1.4F
	1435	1725	1.5E
	2024		
9 F		0014	1.1F
	0406	0620	0.8E
	0846	1224	1.3F
	1514	1812	1.5E
	2114		
10 Sa		0120	1.0F
	0532	0727	0.6E
	0947	1319	1.2F
	1604	1906	1.4E
	2216		
11 Su ◐		0305	1.0F
	0655	0842	0.5E
	1109	1422	1.1F
	1708	2007	1.4E
	2337		
12 M		0506	1.2F
	0804	1000	0.6E
	1234	1531	1.1F
	1824	2114	1.4E
13 Tu	0108	0609	1.4F
	0900	1109	0.8E
	1348	1643	1.2F
	1940	2222	1.4E
14 W	0225	0658	1.7F
	0949	1204	1.1E
	1450	1750	1.3F
	2048	2327	1.6E
15 Th	0328	0743	1.8F
	1035	1251	1.2E
	1545	1850	1.5F
	2149		
16 F		0026	1.7E
	0420	0825	1.9F
	1117	1335	1.4E
	1636	1947	1.6F
	2245		
17 Sa ○		0120	1.8E
	0506	0902	2.0F
	1157	1416	1.5E
	1724	2042	1.7F
	2339		
18 Su		0212	1.8E
	0548	0936	1.9F
	1236	1455	1.5E
	1811	2133	1.6F
19 M	0031	0302	1.7E
	0627	1006	1.9F
	1312	1533	1.6E
	1857	2224	1.6F
20 Tu	0126	0351	1.5E
	0706	1035	1.8F
	1346	1611	1.5E
	1942	2315	1.4F
21 W	0223	0442	1.3E
	0746	1107	1.6F
	1420	1649	1.5E
	2028		
22 Th		0012	1.3F
	0325	0538	1.1E
	0832	1144	1.5F
	1456	1730	1.4E
	2115		
23 F		0122	1.2F
	0434	0642	0.8E
	0925	1228	1.3F
	1536	1815	1.3E
	2206		
24 Sa ◑		0247	1.2F
	0546	0755	0.7E
	1029	1320	1.1F
	1627	1906	1.1E
	2307		
25 Su		0406	1.2F
	0653	0911	0.8E
	1140	1420	1.0F
	1729	2004	1.0E
26 M	0016	0508	1.4F
	0751	1018	0.9E
	1248	1524	1.0F
	1839	2107	1.0E
27 Tu	0121	0558	1.5F
	0841	1113	1.1E
	1347	1629	1.1F
	1944	2213	1.0E
28 W	0218	0641	1.5F
	0925	1159	1.2E
	1439	1728	1.2F
	2041	2312	1.1E
29 Th	0306	0719	1.6F
	1006	1238	1.2E
	1525	1819	1.3F
	2131		
30 F		0001	1.2E
	0348	0750	1.6F
	1042	1310	1.2E
	1605	1905	1.4F
	2217		

Time meridian 120° W. 0000 is midnight. 1200 is noon. Times are not adjusted for Daylight Saving Time.

Oakland, Yerba Buena Island, Calif., 2005

F–Flood, Dir. 167° True E–Ebb, Dir. 338° True

October

Day	Slack h m	Maximum h m	knots
1 Sa		0045	1.3E
	0424	0813	1.5F
	1114	1336	1.2E
	1642	1950	1.4F
	2300		
2 Su		0126	1.3E
	0456	0832	1.5F
	1142	1401	1.2E
	1714	2032	1.4F
	2344		
3 M ●		0207	1.3E
	0525	0854	1.4F
	1205	1427	1.3E
	1742	2114	1.4F
4 Tu	0028	0249	1.2E
	0553	0920	1.4F
	1225	1458	1.4E
	1808	2154	1.3F
5 W	0116	0333	1.1E
	0623	0952	1.4F
	1248	1532	1.6E
	1837	2234	1.3F
6 Th	0209	0420	1.0E
	0658	1028	1.4F
	1316	1610	1.6E
	1912	2319	1.2F
7 F	0310	0512	0.8E
	0742	1111	1.3F
	1354	1653	1.7E
	1955		
8 Sa		0016	1.2F
	0418	0612	0.7E
	0839	1203	1.2F
	1441	1743	1.6E
	2048		
9 Su		0135	1.1F
	0530	0721	0.6E
	0954	1303	1.1F
	1540	1842	1.5E
	2156		
10 M ◐		0315	1.2F
	0638	0834	0.7E
	1116	1410	1.0F
	1654	1948	1.4E
	2324		
11 Tu		0437	1.4F
	0737	0946	0.8E
	1232	1521	1.1F
	1816	2058	1.3E
12 W	0053	0535	1.6F
	0828	1047	1.0E
	1337	1632	1.2F
	1933	2209	1.4E
13 Th	0203	0622	1.7F
	0915	1138	1.2E
	1435	1739	1.4F
	2040	2314	1.5E
14 F	0300	0704	1.8F
	0959	1222	1.4E
	1528	1840	1.5F
	2140		
15 Sa		0012	1.6E
	0348	0743	1.9F
	1039	1303	1.5E
	1618	1939	1.6F
	2237		
16 Su		0105	1.6E
	0432	0818	1.9F
	1117	1342	1.6E
	1706	2036	1.6F
	2333		
17 M ○		0156	1.5E
	0512	0850	1.8F
	1153	1419	1.6E
	1752	2131	1.6F
18 Tu	0028	0246	1.4E
	0552	0920	1.7F
	1226	1455	1.6E
	1835	2224	1.5F
19 W	0125	0337	1.2E
	0634	0952	1.6F
	1258	1531	1.6E
	1915	2317	1.4F
20 Th	0222	0429	1.0E
	0719	1028	1.4F
	1331	1609	1.6E
	1953		
21 F		0012	1.4F
	0321	0525	0.9E
	0810	1109	1.3F
	1407	1649	1.4E
	2030		
22 Sa		0112	1.3F
	0421	0628	0.8E
	0908	1157	1.2F
	1450	1734	1.3E
	2113		
23 Su		0216	1.2F
	0520	0736	0.8E
	1012	1251	1.1F
	1543	1826	1.1E
	2208		
24 M ◑		0321	1.2F
	0617	0842	0.9E
	1117	1350	1.0F
	1650	1926	1.0E
	2315		
25 Tu		0419	1.3F
	0710	0941	1.0E
	1219	1451	1.0F
	1803	2031	0.9E
26 W	0024	0507	1.3F
	0757	1033	1.1E
	1314	1554	1.1F
	1911	2137	1.0E
27 Th	0125	0547	1.4F
	0840	1117	1.1E
	1405	1654	1.2F
	2010	2239	1.1E
28 F	0215	0618	1.4F
	0919	1153	1.2E
	1450	1750	1.3F
	2105	2332	1.1E
29 Sa	0259	0643	1.4F
	0953	1221	1.2E
	1532	1841	1.4F
	2156		
30 Su		0019	1.2E
	0338	0708	1.4F
	1023	1248	1.3E
	1610	1931	1.4F
	2247		
31 M		0104	1.2E
	0414	0737	1.4F
	1048	1316	1.4E
	1644	2021	1.4F
	2338		

November

Day	Slack h m	Maximum h m	knots
1 Tu ●		0148	1.1E
	0449	0809	1.4F
	1112	1347	1.5E
	1714	2110	1.4F
2 W	0030	0234	1.0E
	0524	0844	1.3F
	1137	1423	1.6E
	1743	2157	1.4F
3 Th	0124	0322	0.9E
	0603	0924	1.3F
	1208	1501	1.7E
	1816	2244	1.4F
4 F	0218	0412	0.8E
	0649	1008	1.2F
	1245	1544	1.8E
	1856	2334	1.4F
5 Sa	0314	0506	0.7E
	0744	1056	1.2F
	1331	1631	1.7E
	1944		
6 Su		0033	1.3F
	0412	0605	0.7E
	0847	1151	1.1F
	1426	1725	1.6E
	2042		
7 M		0140	1.3F
	0511	0710	0.8E
	0956	1252	1.1F
	1531	1826	1.5E
	2153		
8 Tu ◐		0250	1.4F
	0608	0816	0.9E
	1107	1357	1.1F
	1647	1933	1.4E
	2315		
9 W		0356	1.5F
	0703	0919	1.0E
	1215	1506	1.2F
	1807	2043	1.3E
10 Th	0030	0452	1.6F
	0753	1016	1.2E
	1319	1619	1.3F
	1922	2154	1.3E
11 F	0132	0539	1.7F
	0839	1107	1.3E
	1418	1730	1.4F
	2030	2300	1.4E
12 Sa	0226	0621	1.8F
	0921	1152	1.5E
	1513	1838	1.5F
	2134	2359	1.4E
13 Su	0313	0658	1.8F
	1001	1232	1.6E
	1606	1942	1.5F
	2234		
14 M		0053	1.3E
	0358	0733	1.7F
	1038	1309	1.6E
	1654	2044	1.6F
	2333		
15 Tu ○		0146	1.2E
	0442	0807	1.6F
	1112	1345	1.7E
	1738	2140	1.6F
16 W	0030	0238	1.1E
	0526	0842	1.5F
	1145	1421	1.7E
	1817	2231	1.6F
17 Th	0124	0329	1.0E
	0613	0920	1.4F
	1218	1458	1.6E
	1850	2318	1.5F
18 F	0216	0420	0.9E
	0703	1000	1.3F
	1253	1536	1.5E
	1921		
19 Sa		0001	1.5F
	0306	0512	0.9E
	0755	1043	1.2F
	1332	1618	1.4E
	1953		
20 Su		0042	1.4F
	0354	0605	0.9E
	0848	1130	1.2F
	1418	1703	1.3E
	2032		
21 M		0122	1.3F
	0442	0659	0.9E
	0943	1221	1.1F
	1510	1755	1.2E
	2121		
22 Tu		0202	1.2F
	0531	0752	0.9E
	1038	1315	1.1F
	1610	1852	1.1E
	2221		
23 W ◑		0246	1.2F
	0619	0844	0.9E
	1134	1412	1.1F
	1718	1954	1.0E
	2325		
24 Th		0332	1.2F
	0705	0932	1.0E
	1229	1512	1.1F
	1828	2057	1.0E
25 F	0025	0416	1.3F
	0747	1015	1.0E
	1321	1615	1.1F
	1936	2200	1.0E
26 Sa	0119	0457	1.3F
	0825	1053	1.1E
	1411	1719	1.2F
	2040	2300	1.0E
27 Su	0208	0535	1.3F
	0858	1127	1.2E
	1457	1823	1.2F
	2142	2354	1.0E
28 M	0253	0613	1.3F
	0928	1200	1.3E
	1539	1926	1.3F
	2241		
29 Tu		0045	0.9E
	0337	0652	1.3F
	0957	1236	1.5E
	1618	2026	1.4F
	2338		
30 W		0136	0.9E
	0422	0734	1.3F
	1028	1314	1.6E
	1653	2121	1.5F

December

Day	Slack h m	Maximum h m	knots
1 Th ●	0032	0226	0.8E
	0509	0820	1.2F
	1104	1355	1.7E
	1729	2209	1.5F
2 F	0123	0317	0.8E
	0559	0907	1.2F
	1146	1440	1.8E
	1809	2254	1.6F
3 Sa	0212	0407	0.9E
	0652	0957	1.3F
	1233	1528	1.8E
	1855	2339	1.6F
4 Su	0259	0457	0.9E
	0746	1047	1.3F
	1325	1618	1.8E
	1947		
5 M		0027	1.6F
	0348	0550	0.9E
	0843	1141	1.3F
	1422	1713	1.7E
	2046		
6 Tu		0117	1.6F
	0439	0646	1.0E
	0942	1238	1.3F
	1525	1813	1.6E
	2149		
7 W		0210	1.6F
	0531	0744	1.0E
	1044	1339	1.2F
	1636	1918	1.4E
	2254		
8 Th ◐		0305	1.6F
	0623	0841	1.1E
	1151	1448	1.2F
	1753	2026	1.3E
	2358		
9 F		0359	1.6F
	0713	0938	1.2E
	1257	1607	1.2F
	1910	2136	1.2E
10 Sa	0056	0449	1.6F
	0759	1031	1.3E
	1402	1732	1.3F
	2024	2246	1.1E
11 Su	0151	0534	1.6F
	0843	1119	1.5E
	1502	1847	1.4F
	2133	2350	1.1E
12 M	0242	0615	1.6F
	0923	1200	1.5E
	1557	1952	1.6F
	2237		
13 Tu		0049	1.0E
	0333	0655	1.5F
	1002	1239	1.6E
	1646	2049	1.7F
	2334		
14 W		0144	1.0E
	0424	0735	1.4F
	1039	1316	1.6E
	1727	2139	1.7F
15 Th ○	0027	0236	1.0E
	0514	0816	1.3F
	1116	1353	1.6E
	1801	2224	1.7F
16 F	0114	0324	1.0E
	0603	0858	1.3F
	1153	1432	1.5E
	1831	2303	1.6F
17 Sa	0157	0408	1.0E
	0651	0940	1.3F
	1232	1512	1.5E
	1858	2334	1.5F
18 Su	0237	0449	1.0E
	0736	1022	1.3F
	1312	1555	1.5E
	1928	2358	1.4F
19 M	0316	0529	1.0E
	0819	1105	1.3F
	1355	1639	1.4E
	2005		
20 Tu		0018	1.4F
	0354	0609	1.0E
	0902	1150	1.2F
	1441	1727	1.3E
	2047		
21 W		0047	1.3F
	0434	0650	0.9E
	0946	1237	1.2F
	1532	1819	1.2E
	2134		
22 Th		0125	1.3F
	0516	0733	0.9E
	1035	1329	1.1F
	1632	1916	1.1E
	2227		
23 F ◑		0209	1.3F
	0558	0817	0.9E
	1128	1425	1.0F
	1744	2017	1.0E
	2322		
24 Sa		0256	1.3F
	0639	0902	1.0E
	1225	1531	1.0F
	1905	2122	0.8E
25 Su	0020	0346	1.2F
	0717	0947	1.1E
	1323	1653	1.0F
	2024	2229	0.7E
26 M	0118	0437	1.2F
	0755	1032	1.2E
	1417	1822	1.1F
	2135	2334	0.7E
27 Tu	0216	0529	1.2F
	0834	1117	1.4E
	1508	1935	1.3F
	2238		
28 W		0034	0.7E
	0314	0620	1.2F
	0915	1202	1.5E
	1555	2032	1.5F
	2332		
29 Th		0129	0.8E
	0411	0712	1.2F
	1000	1248	1.7E
	1640	2120	1.6F
30 F ●	0021	0220	0.9E
	0505	0805	1.3F
	1048	1336	1.8E
	1725	2203	1.7F
31 Sa	0106	0308	1.0E
	0557	0857	1.3F
	1139	1426	1.8E
	1811	2243	1.8F

Time meridian 120° W. 0000 is midnight. 1200 is noon. Times are not adjusted for Daylight Saving Time.

Richmond (Long Wharf), Calif., 2005

F–Flood, Dir. 328° True E–Ebb, Dir. 147° True

January

Day	Slack h m	Maximum h m	knots
1 Sa		0143	1.6F
	0545	0804	1.0E
	1102	1342	1.1F
	1620	1951	1.8E
	2318		
2 Su		0223	1.6F
	0617	0852	1.1E
	1213	1440	1.0F
	1722	2043	1.5E
	2355		
3 M		0306	1.7F
	0650	0945	1.3E
	1329	1546	0.9F
◑	1846	2142	1.1E
4 Tu	0036	0352	1.7F
	0727	1042	1.6E
	1442	1701	0.9F
	2030	2250	0.8E
5 W	0121	0442	1.7F
	0808	1140	2.0E
	1549	1833	1.0F
	2207		
6 Th		0001	0.6E
	0211	0534	1.7F
	0853	1236	2.3E
	1648	2011	1.3F
	2328		
7 F		0110	0.5E
	0307	0629	1.7F
	0942	1329	2.6E
	1743	2115	1.6F
8 Sa	0035	0215	0.5E
	0407	0725	1.8F
	1034	1422	2.9E
	1834	2207	1.9F
9 Su	0129	0314	0.6E
	0509	0821	1.9F
	1129	1515	3.1E
	1923	2253	2.0F
10 M	0216	0409	0.8E
	0611	0916	2.0F
	1224	1606	3.1E
●	2009	2335	2.1F
11 Tu	0258	0500	0.9E
	0711	1011	2.0F
	1319	1655	3.1E
	2052		
12 W		0014	2.1F
	0339	0550	1.1E
	0810	1105	2.0F
	1413	1743	3.0E
	2133		
13 Th		0049	2.0F
	0418	0638	1.3E
	0911	1200	1.8F
	1508	1830	2.7E
	2213		
14 F		0122	2.0F
	0458	0728	1.4E
	1015	1257	1.6F
	1604	1917	2.3E
	2252		
15 Sa		0157	1.9F
	0539	0822	1.5E
	1126	1359	1.4F
	1705	2007	1.9E
	2331		
16 Su		0235	1.8F
	0621	0922	1.6E
	1242	1511	1.1F
◐	1816	2104	1.4E
17 M	0012	0317	1.7F
	0706	1030	1.7E
	1400	1650	1.0F
	1939	2211	1.0E
18 Tu	0058	0403	1.6F
	0752	1137	1.9E
	1513	1828	1.1F
	2110	2329	0.7E
19 W	0150	0453	1.5F
	0838	1237	2.1E
	1616	1940	1.3F
	2234		
20 Th		0044	0.6E
	0247	0546	1.4F
	0924	1329	2.2E
	1710	2037	1.6F
	2341		
21 F		0148	0.6E
	0345	0639	1.3F
	1009	1413	2.3E
	1755	2126	1.7F
22 Sa	0034	0241	0.7E
	0441	0730	1.4F
	1051	1450	2.3E
	1835	2209	1.8F
23 Su	0116	0327	0.8E
	0531	0817	1.4F
	1132	1522	2.3E
	1911	2246	1.8F
24 M	0153	0407	0.9E
	0617	0900	1.5F
	1210	1550	2.3E
	1943	2316	1.7F
25 Tu	0226	0441	0.9E
	0658	0941	1.5F
	1246	1618	2.3E
○	2012	2331	1.7F
26 W	0255	0510	1.0E
	0737	1020	1.6F
	1322	1648	2.3E
	2038	2335	1.7F
27 Th	0321	0538	1.0E
	0814	1100	1.6F
	1357	1722	2.3E
	2102	2353	1.7F
28 F	0345	0607	1.2E
	0853	1141	1.5F
	1434	1758	2.2E
	2127		
29 Sa		0021	1.8F
	0407	0640	1.3E
	0938	1225	1.4F
	1516	1837	2.0E
	2153		
30 Su		0054	1.8F
	0431	0717	1.4E
	1031	1314	1.2F
	1605	1920	1.6E
	2222		
31 M		0133	1.8F
	0459	0800	1.6E
	1137	1411	1.1F
	1711	2010	1.2E
	2254		

February

Day	Slack h m	Maximum h m	knots
1 Tu		0217	1.7F
	0534	0852	1.7E
	1255	1519	0.9F
◑	1847	2111	0.8E
	2334		
2 W		0307	1.6F
	0619	0953	1.8E
	1416	1645	0.9F
	2043	2227	0.5E
3 Th	0026	0404	1.5F
	0714	1102	2.0E
	1530	1908	1.1F
	2224	2354	0.3E
4 F	0139	0507	1.5F
	0819	1213	2.3E
	1633	2019	1.5F
	2333		
5 Sa		0113	0.4E
	0300	0612	1.5F
	0926	1317	2.6E
	1729	2110	1.8F
6 Su	0023	0216	0.6E
	0414	0717	1.7F
	1029	1414	2.8E
	1818	2153	2.0F
7 M	0104	0309	0.9E
	0518	0817	1.9F
	1128	1506	3.0E
	1902	2231	2.1F
8 Tu	0142	0357	1.2E
	0616	0913	2.0F
	1224	1554	3.0E
●	1943	2303	2.1F
9 W	0218	0442	1.4E
	0711	1005	2.1F
	1316	1639	2.9E
	2020	2330	2.1F
10 Th	0252	0525	1.6E
	0805	1055	2.1F
	1407	1722	2.7E
	2055	2355	2.1F
11 F	0327	0607	1.7E
	0900	1145	1.9F
	1458	1804	2.4E
	2128		
12 Sa		0023	2.0F
	0401	0650	1.8E
	0958	1237	1.7F
	1551	1847	2.0E
	2201		
13 Su		0056	1.9F
	0437	0735	1.8E
	1100	1335	1.4F
	1651	1934	1.5E
	2236		
14 M		0134	1.8F
	0514	0824	1.8E
	1210	1445	1.1F
	1803	2028	1.0E
	2315		
15 Tu		0217	1.6F
	0556	0924	1.7E
	1325	1636	1.0F
◐	1933	2139	0.6E
16 W	0005	0308	1.4F
	0646	1040	1.7E
	1440	1815	1.1F
	2111	2313	0.5E
17 Th	0113	0406	1.2F
	0743	1200	1.8E
	1546	1923	1.4F
	2232		
18 F		0038	0.5E
	0230	0513	1.1F
	0845	1303	1.9E
	1641	2017	1.6F
	2326		
19 Sa		0139	0.7E
	0339	0621	1.1F
	0943	1351	2.0E
	1726	2102	1.8F
20 Su	0007	0227	0.9E
	0437	0722	1.2F
	1035	1429	2.1E
	1805	2141	1.8F
21 M	0041	0308	1.0E
	0525	0810	1.4F
	1120	1500	2.2E
	1839	2212	1.8F
22 Tu	0112	0342	1.1E
	0608	0851	1.5F
	1200	1527	2.2E
	1908	2232	1.7F
23 W	0139	0411	1.2E
	0647	0928	1.6F
	1237	1554	2.2E
○	1933	2235	1.7F
24 Th	0203	0436	1.3E
	0723	1005	1.7F
	1313	1624	2.2E
	1956	2245	1.8F
25 F	0223	0501	1.5E
	0800	1043	1.7F
	1350	1657	2.1E
	2018	2308	1.8F
26 Sa	0242	0529	1.6E
	0838	1122	1.6F
	1430	1733	1.9E
	2041	2337	1.9F
27 Su	0301	0601	1.8E
	0921	1206	1.5F
	1516	1812	1.6E
	2105		
28 M		0012	1.9F
	0324	0639	2.0E
	1012	1255	1.4F
	1613	1857	1.3E
	2133		

March

Day	Slack h m	Maximum h m	knots
1 Tu		0051	1.8F
	0355	0722	2.0E
	1114	1353	1.2F
	1730	1949	0.8E
	2205		
2 W		0138	1.6F
	0436	0814	2.0E
	1230	1505	1.0F
	1914	2056	0.5E
	2248		
3 Th		0234	1.5F
	0529	0919	2.0E
	1353	1713	1.0F
◑		2225	*
4 F		0340	1.3F
	0640	1037	2.1E
	1509	1907	1.3F
	2224		
5 Sa		0004	0.3E
	0149	0453	1.3F
	0803	1159	2.2E
	1612	2002	1.6F
	2313		
6 Su		0115	0.7E
	0316	0607	1.4F
	0921	1308	2.5E
	1705	2047	1.9F
	2352		
7 M		0209	1.0E
	0424	0717	1.7F
	1028	1404	2.6E
	1751	2124	2.0F
8 Tu	0027	0255	1.3E
	0523	0817	1.9F
	1126	1452	2.7E
	1831	2155	2.1F
9 W	0100	0338	1.6E
	0617	0910	2.0F
	1220	1536	2.7E
	1907	2218	2.1F
10 Th	0132	0418	1.9E
	0708	0958	2.1F
	1310	1617	2.5E
●	1940	2239	2.1F
11 F	0203	0457	2.0E
	0758	1045	2.0F
	1400	1657	2.3E
	2010	2303	2.1F
12 Sa	0233	0534	2.2E
	0848	1132	1.9F
	1450	1737	1.9E
	2040	2332	2.0F
13 Su	0303	0611	2.2E
	0939	1221	1.6F
	1545	1820	1.5E
	2111		
14 M		0005	1.9F
	0333	0649	2.1E
	1034	1315	1.4F
	1647	1906	1.1E
	2144		
15 Tu		0044	1.7F
	0405	0730	2.0E
	1135	1423	1.2F
	1801	2001	0.7E
	2225		
16 W		0130	1.4F
	0444	0819	1.8E
	1242	1617	1.1F
	1930	2116	0.4E
	2325		
17 Th		0225	1.1F
	0535	0924	1.6E
	1354	1749	1.2F
◐	2100	2259	0.4E
18 F	0052	0330	0.9F
	0644	1055	1.6E
	1500	1853	1.4F
	2206		
19 Sa		0021	0.6E
	0219	0444	0.9F
	0803	1219	1.7E
	1556	1944	1.6F
	2250		
20 Su		0117	0.8E
	0328	0600	1.0F
	0914	1313	1.8E
	1642	2026	1.7F
	2325		
21 M		0201	1.1E
	0424	0705	1.2F
	1011	1352	1.9E
	1720	2100	1.7F
	2355		
22 Tu		0238	1.3E
	0510	0754	1.4F
	1100	1423	1.9E
	1753	2122	1.7F
23 W	0022	0308	1.4E
	0552	0834	1.5F
	1143	1453	2.0E
	1820	2127	1.7F
24 Th	0045	0334	1.6E
	0631	0912	1.6F
	1224	1523	1.9E
	1845	2136	1.8F
25 F	0105	0358	1.8E
	0709	0949	1.7F
	1305	1556	1.8E
○	1908	2157	1.8F
26 Sa	0123	0424	2.0E
	0747	1028	1.7F
	1349	1632	1.7E
	1930	2226	1.9F
27 Su	0141	0455	2.2E
	0827	1110	1.7F
	1438	1712	1.4E
	1954	2259	1.9F
28 M	0204	0530	2.4E
	0911	1155	1.6F
	1534	1755	1.2E
	2022	2336	1.9F
29 Tu	0234	0610	2.4E
	1002	1247	1.5F
	1641	1844	0.8E
	2054		
30 W		0020	1.7F
	0312	0656	2.4E
	1102	1349	1.3F
	1803	1943	0.5E
	2134		
31 Th		0112	1.5F
	0400	0751	2.3E
	1214	1514	1.2F
	1937	2058	0.3E
	2237		

Time meridian 120° W. 0000 is midnight. 1200 is noon. Times are not adjusted for Daylight Saving Time.
* Current weak and variable.

Richmond (Long Wharf), Calif., 2005

F–Flood, Dir. 328° True E–Ebb, Dir. 147° True

April

Day	Slack h m	Maximum h m	knots
1 F ◑		0215	1.3F
	0503	0859	2.1E
	1331	1730	1.3F
	2100	2237	0.3E
2 Sa	0027	0329	1.2F
	0626	1020	2.1E
	1442	1840	1.6F
	2155		
3 Su		0004	0.6E
	0208	0448	1.2F
	0757	1145	2.1E
	1541	1930	1.8F
	2236		
4 M		0104	1.0E
	0325	0608	1.4F
	0917	1251	2.3E
	1630	2011	2.0F
	2312		
5 Tu		0153	1.4E
	0428	0719	1.6F
	1023	1344	2.3E
	1712	2043	2.1F
	2345		
6 W		0236	1.8E
	0524	0818	1.8F
	1120	1430	2.3E
	1749	2108	2.1F
7 Th	0017	0316	2.1E
	0615	0909	1.9F
	1214	1512	2.2E
	1822	2126	2.1F
8 F ●	0047	0353	2.3E
	0704	0956	1.9F
	1306	1552	1.9E
	1853	2148	2.1F
9 Sa	0115	0428	2.4E
	0751	1041	1.9F
	1358	1633	1.6E
	1923	2216	2.0F
10 Su	0142	0501	2.5E
	0836	1125	1.8F
	1451	1715	1.3E
	1954	2248	1.9F
11 M	0208	0534	2.4E
	0922	1212	1.6F
	1549	1759	1.0E
	2027	2325	1.7F
12 Tu	0235	0609	2.3E
	1009	1304	1.4F
	1651	1848	0.8E
	2104		
13 W		0006	1.5F
	0307	0648	2.2E
	1100	1409	1.3F
	1800	1945	0.5E
	2152		
14 Th		0055	1.3F
	0346	0733	1.9E
	1158	1545	1.2F
	1915	2057	0.4E
	2301		
15 F		0152	1.0F
	0437	0829	1.7E
	1301	1706	1.3F
	2025	2231	0.4E

Day	Slack h m	Maximum h m	knots
16 Sa ◐	0033	0258	0.8F
	0548	0940	1.5E
	1403	1807	1.4F
	2120	2347	0.6E
17 Su	0157	0411	0.8F
	0713	1102	1.5E
	1457	1856	1.5F
	2200		
18 M		0041	0.9E
	0305	0525	0.9F
	0832	1208	1.6E
	1543	1933	1.5F
	2233		
19 Tu		0123	1.2E
	0401	0632	1.1F
	0937	1256	1.6E
	1621	1955	1.6F
	2301		
20 W		0158	1.4E
	0449	0726	1.2F
	1032	1335	1.6E
	1653	2002	1.6F
	2325		
21 Th		0226	1.7E
	0532	0813	1.4F
	1123	1412	1.6E
	1722	2017	1.7F
	2347		
22 F		0252	1.9E
	0614	0855	1.5F
	1212	1449	1.5E
	1748	2043	1.8F
23 Sa	0006	0319	2.2E
	0654	0937	1.6F
	1304	1529	1.3E
	1814	2113	1.9F
24 Su ○	0026	0350	2.4E
	0736	1020	1.7F
	1358	1612	1.1E
	1842	2148	1.9F
25 M	0051	0426	2.6E
	0819	1106	1.7F
	1456	1658	0.9E
	1913	2227	1.9F
26 Tu	0123	0506	2.8E
	0905	1156	1.7F
	1559	1747	0.7E
	1949	2311	1.8F
27 W	0202	0551	2.8E
	0957	1253	1.6F
	1706	1843	0.5E
	2033		
28 Th		0001	1.6F
	0249	0641	2.7E
	1055	1403	1.5F
	1817	1947	0.4E
	2134		
29 F		0059	1.4F
	0346	0739	2.5E
	1159	1537	1.5F
	1926	2105	0.4E
	2304		
30 Sa ◑		0207	1.2F
	0457	0846	2.2F
	1305	1658	1.6F
	2024	2233	0.6E

May

Day	Slack h m	Maximum h m	knots
1 Su	0046	0323	1.1F
	0622	1003	2.1E
	1406	1757	1.7F
	2112	2346	1.0E
2 M	0213	0444	1.2F
	0748	1120	2.0E
	1459	1845	1.9F
	2153		
3 Tu		0043	1.4E
	0325	0607	1.3F
	0906	1224	2.0E
	1545	1923	2.0F
	2229		
4 W		0131	1.8E
	0427	0721	1.5F
	1013	1317	1.9E
	1624	1951	2.0F
	2303		
5 Th		0214	2.2E
	0522	0821	1.6F
	1114	1404	1.7E
	1700	2013	2.0F
	2334		
6 F		0253	2.4E
	0613	0915	1.7F
	1212	1448	1.5E
	1734	2037	2.0F
7 Sa	0003	0329	2.5E
	0700	1004	1.8F
	1308	1532	1.3E
	1807	2105	1.9F
8 Su ●	0030	0402	2.6E
	0744	1050	1.7F
	1404	1616	1.1E
	1841	2137	1.9F
9 M	0056	0433	2.6E
	0825	1135	1.7F
	1500	1701	0.9E
	1916	2214	1.7F
10 Tu	0123	0504	2.5E
	0906	1219	1.6F
	1556	1748	0.7E
	1955	2254	1.6F
11 W	0153	0539	2.4E
	0947	1305	1.5F
	1651	1837	0.6E
	2040	2339	1.4F
12 Th	0228	0618	2.3E
	1030	1354	1.4F
	1747	1930	0.5E
	2134		
13 F		0029	1.2F
	0311	0701	2.1E
	1117	1450	1.4F
	1842	2031	0.5E
	2243		
14 Sa		0125	1.0F
	0402	0752	1.9E
	1208	1548	1.3F
	1934	2141	0.5E
15 Su	0003	0227	0.9F
	0504	0849	1.7E
	1259	1638	1.3F
	2020	2251	0.7E

Day	Slack h m	Maximum h m	knots
16 M ◐	0122	0333	0.8F
	0619	0952	1.5E
	1348	1715	1.4F
	2059	2348	1.0E
17 Tu	0231	0441	0.8F
	0739	1056	1.4E
	1432	1744	1.5F
	2131		
18 W		0032	1.3E
	0331	0549	0.9F
	0854	1153	1.4E
	1510	1813	1.6F
	2159		
19 Th		0107	1.6E
	0423	0653	1.1F
	1003	1244	1.3E
	1545	1845	1.7F
	2223		
20 F		0138	1.9E
	0511	0751	1.3F
	1107	1332	1.1E
	1617	1919	1.7F
	2246		
21 Sa		0209	2.2E
	0556	0844	1.4F
	1209	1419	1.0E
	1649	1956	1.8F
	2311		
22 Su		0243	2.5E
	0641	0935	1.6F
	1310	1508	0.8E
	1722	2036	1.9F
	2341		
23 M ○		0322	2.8E
	0726	1025	1.7F
	1412	1558	0.7E
	1800	2119	1.9F
24 Tu	0017	0404	2.9E
	0812	1116	1.8F
	1512	1651	0.6E
	1844	2205	1.8F
25 W	0059	0450	3.0E
	0900	1210	1.8F
	1610	1746	0.6E
	1936	2256	1.8F
26 Th	0148	0540	3.0E
	0950	1309	1.8F
	1706	1844	0.6E
	2038	2351	1.6F
27 F	0244	0633	2.8E
	1043	1411	1.8F
	1758	1946	0.6E
	2152		
28 Sa		0053	1.5F
	0347	0730	2.6E
	1138	1513	1.8F
	1850	2055	0.8E
	2317		
29 Su		0200	1.3F
	0456	0832	2.3E
	1231	1609	1.8F
	1938	2208	1.0E
30 M ◑	0045	0314	1.2F
	0613	0939	2.0E
	1323	1700	1.9F
	2024	2317	1.4E
31 Tu	0206	0436	1.1F
	0734	1047	1.8E
	1410	1743	1.9F
	2106		

June

Day	Slack h m	Maximum h m	knots
1 W		0015	1.8E
	0318	0605	1.2F
	0852	1151	1.6E
	1453	1819	1.9F
	2145		
2 Th		0106	2.1E
	0422	0723	1.3F
	1005	1249	1.4E
	1534	1851	1.9F
	2221		
3 F		0151	2.4E
	0519	0827	1.5F
	1114	1341	1.2E
	1613	1922	1.9F
	2254		
4 Sa		0232	2.5E
	0609	0923	1.6F
	1217	1432	1.0E
	1651	1955	1.8F
	2325		
5 Su		0308	2.6E
	0654	1014	1.7F
	1317	1521	0.8E
	1731	2031	1.7F
	2354		
6 M ●		0341	2.6E
	0735	1102	1.7F
	1413	1609	0.7E
	1812	2110	1.6F
7 Tu	0024	0412	2.5E
	0813	1145	1.7F
	1505	1656	0.6E
	1856	2151	1.6F
8 W	0055	0443	2.5E
	0850	1225	1.7F
	1551	1741	0.6E
	1941	2234	1.5F
9 Th	0130	0518	2.4E
	0926	1257	1.6F
	1634	1824	0.6E
	2029	2320	1.4F
10 F	0209	0555	2.3E
	1002	1322	1.5F
	1715	1907	0.6E
	2121		
11 Sa		0008	1.2F
	0251	0636	2.2E
	1039	1348	1.5F
	1754	1953	0.6E
	2218		
12 Su		0100	1.1F
	0338	0720	2.0E
	1118	1422	1.5F
	1833	2042	0.7E
	2325		
13 M		0155	1.0F
	0430	0808	1.8E
	1157	1500	1.5F
	1911	2135	0.8E
14 Tu ◐	0037	0254	0.9F
	0532	0901	1.6E
	1237	1541	1.5F
	1946	2230	1.1E
15 W	0149	0358	0.8F
	0648	0958	1.3E
	1316	1622	1.0F
	2018	2320	1.4E

Day	Slack h m	Maximum h m	knots
16 Th	0255	0506	0.8F
	0815	1059	1.1E
	1355	1705	1.6F
	2048		
17 F		0006	1.7E
	0355	0620	0.9F
	0942	1200	0.9E
	1434	1748	1.7F
	2119		
18 Sa		0048	2.1E
	0448	0738	1.1F
	1101	1259	0.7E
	1514	1833	1.7F
	2152		
19 Su		0130	2.4E
	0539	0849	1.4F
	1213	1357	0.6E
	1557	1920	1.8F
	2229		
20 M		0214	2.7E
	0627	0947	1.6F
	1319	1455	0.5E
	1645	2009	1.8F
	2312		
21 Tu ○		0301	2.9E
	0715	1040	1.8F
	1416	1551	0.5E
	1739	2100	1.9F
22 W	0000	0350	3.1E
	0803	1128	1.9F
	1507	1647	0.6E
	1838	2153	1.9F
23 Th	0053	0441	3.1E
	0850	1215	1.9F
	1553	1741	0.7E
	1940	2248	1.8F
24 F	0148	0532	3.1E
	0936	1300	2.0F
	1637	1834	0.8E
	2046	2345	1.8F
25 Sa	0246	0623	2.9E
	1022	1343	2.0F
	1720	1930	1.0E
	2156		
26 Su		0045	1.6F
	0346	0716	2.6E
	1106	1426	1.9F
	1803	2029	1.2E
	2312		
27 M		0149	1.4F
	0450	0810	2.3E
	1150	1508	1.9F
	1848	2133	1.4E
28 Tu ◑	0032	0300	1.2F
	0600	0908	1.9E
	1234	1550	1.9F
	1932	2240	1.7E
29 W	0152	0425	1.1F
	0719	1012	1.5E
	1318	1633	1.9F
	2016	2343	1.9E
30 Th	0306	0602	1.1F
	0842	1120	1.1E
	1403	1716	1.8F
	2059		

Time meridian 120° W. 0000 is midnight. 1200 is noon. Times are not adjusted for Daylight Saving Time.

Richmond (Long Wharf), Calif., 2005

F–Flood, Dir. 328° True   E–Ebb, Dir. 147° True

July

Day	Slack h m	Maximum h m	knots
1 F		0039	2.2E
	0412	0723	1.3F
	1004	1227	0.9E
	1448	1759	1.7F
	2139		
2 Sa		0129	2.4E
	0509	0827	1.5F
	1119	1328	0.8E
	1536	1843	1.7F
	2218		
3 Su		0213	2.5E
	0558	0922	1.7F
	1223	1425	0.7E
	1625	1927	1.6F
	2255		
4 M		0252	2.5E
	0642	1010	1.8F
	1319	1517	0.7E
	1714	2010	1.5F
	2331		
5 Tu		0326	2.5E
	0721	1054	1.8F
	1406	1605	0.7E
	1803	2054	1.5F
6 W ●	0007	0357	2.4E
	0757	1134	1.8F
	1448	1648	0.7E
	1850	2137	1.5F
7 Th	0044	0428	2.4E
	0830	1206	1.7F
	1524	1726	0.7E
	1935	2220	1.5F
8 F	0121	0500	2.4E
	0901	1224	1.6F
	1557	1801	0.8E
	2018	2304	1.4F
9 Sa	0159	0535	2.3E
	0930	1233	1.6F
	1628	1835	0.8E
	2103	2348	1.4F
10 Su	0237	0611	2.2E
	0959	1255	1.6F
	1657	1909	0.9E
	2151		
11 M		0034	1.3F
	0319	0650	2.1E
	1028	1326	1.7F
	1726	1947	1.0E
	2247		
12 Tu		0124	1.1F
	0405	0733	1.8E
	1058	1402	1.7F
	1755	2030	1.2E
	2352		
13 W		0219	1.0F
	0501	0820	1.5E
	1131	1443	1.7F
	1825	2118	1.3E
14 Th ◐	0105	0320	0.9F
	0616	0915	1.1E
	1207	1527	1.7F
	1859	2212	1.6E
15 F	0217	0431	0.8F
	0756	1019	0.8E
	1247	1614	1.6F
	1938	2309	1.9E

Day	Slack h m	Maximum h m	knots
16 Sa	0325	0557	0.9F
	0939	1129	0.5E
	1334	1705	1.6F
	2022		
17 Su		0005	2.2E
	0425	0748	1.2F
	1106	1240	0.4E
	1429	1800	1.6F
	2112		
18 M		0100	2.5E
	0520	0856	1.5F
	1215	1347	0.4E
	1531	1856	1.7F
	2205		
19 Tu		0154	2.7E
	0611	0947	1.7F
	1309	1447	0.5E
	1636	1953	1.8F
	2300		
20 W		0247	2.9E
	0700	1032	1.9F
	1355	1542	0.7E
	1740	2050	1.9F
	2357		
21 Th ○		0340	3.1E
	0746	1112	2.0F
	1436	1634	0.9E
	1842	2146	2.0F
22 F	0053	0430	3.1E
	0829	1149	2.0F
	1515	1724	1.1E
	1943	2240	2.0F
23 Sa	0149	0519	3.0E
	0910	1222	2.0F
	1553	1812	1.3E
	2045	2336	1.9F
24 Su	0244	0607	2.8E
	0949	1255	2.0F
	1631	1902	1.5E
	2149		
25 M		0033	1.7F
	0341	0654	2.5E
	1026	1330	2.0F
	1711	1954	1.6E
	2258		
26 Tu		0134	1.5F
	0441	0743	2.0E
	1104	1408	2.0F
	1752	2051	1.7E
27 W ◑	0013	0245	1.2F
	0550	0837	1.5E
	1144	1449	1.9F
	1836	2155	1.8E
28 Th	0131	0416	1.0F
	0710	0941	1.1E
	1228	1536	1.7F
	1923	2304	2.0E
29 F	0246	0558	1.1F
	0840	1057	0.8E
	1320	1626	1.6F
	2011		
30 Sa		0009	2.1E
	0353	0715	1.3F
	1007	1216	0.6E
	1418	1720	1.5F
	2101		
31 Su		0105	2.2E
	0449	0815	1.5F
	1118	1323	0.6E
	1519	1817	1.4F
	2149		

August

Day	Slack h m	Maximum h m	knots
1 M		0154	2.3E
	0538	0906	1.7F
	1213	1419	0.7E
	1617	1911	1.4F
	2235		
2 Tu		0235	2.3E
	0620	0951	1.8F
	1257	1507	0.8E
	1711	2000	1.4F
	2318		
3 W		0311	2.3E
	0657	1030	1.8F
	1335	1550	0.9E
	1759	2045	1.5F
	2359		
4 Th ●		0341	2.3E
	0730	1104	1.7F
	1409	1627	0.9E
	1843	2127	1.5F
5 F	0037	0409	2.3E
	0800	1125	1.7F
	1439	1659	1.0E
	1924	2206	1.5F
6 Sa	0113	0438	2.3E
	0826	1128	1.6F
	1506	1727	1.1E
	2004	2246	1.5F
7 Su	0149	0510	2.2E
	0850	1141	1.7F
	1530	1755	1.2E
	2044	2327	1.5F
8 M	0225	0544	2.1E
	0914	1205	1.7F
	1552	1825	1.3E
	2127		
9 Tu		0010	1.4F
	0305	0621	1.9E
	0938	1237	1.8F
	1613	1900	1.4E
	2216		
10 W		0057	1.2F
	0352	0702	1.6E
	1004	1313	1.8F
	1638	1939	1.6E
	2316		
11 Th		0150	1.1F
	0452	0749	1.2E
	1034	1355	1.7F
	1709	2026	1.7E
12 F ◐	0026	0253	0.9F
	0618	0845	0.8E
	1109	1442	1.6F
	1750	2122	1.8E
13 Sa	0142	0408	0.9F
	0806	0955	0.5E
	1155	1536	1.5F
	1841	2226	2.0E
14 Su	0256	0606	1.0F
	0947	1116	0.3E
	1301	1637	1.5F
	1943	2335	2.2E
15 M	0401	0747	1.3F
	1102	1236	0.4E
	1421	1740	1.5F
	2050		

Day	Slack h m	Maximum h m	knots
16 Tu		0041	2.4E
	0458	0841	1.6F
	1155	1342	0.6E
	1537	1845	1.6F
	2156		
17 W		0141	2.7E
	0549	0925	1.8F
	1237	1437	0.8E
	1644	1946	1.8F
	2258		
18 Th		0235	2.9E
	0635	1003	1.9F
	1315	1527	1.0E
	1745	2044	2.0F
	2356		
19 F ○		0326	2.9E
	0717	1035	2.0F
	1351	1613	1.3E
	1843	2139	2.1F
20 Sa	0051	0413	2.9E
	0756	1103	2.0F
	1426	1659	1.5E
	1940	2231	2.1F
21 Su	0144	0459	2.7E
	0832	1130	2.1F
	1502	1743	1.7E
	2037	2324	1.9F
22 M	0238	0543	2.4E
	0906	1201	2.0F
	1537	1828	1.9E
	2137		
23 Tu		0019	1.7F
	0334	0629	2.0E
	0940	1235	2.0F
	1614	1914	2.0E
	2241		
24 W		0119	1.5F
	0436	0717	1.6E
	1016	1314	1.9F
	1653	2004	2.0E
	2349		
25 Th		0232	1.2F
	0547	0811	1.1E
	1056	1358	1.7F
	1736	2102	1.9E
26 F ◑	0102	0413	1.1F
	0710	0919	0.7E
	1146	1449	1.5F
	1825	2213	1.9E
27 Sa	0215	0546	1.2F
	0841	1046	0.5E
	1252	1547	1.3F
	1922	2332	1.9E
28 Su	0320	0655	1.4F
	1000	1209	0.6E
	1405	1652	1.2F
	2024		
29 M		0037	2.0E
	0417	0751	1.6F
	1057	1312	0.7E
	1514	1800	1.2F
	2124		
30 Tu		0129	2.1E
	0504	0838	1.7F
	1140	1402	0.9E
	1613	1902	1.3F
	2217		
31 W		0210	2.1E
	0545	0919	1.8F
	1217	1446	1.1E
	1704	1952	1.4F
	2304		

September

Day	Slack h m	Maximum h m	knots
1 Th		0244	2.2E
	0620	0953	1.8F
	1250	1523	1.2E
	1749	2035	1.5F
	2346		
2 F		0313	2.1E
	0651	1017	1.7F
	1319	1555	1.3E
	1831	2113	1.6F
3 Sa ●	0025	0341	2.1E
	0718	1022	1.6F
	1344	1622	1.3E
	1910	2150	1.6F
4 Su	0102	0410	2.0E
	0741	1029	1.7F
	1406	1647	1.4E
	1948	2228	1.6F
5 M	0140	0442	1.9E
	0803	1050	1.7F
	1425	1714	1.6E
	2026	2307	1.5F
6 Tu	0220	0517	1.7E
	0825	1119	1.8F
	1442	1744	1.8E
	2108	2350	1.4F
7 W	0305	0556	1.5E
	0848	1152	1.8F
	1504	1819	1.9E
	2155		
8 Th		0037	1.3F
	0401	0639	1.2E
	0914	1231	1.7F
	1532	1900	2.0E
	2251		
9 F		0132	1.2F
	0512	0729	0.8E
	0946	1316	1.6F
	1610	1949	2.0E
	2358		
10 Sa		0238	1.1F
	0644	0831	0.5E
	1028	1409	1.4F
	1700	2047	2.0E
11 Su ◐	0115	0405	1.0F
	0823	0949	0.3E
	1133	1512	1.3F
	1805	2158	2.1E
12 M	0229	0619	1.2F
	0943	1120	0.3E
	1310	1621	1.3F
	1923	2315	2.2E
13 Tu	0334	0724	1.5F
	1037	1235	0.6E
	1437	1732	1.4F
	2043		
14 W		0026	2.4E
	0430	0811	1.7F
	1118	1332	0.9E
	1548	1840	1.6F
	2153		
15 Th		0126	2.5E
	0518	0849	1.9F
	1155	1421	1.2E
	1650	1942	1.8F
	2255		

Day	Slack h m	Maximum h m	knots
16 F		0218	2.6E
	0600	0920	2.0F
	1230	1506	1.6E
	1746	2039	2.0F
	2351		
17 Sa ○		0305	2.6E
	0638	0945	2.0F
	1303	1549	1.8E
	1841	2132	2.0F
18 Su	0045	0350	2.4E
	0713	1010	2.1F
	1336	1631	2.0E
	1935	2223	2.0F
19 M	0139	0434	2.2E
	0746	1038	2.1F
	1408	1712	2.2E
	2029	2315	1.9F
20 Tu	0235	0519	1.9E
	0819	1111	2.0F
	1441	1752	2.3E
	2123		
21 W		0009	1.7F
	0333	0605	1.5E
	0853	1147	1.9F
	1514	1834	2.2E
	2220		
22 Th		0109	1.5F
	0438	0654	1.1E
	0930	1228	1.7F
	1551	1918	2.1E
	2321		
23 F		0226	1.3F
	0550	0752	0.8E
	1015	1316	1.5F
	1634	2010	2.0E
24 Sa ◑	0026	0403	1.2F
	0710	0905	0.5E
	1117	1411	1.2F
	1727	2114	1.8E
25 Su	0133	0521	1.3F
	0829	1035	0.5E
	1237	1516	1.1F
	1834	2237	1.7E
26 M	0236	0624	1.5F
	0932	1152	0.7E
	1357	1627	1.0F
	1947	2354	1.8E
27 Tu	0331	0716	1.6F
	1019	1249	0.9E
	1504	1740	1.1F
	2055		
28 W		0049	1.8E
	0418	0759	1.7F
	1057	1335	1.1E
	1601	1844	1.2F
	2153		
29 Th		0131	1.9E
	0457	0835	1.7F
	1130	1414	1.3E
	1649	1935	1.4F
	2243		
30 F		0205	1.9E
	0531	0900	1.7F
	1158	1448	1.5E
	1734	2018	1.5F
	2327		

Time meridian 120° W. 0000 is midnight. 1200 is noon. Times are not adjusted for Daylight Saving Time.

Richmond (Long Wharf), Calif., 2005

F–Flood, Dir. 328° True E–Ebb, Dir. 147° True

October

Day	Slack h m	Maximum h m	knots
1 Sa		0235	1.9E
	0559	0907	1.6F
	1223	1516	1.6E
	1815	2056	1.5F
2 Su	0010	0306	1.8E
	0625	0915	1.6F
	1244	1541	1.7E
	1854	2134	1.6F
3 M ●	0052	0339	1.6E
	0648	0936	1.7F
	1302	1606	1.9E
	1933	2213	1.6F
4 Tu	0137	0414	1.5E
	0710	1004	1.8F
	1320	1635	2.1E
	2012	2253	1.5F
5 W	0226	0454	1.3E
	0734	1037	1.8F
	1341	1709	2.2E
	2054	2337	1.5F
6 Th	0322	0537	1.0E
	0801	1115	1.7F
	1409	1747	2.3E
	2141		
7 F		0027	1.4F
	0427	0626	0.8E
	0833	1158	1.6F
	1446	1832	2.4E
	2236		
8 Sa		0126	1.3F
	0542	0722	0.5E
	0914	1249	1.5F
	1533	1925	2.3E
	2341		
9 Su		0239	1.2F
	0703	0830	0.4E
	1015	1349	1.3F
	1633	2027	2.2E
10 M ◐	0053	0426	1.3F
	0818	0954	0.4E
	1151	1459	1.2F
	1749	2140	2.1E
11 Tu	0201	0553	1.5F
	0915	1119	0.6E
	1329	1613	1.2F
	1916	2258	2.1E
12 W	0302	0647	1.7F
	0959	1224	1.0E
	1447	1727	1.4F
	2037		
13 Th		0008	2.2E
	0353	0729	1.8F
	1037	1316	1.4E
	1553	1838	1.6F
	2147		
14 F		0105	2.3E
	0437	0802	1.9F
	1112	1401	1.7E
	1652	1941	1.8F
	2249		
15 Sa		0155	2.2E
	0516	0828	2.0F
	1145	1444	2.0E
	1747	2038	1.9F
	2347		
16 Su		0242	2.1E
	0552	0853	2.0F
	1217	1524	2.3E
	1839	2131	1.9F
17 M ○	0043	0326	1.8E
	0626	0921	2.0F
	1248	1603	2.4E
	1930	2223	1.8F
18 Tu	0140	0412	1.6E
	0659	0953	2.0F
	1318	1641	2.5E
	2020	2314	1.7F
19 W	0238	0458	1.3E
	0733	1028	1.9F
	1348	1719	2.5E
	2109		
20 Th		0008	1.6F
	0339	0547	1.0E
	0810	1107	1.7F
	1421	1758	2.4E
	2158		
21 F		0109	1.5F
	0443	0639	0.7E
	0853	1152	1.5F
	1458	1839	2.2E
	2250		
22 Sa		0222	1.4F
	0549	0738	0.6E
	0947	1242	1.3F
	1542	1927	2.0E
	2346		
23 Su		0336	1.4F
	0655	0849	0.5E
	1057	1340	1.1F
	1637	2023	1.8E
24 M ◑	0045	0442	1.4F
	0756	1009	0.6E
	1219	1446	0.9F
	1746	2131	1.7E
25 Tu	0142	0538	1.5F
	0848	1119	0.8E
	1336	1556	0.9F
	1903	2246	1.6E
26 W	0234	0626	1.5F
	0930	1214	1.0E
	1442	1707	1.0F
	2015	2348	1.6E
27 Th	0319	0704	1.6F
	1006	1259	1.3E
	1539	1812	1.1F
	2119		
28 F		0036	1.6E
	0357	0728	1.6F
	1036	1336	1.5E
	1629	1909	1.3F
	2215		
29 Sa		0116	1.6E
	0430	0738	1.6F
	1102	1407	1.7E
	1715	1957	1.4F
	2307		
30 Su		0153	1.5E
	0500	0755	1.6F
	1125	1434	1.9E
	1758	2041	1.4F
	2358		
31 M		0231	1.3E
	0526	0820	1.7F
	1144	1500	2.1E
	1839	2123	1.5F

November

Day	Slack h m	Maximum h m	knots
1 Tu ●	0050	0310	1.2E
	0552	0851	1.7F
	1204	1530	2.3E
	1920	2205	1.5F
2 W	0145	0353	1.0E
	0619	0925	1.8F
	1227	1603	2.5E
	2001	2249	1.6F
3 Th	0243	0438	0.8E
	0649	1004	1.7F
	1257	1642	2.6E
	2045	2336	1.6F
4 F	0344	0528	0.6E
	0726	1048	1.7F
	1335	1727	2.7E
	2133		
5 Sa		0030	1.5F
	0447	0621	0.5E
	0811	1137	1.6F
	1422	1816	2.6E
	2227		
6 Su		0132	1.5F
	0550	0721	0.4E
	0910	1234	1.4F
	1518	1911	2.5E
	2326		
7 M		0247	1.5F
	0651	0830	0.5E
	1031	1339	1.3F
	1626	2014	2.3E
8 Tu ◐	0028	0405	1.6F
	0746	0948	0.6E
	1206	1450	1.2F
	1744	2123	2.2E
9 W	0128	0508	1.7F
	0834	1102	0.9E
	1333	1605	1.2F
	1908	2236	2.1E
10 Th	0221	0556	1.8F
	0916	1204	1.4E
	1448	1723	1.3F
	2027	2343	2.0E
11 F	0308	0636	1.9F
	0955	1255	1.8E
	1553	1839	1.4F
	2139		
12 Sa		0041	1.9E
	0350	0707	2.0F
	1031	1341	2.1E
	1653	1948	1.6F
	2245		
13 Su		0132	1.7E
	0429	0737	2.0F
	1104	1424	2.4E
	1747	2049	1.7F
	2348		
14 M		0221	1.5E
	0505	0808	2.0F
	1136	1503	2.6E
	1838	2144	1.7F
15 Tu ○	0048	0309	1.2E
	0542	0841	1.9F
	1207	1540	2.6E
	1925	2237	1.8F
16 W	0148	0357	1.0E
	0619	0917	1.9F
	1237	1616	2.6E
	2010	2328	1.7F
17 Th	0247	0447	0.8E
	0659	0956	1.7F
	1309	1652	2.6E
	2053		
18 F		0018	1.7F
	0344	0537	0.7E
	0743	1039	1.6F
	1343	1729	2.5E
	2136		
19 Sa		0108	1.6F
	0438	0628	0.6E
	0832	1126	1.4F
	1422	1809	2.3E
	2220		
20 Su		0158	1.5F
	0530	0721	0.6E
	0929	1217	1.3F
	1508	1854	2.1E
	2306		
21 M		0248	1.5F
	0619	0818	0.6E
	1034	1312	1.1F
	1601	1943	1.9E
	2354		
22 Tu		0335	1.4F
	0707	0921	0.7E
	1147	1413	1.0F
	1702	2037	1.7E
23 W ◑	0042	0416	1.4F
	0751	1025	0.8E
	1300	1517	0.9F
	1811	2137	1.6E
24 Th	0128	0450	1.4F
	0831	1123	1.0E
	1409	1623	0.9F
	1925	2237	1.4E
25 F	0211	0519	1.5F
	0906	1210	1.3E
	1510	1730	0.9F
	2038	2334	1.3E
26 Sa	0249	0550	1.5F
	0935	1248	1.6F
	1605	1836	1.1F
	2147		
27 Su		0026	1.2E
	0324	0622	1.6F
	1002	1320	1.8E
	1655	1938	1.2F
	2252		
28 M		0114	1.0E
	0356	0657	1.6F
	1026	1351	2.1E
	1741	2033	1.3F
	2355		
29 Tu		0201	0.9E
	0428	0734	1.7F
	1050	1423	2.4E
	1824	2124	1.5F
30 W	0057	0250	0.7E
	0501	0814	1.7F
	1119	1500	2.6E
	1908	2211	1.6F

December

Day	Slack h m	Maximum h m	knots
1 Th ●	0157	0340	0.6E
	0539	0857	1.7F
	1153	1541	2.7E
	1951	2259	1.7F
2 F	0254	0431	0.5E
	0623	0943	1.7F
	1235	1626	2.9E
	2037	2347	1.7F
3 Sa	0348	0524	0.5E
	0714	1033	1.7F
	1324	1715	2.9E
	2124		
4 Su		0038	1.7F
	0438	0618	0.6E
	0813	1127	1.6F
	1418	1806	2.8E
	2214		
5 M		0130	1.7F
	0526	0714	0.7E
	0922	1225	1.5F
	1518	1901	2.6E
	2305		
6 Tu		0224	1.7F
	0613	0815	0.8E
	1040	1329	1.4F
	1625	1959	2.4E
	2356		
7 W		0317	1.8F
	0700	0923	1.0E
	1204	1439	1.3F
	1737	2101	2.1E
8 Th ◐	0046	0406	1.8F
	0746	1033	1.3E
	1327	1555	1.2F
	1856	2207	1.8E
9 F	0134	0452	1.8F
	0830	1137	1.7E
	1443	1720	1.2F
	2017	2314	1.6E
10 Sa	0220	0535	1.9F
	0911	1233	2.0E
	1552	1849	1.3F
	2136		
11 Su		0017	1.3E
	0303	0615	1.9F
	0950	1322	2.3E
	1652	2002	1.5F
	2250		
12 M		0115	1.1E
	0346	0654	1.9F
	1027	1407	2.5E
	1746	2103	1.6F
	2357		
13 Tu		0210	1.0E
	0428	0733	1.8F
	1103	1448	2.6E
	1834	2157	1.8F
14 W	0059	0303	0.8E
	0512	0813	1.7F
	1137	1525	2.6E
	1918	2246	1.8F
15 Th ○	0156	0354	0.7E
	0558	0855	1.7F
	1211	1600	2.6E
	1958	2331	1.8F
16 F	0246	0443	0.7E
	0645	0937	1.6F
	1247	1634	2.5E
	2036		
17 Sa		0012	1.8F
	0331	0528	0.7E
	0733	1022	1.5F
	1325	1709	2.5E
	2113		
18 Su		0047	1.7F
	0412	0611	0.7E
	0821	1107	1.4F
	1405	1746	2.4E
	2149		
19 M		0112	1.6F
	0451	0653	0.7E
	0911	1155	1.3F
	1448	1825	2.2E
	2224		
20 Tu		0134	1.6F
	0528	0735	0.8E
	1005	1245	1.2F
	1533	1907	2.1E
	2301		
21 W		0202	1.5F
	0605	0820	0.8E
	1106	1338	1.1F
	1623	1952	1.8E
	2338		
22 Th		0237	1.5F
	0642	0909	0.9E
	1214	1435	0.9F
	1721	2042	1.6E
23 F ◑	0016	0316	1.5F
	0717	1002	1.1E
	1326	1537	0.8F
	1833	2137	1.3E
24 Sa	0054	0358	1.5F
	0751	1055	1.3E
	1434	1645	0.8F
	1959	2238	1.0E
25 Su	0133	0441	1.5F
	0823	1144	1.6E
	1536	1803	0.9F
	2128	2341	0.8E
26 M	0214	0525	1.5F
	0856	1228	1.0E
	1631	1933	1.1F
	2250		
27 Tu		0043	0.6E
	0256	0611	1.6F
	0930	1311	2.2E
	1721	2044	1.3F
28 W	0001	0143	0.5E
	0341	0659	1.6F
	1008	1354	2.5E
	1808	2137	1.5F
29 Th	0103	0239	0.5E
	0430	0749	1.7F
	1051	1440	2.7E
	1854	2223	1.7F
30 F ●	0155	0333	0.5E
	0524	0840	1.7F
	1139	1527	2.9E
	1939	2305	1.8F
31 Sa	0241	0425	0.6E
	0621	0932	1.8F
	1231	1616	3.0E
	2024	2344	1.9F

Time meridian 120° W. 0000 is midnight. 1200 is noon. Times are not adjusted for Daylight Saving Time.

Carquinez Strait (West End Bridge), San Pablo Bay, Calif., 2005

F–Flood, Dir. 103° True   E–Ebb, Dir. 283° True

January

Day	Slack h m	Maximum h m	knots
1 Sa	0111	0407	2.2F
	0757	1043	1.4E
	1337	1602	1.5F
	1844	2213	2.6E
2 Su	0148	0444	2.3F
	0827	1128	1.7E
	1443	1708	1.3F
	1950	2308	2.3E
3 M ◑	0226	0524	2.4F
	0859	1215	2.1E
	1555	1817	1.3F
	2105		
4 Tu		0005	1.9E
	0305	0607	2.5F
	0936	1304	2.5E
	1706	1937	1.2F
	2227		
5 W		0103	1.5E
	0344	0651	2.6F
	1014	1353	2.8E
	1812	2103	1.4F
	2347		
6 Th		0204	1.1E
	0426	0738	2.6F
	1055	1444	3.1E
	1912	2213	1.6F
7 F	0107	0308	0.8E
	0509	0828	2.6F
	1139	1534	3.4E
	2008	2311	1.8F
8 Sa	0220	0411	0.7E
	0559	0921	2.6F
	1225	1624	3.5E
	2058		
9 Su		0006	1.9F
	0323	0514	0.7E
	0658	1014	2.6F
	1316	1718	3.6E
	2143		
10 M ●		0056	2.0F
	0417	0616	0.8E
	0802	1108	2.5F
	1410	1814	3.6E
	2228		
11 Tu		0141	2.1F
	0500	0713	1.0E
	0908	1204	2.5F
	1507	1908	3.6E
	2310		
12 W		0222	2.2F
	0540	0804	1.2E
	1011	1302	2.4F
	1604	1959	3.5E
	2350		
13 Th		0259	2.2F
	0619	0855	1.5E
	1117	1401	2.2F
	1702	2047	3.3E
14 F	0030	0336	2.3F
	0659	0948	1.7E
	1223	1502	2.0F
	1801	2138	2.9E
15 Sa	0110	0416	2.3F
	0740	1048	1.9E
	1336	1612	1.7F
	1904	2234	2.5E
16 Su ◐	0150	0458	2.4F
	0825	1151	2.1E
	1456	1735	1.4F
	2013	2335	2.1E
17 M	0232	0542	2.4F
	0911	1255	2.4E
	1620	1901	1.3F
	2130		
18 Tu		0037	1.7E
	0318	0627	2.4F
	0958	1400	2.6E
	1740	2025	1.3F
	2254		
19 W		0143	1.4E
	0404	0714	2.3F
	1042	1501	2.8E
	1848	2135	1.6F
20 Th	0018	0250	1.1E
	0453	0803	2.2F
	1126	1554	3.0E
	1942	2234	1.8F
21 F	0133	0352	1.0E
	0545	0853	2.1F
	1205	1641	3.0E
	2030	2326	2.0F
22 Sa	0239	0449	0.9E
	0639	0939	1.9F
	1242	1721	3.0E
	2110		
23 Su		0014	2.0F
	0332	0542	0.9E
	0731	1021	1.8F
	1320	1754	2.9E
	2142		
24 M		0057	2.1F
	0417	0631	0.9E
	0821	1102	1.8F
	1357	1816	2.8E
	2210		
25 Tu ○		0133	2.0F
	0450	0713	1.0E
	0908	1144	1.8F
	1435	1835	2.9E
	2235		
26 W		0158	2.0F
	0516	0747	1.1E
	0950	1226	1.8F
	1515	1901	2.9E
	2258		
27 Th		0208	2.0F
	0537	0813	1.2E
	1032	1309	1.8F
	1557	1934	3.0E
	2322		
28 F		0216	2.1F
	0555	0836	1.4E
	1118	1354	1.8F
	1641	2012	2.9E
	2350		
29 Sa		0237	2.3F
	0617	0903	1.7E
	1209	1441	1.7F
	1730	2053	2.7E
30 Su	0020	0309	2.4F
	0641	0937	1.9E
	1305	1534	1.6F
	1827	2139	2.4E
31 M	0053	0348	2.5F
	0713	1021	2.2E
	1410	1636	1.4F
	1931	2233	1.9E

February

Day	Slack h m	Maximum h m	knots
1 Tu ◑	0130	0432	2.6F
	0751	1113	2.4E
	1526	1751	1.2F
	2050	2333	1.4E
2 W	0211	0521	2.5F
	0838	1210	2.7E
	1642	1927	1.2F
	2222		
3 Th		0038	1.0E
	0300	0613	2.5F
	0928	1311	2.9E
	1752	2101	1.4F
	2351		
4 F		0152	0.8E
	0357	0709	2.4F
	1021	1417	3.1E
	1853	2207	1.7F
5 Sa	0109	0308	0.7E
	0459	0809	2.3F
	1118	1522	3.2E
	1948	2300	1.9F
6 Su	0210	0413	0.8E
	0605	0911	2.4F
	1215	1622	3.4E
	2037	2347	2.1F
7 M	0259	0512	1.0E
	0710	1010	2.4F
	1313	1718	3.4E
	2120		
8 Tu ●		0030	2.1F
	0340	0606	1.2E
	0811	1105	2.5F
	1410	1810	3.4E
	2200		
9 W		0107	2.2F
	0417	0656	1.5E
	0910	1201	2.5F
	1507	1858	3.4E
	2237		
10 Th		0139	2.2F
	0450	0741	1.7E
	1008	1256	2.4F
	1600	1942	3.2E
	2311		
11 F		0207	2.3F
	0526	0824	1.9E
	1106	1350	2.2F
	1653	2024	3.0E
	2347		
12 Sa		0237	2.3F
	0600	0909	2.1E
	1207	1447	2.0F
	1749	2108	2.6E
13 Su	0021	0311	2.4F
	0639	0959	2.2E
	1315	1552	1.6F
	1849	2158	2.1E
14 M	0059	0352	2.4F
	0720	1059	2.3E
	1433	1718	1.3F
	1959	2259	1.6E
15 Tu ◐	0141	0439	2.3F
	0807	1208	2.4E
	1558	1849	1.2F
	2122		
16 W		0009	1.2E
	0230	0531	2.1F
	0858	1320	2.5E
	1716	2012	1.4F
	2256		
17 Th		0124	1.0E
	0328	0628	2.0F
	0951	1429	2.6E
	1820	2120	1.7F
18 F	0018	0238	1.0E
	0432	0730	1.8F
	1046	1526	2.8E
	1911	2214	2.0F
19 Sa	0121	0341	1.1E
	0538	0834	1.7F
	1137	1614	2.8E
	1956	2301	2.1F
20 Su	0213	0435	1.2E
	0638	0931	1.7F
	1223	1653	2.8E
	2032	2343	2.2F
21 M	0255	0523	1.3E
	0731	1017	1.8F
	1308	1725	2.8E
	2103		
22 Tu		0019	2.1F
	0328	0606	1.4E
	0818	1058	1.8F
	1349	1749	2.7E
	2129		
23 W ○		0046	2.0F
	0352	0642	1.5E
	0900	1137	1.9F
	1429	1811	2.7E
	2151		
24 Th		0058	2.0F
	0411	0710	1.6E
	0939	1215	1.9F
	1509	1837	2.7E
	2212		
25 F		0101	2.0F
	0427	0730	1.7E
	1019	1255	1.9F
	1550	1910	2.7E
	2235		
26 Sa		0119	2.3F
	0443	0748	2.0E
	1100	1337	1.9F
	1634	1946	2.5E
	2300		
27 Su		0147	2.5F
	0506	0814	2.3E
	1149	1422	1.8F
	1723	2026	2.3E
	2329		
28 M		0222	2.6F
	0536	0849	2.6E
	1243	1513	1.6F
	1821	2112	1.9E

March

Day	Slack h m	Maximum h m	knots
1 Tu	0003	0303	2.6F
	0612	0933	2.7E
	1347	1617	1.4F
	1931	2207	1.4E
2 W	0044	0351	2.5F
	0658	1027	2.8E
	1501	1745	1.2F
	2059	2316	1.0E
3 Th ◑	0134	0446	2.3F
	0750	1131	2.8E
	1618	1932	1.3F
	2233		
4 F		0036	0.8E
	0239	0547	2.1F
	0851	1244	2.8E
	1728	2052	1.6F
	2352		
5 Sa		0159	0.8E
	0358	0653	2.0F
	1000	1405	2.9E
	1828	2149	1.9F
6 Su	0052	0312	1.0E
	0511	0804	2.0F
	1109	1520	3.0E
	1919	2236	2.1F
7 M	0140	0410	1.3E
	0619	0912	2.2F
	1213	1618	3.2E
	2005	2316	2.2F
8 Tu	0220	0501	1.6E
	0720	1012	2.3F
	1313	1707	3.2E
	2045	2350	2.2F
9 W	0256	0548	1.8E
	0817	1105	2.4F
	1409	1753	3.1E
	2121		
10 Th ●		0020	2.3F
	0328	0632	2.0E
	0909	1157	2.4F
	1501	1836	3.0E
	2154		
11 F		0045	2.3F
	0358	0713	2.2E
	1001	1249	2.3F
	1552	1917	2.7E
	2226		
12 Sa		0112	2.4F
	0428	0750	2.4E
	1056	1340	2.1F
	1644	1957	2.4E
	2258		
13 Su		0142	2.4F
	0500	0827	2.5E
	1153	1434	1.8F
	1740	2039	2.0E
	2330		
14 M		0218	2.4F
	0536	0905	2.5E
	1256	1540	1.5F
	1842	2128	1.6E
15 Tu	0009	0258	2.3F
	0616	0952	2.5E
	1406	1709	1.3F
	1958	2232	1.1E
16 W	0054	0347	2.1F
	0701	1057	2.4E
	1521	1835	1.4F
	2127	2352	0.9E
17 Th ◐	0152	0445	1.8F
	0756	1220	2.4E
	1632	1949	1.6F
	2252		
18 F		0111	0.8E
	0306	0551	1.6F
	0859	1339	2.4E
	1733	2051	1.9F
	2359		
19 Sa		0221	1.0E
	0423	0704	1.5F
	1006	1443	2.5E
	1822	2142	2.1F
20 Su	0048	0320	1.3E
	0531	0818	1.5F
	1108	1532	2.6E
	1905	2224	2.2F
21 M	0129	0410	1.6E
	0630	0919	1.7F
	1202	1611	2.6E
	1940	2258	2.2F
22 Tu	0201	0453	1.7E
	0720	1007	1.8F
	1251	1642	2.6E
	2010	2325	2.1F
23 W	0228	0530	1.9E
	0805	1048	1.9F
	1337	1708	2.5E
	2035	2339	2.0F
24 Th	0247	0601	2.0E
	0847	1126	1.9F
	1419	1736	2.4E
	2057	2345	2.1F
25 F ○	0301	0623	2.1E
	0927	1204	1.9F
	1502	1807	2.3E
	2118		
26 Sa		0001	2.3F
	0318	0641	2.4E
	1007	1244	1.9F
	1547	1843	2.1E
	2141		
27 Su		0029	2.5F
	0339	0704	2.7E
	1049	1326	1.8F
	1635	1922	1.9E
	2208		
28 M		0104	2.6F
	0408	0736	3.0E
	1137	1412	1.7F
	1730	2005	1.6E
	2240		
29 Tu		0143	2.7F
	0442	0814	3.1E
	1230	1506	1.6F
	1835	2055	1.2E
	2320		
30 W		0228	2.6F
	0526	0900	3.1E
	1331	1618	1.5F
	1951	2158	0.9E
31 Th	0011	0320	2.3F
	0617	0956	3.0E
	1440	1758	1.5F
	2117	2320	0.7E

Time meridian 120° W. 0000 is midnight. 1200 is noon. Times are not adjusted for Daylight Saving Time.

Carquinez Strait (West End Bridge), San Pablo Bay, Calif., 2005

F–Flood, Dir. 103° True   E–Ebb, Dir. 283° True

April

Day	Slack h m	Maximum h m	knots
1 F ◑	0120 0718 1551 2235	0423 1108 1922	2.0F 2.8E 1.6F
2 Sa	 0247 0831 1656 2335	0046 0534 1233 2028	0.8E 1.8F 2.7E 1.9F
3 Su	 0411 0951 1752	0202 0650 1359 2120	1.1E 1.7F 2.8E 2.2F
4 M	0022 0525 1107 1841	0305 0808 1507 2202	1.5E 1.8F 2.9E 2.3F
5 Tu	0103 0628 1211 1923	0357 0918 1600 2236	1.9E 2.0F 2.9E 2.4F
6 W	0139 0725 1310 2001	0444 1015 1645 2304	2.2E 2.2F 2.8E 2.4F
7 Th	0210 0818 1403 2035	0527 1106 1728 2329	2.4E 2.2F 2.6E 2.4F
8 F ●	0239 0909 1457 2106	0608 1156 1809 2353	2.5E 2.2F 2.4E 2.4F
9 Sa	0307 1000 1548 2136	0645 1247 1851	2.7E 2.0F 2.0E
10 Su	 0334 1051 1643 2208	0022 0717 1339 1933	2.4F 2.8E 1.9F 1.7E
11 M	 0404 1143 1744 2241	0055 0747 1435 2017	2.4F 2.8E 1.7F 1.3E
12 Tu	 0439 1238 1851 2324	0134 0819 1543 2110	2.3F 2.8E 1.5F 1.0E
13 W	 0518 1336 2006	0217 0858 1700 2217	2.1F 2.7E 1.5F 0.8E
14 Th	0019 0605 1437 2123	0307 0950 1812 2337	1.8F 2.6E 1.6F 0.7E
15 F	0128 0701 1538 2231	0408 1101 1914	1.5F 2.4E 1.8F
16 Sa ◐	 0250 0809 1632 2322	0051 0520 1225 2009	0.9E 1.3F 2.3E 2.0F
17 Su	 0409 0922 1720	0154 0637 1335 2055	1.2E 1.3F 2.3E 2.1F
18 M	0001 0516 1032 1801	0249 0752 1431 2132	1.5E 1.4F 2.4E 2.2F
19 Tu	0034 0611 1133 1836	0336 0855 1514 2200	1.9E 1.6F 2.4E 2.2F
20 W	0100 0701 1228 1906	0415 0947 1550 2216	2.1E 1.7F 2.3E 2.1F
21 Th	0121 0748 1318 1931	0447 1032 1624 2226	2.3E 1.8F 2.1E 2.2F
22 F	0140 0831 1408 1955	0513 1113 1658 2245	2.5E 1.9F 1.9E 2.3F
23 Sa	0159 0914 1458 2019	0533 1155 1736 2312	2.7E 1.8F 1.6E 2.5F
24 Su ○	0220 0958 1550 2047	0557 1239 1818 2347	3.0E 1.8F 1.4E 2.6F
25 M	0249 1042 1648 2119	0628 1326 1904	3.2E 1.8F 1.2E
26 Tu	 0325 1130 1751 2200	0028 0707 1417 1955	2.7F 3.4E 1.7F 1.0E
27 W	 0407 1221 1857 2252	0114 0750 1518 2052	2.6F 3.5E 1.7F 0.8E
28 Th	 0455 1318 2006	0204 0841 1634 2203	2.4F 3.3E 1.7F 0.7E
29 F	0002 0552 1418 2113	0302 0941 1752 2327	2.1F 3.1E 1.8F 0.8E
30 Sa ◑	0128 0702 1520 2213	0410 1059 1856	1.7F 2.8E 2.0F

May

Day	Slack h m	Maximum h m	knots
1 Su	 0257 0821 1619 2303	0043 0529 1225 1951	1.2E 1.6F 2.7E 2.2F
2 M	 0418 0945 1710 2346	0149 0652 1340 2039	1.6E 1.6F 2.7E 2.4F
3 Tu	 0528 1059 1756	0248 0812 1442 2119	2.0E 1.7F 2.6E 2.5F
4 W	0024 0630 1202 1836	0339 0921 1534 2151	2.4E 1.9F 2.5E 2.5F
5 Th	0058 0728 1302 1911	0425 1018 1619 2218	2.6E 2.0F 2.3E 2.5F
6 F	0129 0822 1400 1946	0508 1110 1702 2242	2.8E 2.0F 2.0E 2.5F
7 Sa	0157 0913 1457 2018	0548 1203 1746 2309	2.9E 1.9F 1.6E 2.4F
8 Su ●	0221 1002 1556 2049	0622 1258 1832 2341	2.9E 1.8F 1.3E 2.4F
9 M	0249 1049 1659 2123	0650 1352 1920	3.0E 1.7F 1.0E
10 Tu	 0319 1133 1801 2204	0018 0715 1446 2010	2.3F 3.0E 1.7F 0.8E
11 W	 0354 1218 1903 2254	0100 0745 1542 2103	2.1F 3.0E 1.7F 0.7E
12 Th	 0435 1302 2002 2358	0147 0824 1640 2204	1.9F 2.9E 1.7F 0.7E
13 F	 0523 1350 2058	0239 0911 1736 2314	1.7F 2.8E 1.8F 0.8E
14 Sa	0110 0620 1439 2148	0339 1011 1825	1.4F 2.6E 1.9F
15 Su	 0227 0727 1528 2229	0018 0449 1119 1909	1.0E 1.2F 2.5E 2.0F
16 M ◐	 0340 0840 1611 2302	0115 0603 1224 1947	1.3E 1.2F 2.4E 2.1F
17 Tu	 0447 0952 1651 2330	0206 0715 1321 2016	1.7E 1.3F 2.3E 2.1F
18 W	 0545 1059 1727 2356	0251 0823 1412 2038	2.0E 1.4F 2.1E 2.2F
19 Th	 0639 1200 1757	0328 0923 1458 2059	2.3E 1.6F 1.9E 2.3F
20 F	0018 0729 1300 1825	0359 1015 1541 2126	2.6E 1.7F 1.6E 2.4F
21 Sa	0040 0818 1400 1852	0424 1104 1624 2158	2.9E 1.7F 1.3E 2.6F
22 Su	0106 0905 1501 1922	0450 1154 1709 2234	3.2E 1.8F 1.0E 2.7F
23 M ○	0137 0950 1605 1959	0523 1246 1801 2315	3.4E 1.8F 0.8E 2.7F
24 Tu	0214 1036 1708 2044	0603 1338 1857	3.6E 1.8F 0.7E
25 W	 0258 1121 1806 2141	0002 0649 1431 1955	2.6F 3.6E 1.9F 0.7E
26 Th	 0347 1209 1900 2250	0054 0739 1525 2055	2.5F 3.6E 1.9F 0.8E
27 F	 0441 1300 1952	0151 0834 1624 2203	2.2F 3.4E 2.0F 0.9E
28 Sa	0008 0545 1351 2044	0253 0936 1723 2316	1.9F 3.1E 2.1F 1.2E
29 Su	0130 0656 1445 2135	0403 1050 1815	1.7F 2.8E 2.2F
30 M ◑	 0252 0812 1537 2221	0024 0524 1204 1903	1.6E 1.6F 2.7E 2.4F
31 Tu	 0412 0931 1624 2305	0126 0648 1310 1948	2.0E 1.5F 2.5E 2.5F

June

Day	Slack h m	Maximum h m	knots
1 W	 0526 1045 1708 2343	0225 0809 1410 2029	2.4E 1.6F 2.3E 2.5F
2 Th	 0632 1152 1748	0319 0920 1506 2103	2.7E 1.7F 2.0E 2.6F
3 F	0019 0733 1259 1825	0408 1021 1556 2134	2.9E 1.7F 1.7E 2.5F
4 Sa	0050 0829 1405 1900	0453 1118 1644 2204	3.1E 1.8F 1.4E 2.4F
5 Su	0119 0919 1511 1938	0534 1214 1735 2236	3.1E 1.8F 1.0E 2.3F
6 M ●	0148 1003 1618 2016	0609 1308 1828 2312	3.0E 1.8F 0.8E 2.2F
7 Tu	0218 1043 1719 2059	0635 1358 1920 2353	3.0E 1.8F 0.7E 2.0F
8 W	0249 1119 1810 2148	0657 1444 2008	3.0E 1.8F 0.6E
9 Th	 0327 1153 1854 2241	0038 0725 1525 2054	1.9F 3.0E 1.9F 0.7E
10 F	 0409 1229 1933 2342	0127 0802 1603 2142	1.8F 3.0E 1.9F 0.8E
11 Sa	 0458 1306 2010	0218 0848 1639 2236	1.6F 2.9E 1.9F 0.9E
12 Su	0047 0552 1344 2046	0313 0936 1712 2330	1.5F 2.8E 2.0F 1.2E
13 M	0152 0653 1424 2119	0415 1031 1742	1.3F 2.6E 2.1F
14 Tu ◐	 0301 0800 1505 2150	0019 0524 1129 1812	1.4E 1.3F 2.4E 2.1F
15 W	 0410 0912 1543 2220	0105 0633 1225 1845	1.8E 1.2F 2.1E 2.2F
16 Th	 0514 1027 1619 2249	0149 0746 1319 1920	2.1E 1.3F 1.8E 2.4F
17 F	 0615 1139 1652 2319	0229 0900 1413 1958	2.5E 1.4F 1.4E 2.5F
18 Sa	 0712 1250 1726 2350	0306 1003 1507 2038	2.9E 1.5F 1.1E 2.6F
19 Su	 0804 1401 1800	0341 1101 1600 2121	3.2E 1.7F 0.8E 2.6F
20 M	0027 0852 1510 1841	0419 1156 1655 2207	3.4E 1.8F 0.6E 2.6F
21 Tu ○	0108 0939 1613 1935	0501 1248 1756 2255	3.6E 1.9F 0.6E 2.6F
22 W	0154 1022 1705 2038	0550 1337 1856 2348	3.6E 1.9F 0.7E 2.5F
23 Th	0246 1108 1750 2144	0643 1422 1952	3.6E 2.0F 0.8E
24 F	 0341 1150 1832 2253	0045 0737 1505 2047	2.4F 3.6E 2.1F 1.1E
25 Sa	 0440 1235 1915	0144 0831 1549 2145	2.2F 3.4E 2.2F 1.3E
26 Su	0006 0612 1319 1959	0245 0928 1635 2249	2.0F 3.1F 2.3F 1.6E
27 M	0120 0648 1404 2046	0353 1030 1721 2353	1.7F 2.8E 2.4F 1.9E
28 Tu ◑	0239 0759 1450 2132	0514 1135 1806	1.5F 2.5E 2.4F
29 W	 0401 0914 1536 2219	0056 0638 1237 1850	2.2E 1.4F 2.2E 2.5F
30 Th	 0520 1031 1620 2300	0159 0803 1339 1934	2.5E 1.4F 1.8E 2.5F

Time meridian 120° W. 0000 is midnight. 1200 is noon. Times are not adjusted for Daylight Saving Time.

Carquinez Strait (West End Bridge), San Pablo Bay, Calif., 2005

F–Flood, Dir. 103° True E–Ebb, Dir. 283° True

July

Day	Slack h m	Maximum h m	knots
1 F		0258	2.8E
	0633	0919	1.5F
	1149	1442	1.5E
	1703	2017	2.4F
	2340		
2 Sa		0352	3.0E
	0735	1022	1.7F
	1306	1541	1.2E
	1747	2058	2.3F
3 Su	0017	0440	3.1E
	0829	1119	1.8F
	1419	1637	1.0E
	1831	2138	2.2F
4 M	0050	0524	3.1E
	0914	1213	1.9F
	1526	1733	0.8E
	1919	2217	2.1F
5 Tu	0124	0602	3.0E
	0953	1302	2.0F
	1622	1827	0.8E
	2008	2258	1.9F
6 W ●	0159	0630	3.0E
	1028	1345	2.0F
	1709	1916	0.8E
	2058	2341	1.8F
7 Th	0235	0650	2.9E
	1057	1421	2.0F
	1746	1958	0.9E
	2147		
8 F		0026	1.8F
	0314	0713	3.0E
	1122	1449	2.0F
	1815	2034	0.9E
	2234		
9 Sa		0112	1.8F
	0357	0745	3.0E
	1150	1507	2.0F
	1841	2108	1.1E
	2324		
10 Su		0159	1.7F
	0441	0823	3.0E
	1220	1521	2.1F
	1906	2142	1.2E
11 M	0018	0248	1.6F
	0531	0905	2.9E
	1251	1546	2.2F
	1931	2220	1.4E
12 Tu	0116	0342	1.5F
	0627	0952	2.6E
	1327	1620	2.3F
	2000	2302	1.7E
13 W	0219	0444	1.3F
	0729	1045	2.2E
	1401	1659	2.4F
	2032	2347	2.0E
14 Th ◐	0331	0554	1.2F
	0841	1141	1.8E
	1439	1741	2.4F
	2108		
15 F		0035	2.4E
	0443	0713	1.2F
	1003	1239	1.4E
	1519	1825	2.5F
	2147		

Day	Slack h m	Maximum h m	knots
16 Sa		0125	2.7E
	0551	0845	1.3F
	1128	1341	1.0E
	1559	1912	2.5F
	2229		
17 Su		0216	3.0E
	0651	0956	1.5F
	1249	1446	0.7E
	1644	2003	2.5F
	2312		
18 M		0308	3.2E
	0747	1054	1.7F
	1402	1550	0.6E
	1735	2057	2.5F
19 Tu	0000	0400	3.4E
	0836	1145	1.9F
	1502	1652	0.6E
	1836	2152	2.5F
20 W	0052	0452	3.5E
	0920	1232	2.0F
	1550	1751	0.8E
	1940	2246	2.5F
21 Th ○	0148	0547	3.5E
	1002	1315	2.1F
	1630	1846	1.0E
	2046	2342	2.5F
22 F	0245	0641	3.5E
	1042	1351	2.1F
	1709	1937	1.3E
	2148		
23 Sa		0039	2.5F
	0341	0732	3.5E
	1121	1425	2.2F
	1746	2025	1.6E
	2250		
24 Su		0136	2.3F
	0438	0820	3.3E
	1200	1459	2.3F
	1824	2115	1.8E
	2355		
25 M		0235	2.1F
	0536	0908	3.0E
	1239	1536	2.4F
	1905	2210	2.0E
26 Tu	0104	0340	1.8F
	0637	1001	2.6E
	1319	1618	2.4F
	1949	2312	2.2E
27 W ◑	0221	0459	1.5F
	0744	1102	2.2E
	1401	1704	2.4F
	2036		
28 Th		0019	2.4E
	0347	0629	1.3F
	0901	1207	1.7E
	1448	1753	2.4F
	2125		
29 F		0128	2.6E
	0511	0757	1.3F
	1029	1315	1.3E
	1537	1844	2.3F
	2214		
30 Sa		0235	2.8E
	0623	0913	1.5F
	1156	1427	1.1E
	1630	1938	2.2F
	2301		
31 Su		0333	3.0E
	0721	1015	1.8F
	1313	1533	1.0E
	1727	2034	2.0F
	2348		

August

Day	Slack h m	Maximum h m	knots
1 M		0424	3.0E
	0811	1107	2.0F
	1418	1631	1.0E
	1824	2126	1.9F
2 Tu	0030	0508	3.0E
	0852	1155	2.1F
	1510	1725	1.1E
	1920	2213	1.9F
3 W	0111	0545	2.9E
	0928	1238	2.1F
	1553	1814	1.1E
	2012	2255	1.8F
4 Th ●	0150	0614	2.9E
	0957	1315	2.1F
	1629	1858	1.2E
	2059	2336	1.8F
5 F	0229	0634	2.8E
	1021	1342	2.0F
	1655	1934	1.3E
	2141		
6 Sa		0017	1.8F
	0308	0655	2.9E
	1043	1355	2.0F
	1716	2002	1.3E
	2222		
7 Su		0058	1.8F
	0348	0723	2.9E
	1107	1401	2.1F
	1734	2023	1.5E
	2305		
8 M		0140	1.8F
	0430	0757	2.8E
	1131	1418	2.2F
	1752	2044	1.7E
	2351		
9 Tu		0225	1.7F
	0516	0835	2.7E
	1200	1447	2.4F
	1817	2114	2.0E
10 W	0043	0313	1.6F
	0609	0918	2.3E
	1230	1523	2.5F
	1846	2152	2.2E
11 Th	0144	0411	1.4F
	0711	1008	1.9E
	1306	1605	2.5F
	1920	2240	2.4E
12 F ◐	0255	0523	1.2F
	0828	1107	1.4E
	1344	1653	2.4F
	2003	2335	2.6E
13 Sa	0412	0656	1.1F
	0959	1213	1.0E
	1430	1745	2.3F
	2052		
14 Su		0035	2.8E
	0523	0836	1.3F
	1130	1327	0.7E
	1528	1841	2.3F
	2149		
15 M		0140	3.0E
	0626	0944	1.6F
	1246	1443	0.7E
	1633	1942	2.2F
	2249		

Day	Slack h m	Maximum h m	knots
16 Tu		0249	3.1E
	0720	1036	1.9F
	1343	1549	0.8E
	1741	2046	2.3F
	2349		
17 W		0352	3.2E
	0809	1121	2.0F
	1430	1646	1.0E
	1848	2147	2.4F
18 Th	0049	0448	3.3E
	0851	1200	2.1F
	1509	1738	1.3E
	1950	2244	2.5F
19 F ○	0148	0541	3.4E
	0931	1235	2.2F
	1543	1827	1.6E
	2049	2338	2.5F
20 Sa	0244	0630	3.3E
	1008	1306	2.2F
	1618	1913	1.9E
	2145		
21 Su		0033	2.5F
	0339	0715	3.2E
	1043	1334	2.3F
	1651	1956	2.1E
	2241		
22 M		0128	2.3F
	0432	0759	3.0E
	1118	1405	2.4F
	1728	2039	2.3E
	2342		
23 Tu		0223	2.1F
	0528	0843	2.6E
	1153	1441	2.5F
	1806	2125	2.4E
24 W	0048	0326	1.7F
	0628	0932	2.1E
	1231	1521	2.4F
	1848	2221	2.4E
25 Th	0202	0449	1.4F
	0737	1032	1.6E
	1313	1608	2.3F
	1933	2332	2.4E
26 F ◑	0327	0623	1.3F
	0900	1144	1.2E
	1402	1702	2.1F
	2027		
27 Sa		0049	2.5E
	0448	0747	1.4F
	1034	1301	1.0E
	1504	1803	1.9F
	2125		
28 Su		0203	2.6E
	0556	0857	1.7F
	1157	1417	1.0E
	1613	1909	1.8F
	2226		
29 M		0306	2.8E
	0650	0954	2.0F
	1300	1522	1.1E
	1722	2020	1.7F
	2321		
30 Tu		0357	2.8E
	0736	1041	2.2F
	1350	1617	1.4E
	1827	2122	1.8F
31 W	0013	0439	2.8E
	0814	1123	2.3F
	1431	1706	1.5E
	1921	2211	1.8F

September

Day	Slack h m	Maximum h m	knots
1 Th	0100	0514	2.8E
	0847	1159	2.2F
	1506	1750	1.6E
	2009	2253	1.8F
2 F	0142	0542	2.7E
	0912	1227	2.1F
	1532	1828	1.7E
	2051	2331	1.9F
3 Sa ●	0222	0604	2.6E
	0934	1243	2.0F
	1551	1859	1.7E
	2130		
4 Su		0007	1.9F
	0300	0627	2.6E
	0955	1247	2.0F
	1608	1919	1.8E
	2209		
5 M		0045	1.9F
	0340	0656	2.5E
	1017	1300	2.2F
	1622	1933	2.0E
	2249		
6 Tu		0124	1.8F
	0421	0730	2.4E
	1040	1326	2.4F
	1641	1954	2.3E
	2331		
7 W		0206	1.8F
	0509	0808	2.1E
	1108	1359	2.5F
	1708	2025	2.6E
8 Th	0021	0253	1.6F
	0606	0851	1.7E
	1139	1438	2.6F
	1741	2104	2.8E
9 F	0119	0351	1.4F
	0712	0943	1.3E
	1217	1523	2.5F
	1823	2153	2.9E
10 Sa	0227	0510	1.3F
	0836	1048	0.9E
	1303	1616	2.3F
	1913	2253	2.8E
11 Su ◐	0340	0655	1.3F
	1008	1207	0.7E
	1407	1717	2.0F
	2013		
12 M		0003	2.8E
	0450	0820	1.6F
	1126	1329	0.7E
	1527	1823	1.9F
	2124		
13 Tu		0120	2.8E
	0552	0920	1.8F
	1223	1443	1.0E
	1646	1934	1.9F
	2238		
14 W		0239	2.9E
	0646	1007	2.1F
	1309	1543	1.3E
	1755	2045	2.1F
	2346		
15 Th		0343	3.1E
	0732	1045	2.2F
	1349	1633	1.6E
	1857	2148	2.3F

Day	Slack h m	Maximum h m	knots
16 F	0048	0435	3.1E
	0813	1118	2.3F
	1423	1719	1.9E
	1953	2243	2.4F
17 Sa ○	0146	0522	3.1E
	0850	1146	2.3F
	1455	1803	2.2E
	2048	2335	2.4F
18 Su	0240	0608	2.9E
	0924	1213	2.4F
	1527	1845	2.4E
	2140		
19 M		0027	2.3F
	0332	0652	2.7E
	0958	1243	2.4F
	1558	1924	2.6E
	2235		
20 Tu		0121	2.2F
	0427	0734	2.3E
	1030	1315	2.5F
	1630	2002	2.7E
	2332		
21 W		0216	1.9F
	0523	0819	1.9E
	1105	1352	2.5F
	1707	2041	2.7E
22 Th	0034	0321	1.6F
	0628	0908	1.5E
	1143	1434	2.4F
	1748	2127	2.7E
23 F	0142	0447	1.4F
	0741	1011	1.1E
	1230	1522	2.1F
	1834	2228	2.5E
24 Sa ◑	0255	0613	1.5F
	0909	1131	0.8E
	1329	1620	1.8F
	1930	2354	2.4E
25 Su	0408	0725	1.6F
	1032	1250	0.9E
	1445	1730	1.6F
	2035		
26 M		0116	2.4E
	0509	0828	1.9F
	1137	1401	1.1E
	1606	1847	1.5F
	2147		
27 Tu		0223	2.5E
	0600	0920	2.1F
	1226	1501	1.4E
	1718	2005	1.5F
	2253		
28 W		0316	2.6E
	0643	1003	2.3F
	1307	1553	1.7E
	1819	2110	1.7F
	2351		
29 Th		0357	2.6E
	0720	1039	2.3F
	1340	1637	1.9E
	1910	2200	1.8F
30 F	0041	0431	2.5E
	0750	1107	2.2F
	1409	1716	2.1E
	1957	2242	1.9F

Time meridian 120° W. 0000 is midnight. 1200 is noon. Times are not adjusted for Daylight Saving Time.

Carquinez Strait (West End Bridge), San Pablo Bay, Calif., 2005

F–Flood, Dir. 103° True E–Ebb, Dir. 283° True

October

Day	Slack h m	Maximum h m	knots
1 Sa	0127	0459	2.4E
	0816	1124	2.1F
	1430	1750	2.1E
	2039	2320	1.9F
2 Su	0210	0525	2.3E
	0838	1130	2.1F
	1446	1814	2.2E
	2118	2357	1.8F
3 M ●	0252	0554	2.1E
	0859	1144	2.2F
	1500	1829	2.4E
	2157		
4 Tu		0035	1.8F
	0336	0627	1.9E
	0921	1209	2.4F
	1518	1847	2.7E
	2237		
5 W		0114	1.8F
	0423	0705	1.7E
	0946	1241	2.5F
	1542	1914	2.9E
	2320		
6 Th		0157	1.7F
	0518	0747	1.4E
	1016	1319	2.6F
	1616	1949	3.2E
7 F	0008	0246	1.6F
	0620	0834	1.1E
	1052	1403	2.5F
	1655	2032	3.2E
8 Sa	0102	0349	1.5F
	0731	0932	0.8E
	1140	1452	2.3F
	1742	2123	3.1E
9 Su	0203	0517	1.5F
	0850	1047	0.7E
	1247	1551	2.0F
	1840	2227	2.9E
10 M ◐	0310	0643	1.6F
	1004	1212	0.7E
	1411	1701	1.7F
	1950	2346	2.7E
11 Tu	0416	0750	1.8F
	1103	1328	1.0E
	1539	1816	1.7F
	2111		
12 W		0108	2.7E
	0513	0845	2.1F
	1150	1433	1.5E
	1656	1933	1.7F
	2231		
13 Th		0224	2.8E
	0603	0928	2.3F
	1230	1528	1.9E
	1800	2047	1.9F
	2340		
14 F		0324	2.8E
	0648	1003	2.4F
	1307	1615	2.2E
	1900	2149	2.1F
15 Sa	0041	0413	2.8E
	0728	1032	2.4F
	1339	1658	2.5E
	1956	2243	2.2F
16 Su	0139	0458	2.6E
	0802	1058	2.5F
	1409	1740	2.7E
	2049	2335	2.2F
17 M ○	0235	0542	2.3E
	0837	1125	2.5F
	1439	1819	2.8E
	2141		
18 Tu		0029	2.1F
	0330	0627	1.9E
	0909	1156	2.5F
	1508	1855	2.9E
	2233		
19 W		0124	1.9F
	0429	0714	1.6E
	0941	1232	2.5F
	1540	1929	3.0E
	2327		
20 Th		0223	1.8F
	0532	0802	1.2E
	1019	1312	2.4F
	1616	2003	3.0E
21 F	0021	0329	1.7F
	0640	0855	0.9E
	1101	1356	2.2F
	1657	2044	2.9E
22 Sa	0118	0442	1.6F
	0752	1000	0.8E
	1157	1447	1.9F
	1745	2134	2.7E
23 Su	0217	0551	1.7F
	0904	1118	0.7E
	1308	1547	1.6F
	1842	2246	2.5E
24 M ◑	0315	0651	1.8F
	1009	1230	0.9E
	1430	1701	1.3F
	1950		
25 Tu		0009	2.4E
	0410	0745	2.0F
	1059	1334	1.3E
	1550	1821	1.3F
	2105		
26 W		0118	2.4E
	0458	0832	2.2F
	1140	1430	1.6E
	1700	1938	1.4F
	2217		
27 Th		0214	2.4E
	0539	0911	2.2F
	1214	1519	2.0E
	1800	2044	1.5F
	2319		
28 F		0300	2.3E
	0614	0941	2.3F
	1242	1601	2.2E
	1851	2138	1.7F
29 Sa	0014	0338	2.2E
	0645	1001	2.2F
	1307	1636	2.4E
	1939	2225	1.8F
30 Su	0106	0411	2.0E
	0711	1012	2.2F
	1325	1705	2.5E
	2023	2308	1.8F
31 M	0157	0445	1.7E
	0736	1028	2.3F
	1341	1724	2.7E
	2107	2350	1.8F

November

Day	Slack h m	Maximum h m	knots
1 Tu ●	0249	0521	1.5E
	0759	1054	2.4F
	1401	1742	2.9E
	2148		
2 W		0033	1.7F
	0341	0601	1.2E
	0824	1126	2.5F
	1428	1808	3.2E
	2229		
3 Th		0118	1.7F
	0439	0647	1.0E
	0855	1205	2.6F
	1500	1844	3.4E
	2311		
4 F		0204	1.7F
	0539	0737	0.8E
	0933	1249	2.5F
	1539	1926	3.5E
	2358		
5 Sa		0256	1.7F
	0640	0831	0.7E
	1024	1339	2.3F
	1626	2013	3.4E
6 Su	0048	0358	1.7F
	0741	0933	0.7E
	1130	1434	2.1F
	1720	2109	3.2E
7 M	0142	0511	1.8F
	0841	1050	0.8E
	1252	1538	1.8F
	1825	2216	2.9E
8 Tu ◐	0240	0614	1.9F
	0939	1207	1.1E
	1419	1652	1.6F
	1941	2336	2.7E
9 W	0338	0709	2.1F
	1029	1313	1.5E
	1542	1812	1.5F
	2102		
10 Th		0051	2.6E
	0430	0758	2.3F
	1111	1414	1.9E
	1657	1933	1.6F
	2221		
11 F		0159	2.6E
	0518	0840	2.4F
	1151	1508	2.3E
	1802	2048	1.8F
	2330		
12 Sa		0257	2.5E
	0600	0916	2.5F
	1227	1556	2.6E
	1903	2152	1.9F
13 Su	0034	0348	2.3E
	0639	0946	2.6F
	1259	1640	2.9E
	2000	2248	1.9F
14 M	0136	0435	1.9E
	0716	1015	2.6F
	1329	1722	3.0E
	2054	2344	1.9F
15 Tu ○	0238	0522	1.6E
	0750	1045	2.5F
	1359	1800	3.1E
	2146		
16 W		0042	1.8F
	0341	0612	1.2E
	0825	1119	2.4F
	1429	1834	3.1E
	2234		
17 Th		0139	1.8F
	0448	0705	0.9E
	0901	1158	2.3F
	1500	1904	3.1E
	2320		
18 F		0234	1.8F
	0552	0757	0.8E
	0944	1241	2.2F
	1539	1937	3.1E
19 Sa	0004	0328	1.8F
	0651	0850	0.7E
	1036	1330	2.0F
	1620	2015	3.0E
20 Su	0049	0423	1.8F
	0746	0948	0.7E
	1139	1422	1.7F
	1710	2101	2.8E
21 M	0133	0515	1.9F
	0837	1055	0.8E
	1250	1522	1.5F
	1807	2158	2.7E
22 Tu	0220	0603	2.0F
	0923	1157	1.1E
	1407	1631	1.3F
	1911	2303	2.5E
23 W ◑	0306	0645	2.1F
	1005	1254	1.4E
	1521	1746	1.2F
	2021		
24 Th		0007	2.4E
	0349	0723	2.1F
	1041	1346	1.7E
	1630	1859	1.3F
	2133		
25 F		0103	2.2E
	0429	0755	2.2F
	1111	1434	2.1E
	1731	2009	1.4F
	2241		
26 Sa		0155	2.0E
	0505	0819	2.2F
	1130	1510	2.4E
	1828	2113	1.5F
	2346		
27 Su		0242	1.8E
	0537	0842	2.3F
	1201	1549	2.6E
	1920	2207	1.6F
28 M	0047	0326	1.5E
	0606	0908	2.3F
	1223	1616	2.9E
	2009	2258	1.7F
29 Tu	0149	0410	1.2E
	0632	0939	2.4F
	1248	1638	3.1E
	2054	2349	1.7F
30 W	0252	0455	0.9E
	0700	1014	2.5F
	1317	1706	3.3E
	2138		

December

Day	Slack h m	Maximum h m	knots
1 Th ●		0039	1.7F
	0357	0546	0.7E
	0735	1054	2.5F
	1351	1742	3.5E
	2219		
2 F		0128	1.8F
	0458	0641	0.6E
	0819	1140	2.5F
	1433	1826	3.6E
	2301		
3 Sa		0214	1.8F
	0550	0735	0.6E
	0916	1230	2.4F
	1521	1914	3.6E
	2345		
4 Su		0259	1.9F
	0637	0830	0.7E
	1021	1326	2.2F
	1614	2007	3.4E
5 M	0030	0347	2.0F
	0722	0929	0.9E
	1137	1425	2.0F
	1714	2104	3.2E
6 Tu	0118	0439	2.1F
	0809	1037	1.2E
	1255	1530	1.8F
	1821	2208	3.0E
7 W	0208	0531	2.2F
	0857	1145	1.5E
	1416	1644	1.6F
	1934	2319	2.7E
8 Th ◐	0257	0619	2.3F
	0944	1248	1.9E
	1536	1806	1.5F
	2050		
9 F		0026	2.5E
	0346	0704	2.5F
	1029	1349	2.3E
	1652	1929	1.5F
	2208		
10 Sa		0130	2.3E
	0431	0747	2.5F
	1110	1447	2.6E
	1803	2049	1.5F
	2320		
11 Su		0230	2.0E
	0514	0827	2.6F
	1149	1540	2.9E
	1909	2156	1.7F
12 M	0032	0327	1.7E
	0555	0904	2.5F
	1222	1628	3.1E
	2008	2257	1.7F
13 Tu	0143	0420	1.3E
	0634	0940	2.5F
	1256	1712	3.2E
	2100	2355	1.8F
14 W	0254	0514	1.0E
	0714	1016	2.4F
	1328	1752	3.1E
	2148		
15 Th ○		0051	1.9F
	0403	0611	0.8E
	0757	1054	2.2F
	1400	1827	3.1E
	2230		
16 F		0143	1.9F
	0505	0706	0.7E
	0842	1137	2.1F
	1438	1855	3.1E
	2308		
17 Sa		0229	1.9F
	0555	0755	0.7E
	0933	1223	1.9F
	1517	1923	3.0E
	2341		
18 Su		0309	1.9F
	0637	0841	0.8E
	1028	1313	1.8F
	1600	1957	3.0E
19 M	0016	0346	2.0F
	0713	0927	0.9E
	1128	1404	1.7F
	1648	2037	3.0E
20 Tu	0050	0419	2.0F
	0748	1017	1.0E
	1229	1458	1.6F
	1740	2122	2.8E
21 W	0127	0450	2.0F
	0821	1109	1.2E
	1333	1557	1.4F
	1838	2214	2.6E
22 Th	0204	0519	2.1F
	0855	1158	1.5E
	1440	1705	1.3F
	1942	2310	2.4E
23 F ◑	0242	0550	2.2F
	0929	1245	1.8E
	1550	1815	1.2F
	2051		
24 Sa		0005	2.1E
	0320	0623	2.2F
	0959	1331	2.1E
	1658	1930	1.2F
	2207		
25 Su		0100	1.7E
	0358	0659	2.3F
	1029	1414	2.4E
	1801	2047	1.3F
	2321		
26 M		0154	1.3E
	0431	0736	2.1F
	1059	1452	2.8E
	1859	2154	1.5F
27 Tu	0037	0250	1.0E
	0506	0817	2.4F
	1130	1526	3.0E
	1951	2251	1.6F
28 W	0150	0345	0.7E
	0539	0900	2.5F
	1207	1602	3.3E
	2039	2344	1.7F
29 Th	0259	0441	0.5E
	0620	0946	2.5F
	1247	1642	3.4E
	2121		
30 F ●		0034	1.8F
	0359	0539	0.5E
	0713	1035	2.5F
	1332	1728	3.5E
	2203		
31 Sa		0119	1.9F
	0446	0635	0.6E
	0816	1126	2.4F
	1423	1819	3.6E
	2244		

Time meridian 120° W. 0000 is midnight. 1200 is noon. Times are not adjusted for Daylight Saving Time.

Benicia Bridge, Suisun Bay, Calif., 2005

F–Flood, Dir. 047° True   E–Ebb, Dir. 230° True

January

Day	Slack h m	Maximum h m	knots
1 Sa	0103	0404	1.6F
	0754	1027	1.0E
	1322	1547	1.0F
	1821	2158	1.8E
2 Su	0136	0437	1.6F
	0824	1111	1.1E
	1432	1648	0.8F
	1924	2248	1.5E
3 M ◑	0211	0516	1.6F
	0856	1200	1.3E
	1549	1803	0.7F
	2045	2345	1.1E
4 Tu	0250	0601	1.5F
	0932	1256	1.5E
	1706	1935	0.8F
	2220		
5 W		0049	0.8E
	0335	0652	1.5F
	1013	1358	1.7E
	1816	2100	1.0F
	2351		
6 Th		0159	0.6E
	0428	0747	1.5F
	1059	1503	1.9E
	1917	2207	1.2F
7 F	0109	0310	0.6E
	0526	0843	1.6F
	1150	1607	2.2E
	2010	2302	1.4F
8 Sa	0214	0417	0.6E
	0625	0938	1.7F
	1242	1705	2.4E
	2100	2352	1.6F
9 Su	0309	0518	0.7E
	0725	1032	1.8F
	1334	1757	2.5E
	2146		
10 M ●		0038	1.7F
	0357	0614	0.8E
	0824	1125	1.9F
	1426	1843	2.6E
	2229		
11 Tu		0123	1.9F
	0443	0707	0.9E
	0922	1217	1.9F
	1517	1926	2.6E
	2310		
12 W		0205	1.9F
	0526	0758	1.1E
	1021	1309	1.8F
	1607	2006	2.4E
	2349		
13 Th		0247	2.0F
	0609	0849	1.2E
	1122	1402	1.6F
	1657	2046	2.2E
14 F	0026	0328	2.0F
	0652	0941	1.3E
	1228	1459	1.3F
	1750	2128	1.9E
15 Sa	0104	0410	1.9F
	0735	1036	1.4E
	1340	1604	1.0F
	1849	2215	1.6E
16 Su ◐	0143	0454	1.8F
	0820	1139	1.4E
	1501	1726	0.8F
	2002	2312	1.2E
17 M	0226	0544	1.6F
	0907	1252	1.5E
	1628	1904	0.8F
	2132		
18 Tu		0026	0.8E
	0316	0638	1.5F
	0955	1407	1.6E
	1747	2028	0.9F
	2308		
19 W		0155	0.6E
	0414	0736	1.3F
	1044	1514	1.8E
	1849	2134	1.2F
20 Th	0035	0312	0.6E
	0516	0832	1.3F
	1132	1609	2.0E
	1939	2229	1.4F
21 F	0142	0412	0.7E
	0614	0922	1.3F
	1216	1655	2.1E
	2022	2316	1.6F
22 Sa	0233	0502	0.8E
	0705	1005	1.3F
	1257	1734	2.2E
	2100	2358	1.7F
23 Su	0316	0545	0.8E
	0750	1043	1.4F
	1334	1804	2.2E
	2133		
24 M		0036	1.8F
	0354	0623	0.9E
	0831	1118	1.4F
	1408	1823	2.2E
	2202		
25 Tu ○		0109	1.8F
	0429	0658	0.9E
	0911	1153	1.4F
	1441	1838	2.2E
	2229		
26 W		0137	1.8F
	0500	0729	0.9E
	0949	1229	1.5F
	1516	1900	2.3E
	2254		
27 Th		0157	1.8F
	0527	0758	1.0E
	1029	1307	1.5F
	1552	1930	2.3E
	2319		
28 F		0215	1.8F
	0552	0827	1.1E
	1112	1348	1.4F
	1632	2005	2.2E
	2345		
29 Sa		0237	1.8F
	0615	0858	1.2E
	1159	1433	1.3F
	1718	2044	2.0E
30 Su	0013	0306	1.8F
	0638	0933	1.3E
	1254	1524	1.2F
	1811	2128	1.6E
31 M	0044	0342	1.7F
	0707	1014	1.4E
	1359	1624	1.0F
	1918	2219	1.3E

February

Day	Slack h m	Maximum h m	knots
1 Tu ◑	0119	0424	1.6F
	0743	1104	1.5E
	1518	1743	0.8F
	2045	2319	0.9E
2 W	0201	0514	1.5F
	0830	1205	1.6E
	1643	1929	0.8F
	2224		
3 Th		0031	0.6E
	0256	0615	1.4F
	0927	1325	1.7E
	1800	2056	1.0F
	2355		
4 F		0152	0.5E
	0405	0721	1.4F
	1032	1502	1.9E
	1903	2159	1.3F
5 Sa	0106	0311	0.5E
	0518	0828	1.5F
	1135	1612	2.2E
	1955	2250	1.5F
6 Su	0200	0417	0.7E
	0625	0930	1.7F
	1234	1705	2.4E
	2041	2334	1.7F
7 M	0245	0512	0.9E
	0726	1025	1.8F
	1328	1749	2.5E
	2122		
8 Tu ●		0014	1.9F
	0326	0602	1.1E
	0823	1117	1.9F
	1419	1827	2.5E
	2159		
9 W		0052	2.0F
	0404	0648	1.3E
	0919	1207	1.9F
	1507	1903	2.4E
	2234		
10 Th		0127	2.0F
	0442	0733	1.4E
	1015	1257	1.8F
	1555	1938	2.2E
	2308		
11 F		0202	2.0F
	0518	0818	1.5E
	1112	1349	1.6F
	1645	2014	1.9E
	2341		
12 Sa		0237	1.9F
	0556	0902	1.6E
	1213	1444	1.3F
	1739	2055	1.6E
13 Su	0015	0313	1.8F
	0634	0949	1.6E
	1319	1548	1.0F
	1842	2142	1.2E
14 M	0053	0354	1.6F
	0716	1044	1.5E
	1435	1714	0.8F
	2000	2241	0.9E
15 Tu ◐	0137	0442	1.3F
	0803	1200	1.5E
	1601	1854	0.8F
	2135		
16 W		0005	0.6E
	0234	0542	1.1F
	0859	1337	1.5E
	1721	2015	1.0F
	2312		
17 Th		0143	0.5E
	0347	0657	1.0F
	1000	1451	1.6E
	1824	2117	1.3F
18 F	0029	0259	0.6E
	0502	0809	1.0F
	1101	1547	1.8E
	1912	2208	1.5F
19 Sa	0123	0355	0.8E
	0605	0906	1.1F
	1153	1631	2.0E
	1952	2251	1.7F
20 Su	0205	0442	0.9E
	0657	0951	1.2F
	1237	1706	2.1E
	2026	2329	1.8F
21 M	0241	0521	1.0E
	0741	1029	1.3F
	1315	1731	2.1E
	2055		
22 Tu		0001	1.9F
	0313	0557	1.1E
	0820	1103	1.4F
	1350	1747	2.1E
	2120		
23 W ○		0027	1.8F
	0341	0627	1.2E
	0858	1137	1.5F
	1425	1804	2.1E
	2144		
24 Th		0046	1.8F
	0404	0654	1.3E
	0935	1212	1.5F
	1501	1829	2.1E
	2208		
25 F		0101	1.8F
	0425	0718	1.4E
	1014	1250	1.6F
	1541	1901	2.0E
	2232		
26 Sa		0121	1.9F
	0444	0743	1.5E
	1055	1332	1.5F
	1626	1939	1.9E
	2259		
27 Su		0148	1.9F
	0505	0813	1.7E
	1141	1417	1.5F
	1717	2021	1.6E
	2329		
28 M		0221	1.8F
	0530	0848	1.8E
	1234	1510	1.3F
	1816	2108	1.3E

March

Day	Slack h m	Maximum h m	knots
1 Tu	0002	0301	1.7F
	0604	0931	1.8E
	1338	1613	1.1F
	1929	2203	1.0E
2 W	0042	0347	1.5F
	0648	1024	1.7E
	1456	1740	1.0F
	2059	2309	0.6E
3 Th ◑	0133	0444	1.4F
	0746	1134	1.7E
	1622	1927	1.0F
	2232		
4 F		0030	0.5E
	0245	0552	1.3F
	0859	1317	1.7E
	1738	2043	1.2F
	2350		
5 Sa		0159	0.5E
	0409	0709	1.3F
	1016	1459	1.9E
	1839	2139	1.5F
6 Su	0048	0313	0.7E
	0525	0823	1.4F
	1126	1600	2.1E
	1928	2225	1.7F
7 M	0132	0411	1.0E
	0631	0926	1.6F
	1227	1646	2.3E
	2009	2304	1.9F
8 Tu	0211	0500	1.3E
	0730	1021	1.8F
	1320	1726	2.3E
	2046	2340	2.0F
9 W	0246	0544	1.5E
	0824	1111	1.8F
	1410	1801	2.2E
	2119		
10 Th ●		0012	2.0F
	0319	0626	1.6E
	0917	1200	1.8F
	1459	1834	2.0E
	2151		
11 F		0043	2.0F
	0351	0705	1.8E
	1009	1249	1.7F
	1548	1908	1.8E
	2222		
12 Sa		0114	1.9F
	0423	0742	1.8E
	1102	1339	1.5F
	1641	1946	1.5E
	2254		
13 Su		0146	1.8F
	0456	0818	1.8E
	1157	1434	1.3F
	1739	2029	1.2E
	2328		
14 M		0221	1.6F
	0530	0858	1.7E
	1257	1540	1.1F
	1847	2120	0.9E
15 Tu	0007	0301	1.4F
	0610	0944	1.6E
	1405	1706	1.0F
	2007	2225	0.6E
16 W	0056	0349	1.1F
	0657	1046	1.4E
	1523	1837	1.0F
	2138	2353	0.4E
17 Th ◐	0204	0451	0.9F
	0758	1244	1.4E
	1639	1951	1.1F
	2301		
18 F		0125	0.5E
	0328	0614	0.8F
	0910	1412	1.5E
	1741	2049	1.4F
19 Sa	0003	0235	0.7E
	0448	0740	0.8F
	1020	1508	1.6E
	1829	2136	1.6F
20 Su	0048	0329	0.9E
	0553	0843	1.0F
	1119	1549	1.7E
	1906	2216	1.7F
21 M	0125	0414	1.1E
	0644	0930	1.1F
	1207	1620	1.8E
	1937	2249	1.8F
22 Tu	0156	0452	1.3E
	0727	1009	1.3F
	1249	1642	1.9E
	2004	2315	1.8F
23 W	0223	0524	1.4E
	0807	1045	1.4F
	1329	1701	1.9E
	2028	2333	1.8F
24 Th	0244	0551	1.5E
	0844	1121	1.5F
	1409	1726	1.8E
	2052	2348	1.8F
25 F ○	0303	0614	1.7E
	0921	1158	1.6F
	1452	1758	1.8E
	2117		
26 Sa		0008	1.9F
	0321	0636	1.8E
	1000	1238	1.6F
	1539	1836	1.6E
	2145		
27 Su		0034	1.9F
	0341	0703	2.0E
	1042	1322	1.6F
	1630	1918	1.4E
	2216		
28 M		0107	1.9F
	0407	0736	2.1E
	1129	1410	1.5F
	1728	2005	1.2E
	2250		
29 Tu		0146	1.8F
	0440	0815	2.1E
	1223	1506	1.4F
	1833	2057	0.9E
	2331		
30 W		0231	1.6F
	0522	0903	2.0E
	1327	1616	1.3F
	1948	2158	0.7E
31 Th	0021	0323	1.5F
	0615	1003	1.9E
	1441	1746	1.2F
	2110	2311	0.6E

Time meridian 120° W. 0000 is midnight. 1200 is noon. Times are not adjusted for Daylight Saving Time.

Benicia Bridge, Suisun Bay, Calif., 2005

F–Flood, Dir. 047° True E–Ebb, Dir. 230° True

April

Day	Slack (h m)	Maximum (h m)	knots
1 F	0128 0721 1557 2227	0425 1122 1912	1.3F 1.8E 1.3F
2 Sa	0252 0840 1706 2329	0037 0540 1309 2018	0.6E 1.1F 1.8E 1.5F
3 Su	0418 1002 1801	0159 0703 1434 2109	0.8E 1.2F 1.9E 1.7F
4 M	0017 0533 1114 1847	0304 0820 1531 2151	1.1E 1.3F 2.0E 1.9F
5 Tu	0057 0637 1216 1926	0358 0924 1616 2228	1.4E 1.5F 2.0E 2.0F
6 W	0132 0734 1311 2001	0444 1018 1655 2300	1.7E 1.6F 1.9E 2.0F
7 Th	0204 0826 1404 2034	0525 1109 1732 2330	1.9E 1.7F 1.7E 2.0F
8 F	0233 0916 1457 2105	0603 1157 1807 2359	2.0E 1.6F 1.5E 1.9F
9 Sa	0302 1004 1551 2137	0636 1246 1845	2.0E 1.6F 1.2E
10 Su	0331 1053 1649 2210	0029 0707 1337 1927	1.8F 2.0E 1.5F 1.0E
11 M	0401 1143 1751 2247	0102 0739 1433 2015	1.6F 2.0E 1.3F 0.8E
12 Tu	0435 1236 1859 2330	0139 0816 1538 2110	1.4F 1.9E 1.2F 0.6E
13 W	0514 1335 2013	0221 0859 1654 2217	1.2F 1.7E 1.1F 0.5E
14 Th	0026 0601 1439 2127	0310 0953 1809 2336	1.0F 1.6E 1.2F 0.4E
15 F	0139 0659 1543 2232	0410 1104 1914	0.8F 1.4E 1.3F
16 Sa	0304 0810 1639 2323	0055 0526 1239 2008	0.5E 0.6F 1.4E 1.4F
17 Su	0426 0924 1725	0201 0654 1350 2052	0.7E 0.6F 1.4E 1.6F
18 M	0004 0533 1031 1802	0254 0805 1436 2128	1.0E 0.8F 1.5E 1.7F
19 Tu	0036 0626 1128 1834	0339 0900 1510 2156	1.2E 0.9F 1.5E 1.7F
20 W	0103 0711 1220 1902	0415 0945 1541 2216	1.5E 1.1F 1.5E 1.8F
21 Th	0125 0751 1309 1929	0446 1026 1614 2232	1.6E 1.3F 1.5E 1.8F
22 F	0144 0830 1359 1957	0510 1106 1650 2253	1.8E 1.4F 1.4E 1.8F
23 Sa	0203 0909 1450 2026	0533 1148 1731 2321	2.0E 1.5F 1.3E 1.8F
24 Su	0224 0950 1545 2100	0559 1232 1816 2356	2.2E 1.6F 1.1E 1.8F
25 M	0252 1035 1642 2138	0631 1319 1904	2.3E 1.6F 1.0E
26 Tu	0327 1124 1744 2221	0036 0711 1412 1958	1.8F 2.3E 1.6F 0.9E
27 W	0409 1218 1848 2313	0121 0758 1512 2056	1.7F 2.3E 1.5F 0.7E
28 Th	0459 1318 1956	0211 0853 1621 2201	1.6F 2.2E 1.5F 0.7E
29 F	0015 0557 1422 2103	0308 0957 1735 2314	1.4F 2.0E 1.5F 0.7E
30 Sa	0132 0705 1525 2204	0413 1113 1844	1.2F 1.9E 1.6F

May

Day	Slack (h m)	Maximum (h m)	knots
1 Su	0258 0823 1623 2257	0033 0531 1239 1941	0.8E 1.1F 1.8E 1.7F
2 M	0423 0943 1714 2341	0146 0656 1353 2029	1.1E 1.0F 1.8E 1.9F
3 Tu	0538 1057 1758	0247 0816 1452 2111	1.4E 1.1F 1.7E 2.0F
4 W	0019 0642 1204 1837	0340 0921 1541 2147	1.7E 1.3F 1.6E 2.0F
5 Th	0052 0737 1307 1914	0427 1018 1626 2220	2.0E 1.4F 1.4E 2.0F
6 F	0123 0828 1406 1948	0508 1110 1709 2250	2.1E 1.5F 1.2E 1.9F
7 Sa	0151 0915 1505 2022	0544 1200 1751 2321	2.2E 1.5F 1.0E 1.7F
8 Su	0219 1000 1604 2057	0616 1250 1835 2353	2.2E 1.5F 0.8E 1.6F
9 M	0248 1045 1704 2135	0644 1341 1922	2.2E 1.5F 0.6E
10 Tu	0320 1131 1804 2217	0029 0714 1435 2012	1.5F 2.1E 1.4F 0.5E
11 W	0356 1217 1903 2306	0108 0749 1532 2105	1.3F 2.0E 1.4F 0.5E
12 Th	0436 1304 2001	0152 0831 1630 2202	1.2F 1.9E 1.4F 0.5E
13 F	0004 0521 1352 2056	0241 0918 1729 2305	1.0F 1.8E 1.4F 0.5E
14 Sa	0112 0611 1440 2147	0336 1010 1823	0.8F 1.7E 1.4F
15 Su	0231 0710 1525 2231	0011 0441 1106 1910	0.6E 0.6F 1.5E 1.5F
16 M	0351 0820 1606 2307	0113 0557 1204 1950	0.8E 0.5F 1.4E 1.5F
17 Tu	0503 0934 1644 2337	0208 0716 1301 2021	1.0E 0.6F 1.3E 1.6F
18 W	0602 1046 1719	0253 0824 1354 2044	1.3E 0.7F 1.3E 1.6F
19 Th	0002 0650 1153 1753	0331 0920 1443 2107	1.6E 0.9F 1.2E 1.7F
20 F	0023 0733 1255 1826	0402 1010 1531 2134	1.8E 1.2F 1.1E 1.7F
21 Sa	0045 0815 1356 1901	0429 1057 1620 2206	2.0E 1.4F 0.9E 1.8F
22 Su	0111 0858 1455 1939	0458 1143 1710 2245	2.2E 1.5F 0.8E 1.8F
23 M	0142 0943 1554 2022	0533 1231 1802 2327	2.4E 1.6F 0.8E 1.8F
24 Tu	0220 1030 1653 2110	0614 1322 1858	2.5E 1.7F 0.7E
25 W	0305 1120 1750 2204	0014 0702 1415 1955	1.8F 2.5E 1.7F 0.7E
26 Th	0354 1211 1847 2305	0104 0755 1511 2055	1.7F 2.5E 1.7F 0.7E
27 F	0448 1303 1943	0158 0851 1610 2158	1.6F 2.4E 1.7F 0.8E
28 Sa	0014 0547 1354 2037	0257 0949 1708 2305	1.4F 2.2E 1.8F 0.9E
29 Su	0132 0651 1445 2129	0403 1052 1804	1.2F 2.0E 1.8F
30 M	0256 0802 1534 2217	0015 0520 1158 1856	1.1E 1.0F 1.8E 1.9F
31 Tu	0421 0921 1621 2300	0123 0647 1305 1944	1.4E 0.9F 1.5E 1.9F

June

Day	Slack (h m)	Maximum (h m)	knots
1 W	0539 1042 1706 2338	0226 0810 1410 2027	1.7E 1.0F 1.3E 1.9F
2 Th	0643 1159 1749	0322 0920 1510 2107	1.9E 1.1F 1.1E 1.9F
3 F	0013 0738 1311 1831	0411 1019 1606 2143	2.1E 1.3F 0.9E 1.8F
4 Sa	0046 0827 1418 1911	0456 1113 1658 2218	2.2E 1.5F 0.8E 1.7F
5 Su	0117 0912 1520 1951	0535 1203 1747 2252	2.3E 1.5F 0.7E 1.6F
6 M	0148 0955 1618 2032	0610 1252 1835 2328	2.2E 1.6F 0.6E 1.5F
7 Tu	0221 1036 1712 2115	0639 1340 1921	2.2E 1.6F 0.5E
8 W	0256 1116 1801 2200	0007 0705 1426 2006	1.4F 2.2E 1.6F 0.5E
9 Th	0333 1154 1846 2250	0047 0735 1511 2051	1.3F 2.1E 1.6F 0.5E
10 F	0412 1230 1929 2344	0131 0809 1554 2137	1.2F 2.1E 1.6F 0.6E
11 Sa	0453 1304 2010	0217 0848 1633 2225	1.1F 2.0E 1.5F 0.7E
12 Su	0044 0537 1339 2049	0306 0930 1709 2316	0.9F 1.9E 1.5F 0.8E
13 M	0152 0627 1413 2125	0402 1016 1742	0.7F 1.7E 1.5F
14 Tu	0306 0728 1449 2157	0008 0507 1106 1813	0.9E 0.6F 1.5E 1.5F
15 W	0421 0845 1526 2226	0101 0624 1201 1847	1.1E 0.5F 1.3E 1.5F
16 Th	0528 1011 1605 2253	0149 0746 1259 1923	1.3E 0.6F 1.0E 1.5F
17 F	0625 1134 1645 2323	0233 0858 1359 2003	1.6E 0.9F 0.8E 1.6F
18 Sa	0715 1249 1728 2355	0313 0958 1459 2046	1.9E 1.1F 0.7E 1.6F
19 Su	0803 1357 1814	0353 1051 1558 2131	2.1E 1.3F 0.6E 1.7F
20 M	0034 0850 1459 1904	0436 1141 1656 2219	2.3E 1.5F 0.6E 1.8F
21 Tu	0117 0936 1555 1958	0523 1229 1754 2308	2.4E 1.6F 0.6E 1.8F
22 W	0204 1023 1647 2056	0613 1318 1852	2.5E 1.7F 0.7E
23 Th	0255 1109 1737 2157	0000 0703 1406 1948	1.9F 2.6E 1.8F 0.8E
24 F	0347 1153 1824 2302	0053 0753 1453 2045	1.8F 2.5E 1.9F 0.9E
25 Sa	0440 1236 1911	0148 0842 1540 2142	1.6F 2.4E 1.9F 1.0E
26 Su	0011 0536 1318 1958	0246 0930 1628 2242	1.4F 2.2E 2.0F 1.2E
27 M	0126 0636 1400 2044	0351 1021 1716 2346	1.2F 1.9E 2.0F 1.4E
28 Tu	0247 0744 1444 2130	0507 1118 1805	0.9F 1.6E 1.9F
29 W	0413 0905 1530 2214	0054 0638 1223 1854	1.5E 0.8F 1.2E 1.8F
30 Th	0533 1035 1618 2257	0201 0805 1337 1943	1.8E 0.9F 0.9E 1.7F

Time meridian 120° W. 0000 is midnight. 1200 is noon. Times are not adjusted for Daylight Saving Time.

Benicia Bridge, Suisun Bay, Calif., 2005

F–Flood, Dir. 047° True   E–Ebb, Dir. 230° True

July

Day	Slack h m	Maximum h m	knots
1 F		0303	2.0E
	0638	0917	1.1F
	1202	1451	0.8E
	1709	2031	1.6F
	2337		
2 Sa		0358	2.1E
	0733	1017	1.3F
	1320	1556	0.7E
	1800	2115	1.6F
3 Su	0016	0447	2.2E
	0820	1110	1.5F
	1427	1653	0.6E
	1849	2156	1.5F
4 M	0053	0530	2.3E
	0903	1158	1.6F
	1523	1743	0.6E
	1936	2236	1.4F
5 Tu	0129	0607	2.3E
	0943	1242	1.7F
	1611	1828	0.6E
	2021	2314	1.4F
6 W	0205	0636	2.2E
	1019	1324	1.7F
	1654	1909	0.6E
●	2105	2353	1.4F
7 Th	0241	0656	2.2E
	1052	1402	1.7F
	1733	1948	0.7E
	2150		
8 F		0032	1.3F
	0317	0717	2.2E
	1122	1436	1.7F
	1808	2026	0.7E
	2236		
9 Sa		0112	1.3F
	0354	0745	2.2E
	1149	1503	1.7F
	1841	2102	0.8E
	2324		
10 Su		0155	1.2F
	0432	0819	2.1E
	1216	1526	1.7F
	1911	2138	0.9E
11 M	0016	0241	1.0F
	0514	0857	2.0E
	1244	1549	1.7F
	1939	2216	1.0E
12 Tu	0115	0331	0.9F
	0602	0939	1.8E
	1313	1617	1.6F
	2007	2257	1.1E
13 W	0221	0430	0.7F
	0702	1027	1.5E
	1345	1652	1.6F
	2036	2342	1.3E
14 Th	0334	0543	0.6F
	0820	1121	1.1E
	1421	1734	1.5F
◐	2109		
15 F		0033	1.5E
	0449	0714	0.7F
	0953	1222	0.8E
	1502	1823	1.5F
	2147		
16 Sa		0130	1.7E
	0557	0841	0.9F
	1126	1329	0.6E
	1551	1916	1.5F
	2232		
17 Su		0231	1.9E
	0657	0947	1.1F
	1247	1439	0.5E
	1647	2012	1.6F
	2321		
18 M		0334	2.1E
	0749	1042	1.3F
	1353	1546	0.5E
	1748	2108	1.7F
19 Tu	0012	0432	2.3E
	0838	1131	1.5F
	1448	1648	0.6E
	1850	2203	1.8F
20 W	0105	0525	2.5E
	0923	1216	1.7F
	1536	1746	0.7E
	1951	2257	1.9F
21 Th	0157	0612	2.6E
	1005	1259	1.8F
	1620	1840	0.9E
○	2053	2350	1.9F
22 F	0249	0657	2.6E
	1045	1340	1.9F
	1703	1933	1.0E
	2154		
23 Sa		0043	1.8F
	0340	0739	2.5E
	1123	1420	2.0F
	1744	2024	1.2E
	2257		
24 Su		0137	1.6F
	0432	0821	2.3E
	1200	1500	2.0F
	1826	2116	1.4E
25 M	0003	0235	1.4F
	0525	0904	2.0E
	1237	1540	2.0F
	1908	2210	1.5E
26 Tu	0114	0338	1.1F
	0625	0951	1.7E
	1315	1623	1.9F
	1951	2309	1.5E
27 W	0232	0456	0.9F
	0736	1045	1.3E
	1357	1710	1.7F
◑	2036		
28 Th		0018	1.6E
	0356	0630	0.8F
	0901	1153	0.9E
	1444	1803	1.6F
	2125		
29 F		0133	1.7E
	0516	0758	0.9F
	1037	1318	0.7E
	1541	1902	1.4F
	2215		
30 Sa		0244	1.9E
	0623	0908	1.2F
	1207	1440	0.6E
	1643	2001	1.4F
	2305		
31 Su		0344	2.0E
	0717	1006	1.4F
	1319	1547	0.6E
	1745	2056	1.3F
	2353		

August

Day	Slack h m	Maximum h m	knots
1 M		0434	2.1E
	0803	1055	1.6F
	1415	1641	0.7E
	1841	2143	1.4F
2 Tu	0036	0516	2.2E
	0843	1139	1.7F
	1501	1728	0.8E
	1930	2225	1.4F
3 W	0116	0550	2.2E
	0918	1219	1.8F
	1541	1809	0.8E
	2015	2303	1.4F
4 Th	0152	0615	2.2E
	0949	1255	1.8F
	1616	1847	0.9E
●	2058	2340	1.4F
5 F	0227	0631	2.2E
	1016	1324	1.8F
	1648	1921	0.9E
	2139		
6 Sa		0017	1.4F
	0302	0649	2.2E
	1041	1347	1.8F
	1715	1952	1.0E
	2222		
7 Su		0055	1.3F
	0339	0716	2.2E
	1104	1404	1.8F
	1740	2020	1.1E
	2305		
8 M		0136	1.3F
	0418	0750	2.1E
	1129	1424	1.8F
	1802	2049	1.2E
	2352		
9 Tu		0220	1.2F
	0503	0828	1.9E
	1155	1449	1.8F
	1824	2120	1.4E
10 W	0044	0308	1.1F
	0555	0911	1.6E
	1224	1522	1.7F
	1849	2157	1.5E
11 Th	0143	0406	0.9F
	0658	0959	1.3E
	1256	1601	1.6F
	1920	2241	1.6E
12 F	0254	0518	0.8F
	0819	1055	0.9E
	1334	1647	1.5F
◐	2002	2336	1.6E
13 Sa	0413	0655	0.8F
	0952	1200	0.6E
	1422	1743	1.4F
	2055		
14 Su		0043	1.7E
	0530	0827	1.0F
	1123	1314	0.5E
	1523	1846	1.4F
	2157		
15 M		0207	1.9E
	0635	0932	1.2F
	1237	1431	0.5E
	1635	1952	1.5F
	2300		
16 Tu		0327	2.1E
	0729	1024	1.4F
	1333	1540	0.6E
	1746	2056	1.6F
17 W	0001	0427	2.3E
	0815	1109	1.6F
	1420	1639	0.8E
	1851	2154	1.8F
18 Th	0057	0515	2.4E
	0856	1149	1.8F
	1501	1732	1.0E
	1953	2249	1.9F
19 F	0150	0557	2.4E
	0934	1226	1.9F
	1539	1822	1.2E
○	2053	2342	1.8F
20 Sa	0241	0636	2.4E
	1010	1302	2.0F
	1617	1910	1.4E
	2152		
21 Su		0034	1.8F
	0332	0715	2.2E
	1044	1337	2.0F
	1654	1957	1.6E
	2252		
22 M		0128	1.6F
	0425	0755	1.9E
	1118	1413	2.0F
	1732	2043	1.7E
	2353		
23 Tu		0225	1.4F
	0521	0838	1.6E
	1153	1450	1.9F
	1811	2131	1.7E
24 W	0059	0330	1.1F
	0624	0926	1.3E
	1231	1531	1.7F
	1852	2225	1.7E
25 Th	0210	0449	1.0F
	0739	1024	0.9E
	1315	1619	1.5F
	1939	2333	1.6E
26 F	0330	0622	0.9F
	0906	1138	0.7E
	1409	1715	1.3F
◑	2033		
27 Sa		0100	1.6E
	0448	0743	1.0F
	1038	1308	0.6E
	1516	1824	1.1F
	2132		
28 Su		0219	1.7E
	0554	0849	1.3F
	1158	1427	0.6E
	1630	1937	1.1F
	2233		
29 M		0320	1.9E
	0648	0943	1.5F
	1258	1529	0.7E
	1737	2039	1.2F
	2328		
30 Tu		0408	2.0E
	0731	1029	1.7F
	1344	1620	0.9E
	1834	2130	1.2F
31 W	0016	0447	2.1E
	0808	1109	1.8F
	1423	1704	1.0E
	1923	2212	1.3F

September

Day	Slack h m	Maximum h m	knots
1 Th	0057	0516	2.1E
	0839	1144	1.9F
	1457	1743	1.1E
	2007	2250	1.4F
2 F	0135	0537	2.0E
	0905	1213	1.8F
	1526	1818	1.2E
	2048	2326	1.4F
3 Sa	0211	0553	2.0E
	0929	1234	1.8F
	1552	1847	1.3E
●	2128		
4 Su		0002	1.4F
	0248	0616	2.0E
	0952	1250	1.8F
	1613	1912	1.4E
	2207		
5 M		0040	1.4F
	0329	0647	1.9E
	1016	1307	1.8F
	1632	1935	1.5E
	2248		
6 Tu		0121	1.4F
	0414	0724	1.7E
	1042	1332	1.8F
	1651	2001	1.6E
	2332		
7 W		0205	1.3F
	0504	0805	1.5E
	1111	1403	1.8F
	1714	2033	1.7E
8 Th	0021	0255	1.3F
	0602	0851	1.3E
	1143	1440	1.7F
	1745	2113	1.8E
9 F	0118	0353	1.1F
	0710	0943	1.0E
	1221	1524	1.6F
	1825	2201	1.8E
10 Sa	0227	0509	1.0F
	0830	1043	0.7E
	1306	1616	1.5F
	1917	2301	1.8E
11 Su	0345	0645	1.0F
	0955	1153	0.5E
	1406	1718	1.4F
◐	2022		
12 M		0018	1.8E
	0500	0807	1.1F
	1112	1311	0.5E
	1522	1828	1.3F
	2135		
13 Tu		0152	1.9E
	0604	0907	1.4F
	1213	1427	0.6E
	1641	1941	1.4F
	2246		
14 W		0310	2.0E
	0656	0955	1.6F
	1301	1531	0.9E
	1752	2049	1.6F
	2350		
15 Th		0405	2.2E
	0739	1036	1.7F
	1342	1626	1.1E
	1857	2149	1.7F
16 F	0048	0450	2.2E
	0817	1112	1.9F
	1418	1715	1.4E
	1957	2244	1.7F
17 Sa	0142	0530	2.1E
	0853	1146	1.9F
	1453	1801	1.6E
○	2054	2337	1.7F
18 Su	0235	0609	1.9E
	0926	1218	2.0F
	1527	1844	1.8E
	2150		
19 M		0029	1.6F
	0329	0649	1.7E
	1000	1252	1.9F
	1601	1926	1.9E
	2246		
20 Tu		0123	1.5F
	0425	0731	1.4E
	1034	1326	1.8F
	1636	2007	1.9E
	2342		
21 W		0221	1.4F
	0526	0818	1.2E
	1111	1404	1.7F
	1713	2050	1.8E
22 Th	0042	0327	1.2F
	0634	0911	0.9E
	1152	1446	1.5F
	1755	2138	1.7E
23 F	0146	0445	1.1F
	0749	1014	0.7E
	1242	1535	1.3F
	1843	2240	1.6E
24 Sa	0257	0607	1.1F
	0909	1130	0.6E
	1344	1634	1.1F
◑	1940		
25 Su		0010	1.5E
	0408	0719	1.2F
	1027	1252	0.6E
	1458	1747	0.9F
	2046		
26 M		0136	1.6E
	0511	0820	1.4F
	1131	1403	0.7E
	1616	1907	0.9F
	2153		
27 Tu		0237	1.7E
	0602	0910	1.6F
	1222	1502	0.9E
	1725	2015	1.0F
	2254		
28 W		0324	1.7E
	0643	0953	1.7F
	1303	1552	1.1E
	1823	2109	1.1F
	2346		
29 Th		0359	1.8E
	0717	1029	1.8F
	1337	1635	1.3E
	1912	2154	1.2F
30 F	0031	0425	1.8E
	0745	1059	1.8F
	1406	1712	1.4E
	1956	2234	1.3F

Time meridian 120° W. 0000 is midnight. 1200 is noon. Times are not adjusted for Daylight Saving Time.

Benicia Bridge, Suisun Bay, Calif., 2005

F–Flood, Dir. 047° True E–Ebb, Dir. 230° True

October

Day	Slack h m	Maximum h m	knots
1 Sa	0113	0447	1.7E
	0811	1120	1.8F
	1430	1743	1.5E
	2036	2312	1.3F
2 Su	0155	0512	1.7E
	0835	1136	1.7F
	1450	1808	1.6E
	2115	2349	1.4F
3 M	0240	0544	1.6E
	0901	1153	1.7F
	1508	1829	1.7E
●	2153		
4 Tu		0029	1.4F
	0327	0622	1.4E
	0928	1218	1.7F
	1527	1852	1.8E
	2234		
5 W		0112	1.5F
	0419	0704	1.3E
	0959	1250	1.7F
	1551	1922	2.0E
	2318		
6 Th		0158	1.5F
	0516	0750	1.1E
	1033	1327	1.7F
	1622	2000	2.1E
7 F	0007	0251	1.4F
	0617	0841	0.9E
	1112	1410	1.7F
	1702	2045	2.1E
8 Sa	0104	0353	1.3F
	0725	0937	0.8E
	1159	1500	1.5F
	1751	2139	2.0E
9 Su	0209	0508	1.2F
	0836	1040	0.6E
	1256	1556	1.4F
	1851	2244	1.9E
10 M	0319	0629	1.3F
	0947	1151	0.6E
	1407	1702	1.3F
◐	2001		
11 Tu		0002	1.9E
	0425	0737	1.4F
	1049	1307	0.7E
	1529	1816	1.2F
	2117		
12 W		0127	1.9E
	0523	0832	1.6F
	1141	1418	0.9E
	1648	1933	1.3F
	2230		
13 Th		0238	1.9E
	0611	0917	1.7F
	1224	1518	1.2E
	1759	2044	1.4F
	2337		
14 F		0332	1.9E
	0653	0956	1.9F
	1301	1611	1.5E
	1903	2146	1.5F
15 Sa	0039	0419	1.8E
	0731	1031	1.9F
	1336	1658	1.8E
	2001	2243	1.6F

Day	Slack h m	Maximum h m	knots
16 Su	0137	0502	1.6E
	0806	1104	1.9F
	1408	1741	1.9E
	2056	2337	1.6F
17 M	0235	0545	1.4E
	0841	1136	1.9F
	1440	1820	2.0E
○	2148		
18 Tu		0030	1.5F
	0335	0630	1.2E
	0916	1210	1.8F
	1512	1858	2.1E
	2240		
19 W		0125	1.5F
	0436	0718	1.0E
	0954	1246	1.7F
	1546	1936	2.0E
	2331		
20 Th		0223	1.4F
	0540	0809	0.8E
	1035	1326	1.5F
	1624	2015	2.0E
21 F	0025	0326	1.3F
	0645	0904	0.7E
	1121	1410	1.3F
	1707	2100	1.8E
22 Sa	0121	0434	1.3F
	0752	1005	0.6E
	1215	1500	1.1F
	1755	2153	1.7E
23 Su	0219	0541	1.3F
	0857	1112	0.5E
	1320	1557	0.9F
	1851	2257	1.6E
24 M	0318	0644	1.4F
	0959	1223	0.6E
	1435	1706	0.8F
◑	1954		
25 Tu		0009	1.5E
	0412	0738	1.5F
	1052	1329	0.8E
	1554	1825	0.7F
	2101		
26 W		0115	1.5E
	0458	0825	1.6F
	1136	1428	1.0E
	1706	1939	0.8F
	2206		
27 Th		0206	1.5E
	0538	0905	1.7F
	1214	1517	1.2E
	1807	2040	0.9F
	2306		
28 F		0245	1.5E
	0611	0937	1.7F
	1244	1600	1.4E
	1858	2131	1.0F
29 Sa	0001	0321	1.4E
	0641	1001	1.7F
	1309	1635	1.6E
	1942	2217	1.1F
30 Su	0053	0356	1.3E
	0710	1019	1.7F
	1330	1704	1.7E
	2022	2259	1.2F
31 M	0146	0435	1.2E
	0739	1038	1.7F
	1349	1727	1.9E
	2101	2341	1.4F

November

Day	Slack h m	Maximum h m	knots
1 Tu	0239	0516	1.1E
	0809	1105	1.7F
	1410	1750	2.0E
●	2141		
2 W		0024	1.5F
	0334	0602	1.0E
	0843	1138	1.7F
	1436	1819	2.1E
	2223		
3 Th		0109	1.5F
	0430	0651	0.9E
	0921	1218	1.7F
	1510	1857	2.2E
	2309		
4 F		0159	1.6F
	0528	0742	0.8E
	1004	1302	1.7F
	1551	1942	2.3E
	2359		
5 Sa		0253	1.5F
	0627	0836	0.8E
	1052	1350	1.7F
	1639	2033	2.3E
6 Su	0053	0352	1.5F
	0727	0933	0.7E
	1148	1443	1.6F
	1734	2129	2.2E
7 M	0150	0456	1.5F
	0826	1035	0.7E
	1253	1543	1.4F
	1835	2232	2.1E
8 Tu	0248	0600	1.6F
	0924	1143	0.8E
	1410	1650	1.2F
◐	1943	2340	1.9E
9 W	0343	0658	1.7F
	1016	1254	1.0E
	1532	1806	1.1F
	2057		
10 Th		0051	1.8E
	0435	0749	1.8F
	1103	1402	1.3E
	1653	1927	1.1F
	2213		
11 F		0158	1.7E
	0521	0834	1.9F
	1145	1502	1.6E
	1806	2043	1.2F
	2327		
12 Sa		0258	1.5E
	0604	0914	1.9F
	1222	1556	1.8E
	1909	2149	1.3F
13 Su	0036	0352	1.3E
	0644	0951	1.9F
	1256	1644	2.0E
	2005	2247	1.4F
14 M	0142	0444	1.1E
	0723	1026	1.8F
	1328	1727	2.1E
	2057	2342	1.5F
15 Tu	0247	0535	0.9E
	0802	1101	1.7F
	1401	1807	2.2E
○	2146		

Day	Slack h m	Maximum h m	knots
16 W		0036	1.5F
	0350	0625	0.8E
	0841	1138	1.6F
	1434	1844	2.2E
	2233		
17 Th		0129	1.5F
	0451	0715	0.7E
	0923	1217	1.5F
	1511	1919	2.1E
	2320		
18 F		0223	1.5F
	0549	0805	0.6E
	1008	1258	1.4F
	1550	1954	2.1E
19 Sa	0006	0316	1.5F
	0644	0855	0.6E
	1058	1343	1.3F
	1633	2033	2.0E
20 Su	0052	0410	1.5F
	0736	0947	0.6E
	1152	1431	1.1F
	1719	2115	1.9E
21 M	0137	0503	1.5F
	0827	1042	0.6E
	1254	1524	1.0F
	1808	2201	1.8E
22 Tu	0220	0554	1.5F
	0915	1141	0.7E
	1404	1624	0.8F
	1901	2249	1.6E
23 W	0302	0641	1.5F
	1000	1242	0.8E
	1521	1733	0.6F
◑	2002	2341	1.5E
24 Th	0342	0723	1.5F
	1040	1341	1.0E
	1637	1851	0.6F
	2111		
25 F		0035	1.3E
	0420	0758	1.5F
	1114	1433	1.2E
	1744	2006	0.7F
	2223		
26 Sa		0129	1.2E
	0456	0825	1.5F
	1140	1517	1.5E
	1839	2108	0.8F
	2334		
27 Su		0222	1.0E
	0532	0850	1.5F
	1208	1553	1.7E
	1925	2202	1.0F
28 M	0040	0314	0.9E
	0607	0917	1.5F
	1231	1623	1.8E
	2007	2250	1.2F
29 Tu	0143	0405	0.8E
	0644	0950	1.6F
	1256	1651	2.0E
	2048	2336	1.4F
30 W	0243	0457	0.7E
	0724	1027	1.6F
	1327	1723	2.2E
	2131		

December

Day	Slack h m	Maximum h m	knots
1 Th		0022	1.5F
	0341	0549	0.7E
	0807	1110	1.7F
●	1404	1802	2.3E
	2215		
2 F		0109	1.6F
	0435	0642	0.7E
	0854	1155	1.8F
	1448	1847	2.4E
	2301		
3 Sa		0157	1.7F
	0528	0735	0.7E
	0945	1244	1.8F
	1536	1936	2.5E
	2348		
4 Su		0246	1.7F
	0619	0828	0.8E
	1041	1336	1.7F
	1628	2027	2.5E
5 M	0036	0336	1.8F
	0708	0923	0.9E
	1142	1431	1.6F
	1723	2119	2.4E
6 Tu	0123	0428	1.8F
	0758	1022	1.0E
	1251	1530	1.4F
	1821	2213	2.2E
7 W	0210	0520	1.8F
	0847	1125	1.1E
	1407	1637	1.2F
	1925	2310	1.9E
8 Th	0257	0611	1.8F
	0935	1232	1.3E
	1531	1756	1.0F
◐	2039		
9 F		0013	1.6E
	0344	0702	1.8F
	1021	1341	1.5E
	1655	1924	0.9F
	2200		
10 Sa		0121	1.3E
	0431	0750	1.8F
	1104	1446	1.8E
	1810	2046	1.0F
	2324		
11 Su		0232	1.1E
	0519	0835	1.8F
	1145	1544	2.0E
	1913	2154	1.2F
12 M	0044	0339	0.9E
	0605	0918	1.7F
	1222	1636	2.1E
	2007	2253	1.4F
13 Tu	0157	0440	0.8E
	0652	0958	1.6F
	1259	1723	2.2E
	2055	2346	1.5F
14 W	0302	0535	0.7E
	0737	1038	1.6F
	1335	1806	2.2E
	2141		
15 Th		0036	1.6F
	0400	0624	0.7E
	0822	1118	1.5F
○	1412	1843	2.2E
	2223		

Day	Slack h m	Maximum h m	knots
16 F		0124	1.6F
	0452	0710	0.6E
	0907	1158	1.4F
	1450	1913	2.2E
	2304		
17 Sa		0209	1.7F
	0538	0754	0.6E
	0952	1239	1.4F
	1529	1939	2.2E
	2341		
18 Su		0252	1.6F
	0620	0836	0.7E
	1040	1322	1.3F
	1610	2007	2.1E
19 M	0016	0332	1.6F
	0700	0918	0.7E
	1130	1406	1.2F
	1650	2040	2.1E
20 Tu	0049	0409	1.6F
	0739	1001	0.8E
	1225	1454	1.0F
	1732	2117	1.9E
21 W	0120	0442	1.6F
	0817	1047	0.8E
	1328	1545	0.8F
	1818	2158	1.8E
22 Th	0152	0513	1.5F
	0854	1137	0.9E
	1438	1645	0.7F
	1913	2245	1.5E
23 F	0225	0544	1.5F
	0929	1230	1.1E
	1555	1759	0.5F
◑	2023	2337	1.2E
24 Sa	0301	0619	1.4F
	1001	1325	1.2E
	1709	1927	0.6F
	2149		
25 Su		0035	0.9E
	0341	0659	1.4F
	1033	1417	1.4E
	1812	2047	0.7F
	2316		
26 M		0138	0.7E
	0425	0741	1.4F
	1104	1504	1.7E
	1904	2150	1.0F
27 Tu	0036	0243	0.6E
	0512	0827	1.4F
	1139	1547	1.9E
	1951	2242	1.2F
28 W	0144	0346	0.6E
	0602	0914	1.5F
	1218	1629	2.1E
	2036	2330	1.4F
29 Th	0243	0445	0.6E
	0653	1002	1.6F
	1301	1713	2.3E
	2120		
30 F		0015	1.6F
	0335	0539	0.6E
	0745	1051	1.8F
●	1348	1759	2.5E
	2203		
31 Sa		0058	1.7F
	0422	0631	0.7E
	0839	1141	1.9F
	1438	1844	2.6E
	2246		

Time meridian 120° W. 0000 is midnight. 1200 is noon. Times are not adjusted for Daylight Saving Time.

Grays Harbor Entrance, Washington, 2005

F–Flood, Dir. 060° True E–Ebb, Dir. 240° True

January

Day	Slack h m	Maximum h m	knots
1 Sa		0128	2.0F
	0511	0805	1.9E
	1110	1321	1.1F
	1609	1951	2.6E
	2321		
2 Su		0206	2.0F
	0550	0853	2.1E
	1210	1417	1.0F
	1710	2041	2.2E
3 M ◑	0003	0249	1.9F
	0630	0947	2.3E
	1316	1524	1.0F
	1830	2139	1.9E
4 Tu	0051	0337	1.8F
	0714	1048	2.6E
	1424	1644	1.0F
	1958	2250	1.6E
5 W	0144	0433	1.7F
	0803	1150	2.9E
	1529	1822	1.3F
	2120		
6 Th		0009	1.5E
	0242	0534	1.7F
	0856	1250	3.4E
	1629	1946	1.7F
	2233		
7 F		0120	1.5E
	0343	0638	1.8F
	0951	1346	3.8E
	1724	2045	2.1F
	2337		
8 Sa		0221	1.7E
	0444	0741	2.0F
	1046	1440	4.2E
	1815	2137	2.5F
9 Su	0033	0317	1.9E
	0543	0841	2.1F
	1140	1531	4.5E
	1904	2224	2.9F
10 M ●	0124	0409	2.2E
	0640	0937	2.3F
	1233	1620	4.6E
	1950	2309	3.1F
11 Tu	0211	0459	2.4E
	0735	1030	2.3F
	1325	1709	4.6E
	2036	2352	3.1F
12 W	0256	0549	2.6E
	0830	1123	2.3F
	1417	1756	4.3E
	2119		
13 Th		0035	3.1F
	0341	0638	2.7E
	0927	1216	2.1F
	1509	1844	3.9E
	2202		
14 F		0117	2.9F
	0424	0728	2.8E
	1026	1311	1.9F
	1602	1933	3.4E
	2245		
15 Sa		0158	2.6F
	0508	0820	2.8E
	1129	1413	1.5F
	1659	2024	2.8E
	2327		
16 Su ◐		0241	2.3F
	0552	0915	2.8E
	1238	1532	1.3F
	1802	2120	2.2E
17 M	0012	0327	1.9F
	0638	1012	2.8E
	1352	1702	1.1F
	1912	2222	1.7E
18 Tu	0100	0421	1.6F
	0726	1111	2.8E
	1509	1819	1.2F
	2029	2329	1.3E
19 W	0154	0533	1.3F
	0817	1209	2.8E
	1617	1923	1.3F
	2148		
20 Th		0034	1.2E
	0255	0648	1.2F
	0909	1304	2.9E
	1713	2019	1.6F
	2257		
21 F		0133	1.2E
	0359	0747	1.2F
	1000	1356	3.1E
	1757	2108	1.8F
	2351		
22 Sa		0227	1.3E
	0457	0836	1.3F
	1048	1442	3.2E
	1834	2151	2.0F
23 Su	0035	0314	1.5E
	0547	0914	1.4F
	1133	1525	3.4E
	1906	2230	2.1F
24 M	0113	0357	1.7E
	0631	0942	1.4F
	1214	1604	3.5E
	1936	2303	2.2F
25 Tu ○	0147	0436	1.8E
	0712	1006	1.5F
	1252	1639	3.5E
	2005	2329	2.2F
26 W	0219	0513	2.0E
	0750	1033	1.6F
	1328	1711	3.5E
	2033	2344	2.3F
27 Th	0251	0547	2.1E
	0829	1105	1.6F
	1404	1741	3.4E
	2103	2359	2.3F
28 F	0321	0618	2.3E
	0910	1140	1.7F
	1440	1810	3.2E
	2133		
29 Sa		0022	2.3F
	0352	0649	2.4E
	0953	1218	1.7F
	1519	1843	2.9E
	2206		
30 Su		0051	2.3F
	0422	0723	2.5E
	1041	1302	1.6F
	1604	1922	2.6E
	2241		
31 M		0124	2.2F
	0454	0805	2.6E
	1134	1351	1.5F
	1658	2007	2.2E
	2320		

February

Day	Slack h m	Maximum h m	knots
1 Tu ◑		0203	2.1F
	0530	0854	2.7E
	1234	1450	1.3F
	1808	2101	1.8E
2 W	0006	0250	1.9F
	0615	0954	2.8E
	1343	1604	1.2F
	1933	2209	1.4E
3 Th	0102	0346	1.7F
	0712	1107	3.0E
	1456	1749	1.2F
	2059	2341	1.2E
4 F	0209	0456	1.5F
	0819	1222	3.2E
	1606	1936	1.6F
	2218		
5 Sa		0104	1.3E
	0322	0616	1.6F
	0928	1328	3.6E
	1708	2037	2.1F
	2323		
6 Su		0209	1.7E
	0434	0737	1.8F
	1033	1427	4.0E
	1802	2128	2.5F
7 M	0017	0306	2.1E
	0539	0846	2.0F
	1133	1520	4.3E
	1850	2212	2.9F
8 Tu ●	0104	0356	2.5E
	0637	0943	2.3F
	1228	1609	4.4E
	1934	2253	3.1F
9 W	0147	0444	2.9E
	0731	1034	2.5F
	1321	1655	4.4E
	2015	2331	3.2F
10 Th	0227	0529	3.1E
	0823	1123	2.5F
	1411	1740	4.1E
	2054		
11 F		0007	3.1F
	0306	0614	3.3E
	0914	1209	2.4F
	1500	1823	3.7E
	2132		
12 Sa		0040	2.8F
	0344	0658	3.3E
	1005	1256	2.1F
	1550	1907	3.1E
	2209		
13 Su		0112	2.5F
	0422	0743	3.2E
	1058	1346	1.8F
	1641	1951	2.5E
	2246		
14 M		0144	2.2F
	0501	0830	3.0E
	1156	1444	1.4F
	1736	2040	1.9E
	2327		
15 Tu ◐		0220	1.8F
	0543	0923	2.8E
	1301	1610	1.1F
	1839	2139	1.4E
16 W	0012	0301	1.4F
	0630	1023	2.6E
	1416	1743	1.0F
	1953	2250	1.1E
17 Th	0108	0357	1.0F
	0725	1128	2.5E
	1535	1855	1.1F
	2115		
18 F		0003	0.9E
	0219	0605	0.8F
	0827	1232	2.5E
	1642	1954	1.4F
	2229		
19 Sa		0109	1.0E
	0339	0732	0.9F
	0929	1329	2.7E
	1730	2044	1.6F
	2325		
20 Su		0205	1.3E
	0448	0826	1.1F
	1026	1419	2.9E
	1808	2127	1.9F
21 M	0007	0253	1.6E
	0541	0908	1.3F
	1115	1503	3.1E
	1839	2203	2.1F
22 Tu	0042	0335	1.9E
	0623	0940	1.5F
	1200	1542	3.3E
	1908	2234	2.2F
23 W ○	0113	0412	2.2E
	0700	1004	1.7F
	1240	1618	3.4E
	1935	2254	2.3F
24 Th	0143	0446	2.4E
	0736	1027	1.8F
	1318	1650	3.3E
	2002	2306	2.3F
25 F	0211	0517	2.7E
	0812	1055	1.9F
	1356	1719	3.2E
	2030	2322	2.4F
26 Sa	0238	0545	2.9E
	0850	1127	2.0F
	1435	1748	3.1E
	2100	2345	2.4F
27 Su	0305	0614	3.0E
	0930	1204	2.1F
	1516	1820	2.8E
	2132		
28 M		0014	2.4F
	0333	0648	3.2E
	1014	1245	2.0F
	1601	1858	2.5E
	2207		

March

Day	Slack h m	Maximum h m	knots
1 Tu		0048	2.3F
	0404	0729	3.2E
	1105	1332	1.8F
	1654	1943	2.0E
	2248		
2 W		0128	2.1F
	0441	0817	3.1E
	1203	1428	1.6F
	1800	2037	1.6E
	2336		
3 Th ◑		0215	1.8F
	0530	0918	3.0E
	1311	1541	1.3F
	1921	2151	1.3E
4 F	0038	0316	1.5F
	0635	1038	2.9E
	1428	1745	1.3F
	2046	2333	1.2E
5 Sa	0155	0436	1.3F
	0756	1203	3.1E
	1544	1923	1.7F
	2202		
6 Su		0054	1.4E
	0320	0619	1.3F
	0915	1313	3.4E
	1649	2021	2.2F
	2303		
7 M		0157	1.9E
	0438	0754	1.7F
	1026	1413	3.7E
	1742	2108	2.6F
	2353		
8 Tu		0250	2.4E
	0541	0857	2.1F
	1128	1505	3.9E
	1828	2150	2.9F
9 W	0035	0338	2.9E
	0635	0948	2.4F
	1223	1553	4.0E
	1909	2228	3.0F
10 Th ●	0114	0423	3.4E
	0724	1034	2.6F
	1314	1637	3.9E
	1947	2301	3.0F
11 F	0151	0505	3.6E
	0810	1117	2.7F
	1403	1719	3.6E
	2023	2331	2.9F
12 Sa	0226	0545	3.7E
	0855	1157	2.5F
	1449	1800	3.2E
	2057	2358	2.6F
13 Su	0301	0624	3.7E
	0940	1237	2.3F
	1535	1840	2.7E
	2132		
14 M		0026	2.3F
	0335	0704	3.5E
	1026	1317	2.0F
	1622	1920	2.2E
	2208		
15 Tu		0055	1.9F
	0410	0745	3.2E
	1115	1401	1.6F
	1712	2004	1.7E
	2248		
16 W		0127	1.6F
	0448	0832	2.8E
	1210	1500	1.2F
	1810	2059	1.3E
	2335		
17 Th ◐		0206	1.2F
	0532	0931	2.5E
	1317	1659	1.0F
	1919	2214	1.0E
18 F	0035	0259	0.8F
	0630	1043	2.3E
	1434	1820	1.0F
	2036	2333	0.9E
19 Sa	0156	0426	0.6F
	0742	1154	2.2E
	1548	1920	1.3F
	2148		
20 Su		0040	1.1E
	0328	0710	0.7F
	0855	1255	2.4E
	1642	2010	1.5F
	2242		
21 M		0137	1.5E
	0439	0805	1.0F
	0959	1348	2.6E
	1723	2051	1.8F
	2323		
22 Tu		0224	1.9E
	0527	0848	1.3F
	1053	1433	2.8E
	1756	2125	2.0F
	2357		
23 W		0304	2.3E
	0606	0922	1.6F
	1141	1513	3.0E
	1826	2151	2.1F
24 Th	0028	0341	2.6E
	0641	0948	1.8F
	1225	1550	3.1E
	1855	2206	2.2F
25 F ○	0057	0413	3.0E
	0716	1012	2.1F
	1306	1623	3.0E
	1924	2220	2.3F
26 Sa	0125	0443	3.3E
	0751	1040	2.2F
	1347	1655	2.9E
	1955	2242	2.4F
27 Su	0152	0512	3.5E
	0828	1113	2.4F
	1428	1727	2.8E
	2027	2310	2.4F
28 M	0221	0544	3.6E
	0909	1150	2.4F
	1512	1802	2.5E
	2102	2342	2.3F
29 Tu	0251	0620	3.7E
	0953	1232	2.3F
	1600	1842	2.2E
	2140		
30 W		0019	2.2F
	0326	0703	3.6E
	1044	1320	2.1F
	1655	1930	1.8E
	2226		
31 Th		0102	1.9F
	0408	0754	3.3E
	1142	1418	1.8F
	1800	2031	1.5E
	2322		

Time meridian 120° W. 0000 is midnight. 1200 is noon. Times are not adjusted for Daylight Saving Time.

Grays Harbor Entrance, Washington, 2005

F–Flood, Dir. 060° True E–Ebb, Dir. 240° True

April

Day	Slack h m	Maximum h m	knots
1 F ◑	 0502 1249 1915	0155 0859 1537 2159	1.6F 3.1E 1.5F 1.3E
2 Sa	0034 0617 1404 2032	0303 1023 1742 2329	1.2F 2.9E 1.6F 1.4E
3 Su	0202 0745 1518 2139	0441 1146 1859	1.0F 2.9E 1.9F
4 M	 0332 0908 1620 2234	0040 0649 1255 1954	1.8E 1.3F 3.1E 2.2F
5 Tu	 0444 1019 1712 2319	0139 0801 1354 2040	2.4E 1.7F 3.2E 2.5F
6 W	 0541 1121 1757 2359	0230 0856 1445 2120	2.9E 2.2F 3.4E 2.7F
7 Th	 0629 1216 1836	0315 0943 1532 2155	3.4E 2.5F 3.3E 2.7F
8 F ●	0036 0713 1306 1913	0357 1027 1616 2226	3.8E 2.7F 3.2E 2.7F
9 Sa	0111 0755 1353 1948	0437 1106 1657 2252	3.9E 2.7F 2.9E 2.5F
10 Su	0145 0835 1437 2022	0515 1143 1737 2317	4.0E 2.5F 2.6E 2.3F
11 M	0218 0915 1521 2057	0551 1218 1815 2343	3.8E 2.3F 2.3E 2.0F
12 Tu	0250 0956 1605 2135	0627 1252 1854	3.6E 2.0F 1.9E
13 W	 0323 1040 1652 2217	0013 0705 1329 1937	1.7F 3.2E 1.7F 1.5E
14 Th	 0358 1129 1746 2307	0047 0747 1416 2032	1.4F 2.8E 1.4F 1.2E
15 F	 0438 1225 1849	0127 0840 1540 2145	1.0F 2.5E 1.1F 1.0E
16 Sa ◐	0013 0533 1330 1956	0221 0951 1731 2300	0.7F 2.2E 1.1F 1.1E
17 Su	0137 0653 1436 2058	0341 1107 1834	0.5F 2.1E 1.3F
18 M	 0308 0814 1533 2149	0005 0632 1212 1923	1.3E 0.6F 2.2E 1.5F
19 Tu	 0415 0925 1620 2230	0100 0732 1307 2002	1.7E 0.9F 2.3E 1.7F
20 W	 0501 1025 1700 2305	0146 0817 1355 2031	2.2E 1.2F 2.4E 1.9F
21 Th	 0540 1118 1735 2337	0227 0853 1438 2050	2.6E 1.6F 2.5E 2.0F
22 F	 0616 1206 1809	0303 0924 1517 2109	3.1E 1.9F 2.6E 2.1F
23 Sa	0008 0652 1252 1843	0337 0953 1555 2134	3.4E 2.2F 2.6E 2.3F
24 Su ○	0039 0729 1336 1919	0410 1025 1631 2204	3.8E 2.5F 2.5E 2.3F
25 M	0110 0809 1422 1956	0443 1101 1709 2239	4.0E 2.6F 2.4E 2.3F
26 Tu	0144 0852 1509 2037	0520 1141 1750 2317	4.1E 2.6F 2.2E 2.2F
27 W	0220 0938 1600 2123	0601 1226 1836	4.0E 2.5F 2.0E
28 Th	 0302 1030 1656 2216	0000 0648 1318 1932	2.0F 3.8E 2.3F 1.7E
29 F	 0352 1127 1759 2322	0050 0744 1421 2043	1.7F 3.5E 2.0F 1.6E
30 Sa ◑	 0454 1230 1905	0150 0851 1544 2202	1.4F 3.1E 1.9F 1.6E

May

Day	Slack h m	Maximum h m	knots
1 Su	0043 0612 1337 2010	0309 1010 1718 2316	1.1F 2.9E 1.9F 1.9E
2 M	0214 0737 1443 2107	0512 1126 1826	1.0F 2.7E 2.1F
3 Tu	 0337 0858 1542 2157	0020 0650 1231 1919	2.3E 1.3F 2.7E 2.3F
4 W	 0442 1010 1633 2240	0115 0753 1330 2005	2.8E 1.8F 2.7E 2.4F
5 Th	 0534 1112 1719 2320	0205 0845 1422 2045	3.3E 2.1F 2.7E 2.4F
6 F	 0619 1208 1759 2357	0249 0933 1509 2120	3.7E 2.4F 2.6E 2.4F
7 Sa	 0700 1257 1837	0331 1016 1553 2149	3.9E 2.5F 2.5E 2.2F
8 Su ●	0032 0738 1342 1914	0410 1056 1635 2214	4.0E 2.5F 2.3E 2.1F
9 M	0106 0816 1425 1950	0448 1132 1715 2240	3.9E 2.4F 2.1E 1.9F
10 Tu	0140 0853 1507 2028	0523 1205 1755 2308	3.8E 2.2F 1.9E 1.7F
11 W	0212 0931 1550 2108	0559 1236 1835 2340	3.5E 2.0F 1.6E 1.5F
12 Th	0245 1011 1635 2154	0635 1310 1919	3.2E 1.8F 1.4E
13 F	 0319 1055 1725 2248	0017 0714 1350 2012	1.2F 2.9E 1.6F 1.3E
14 Sa	 0357 1143 1818 2353	0101 0801 1441 2115	1.0F 2.6E 1.4F 1.3E
15 Su	 0447 1235 1913	0155 0859 1548 2221	0.7F 2.3E 1.4F 1.4E
16 M ◐	0110 0601 1329 2005	0305 1007 1705 2321	0.5F 2.1E 1.4F 1.6E
17 Tu	0229 0726 1421 2051	0442 1115 1759	0.5F 2.0E 1.5F
18 W	 0334 0844 1511 2131	0015 0639 1216 1835	2.0E 0.8F 2.0E 1.6F
19 Th	 0424 0951 1557 2209	0102 0735 1309 1907	2.4E 1.2F 2.0E 1.7F
20 F	 0507 1051 1640 2245	0144 0818 1358 1940	2.9E 1.6F 2.1E 1.9F
21 Sa	 0548 1146 1722 2321	0224 0857 1444 2015	3.4E 1.9F 2.1E 2.0F
22 Su	 0628 1237 1804 2358	0302 0934 1528 2053	3.8E 2.3F 2.1E 2.2F
23 M ○	 0710 1326 1847	0341 1013 1611 2133	4.1E 2.5F 2.1E 2.2F
24 Tu	0037 0753 1414 1932	0421 1054 1656 2216	4.3E 2.7F 2.1E 2.2F
25 W	0119 0839 1504 2020	0504 1139 1744 2302	4.3E 2.7F 2.0E 2.1F
26 Th	0204 0927 1556 2114	0551 1228 1837 2352	4.2E 2.6F 2.0E 1.9F
27 F	0253 1018 1650 2215	0642 1322 1936	4.0E 2.5F 1.9E
28 Sa	 0349 1111 1746 2325	0048 0739 1423 2041	1.7F 3.6E 2.4F 2.0E
29 Su	 0453 1207 1842	0155 0842 1531 2148	1.3F 3.2E 2.2F 2.1E
30 M ◑	0046 0607 1304 1937	0322 0951 1641 2252	1.1F 2.8E 2.2F 2.4E
31 Tu	0211 0726 1402 2028	0515 1059 1744 2352	1.2F 2.5E 2.1F 2.8E

June

Day	Slack h m	Maximum h m	knots
1 W	0327 0844 1457 2116	0634 1203 1839	1.4F 2.3E 2.1F
2 Th	 0430 0957 1550 2200	0047 0736 1302 1927	3.1E 1.7F 2.1E 2.1F
3 F	 0521 1102 1638 2242	0137 0831 1357 2010	3.5E 2.0F 2.0E 2.0F
4 Sa	 0606 1158 1723 2321	0223 0920 1446 2047	3.7E 2.2F 1.9E 1.9F
5 Su	 0647 1248 1805 2359	0306 1005 1533 2118	3.8E 2.3F 1.9E 1.8F
6 M ●	 0724 1332 1845	0347 1046 1616 2145	3.8E 2.3F 1.8E 1.7F
7 Tu	0035 0800 1413 1925	0425 1124 1658 2213	3.8E 2.3F 1.7E 1.6F
8 W	0111 0835 1453 2006	0502 1158 1738 2245	3.6E 2.2F 1.6E 1.5F
9 Th	0146 0910 1533 2049	0538 1227 1819 2319	3.5E 2.0F 1.6E 1.4F
10 F	0220 0947 1615 2136	0613 1254 1902 2358	3.3E 1.9F 1.5E 1.2F
11 Sa	0255 1025 1657 2228	0649 1324 1948	3.0E 1.8F 1.5E
12 Su	 0333 1105 1741 2327	0042 0729 1400 2038	1.0F 2.8E 1.8F 1.6E
13 M	 0420 1147 1824	0133 0815 1440 2132	0.9F 2.5E 1.7F 1.7E
14 Tu ◐	0032 0522 1232 1907	0232 0907 1524 2227	0.7F 2.2E 1.7F 1.9E
15 W	0140 0640 1319 1950	0342 1007 1612 2320	0.7F 1.9E 1.6F 2.3E
16 Th	0244 0801 1408 2032	0505 1113 1701	0.8F 1.7E 1.7F
17 F	 0341 0917 1458 2114	0011 0633 1218 1752	2.7E 1.1F 1.6E 1.7F
18 Sa	 0432 1025 1550 2157	0059 0740 1318 1843	3.1E 1.5F 1.6E 1.8F
19 Su	 0520 1126 1641 2242	0146 0833 1413 1933	3.5E 1.9F 1.7E 1.9F
20 M	 0607 1222 1732 2329	0233 0921 1505 2023	3.9E 2.3F 1.8E 2.1F
21 Tu ○	 0654 1314 1824	0319 1006 1556 2114	4.2E 2.5F 1.9E 2.2F
22 W	0017 0740 1404 1916	0407 1052 1646 2205	4.4E 2.7F 2.1E 2.2F
23 Th	0106 0827 1453 2011	0455 1139 1737 2257	4.5E 2.9F 2.2E 2.2F
24 F	0157 0914 1541 2109	0544 1226 1830 2352	4.4E 2.9F 2.3E 2.0F
25 Sa	0251 1002 1630 2211	0635 1314 1925	4.1E 2.8F 2.4E
26 Su	 0347 1049 1719 2319	0051 0728 1405 2022	1.8F 3.7E 2.7F 2.5E
27 M	 0449 1138 1807	0157 0824 1457 2121	1.5F 3.2E 2.5F 2.6E
28 Tu ◑	0033 0556 1227 1856	0319 0925 1553 2221	1.3F 2.7E 2.2F 2.8E
29 W	0150 0709 1318 1945	0454 1028 1652 2319	1.3F 2.2E 2.0F 3.0E
30 Th	0304 0825 1412 2033	0612 1133 1752	1.4F 1.8E 1.8F

Time meridian 120° W. 0000 is midnight. 1200 is noon. Times are not adjusted for Daylight Saving Time.

Grays Harbor Entrance, Washington, 2005

F–Flood, Dir. 060° True   E–Ebb, Dir. 240° True

July

Day	Slack h m	Maximum h m	knots
1 F		0015	3.2E
	0410	0717	1.6F
	0940	1235	1.6E
	1507	1849	1.7F
	2121		
2 Sa		0108	3.3E
	0506	0814	1.8F
	1049	1332	1.5E
	1601	1941	1.6F
	2207		
3 Su		0158	3.5E
	0553	0905	2.0F
	1148	1426	1.5E
	1654	2027	1.5F
	2252		
4 M		0244	3.5E
	0634	0951	2.1F
	1237	1514	1.5E
	1742	2105	1.5F
	2334		
5 Tu		0327	3.6E
	0711	1033	2.2F
	1319	1559	1.6E
	1827	2136	1.5F
6 W ●	0015	0407	3.6E
	0745	1111	2.2F
	1357	1641	1.6E
	1910	2204	1.5F
7 Th	0053	0445	3.5E
	0817	1144	2.2F
	1434	1721	1.7E
	1952	2234	1.4F
8 F	0130	0520	3.4E
	0848	1210	2.1F
	1509	1800	1.8E
	2034	2307	1.4F
9 Sa	0205	0553	3.3E
	0920	1228	2.1F
	1545	1837	1.8E
	2118	2344	1.4F
10 Su	0241	0625	3.1E
	0953	1249	2.1F
	1620	1915	1.9E
	2205		
11 M		0024	1.3F
	0319	0659	2.9E
	1027	1317	2.0F
	1655	1953	2.0E
	2256		
12 Tu		0109	1.2F
	0403	0737	2.6E
	1103	1350	2.0F
	1730	2035	2.1E
	2351		
13 W		0200	1.1F
	0456	0822	2.2E
	1142	1427	1.9F
	1807	2123	2.3E
14 Th ◐	0050	0259	1.0F
	0605	0913	1.9E
	1226	1511	1.8F
	1847	2217	2.5E
15 F	0154	0408	1.0F
	0725	1015	1.6E
	1315	1601	1.7F
	1932	2316	2.8E
16 Sa	0258	0532	1.1F
	0846	1130	1.4E
	1410	1658	1.7F
	2023		
17 Su		0018	3.1E
	0400	0709	1.4F
	1001	1246	1.4E
	1510	1801	1.7F
	2119		
18 M		0117	3.5E
	0457	0818	1.8F
	1108	1352	1.5E
	1612	1906	1.8F
	2216		
19 Tu		0212	3.9E
	0550	0912	2.2F
	1206	1450	1.7E
	1713	2009	2.0F
	2312		
20 W		0305	4.2E
	0639	0959	2.6F
	1258	1543	2.0E
	1812	2108	2.2F
21 Th ○	0007	0355	4.4E
	0726	1044	2.9F
	1345	1634	2.3E
	1909	2204	2.3F
22 F	0101	0444	4.5E
	0811	1127	3.0F
	1430	1724	2.6E
	2005	2258	2.3F
23 Sa	0154	0532	4.4E
	0855	1208	3.1F
	1514	1813	2.8E
	2102	2352	2.3F
24 Su	0247	0620	4.0E
	0938	1249	3.0F
	1557	1902	3.0E
	2200		
25 M		0047	2.1F
	0341	0709	3.6E
	1020	1331	2.8F
	1641	1953	3.0E
	2301		
26 Tu		0147	1.8F
	0438	0800	3.0E
	1103	1413	2.5F
	1725	2046	3.0E
27 W ◑	0006	0257	1.5F
	0539	0854	2.4E
	1148	1458	2.1F
	1810	2143	3.0E
28 Th	0116	0424	1.3F
	0647	0955	1.8E
	1236	1549	1.7F
	1858	2241	3.0E
29 F	0230	0546	1.3F
	0802	1101	1.4E
	1329	1654	1.4F
	1950	2341	2.9E
30 Sa	0343	0655	1.4F
	0920	1208	1.2E
	1431	1817	1.3F
	2044		
31 Su		0040	3.0E
	0446	0755	1.6F
	1033	1310	1.2E
	1536	1926	1.2F
	2138		

August

Day	Slack h m	Maximum h m	knots
1 M		0134	3.1E
	0537	0847	1.8F
	1132	1406	1.3E
	1639	2020	1.3F
	2230		
2 Tu		0224	3.2E
	0618	0933	2.0F
	1218	1456	1.5E
	1733	2105	1.4F
	2317		
3 W		0308	3.3E
	0653	1013	2.1F
	1256	1540	1.7E
	1819	2140	1.4F
4 Th ●	0000	0349	3.4E
	0723	1049	2.2F
	1331	1621	1.9E
	1901	2206	1.5F
5 F	0040	0426	3.4E
	0752	1118	2.2F
	1403	1658	2.0E
	1940	2229	1.6F
6 Sa	0118	0500	3.4E
	0820	1136	2.2F
	1434	1733	2.2E
	2018	2257	1.6F
7 Su	0154	0530	3.3E
	0848	1149	2.2F
	1504	1805	2.3E
	2057	2329	1.6F
8 M	0231	0559	3.1E
	0918	1208	2.2F
	1533	1835	2.4E
	2138		
9 Tu		0005	1.6F
	0309	0629	2.8E
	0949	1233	2.2F
	1602	1907	2.5E
	2223		
10 W		0045	1.6F
	0351	0705	2.5E
	1022	1304	2.1F
	1632	1943	2.6E
	2312		
11 Th		0131	1.5F
	0440	0746	2.2E
	1100	1340	2.0F
	1705	2028	2.7E
12 F ◐	0007	0224	1.3F
	0542	0835	1.8E
	1142	1423	1.8F
	1745	2122	2.7E
13 Sa	0111	0329	1.2F
	0700	0936	1.4E
	1234	1515	1.6F
	1837	2229	2.8E
14 Su	0221	0455	1.2F
	0824	1059	1.2E
	1337	1619	1.5F
	1942	2345	3.0E
15 M	0332	0655	1.4F
	0943	1229	1.2E
	1449	1736	1.5F
	2053		
16 Tu		0056	3.4E
	0436	0806	1.9F
	1051	1339	1.5E
	1602	1858	1.6F
	2201		
17 W		0157	3.7E
	0532	0858	2.3F
	1146	1437	2.0E
	1709	2011	1.9F
	2304		
18 Th		0252	4.1E
	0621	0943	2.7F
	1234	1529	2.4E
	1809	2113	2.2F
19 F ○	0002	0342	4.3E
	0706	1024	3.0F
	1318	1617	2.9E
	1904	2207	2.5F
20 Sa	0056	0430	4.3E
	0748	1102	3.1F
	1359	1703	3.2E
	1957	2257	2.6F
21 Su	0148	0515	4.1E
	0828	1139	3.1F
	1438	1748	3.4E
	2049	2346	2.6F
22 M	0239	0600	3.8E
	0907	1214	2.9F
	1517	1833	3.5E
	2140		
23 Tu		0035	2.4F
	0331	0645	3.2E
	0946	1248	2.6F
	1557	1918	3.5E
	2234		
24 W		0126	2.1F
	0424	0732	2.7E
	1026	1324	2.3F
	1637	2007	3.3E
	2331		
25 Th		0225	1.7F
	0520	0823	2.1E
	1108	1402	1.8F
	1720	2100	3.0E
26 F ◑	0035	0346	1.4F
	0623	0922	1.5E
	1156	1446	1.4F
	1808	2200	2.8E
27 Sa	0148	0516	1.2F
	0736	1031	1.2E
	1253	1546	1.1F
	1904	2306	2.6E
28 Su	0307	0630	1.3F
	0855	1143	1.0E
	1406	1756	0.9F
	2007		
29 M		0010	2.6E
	0417	0731	1.4F
	1008	1249	1.1E
	1528	1916	1.0F
	2111		
30 Tu		0109	2.7E
	0510	0822	1.7F
	1104	1346	1.4E
	1638	2011	1.2F
	2210		
31 W		0200	2.9E
	0550	0906	1.9F
	1146	1434	1.7E
	1730	2055	1.4F
	2301		

September

Day	Slack h m	Maximum h m	knots
1 Th		0245	3.1E
	0622	0944	2.1F
	1221	1517	2.0E
	1812	2132	1.5F
	2346		
2 F		0325	3.2E
	0651	1016	2.2F
	1253	1555	2.3E
	1849	2159	1.7F
3 Sa ●	0027	0402	3.2E
	0718	1039	2.2F
	1322	1629	2.5E
	1924	2220	1.8F
4 Su	0106	0435	3.2E
	0744	1051	2.3F
	1350	1700	2.7E
	1958	2244	1.9F
5 M	0143	0504	3.1E
	0812	1104	2.3F
	1417	1728	2.9E
	2034	2312	2.0F
6 Tu	0221	0532	2.9E
	0841	1125	2.3F
	1443	1755	3.0E
	2111	2346	2.0F
7 W	0300	0602	2.7E
	0912	1152	2.2F
	1509	1826	3.1E
	2153		
8 Th		0024	2.0F
	0342	0637	2.3E
	0945	1224	2.1F
	1537	1903	3.1E
	2239		
9 F		0107	1.8F
	0432	0719	2.0E
	1024	1301	2.0F
	1610	1949	3.0E
	2334		
10 Sa		0159	1.6F
	0533	0810	1.6E
	1110	1346	1.7F
	1654	2045	2.9E
11 Su ◐	0038	0304	1.4F
	0648	0916	1.3E
	1209	1442	1.4F
	1754	2157	2.8E
12 M	0151	0438	1.3F
	0811	1054	1.1E
	1324	1557	1.2F
	1916	2325	2.9E
13 Tu	0306	0644	1.5F
	0927	1221	1.4E
	1448	1733	1.2F
	2040		
14 W		0040	3.2E
	0413	0747	2.0F
	1029	1326	1.8E
	1606	1913	1.5F
	2154		
15 Th		0142	3.5E
	0508	0836	2.4F
	1120	1421	2.4E
	1711	2024	2.0F
	2259		
16 F		0236	3.8E
	0556	0918	2.8F
	1204	1510	3.0E
	1807	2118	2.4F
	2357		
17 Sa ○		0325	3.9E
	0638	0956	3.0F
	1244	1555	3.5E
	1857	2206	2.7F
18 Su	0050	0411	3.8E
	0718	1030	3.0F
	1322	1638	3.8E
	1945	2251	2.8F
19 M	0141	0455	3.6E
	0756	1103	2.9F
	1359	1719	3.9E
	2031	2335	2.7F
20 Tu	0230	0538	3.3E
	0834	1134	2.7F
	1435	1800	3.9E
	2118		
21 W		0018	2.5F
	0318	0620	2.8E
	0911	1204	2.4F
	1511	1842	3.7E
	2205		
22 Th		0102	2.2F
	0408	0704	2.3E
	0950	1237	2.0F
	1548	1926	3.4E
	2256		
23 F		0152	1.8F
	0500	0752	1.8E
	1032	1312	1.6F
	1628	2015	3.0E
	2352		
24 Sa ◑		0301	1.4F
	0559	0851	1.3E
	1121	1354	1.2F
	1715	2115	2.6E
25 Su	0058	0442	1.2F
	0708	1003	1.0E
	1225	1451	0.8F
	1815	2226	2.3E
26 M	0215	0558	1.2F
	0823	1117	1.0E
	1351	1735	0.6F
	1927	2335	2.3E
27 Tu	0327	0659	1.4F
	0931	1223	1.2E
	1526	1857	0.8F
	2041		
28 W		0037	2.4E
	0422	0749	1.6F
	1023	1319	1.6E
	1633	1951	1.1F
	2145		
29 Th		0130	2.6E
	0503	0830	1.9F
	1103	1406	2.0E
	1719	2035	1.3F
	2240		
30 F		0215	2.7E
	0537	0906	2.0F
	1137	1446	2.4E
	1756	2112	1.6F
	2327		

Time meridian 120° W. 0000 is midnight. 1200 is noon. Times are not adjusted for Daylight Saving Time.

Grays Harbor Entrance, Washington, 2005

F–Flood, Dir. 060° True E–Ebb, Dir. 240° True

October

Day	Slack h m	Maximum h m	knots
1 Sa		0256	2.9E
	0607	0933	2.1F
	1208	1523	2.7E
	1830	2141	1.8F
2 Su	0011	0333	2.9E
	0636	0949	2.2F
	1236	1556	3.0E
	1902	2204	2.0F
3 M ●	0051	0406	2.9E
	0704	1001	2.2F
	1304	1625	3.3E
	1935	2227	2.2F
4 Tu	0131	0437	2.8E
	0734	1021	2.3F
	1330	1653	3.4E
	2010	2255	2.3F
5 W	0211	0507	2.6E
	0805	1046	2.3F
	1356	1721	3.6E
	2048	2329	2.3F
6 Th	0253	0540	2.4E
	0838	1116	2.2F
	1423	1755	3.6E
	2129		
7 F		0007	2.2F
	0338	0617	2.1E
	0915	1152	2.1F
	1454	1835	3.5E
	2216		
8 Sa		0052	2.0F
	0429	0702	1.8E
	0958	1233	1.9F
	1532	1923	3.3E
	2311		
9 Su		0146	1.8F
	0531	0758	1.5E
	1052	1322	1.5F
	1622	2022	3.1E
10 M ◐	0014	0255	1.6F
	0643	0918	1.3E
	1201	1426	1.2F
	1731	2139	2.8E
11 Tu	0125	0439	1.5F
	0758	1051	1.3E
	1327	1554	1.0F
	1903	2308	2.8E
12 W	0238	0621	1.8F
	0905	1209	1.8E
	1457	1757	1.1F
	2032		
13 Th		0021	2.9E
	0342	0719	2.1F
	1000	1309	2.3E
	1612	1927	1.6F
	2147		
14 F		0123	3.1E
	0436	0806	2.5F
	1047	1401	2.9E
	1712	2026	2.1F
	2253		
15 Sa		0216	3.3E
	0524	0847	2.7F
	1129	1448	3.5E
	1802	2116	2.5F
	2350		
16 Su		0305	3.3E
	0606	0924	2.8F
	1207	1531	3.9E
	1848	2201	2.7F
17 M ○	0043	0351	3.2E
	0645	0956	2.7F
	1244	1612	4.2E
	1931	2243	2.8F
18 Tu	0132	0434	3.0E
	0723	1026	2.6F
	1320	1652	4.2E
	2013	2324	2.7F
19 W	0219	0516	2.7E
	0800	1055	2.4F
	1355	1731	4.1E
	2055		
20 Th		0003	2.5F
	0305	0558	2.3E
	0838	1125	2.1F
	1429	1809	3.8E
	2138		
21 F		0042	2.2F
	0352	0640	1.9E
	0917	1157	1.7F
	1504	1849	3.4E
	2224		
22 Sa		0124	1.8F
	0441	0727	1.5E
	1001	1232	1.4F
	1541	1934	3.0E
	2313		
23 Su		0219	1.5F
	0535	0824	1.3E
	1054	1314	1.0F
	1624	2028	2.6E
24 M ◑	0010	0352	1.3F
	0637	0933	1.1E
	1202	1410	0.7F
	1721	2137	2.3E
25 Tu	0113	0513	1.2F
	0742	1045	1.2E
	1332	1541	0.4F
	1839	2250	2.1E
26 W	0217	0614	1.4F
	0842	1149	1.4E
	1508	1824	0.6F
	1959	2354	2.1E
27 Th	0314	0703	1.6F
	0931	1243	1.8E
	1612	1921	0.9F
	2110		
28 F		0049	2.2E
	0401	0744	1.7F
	1011	1330	2.3E
	1655	2007	1.3F
	2211		
29 Sa		0138	2.3E
	0440	0815	1.9F
	1046	1410	2.7E
	1731	2045	1.6F
	2304		
30 Su		0221	2.4E
	0516	0837	2.0F
	1118	1447	3.1E
	1805	2117	1.9F
	2351		
31 M		0301	2.4E
	0549	0853	2.0F
	1149	1521	3.4E
	1838	2144	2.1F

November

Day	Slack h m	Maximum h m	knots
1 Tu ●	0036	0338	2.4E
	0623	0914	2.1F
	1218	1552	3.7E
	1913	2211	2.3F
2 W	0119	0413	2.4E
	0657	0942	2.2F
	1248	1623	3.9E
	1950	2242	2.5F
3 Th	0202	0448	2.2E
	0733	1014	2.2F
	1319	1656	4.0E
	2030	2318	2.5F
4 F	0247	0525	2.1E
	0811	1050	2.2F
	1352	1734	4.0E
	2113		
5 Sa		0000	2.4F
	0335	0608	1.9E
	0855	1131	2.0F
	1430	1818	3.8E
	2201		
6 Su		0047	2.3F
	0428	0659	1.7E
	0945	1218	1.8F
	1516	1909	3.5E
	2255		
7 M		0143	2.1F
	0527	0804	1.5E
	1048	1314	1.4F
	1612	2010	3.2E
	2354		
8 Tu ◐		0253	1.9F
	0631	0924	1.5E
	1205	1425	1.1F
	1727	2126	2.9E
9 W	0058	0423	1.9F
	0735	1041	1.8E
	1334	1605	0.9F
	1856	2247	2.7E
10 Th	0203	0544	2.0F
	0833	1148	2.3E
	1501	1810	1.2F
	2022	2357	2.6E
11 F	0304	0643	2.2F
	0925	1246	2.8E
	1611	1922	1.7F
	2138		
12 Sa		0059	2.6E
	0358	0732	2.3F
	1010	1337	3.3E
	1706	2018	2.1F
	2245		
13 Su		0154	2.6E
	0447	0814	2.4F
	1052	1424	3.8E
	1754	2108	2.4F
	2344		
14 M		0244	2.6E
	0532	0852	2.4F
	1132	1507	4.1E
	1837	2154	2.6F
15 Tu ○	0037	0331	2.5E
	0613	0925	2.3F
	1209	1549	4.2E
	1918	2236	2.7F
16 W	0125	0415	2.3E
	0653	0955	2.2F
	1246	1628	4.2E
	1958	2316	2.6F
17 Th	0210	0458	2.1E
	0731	1025	2.0F
	1321	1706	4.0E
	2037	2354	2.4F
18 F	0253	0539	1.9E
	0810	1055	1.8F
	1356	1744	3.7E
	2116		
19 Sa		0030	2.2F
	0336	0622	1.7E
	0852	1128	1.5F
	1431	1822	3.4E
	2157		
20 Su		0105	1.9F
	0421	0707	1.5E
	0938	1205	1.3F
	1507	1902	3.1E
	2240		
21 M		0145	1.7F
	0510	0759	1.4E
	1032	1248	1.0F
	1546	1948	2.7E
	2326		
22 Tu		0235	1.5F
	0601	0859	1.3E
	1138	1340	0.7F
	1634	2042	2.3E
23 W ◑	0015	0340	1.4F
	0654	1002	1.4E
	1256	1450	0.5F
	1744	2148	2.1E
24 Th	0107	0452	1.4F
	0745	1103	1.6E
	1420	1649	0.4F
	1907	2256	1.9E
25 F	0159	0546	1.5F
	0831	1157	2.0E
	1528	1836	0.7F
	2026	2358	1.8E
26 Sa	0250	0625	1.5F
	0913	1246	2.4E
	1617	1930	1.1F
	2136		
27 Su		0053	1.8E
	0337	0656	1.6F
	0951	1329	2.8E
	1658	2014	1.5F
	2237		
28 M		0142	1.9E
	0421	0726	1.8F
	1028	1409	3.3E
	1737	2052	1.8F
	2331		
29 Tu		0228	1.9E
	0503	0758	1.9F
	1103	1447	3.6E
	1814	2126	2.1F
30 W	0020	0311	2.0E
	0544	0834	2.0F
	1139	1524	3.9E
	1853	2200	2.4F

December

Day	Slack h m	Maximum h m	knots
1 Th ●	0108	0353	2.0E
	0625	0912	2.1F
	1216	1602	4.1E
	1934	2236	2.5F
2 F	0154	0435	2.0E
	0708	0952	2.2F
	1255	1642	4.2E
	2017	2316	2.6F
3 Sa	0241	0519	2.0E
	0754	1036	2.1F
	1338	1725	4.2E
	2102	2359	2.6F
4 Su	0329	0607	2.0E
	0845	1123	2.0F
	1424	1812	4.0E
	2149		
5 M		0048	2.5F
	0419	0702	1.9E
	0942	1216	1.8F
	1515	1904	3.7E
	2239		
6 Tu		0141	2.4F
	0512	0803	2.0E
	1048	1316	1.5F
	1615	2002	3.3E
	2331		
7 W		0240	2.3F
	0606	0910	2.1E
	1204	1430	1.2F
	1726	2108	2.9E
8 Th ◐	0026	0346	2.2F
	0701	1016	2.3E
	1328	1611	1.1F
	1846	2220	2.5E
9 F	0123	0455	2.1F
	0754	1119	2.7E
	1449	1757	1.3F
	2008	2329	2.2E
10 Sa	0220	0559	2.1F
	0844	1217	3.1E
	1558	1908	1.6F
	2126		
11 Su		0033	2.1E
	0317	0654	2.1F
	0932	1311	3.5E
	1656	2006	2.0F
	2237		
12 M		0132	2.0E
	0411	0744	2.0F
	1017	1400	3.8E
	1745	2058	2.3F
	2338		
13 Tu		0225	1.9E
	0500	0827	2.0F
	1100	1446	3.9E
	1828	2146	2.4F
14 W	0031	0314	1.9E
	0547	0905	1.9F
	1141	1529	4.0E
	1909	2229	2.5F
15 Th ○	0118	0400	1.9E
	0630	0938	1.8F
	1221	1610	4.0E
	1946	2310	2.5F
16 F	0200	0443	1.8E
	0712	1008	1.7F
	1259	1649	3.9E
	2022	2347	2.4F
17 Sa	0240	0524	1.8E
	0753	1039	1.6F
	1335	1726	3.7E
	2057		
18 Su		0020	2.2F
	0318	0605	1.7E
	0836	1112	1.5F
	1411	1802	3.4E
	2132		
19 M		0047	2.1F
	0358	0647	1.7E
	0921	1149	1.3F
	1447	1837	3.2E
	2208		
20 Tu		0112	2.0F
	0438	0730	1.7E
	1012	1229	1.1F
	1524	1914	2.9E
	2246		
21 W		0143	1.9F
	0519	0817	1.7E
	1108	1316	0.9F
	1607	1955	2.5E
	2325		
22 Th		0218	1.8F
	0601	0908	1.8E
	1210	1411	0.8F
	1701	2043	2.2E
23 F ◑	0008	0259	1.7F
	0643	1003	1.9E
	1318	1516	0.6F
	1814	2140	1.8E
24 Sa	0053	0344	1.6F
	0726	1058	2.2E
	1426	1639	0.7F
	1936	2247	1.6E
25 Su	0142	0434	1.5F
	0809	1152	2.5E
	1527	1830	0.9F
	2055	2358	1.4E
26 M	0234	0527	1.5F
	0853	1242	2.8E
	1620	1938	1.3F
	2207		
27 Tu		0101	1.4E
	0328	0620	1.6F
	0937	1330	3.2E
	1707	2029	1.7F
	2309		
28 W		0157	1.5E
	0420	0713	1.7F
	1023	1417	3.6E
	1753	2113	2.0F
29 Th	0005	0249	1.7E
	0512	0804	1.9F
	1109	1502	4.0E
	1837	2154	2.4F
30 F ●	0055	0338	1.8E
	0603	0853	2.0F
	1157	1547	4.3E
	1921	2234	2.6F
31 Sa	0142	0425	2.0E
	0654	0943	2.2F
	1245	1633	4.4E
	2005	2315	2.8F

Time meridian 120° W. 0000 is midnight. 1200 is noon. Times are not adjusted for Daylight Saving Time.

Strait of Juan de Fuca Entrance, 2005

F–Flood, Dir. 115° True E–Ebb, Dir. 290° True

January

Day	Slack h m	Maximum h m	knots
1 Sa	0247	0529	0.8F
	0813	1214	1.3E
		1757	*
		2330	1.2E
2 Su	0334	0607	0.6F
	0837	1255	1.4E
		1900	*
3 M ◑		0030	1.0E
	0430	0648	0.5F
	0901	1338	1.5E
		2008	*
4 Tu		0141	0.9E
	0537	0735	0.3F
	0925	1425	1.7E
	1924	2116	0.3F
	2319		
5 W		0301	0.8E
		0827	*
		1514	1.8E
	1957	2220	0.6F
6 Th	0059	0420	0.8E
		0925	*
		1605	2.0E
	2034	2317	0.9F
7 F	0219	0530	0.9E
		1025	*
		1656	2.1E
	2115		
8 Sa		0009	1.1F
	0322	0631	1.0E
		1125	*
		1747	2.3E
	2158		
9 Su		0058	1.4F
	0416	0724	1.2E
		1223	*
		1838	2.3E
	2243		
10 M ●		0145	1.5F
	0503	0814	1.3E
		1320	*
		1929	2.4E
	2329		
11 Tu		0231	1.6F
	0547	0901	1.5E
		1416	*
		2020	2.3E
12 W	0015	0316	1.6F
	0627	0946	1.6E
		1512	*
		2112	2.1E
13 Th	0104	0400	1.4F
	0704	1032	1.7E
		1610	*
		2204	1.9E
14 F	0153	0445	1.3F
	0740	1119	1.7E
	1544	1709	0.3F
	1836	2259	1.7E
15 Sa	0246	0530	1.0F
	0813	1207	1.8E
	1641	1813	0.3F
	1949	2359	1.4E
16 Su	0344	0616	0.8F
	0844	1256	1.8E
	1740	1922	0.3F
◐	2110		
17 M		0105	1.1E
	0452	0704	0.5F
	0911	1348	1.8E
	1837	2034	0.4F
	2242		
18 Tu		0221	0.9E
	0614	0757	0.3F
	0934	1442	1.8E
	1930	2145	0.5F
19 W	0016	0344	0.8E
		0854	*
		1536	1.8E
	2018	2249	0.6F
20 Th	0138	0502	0.8E
		0954	*
		1628	1.8E
	2102	2344	0.8F
21 F	0244	0607	0.9E
		1053	*
		1716	1.8E
	2142		
22 Sa		0030	0.9F
	0335	0658	0.9E
		1147	*
		1800	1.8E
	2218		
23 Su		0110	1.0F
	0416	0740	1.0E
		1234	*
		1840	1.9E
	2253		
24 M		0145	1.1F
	0451	0817	1.1E
		1317	*
		1918	1.8E
	2326		
25 Tu ○		0218	1.1F
	0522	0849	1.2E
		1357	*
		1954	1.8E
	2359		
26 W		0249	1.1F
	0549	0919	1.3E
		1435	*
		2030	1.8E
27 Th	0032	0319	1.0F
	0614	0949	1.3E
		1513	*
		2105	1.7E
28 F	0106	0348	0.9F
	0637	1018	1.4E
		1553	*
		2143	1.6E
29 Sa	0142	0418	0.8F
	0657	1048	1.4E
		1635	*
		2223	1.4E
30 Su	0221	0449	0.7F
	0716	1120	1.5E
		1724	*
		2309	1.2E
31 M	0305	0521	0.5F
	0734	1157	1.6E
		1820	*

February

Day	Slack h m	Maximum h m	knots
1 Tu		0003	1.0E
	0359	0558	0.4F
	0753	1240	1.7E
◑	1734	1925	0.3F
	2126		
2 W		0111	0.8E
		0643	*
		1330	1.7E
	1824	2037	0.5F
	2307		
3 Th		0235	0.7E
		0740	*
		1428	1.8E
	1915	2150	0.7F
4 F	0045	0404	0.7E
		0850	*
		1532	1.9E
	2006	2255	0.9F
5 Sa	0204	0520	0.9E
		1007	*
		1635	2.0E
	2055	2351	1.1F
6 Su	0304	0620	1.1E
		1118	*
		1735	2.1E
	2143		
7 M		0042	1.3F
	0353	0710	1.3E
		1221	*
		1831	2.2E
	2231		
8 Tu		0128	1.4F
	0436	0754	1.5E
	1158	1319	0.3F
●	1438	1924	2.2E
	2318		
9 W		0212	1.4F
	0514	0836	1.6E
	1234	1413	0.4F
	1549	2015	2.2E
10 Th	0004	0254	1.4F
	0549	0917	1.8E
	1315	1504	0.5F
	1653	2105	2.0E
11 F	0051	0334	1.2F
	0621	0958	1.9E
	1400	1556	0.6F
	1754	2154	1.8E
12 Sa	0139	0414	1.0F
	0651	1039	1.9E
	1448	1649	0.6F
	1853	2245	1.6E
13 Su	0229	0454	0.8F
	0718	1122	1.9E
	1541	1744	0.5F
	1953	2338	1.3E
14 M	0326	0535	0.5F
	0741	1207	1.8E
	1640	1845	0.5F
	2100		
15 Tu		0039	1.0E
	0436	0619	0.3F
	0758	1256	1.7E
◐	1742	1952	0.5F
	2216		
16 W		0152	0.8E
		0709	*
		1350	1.7E
	1845	2104	0.5F
	2341		
17 Th		0318	0.7E
		0811	*
		1451	1.6E
	1942	2213	0.6F
18 F	0102	0442	0.8E
		0925	*
		1552	1.6E
	2032	2312	0.7F
19 Sa	0206	0546	0.9E
		1036	*
		1649	1.6E
	2116		
20 Su		0000	0.8F
	0255	0634	1.0E
		1136	*
		1739	1.7E
	2155		
21 M		0040	0.9F
	0334	0711	1.1E
		1224	*
		1823	1.7E
	2230		
22 Tu		0115	0.9F
	0407	0743	1.3E
		1306	*
		1903	1.7E
	2305		
23 W		0147	1.0F
	0436	0812	1.4E
		1344	*
○		1940	1.7E
	2339		
24 Th		0217	0.9F
	0500	0839	1.4E
		1419	*
		2016	1.7E
25 F	0013	0246	0.9F
	0522	0905	1.5E
	1317	1455	0.3F
	1633	2052	1.6E
26 Sa	0048	0313	0.8F
	0541	0932	1.6E
	1342	1532	0.4F
	1723	2130	1.5E
27 Su	0124	0341	0.7F
	0558	1000	1.7E
	1413	1612	0.5F
	1814	2210	1.4E
28 M	0205	0411	0.5F
	0614	1032	1.7E
	1451	1657	0.5F
	1910	2255	1.2E

March

Day	Slack h m	Maximum h m	knots
1 Tu	0253	0443	0.4F
	0630	1108	1.8E
	1536	1750	0.6F
	2014	2350	1.0E
2 W		0520	*
		1153	1.8E
	1631	1853	0.6F
	2131		
3 Th		0059	0.8E
		0608	*
		1247	1.8E
◑	1733	2006	0.6F
	2258		
4 F		0226	0.8E
		0715	*
		1354	1.7E
	1837	2122	0.7F
5 Sa	0025	0356	0.8E
		0841	*
		1509	1.8E
	1939	2230	0.9F
6 Su	0135	0507	1.0E
		1009	*
		1622	1.8E
	2036	2329	1.0F
7 M	0230	0600	1.2E
		1122	*
		1728	1.9E
	2129		
8 Tu		0019	1.1F
	0315	0645	1.5E
	1053	1222	0.3F
	1350	1826	2.0E
	2219		
9 W		0104	1.2F
	0354	0727	1.7E
	1123	1315	0.5F
	1507	1919	2.0E
	2307		
10 Th		0146	1.2F
	0428	0805	1.8E
	1158	1405	0.7F
●	1612	2008	1.9E
	2354		
11 F		0226	1.1F
	0459	0843	2.0E
	1237	1452	0.8F
	1709	2056	1.8E
12 Sa	0040	0304	0.9F
	0527	0921	2.0E
	1318	1538	0.9F
	1804	2142	1.6E
13 Su	0128	0341	0.7F
	0552	0958	2.0E
	1402	1625	0.8F
	1856	2230	1.4E
14 M	0220	0418	0.5F
	0613	1037	1.9E
	1451	1714	0.8F
	1949	2320	1.2E
15 Tu	0321	0456	0.3F
	0628	1118	1.8E
	1544	1808	0.7F
	2045		
16 W		0017	1.0E
		0537	*
		1203	1.7E
	1644	1908	0.6F
	2148		
17 Th		0127	0.8E
		0629	*
		1257	1.5E
◐	1749	2016	0.5F
	2258		
18 F		0251	0.8E
		0739	0.3E
		1401	1.4E
	1853	2125	0.5F
19 Sa	0008	0411	0.9E
		0905	0.3E
		1512	1.4E
	1950	2226	0.6F
20 Su	0108	0509	1.0E
		1022	0.3E
		1617	1.4E
	2040	2316	0.6F
21 M	0156	0553	1.2E
		1120	*
		1713	1.4E
	2123	2358	0.7F
22 Tu	0235	0627	1.3E
		1207	*
		1801	1.5E
	2202		
23 W		0034	0.8F
	0307	0658	1.4E
		1246	*
		1843	1.6E
	2240		
24 Th		0107	0.8F
	0334	0726	1.5E
	1136	1323	0.4F
	1510	1922	1.6E
	2317		
25 F		0137	0.7F
	0358	0752	1.6E
	1155	1358	0.6F
○	1602	2000	1.6E
	2354		
26 Sa		0207	0.7F
	0418	0819	1.7E
	1219	1434	0.7F
	1651	2039	1.5E
27 Su	0033	0235	0.6F
	0436	0847	1.8E
	1248	1511	0.8F
	1740	2118	1.4E
28 M	0115	0305	0.4F
	0453	0917	1.9E
	1322	1552	0.9F
	1831	2202	1.3E
29 Tu	0202	0337	0.3F
	0509	0952	1.9E
	1403	1638	0.9F
	1926	2251	1.1E
30 W		0413	*
		1031	1.9E
	1451	1730	0.9F
	2026	2349	1.0E
31 Th		0457	*
		1119	1.8E
	1548	1831	0.8F
	2134		

Time meridian 120° W. 0000 is midnight. 1200 is noon. Times are not adjusted for Daylight Saving Time.
If three consecutive entries are marked (E) the middle one is not a true maximum but an intermediate value to show the current pattern.
* Current weak and variable.

Strait of Juan de Fuca Entrance, 2005

F–Flood, Dir. 115° True E–Ebb, Dir. 290° True

April

Day	Slack h m	Maximum h m	knots
1 F		0101	0.9E
		0556	*
		1219	1.7E
◑	1653	1941	0.8F
	2246		
2 Sa		0224	0.9E
		0718	0.3E
		1333	1.6E
	1804	2054	0.8F
	2355		
3 Su		0341	1.1E
		0853	*
		1456	1.6E
	1913	2201	0.9F
4 M	0054	0442	1.3E
		1016	*
		1614	1.6E
	2017	2259	0.9F
5 Tu	0143	0531	1.5E
	0954	1123	0.3F
	1251	1721	1.7E
	2114	2350	0.9F
6 W	0225	0614	1.7E
	1019	1218	0.5F
	1418	1819	1.7E
	2207		
7 Th		0034	0.9F
	0301	0654	1.9E
	1051	1307	0.8F
	1526	1912	1.7E
	2257		
8 F		0116	0.8F
	0332	0731	2.0E
	1126	1353	1.0F
●	1625	2000	1.6E
	2346		
9 Sa		0154	0.7F
	0400	0808	2.1E
	1204	1437	1.1F
	1718	2046	1.5E
10 Su	0035	0231	0.5F
	0424	0843	2.1E
	1243	1520	1.1F
	1807	2132	1.4E
11 M	0127	0306	0.4F
	0444	0919	2.0E
	1324	1603	1.0F
	1854	2218	1.2E
12 Tu		0342	*
		0955	1.9E
	1408	1647	0.9F
	1941	2307	1.1E
13 W		0420	*
		1033	1.8E
	1456	1735	0.8F
	2029		
14 Th		0002	1.0E
		0504	*
		1116	1.6E
	1549	1827	0.7F
	2120		
15 F		0107	0.9F
		0601	0.3E
		1207	1.4E
	1648	1926	0.6F
	2214		
16 Sa		0219	0.9E
		0717	0.4E
		1311	1.3E
◐	1751	2028	0.5F
	2309		
17 Su		0326	1.0E
		0844	0.4E
		1425	1.2E
	1854	2127	0.5F
18 M	0000	0419	1.2E
		0958	*
		1537	1.2E
	1951	2219	0.5F
19 Tu	0045	0500	1.3E
		1054	*
		1638	1.2E
	2042	2303	0.6F
20 W	0122	0535	1.5E
		1140	*
		1731	1.3E
	2128	2342	0.6F
21 Th	0154	0606	1.6E
	1026	1220	0.4F
	1417	1817	1.4E
	2212		
22 F		0018	0.5F
	0221	0636	1.7E
	1044	1258	0.6F
	1516	1901	1.4E
	2255		
23 Sa		0051	0.5F
	0245	0705	1.8E
	1108	1335	0.8F
	1609	1943	1.4E
	2339		
24 Su		0124	0.4F
	0306	0735	2.0E
	1136	1413	1.0F
○	1659	2026	1.4E
25 M	0025	0157	0.3F
	0327	0807	2.0E
	1209	1454	1.1F
	1750	2110	1.3E
26 Tu		0232	*
		0843	2.1E
	1247	1537	1.2F
	1841	2158	1.2E
27 W		0311	*
		0922	2.1E
	1331	1624	1.2F
	1934	2252	1.1E
28 Th		0356	*
		1007	2.0E
	1421	1716	1.1F
	2030	2353	1.1E
29 F		0452	*
		1100	1.8E
	1518	1815	1.0F
	2127		
30 Sa		0101	1.1E
		0603	*
		1205	1.7E
◑	1623	1919	0.9F
	2224		

May

Day	Slack h m	Maximum h m	knots
1 Su		0211	1.2E
		0731	*
		1322	1.5E
	1734	2024	0.9F
	2317		
2 M		0315	1.4E
		0859	*
		1445	1.4E
	1846	2127	0.8F
3 Tu	0006	0409	1.6E
		1014	*
		1603	1.4E
	1954	2223	0.7F
4 W	0049	0457	1.7E
	0918	1115	0.5F
	1317	1711	1.4E
	2057	2314	0.7F
5 Th	0126	0540	1.9E
	0949	1208	0.8F
	1434	1811	1.4E
	2156		
6 F		0000	0.6F
	0159	0620	2.0E
	1024	1256	1.0F
	1537	1904	1.4E
	2252		
7 Sa		0042	0.4F
	0228	0658	2.1E
	1100	1340	1.1F
	1632	1953	1.3E
	2347		
8 Su		0121	0.3F
	0252	0734	2.1E
	1137	1423	1.2F
●	1721	2039	1.3E
9 M		0158	*
		0809	2.1E
	1215	1503	1.2F
	1807	2124	1.2E
10 Tu		0235	*
		0844	2.0E
	1253	1543	1.1F
	1849	2210	1.1E
11 W		0312	*
		0919	1.9E
	1333	1624	1.0F
	1930	2257	1.1E
12 Th		0354	*
		0957	1.7E
	1416	1706	0.9F
	2011	2348	1.0E
13 F		0442	0.3E
		1038	1.6E
	1502	1751	0.8F
	2051		
14 Sa		0043	1.0E
		0541	0.4E
		1127	1.4E
	1553	1839	0.7F
	2133		
15 Su		0139	1.1E
		0653	0.4E
		1226	1.2E
	1650	1931	0.6F
	2214		
16 M		0233	1.2E
		0811	0.3E
		1335	1.1E
◐	1751	2024	0.5F
	2253		
17 Tu		0320	1.3E
		0920	*
		1448	1.0E
	1853	2114	0.5F
	2330		
18 W		0401	1.4E
		1018	*
		1555	1.0E
	1953	2201	0.4F
19 Th	0003	0438	1.6E
	0920	1106	0.3F
	1258	1655	1.1E
	2050	2244	0.4F
20 F	0033	0513	1.7E
	0936	1150	0.6F
	1412	1749	1.1E
	2144	2325	0.3F
21 Sa	0101	0547	1.9E
	1000	1232	0.8F
	1514	1839	1.2E
22 Su		0004	*
		0621	2.0E
	1030	1313	1.1F
	1609	1927	1.2E
23 M		0043	*
		0658	2.2E
	1103	1355	1.2F
○	1702	2015	1.2E
24 Tu		0124	*
		0736	2.2E
	1141	1438	1.4F
	1753	2104	1.2E
25 W		0207	*
		0818	2.2E
	1223	1521	1.4F
	1842	2155	1.2E
26 Th		0256	*
		0904	2.2E
	1309	1612	1.4F
	1931	2249	1.3E
27 F		0351	*
		0954	2.0E
	1400	1703	1.3F
	2019	2346	1.3E
28 Sa		0456	*
		1051	1.8E
	1456	1757	1.2F
	2106		
29 Su		0045	1.4E
		0610	*
		1157	1.6E
	1558	1853	1.0F
	2151		
30 M		0144	1.5E
		0731	*
		1311	1.4E
◑	1706	1951	0.9F
	2234		
31 Tu		0240	1.6E
		0850	*
		1431	1.2E
	1818	2049	0.7F
	2313		

June

Day	Slack h m	Maximum h m	knots
1 W		0332	1.8E
	0810	1000	0.4F
	1157	1548	1.1E
	1931	2144	0.5F
	2350		
2 Th		0421	1.9E
	0846	1101	0.6F
	1328	1659	1.1E
	2043	2236	0.4F
3 F	0023	0506	2.0E
	0923	1155	0.9F
	1441	1801	1.1E
	2153	2324	0.3F
4 Sa	0051	0548	2.1E
	1001	1243	1.0F
	1542	1857	1.1E
5 Su		0009	*
		0627	2.1E
	1038	1328	1.1F
	1634	1948	1.1E
6 M		0051	*
		0705	2.1E
	1115	1409	1.2F
●	1720	2034	1.1E
7 Tu		0131	*
		0741	2.0E
	1152	1448	1.2F
	1800	2118	1.1E
8 W		0211	*
		0817	1.9E
	1229	1525	1.1F
	1837	2200	1.1E
9 Th		0252	*
		0853	1.8E
	1306	1602	1.1F
	1912	2242	1.1E
10 F		0336	0.3E
		0930	1.7E
	1344	1639	1.0F
	1945	2324	1.1E
11 Sa		0425	0.3E
		1011	1.5E
	1424	1717	0.9F
	2017		
12 Su		0007	1.2E
		0520	0.3E
		1056	1.4E
	1509	1757	0.8F
	2048		
13 M		0051	1.2E
		0621	0.3E
		1149	1.2E
	1558	1839	0.7F
	2118		
14 Tu		0135	1.3E
		0727	*
		1250	1.1E
◐	1654	1923	0.5F
	2148		
15 W		0218	1.4E
		0833	*
		1358	0.9E
	1756	2010	0.4F
	2216		
16 Th		0300	1.6E
		0934	*
		1510	0.9E
	1904	2057	0.3F
	2243		
17 F		0341	1.7E
	0827	1029	0.4F
	1241	1619	0.9E
		2145	*
18 Sa		0422	1.8E
	0852	1119	0.7F
	1401	1722	0.9E
		2233	*
19 Su		0503	2.0E
	0924	1207	0.9F
	1507	1820	1.0E
		2321	*
20 M		0546	2.1E
	1000	1253	1.2F
	1605	1914	1.1E
21 Tu		0011	*
		0630	2.3E
	1039	1338	1.4F
○	1656	2005	1.2E
22 W		0102	*
		0716	2.3E
	1121	1424	1.5F
	1744	2054	1.3E
23 Th		0155	*
		0804	2.3E
	1206	1510	1.5F
	1829	2144	1.3E
24 F		0251	*
		0854	2.2E
	1254	1556	1.5F
	1912	2233	1.4E
25 Sa		0350	*
		0947	2.0E
	1344	1644	1.4F
	1952	2323	1.5E
26 Su		0454	*
		1045	1.8E
	1438	1733	1.2F
	2031		
27 M		0015	1.6E
		0602	*
		1147	1.5E
	1537	1823	1.0F
	2109		
28 Tu		0107	1.7E
		0715	*
		1256	1.3E
◑	1641	1915	0.7F
	2144		
29 W		0200	1.8E
	0644	0828	0.3F
	1019	1412	1.1E
	1754	2009	0.5F
	2217		
30 Th		0253	1.9E
	0731	0939	0.5F
	1158	1530	1.0E
	1916	2104	0.3F
	2246		

Time meridian 120° W. 0000 is midnight. 1200 is noon. Times are not adjusted for Daylight Saving Time.
If three consecutive entries are marked (E) the middle one is not a true maximum but an intermediate value to show the current pattern.
* Current weak and variable.

Strait of Juan de Fuca Entrance, 2005

F–Flood, Dir. 115° True E–Ebb, Dir. 290° True

July

Day	Slack h m	Maximum h m	knots
1 F		0344	1.9E
	0816	1043	0.7F
	1325	1645	0.9E
		2159	*
2 Sa		0433	2.0E
	0859	1139	0.9F
	1438	1753	0.9E
		2252	*
3 Su		0519	2.0E
	0941	1229	1.0F
	1536	1851	1.0E
		2343	*
4 M		0602	2.0E
	1020	1313	1.1F
	1625	1941	1.0E
5 Tu		0030	*
		0643	2.0E
	1057	1353	1.1F
	1706	2024	1.1E
6 W ●		0115	*
		0721	1.9E
	1133	1429	1.1F
	1741	2103	1.1E
7 Th		0157	*
		0758	1.9E
	1208	1503	1.1F
	1812	2139	1.2E
8 F		0239	*
		0835	1.8E
	1243	1536	1.1F
	1841	2213	1.2E
9 Sa		0321	*
		0912	1.7E
	1318	1609	1.0F
	1907	2247	1.3E
10 Su		0405	*
		0950	1.5E
	1355	1641	0.9F
	1932	2322	1.3E
11 M		0452	*
		1032	1.4E
	1435	1715	0.8F
	1956	2357	1.4E
12 Tu		0543	*
		1118	1.2E
	1519	1750	0.6F
	2019		
13 W		0035	1.4E
		0640	*
		1212	1.1E
	1611	1828	0.5F
	2040		
14 Th ◐		0115	1.5E
		0743	*
		1316	0.9E
	1713	1910	0.3F
	2101		
15 F		0159	1.6E
	0702	0848	0.3F
	1043	1430	0.8E
		1958	*
16 Sa		0246	1.7E
	0736	0952	0.5F
	1223	1549	0.7E
		2053	*
17 Su		0336	1.9E
	0813	1050	0.7F
	1347	1702	0.8E
		2153	*
18 M		0428	2.0E
	0854	1144	1.0F
	1454	1805	0.9E
		2254	*
19 Tu		0520	2.1E
	0937	1234	1.2F
	1550	1900	1.1E
		2355	*
20 W		0612	2.2E
	1021	1321	1.4F
	1637	1949	1.2E
21 Th ○		0053	*
		0704	2.3E
	1106	1406	1.5F
	1720	2035	1.4E
22 F		0150	*
		0756	2.3E
	1152	1451	1.5F
	1800	2120	1.5E
23 Sa		0246	*
		0848	2.2E
	1240	1535	1.4F
	1837	2205	1.7E
24 Su	0215	0343	0.3F
	0512	0941	2.0E
	1329	1619	1.3F
	1913	2250	1.8E
25 M	0304	0442	0.4F
	0620	1035	1.7E
	1422	1704	1.1F
	1946	2336	1.8E
26 Tu	0358	0543	0.4F
	0732	1134	1.5E
	1518	1749	0.8F
	2018		
27 W ◑		0025	1.8E
	0455	0649	0.4F
	0849	1238	1.2E
	1623	1837	0.6F
	2046		
28 Th		0117	1.8E
	0554	0759	0.5F
	1015	1351	1.0E
	1742	1929	0.3F
	2111		
29 F		0211	1.8E
	0652	0911	0.6F
	1146	1512	0.8E
		2026	*
30 Sa		0307	1.8E
	0746	1019	0.7F
	1310	1633	0.8E
		2129	*
31 Su		0402	1.8E
	0835	1118	0.8F
	1420	1743	0.9E
		2232	*

August

Day	Slack h m	Maximum h m	knots
1 M		0455	1.8E
	0920	1209	0.9F
	1514	1839	1.0E
		2329	*
2 Tu		0543	1.8E
	1001	1252	1.0F
	1558	1923	1.1E
3 W		0021	*
		0626	1.8E
	1039	1330	1.0F
	1634	2001	1.1E
4 Th ●		0106	*
		0707	1.8E
	1114	1404	1.0F
	1704	2034	1.2E
5 F		0147	*
		0744	1.8E
	1148	1435	1.0F
	1731	2104	1.3E
6 Sa		0225	*
		0820	1.7E
	1222	1505	1.0F
	1755	2133	1.4E
7 Su		0303	*
		0856	1.6E
	1256	1534	0.9F
	1817	2201	1.4E
8 M		0341	*
		0932	1.5E
	1331	1603	0.8F
	1837	2230	1.5E
9 Tu		0422	*
		1011	1.4E
	1410	1633	0.6F
	1855	2301	1.5E
10 W		0507	*
		1054	1.2E
	1453	1704	0.5F
	1912	2336	1.6E
11 Th	0422	0558	0.3F
	0740	1145	1.0E
	1545	1738	0.3F
	1927		
12 F ◐		0015	1.6E
	0507	0658	0.3F
	0859	1247	0.9E
		1819	*
13 Sa		0102	1.7E
	0556	0806	0.4F
	1031	1405	0.7E
		1911	*
14 Su		0158	1.7E
	0648	0917	0.6F
	1206	1531	0.7E
		2018	*
15 M		0300	1.8E
	0739	1023	0.8F
	1327	1649	0.8E
		2134	*
16 Tu		0404	1.9E
	0829	1121	1.0F
	1430	1750	1.0E
		2247	*
17 W		0505	2.0E
	0917	1213	1.2F
	1521	1840	1.2E
		2352	*
18 Th		0603	2.1E
	1005	1300	1.3F
	1604	1925	1.4E
	2329		
19 F ○		0051	0.3F
	0212	0657	2.2E
	1052	1344	1.4F
	1642	2007	1.6E
20 Sa	0003	0146	0.4F
	0326	0750	2.1E
	1140	1426	1.3F
	1718	2048	1.8E
21 Su	0042	0238	0.6F
	0433	0840	2.0E
	1227	1508	1.2F
	1751	2129	1.9E
22 M	0126	0330	0.7F
	0535	0931	1.9E
	1316	1549	1.0F
	1822	2211	1.9E
23 Tu	0213	0423	0.7F
	0636	1023	1.6E
	1408	1630	0.8F
	1851	2254	1.9E
24 W	0305	0518	0.7F
	0739	1118	1.4E
	1506	1713	0.6F
	1917	2340	1.9E
25 Th	0402	0618	0.6F
	0845	1220	1.1E
	1617	1758	0.3F
	1937		
26 F ◑		0030	1.8E
	0504	0725	0.6F
	1000	1332	0.9E
		1851	*
27 Sa		0125	1.7E
	0609	0836	0.6F
	1121	1456	0.8E
		1955	*
28 Su		0227	1.6E
	0711	0947	0.6F
	1238	1618	0.8E
		2109	*
29 M		0331	1.6E
	0807	1048	0.7F
	1342	1724	0.9E
		2222	*
30 Tu		0431	1.6E
	0855	1139	0.8F
	1432	1813	1.1E
		2323	*
31 W		0524	1.6E
	0937	1221	0.8F
	1511	1851	1.2E

September

Day	Slack h m	Maximum h m	knots
1 Th		0012	*
		0610	1.7E
	1016	1257	0.9F
	1544	1923	1.3E
2 F		0054	*
		0651	1.7E
	1052	1330	0.9F
	1612	1953	1.4E
3 Sa ●		0131	*
		0728	1.7E
	1126	1400	0.9F
	1637	2020	1.5E
4 Su	0035	0207	0.3F
	0339	0804	1.6E
	1201	1428	0.8F
	1658	2046	1.5E
5 M	0057	0241	0.4F
	0426	0840	1.6E
	1236	1456	0.7F
	1716	2112	1.6E
6 Tu	0122	0317	0.5F
	0513	0916	1.5E
	1313	1523	0.6F
	1732	2139	1.6E
7 W	0152	0355	0.5F
	0601	0955	1.3E
	1354	1551	0.4F
	1746	2209	1.7E
8 Th	0228	0437	0.5F
	0653	1038	1.2E
	1442	1621	0.3F
	1758	2243	1.7E
9 F	0311	0526	0.6F
	0752	1129	1.0E
		1656	*
		2324	1.7E
10 Sa	0402	0624	0.6F
	0902	1234	0.8E
		1741	*
11 Su ◐		0016	1.7E
	0501	0733	0.6F
	1023	1356	0.8E
		1844	*
12 M		0120	1.7E
	0605	0846	0.7F
	1146	1523	0.8E
		2008	*
13 Tu		0234	1.7E
	0707	0956	0.8F
	1257	1634	1.0E
		2137	*
14 W		0349	1.7E
	0806	1055	0.9F
	1353	1728	1.2E
		2251	*
15 Th		0457	1.8E
	0900	1147	1.1F
	1439	1813	1.5E
	2223	2353	0.3F
16 F	0122	0557	1.9E
	0951	1233	1.1F
	1518	1855	1.7E
	2251		
17 Sa ○		0047	0.5F
	0242	0651	2.0E
	1039	1316	1.1F
	1554	1934	1.9E
	2326		
18 Su		0138	0.8F
	0350	0743	1.9E
	1127	1357	1.0F
	1626	2013	2.0E
19 M	0005	0226	0.9F
	0450	0832	1.8E
	1216	1437	0.9F
	1656	2052	2.1E
20 Tu	0046	0314	1.0F
	0546	0921	1.7E
	1306	1516	0.7F
	1723	2131	2.1E
21 W	0131	0402	1.0F
	0641	1011	1.5E
	1401	1555	0.5F
	1747	2212	2.0E
22 Th	0220	0453	0.9F
	0737	1104	1.2E
	1507	1636	0.3F
	1803	2254	1.9E
23 F	0314	0547	0.8F
	0835	1203	1.0E
		1721	*
		2342	1.7E
24 Sa ◑	0414	0648	0.7F
	0938	1315	0.9E
		1817	*
25 Su		0037	1.6E
	0519	0755	0.6F
	1045	1437	0.9E
		1931	0.3E
26 M		0144	1.4E
	0626	0903	0.6F
	1151	1553	0.9E
		2057	0.3E
27 Tu		0256	1.4E
	0727	1005	0.6F
	1247	1650	1.1E
		2212	*
28 W		0403	1.4E
	0820	1055	0.7F
	1333	1733	1.2E
		2310	*
29 Th		0500	1.4E
	0905	1138	0.7F
	1411	1808	1.4E
		2355	*
30 F		0548	1.5E
	0947	1214	0.7F
	1442	1838	1.5E
	2301		

Time meridian 120° W. 0000 is midnight. 1200 is noon. Times are not adjusted for Daylight Saving Time.
If three consecutive entries are marked (E) the middle one is not a true maximum but an intermediate value to show the current pattern.
* Current weak and variable.

Strait of Juan de Fuca Entrance, 2005

F–Flood, Dir. 115° True E–Ebb, Dir. 290° True

October

Day	Slack h m	Maximum h m	knots
1 Sa		0034	0.3F
	0209	0630	1.5E
	1025	1247	0.7F
	1509	1906	1.6E
	2317		
2 Su		0110	0.4F
	0304	0709	1.5E
	1103	1318	0.7F
	1532	1933	1.7E
	2338		
3 M ●		0144	0.6F
	0353	0747	1.5E
	1141	1347	0.6F
	1551	1959	1.7E
4 Tu	0001	0218	0.7F
	0440	0824	1.4E
	1220	1415	0.5F
	1608	2026	1.8E
5 W	0029	0254	0.8F
	0526	0902	1.4E
	1303	1443	0.3F
	1621	2054	1.9E
6 Th	0100	0332	0.9F
	0614	0944	1.2E
		1513	*
		2126	1.9E
7 F	0138	0415	0.9F
	0705	1030	1.1E
		1547	*
		2203	1.9E
8 Sa	0223	0504	0.9F
	0801	1126	1.0E
		1628	*
		2247	1.8E
9 Su	0316	0601	0.8F
	0904	1233	0.9E
		1723	*
		2343	1.7E
10 M ◐	0418	0706	0.8F
	1011	1352	0.9E
		1842	0.3E
11 Tu		0055	1.6E
	0527	0817	0.8F
	1118	1508	1.1E
		2018	0.3E
12 W		0218	1.5E
	0637	0924	0.8F
	1216	1609	1.3E
		2144	*
13 Th		0339	1.5E
	0742	1024	0.9F
	1306	1659	1.5E
		2253	*
14 F		0449	1.6E
	0841	1116	0.9F
	1348	1742	1.7E
	2150	2350	0.5F
15 Sa	0151	0550	1.6E
	0936	1202	0.9F
	1425	1823	1.9E
	2221		

Day	Slack h m	Maximum h m	knots
16 Su		0040	0.8F
	0303	0645	1.7E
	1028	1245	0.8F
	1459	1902	2.1E
	2257		
17 M ○		0127	1.0F
	0404	0735	1.6E
	1119	1326	0.7F
	1529	1940	2.2E
	2335		
18 Tu		0213	1.2F
	0500	0824	1.6E
	1210	1404	0.5F
	1556	2017	2.2E
19 W	0015	0258	1.2F
	0551	0912	1.4E
	1305	1443	0.4F
	1618	2055	2.1E
20 Th	0057	0343	1.2F
	0641	1001	1.3E
		1522	*
		2133	2.0E
21 F	0142	0429	1.1F
	0730	1052	1.1E
		1603	*
		2213	1.8E
22 Sa	0231	0517	0.9F
	0820	1150	1.0E
		1650	*
		2258	1.6E
23 Su	0325	0610	0.8F
	0911	1255	1.0E
		1750	0.3E
		2350	1.4E
24 M ◑	0424	0708	0.7F
	1003	1406	1.0E
		1909	0.4E
25 Tu		0055	1.3E
	0529	0809	0.6F
	1054	1510	1.1E
		2036	0.3E
26 W		0210	1.2E
	0633	0908	0.5F
	1141	1602	1.2E
		2149	*
27 Th		0323	1.1E
	0732	0959	0.5F
	1223	1643	1.4E
		2245	*
28 F		0425	1.2E
	0825	1044	0.5F
	1300	1718	1.5E
		2330	*
29 Sa		0518	1.2E
	0913	1124	0.5F
	1331	1749	1.6E
	2211		
30 Su		0009	0.5F
	0210	0605	1.3E
	0958	1159	0.5F
	1357	1819	1.8E
	2230		
31 M		0046	0.7F
	0307	0648	1.3E
	1042	1232	0.4F
	1420	1847	1.9E
	2253		

November

Day	Slack h m	Maximum h m	knots
1 Tu ●		0121	0.8F
	0357	0729	1.3E
	1126	1304	0.3F
	1439	1916	2.0E
	2320		
2 W		0157	1.0F
	0445	0810	1.3E
		1336	*
		1946	2.0E
	2350		
3 Th		0235	1.1F
	0533	0853	1.2E
		1408	*
		2019	2.1E
4 F	0025	0316	1.2F
	0621	0938	1.2E
		1444	*
		2055	2.1E
5 Sa	0105	0359	1.2F
	0711	1028	1.1E
		1526	*
		2137	2.0E
6 Su	0152	0448	1.1F
	0802	1125	1.1E
		1618	*
		2226	1.8E
7 M	0245	0542	1.0F
	0856	1229	1.1E
		1726	0.3E
		2327	1.7E
8 Tu ◐	0346	0642	0.9F
	0949	1336	1.2E
		1851	0.3E
9 W		0041	1.5E
	0454	0746	0.9F
	1041	1440	1.3E
		2022	*
10 Th		0205	1.4E
	0607	0849	0.8F
	1129	1536	1.5E
		2141	*
11 F		0327	1.3E
	0717	0948	0.7F
	1213	1625	1.7E
	2051	2246	0.5F
12 Sa	0045	0439	1.3E
	0824	1041	0.6F
	1252	1710	1.9E
	2121	2341	0.8F
13 Su	0208	0542	1.3E
	0925	1129	0.6F
	1327	1751	2.1E
	2156		
14 M		0031	1.0F
	0316	0639	1.4E
	1023	1213	0.5F
	1359	1831	2.2E
	2234		
15 Tu ○		0117	1.2F
	0413	0730	1.3E
	1121	1255	0.3F
	1427	1910	2.2E
	2312		

Day	Slack h m	Maximum h m	knots
16 W		0201	1.3F
	0505	0819	1.3E
		1335	*
		1947	2.2E
	2351		
17 Th		0244	1.3F
	0553	0907	1.2E
		1415	*
		2024	2.1E
18 F	0031	0326	1.2F
	0637	0954	1.2E
		1455	*
		2102	2.0E
19 Sa	0113	0407	1.1F
	0719	1042	1.1E
		1538	*
		2141	1.8E
20 Su	0156	0450	1.0F
	0759	1132	1.1E
		1628	0.3E
		2223	1.6E
21 M	0242	0534	0.9F
	0838	1226	1.1E
		1727	0.4E
		2311	1.4E
22 Tu	0332	0621	0.7F
	0917	1321	1.1E
		1839	0.4E
23 W ◑		0008	1.2E
	0429	0711	0.6F
	0955	1414	1.2E
		1957	0.3E
24 Th		0117	1.1E
	0530	0803	0.5F
	1032	1502	1.3E
		2108	*
25 F		0231	1.0E
	0634	0854	0.4F
	1108	1545	1.5E
		2207	*
26 Sa		0340	1.0E
	0737	0942	0.4F
	1141	1023	1.6E
	2111	2256	0.3F
27 Su	0047	0442	1.0E
	0836	1026	0.3F
	1210	1658	1.7E
	2127	2339	0.6F
28 M	0201	0537	1.0E
		1107	*
		1732	1.9E
	2150		
29 Tu		0020	0.8F
	0302	0627	1.1E
		1146	*
		1805	2.0E
	2217		
30 W		0059	1.0F
	0356	0714	1.1E
		1224	*
		1840	2.1E
	2248		

December

Day	Slack h m	Maximum h m	knots
1 Th ●		0139	1.2F
	0446	0759	1.2E
		1303	*
		1916	2.2E
	2323		
2 F		0220	1.3F
	0534	0845	1.2E
		1344	*
		1955	2.2E
3 Sa	0001	0302	1.4F
	0620	0933	1.2E
		1430	*
		2038	2.2E
4 Su	0044	0346	1.4F
	0705	1022	1.2E
		1521	*
		2125	2.1E
5 M	0131	0433	1.3F
	0750	1114	1.3E
		1621	*
		2218	1.9E
6 Tu	0223	0523	1.2F
	0833	1209	1.3E
		1730	*
		2319	1.6E
7 W	0321	0617	1.0F
	0916	1306	1.5E
		1849	*
8 Th ◐		0030	1.4E
	0426	0713	0.9F
	0957	1403	1.6E
		2010	*
9 F		0149	1.2E
	0538	0811	0.7F
	1037	1457	1.7E
	1941	2126	0.3F
	2316		
10 Sa		0311	1.1E
	0655	0908	0.5F
	1115	1549	1.9E
	2018	2232	0.6F
11 Su	0057	0428	1.1E
	0811	1004	0.4F
	1150	1637	2.0E
	2057	2330	0.9F
12 M	0217	0536	1.1E
	0926	1056	0.3F
	1223	1723	2.1E
	2136		
13 Tu		0021	1.1F
	0322	0636	1.1E
		1145	*
		1805	2.2E
	2216		
14 W		0108	1.2F
	0418	0729	1.1E
		1231	*
		1846	2.2E
	2255		
15 Th ○		0151	1.3F
	0505	0817	1.1E
		1315	*
		1925	2.1E
	2334		

Day	Slack h m	Maximum h m	knots
16 F		0232	1.3F
	0547	0902	1.2E
		1357	*
		2003	2.0E
17 Sa	0012	0310	1.2F
	0624	0944	1.2E
		1440	*
		2041	1.9E
18 Su	0049	0347	1.2F
	0658	1024	1.2E
		1524	*
		2118	1.8E
19 M	0128	0423	1.0F
	0729	1105	1.2E
		1611	*
		2158	1.6E
20 Tu	0207	0459	0.9F
	0759	1146	1.2E
		1703	0.3E
		2241	1.4E
21 W	0250	0537	0.8F
	0828	1228	1.3E
		1801	*
		2330	1.2E
22 Th	0337	0617	0.7F
	0856	1311	1.3E
		1905	*
23 F ◑		0027	1.0E
	0432	0700	0.5F
	0923	1355	1.4E
		2012	*
24 Sa		0135	0.9E
	0535	0746	0.4F
	0949	1439	1.5E
		2117	*
25 Su		0249	0.8E
		0834	*
		1522	1.6E
	2020	2215	0.3F
26 M	0021	0403	0.8E
		0924	*
		1605	1.8E
	2044	2307	0.6F
27 Tu	0145	0509	0.8E
		1014	*
		1647	1.9E
	2114	2354	0.8F
28 W	0252	0608	0.9E
		1104	*
		1729	2.0E
	2147		
29 Th		0039	1.1F
	0349	0700	1.0E
		1153	*
		1812	2.2E
	2224		
30 F ●		0122	1.3F
	0438	0748	1.1E
		1243	*
		1856	2.2E
	2303		
31 Sa		0204	1.4F
	0523	0834	1.2E
		1333	*
		1942	2.3E
	2345		

Time meridian 120° W. 0000 is midnight. 1200 is noon. Times are not adjusted for Daylight Saving Time.
If three consecutive entries are marked (E) the middle one is not a true maximum but an intermediate value to show the current pattern.
* Current weak and variable.

Race Rocks, Strait of Juan de Fuca, B.C., 2005

F–Flood, Dir. 091° True E–Ebb, Dir. 271° True

January

Day	Slack h m	Maximum h m	knots
1 Sa	0255	0620	1.9F
	0950	1245	1.1E
		1815	0.5E
2 Su		0010	1.2E
	0330	0655	1.6F
	1010	1325	1.3E
		1930	0.4E
3 M		0115	0.8E
	0410	0730	1.4F
	1025	1415	1.6E
		2210	*
4 Tu		0235	0.5E
	0455	0805	1.1F
	1045	1505	1.9E
	2025	2255	0.6F
5 W	0220	0400	0.3E
	0555	0845	0.7F
	1100	1555	2.2E
	2040	2355	1.2F
6 Th		0610	*
		0940	0.3F
	1120	1645	2.4E
	2110		
7 F		0040	1.9F
	0500	0705	0.4E
		1030	*
		1740	2.7E
	2150		
8 Sa		0120	2.6F
	0550	0810	0.7E
		1200	*
		1835	2.9E
	2230		
9 Su		0205	3.0F
	0635	0855	0.9E
		1310	*
		1935	3.1E
	2315		
10 M		0245	3.3F
	0710	0930	1.1E
		1405	*
		2030	3.2E
11 Tu	0000	0330	3.4F
	0750	1005	1.2E
	1335	1455	0.3F
	1605	2125	3.1E
12 W	0050	0410	3.4F
	0820	1040	1.3E
	1430	1550	0.3F
	1705	2215	2.9E
13 Th	0140	0455	3.1F
	0850	1120	1.4E
		1645	*
		2305	2.5E
14 F	0225	0535	2.8F
	0920	1205	1.5E
		1750	*
15 Sa		0000	1.9E
	0315	0620	2.4F
	0945	1255	1.7E
		1940	*
16 Su		0105	1.4E
	0405	0700	1.9F
	1005	1355	1.9E
	1905	2100	0.4F
	2330		
17 M		0225	0.9E
	0455	0740	1.3F
	1025	1455	2.0E
	1950	2235	0.8F
18 Tu	0140	0345	0.6E
	0605	0825	0.7F
	1035	1545	2.1E
	2030	2345	1.2F
19 W	0310	0510	0.5E
		0855	*
		1635	2.2E
	2105		
20 Th		0035	1.6F
	0425	0645	0.5E
		1015	*
		1720	2.2E
	2140		
21 F		0110	1.9F
	0515	0820	0.7E
		1125	0.5E
		1810	2.2E
	2210		
22 Sa		0145	2.1F
	0600	0905	0.9E
		1230	0.5E
		1855	2.2E
	2245		
23 Su		0215	2.2F
	0630	0920	1.0E
		1320	0.4E
		1940	2.3E
	2315		
24 M		0240	2.3F
	0655	0930	1.0E
		1400	0.3E
		2020	2.3E
	2350		
25 Tu		0310	2.4F
	0715	0950	1.1E
		1440	*
		2055	2.4E
26 W	0020	0340	2.4F
	0735	1010	1.2E
		1520	*
		2125	2.3E
27 Th	0055	0410	2.4F
	0750	1030	1.2E
		1555	*
		2200	2.2E
28 F	0125	0445	2.3F
	0805	1055	1.4E
		1625	*
		2230	2.0E
29 Sa	0155	0515	2.2F
	0825	1120	1.6E
		1705	*
		2305	1.7E
30 Su	0225	0545	2.0F
	0840	1155	1.8E
		1805	*
31 M		0000	1.3E
	0300	0610	1.7F
	0855	1240	2.0E
		1905	*

February

Day	Slack h m	Maximum h m	knots
1 Tu		0105	0.9E
	0340	0645	1.3F
	0915	1325	2.2E
	1825	2035	0.5F
	2325		
2 W		0220	0.6E
	0425	0720	1.0F
	0935	1415	2.4E
	1900	2150	1.0F
3 Th	0155	0335	0.5E
	0530	0800	0.6F
	0955	1515	2.5E
	1945	2300	1.6F
4 F	0315	0455	0.4E
		0845	*
		1615	2.7E
	2030		
5 Sa		0000	2.2F
	0415	0620	0.6E
		1005	*
		1715	2.9E
	2120		
6 Su		0050	2.7F
	0505	0720	0.8E
		1125	*
		1820	3.1E
	2210		
7 M		0135	3.0F
	0545	0800	1.0E
	1110	1250	0.4F
	1405	1920	3.3E
	2300		
8 Tu		0220	3.3F
	0620	0840	1.3E
	1150	1345	0.6F
	1525	2015	3.3E
	2345		
9 W		0305	3.4F
	0655	0915	1.5E
	1235	1440	0.8F
	1630	2105	3.2E
10 Th	0035	0345	3.3F
	0720	0950	1.7E
	1325	1535	0.9F
	1735	2155	2.9E
11 F	0120	0425	3.1F
	0745	1035	2.0E
	1425	1630	0.9F
	1840	2250	2.4E
12 Sa	0200	0505	2.7F
	0810	1115	2.1E
	1520	1730	0.9F
	1950	2345	1.9E
13 Su	0245	0540	2.2F
	0830	1200	2.3E
	1625	1835	0.9F
	2115		
14 M		0050	1.4E
	0335	0615	1.6F
	0850	1250	2.3E
	1725	1950	0.9F
	2325		
15 Tu		0155	1.0E
	0430	0655	1.1F
	0905	1340	2.3E
	1825	2115	1.1F
16 W	0100	0305	0.8E
	0550	0740	0.5F
	0910	1435	2.2E
	1915	2245	1.3F
17 Th	0215	0425	0.7E
		0820	*
		1530	2.1E
	2005	2340	1.5F
18 F	0315	0600	0.7E
		0925	0.3E
		1635	2.1E
	2050		
19 Sa		0025	1.7F
	0405	0700	0.9E
		1045	0.5E
		1740	2.1E
	2135		
20 Su		0105	1.9F
	0445	0735	1.0E
		1200	0.4E
		1835	2.2E
	2215		
21 M		0140	2.0F
	0520	0805	1.2E
		1300	*
		1920	2.4E
	2255		
22 Tu		0210	2.1F
	0540	0830	1.3E
		1355	*
		2000	2.5E
	2330		
23 W		0240	2.3F
	0605	0855	1.5E
		1415	*
		2035	2.5E
24 Th	0005	0310	2.3F
	0620	0915	1.6E
	1330	1455	0.3F
	1620	2105	2.5E
25 F	0035	0340	2.3F
	0640	0940	1.9E
	1350	1535	0.5F
	1710	2140	2.3E
26 Sa	0110	0410	2.2F
	0655	1010	2.1E
	1420	1615	0.6F
	1800	2225	2.0E
27 Su	0140	0440	2.0F
	0715	1040	2.3E
	1455	1700	0.7F
	1855	2315	1.7E
28 M	0220	0510	1.7F
	0730	1120	2.5E
	1535	1750	0.8F
	2005		

March

Day	Slack h m	Maximum h m	knots
1 Tu		0015	1.4E
	0305	0540	1.3F
	0750	1200	2.6E
	1620	1845	1.0F
	2135		
2 W		0115	1.2E
	0355	0615	0.9F
	0810	1250	2.7E
	1710	1955	1.2F
	2340		
3 Th		0215	1.0E
	0505	0655	0.6F
	0830	1345	2.7E
	1805	2105	1.5F
4 F	0110	0315	0.9E
		0750	*
		1445	2.8E
	1900	2215	1.9F
5 Sa	0215	0420	1.0E
		0850	*
		1550	2.8E
	2000	2320	2.2F
6 Su	0310	0530	1.1E
		0950	*
		1700	2.9E
	2055		
7 M		0015	2.5F
	0400	0630	1.3E
	1000	1145	0.4F
	1245	1805	3.1E
	2150		
8 Tu		0105	2.8F
	0445	0715	1.6E
	1040	1240	0.6F
	1425	1905	3.2E
	2240		
9 W		0155	2.9F
	0520	0755	1.8E
	1125	1340	0.9F
	1545	1955	3.1E
	2330		
10 Th		0235	2.9F
	0545	0830	2.2E
	1205	1435	1.3F
	1655	2050	2.9E
11 F	0015	0315	2.8F
	0610	0910	2.4E
	1250	1530	1.5F
	1755	2145	2.6E
12 Sa	0100	0350	2.5F
	0635	0950	2.7E
	1335	1620	1.7F
	1900	2240	2.2E
13 Su	0145	0425	2.1F
	0655	1035	2.8E
	1425	1705	1.7F
	2005	2335	1.8E
14 M	0235	0505	1.6F
	0715	1115	2.8E
	1510	1800	1.7F
	2135		
15 Tu		0035	1.5E
	0330	0540	1.1F
	0730	1200	2.7E
	1605	1855	1.6F
	2300		
16 W		0130	1.3E
	0450	0620	0.6F
	0745	1250	2.5E
	1700	1955	1.4F
17 Th	0005	0230	1.1E
		0655	*
		1340	2.2E
	1800	2110	1.4F
18 F	0100	0335	1.1E
		0755	*
		1440	2.0E
	1900	2230	1.4F
19 Sa	0150	0500	1.1E
		0900	0.5E
		1545	1.9E
	2000	2325	1.4F
20 Su	0240	0600	1.3E
		1025	0.5E
		1700	2.0E
	2055		
21 M		0015	1.5F
	0320	0640	1.5E
		1145	0.3E
		1805	2.1E
	2145		
22 Tu		0055	1.7F
	0355	0710	1.6E
		1250	*
		1850	2.3E
	2230		
23 W		0130	1.8F
	0425	0740	1.8E
	1205	1330	0.3F
	1455	1935	2.4E
	2305		
24 Th		0205	1.9F
	0445	0805	2.1E
	1215	1410	0.6F
	1600	2015	2.4E
	2345		
25 F		0235	1.9F
	0510	0830	2.3E
	1235	1450	0.9F
	1650	2055	2.3E
26 Sa	0020	0305	1.8F
	0525	0855	2.6E
	1300	1530	1.2F
	1745	2140	2.1E
27 Su	0100	0335	1.6F
	0545	0925	2.9E
	1330	1610	1.4F
	1835	2235	1.9E
28 M	0140	0405	1.3F
	0605	1005	3.0E
	1405	1650	1.6F
	1935	2325	1.7E
29 Tu	0230	0435	1.0F
	0620	1045	3.1E
	1445	1735	1.7F
	2040		
30 W		0020	1.6E
	0330	0510	0.6F
	0640	1135	3.1E
	1530	1825	1.8F
	2210		
31 Th		0110	1.5E
	0450	0550	0.3F
	0655	1225	3.0E
	1625	1925	1.8F
	2325		

Time meridian 120° W. 0000 is midnight. 1200 is noon. Times are not adjusted for Daylight Saving Time.
If three consecutive entries are marked (E) the middle one is not a true maximum but an intermediate value to show the current pattern.
* Current weak and variable.

Race Rocks, Strait of Juan de Fuca, B.C., 2005

F–Flood, Dir. 091° True E–Ebb, Dir. 271° True

April

Day	Slack h m	Maximum h m	knots
1 F		0200	1.5E
		0650	*
		1325	2.9E
◑	1725	2030	1.9F
2 Sa	0025	0255	1.4E
		0740	*
		1425	2.7E
	1830	2135	1.9F
3 Su	0120	0400	1.5E
		0840	*
		1530	2.7E
	1935	2240	2.0F
4 M	0215	0505	1.6E
		0950	*
		1640	2.7E
	2035	2340	2.1F
5 Tu	0300	0555	1.9E
	1010	1145	0.4F
	1305	1750	2.7E
	2130		
6 W		0035	2.1F
	0340	0640	2.2E
	1040	1245	0.7F
	1445	1850	2.6E
	2225		
7 Th		0120	2.1F
	0410	0720	2.5E
	1115	1345	1.2F
	1605	1945	2.5E
	2315		
8 F		0200	2.0F
	0435	0755	2.8E
	1150	1430	1.6F
●	1710	2045	2.3E
9 Sa	0005	0240	1.8F
	0455	0835	3.0E
	1225	1515	2.0F
	1815	2140	2.1E
10 Su	0055	0315	1.5F
	0515	0915	3.2E
	1300	1600	2.2F
	1915	2235	1.9E
11 M	0145	0350	1.2F
	0540	0955	3.2E
	1340	1640	2.3F
	2025	2330	1.7E
12 Tu	0245	0425	0.8F
	0555	1035	3.0E
	1420	1725	2.1F
	2130		
13 W		0020	1.6E
	0400	0505	0.4F
	0610	1120	2.8E
	1505	1810	1.9F
	2220		
14 Th		0105	1.5E
		0555	*
		1205	2.5E
	1600	1900	1.6F
	2310		
15 F		0155	1.4E
		0635	0.4E
		1255	2.1E
	1700	2000	1.4F
	2355		

Day	Slack h m	Maximum h m	knots
16 Sa		0255	1.4E
		0735	0.6E
		1350	1.9E
◐	1810	2110	1.2F
17 Su	0040	0425	1.4E
		0845	0.7E
		1455	1.7E
	1910	2215	1.2F
18 M	0120	0520	1.6E
		1040	0.6E
		1610	1.7E
	2010	2310	1.2F
19 Tu	0200	0600	1.8E
		1155	*
		1725	1.8E
	2105	2355	1.2F
20 W	0235	0630	2.0E
		1240	*
		1820	1.9E
	2155		
21 Th		0040	1.2F
	0305	0650	2.3E
	1130	1320	0.5F
	1505	1905	1.9E
	2240		
22 F		0115	1.2F
	0330	0715	2.6E
	1140	1400	0.9F
	1610	2000	1.8E
	2330		
23 Sa		0150	1.1F
	0350	0740	2.9E
	1200	1435	1.4F
	1710	2055	1.8E
24 Su	0015	0225	1.0F
	0410	0815	3.2E
	1225	1515	1.8F
○	1805	2150	1.7E
25 M	0105	0300	0.8F
	0430	0850	3.3E
	1255	1555	2.1F
	1905	2240	1.7E
26 Tu	0200	0330	0.5F
	0450	0930	3.4E
	1335	1635	2.2F
	2010	2325	1.7E
27 W		0400	*
		1020	3.4E
	1415	1720	2.3F
	2105		
28 Th		0010	1.7E
		0455	*
		1115	3.2E
	1505	1805	2.2F
	2205		
29 F		0055	1.7E
		0540	*
		1210	3.0E
	1600	1900	2.1F
	2255		
30 Sa		0145	1.7E
		0630	0.4E
		1305	2.8E
◑	1705	2000	1.9F
	2345		

May

Day	Slack h m	Maximum h m	knots
1 Su		0240	1.7E
		0735	0.5E
		1405	2.6E
	1810	2105	1.8F
2 M	0035	0350	1.8E
		0850	0.5E
		1515	2.3E
	1910	2205	1.6F
3 Tu	0115	0455	2.0E
		1035	*
		1630	2.1E
	2015	2300	1.5F
4 W	0155	0540	2.3E
		1200	*
		1740	2.0E
	2115	2355	1.3F
5 Th	0225	0615	2.6E
	1050	1255	0.8F
	1510	1845	1.9E
	2215		
6 F		0040	1.1F
	0250	0650	2.9E
	1110	1345	1.3F
	1630	1945	1.7E
	2320		
7 Sa		0125	0.9F
	0315	0725	3.1E
	1140	1425	1.8F
	1735	2045	1.6E
8 Su	0015	0205	0.7F
	0335	0800	3.2E
	1205	1505	2.2F
●	1840	2140	1.6E
9 M	0115	0240	0.5F
	0355	0840	3.3E
	1235	1540	2.3F
	1935	2230	1.6E
10 Tu	0220	0310	0.3F
	0415	0920	3.1E
	1310	1615	2.3F
	2020	2315	1.6E
11 W		0400	*
		1000	2.9E
	1350	1655	2.2F
	2100	2355	1.5E
12 Th		0435	0.3E
		1045	2.6E
	1430	1735	1.9F
	2140		
13 F		0035	1.5E
		0515	0.5E
		1135	2.4E
	1520	1820	1.6F
	2215		
14 Sa		0120	1.4E
		0605	0.7E
		1220	2.1E
	1620	1910	1.4F
	2255		
15 Su		0215	1.4E
		0655	0.9E
		1305	1.8E
	1720	2005	1.2F
	2330		

Day	Slack h m	Maximum h m	knots
16 M		0415	1.4E
		0810	0.9E
		1400	1.6E
◐	1820	2100	1.0F
17 Tu	0005	0455	1.6E
		1035	0.8E
		1510	1.5E
	1915	2150	0.9F
18 W	0035	0520	1.8E
		1140	0.4E
		1625	1.4E
	2010	2240	0.8F
19 Th	0105	0540	2.1E
		1220	*
		1735	1.3E
	2110	2330	0.6F
20 F	0130	0600	2.5E
	1100	1305	0.6F
	1510	1840	1.3E
	2220		
21 Sa		0020	0.5F
	0150	0625	2.8E
	1110	1340	1.2F
	1625	1950	1.3E
	2335		
22 Su		0105	0.3F
	0215	0700	3.1E
	1130	1420	1.7F
	1730	2055	1.3E
23 M		0135	*
		0740	3.3E
	1200	1500	2.2F
○	1835	2150	1.4E
24 Tu		0215	*
		0825	3.5E
	1235	1540	2.4F
	1930	2230	1.5E
25 W		0305	*
		0915	3.5E
	1315	1620	2.6F
	2015	2310	1.6E
26 Th		0350	*
		1010	3.4E
	1400	1705	2.5F
	2100	2350	1.6E
27 F		0430	0.3E
		1100	3.3E
	1450	1750	2.3F
	2145		
28 Sa		0030	1.6E
		0520	0.5E
		1155	3.0E
	1550	1840	2.1F
	2225		
29 Su		0120	1.6E
		0620	0.6E
		1245	2.6E
	1650	1935	1.8F
	2305		
30 M		0230	1.7E
		0735	0.7E
		1350	2.2E
◑	1750	2030	1.5F
	2340		
31 Tu		0400	1.9E
		0940	0.6E
		1505	1.8E
	1850	2125	1.2F

June

Day	Slack h m	Maximum h m	knots
1 W	0010	0445	2.2E
		1120	*
		1630	1.5E
	1955	2215	0.9F
2 Th	0035	0520	2.5E
	1035	1215	0.4F
	1415	1740	1.3E
	2115	2310	0.5F
3 F	0055	0550	2.7E
	1045	1305	1.0F
	1555	1850	1.2E
		2350	*
4 Sa		0625	2.9E
	1105	1345	1.4F
	1705	2000	1.2E
5 Su		0045	*
		0700	3.0E
	1125	1420	1.8F
	1805	2100	1.2E
6 M		0135	0.3E
		0740	3.0E
	1150	1450	2.1F
●	1850	2145	1.3E
7 Tu		0215	0.4E
		0820	3.0E
	1220	1525	2.2F
	1930	2220	1.3E
8 W		0250	0.4E
		0905	2.9E
	1255	1555	2.1F
	2000	2255	1.3E
9 Th		0325	0.5E
		0945	2.7E
	1330	1635	2.0F
	2030	2325	1.3E
10 F		0400	0.6E
		1030	2.5E
	1415	1710	1.8F
	2055		
11 Sa		0000	1.3E
		0440	0.7E
		1105	2.3E
	1500	1755	1.6F
	2125		
12 Su		0035	1.2E
		0520	0.8E
		1145	2.1E
	1545	1835	1.4F
	2155		
13 M		0110	1.3E
		0610	0.9E
		1220	1.8E
	1635	1915	1.2F
	2220		
14 Tu		0200	1.4E
		0715	1.0E
		1310	1.5E
◐	1720	1955	1.0F
	2240		
15 W		0300	1.6E
		0930	0.8E
		1415	1.2E
	1810	2030	0.8F
	2300		

Day	Slack h m	Maximum h m	knots
16 Th		0340	1.8E
		1140	0.4E
		1535	1.0E
	1905	2110	0.5F
	2315		
17 F		0420	2.2E
		1210	*
		1700	0.8E
		2150	*
18 Sa		0500	2.5E
	1015	1240	0.9F
	1535	1830	0.8E
		2240	*
19 Su		0540	2.9E
	1030	1320	1.5F
	1650	1950	0.9E
		2350	*
20 M		0630	3.1E
	1100	1400	2.1F
	1750	2045	1.1E
21 Tu		0100	*
		0720	3.3E
	1135	1440	2.5F
○	1840	2130	1.3E
22 W		0150	*
		0815	3.5E
	1215	1525	2.7F
	1925	2205	1.4E
23 Th		0240	*
		0910	3.5E
	1300	1605	2.8F
	2005	2240	1.5E
24 F		0325	*
		1000	3.5E
	1350	1650	2.7F
	2040	2320	1.5E
25 Sa		0415	*
		1050	3.2E
	1440	1735	2.4F
	2115		
26 Su		0000	1.6E
		0510	0.3E
		1140	2.9E
	1535	1820	2.1F
	2150		
27 M		0050	1.6E
		0620	0.5E
		1235	2.3E
	1625	1910	1.8F
	2215		
28 Tu		0200	1.8E
		0810	0.5E
		1345	1.8E
◑	1725	1955	1.3F
	2240		
29 W		0320	2.1E
		1010	*
		1510	1.4E
	1825	2040	0.9F
	2255		
30 Th		0405	2.3E
	0945	1135	0.3F
	1325	1630	1.1E
	1950	2130	0.4F
	2305		

Time meridian 120° W. 0000 is midnight. 1200 is noon. Times are not adjusted for Daylight Saving Time.
If three consecutive entries are marked (E) the middle one is not a true maximum but an intermediate value to show the current pattern.
* Current weak and variable.

Race Rocks, Strait of Juan de Fuca, B.C., 2005

F–Flood, Dir. 091° True E–Ebb, Dir. 271° True

July

Day	Slack (h m)	Maximum (h m)	knots
1 F		0445	2.5E
	0955	1230	0.8F
	1510	1750	0.9E
		2220	*
2 Sa		0525	2.6E
	1015	1310	1.3F
	1625	1915	0.9E
		2320	0.4E
3 Su		0605	2.7E
	1040	1340	1.6F
	1725	2035	1.0E
4 M		0025	0.6E
		0650	2.7E
	1110	1410	1.8F
	1810	2115	1.1E
5 Tu		0115	0.6E
		0735	2.7E
	1140	1440	2.0F
	1845	2135	1.2E
6 W ●		0155	0.6E
		0820	2.7E
	1210	1510	2.0F
	1915	2155	1.2E
7 Th		0230	0.5E
		0900	2.7E
	1245	1545	2.0F
	1935	2225	1.2E
8 F		0305	0.4E
		0940	2.6E
	1320	1620	1.9F
	1955	2250	1.2E
9 Sa		0345	0.4E
		1015	2.5E
	1400	1655	1.9F
	2015	2315	1.3E
10 Su		0420	0.5E
		1045	2.3E
	1435	1730	1.7F
	2040	2340	1.3E
11 M		0500	0.6E
		1115	2.0E
	1510	1800	1.6F
	2100		
12 Tu		0015	1.5E
		0550	0.6E
		1155	1.7E
	1545	1830	1.4F
	2115		
13 W		0055	1.7E
		0655	0.6E
		1245	1.3E
	1625	1900	1.1F
	2130		
14 Th ◐		0140	1.9E
		0835	0.4E
		1355	1.0E
	1710	1930	0.8F
	2145		
15 F		0230	2.1E
		1025	*
		1520	0.7E
	1815	2000	0.4F
	2155		
16 Sa		0320	2.3E
	0850	1120	0.7F
	1435	1645	0.6E
		2100	*
17 Su		0415	2.6E
	0910	1215	1.3F
	1555	1820	0.7E
		2220	*
18 M		0515	2.8E
	0945	1300	1.9F
	1650	1935	0.8E
		2335	*
19 Tu		0615	3.1E
	1025	1340	2.4F
	1740	2020	1.0E
20 W		0045	*
		0715	3.3E
	1110	1425	2.8F
	1825	2055	1.2E
21 Th ○		0130	*
		0810	3.5E
	1155	1510	3.0F
	1900	2130	1.4E
22 F		0225	*
		0900	3.5E
	1245	1550	3.0F
	1935	2205	1.6E
23 Sa		0320	*
		0950	3.4E
	1330	1635	2.9F
	2005	2245	1.7E
24 Su		0415	*
		1040	3.0E
	1420	1715	2.6F
	2035	2330	1.8E
25 M		0510	*
		1135	2.5E
	1505	1755	2.2F
	2100		
26 Tu		0020	2.0E
		0645	*
		1235	2.0E
	1600	1835	1.8F
	2120		
27 W ◑		0115	2.2E
		0800	*
		1350	1.5E
	1655	1915	1.2F
	2140		
28 Th		0210	2.3E
	0730	0955	0.6F
	1225	1505	1.1E
	1805	2000	0.6F
	2150		
29 F		0310	2.4E
	0810	1120	1.0F
	1400	1630	0.9E
		2050	*
30 Sa		0405	2.4E
	0850	1205	1.4F
	1515	1800	0.9E
		2155	0.3E
31 Su		0500	2.4E
	0930	1250	1.6F
	1615	1925	1.0E
		2310	0.5E

August

Day	Slack (h m)	Maximum (h m)	knots
1 M		0555	2.4E
	1005	1325	1.8F
	1705	2020	1.2E
2 Tu		0020	0.6E
		0645	2.4E
	1045	1400	2.0F
	1740	2040	1.2E
3 W		0110	0.4E
		0730	2.5E
	1120	1430	2.1F
	1815	2100	1.3E
4 Th ●		0155	0.3E
		0815	2.6E
	1155	1500	2.1F
	1835	2120	1.3E
5 F		0230	*
		0850	2.6E
	1230	1530	2.1F
	1900	2150	1.4E
6 Sa		0310	*
		0925	2.5E
	1300	1600	2.1F
	1915	2215	1.5E
7 Su		0345	*
		0955	2.4E
	1335	1630	2.1F
	1935	2235	1.6E
8 M		0425	*
		1030	2.1E
	1405	1700	1.9F
	1950	2305	1.8E
9 Tu		0510	*
		1110	1.8E
	1435	1730	1.6F
	2005	2340	1.9E
10 W		0555	*
		1155	1.4E
	1515	1755	1.3F
	2020		
11 Th		0015	2.1E
		0655	*
		1300	1.1E
	1550	1825	1.0F
	2035		
12 F ◐		0100	2.2E
	0615	0810	0.4F
	1040	1410	0.8E
	1655	1850	0.6F
	2050		
13 Sa		0150	2.3E
	0650	0930	0.8F
	1310	1515	0.7E
	1825	1905	0.3F
	2100		
14 Su		0250	2.4E
	0730	1045	1.3F
	1435	1630	0.7E
		2050	*
15 M		0355	2.6E
	0820	1140	1.9F
	1535	1800	0.8E
		2205	*
16 Tu		0500	2.8E
	0910	1235	2.3F
	1630	1900	1.0E
		2315	*
17 W		0605	3.0E
	1000	1320	2.7F
	1710	1940	1.2E
18 Th		0015	*
		0705	3.3E
	1045	1405	3.0F
	1750	2015	1.5E
	2345		
19 F ○		0135	0.5F
	0305	0755	3.4E
	1135	1450	3.2F
	1825	2055	1.7E
20 Sa	0025	0230	0.8F
	0420	0850	3.3E
	1220	1530	3.2F
	1855	2130	1.9E
21 Su	0110	0325	1.0F
	0525	0940	3.1E
	1310	1610	3.0F
	1920	2215	2.2E
22 M	0205	0420	1.1F
	0635	1035	2.6E
	1355	1645	2.6F
	1945	2255	2.4E
23 Tu	0300	0520	1.1F
	0745	1135	2.1E
	1440	1725	2.1F
	2005	2345	2.5E
24 W	0355	0620	1.2F
	0915	1240	1.7E
	1535	1805	1.6F
	2025		
25 Th		0035	2.5E
	0455	0730	1.2F
	1115	1345	1.3E
	1640	1845	1.0F
	2040		
26 F ◑		0125	2.5E
	0555	0905	1.3F
	1235	1450	1.1E
	1810	1935	0.4F
	2045		
27 Sa		0220	2.3E
	0650	1030	1.6F
	1345	1610	1.0E
		2020	*
28 Su		0325	2.2E
	0745	1125	1.7F
	1445	1745	1.1E
		2135	0.4E
29 M		0430	2.1E
	0840	1210	1.9F
	1540	1845	1.2E
		2300	0.4E
30 Tu		0540	2.1E
	0930	1255	2.0F
	1625	1925	1.4E
31 W		0005	0.3E
		0635	2.3E
	1015	1335	2.0F
	1700	1955	1.5E

September

Day	Slack (h m)	Maximum (h m)	knots
1 Th		0110	*
		0720	2.4E
	1055	1410	2.1F
	1730	2020	1.5E
2 F		0155	*
		0800	2.5E
	1130	1440	2.2F
	1750	2045	1.7E
3 Sa ●	0110	0240	0.4F
	0405	0835	2.4E
	1205	1510	2.2F
	1810	2115	1.8E
4 Su	0130	0310	0.5F
	0450	0915	2.3E
	1240	1535	2.1F
	1825	2140	1.9E
5 M	0150	0350	0.6F
	0535	0955	2.1E
	1310	1605	2.0F
	1840	2205	2.1E
6 Tu	0215	0425	0.7F
	0620	1035	1.8E
	1345	1630	1.7F
	1855	2230	2.2E
7 W	0245	0505	0.9F
	0715	1125	1.5E
	1420	1700	1.4F
	1910	2305	2.3E
8 Th	0320	0550	1.0F
	0825	1220	1.3E
	1510	1730	1.0F
	1925	2345	2.4E
9 F	0400	0640	1.1F
	1015	1315	1.1E
	1605	1800	0.6F
	1935		
10 Sa		0035	2.4E
	0450	0745	1.3F
	1200	1405	1.0E
		1835	*
11 Su ◐		0130	2.4E
	0545	0855	1.5F
	1305	1505	0.9E
		1935	*
12 M		0230	2.4E
	0640	1005	1.8F
	1405	1610	1.0E
		2030	*
13 Tu		0335	2.5E
	0740	1110	2.2F
	1500	1730	1.1E
		2135	*
14 W		0445	2.6E
	0835	1205	2.5F
	1550	1825	1.3E
	2200	2355	0.4F
15 Th	0040	0550	2.8E
	0930	1255	2.7F
	1630	1905	1.5E
	2235		
16 F		0040	0.7F
	0230	0650	2.9E
	1025	1340	2.9F
	1710	1945	1.8E
	2310		
17 Sa ○		0140	1.0F
	0350	0745	2.9E
	1115	1425	2.9F
	1735	2020	2.1E
	2355		
18 Su		0235	1.4F
	0500	0840	2.8E
	1200	1505	2.8F
	1805	2100	2.4E
19 M	0035	0325	1.8F
	0605	0935	2.5E
	1250	1540	2.6F
	1825	2140	2.6E
20 Tu	0120	0415	2.0F
	0715	1035	2.1E
	1335	1615	2.2F
	1845	2225	2.8E
21 W	0205	0505	2.1F
	0835	1135	1.8E
	1430	1655	1.7F
	1910	2310	2.7E
22 Th	0255	0555	2.1F
	1000	1230	1.6E
	1530	1735	1.1F
	1930	2355	2.6E
23 F	0345	0650	2.0F
	1110	1325	1.4E
	1650	1820	0.6F
	1950		
24 Sa ◑		0050	2.3E
	0445	0805	1.8F
	1210	1425	1.3E
		1900	*
25 Su		0145	2.0E
	0545	0935	1.8F
	1305	1545	1.2E
		2000	*
26 M		0250	1.8E
	0650	1040	1.8F
	1400	1720	1.3E
		2115	0.4E
27 Tu		0405	1.7E
	0755	1135	1.8F
	1450	1810	1.4E
		2350	0.4E
28 W		0525	1.8E
	0850	1220	1.8F
	1535	1845	1.6E
	2315		
29 Th		0040	0.3F
	0150	0620	1.9E
	0940	1300	1.9F
	1605	1915	1.7E
	2335		
30 F		0125	0.5F
	0255	0705	2.0E
	1025	1335	1.9F
	1635	1945	1.8E
	2350		

Time meridian 120° W. 0000 is midnight. 1200 is noon. Times are not adjusted for Daylight Saving Time.
If three consecutive entries are marked (E) the middle one is not a true maximum but an intermediate value to show the current pattern.
* Current weak and variable.

Race Rocks, Strait of Juan de Fuca, B.C., 2005

F–Flood, Dir. 091° True   E–Ebb, Dir. 271° True

October

Day	Slack h m	Maximum h m	knots
1 Sa		0200	0.7F
	0350	0745	2.1E
	1105	1405	1.9F
	1655	2015	1.9E
2 Su	0010	0230	0.9F
	0440	0830	2.0E
	1145	1435	1.8F
	1715	2040	2.1E
3 M	0030	0305	1.2F
	0525	0915	1.9E
	1220	1505	1.7F
●	1730	2100	2.3E
4 Tu	0055	0335	1.4F
	0615	1000	1.7E
	1300	1535	1.4F
	1740	2130	2.4E
5 W	0120	0410	1.6F
	0710	1050	1.5E
	1345	1605	1.1F
	1755	2200	2.5E
6 Th	0150	0450	1.8F
	0820	1135	1.4E
	1430	1635	0.8F
	1810	2240	2.5E
7 F	0225	0530	1.9F
	0935	1225	1.3E
	1535	1710	0.4F
	1825	2325	2.4E
8 Sa	0305	0620	1.9F
	1045	1305	1.3E
		1740	*
9 Su		0020	2.3E
	0400	0715	1.9F
	1140	1355	1.2E
		1825	*
10 M		0115	2.3E
	0500	0820	1.9F
	1235	1450	1.2E
◐		1920	*
11 Tu		0215	2.2F
	0605	0930	2.0F
	1330	1600	1.2E
		2020	*
12 W		0320	2.2E
	0710	1035	2.1F
	1420	1715	1.3E
	2125	2220	0.3F
	2320		
13 Th		0435	2.2E
	0810	1130	2.2F
	1505	1800	1.6E
	2145	2345	0.5F
14 F	0140	0545	2.3E
	0910	1225	2.3F
	1545	1840	1.8E
	2215		
15 Sa		0045	1.0F
	0310	0645	2.3E
	1005	1310	2.3F
	1615	1915	2.1E
	2250		
16 Su		0140	1.5F
	0425	0745	2.2E
	1100	1355	2.2F
	1640	1950	2.4E
	2330		
17 M		0230	2.0F
	0535	0845	2.0E
	1150	1435	1.9F
○	1705	2030	2.7E
18 Tu	0005	0315	2.4F
	0645	0940	1.9E
	1245	1510	1.7F
	1725	2110	2.8E
19 W	0045	0355	2.7F
	0755	1040	1.7E
	1340	1550	1.3F
	1745	2150	2.7E
20 Th	0120	0440	2.7F
	0900	1130	1.6E
	1435	1625	0.9F
	1805	2235	2.6E
21 F	0205	0525	2.6F
	0955	1215	1.5E
	1545	1710	0.5F
	1825	2325	2.3E
22 Sa	0250	0610	2.3F
	1045	1305	1.4E
		1800	*
23 Su		0015	1.9E
	0345	0710	2.0F
	1135	1400	1.3E
		1845	*
24 M		0110	1.6E
	0450	0825	1.7F
	1225	1540	1.3E
◑		1955	0.4E
25 Tu		0210	1.4E
	0600	0940	1.5F
	1310	1700	1.4E
		2305	*
26 W		0335	1.3E
	0705	1040	1.5F
	1400	1740	1.5E
		2340	*
27 Th		0505	1.3E
	0805	1130	1.4F
	1435	1815	1.6E
	2240		
28 F		0030	0.4F
	0205	0600	1.4E
	0900	1215	1.4F
	1505	1845	1.7E
	2255		
29 Sa		0110	0.7F
	0315	0645	1.4E
	0950	1250	1.3F
	1530	1910	1.9E
	2310		
30 Su		0145	1.0F
	0410	0735	1.4E
	1045	1325	1.2F
	1545	1935	2.0E
	2325		
31 M		0215	1.4F
	0505	0825	1.4E
	1135	1400	1.0F
	1600	1955	2.2E
	2345		

November

Day	Slack h m	Maximum h m	knots
1 Tu		0245	1.8F
	0605	0920	1.4E
	1230	1430	0.8F
●	1615	2020	2.4E
2 W	0005	0320	2.1F
	0705	1010	1.4E
	1320	1505	0.6F
	1630	2050	2.5E
3 Th	0035	0355	2.3F
	0800	1050	1.4E
	1415	1535	0.4F
	1645	2130	2.5E
4 F	0105	0430	2.4F
	0850	1130	1.4E
		1610	*
		2215	2.5E
5 Sa	0145	0510	2.4F
	0940	1210	1.4E
		1650	*
		2310	2.4E
6 Su	0235	0555	2.3F
	1025	1250	1.3E
		1730	*
7 M		0000	2.2E
	0330	0650	2.2F
	1115	1335	1.3E
		1820	0.3E
8 Tu		0055	2.1E
	0430	0750	2.1F
	1200	1435	1.2E
◐		1925	0.3E
9 W		0155	1.9E
	0535	0850	2.0F
	1250	1615	1.3E
		2120	*
10 Th		0305	1.7E
	0640	0950	1.8F
	1335	1705	1.5E
		2250	*
11 F		0430	1.6E
	0740	1045	1.7F
	1410	1740	1.8E
	2140		
12 Sa		0005	0.7F
	0230	0540	1.5E
	0845	1140	1.5F
	1440	1810	2.1E
	2205		
13 Su		0055	1.3F
	0355	0650	1.4E
	0950	1235	1.3F
	1505	1845	2.3E
	2235		
14 M		0145	1.9F
	0510	0755	1.4E
	1105	1320	1.0F
	1525	1920	2.5E
	2305		
15 Tu		0225	2.4F
	0615	0855	1.4F
	1210	1400	0.8F
○	1545	1955	2.6E
	2335		
16 W		0300	2.8F
	0715	0950	1.4E
	1305	1440	0.6F
	1610	2040	2.6E
17 Th	0010	0335	2.9F
	0805	1035	1.5E
	1405	1520	0.4F
	1630	2120	2.5E
18 F	0045	0415	2.9F
	0850	1115	1.5E
		1600	*
		2210	2.2E
19 Sa	0120	0455	2.7F
	0935	1155	1.4E
		1645	*
		2300	2.0E
20 Su	0205	0535	2.3F
	1015	1240	1.3E
		1730	*
		2345	1.7E
21 M	0255	0625	2.0F
	1050	1330	1.2E
		1825	0.4E
22 Tu		0035	1.4E
	0355	0720	1.7F
	1130	1550	1.2E
		2140	0.4E
23 W		0125	1.1E
	0455	0815	1.4F
	1210	1640	1.3E
◑		2240	*
24 Th		0240	0.8E
	0555	0905	1.2F
	1240	1710	1.4E
		2320	*
25 F		0420	0.7E
	0650	0950	1.0F
	1305	1735	1.5E
	2215		
26 Sa		0010	0.5F
	0215	0525	0.7E
	0750	1035	0.8F
	1325	1800	1.7E
	2215		
27 Su		0055	0.8F
	0340	0625	0.7E
	0910	1125	0.5F
	1340	1820	1.8E
	2225		
28 M		0130	1.3F
	0450	0730	0.8E
	1050	1225	0.3F
	1350	1835	2.1E
	2240		
29 Tu		0155	1.7F
	0545	0830	0.9E
		1255	*
		1900	2.3E
	2300		
30 W		0225	2.2F
	0635	0920	1.1E
		1340	*
		1930	2.4E
	2330		

December

Day	Slack h m	Maximum h m	knots
1 Th		0255	2.5F
	0720	1005	1.2E
		1410	*
●		2015	2.5E
2 F	0000	0330	2.8F
	0800	1040	1.3E
		1500	*
		2110	2.6E
3 Sa	0040	0410	2.9F
	0840	1115	1.3E
		1540	*
		2200	2.6E
4 Su	0125	0450	2.8F
	0920	1145	1.3E
		1625	*
		2250	2.5E
5 M	0210	0535	2.7F
	1000	1220	1.3E
		1710	*
		2340	2.3E
6 Tu	0305	0625	2.4F
	1035	1305	1.2E
		1810	0.3E
7 W		0030	2.0E
	0400	0715	2.2F
	1115	1420	1.3E
		1940	0.3E
8 Th		0135	1.6E
	0500	0805	1.9F
	1145	1550	1.5E
◐		2155	*
9 F		0255	1.2E
	0600	0855	1.5F
	1215	1630	1.8E
	2110	2315	0.5F
10 Sa	0130	0425	1.0E
	0700	0945	1.1F
	1235	1705	2.0E
	2125		
11 Su		0015	1.1F
	0320	0540	0.8E
	0815	1040	0.7F
	1255	1735	2.3E
	2150		
12 M		0100	1.7F
	0440	0700	0.8E
	1010	1145	0.4F
	1310	1810	2.4E
	2215		
13 Tu		0140	2.2F
	0545	0820	0.9E
		1235	*
		1850	2.5E
	2245		
14 W		0215	2.5F
	0635	0920	1.0E
		1325	*
		1930	2.5E
	2315		
15 Th		0245	2.8F
	0720	0955	1.2E
		1405	*
○		2015	2.4E
	2345		
16 F		0320	2.8F
	0755	1025	1.2E
		1450	*
		2100	2.3E
17 Sa	0020	0350	2.7F
	0830	1055	1.2E
		1530	*
		2145	2.1E
18 Su	0100	0430	2.5F
	0900	1125	1.2E
		1610	*
		2230	1.9E
19 M	0135	0505	2.3F
	0930	1200	1.1E
		1655	0.3E
		2310	1.6E
20 Tu	0220	0545	2.0F
	0955	1235	1.0E
		1745	0.4E
		2350	1.3E
21 W	0300	0625	1.8F
	1015	1420	1.0E
		1845	0.5E
22 Th		0030	1.0E
	0345	0705	1.5F
	1035	1535	1.1E
		2240	0.5E
23 F		0125	0.7E
	0425	0740	1.3F
	1050	1545	1.3E
◑		2355	*
24 Sa		0255	0.4E
	0510	0815	1.0F
	1100	1550	1.5E
	2135	2330	0.4F
25 Su		0400	*
		0850	0.6F
	1110	1610	1.7E
	2120		
26 M		0030	0.9F
		0540	*
		0935	0.3F
	1110	1640	2.0E
	2130		
27 Tu		0100	1.4F
	0510	0710	0.3E
		1000	*
		1720	2.2E
	2150		
28 W		0125	2.0F
	0555	0820	0.6E
		1150	*
		1810	2.4E
	2220		
29 Th		0200	2.5F
	0630	0905	0.8E
		1250	*
		1905	2.6E
	2300		
30 F		0235	2.8F
	0705	0935	1.0E
		1345	*
●		2000	2.8E
	2340		
31 Sa		0310	3.1F
	0740	1005	1.1E
		1430	*
		2055	2.9E

Time meridian 120° W. 0000 is midnight. 1200 is noon. Times are not adjusted for Daylight Saving Time.
If three consecutive entries are marked (E) the middle one is not a true maximum but an intermediate value to show the current pattern.
* Current weak and variable.

Admiralty Inlet (off Bush Pt.), Washington, 2005

F–Flood, Dir. 180° True E–Ebb, Dir. 005° True

January

Day	Slack h m	Maximum h m	knots
1 Sa	0245	0559	2.1F
	0916	1246	2.2E
		1826	*
		2358	1.9E
2 Su	0327	0636	1.8F
	0941	1325	2.3E
	1758	1930	0.4F
	2106		
3 M ◑		0058	1.6E
	0416	0717	1.4F
	1007	1408	2.6E
	1834	2038	0.7F
	2254		
4 Tu		0210	1.2E
	0517	0803	1.1F
	1034	1453	2.8E
	1912	2145	1.2F
5 W	0040	0332	1.1E
	0631	0856	0.8F
	1105	1542	3.1E
	1953	2248	1.7F
6 Th	0210	0451	1.1E
	0753	0955	0.6F
	1143	1633	3.4E
	2036	2345	2.3F
7 F	0322	0601	1.3E
	0912	1055	0.5F
	1230	1725	3.7E
	2122		
8 Sa		0038	2.8F
	0420	0701	1.5E
	1019	1155	0.5F
	1325	1817	4.0E
	2209		
9 Su		0128	3.2F
	0510	0754	1.8E
	1116	1253	0.6F
	1424	1909	4.2E
	2256		
10 M ●		0215	3.5F
	0556	0843	2.1E
	1208	1349	0.7F
	1525	2000	4.2E
	2343		
11 Tu		0302	3.6F
	0638	0930	2.3E
	1259	1445	0.8F
	1627	2051	4.1E
12 W	0030	0347	3.6F
	0718	1017	2.6E
	1350	1541	0.9F
	1729	2142	3.8E
13 Th	0117	0431	3.4F
	0756	1103	2.8E
	1444	1639	1.0F
	1833	2234	3.4E
14 F	0204	0516	3.0F
	0832	1150	2.9E
	1539	1740	1.0F
	1942	2329	2.8E
15 Sa	0253	0600	2.5F
	0906	1237	3.0E
	1636	1844	1.1F
	2059		
16 Su ◐		0029	2.2E
	0345	0645	2.0F
	0939	1327	3.1E
	1734	1954	1.2F
	2226		
17 M		0137	1.6E
	0444	0733	1.5F
	1011	1418	3.1E
	1832	2106	1.3F
18 Tu	0002	0254	1.2E
	0555	0826	1.0F
	1042	1511	3.1E
	1926	2216	1.6F
19 W	0135	0417	1.1E
	0721	0924	0.6F
	1115	1605	3.1E
	2017	2319	1.8F
20 Th	0252	0534	1.1E
	0853	1026	0.3F
	1151	1658	3.1E
	2104		
21 F		0013	2.1F
	0350	0637	1.2E
		1125	*
		1747	3.2E
	2146		
22 Sa		0100	2.3F
	0437	0727	1.4E
		1218	*
		1832	3.2E
	2225		
23 Su		0140	2.4F
	0515	0809	1.6E
		1306	*
		1912	3.2E
	2301		
24 M		0216	2.6F
	0548	0845	1.8E
	1233	1348	0.3F
	1500	1950	3.2E
	2335		
25 Tu ○		0249	2.6F
	0618	0918	1.9E
	1307	1427	0.4F
	1545	2025	3.2E
26 W	0007	0320	2.6F
	0646	0949	2.0E
	1339	1505	0.4F
	1629	2100	3.1E
27 Th	0038	0350	2.6F
	0711	1019	2.2E
	1411	1543	0.5F
	1714	2135	2.9E
28 F	0110	0419	2.4F
	0734	1049	2.3E
	1443	1623	0.6F
	1803	2212	2.6E
29 Sa	0143	0448	2.2F
	0756	1119	2.4E
	1516	1706	0.7F
	1858	2252	2.3E
30 Su	0218	0518	1.9F
	0816	1151	2.6E
	1553	1755	0.8F
	2002	2338	1.9E
31 M	0258	0551	1.6F
	0837	1227	2.7E
	1634	1850	1.0F
	2116		

February

Day	Slack h m	Maximum h m	knots
1 Tu ◑		0033	1.5E
	0346	0628	1.2F
	0859	1309	2.8E
	1722	1954	1.2F
	2245		
2 W		0142	1.2E
	0447	0713	0.9F
	0925	1358	2.9E
	1815	2105	1.5F
3 Th	0023	0305	1.0E
	0610	0810	0.5F
	0959	1457	3.1E
	1911	2217	1.8F
4 F	0154	0433	1.0E
	0750	0921	0.3F
	1045	1600	3.3E
	2008	2322	2.3F
5 Sa	0306	0549	1.3E
	0917	1037	0.3F
	1152	1704	3.5E
	2102		
6 Su		0020	2.7F
	0401	0649	1.6E
	1017	1147	0.4F
	1312	1805	3.8E
	2154		
7 M		0112	3.1F
	0448	0739	2.0E
	1105	1250	0.7F
	1429	1901	3.9E
	2243		
8 Tu ●		0159	3.3F
	0528	0824	2.4E
	1150	1347	1.0F
	1540	1954	4.0E
	2331		
9 W		0242	3.4F
	0605	0907	2.7E
	1234	1441	1.3F
	1645	2044	3.9E
10 Th	0016	0324	3.3F
	0640	0948	3.0E
	1318	1534	1.5F
	1747	2134	3.6E
11 F	0102	0405	3.0F
	0712	1029	3.2E
	1404	1626	1.6F
	1849	2224	3.1E
12 Sa	0147	0444	2.6F
	0742	1110	3.3E
	1452	1719	1.7F
	1952	2315	2.6E
13 Su	0233	0524	2.1F
	0811	1152	3.2E
	1543	1815	1.6F
	2058		
14 M		0009	2.0E
	0324	0604	1.5F
	0838	1236	3.1E
	1637	1916	1.5F
	2212		
15 Tu ◐		0111	1.5E
	0422	0648	1.0F
	0904	1324	3.0E
	1736	2023	1.5F
	2335		
16 W		0225	1.1E
	0539	0739	0.5F
	0929	1420	2.8E
	1837	2134	1.5F
17 Th	0102	0351	1.0E
		0843	*
		1521	2.7E
	1937	2244	1.6F
18 F	0218	0513	1.0E
		0957	*
		1624	2.7E
	2032	2343	1.8F
19 Sa	0316	0616	1.3E
		1108	*
		1722	2.7E
	2120		
20 Su		0031	2.0F
	0400	0703	1.5E
		1206	*
		1812	2.8E
	2201		
21 M		0112	2.2F
	0436	0740	1.8E
	1130	1254	0.4F
	1416	1855	2.9E
	2238		
22 Tu		0147	2.3F
	0506	0813	2.0E
	1156	1336	0.6F
	1513	1934	3.0E
	2312		
23 W ○		0218	2.4F
	0533	0842	2.2E
	1222	1414	0.8F
	1603	2011	3.0E
	2345		
24 Th		0248	2.4F
	0557	0910	2.4E
	1248	1450	1.0F
	1651	2046	2.9E
25 F	0017	0316	2.3F
	0618	0936	2.6E
	1314	1526	1.2F
	1738	2122	2.8E
26 Sa	0050	0343	2.1F
	0637	1003	2.7E
	1343	1603	1.3F
	1826	2159	2.5E
27 Su	0124	0411	1.9F
	0656	1031	2.9E
	1414	1643	1.5F
	1918	2240	2.2E
28 M	0202	0440	1.6F
	0714	1102	3.0E
	1451	1728	1.6F
	2016	2326	1.9E

March

Day	Slack h m	Maximum h m	knots
1 Tu	0244	0513	1.2F
	0734	1138	3.0E
	1535	1820	1.6F
	2122		
2 W		0021	1.5E
	0337	0552	0.9F
	0758	1222	3.0E
	1628	1921	1.7F
	2240		
3 Th ◑		0129	1.2E
	0447	0641	0.5F
	0825	1317	3.0E
	1729	2033	1.8F
4 F	0008	0255	1.1E
		0747	*
		1424	3.0E
	1836	2149	2.0F
5 Sa	0131	0423	1.2E
		0912	*
		1539	3.0E
	1942	2259	2.3F
6 Su	0236	0535	1.5E
	0924	1038	0.3F
	1149	1652	3.2E
	2043	2358	2.6F
7 M	0327	0630	2.0E
	1007	1150	0.6F
	1328	1757	3.4E
	2138		
8 Tu		0049	2.8F
	0409	0716	2.4E
	1046	1251	1.1F
	1450	1855	3.5E
	2229		
9 W		0135	2.9F
	0446	0757	2.8E
	1126	1344	1.5F
	1600	1947	3.5E
	2316		
10 Th ●		0216	2.9F
	0520	0836	3.2E
	1205	1434	1.9F
	1703	2037	3.4E
11 F	0002	0256	2.7F
	0550	0914	3.4E
	1245	1522	2.1F
	1801	2125	3.1E
12 Sa	0047	0333	2.3F
	0618	0950	3.5E
	1326	1608	2.2F
	1858	2212	2.7E
13 Su	0132	0410	1.9F
	0645	1027	3.5E
	1408	1655	2.2F
	1953	2300	2.3E
14 M	0220	0447	1.4F
	0709	1105	3.3E
	1453	1744	2.1F
	2051	2351	1.8E
15 Tu	0312	0525	1.0F
	0732	1146	3.1E
	1543	1837	1.8F
	2152		
16 W		0049	1.4E
	0417	0607	0.5F
	0752	1232	2.8E
	1639	1937	1.6F
	2301		
17 Th ◐		0159	1.2E
		0700	*
		1326	2.5E
	1741	2045	1.5F
18 F	0016	0321	1.1E
		0810	*
		1433	2.3E
	1846	2156	1.5F
19 Sa	0125	0441	1.2E
		0935	*
		1545	2.2E
	1948	2258	1.6F
20 Su	0221	0540	1.5E
		1051	*
		1650	2.3E
	2040	2349	1.8F
21 M	0304	0623	1.8E
	1031	1149	0.3F
	1307	1745	2.4E
	2126		
22 Tu		0030	1.9F
	0338	0659	2.1E
	1050	1236	0.6F
	1422	1832	2.5E
	2206		
23 W		0106	2.0F
	0407	0729	2.3E
	1111	1317	0.9F
	1522	1913	2.6E
	2243		
24 Th		0137	2.0F
	0432	0757	2.6E
	1135	1354	1.3F
	1614	1953	2.7E
	2320		
25 F ○		0207	2.0F
	0454	0823	2.8E
	1159	1430	1.6F
	1703	2031	2.6E
	2356		
26 Sa		0236	1.8F
	0514	0850	3.0E
	1225	1506	1.9F
	1751	2109	2.5E
27 Su	0033	0305	1.6F
	0533	0917	3.2E
	1255	1543	2.1F
	1839	2150	2.3E
28 M	0112	0335	1.4F
	0553	0947	3.3E
	1329	1623	2.3F
	1930	2233	2.1E
29 Tu	0155	0407	1.1F
	0614	1022	3.3E
	1408	1708	2.3F
	2025	2322	1.8E
30 W	0246	0445	0.8F
	0638	1102	3.3E
	1455	1800	2.2F
	2126		
31 Th		0019	1.5E
	0350	0530	0.5F
	0704	1150	3.1E
	1550	1900	2.1F
	2236		

Time meridian 120° W. 0000 is midnight. 1200 is noon. Times are not adjusted for Daylight Saving Time.
* Current weak and variable.

Admiralty Inlet (off Bush Pt.), Washington, 2005

F–Flood, Dir. 180° True E–Ebb, Dir. 005° True

April

Day	Slack h m	Maximum h m	knots
1 F ◑		0130	1.4E
		0629	*
		1251	2.9E
	1655	2010	2.1F
	2350		
2 Sa		0252	1.4E
		0749	*
		1405	2.7E
	1805	2124	2.1F
3 Su	0058	0409	1.6E
		0922	*
		1527	2.7E
	1915	2231	2.2F
4 M	0154	0511	2.0E
	0910	1045	0.5F
	1218	1643	2.7E
	2020	2329	2.3F
5 Tu	0240	0602	2.5E
	0946	1151	1.0F
	1356	1750	2.8E
	2118		
6 W		0019	2.4F
	0319	0645	2.9E
	1022	1247	1.5F
	1513	1848	2.9E
	2211		
7 Th		0104	2.3F
	0354	0724	3.3E
	1059	1337	2.0F
	1619	1940	2.9E
	2301		
8 F ●		0145	2.2F
	0425	0801	3.6E
	1136	1423	2.4F
	1717	2029	2.8E
	2349		
9 Sa		0223	1.9F
	0453	0837	3.7E
	1213	1507	2.6F
	1811	2116	2.6E
10 Su	0036	0300	1.6F
	0519	0912	3.7E
	1251	1549	2.7F
	1901	2202	2.3E
11 M	0124	0336	1.2F
	0543	0947	3.5E
	1330	1632	2.6F
	1951	2248	2.0E
12 Tu	0216	0412	0.8F
	0604	1023	3.3E
	1411	1716	2.4F
	2040	2337	1.7E
13 W	0315	0451	0.5F
	0623	1101	3.0E
	1456	1803	2.1F
	2132		
14 Th		0032	1.5E
		0535	*
		1145	2.7E
	1546	1856	1.8F
	2228		
15 F		0136	1.3E
		0631	*
		1237	2.3E
	1643	1956	1.6F
	2328		

Day	Slack h m	Maximum h m	knots
16 Sa ◐		0248	1.3E
		0745	0.3E
		1342	2.1E
	1745	2100	1.5F
17 Su	0024	0357	1.5E
		0911	*
		1456	1.9E
	1848	2200	1.5F
18 M	0113	0450	1.8E
		1026	*
		1607	1.9E
	1946	2251	1.6F
19 Tu	0154	0532	2.1E
	0947	1124	0.4F
	1302	1708	2.0E
	2038	2335	1.6F
20 W	0227	0607	2.4E
	1003	1211	0.9F
	1421	1801	2.1E
	2125		
21 Th		0013	1.6F
	0256	0637	2.7E
	1024	1251	1.4F
	1523	1848	2.2E
	2210		
22 F		0048	1.5F
	0320	0706	3.0E
	1048	1330	1.8F
	1618	1932	2.2E
	2253		
23 Sa		0121	1.4F
	0343	0735	3.2E
	1115	1407	2.2F
	1708	2015	2.2E
	2336		
24 Su ○		0153	1.3F
	0405	0805	3.5E
	1144	1445	2.5F
	1757	2058	2.2E
25 M	0020	0227	1.1F
	0428	0837	3.6E
	1218	1524	2.8F
	1846	2142	2.1E
26 Tu	0106	0303	0.9F
	0453	0913	3.7E
	1257	1607	2.9F
	1936	2229	2.0E
27 W	0158	0342	0.7F
	0521	0953	3.6E
	1340	1654	2.8F
	2030	2322	1.8E
28 Th	0300	0428	0.4F
	0552	1039	3.5E
	1429	1746	2.7F
	2126		
29 F		0022	1.7E
		0524	*
		1132	3.2E
	1525	1845	2.5F
	2225		
30 Sa ◑		0130	1.7E
		0635	*
		1237	2.8E
	1628	1949	2.4F
	2324		

May

Day	Slack h m	Maximum h m	knots
1 Su		0240	1.9E
		0801	*
		1353	2.5E
	1736	2055	2.2F
2 M	0017	0345	2.2E
	0803	0928	0.4F
	1054	1515	2.3E
	1846	2157	2.1F
3 Tu	0105	0440	2.6E
	0842	1044	0.9F
	1248	1632	2.2E
	1953	2253	2.0F
4 W	0146	0528	3.0E
	0919	1146	1.5F
	1418	1740	2.3E
	2055	2343	1.9F
5 Th	0223	0610	3.4E
	0956	1239	2.0F
	1531	1839	2.3E
	2153		
6 F		0028	1.7F
	0255	0650	3.6E
	1033	1326	2.4F
	1632	1933	2.2E
	2248		
7 Sa		0110	1.4F
	0325	0727	3.8E
	1109	1410	2.7F
	1726	2023	2.2E
	2341		
8 Su ●		0150	1.1F
	0353	0803	3.8E
	1145	1451	2.8F
	1815	2109	2.1E
9 M	0033	0228	0.9F
	0417	0837	3.7E
	1222	1532	2.8F
	1901	2154	1.9E
10 Tu	0126	0305	0.6F
	0440	0912	3.5E
	1259	1611	2.7F
	1945	2239	1.8E
11 W	0224	0343	0.3F
	0459	0948	3.2E
	1337	1652	2.5F
	2028	2326	1.7E
12 Th		0424	*
		1026	2.9E
	1418	1735	2.3F
	2112		
13 F		0016	1.6E
		0511	*
		1108	2.6E
	1502	1820	2.1F
	2156		
14 Sa		0111	1.6E
		0609	*
		1157	2.3E
	1550	1910	1.9F
	2241		
15 Su		0209	1.6E
		0720	0.3E
		1255	2.0E
	1644	2003	1.7F
	2324		

Day	Slack h m	Maximum h m	knots
16 M ◐		0304	1.8E
		0838	*
		1404	1.7E
	1741	2055	1.6F
17 Tu	0004	0352	2.1E
		0949	*
		1516	1.6E
	1841	2145	1.5F
18 W	0038	0432	2.4E
	0853	1048	0.6F
	1249	1625	1.6E
	1941	2231	1.3F
19 Th	0109	0509	2.7E
	0913	1137	1.1F
	1412	1726	1.6E
	2039	2313	1.2F
20 F	0137	0543	3.0E
	0937	1221	1.7F
	1519	1821	1.7E
	2134	2354	1.1F
21 Sa	0204	0616	3.3E
	1005	1303	2.2F
	1616	1911	1.8E
	2227		
22 Su		0033	0.9F
	0231	0651	3.6E
	1037	1344	2.7F
	1708	1959	1.9E
	2319		
23 M ○		0113	0.8F
	0300	0727	3.8E
	1113	1425	3.0F
	1758	2047	1.9E
24 Tu	0010	0154	0.7F
	0332	0806	3.9E
	1152	1509	3.2F
	1847	2135	2.0E
25 W	0104	0238	0.6F
	0408	0849	3.9E
	1235	1554	3.3F
	1936	2225	2.0E
26 Th	0202	0327	0.4F
	0448	0935	3.8E
	1322	1643	3.2F
	2025	2319	2.0E
27 F	0308	0422	0.3F
	0534	1026	3.6E
	1412	1734	3.1F
	2114		
28 Sa		0016	2.1E
		0526	*
		1123	3.2E
	1506	1828	2.8F
	2202		
29 Su		0115	2.2E
		0640	*
		1228	2.7E
	1605	1924	2.5F
	2248		
30 M ◑		0214	2.5E
	0634	0801	0.4F
	0930	1342	2.3E
	1707	2022	2.2F
	2331		
31 Tu		0311	2.8E
	0724	0921	0.8F
	1123	1500	2.0E
	1814	2118	1.9F

June

Day	Slack h m	Maximum h m	knots
1 W	0011	0403	3.1E
	0808	1032	1.3F
	1306	1618	1.8E
	1924	2213	1.6F
2 Th	0049	0451	3.4E
	0850	1133	1.8F
	1432	1728	1.7E
	2033	2304	1.3F
3 F	0124	0535	3.6E
	0930	1226	2.3F
	1541	1831	1.7E
	2139	2353	1.0F
4 Sa	0156	0617	3.7E
	1008	1313	2.6F
	1640	1927	1.7E
	2242		
5 Su		0038	0.8F
	0226	0656	3.7E
	1046	1356	2.8F
	1730	2017	1.8E
	2342		
6 M ●		0121	0.6F
	0255	0734	3.7E
	1123	1437	2.9F
	1814	2103	1.8E
7 Tu	0039	0202	0.4F
	0322	0811	3.5E
	1159	1516	2.8F
	1855	2146	1.7E
8 W		0242	*
		0847	3.4E
	1235	1553	2.7F
	1933	2228	1.7E
9 Th		0322	*
		0923	3.1E
	1312	1631	2.6F
	2009	2310	1.7E
10 F		0405	*
		1000	2.9E
	1348	1708	2.4F
	2045	2353	1.8E
11 Sa		0453	*
		1040	2.6E
	1427	1747	2.3F
	2120		
12 Su		0037	1.8E
		0546	*
		1125	2.3E
	1507	1827	2.1F
	2153		
13 M		0122	2.0E
		0648	*
		1216	2.0E
	1552	1909	1.8F
	2225		
14 Tu ◐		0206	2.1E
		0755	*
		1317	1.6E
	1642	1953	1.6F
	2255		
15 W		0248	2.3E
	0729	0902	0.4F
	1042	1426	1.4E
	1740	2038	1.3F
	2323		

Day	Slack h m	Maximum h m	knots
16 Th		0329	2.6E
	0752	1004	0.8F
	1229	1540	1.2E
	1844	2125	1.1F
	2351		
17 F		0409	2.9E
	0820	1059	1.4F
	1358	1650	1.2E
	1954	2214	0.9F
18 Sa	0020	0450	3.2E
	0852	1149	1.9F
	1509	1754	1.3E
	2102	2302	0.7F
19 Su	0052	0531	3.5E
	0928	1236	2.4F
	1609	1852	1.5E
	2206	2352	0.6F
20 M	0129	0615	3.8E
	1008	1323	2.9F
	1702	1945	1.6E
	2305		
21 Tu ○		0041	0.5F
	0211	0659	4.0E
	1050	1408	3.2F
	1751	2035	1.8E
22 W	0000	0132	0.5F
	0259	0746	4.1E
	1134	1454	3.4F
	1837	2124	2.0E
23 Th	0054	0225	0.5F
	0351	0835	4.1E
	1220	1541	3.5F
	1922	2213	2.2E
24 F	0150	0320	0.5F
	0447	0925	3.9E
	1308	1628	3.4F
	2005	2303	2.3E
25 Sa	0248	0420	0.6F
	0549	1019	3.6E
	1357	1716	3.2F
	2046	2354	2.5E
26 Su	0348	0524	0.6F
	0659	1116	3.1E
	1448	1804	2.9F
	2125		
27 M		0046	2.7E
	0448	0633	0.7F
	0820	1218	2.6E
	1542	1854	2.5F
	2203		
28 Tu ◑		0138	3.0E
	0546	0746	0.9F
	0953	1327	2.1E
	1641	1945	2.0F
	2240		
29 W		0231	3.1E
	0640	0900	1.2F
	1134	1443	1.6E
	1746	2038	1.5F
	2316		
30 Th		0323	3.3E
	0731	1010	1.6F
	1311	1601	1.4E
	1900	2133	1.1F
	2351		

Time meridian 120° W. 0000 is midnight. 1200 is noon. Times are not adjusted for Daylight Saving Time.
* Current weak and variable.

Admiralty Inlet (off Bush Pt.), Washington, 2005

F–Flood, Dir. 180° True E–Ebb, Dir. 005° True

July

Day	Slack h m	Maximum h m	knots
1 F		0413	3.4E
	0819	1113	2.0F
	1434	1717	1.3E
	2019	2229	0.8F
2 Sa	0025	0502	3.5E
	0904	1209	2.3F
	1541	1823	1.4E
	2137	2323	0.5F
3 Su	0100	0549	3.5E
	0947	1258	2.5F
	1636	1920	1.5E
	2247		
4 M		0014	0.4F
	0136	0632	3.5E
	1027	1342	2.7F
	1722	2009	1.6E
	2348		
5 Tu		0102	0.3F
	0213	0714	3.4E
	1105	1422	2.7F
	1801	2052	1.7E
6 W ●		0146	*
		0752	3.4E
	1141	1459	2.7F
	1836	2131	1.7E
7 Th		0228	*
		0829	3.2E
	1216	1533	2.7F
	1908	2208	1.8E
8 F		0309	*
		0905	3.1E
	1249	1606	2.6F
	1938	2243	1.9E
9 Sa		0350	*
		0941	2.9E
	1323	1639	2.5F
	2007	2318	2.0E
10 Su		0434	*
		1019	2.6E
	1357	1711	2.3F
	2033	2353	2.1E
11 M		0521	*
		1100	2.3E
	1433	1745	2.1F
	2058		
12 Tu		0028	2.2E
	0455	0613	0.3F
	0732	1146	2.0E
	1513	1819	1.8F
	2122		
13 W		0105	2.4E
	0531	0710	0.5F
	0854	1240	1.6E
	1558	1857	1.5F
	2145		
14 Th ◐		0144	2.6E
	0608	0812	0.7F
	1028	1346	1.3E
	1653	1939	1.1F
	2209		
15 F		0227	2.7E
	0647	0917	1.1F
	1208	1501	1.1E
	1802	2027	0.8F
	2237		
16 Sa		0314	3.0E
	0728	1020	1.5F
	1339	1620	1.0E
	1922	2123	0.6F
	2310		
17 Su		0404	3.2E
	0813	1119	2.0F
	1455	1733	1.1E
	2044	2223	0.4F
	2353		
18 M		0457	3.5E
	0859	1213	2.5F
	1555	1835	1.3E
	2155	2325	0.4F
19 Tu	0048	0550	3.8E
	0945	1303	2.9F
	1647	1930	1.6E
	2253		
20 W		0025	0.5F
	0151	0643	4.0E
	1032	1351	3.2F
	1732	2019	1.9E
	2343		
21 Th ○		0122	0.6F
	0256	0735	4.1E
	1120	1437	3.4F
	1813	2105	2.2E
22 F	0032	0219	0.8F
	0401	0826	4.1E
	1206	1522	3.5F
	1852	2150	2.5E
23 Sa	0121	0315	1.0F
	0506	0918	3.8E
	1253	1606	3.3F
	1929	2235	2.8E
24 Su	0211	0413	1.1F
	0612	1011	3.5E
	1341	1650	3.0F
	2004	2321	3.0E
25 M	0304	0512	1.2F
	0722	1105	3.0E
	1429	1734	2.6F
	2038		
26 Tu		0007	3.1E
	0358	0614	1.3F
	0836	1204	2.4E
	1521	1819	2.1F
	2111		
27 W ◑		0055	3.2E
	0455	0721	1.4F
	0959	1309	1.8E
	1619	1906	1.6F
	2144		
28 Th		0146	3.2E
	0553	0831	1.5F
	1129	1423	1.4E
	1727	1959	1.1F
	2217		
29 F		0240	3.2E
	0650	0942	1.7F
	1300	1545	1.2E
	1851	2057	0.7F
	2251		
30 Sa		0336	3.2E
	0746	1049	1.9F
	1421	1705	1.1E
	2024	2200	0.4F
	2329		
31 Su		0432	3.2E
	0838	1148	2.1F
	1525	1813	1.3E
		2303	*

August

Day	Slack h m	Maximum h m	knots
1 M		0526	3.2E
	0925	1239	2.3F
	1616	1907	1.4E
2 Tu		0001	*
		0615	3.2E
	1008	1322	2.4F
	1656	1951	1.6E
	2338		
3 W		0051	0.3F
	0202	0658	3.2E
	1047	1400	2.5F
	1731	2029	1.8E
4 Th ●	0015	0136	0.3F
	0254	0738	3.2E
	1122	1435	2.6F
	1801	2103	1.9E
5 F	0049	0217	0.4F
	0341	0815	3.1E
	1155	1506	2.5F
	1828	2134	2.1E
6 Sa	0121	0255	0.5F
	0427	0850	3.0E
	1227	1535	2.5F
	1853	2204	2.2E
7 Su	0153	0333	0.6F
	0512	0925	2.8E
	1259	1604	2.3F
	1915	2232	2.3E
8 M	0224	0412	0.7F
	0600	1001	2.6E
	1331	1633	2.1F
	1936	2301	2.4E
9 Tu	0256	0453	0.8F
	0652	1040	2.3E
	1406	1702	1.8F
	1956	2332	2.5E
10 W	0331	0538	0.9F
	0750	1124	1.9E
	1445	1733	1.5F
	2015		
11 Th		0005	2.6E
	0410	0628	1.0F
	0858	1215	1.6E
	1531	1808	1.1F
	2035		
12 F ◐		0044	2.7E
	0455	0727	1.2F
	1017	1318	1.2E
	1628	1849	0.8F
	2058		
13 Sa		0130	2.8E
	0546	0834	1.4F
	1148	1435	1.0E
	1745	1942	0.5F
	2127		
14 Su		0226	2.9E
	0642	0944	1.7F
	1318	1600	1.0E
	1922	2049	0.3F
	2209		
15 M		0328	3.1E
	0738	1051	2.1F
	1432	1717	1.2E
		2204	*
16 Tu		0433	3.3E
	0834	1150	2.5F
	1530	1819	1.5E
	2150	2316	0.4F
17 W	0037	0535	3.6E
	0926	1243	2.8F
	1617	1910	1.9E
	2237		
18 Th		0021	0.7F
	0159	0633	3.8E
	1017	1330	3.1F
	1658	1955	2.3E
	2320		
19 F ○		0119	1.0F
	0313	0727	3.9E
	1105	1415	3.2F
	1735	2037	2.7E
20 Sa	0003	0214	1.3F
	0422	0819	3.8E
	1151	1457	3.2F
	1809	2119	3.0E
21 Su	0047	0307	1.6F
	0526	0910	3.6E
	1238	1538	2.9F
	1842	2200	3.3E
22 M	0132	0400	1.8F
	0629	1001	3.2E
	1324	1619	2.6F
	1913	2241	3.4E
23 Tu	0220	0453	1.9F
	0733	1053	2.7E
	1413	1700	2.1F
	1943	2324	3.4E
24 W	0310	0549	1.9F
	0840	1149	2.2E
	1506	1742	1.6F
	2012		
25 Th		0009	3.3E
	0404	0649	1.8F
	0952	1252	1.7E
	1607	1828	1.0F
	2041		
26 F ◑		0059	3.1E
	0503	0755	1.7F
	1112	1405	1.3E
	1725	1921	0.6F
	2110		
27 Sa		0155	2.9E
	0606	0907	1.7F
	1235	1528	1.1E
		2027	*
28 Su		0258	2.7E
	0709	1017	1.7F
	1351	1649	1.2E
		2142	*
29 M		0403	2.7E
	0808	1119	1.9F
	1450	1753	1.4E
		2253	*
30 Tu		0504	2.7E
	0859	1210	2.0F
	1536	1842	1.6E
	2239	2353	0.3F
31 W	0105	0556	2.8E
	0944	1252	2.2F
	1612	1920	1.9E
	2308		

September

Day	Slack h m	Maximum h m	knots
1 Th		0041	0.5F
	0212	0641	2.9E
	1022	1329	2.3F
	1643	1953	2.1E
	2336		
2 F		0124	0.7F
	0309	0721	2.9E
	1058	1401	2.3F
	1710	2023	2.3E
3 Sa ●	0003	0202	0.9F
	0359	0758	2.9E
	1131	1430	2.3F
	1734	2051	2.5E
4 Su	0029	0237	1.1F
	0445	0834	2.8E
	1203	1458	2.1F
	1755	2117	2.6E
5 M	0056	0313	1.2F
	0531	0909	2.6E
	1236	1525	1.9F
	1814	2143	2.7E
6 Tu	0123	0348	1.4F
	0617	0946	2.4E
	1311	1552	1.7F
	1831	2210	2.8E
7 W	0153	0426	1.5F
	0705	1025	2.1E
	1348	1620	1.4F
	1848	2239	2.9E
8 Th	0227	0507	1.6F
	0759	1109	1.8E
	1431	1651	1.1F
	1906	2313	2.9E
9 F	0308	0555	1.7F
	0900	1200	1.5E
	1522	1728	0.7F
	1926	2354	2.9E
10 Sa	0358	0652	1.7F
	1011	1304	1.2E
	1631	1813	0.4F
	1950		
11 Su ◐		0045	2.8E
	0456	0800	1.7F
	1133	1424	1.1E
		1916	*
12 M		0149	2.8E
	0601	0914	1.8F
	1254	1550	1.2E
		2039	*
13 Tu		0304	2.8E
	0708	1025	2.1F
	1400	1702	1.5E
		2206	*
14 W		0419	3.0E
	0810	1125	2.4F
	1452	1757	1.9E
	2139	2320	0.6F
15 Th	0056	0526	3.2E
	0907	1217	2.6F
	1534	1844	2.4E
	2216		
16 F		0021	1.0F
	0223	0625	3.4E
	0959	1303	2.8F
	1612	1925	2.8E
	2254		
17 Sa ○		0116	1.5F
	0336	0720	3.4E
	1048	1346	2.8F
	1646	2005	3.2E
	2334		
18 Su		0207	2.0F
	0441	0811	3.3E
	1136	1427	2.6F
	1717	2043	3.5E
19 M	0014	0256	2.3F
	0541	0901	3.1E
	1223	1506	2.3F
	1747	2122	3.7E
20 Tu	0056	0344	2.5F
	0639	0951	2.8E
	1311	1545	1.9F
	1815	2200	3.7E
21 W	0139	0432	2.5F
	0737	1041	2.4E
	1402	1624	1.4F
	1842	2240	3.5E
22 Th	0225	0522	2.4F
	0836	1135	2.0E
	1458	1705	0.9F
	1907	2323	3.2E
23 F	0316	0616	2.1F
	0939	1235	1.6E
	1609	1751	0.5F
	1930		
24 Sa ◑		0011	2.9E
	0412	0717	1.9F
	1047	1346	1.3E
		1848	*
25 Su		0107	2.6E
	0515	0825	1.7F
	1158	1507	1.2E
		2002	*
26 M		0215	2.3E
	0622	0935	1.6F
	1304	1623	1.4E
		2127	*
27 Tu		0329	2.2E
	0726	1037	1.7F
	1357	1720	1.6E
		2241	*
28 W		0435	2.2E
	0821	1128	1.8F
	1439	1804	1.9E
	2209	2339	0.4F
29 Th	0108	0531	2.3E
	0908	1210	1.9F
	1514	1839	2.2E
	2231		
30 F		0025	0.7F
	0219	0618	2.4E
	0949	1246	1.9F
	1542	1910	2.5E
	2254		

Time meridian 120° W. 0000 is midnight. 1200 is noon. Times are not adjusted for Daylight Saving Time.

* Current weak and variable.

Admiralty Inlet (off Bush Pt.), Washington, 2005

F–Flood, Dir. 180° True   E–Ebb, Dir. 005° True

October

Day	Slack h m	Maximum h m	knots
1 Sa		0105	1.1F
	0317	0700	2.5E
	1027	1318	1.9F
	1607	1937	2.7E
	2317		
2 Su		0141	1.4F
	0408	0739	2.5E
	1104	1348	1.8F
	1629	2004	2.9E
	2342		
3 M ●		0216	1.7F
	0455	0817	2.5E
	1141	1416	1.7F
	1649	2030	3.0E
4 Tu	0007	0250	1.9F
	0540	0855	2.4E
	1218	1444	1.4F
	1707	2056	3.2E
5 W	0035	0325	2.1F
	0626	0933	2.2E
	1257	1513	1.2F
	1724	2124	3.2E
6 Th	0106	0403	2.2F
	0713	1015	2.0E
	1340	1544	0.9F
	1743	2156	3.2E
7 F	0143	0445	2.3F
	0804	1101	1.7E
	1430	1619	0.6F
	1802	2233	3.2E
8 Sa	0226	0533	2.2F
	0902	1155	1.5E
	1534	1701	0.4F
	1824	2318	3.1E
9 Su	0318	0629	2.1F
	1007	1302	1.4E
		1756	*
10 M ◐		0015	2.8E
	0419	0735	2.0F
	1116	1420	1.4E
		1913	*
11 Tu		0126	2.6E
	0527	0847	2.0F
	1222	1636	1.6E
		2047	*
12 W		0248	2.5E
	0637	0955	2.1F
	1318	1639	2.0E
	2046	2212	0.4F
	2338		
13 Th		0408	2.6E
	0744	1054	2.2F
	1404	1729	2.5E
	2118	2321	0.9F
14 F	0124	0517	2.7E
	0844	1145	2.3F
	1444	1813	3.0E
	2153		
15 Sa		0019	1.6F
	0247	0618	2.8E
	0940	1231	2.3F
	1519	1853	3.4E
	2229		

Day	Slack h m	Maximum h m	knots
16 Su		0110	2.1F
	0355	0713	2.8E
	1032	1314	2.1F
	1551	1931	3.7E
	2307		
17 M ○		0157	2.6F
	0456	0805	2.8E
	1122	1354	1.9F
	1621	2009	3.9E
	2345		
18 Tu		0243	2.8F
	0551	0854	2.6E
	1212	1434	1.5F
	1649	2046	3.9E
19 W	0024	0327	2.9F
	0644	0942	2.4E
	1304	1512	1.2F
	1716	2123	3.8E
20 Th	0105	0411	2.8F
	0736	1031	2.1E
	1359	1552	0.8F
	1740	2201	3.5E
21 F	0148	0457	2.6F
	0827	1123	1.8E
	1503	1633	0.4F
	1800	2242	3.2E
22 Sa	0234	0546	2.3F
	0921	1219	1.6E
		1720	*
		2327	2.8E
23 Su	0325	0640	2.0F
	1016	1325	1.4E
		1819	*
24 M ◑		0021	2.4E
	0422	0739	1.8F
	1113	1436	1.5E
		1937	0.3E
25 Tu		0126	2.1E
	0524	0842	1.6F
	1206	1541	1.6E
		2104	*
26 W		0241	1.9E
	0627	0940	1.6F
	1253	1634	1.9E
		2218	*
27 Th		0354	1.8E
	0726	1032	1.6F
	1332	1715	2.2E
	2131	2315	0.5F
28 F	0101	0456	1.9E
	0819	1115	1.6F
	1404	1749	2.5E
	2149		
29 Sa		0001	1.0F
	0217	0549	1.9E
	0908	1154	1.5F
	1433	1820	2.8E
	2211		
30 Su		0041	1.4F
	0318	0636	2.0E
	0953	1229	1.4F
	1457	1849	3.1E
	2234		
31 M		0117	1.8F
	0410	0719	2.1E
	1037	1302	1.3F
	1519	1917	3.3E
	2300		

November

Day	Slack h m	Maximum h m	knots
1 Tu ●		0153	2.2F
	0458	0801	2.1E
	1121	1334	1.1F
	1540	1946	3.4E
	2328		
2 W		0229	2.5F
	0544	0842	2.1E
	1204	1406	0.9F
	1601	2016	3.6E
	2359		
3 Th		0306	2.7F
	0630	0924	2.0E
	1250	1439	0.7F
	1624	2049	3.6E
4 F	0034	0346	2.8F
	0717	1009	1.9E
	1340	1516	0.5F
	1648	2126	3.6E
5 Sa	0115	0429	2.8F
	0807	1058	1.7E
	1439	1559	0.3F
	1715	2209	3.4E
6 Su	0200	0518	2.7F
	0900	1154	1.7E
		1650	*
		2259	3.2E
7 M	0252	0613	2.5F
	0955	1258	1.7E
		1757	*
		2359	2.8E
8 Tu ◐	0351	0713	2.4F
	1050	1406	1.8E
		1921	*
9 W		0112	2.5E
	0456	0817	2.2F
	1142	1510	2.2E
		2051	*
10 Th		0234	2.2E
	0605	0919	2.1F
	1229	1606	2.6E
	2015	2211	0.7F
11 F	0009	0355	2.1E
	0714	1017	2.0F
	1311	1656	3.0E
	2051	2316	1.4F
12 Sa	0148	0508	2.1E
	0820	1109	1.8F
	1349	1740	3.4E
	2128		
13 Su		0012	2.0F
	0306	0611	2.2E
	0922	1157	1.6F
	1423	1821	3.7E
	2206		
14 M		0101	2.5F
	0410	0708	2.2E
	1020	1242	1.4F
	1456	1900	3.9E
	2243		
15 Tu ○		0147	2.9F
	0507	0800	2.2E
	1116	1325	1.2F
	1526	1938	4.0E
	2321		

Day	Slack h m	Maximum h m	knots
16 W		0230	3.1F
	0558	0849	2.1E
	1211	1405	0.9F
	1554	2016	3.9E
17 Th	0000	0312	3.1F
	0645	0936	2.0E
	1306	1445	0.6F
	1621	2053	3.7E
18 F	0039	0354	3.0F
	0731	1023	1.9E
	1406	1526	0.4F
	1644	2131	3.4E
19 Sa	0119	0436	2.8F
	0815	1110	1.8E
		1609	*
		2210	3.1E
20 Su	0201	0519	2.5F
	0858	1201	1.7E
		1657	*
		2252	2.7E
21 M	0245	0604	2.2F
	0941	1255	1.7E
		1754	*
		2340	2.3E
22 Tu	0332	0652	2.0F
	1024	1352	1.7E
		1905	0.3E
23 W ◑		0038	2.0E
	0423	0743	1.8F
	1105	1446	1.9E
		2024	*
24 Th		0145	1.7E
	0519	0834	1.6F
	1142	1534	2.1E
		2137	*
25 F		0259	1.5E
	0620	0924	1.4F
	1217	1616	2.4E
	2043	2238	0.6F
26 Sa	0041	0410	1.4E
	0722	1011	1.3F
	1247	1653	2.7E
	2102	2327	1.1F
27 Su	0205	0513	1.4E
	0822	1055	1.1F
	1316	1727	3.0E
	2126		
28 M		0011	1.6F
	0311	0609	1.5E
	0920	1136	1.0F
	1342	1801	3.3E
	2153		
29 Tu		0051	2.1F
	0406	0659	1.7E
	1014	1216	0.8F
	1409	1834	3.5E
	2224		
30 W		0130	2.5F
	0456	0746	1.7E
	1105	1254	0.7F
	1437	1909	3.7E
	2257		

December

Day	Slack h m	Maximum h m	knots
1 Th ●		0209	2.9F
	0543	0831	1.8E
	1155	1334	0.6F
	1507	1946	3.8E
	2334		
2 F		0250	3.1F
	0629	0916	1.9E
	1245	1416	0.5F
	1541	2026	3.9E
3 Sa	0013	0332	3.2F
	0714	1002	1.9E
	1338	1501	0.4F
	1620	2110	3.8E
4 Su	0057	0417	3.2F
	0759	1051	1.9E
	1438	1552	0.3F
	1704	2157	3.6E
5 M	0143	0505	3.1F
	0845	1144	2.0E
		1651	*
		2250	3.3E
6 Tu	0234	0555	2.9F
	0929	1240	2.2E
		1800	*
		2350	2.8E
7 W	0328	0648	2.6F
	1013	1337	2.4E
	1758	1919	0.4F
	2041		
8 Th ◐		0100	2.3E
	0428	0743	2.3F
	1055	1434	2.7E
	1851	2041	0.7F
	2237		
9 F		0219	1.9E
	0534	0840	1.9F
	1136	1528	3.0E
	1937	2157	1.2F
10 Sa	0028	0340	1.7E
	0646	0937	1.6F
	1214	1619	3.4E
	2021	2303	1.7F
11 Su	0202	0457	1.6E
	0800	1033	1.3F
	1252	1707	3.6E
	2103		
12 M		0000	2.3F
	0318	0606	1.6E
	0912	1125	1.0F
	1328	1752	3.8E
	2144		
13 Tu		0051	2.7F
	0420	0705	1.7E
	1019	1215	0.8F
	1403	1835	3.9E
	2225		
14 W		0137	2.9F
	0512	0758	1.8E
	1120	1302	0.6F
	1437	1916	3.8E
	2304		
15 Th ○		0219	3.0F
	0558	0846	1.8E
	1218	1340	0.5F
	1510	1955	3.7E
	2343		

Day	Slack h m	Maximum h m	knots
16 F		0300	3.0F
	0640	0930	1.8E
	1313	1428	0.3F
	1541	2033	3.6E
17 Sa	0020	0338	2.9F
	0718	1012	1.8E
		1510	*
		2111	3.3E
18 Su	0057	0416	2.8F
	0754	1053	1.8E
		1553	*
		2148	3.0E
19 M	0134	0453	2.6F
	0828	1134	1.9E
		1639	*
		2227	2.7E
20 Tu	0211	0530	2.4F
	0900	1216	1.9E
		1730	*
		2309	2.4E
21 W	0249	0607	2.1F
	0931	1259	2.0E
		1828	*
		2358	2.0E
22 Th	0332	0647	1.8F
	1001	1343	2.2E
		1934	*
23 F ◑		0055	1.6E
	0419	0729	1.6F
	1030	1426	2.3E
	1912	2042	0.3F
	2219		
24 Sa		0203	1.3E
	0516	0814	1.2F
	1058	1509	2.5E
	1939	2147	0.7F
25 Su	0010	0319	1.1E
	0622	0902	1.0F
	1125	1551	2.8E
	2009	2245	1.2F
26 M	0144	0435	1.1E
	0737	0953	0.7F
	1154	1633	3.0E
	2041	2337	1.7F
27 Tu	0258	0542	1.2E
	0851	1044	0.5F
	1226	1715	3.3E
	2116		
28 W		0024	2.2F
	0357	0640	1.3E
	0958	1134	0.4F
	1303	1758	3.6E
	2154		
29 Th		0108	2.7F
	0448	0731	1.5E
	1055	1224	0.4F
	1347	1842	3.8E
	2234		
30 F ●		0152	3.0F
	0534	0818	1.7E
	1145	1313	0.5F
	1436	1927	4.0E
	2315		
31 Sa		0235	3.3F
	0617	0903	1.9E
	1232	1403	0.5F
	1529	2013	4.0E
	2359		

Time meridian 120° W. 0000 is midnight. 1200 is noon. Times are not adjusted for Daylight Saving Time.
* Current weak and variable.

The Narrows (north end), Puget Sound, Washington, 2005

F–Flood, Dir. 135° True E–Ebb, Dir. 335° True

January

Day	Slack h m	Maximum h m	knots
1 Sa	0303	0624	3.8F
	1003	1249	2.1E
	1627	1840	1.6F
	2058		
2 Su		0016	2.5E
	0342	0706	3.4F
	1031	1326	2.2E
	1713	1934	1.8F
	2207		
3 M ◑		0108	2.0E
	0427	0750	3.0F
	1100	1407	2.5E
	1800	2032	2.1F
	2330		
4 Tu		0211	1.6E
	0522	0838	2.6F
	1131	1452	2.7E
	1847	2132	2.6F
5 W	0102	0325	1.2E
	0632	0929	2.1F
	1206	1542	3.1E
	1937	2234	3.1F
6 Th	0231	0451	1.1E
	0750	1024	1.8F
	1246	1635	3.4E
	2027	2335	3.8F
7 F	0346	0620	1.3E
	0905	1121	1.7F
	1332	1731	3.8E
	2119		
8 Sa		0033	4.4F
	0448	0730	1.6E
	1009	1217	1.7F
	1425	1827	4.3E
	2210		
9 Su		0128	5.0F
	0542	0824	1.9E
	1106	1313	1.8F
	1522	1922	4.7E
	2302		
10 M ●		0220	5.5F
	0630	0912	2.2E
	1157	1407	2.1F
	1620	2016	4.9E
	2353		
11 Tu		0309	5.7F
	0714	0957	2.5E
	1246	1500	2.3F
	1719	2108	5.0E
12 W	0043	0357	5.8F
	0755	1041	2.7E
	1335	1553	2.5F
	1819	2200	4.8E
13 Th	0132	0443	5.5F
	0834	1124	3.0E
	1425	1647	2.7F
	1921	2253	4.3E
14 F	0220	0529	5.1F
	0911	1208	3.1E
	1517	1742	2.8F
	2027	2346	3.7E
15 Sa	0309	0613	4.5F
	0946	1253	3.2E
	1610	1839	2.8F
	2137		
16 Su ◐		0044	2.9E
	0359	0659	3.9F
	1022	1340	3.3E
	1705	1939	2.8F
	2256		
17 M		0148	2.1E
	0453	0746	3.1F
	1058	1430	3.3E
	1801	2042	2.9F
18 Tu	0023	0310	1.5E
	0553	0836	2.5F
	1135	1524	3.2E
	1857	2149	2.9F
19 W	0156	0447	1.2E
	0701	0929	1.9F
	1216	1623	3.2E
	1953	2256	3.1F
20 Th	0321	0608	1.2E
	0813	1027	1.5F
	1301	1722	3.2E
	2046	2359	3.3F
21 F	0428	0711	1.4E
	0922	1125	1.3F
	1351	1817	3.2E
	2136		
22 Sa		0054	3.6F
	0519	0804	1.6E
	1021	1221	1.3F
	1443	1905	3.3E
	2222		
23 Su		0140	3.8F
	0559	0849	1.8E
	1111	1312	1.4F
	1533	1946	3.4E
	2306		
24 M		0220	4.0F
	0634	0928	1.9E
	1154	1357	1.5F
	1620	2022	3.5E
	2346		
25 Tu ○		0257	4.2F
	0705	1001	2.0E
	1234	1440	1.7F
	1703	2055	3.6E
26 W	0023	0331	4.3F
	0732	1029	2.1E
	1311	1521	1.9F
	1746	2127	3.5E
27 Th	0058	0405	4.3F
	0758	1050	2.2E
	1348	1602	2.0F
	1829	2200	3.4E
28 F	0132	0439	4.2F
	0822	1110	2.3E
	1423	1644	2.2F
	1913	2235	3.2E
29 Sa	0205	0514	4.0F
	0845	1133	2.5E
	1459	1727	2.4F
	2002	2314	2.9E
30 Su	0239	0550	3.7F
	0908	1201	2.7E
	1535	1813	2.5F
	2056	2356	2.5E
31 M	0315	0629	3.3F
	0931	1234	2.8E
	1615	1902	2.7F
	2159		

February

Day	Slack h m	Maximum h m	knots
1 Tu ◑		0045	2.1E
	0356	0710	2.8F
	0957	1313	3.0E
	1700	1956	2.9F
	2314		
2 W		0142	1.6E
	0448	0757	2.2F
	1028	1358	3.1E
	1752	2056	3.1F
3 Th	0042	0253	1.1E
	0559	0850	1.8F
	1106	1452	3.2E
	1852	2201	3.4F
4 F	0215	0426	0.9E
	0726	0950	1.4F
	1155	1555	3.4E
	1955	2308	3.8F
5 Sa	0335	0616	1.1E
	0847	1054	1.4F
	1255	1704	3.7E
	2057		
6 Su		0013	4.3F
	0437	0725	1.5E
	0954	1158	1.5F
	1405	1813	4.1E
	2156		
7 M		0112	4.8F
	0526	0815	2.0E
	1050	1259	1.9F
	1517	1916	4.4E
	2251		
8 Tu ●		0205	5.2F
	0608	0858	2.4E
	1139	1356	2.4F
	1625	2013	4.7E
	2342		
9 W		0253	5.4F
	0646	0938	2.9E
	1226	1451	2.8F
	1728	2106	4.7E
10 Th	0031	0337	5.3F
	0720	1016	3.2E
	1311	1543	3.2F
	1830	2157	4.5E
11 F	0118	0420	5.1F
	0753	1053	3.5E
	1357	1633	3.5F
	1930	2246	4.0E
12 Sa	0204	0501	4.6F
	0825	1130	3.7E
	1443	1724	3.6F
	2030	2335	3.4E
13 Su	0249	0542	4.0F
	0855	1207	3.7E
	1531	1815	3.6F
	2133		
14 M		0026	2.7E
	0336	0624	3.3F
	0927	1247	3.5E
	1620	1907	3.4F
	2241		
15 Tu ◐		0123	1.9E
	0426	0707	2.6F
	0959	1330	3.3E
	1713	2003	3.2F
	2358		
16 W		0236	1.3E
	0525	0755	2.0F
	1034	1420	3.0E
	1810	2105	3.0F
17 Th	0125	0419	1.0E
	0634	0849	1.4F
	1116	1521	2.7E
	1910	2212	2.9F
18 F	0254	0546	1.0E
	0750	0950	1.1F
	1210	1637	2.6E
	2010	2322	3.0F
19 Sa	0403	0650	1.3E
	0901	1055	1.0F
	1315	1751	2.7E
	2107		
20 Su		0025	3.2F
	0452	0741	1.5E
	0959	1158	1.1F
	1423	1849	2.9E
	2158		
21 M		0114	3.5F
	0528	0823	1.8E
	1047	1253	1.4F
	1524	1934	3.1E
	2244		
22 Tu		0155	3.7F
	0558	0859	2.0E
	1127	1340	1.7F
	1616	2010	3.3E
	2324		
23 W ○		0230	3.9F
	0624	0927	2.3E
	1204	1423	2.1F
	1704	2043	3.4E
24 Th	0002	0303	4.0F
	0648	0949	2.4E
	1238	1503	2.5F
	1748	2115	3.4E
25 F	0037	0335	4.1F
	0709	1005	2.6E
	1311	1543	2.8F
	1832	2147	3.3E
26 Sa	0111	0408	3.9F
	0730	1024	2.9E
	1342	1622	3.1F
	1917	2222	3.1E
27 Su	0146	0442	3.7F
	0750	1048	3.1E
	1414	1703	3.3F
	2004	2300	2.9E
28 M	0221	0517	3.3F
	0811	1117	3.3E
	1447	1747	3.5F
	2056	2341	2.5E

March

Day	Slack h m	Maximum h m	knots
1 Tu	0259	0555	2.8F
	0835	1151	3.4E
	1526	1834	3.6F
	2155		
2 W		0029	2.0E
	0342	0637	2.3F
	0903	1230	3.4E
	1613	1927	3.6F
	2305		
3 Th ◑		0126	1.5E
	0437	0725	1.8F
	0938	1318	3.4E
	1710	2028	3.6F
4 F	0028	0238	1.1E
	0552	0821	1.4F
	1023	1418	3.3E
	1818	2135	3.6F
5 Sa	0158	0424	1.0E
	0720	0927	1.2F
	1124	1530	3.3E
	1930	2245	3.8F
6 Su	0314	0610	1.3E
	0838	1038	1.3F
	1243	1651	3.4E
	2039	2353	4.1F
7 M	0410	0708	1.8E
	0940	1147	1.7F
	1410	1810	3.7E
	2141		
8 Tu		0053	4.4F
	0454	0753	2.3E
	1032	1251	2.2F
	1529	1916	4.0E
	2237		
9 W		0144	4.7F
	0531	0832	2.9E
	1118	1349	2.9F
	1638	2013	4.2E
	2328		
10 Th ●		0229	4.7F
	0604	0908	3.4E
	1202	1441	3.5F
	1741	2104	4.2E
11 F	0016	0311	4.6F
	0635	0942	3.7E
	1244	1529	4.0F
	1838	2152	3.9E
12 Sa	0102	0351	4.3F
	0705	1015	3.9E
	1326	1616	4.2F
	1934	2237	3.5E
13 Su	0146	0430	3.8F
	0734	1048	4.0E
	1408	1701	4.3F
	2028	2323	3.0E
14 M	0230	0509	3.3F
	0802	1122	3.8E
	1451	1747	4.1F
	2123		
15 Tu		0009	2.4E
	0316	0549	2.7F
	0832	1158	3.6E
	1536	1835	3.8F
	2222		
16 W		0100	1.8E
	0406	0632	2.1F
	0903	1238	3.2E
	1625	1926	3.5F
	2328		
17 Th ◐		0205	1.3E
	0505	0720	1.5F
	0939	1324	2.8E
	1720	2023	3.1F
18 F	0045	0348	1.0E
	0614	0815	1.1F
	1024	1423	2.4E
	1823	2127	2.8F
19 Sa	0207	0517	1.0E
	0729	0918	0.9F
	1126	1542	2.2E
	1929	2236	2.8F
20 Su	0314	0619	1.3E
	0836	1027	0.9F
	1247	1715	2.3E
	2031	2341	2.9F
21 M	0401	0707	1.6E
	0931	1134	1.2F
	1408	1822	2.5E
	2125		
22 Tu		0033	3.1F
	0435	0746	2.0E
	1015	1230	1.6F
	1516	1910	2.8E
	2213		
23 W		0115	3.4F
	0503	0816	2.3E
	1054	1318	2.1F
	1612	1950	3.0E
	2255		
24 Th		0151	3.5F
	0527	0839	2.6E
	1128	1401	2.7F
	1702	2025	3.1E
	2335		
25 F ○		0226	3.6F
	0550	0857	2.9E
	1200	1441	3.2F
	1748	2059	3.1E
26 Sa	0013	0259	3.5F
	0611	0916	3.2E
	1231	1520	3.6F
	1833	2133	3.1E
27 Su	0051	0334	3.3F
	0631	0939	3.4E
	1302	1559	4.0F
	1918	2209	2.9E
28 M	0129	0409	3.1F
	0653	1007	3.7E
	1334	1640	4.3F
	2006	2249	2.7E
29 Tu	0208	0447	2.7F
	0717	1039	3.8E
	1410	1724	4.4F
	2057	2332	2.3E
30 W	0251	0527	2.3F
	0746	1116	3.8E
	1452	1813	4.3F
	2155		
31 Th		0022	1.9E
	0341	0612	1.9F
	0821	1200	3.7E
	1542	1906	4.2F
	2302		

Time meridian 120° W. 0000 is midnight. 1200 is noon. Times are not adjusted for Daylight Saving Time.
NOTE–These predictions are for midstream. On the west side the current floods most of the time while on the east side it ebbs most of the time.

The Narrows (north end), Puget Sound, Washington, 2005

F–Flood, Dir. 135° True E–Ebb, Dir. 335° True

April

Day	Slack h m	Maximum h m	knots
1 F ◑		0121	1.5E
	0443	0704	1.5F
	0904	1253	3.5E
	1643	2007	4.0F
2 Sa	0017	0240	1.2E
	0600	0805	1.2F
	1000	1358	3.2E
	1754	2114	3.8F
3 Su	0133	0429	1.3E
	0718	0916	1.2F
	1118	1518	3.0E
	1909	2223	3.8F
4 M	0237	0547	1.7E
	0826	1030	1.5F
	1255	1649	3.0E
	2019	2328	3.9F
5 Tu	0326	0639	2.3E
	0922	1142	2.1F
	1428	1811	3.2E
	2122		
6 W		0025	3.9F
	0406	0722	2.9E
	1010	1245	2.8F
	1546	1916	3.4E
	2219		
7 Th		0115	4.0F
	0441	0759	3.4E
	1055	1339	3.5F
	1653	2011	3.5E
	2310		
8 F ●		0159	3.9F
	0513	0833	3.8E
	1136	1428	4.1F
	1751	2101	3.5E
	2359		
9 Sa		0240	3.6F
	0544	0906	4.1E
	1216	1513	4.5F
	1844	2146	3.3E
10 Su	0045	0319	3.3F
	0613	0937	4.2E
	1256	1556	4.7F
	1934	2230	2.9E
11 M	0130	0358	2.9F
	0642	1009	4.1E
	1335	1638	4.7F
	2023	2312	2.6E
12 Tu	0215	0437	2.5F
	0710	1041	3.8E
	1414	1720	4.4F
	2112	2356	2.1E
13 W	0301	0518	2.0F
	0741	1116	3.5E
	1456	1805	4.1F
	2204		
14 Th		0043	1.7E
	0352	0601	1.6F
	0814	1155	3.1E
	1542	1853	3.7F
	2301		
15 F		0143	1.3E
	0450	0650	1.2F
	0853	1241	2.7E
	1635	1946	3.3F
16 Sa ◐	0004	0313	1.1E
	0556	0746	1.0F
	0943	1337	2.3E
	1735	2045	3.0F
17 Su	0110	0436	1.2E
	0705	0849	0.9F
	1052	1449	2.0E
	1840	2147	2.8F
18 M	0206	0535	1.5E
	0806	0958	1.0F
	1222	1617	1.9E
	1943	2247	2.8F
19 Tu	0250	0619	1.8E
	0856	1104	1.4F
	1351	1737	2.1E
	2041	2339	2.9F
20 W	0325	0653	2.2E
	0937	1201	1.9F
	1504	1835	2.3E
	2133		
21 Th		0024	3.0F
	0353	0719	2.5E
	1014	1250	2.6F
	1604	1921	2.4E
	2220		
22 F		0105	3.0F
	0419	0739	2.9E
	1048	1334	3.2F
	1656	2002	2.6E
	2305		
23 Sa		0143	3.0F
	0443	0801	3.3E
	1120	1415	3.9F
	1744	2040	2.7E
	2348		
24 Su ○		0221	2.8F
	0507	0827	3.7E
	1152	1456	4.4F
	1831	2119	2.7E
25 M	0031	0259	2.7F
	0532	0857	4.0E
	1226	1537	4.8F
	1917	2159	2.6F
26 Tu	0114	0338	2.4F
	0600	0932	4.2E
	1303	1620	5.0F
	2006	2242	2.4E
27 W	0159	0419	2.2F
	0632	1010	4.2E
	1344	1706	5.1F
	2058	2329	2.2E
28 Th	0248	0504	1.9F
	0709	1053	4.2E
	1432	1756	4.9F
	2155		
29 F		0023	1.9E
	0343	0554	1.6F
	0754	1143	3.9E
	1526	1851	4.6F
	2256		
30 Sa ◑		0127	1.7E
	0448	0651	1.4F
	0849	1241	3.5E
	1627	1950	4.3F
	2358		

May

Day	Slack h m	Maximum h m	knots
1 Su		0246	1.7E
	0558	0757	1.3F
	1000	1350	3.0E
	1735	2052	4.0F
2 M	0057	0408	2.0E
	0706	0909	1.5F
	1134	1514	2.7E
	1846	2156	3.7F
3 Tu	0149	0511	2.4E
	0806	1024	1.9F
	1316	1647	2.6E
	1955	2256	3.5F
4 W	0233	0601	2.9E
	0859	1134	2.6F
	1447	1808	2.6E
	2058	2351	3.4F
5 Th	0311	0644	3.4E
	0946	1235	3.3F
	1601	1913	2.7E
	2157		
6 F		0040	3.2F
	0347	0722	3.8E
	1029	1327	4.0F
	1703	2008	2.8E
	2251		
7 Sa		0125	2.9F
	0420	0757	4.1E
	1110	1413	4.5F
	1758	2058	2.7E
	2341		
8 Su ●		0207	2.7F
	0452	0830	4.2E
	1150	1456	4.8F
	1847	2143	2.6E
9 M	0029	0248	2.4F
	0523	0902	4.1E
	1228	1536	4.8F
	1932	2226	2.4E
10 Tu	0115	0328	2.1F
	0553	0934	4.0E
	1306	1616	4.7F
	2016	2307	2.2E
11 W	0201	0408	1.8F
	0625	1008	3.7E
	1345	1657	4.5F
	2101	2348	1.9E
12 Th	0248	0450	1.6F
	0658	1044	3.4E
	1425	1740	4.2F
	2146		
13 F		0032	1.7E
	0338	0535	1.3F
	0736	1123	3.1E
	1509	1825	3.9F
	2234		
14 Sa		0124	1.5E
	0432	0624	1.1F
	0820	1208	2.7E
	1556	1914	3.5F
	2324		
15 Su		0227	1.4E
	0532	0719	1.0F
	0915	1301	2.3E
	1649	2005	3.2F
16 M ◐	0012	0333	1.5E
	0632	0820	1.0F
	1027	1404	2.0E
	1747	2059	3.0F
17 Tu	0055	0427	1.7E
	0726	0925	1.2F
	1157	1517	1.7E
	1847	2152	2.8F
18 W	0133	0507	2.0E
	0812	1028	1.7F
	1328	1636	1.7E
	1948	2243	2.7F
19 Th	0207	0538	2.4E
	0853	1125	2.3F
	1446	1748	1.7E
	2046	2330	2.6F
20 F	0237	0606	2.8E
	0930	1216	3.0F
	1551	1847	1.9E
	2141		
21 Sa		0016	2.5F
	0306	0637	3.3E
	1006	1303	3.7F
	1647	1937	2.1E
	2234		
22 Su		0100	2.3F
	0335	0709	3.7E
	1042	1348	4.4F
	1738	2023	2.2E
	2324		
23 M ○		0143	2.2F
	0406	0745	4.1E
	1119	1432	4.9F
	1827	2108	2.3E
24 Tu	0012	0226	2.1F
	0441	0824	4.4E
	1159	1517	5.3F
	1915	2152	2.3E
25 W	0100	0311	2.0F
	0519	0906	4.5E
	1243	1604	5.5F
	2006	2239	2.3E
26 Th	0148	0357	1.9F
	0602	0951	4.6E
	1330	1652	5.5F
	2055	2329	2.2E
27 F	0240	0447	1.8F
	0650	1040	4.4E
	1420	1743	5.2F
	2146		
28 Sa		0024	2.1E
	0336	0542	1.7F
	0746	1134	4.0E
	1515	1836	4.9F
	2237		
29 Su		0124	2.2E
	0437	0642	1.7F
	0853	1235	3.5E
	1613	1930	4.5F
	2326		
30 M ◑		0228	2.3E
	0540	0748	1.8F
	1016	1345	2.9E
	1715	2026	4.0F
31 Tu	0012	0331	2.7E
	0642	0859	2.0F
	1152	1507	2.4E
	1820	2123	3.5F

June

Day	Slack h m	Maximum h m	knots
1 W	0056	0428	3.0E
	0738	1012	2.5F
	1330	1639	2.1E
	1927	2219	3.1F
2 Th	0136	0519	3.4E
	0830	1119	3.1F
	1457	1800	2.0E
	2032	2313	2.7F
3 F	0215	0605	3.7E
	0918	1219	3.7F
	1609	1907	2.1E
	2134		
4 Sa		0004	2.4F
	0253	0647	3.9E
	1003	1311	4.1F
	1710	2003	2.1E
	2232		
5 Su		0052	2.1F
	0330	0725	4.0E
	1046	1356	4.5F
	1801	2054	2.2E
	2325		
6 M ●		0137	1.9F
	0406	0801	4.0E
	1126	1438	4.6F
	1846	2140	2.2E
7 Tu	0015	0220	1.8F
	0441	0836	3.9E
	1206	1518	4.7F
	1928	2223	2.1E
8 W	0101	0302	1.6F
	0517	0910	3.8E
	1245	1557	4.6F
	2008	2302	2.0E
9 Th	0146	0344	1.5F
	0553	0945	3.6E
	1323	1637	4.5F
	2047	2340	1.8E
10 F	0231	0427	1.4F
	0631	1021	3.4E
	1402	1717	4.3F
	2125		
11 Sa		0017	1.7E
	0317	0512	1.3F
	0713	1100	3.1E
	1442	1759	4.0F
	2202		
12 Su		0054	1.7E
	0406	0600	1.3F
	0801	1143	2.8E
	1523	1842	3.8F
	2239		
13 M		0132	1.7E
	0456	0653	1.3F
	0859	1231	2.4E
	1607	1927	3.4F
	2313		
14 Tu ◐		0213	1.9E
	0547	0749	1.4F
	1008	1325	2.0E
	1655	2013	3.1F
	2346		
15 W		0254	2.1E
	0635	0848	1.6F
	1132	1428	1.6E
	1750	2101	2.8F
16 Th	0018	0335	2.4E
	0720	0947	2.1F
	1301	1540	1.4E
	1852	2150	2.5F
17 F	0049	0417	2.7E
	0803	1045	2.6F
	1424	1657	1.3E
	1958	2240	2.2F
18 Sa	0122	0459	3.1E
	0844	1140	3.3F
	1535	1812	1.4E
	2103	2331	2.0F
19 Su	0158	0542	3.5E
	0926	1233	4.0F
	1636	1915	1.6E
	2204		
20 M		0020	1.8F
	0236	0627	3.9E
	1010	1323	4.6F
	1730	2010	1.8E
	2300		
21 Tu ○		0110	1.8F
	0319	0713	4.3E
	1055	1412	5.1F
	1820	2059	2.0E
	2352		
22 W		0200	1.8F
	0406	0801	4.6E
	1142	1501	5.5F
	1908	2146	2.2E
23 Th	0042	0250	1.9F
	0456	0850	4.8E
	1230	1549	5.6F
	1955	2233	2.3E
24 F	0132	0341	2.0F
	0550	0941	4.7E
	1319	1638	5.6F
	2039	2321	2.4E
25 Sa	0223	0435	2.1F
	0649	1034	4.5E
	1410	1726	5.4F
	2122		
26 Su		0009	2.6E
	0316	0531	2.2F
	0753	1129	4.0E
	1502	1815	4.9F
	2203		
27 M		0100	2.8E
	0413	0631	2.3F
	0906	1229	3.4E
	1556	1905	4.4F
	2243		
28 Tu ◑		0152	3.0E
	0510	0734	2.5F
	1028	1336	2.7E
	1652	1955	3.8F
	2322		
29 W		0245	3.2E
	0608	0841	2.7F
	1159	1455	2.0E
	1753	2047	3.2F
30 Th	0001	0340	3.4E
	0704	0949	3.0F
	1332	1626	1.7E
	1859	2140	2.6F

Time meridian 120° W. 0000 is midnight. 1200 is noon. Times are not adjusted for Daylight Saving Time.
NOTE–These predictions are for midstream. On the west side the current floods most of the time while on the east side it ebbs most of the time.

The Narrows (north end), Puget Sound, Washington, 2005

F–Flood, Dir. 135° True E–Ebb, Dir. 335° True

July

Day	Slack h m	Maximum h m	knots
1 F	0041	0434	3.6E
	0758	1056	3.3F
	1458	1749	1.6E
	2007	2235	2.1F
2 Sa	0122	0527	3.7E
	0850	1158	3.7F
	1611	1858	1.6E
	2113	2330	1.8F
3 Su	0205	0616	3.7E
	0938	1252	4.0F
	1709	1956	1.8E
	2215		
4 M		0023	1.6F
	0249	0702	3.7E
	1024	1340	4.2F
	1758	2046	1.9E
	2309		
5 Tu		0112	1.5F
	0333	0743	3.7E
	1107	1423	4.3F
	1839	2131	1.9E
	2358		
6 W ●		0158	1.5F
	0416	0821	3.7E
	1148	1502	4.4F
	1916	2211	2.0E
7 Th	0042	0242	1.5F
	0458	0856	3.6E
	1228	1540	4.4F
	1950	2247	2.0E
8 F	0124	0325	1.6F
	0539	0930	3.5E
	1305	1616	4.4F
	2022	2318	1.9E
9 Sa	0205	0407	1.6F
	0621	1005	3.4E
	1342	1653	4.3F
	2052	2344	2.0E
10 Su	0246	0451	1.7F
	0706	1042	3.2E
	1418	1730	4.1F
	2120		
11 M		0009	2.1E
	0327	0536	1.7F
	0755	1121	2.9E
	1453	1808	3.8F
	2147		
12 Tu		0036	2.2E
	0409	0624	1.8F
	0850	1205	2.5E
	1531	1848	3.4F
	2213		
13 W		0109	2.3E
	0452	0715	2.0F
	0954	1253	2.0E
	1612	1930	3.0F
	2239		
14 Th ◐		0146	2.5E
	0536	0809	2.2F
	1109	1350	1.6E
	1701	2015	2.6F
	2307		
15 F		0227	2.7E
	0622	0907	2.5F
	1233	1456	1.2E
	1803	2104	2.1F
	2340		
16 Sa		0313	3.0E
	0710	1006	3.0F
	1400	1615	1.0E
	1918	2157	1.8F
17 Su	0017	0405	3.2E
	0801	1106	3.5F
	1519	1744	1.1E
	2033	2253	1.6F
18 M	0102	0500	3.6E
	0853	1205	4.1F
	1624	1902	1.3E
	2141	2349	1.5F
19 Tu	0154	0557	4.0E
	0945	1301	4.6F
	1719	2000	1.6E
	2239		
20 W		0046	1.7F
	0252	0654	4.3E
	1037	1354	5.1F
	1807	2049	2.0E
	2332		
21 Th ○		0141	1.9F
	0352	0749	4.7E
	1129	1445	5.4F
	1850	2133	2.3E
22 F	0020	0235	2.2F
	0453	0843	4.8E
	1219	1533	5.6F
	1930	2215	2.6E
23 Sa	0108	0328	2.5F
	0555	0936	4.7E
	1308	1619	5.5F
	2008	2256	2.9E
24 Su	0157	0422	2.8F
	0658	1029	4.4E
	1357	1704	5.2F
	2043	2339	3.2E
25 M	0247	0517	3.0F
	0804	1123	3.9E
	1446	1749	4.7F
	2118		
26 Tu		0022	3.4E
	0339	0613	3.2F
	0913	1219	3.2E
	1536	1834	4.0F
	2153		
27 W ◑		0107	3.5E
	0433	0712	3.2F
	1029	1322	2.4E
	1629	1921	3.3F
	2229		
28 Th		0156	3.5E
	0528	0813	3.2F
	1152	1437	1.8E
	1728	2010	2.6F
	2306		
29 F		0249	3.4E
	0626	0918	3.2F
	1321	1610	1.4E
	1835	2104	2.0F
	2348		
30 Sa		0348	3.3E
	0723	1026	3.3F
	1448	1737	1.3E
	1946	2201	1.6F
31 Su	0035	0451	3.3E
	0820	1132	3.4F
	1601	1845	1.4E
	2057	2301	1.3F

August

Day	Slack h m	Maximum h m	knots
1 M	0128	0553	3.3E
	0913	1232	3.6F
	1657	1942	1.6E
	2159		
2 Tu		0000	1.3F
	0224	0648	3.3E
	1003	1322	3.8F
	1741	2029	1.8E
	2251		
3 W		0054	1.4F
	0318	0734	3.4E
	1049	1405	4.0F
	1817	2111	2.0E
	2336		
4 Th ●		0142	1.5F
	0408	0813	3.5E
	1131	1442	4.1F
	1849	2146	2.1E
5 F	0017	0226	1.7F
	0454	0847	3.5E
	1209	1517	4.2F
	1916	2216	2.2E
6 Sa	0055	0307	1.9F
	0538	0919	3.5E
	1246	1550	4.2F
	1941	2238	2.2E
7 Su	0131	0348	2.1F
	0621	0951	3.3E
	1320	1623	4.1F
	2005	2257	2.4E
8 M	0206	0429	2.3F
	0706	1025	3.1E
	1354	1657	3.9F
	2027	2318	2.5E
9 Tu	0241	0511	2.5F
	0753	1102	2.8E
	1428	1733	3.6F
	2048	2344	2.7E
10 W	0316	0555	2.6F
	0844	1142	2.5E
	1503	1810	3.2F
	2111		
11 Th		0014	2.8E
	0353	0642	2.7F
	0943	1227	2.0E
	1542	1850	2.7F
	2135		
12 F ◐		0050	2.9E
	0435	0733	2.9F
	1051	1320	1.6E
	1630	1935	2.2F
	2204		
13 Sa		0132	3.0E
	0524	0830	3.0F
	1211	1425	1.2E
	1733	2025	1.7F
	2240		
14 Su		0223	3.1E
	0622	0932	3.2F
	1339	1547	0.9E
	1855	2122	1.4F
	2326		
15 M		0323	3.2E
	0724	1037	3.6F
	1502	1734	1.0E
	2017	2225	1.3F
16 Tu	0025	0431	3.4E
	0827	1142	4.0F
	1607	1853	1.3E
	2125	2329	1.4F
17 W	0134	0540	3.8E
	0927	1242	4.5F
	1657	1945	1.8E
	2221		
18 Th		0031	1.8F
	0247	0645	4.2E
	1023	1336	4.9F
	1739	2028	2.2E
	2310		
19 F ○		0129	2.3F
	0357	0744	4.5E
	1116	1425	5.1F
	1816	2107	2.7E
	2357		
20 Sa		0224	2.9F
	0503	0839	4.6E
	1206	1510	5.2F
	1851	2145	3.2E
21 Su	0042	0316	3.4F
	0606	0931	4.5E
	1254	1554	5.0F
	1924	2222	3.5E
22 M	0128	0408	3.7F
	0707	1022	4.1E
	1341	1636	4.6F
	1955	2300	3.8E
23 Tu	0214	0459	3.9F
	0809	1113	3.6E
	1428	1718	4.0F
	2027	2339	3.9E
24 W	0302	0551	4.0F
	0913	1206	2.9E
	1516	1801	3.4F
	2059		
25 Th		0020	3.8E
	0352	0644	3.8F
	1020	1305	2.2E
	1608	1846	2.7F
	2134		
26 F ◑		0105	3.5E
	0445	0741	3.6F
	1135	1417	1.6E
	1708	1935	2.0F
	2212		
27 Sa		0156	3.2E
	0543	0842	3.3F
	1259	1553	1.2E
	1817	2030	1.5F
	2256		
28 Su		0258	2.9E
	0645	0949	3.1F
	1424	1720	1.2E
	1931	2131	1.2F
	2353		
29 M		0416	2.7E
	0747	1100	3.1F
	1535	1826	1.4E
	2041	2238	1.1F
30 Tu	0102	0534	2.8E
	0846	1204	3.2F
	1626	1919	1.7E
	2140	2342	1.2F
31 W	0212	0635	2.9E
	0939	1256	3.4F
	1705	2002	1.9E
	2228		

September

Day	Slack h m	Maximum h m	knots
1 Th		0038	1.5F
	0314	0722	3.1E
	1026	1337	3.6F
	1736	2039	2.2E
	2309		
2 F		0126	1.8F
	0408	0801	3.2E
	1108	1413	3.8F
	1803	2109	2.3E
	2346		
3 Sa ●		0208	2.2F
	0455	0834	3.3E
	1146	1445	3.9F
	1826	2131	2.5E
4 Su	0020	0248	2.6F
	0540	0905	3.3E
	1222	1517	3.8F
	1848	2148	2.7E
5 M	0052	0327	2.9F
	0623	0936	3.2E
	1257	1549	3.7F
	1908	2206	2.9E
6 Tu	0123	0405	3.2F
	0706	1009	3.0E
	1331	1623	3.4F
	1928	2229	3.0E
7 W	0154	0445	3.4F
	0751	1044	2.7E
	1406	1657	3.1F
	1948	2256	3.2E
8 Th	0226	0527	3.5F
	0840	1124	2.4E
	1443	1735	2.6F
	2011	2327	3.3E
9 F	0302	0612	3.6F
	0935	1208	2.0E
	1525	1815	2.2F
	2038		
10 Sa		0005	3.3E
	0345	0702	3.5F
	1039	1301	1.5E
	1617	1901	1.7F
	2111		
11 Su ◐		0050	3.2E
	0438	0800	3.5F
	1156	1407	1.1E
	1727	1955	1.3F
	2154		
12 M		0146	3.1E
	0543	0904	3.5F
	1321	1539	0.9E
	1853	2059	1.1F
	2252		
13 Tu		0255	3.1E
	0655	1013	3.6F
	1438	1732	1.2E
	2009	2208	1.2F
14 W	0009	0414	3.2E
	0806	1120	3.9F
	1536	1835	1.7E
	2111	2318	1.6F
15 Th	0137	0534	3.5E
	0909	1220	4.2F
	1620	1920	2.2E
	2202		
16 F		0022	2.2F
	0259	0643	3.8E
	1007	1312	4.5F
	1657	1959	2.8E
	2248		
17 Sa ○		0120	2.9F
	0411	0743	4.0E
	1100	1359	4.6F
	1731	2035	3.4E
	2332		
18 Su		0213	3.6F
	0516	0836	4.1E
	1150	1443	4.5F
	1803	2110	3.8E
19 M	0015	0303	4.2F
	0616	0926	4.0E
	1237	1524	4.2F
	1833	2145	4.1E
20 Tu	0058	0351	4.6F
	0713	1015	3.6E
	1324	1605	3.8F
	1903	2220	4.2E
21 W	0141	0438	4.7F
	0809	1103	3.1E
	1410	1645	3.2F
	1934	2257	4.1E
22 Th	0225	0526	4.5F
	0906	1152	2.5E
	1459	1728	2.6F
	2006	2335	3.8E
23 F	0311	0615	4.2F
	1006	1248	2.0E
	1552	1812	2.1F
	2040		
24 Sa ◑		0018	3.4E
	0402	0707	3.8F
	1112	1357	1.5E
	1652	1902	1.5F
	2119		
25 Su		0107	2.9E
	0459	0804	3.3F
	1227	1533	1.2E
	1802	1958	1.1F
	2208		
26 M		0208	2.5E
	0602	0909	3.0F
	1344	1656	1.3E
	1916	2103	0.9F
	2315		
27 Tu		0332	2.3E
	0708	1018	2.9F
	1449	1758	1.5E
	2022	2214	1.0F
28 W	0040	0506	2.3E
	0811	1122	3.0F
	1536	1846	1.8E
	2115	2322	1.3F
29 Th	0202	0611	2.5E
	0906	1214	3.1F
	1611	1925	2.1E
	2158		
30 F		0018	1.7F
	0309	0700	2.7E
	0955	1256	3.3F
	1639	1957	2.4E
	2236		

Time meridian 120° W. 0000 is midnight. 1200 is noon. Times are not adjusted for Daylight Saving Time.
NOTE–These predictions are for midstream. On the west side the current floods most of the time while on the east side it ebbs most of the time.

The Narrows (north end), Puget Sound, Washington, 2005

F–Flood, Dir. 135° True E–Ebb, Dir. 335° True

October

Day	Slack h m	Maximum h m	knots
1 Sa		0105	2.2F
	0405	0741	2.9E
	1038	1332	3.4F
	1704	2021	2.7E
	2310		
2 Su		0146	2.8F
	0454	0815	2.9E
	1119	1406	3.4F
	1726	2039	3.0E
	2342		
3 M ●		0225	3.3F
	0539	0848	2.9E
	1157	1440	3.3F
	1747	2057	3.2E
4 Tu	0013	0303	3.7F
	0622	0920	2.9E
	1234	1513	3.1F
	1807	2119	3.4E
5 W	0043	0341	4.0F
	0705	0954	2.7E
	1312	1548	2.8F
	1829	2145	3.6E
6 Th	0113	0420	4.2F
	0749	1031	2.5E
	1351	1625	2.5F
	1852	2215	3.7E
7 F	0146	0502	4.3F
	0837	1111	2.2E
	1432	1704	2.1F
	1919	2250	3.7E
8 Sa	0225	0548	4.3F
	0931	1157	1.9E
	1519	1747	1.7F
	1952	2331	3.6E
9 Su	0311	0639	4.1F
	1033	1253	1.5E
	1619	1837	1.4F
	2032		
10 M ◐		0021	3.4E
	0408	0737	3.9F
	1145	1404	1.2E
	1733	1936	1.1F
	2126		
11 Tu		0123	3.1E
	0517	0841	3.8F
	1258	1544	1.3E
	1851	2045	1.1F
	2240		
12 W		0239	2.9E
	0631	0949	3.7F
	1401	1711	1.7E
	1958	2159	1.4F
13 Th	0017	0407	2.8E
	0744	1054	3.8F
	1451	1805	2.2E
	2053	2311	2.0F
14 F	0155	0532	3.0E
	0849	1152	3.9F
	1531	1847	2.8E
	2141		
15 Sa		0015	2.8F
	0318	0643	3.2E
	0948	1243	3.9F
	1607	1925	3.4E
	2226		
16 Su		0111	3.6F
	0427	0742	3.4E
	1042	1329	3.8F
	1640	2000	3.9E
	2308		
17 M ○		0202	4.3F
	0527	0834	3.4E
	1132	1412	3.6F
	1711	2035	4.3E
	2349		
18 Tu		0249	4.8F
	0623	0923	3.3E
	1221	1453	3.3F
	1742	2109	4.4E
19 W	0030	0334	5.1F
	0715	1009	3.0E
	1308	1534	2.9F
	1813	2144	4.4E
20 Th	0111	0417	5.0F
	0806	1055	2.7E
	1355	1615	2.5F
	1845	2219	4.1E
21 F	0153	0501	4.8F
	0857	1142	2.2E
	1444	1657	2.0F
	1918	2257	3.8E
22 Sa	0236	0547	4.4F
	0950	1234	1.8E
	1537	1743	1.6F
	1953	2338	3.3E
23 Su	0324	0636	3.9F
	1047	1339	1.5E
	1637	1832	1.2F
	2034		
24 M ◑		0025	2.8E
	0417	0729	3.5F
	1149	1503	1.3E
	1744	1929	0.9F
	2127		
25 Tu		0122	2.3E
	0516	0827	3.1F
	1252	1619	1.4E
	1852	2034	0.8F
	2240		
26 W		0237	2.0E
	0621	0929	2.9F
	1346	1716	1.6E
	1952	2145	1.0F
27 Th	0015	0411	1.9E
	0724	1028	2.8F
	1428	1801	2.0E
	2041	2252	1.4F
28 F	0145	0531	2.0E
	0822	1120	2.8F
	1503	1836	2.3E
	2122	2349	2.0F
29 Sa	0258	0628	2.1E
	0915	1205	2.9F
	1531	1903	2.7E
	2158		
30 Su		0037	2.6F
	0357	0714	2.3E
	1003	1245	2.9F
	1557	1924	3.0E
	2232		
31 M		0120	3.3F
	0448	0754	2.4E
	1048	1323	2.8F
	1621	1945	3.3E
	2304		

November

Day	Slack h m	Maximum h m	knots
1 Tu ●		0200	3.8F
	0534	0830	2.5E
	1131	1401	2.7F
	1645	2009	3.6E
	2336		
2 W		0239	4.3F
	0618	0906	2.5E
	1213	1438	2.5F
	1709	2038	3.9E
3 Th	0008	0318	4.7F
	0702	0943	2.4E
	1255	1516	2.3F
	1736	2110	4.1E
4 F	0043	0359	4.9F
	0748	1022	2.3E
	1338	1556	2.0F
	1806	2146	4.1E
5 Sa	0121	0443	4.9F
	0836	1105	2.1E
	1424	1639	1.8F
	1841	2226	4.1E
6 Su	0204	0531	4.8F
	0929	1154	1.9E
	1517	1727	1.5F
	1923	2312	3.9E
7 M	0254	0622	4.6F
	1026	1253	1.7E
	1618	1822	1.3F
	2014		
8 Tu ◐		0006	3.5E
	0351	0719	4.3F
	1125	1403	1.6E
	1727	1925	1.3F
	2121		
9 W		0111	3.1E
	0457	0819	4.0F
	1222	1524	1.9E
	1835	2036	1.4F
	2251		
10 Th		0230	2.6E
	0607	0921	3.7F
	1313	1632	2.3E
	1936	2150	1.8F
11 F	0036	0401	2.4E
	0718	1021	3.5F
	1357	1725	2.9E
	2029	2302	2.5F
12 Sa	0214	0530	2.4E
	0825	1117	3.3F
	1437	1810	3.4E
	2117		
13 Su		0005	3.3F
	0333	0642	2.6E
	0927	1209	3.1F
	1513	1850	3.9E
	2202		
14 M		0100	4.1F
	0439	0741	2.7E
	1024	1257	2.9F
	1549	1928	4.2E
	2244		
15 Tu ○		0149	4.7F
	0537	0834	2.7E
	1117	1342	2.7F
	1623	2004	4.4E
	2326		
16 W		0234	5.0F
	0628	0922	2.7E
	1207	1425	2.4F
	1657	2040	4.4E
17 Th	0007	0317	5.1F
	0716	1008	2.5E
	1255	1507	2.2F
	1731	2115	4.3E
18 F	0047	0359	5.0F
	0801	1052	2.3E
	1342	1549	1.9F
	1805	2151	4.0E
19 Sa	0128	0440	4.8F
	0846	1135	2.0E
	1430	1632	1.6F
	1841	2229	3.7E
20 Su	0209	0523	4.5F
	0932	1222	1.8E
	1520	1717	1.4F
	1920	2309	3.3E
21 M	0253	0608	4.1F
	1019	1314	1.6E
	1615	1806	1.2F
	2004	2353	2.8E
22 Tu	0339	0656	3.7F
	1106	1414	1.5E
	1714	1901	1.0F
	2059		
23 W ◑		0044	2.4E
	0430	0746	3.3F
	1152	1516	1.6E
	1814	2002	1.0F
	2211		
24 Th		0146	1.9E
	0526	0838	3.0F
	1234	1610	1.8E
	1908	2107	1.2F
	2344		
25 F		0300	1.6E
	0627	0931	2.8F
	1312	1652	2.1E
	1956	2212	1.6F
26 Sa	0118	0425	1.5E
	0728	1022	2.6F
	1345	1725	2.4E
	2038	2311	2.2F
27 Su	0239	0542	1.5E
	0828	1110	2.4F
	1416	1754	2.8E
	2116		
28 M		0002	2.9F
	0344	0642	1.7E
	0924	1157	2.3F
	1446	1823	3.2E
	2152		
29 Tu		0049	3.6F
	0438	0731	1.9E
	1017	1241	2.2F
	1516	1854	3.6E
	2228		
30 W		0133	4.2F
	0527	0815	2.0E
	1107	1324	2.1F
	1547	1928	3.9E
	2304		

December

Day	Slack h m	Maximum h m	knots
1 Th ●		0215	4.7F
	0613	0855	2.1E
	1154	1406	2.0F
	1620	2005	4.2E
	2342		
2 F		0258	5.1F
	0658	0936	2.2E
	1240	1449	1.9F
	1657	2045	4.4E
3 Sa	0023	0343	5.3F
	0744	1018	2.1E
	1325	1534	1.9F
	1738	2127	4.5E
4 Su	0106	0429	5.4F
	0831	1103	2.1E
	1414	1622	1.8F
	1824	2214	4.4E
5 M	0153	0516	5.2F
	0918	1152	2.1E
	1506	1714	1.7F
	1916	2304	4.1E
6 Tu	0244	0606	5.0F
	1005	1245	2.1E
	1603	1811	1.7F
	2019		
7 W		0000	3.6E
	0338	0658	4.6F
	1051	1343	2.3E
	1705	1914	1.8F
	2135		
8 Th ◐		0105	3.0E
	0437	0752	4.1F
	1136	1444	2.6E
	1806	2023	2.0F
	2308		
9 F		0220	2.4E
	0542	0848	3.6F
	1219	1544	2.9E
	1905	2135	2.5F
10 Sa	0050	0351	2.0E
	0650	0944	3.1F
	1300	1640	3.3E
	1959	2246	3.0F
11 Su	0225	0524	1.9E
	0759	1040	2.7F
	1341	1731	3.7E
	2050	2350	3.7F
12 M	0344	0639	1.9E
	0906	1135	2.4F
	1422	1818	4.0E
	2138		
13 Tu		0046	4.2F
	0449	0741	2.1E
	1008	1227	2.1F
	1503	1901	4.2E
	2223		
14 W		0136	4.6F
	0543	0834	2.2E
	1104	1316	1.9F
	1543	1942	4.2E
	2307		
15 Th ○		0221	4.8F
	0630	0922	2.2E
	1155	1402	1.8F
	1623	2021	4.2E
	2349		
16 F		0303	4.9F
	0713	1006	2.2E
	1243	1446	1.7F
	1703	2058	4.0E
17 Sa	0030	0343	4.8F
	0753	1047	2.1E
	1328	1529	1.6F
	1742	2134	3.8E
18 Su	0109	0422	4.7F
	0831	1125	2.0E
	1412	1612	1.6F
	1821	2210	3.6E
19 M	0148	0502	4.5F
	0908	1202	1.9E
	1457	1656	1.5F
	1903	2248	3.3E
20 Tu	0227	0542	4.2F
	0943	1236	1.9E
	1544	1743	1.4F
	1950	2329	2.9E
21 W	0307	0623	3.9F
	1017	1312	1.9E
	1632	1833	1.4F
	2045		
22 Th		0014	2.5E
	0349	0706	3.5F
	1050	1349	2.0E
	1722	1927	1.5F
	2151		
23 F ◑		0105	2.0E
	0434	0750	3.1F
	1122	1429	2.1E
	1812	2025	1.6F
	2312		
24 Sa		0205	1.6E
	0526	0837	2.7F
	1153	1511	2.3E
	1859	2126	2.0F
25 Su	0043	0317	1.2E
	0628	0927	2.3F
	1225	1555	2.6E
	1945	2225	2.5F
26 M	0210	0440	1.1E
	0737	1018	2.0F
	1259	1640	2.9E
	2028	2323	3.0F
27 Tu	0325	0605	1.2E
	0845	1110	1.8F
	1335	1725	3.2E
	2111		
28 W		0016	3.7F
	0426	0711	1.4E
	0948	1201	1.6F
	1416	1810	3.6E
	2155		
29 Th		0106	4.3F
	0518	0803	1.6E
	1044	1251	1.6F
	1459	1856	4.0E
	2239		
30 F ●		0154	4.8F
	0606	0848	1.8E
	1134	1340	1.7F
	1546	1943	4.4E
	2324		
31 Sa		0241	5.2F
	0650	0929	2.0E
	1221	1429	1.9F
	1636	2030	4.6E

Time meridian 120° W. 0000 is midnight. 1200 is noon. Times are not adjusted for Daylight Saving Time.
NOTE–These predictions are for midstream. On the west side the current floods most of the time while on the east side it ebbs most of the time.

Deception Pass (Narrows), Washington, 2005

F–Flood, Dir. 090° True   E–Ebb, Dir. 270° True

January

Day	Slack (h m)	Maximum (h m)	knots
1 Sa	0028	0410	5.7F
	0734	1031	6.5E
	1415	1644	4.0F
	1920	2231	5.8E
2 Su	0113	0455	5.4F
	0809	1115	6.5E
	1454	1734	4.2F
	2023	2324	5.6E
3 M	0203	0542	5.1F
	0844	1201	6.5E
	1534	1826	4.5F
	2132		
4 Tu		0020	5.4E
	0300	0632	4.7F
	0921	1247	6.6E
	1616	1919	4.9F
	2243		
5 W		0118	5.4E
	0407	0724	4.4F
	1001	1336	6.7E
	1701	2014	5.3F
	2352		
6 Th		0219	5.4E
	0524	0819	4.1F
	1046	1427	6.8E
	1748	2109	5.7F
7 F	0055	0320	5.6E
	0642	0916	3.9F
	1135	1519	7.0E
	1837	2204	6.2F
8 Sa	0153	0420	5.9E
	0752	1013	3.9F
	1230	1612	7.2E
	1928	2257	6.6F
9 Su	0246	0517	6.3E
	0854	1109	4.1F
	1327	1705	7.4E
	2019	2348	6.9F
10 M	0336	0611	6.7E
	0948	1203	4.3F
	1426	1758	7.6E
	2110		
11 Tu		0038	7.1F
	0422	0701	7.0E
	1038	1257	4.6F
	1526	1850	7.6E
	2200		
12 W		0127	7.2F
	0507	0750	7.3E
	1125	1349	4.9F
	1625	1942	7.6E
	2251		
13 Th		0215	7.2F
	0550	0837	7.5E
	1212	1440	5.2F
	1724	2034	7.5E
	2341		
14 F		0303	7.0F
	0633	0923	7.5E
	1258	1532	5.3F
	1824	2126	7.2E
15 Sa	0032	0350	6.6F
	0714	1009	7.5E
	1344	1624	5.4F
	1926	2219	6.9E
16 Su	0125	0439	6.1F
	0755	1056	7.4E
	1433	1717	5.4F
	2032	2314	6.4E
17 M	0221	0528	5.6F
	0837	1144	7.2E
	1524	1813	5.4F
	2142		
18 Tu		0012	5.9E
	0324	0621	4.9F
	0921	1234	6.9E
	1617	1912	5.3F
	2256		
19 W		0114	5.4E
	0439	0717	4.2F
	1008	1327	6.7E
	1713	2014	5.3F
20 Th	0008	0223	5.1E
	0600	0817	3.7F
	1101	1423	6.5E
	1809	2117	5.4F
21 F	0114	0336	5.1E
	0717	0921	3.3F
	1200	1519	6.4E
	1903	2216	5.6F
22 Sa	0211	0448	5.4E
	0822	1022	3.3F
	1301	1615	6.4E
	1953	2308	5.9F
23 Su	0258	0542	5.7E
	0915	1118	3.5F
	1357	1706	6.5E
	2038	2353	6.1F
24 M	0340	0622	6.1E
	0959	1206	3.8F
	1447	1753	6.7E
	2119		
25 Tu		0033	6.3F
	0417	0656	6.4E
	1037	1248	4.2F
	1532	1836	6.8E
	2156		
26 W		0111	6.4F
	0451	0730	6.7E
	1111	1328	4.5F
	1613	1917	6.9E
	2230		
27 Th		0147	6.4F
	0522	0804	6.9E
	1142	1406	4.7F
	1651	1957	7.0E
	2303		
28 F		0223	6.4F
	0551	0838	7.0E
	1211	1444	4.9F
	1729	2036	6.9E
	2336		
29 Sa		0259	6.2F
	0618	0913	7.1E
	1238	1522	5.0F
	1808	2116	6.8E
30 Su	0009	0335	6.0F
	0643	0949	7.0E
	1305	1602	5.1F
	1850	2158	6.5E
31 M	0044	0413	5.6F
	0709	1026	6.9E
	1336	1645	5.1F
	1939	2244	6.1E

February

Day	Slack (h m)	Maximum (h m)	knots
1 Tu	0124	0454	5.0F
	0737	1107	6.8E
	1412	1733	5.1F
	2037	2335	5.7E
2 W	0211	0539	4.4F
	0810	1152	6.6E
	1457	1828	5.1F
	2151		
3 Th		0035	5.2E
	0313	0634	3.7F
	0849	1245	6.4E
	1553	1931	5.1F
	2318		
4 F		0144	4.9E
	0451	0741	3.1F
	0942	1347	6.3E
	1701	2039	5.4F
5 Sa	0040	0259	5.1E
	0647	0853	3.0F
	1054	1453	6.4E
	1814	2146	5.8F
6 Su	0146	0411	5.6E
	0803	1003	3.3F
	1221	1558	6.7E
	1921	2247	6.3F
7 M	0238	0513	6.3E
	0857	1106	4.0F
	1339	1659	7.1E
	2020	2341	6.8F
8 Tu	0324	0604	6.9E
	0941	1201	4.7F
	1443	1754	7.5E
	2112		
9 W		0029	7.1F
	0405	0649	7.4E
	1021	1250	5.4F
	1539	1845	7.8E
	2201		
10 Th		0114	7.2F
	0444	0730	7.8E
	1059	1336	5.9F
	1631	1932	7.9E
	2246		
11 F		0157	7.2F
	0520	0810	8.0E
	1137	1420	6.2F
	1720	2018	7.9E
	2330		
12 Sa		0238	7.0F
	0555	0850	8.0E
	1214	1504	6.2F
	1809	2104	7.6E
13 Su	0013	0319	6.5F
	0629	0930	7.8E
	1253	1549	6.2F
	1900	2150	7.1E
14 M	0057	0401	5.9F
	0702	1011	7.5E
	1334	1635	5.9F
	1954	2238	6.5E
15 Tu	0143	0446	5.2F
	0736	1055	7.1E
	1418	1726	5.5F
	2058	2331	5.7E
16 W	0239	0535	4.3F
	0812	1143	6.6E
	1512	1824	5.1F
	2215		
17 Th		0032	5.0E
	0402	0633	3.3F
	0856	1239	6.0E
	1618	1931	4.8F
	2340		
18 F		0147	4.5E
	0553	0744	2.6F
	1005	1345	5.6E
	1735	2049	4.8F
19 Sa	0056	0341	4.6E
	0721	0907	2.4F
	1147	1457	5.6E
	1845	2204	5.1F
20 Su	0154	0519	5.3E
	0820	1026	3.0F
	1308	1606	5.8E
	1942	2300	5.5F
21 M	0238	0548	5.9E
	0902	1120	3.7F
	1406	1701	6.2E
	2029	2341	5.9F
22 Tu	0315	0609	6.4E
	0936	1159	4.4F
	1452	1745	6.7E
	2108		
23 W		0015	6.2F
	0348	0634	6.8E
	1007	1233	4.9F
	1531	1823	7.0E
	2143		
24 Th		0048	6.4F
	0417	0701	7.2E
	1034	1306	5.4F
	1605	1900	7.3E
	2215		
25 F		0121	6.5F
	0444	0731	7.4E
	1058	1339	5.7F
	1638	1936	7.4E
	2246		
26 Sa		0154	6.5F
	0508	0802	7.5E
	1120	1413	5.9F
	1711	2012	7.4E
	2315		
27 Su		0226	6.3F
	0530	0833	7.5E
	1143	1447	5.9F
	1745	2049	7.2E
	2345		
28 M		0300	5.9F
	0552	0906	7.4E
	1207	1524	5.9F
	1823	2128	6.9E

March

Day	Slack (h m)	Maximum (h m)	knots
1 Tu	0018	0334	5.4F
	0615	0940	7.2E
	1237	1604	5.8F
	1907	2211	6.4E
2 W	0055	0412	4.7F
	0641	1018	6.9E
	1314	1651	5.5F
	2003	2302	5.7E
3 Th	0142	0457	3.9F
	0713	1104	6.4E
	1400	1749	5.2F
	2119		
4 F		0006	5.1E
	0252	0600	2.9F
	0755	1205	6.0E
	1503	1901	5.0F
	2301		
5 Sa		0125	4.7E
	0526	0723	2.2F
	0902	1322	5.7E
	1632	2020	5.2F
6 Su	0028	0252	5.1E
	0705	0850	2.6F
	1101	1442	5.9E
	1808	2135	5.6F
7 M	0130	0409	5.8E
	0759	1005	3.6F
	1248	1555	6.4E
	1920	2237	6.2F
8 Tu	0218	0503	6.7E
	0840	1104	4.7F
	1357	1655	7.1E
	2017	2327	6.6F
9 W	0258	0546	7.3E
	0916	1151	5.6F
	1451	1747	7.6E
	2106		
10 Th		0011	6.9F
	0335	0624	7.8E
	0951	1233	6.2F
	1538	1833	7.9E
	2151		
11 F		0051	7.0F
	0409	0701	8.1E
	1024	1314	6.6F
	1623	1915	8.0E
	2232		
12 Sa		0130	6.9F
	0442	0737	8.2E
	1057	1353	6.8F
	1706	1957	7.9E
	2312		
13 Su		0208	6.6F
	0512	0813	8.1E
	1130	1433	6.8F
	1749	2038	7.6E
	2350		
14 M		0246	6.1F
	0541	0850	7.9E
	1204	1513	6.5F
	1833	2120	7.1E
15 Tu	0029	0325	5.4F
	0609	0928	7.4E
	1238	1556	6.1F
	1921	2205	6.4E
16 W	0112	0407	4.5F
	0637	1009	6.8E
	1317	1644	5.5F
	2020	2256	5.5E
17 Th	0208	0456	3.5F
	0709	1057	6.1E
	1404	1742	4.9F
	2138	2359	4.7E
18 F	0358	0600	2.4F
	0752	1158	5.3E
	1516	1854	4.5F
	2310		
19 Sa		0122	4.3E
	0601	0726	1.8F
	0935	1315	4.8E
	1701	2019	4.4F
20 Su	0025	0419	4.8E
	0713	0912	2.4F
	1157	1441	4.9E
	1823	2140	4.8F
21 M	0119	0452	5.6E
	0756	1026	3.4F
	1308	1554	5.5E
	1920	2232	5.3F
22 Tu	0200	0508	6.2E
	0829	1102	4.3F
	1358	1644	6.2E
	2006	2309	5.8F
23 W	0234	0528	6.8E
	0858	1133	5.1F
	1437	1723	6.7E
	2044	2343	6.1F
24 Th	0305	0553	7.2E
	0924	1203	5.6F
	1512	1759	7.2E
	2119		
25 F		0015	6.3F
	0332	0622	7.5E
	0948	1234	6.1F
	1544	1834	7.4E
	2150		
26 Sa		0047	6.3F
	0356	0651	7.7E
	1009	1306	6.4F
	1616	1909	7.6E
	2221		
27 Su		0119	6.2F
	0418	0722	7.8E
	1031	1339	6.5F
	1649	1945	7.5E
	2251		
28 M		0152	5.9F
	0440	0754	7.7E
	1054	1414	6.6F
	1724	2022	7.3E
	2323		
29 Tu		0226	5.5F
	0502	0826	7.5E
	1122	1451	6.4F
	1803	2103	6.9E
	2358		
30 W		0302	4.9F
	0527	0901	7.2E
	1154	1533	6.2F
	1850	2149	6.3E
31 Th	0041	0343	4.1F
	0557	0942	6.8E
	1235	1623	5.8F
	1950	2245	5.7E

Time meridian 120° W. 0000 is midnight. 1200 is noon. Times are not adjusted for Daylight Saving Time.

Deception Pass (Narrows), Washington, 2005

F–Flood, Dir. 090° True   E–Ebb, Dir. 270° True

April

Day	Slack h m	Maximum h m	knots
1 F ◑	0142	0437	3.1F
	0634	1034	6.2E
	1326	1727	5.4F
	2112	2355	5.1E
2 Sa	0347	0554	2.2F
	0729	1148	5.6E
	1438	1844	5.1F
	2248		
3 Su		0119	5.1E
	0549	0725	2.2F
	0923	1314	5.4E
	1627	2006	5.2F
4 M	0003	0245	5.7E
	0650	0853	3.3F
	1144	1438	5.8E
	1803	2119	5.6F
5 Tu	0058	0350	6.5E
	0732	0959	4.5F
	1300	1548	6.5E
	1910	2216	6.1F
6 W	0143	0436	7.2E
	0808	1048	5.5F
	1356	1644	7.1E
	2004	2302	6.4F
7 Th	0221	0514	7.7E
	0843	1130	6.3F
	1443	1731	7.6E
	2051	2344	6.6F
8 F ●	0256	0551	8.0E
	0915	1209	6.8F
	1527	1813	7.8E
	2133		
9 Sa		0022	6.5F
	0328	0626	8.2E
	0947	1247	7.1F
	1608	1853	7.8E
	2213		
10 Su		0059	6.3F
	0358	0701	8.2E
	1019	1325	7.1F
	1649	1933	7.0E
	2252		
11 M		0136	5.9F
	0426	0737	8.0F
	1049	1403	6.9F
	1729	2013	7.2E
	2330		
12 Tu		0214	5.3F
	0453	0813	7.6E
	1120	1442	6.6F
	1812	2054	6.7E
13 W	0010	0253	4.6F
	0519	0851	7.1E
	1153	1525	6.1F
	1858	2140	6.1E
14 Th	0059	0337	3.7F
	0548	0933	6.4E
	1229	1613	5.5F
	1956	2233	5.4E
15 F	0216	0432	2.7F
	0624	1024	5.6E
	1314	1711	4.9F
	2109	2338	4.9E
16 Sa ◐	0413	0544	1.9F
	0721	1130	4.8E
	1424	1822	4.5F
	2230		
17 Su		0059	4.8E
	0543	0713	1.9F
	0959	1251	4.5E
	1617	1941	4.5F
	2336		
18 M		0234	5.3E
	0635	0849	2.9F
	1146	1413	4.8E
	1743	2050	4.8F
19 Tu	0027	0330	5.9E
	0712	0944	3.9F
	1246	1520	5.4E
	1842	2142	5.2F
20 W	0107	0403	6.5E
	0742	1020	4.8F
	1331	1608	6.1E
	1929	2223	5.5F
21 Th	0141	0434	7.0E
	0810	1053	5.5F
	1409	1649	6.7E
	2009	2259	5.8F
22 F	0211	0506	7.4E
	0835	1125	6.1F
	1444	1726	7.1E
	2045	2334	5.9F
23 Sa	0238	0538	7.7E
	0858	1159	6.5F
	1518	1803	7.3E
	2120		
24 Su ○		0009	5.8F
	0303	0610	7.8E
	0922	1233	6.8F
	1553	1841	7.4E
	2155		
25 M		0044	5.7F
	0326	0643	7.8E
	0947	1308	6.9F
	1029	1920	7.3E
	2230		
26 Tu		0120	5.3F
	0351	0717	7.7E
	1015	1346	6.9F
	1709	2001	7.1E
	2309		
27 W		0158	4.8F
	0419	0754	7.5E
	1049	1428	6.7F
	1754	2047	6.7E
	2355		
28 Th		0241	4.2F
	0451	0835	7.1E
	1128	1515	6.4F
	1847	2139	6.3E
29 F	0055	0333	3.5F
	0531	0924	6.6E
	1214	1611	6.0F
	1952	2240	5.9E
30 Sa ◑	0228	0439	2.8F
	0626	1028	6.0E
	1315	1718	5.6F
	2109	2351	5.7E

May

Day	Slack h m	Maximum h m	knots
1 Su	0411	0559	2.6F
	0759	1145	5.5E
	1439	1831	5.4F
	2223		
2 M		0108	5.9E
	0521	0723	3.3F
	1015	1307	5.5E
	1622	1945	5.5F
	2324		
3 Tu		0217	6.4E
	0611	0837	4.3F
	1145	1424	5.9E
	1744	2049	5.7F
4 W	0015	0312	7.0E
	0653	0934	5.3F
	1249	1529	6.5E
	1848	2143	5.8F
5 Th	0059	0357	7.5E
	0730	1021	6.1F
	1341	1622	6.9E
	1942	2229	5.9F
6 F	0137	0437	7.8E
	0806	1103	6.6F
	1428	1708	7.2E
	2029	2311	5.9F
7 Sa	0212	0514	8.0E
	0840	1142	6.9F
	1512	1751	7.3E
	2114	2350	5.7F
8 Su ●	0244	0551	8.0E
	0913	1221	7.1F
	1554	1831	7.3E
	2156		
9 M		0029	5.4F
	0314	0627	7.9E
	0945	1259	7.0F
	1635	1912	7.0E
	2237		
10 Tu		0108	5.0F
	0343	0704	7.6E
	1016	1338	6.8F
	1717	1953	6.7E
	2321		
11 W		0148	4.4F
	0411	0742	7.2E
	1048	1419	6.4F
	1800	2037	6.3E
12 Th	0010	0232	3.8F
	0442	0823	6.7E
	1122	1503	6.0F
	1847	2125	5.9E
13 F	0112	0321	3.1F
	0519	0910	6.1E
	1201	1552	5.5F
	1941	2219	5.5E
14 Sa	0232	0421	2.5F
	0611	1004	5.4E
	1249	1649	5.1F
	2041	2320	5.4E
15 Su	0351	0530	2.3F
	0737	1110	4.9E
	1355	1751	4.8F
	2142		
16 M ◐		0025	5.5E
	0453	0644	2.7F
	0942	1221	4.7E
	1523	1855	4.7F
	2236		
17 Tu		0126	5.9E
	0538	0752	3.5F
	1106	1330	4.9E
	1645	1954	4.8F
	2323		
18 W		0218	6.3E
	0615	0845	4.3F
	1204	1430	5.4E
	1748	2045	5.0F
19 Th	0003	0301	6.7E
	0646	0928	5.1F
	1252	1522	5.9E
	1840	2130	5.2F
20 F	0038	0340	7.1E
	0715	1007	5.7F
	1334	1607	6.4E
	1926	2212	5.2F
21 Sa	0110	0417	7.4E
	0743	1045	6.2F
	1414	1651	6.7E
	2009	2252	5.2F
22 Su	0139	0454	7.6E
	0810	1123	6.6F
	1454	1733	6.9E
	2051	2332	5.1F
23 M ○	0208	0531	7.7E
	0840	1203	6.9F
	1535	1816	7.0E
	2134		
24 Tu		0012	4.9F
	0239	0609	7.7E
	0913	1244	7.0F
	1618	1901	7.0E
	2220		
25 W		0055	4.6F
	0312	0650	7.6E
	0949	1327	7.0F
	1704	1948	6.8E
	2311		
26 Th		0141	4.2F
	0351	0734	7.4E
	1031	1415	6.8F
	1754	2039	6.7E
27 F	0011	0233	3.8F
	0437	0824	7.0E
	1118	1507	6.6F
	1849	2134	6.5E
28 Sa	0120	0333	3.5F
	0536	0921	6.6E
	1213	1604	6.3F
	1948	2234	6.4E
29 Su	0233	0439	3.5F
	0655	1026	6.2E
	1320	1706	6.0F
	2047	2337	6.5E
30 M ◑	0339	0550	3.8F
	0835	1137	5.9E
	1438	1810	5.8F
	2145		
31 Tu		0039	6.7E
	0436	0700	4.4F
	1009	1248	5.9E
	1601	1912	5.6F
	2238		

June

Day	Slack h m	Maximum h m	knots
1 W		0136	7.0E
	0525	0804	5.1F
	1124	1357	6.1E
	1714	2010	5.5F
	2326		
2 Th		0228	7.3E
	0609	0859	5.7F
	1227	1459	6.3E
	1819	2104	5.4F
3 F	0010	0314	7.5E
	0651	0949	6.2F
	1322	1554	6.5E
	1916	2153	5.2F
4 Sa	0050	0358	7.6E
	0730	1034	6.5F
	1412	1644	6.6E
	2009	2238	5.0F
5 Su	0128	0439	7.7E
	0808	1117	6.7F
	1459	1730	6.6E
	2059	2322	4.7F
6 M ●	0203	0520	7.6E
	0845	1159	6.8F
	1544	1814	6.5E
	2148		
7 Tu		0005	4.4F
	0238	0600	7.4E
	0921	1240	6.7F
	1627	1858	6.4E
	2236		
8 W		0049	4.1F
	0313	0641	7.1E
	0956	1322	6.5F
	1710	1941	6.3E
	2326		
9 Th		0134	3.7F
	0350	0724	6.8E
	1032	1404	6.3F
	1753	2026	6.2E
10 F	0018	0221	3.4F
	0432	0809	6.4E
	1110	1449	6.0F
	1836	2113	6.1F
11 Sa	0113	0312	3.2F
	0523	0857	6.0E
	1151	1536	5.7F
	1920	2202	6.0E
12 Su	0208	0407	3.2F
	0624	0950	5.7E
	1238	1625	5.5F
	2004	2252	6.1E
13 M	0259	0503	3.3F
	0738	1045	5.4E
	1331	1716	5.2F
	2048	2342	6.2E
14 Tu ◐	0346	0600	3.6F
	0856	1144	5.3E
	1431	1808	5.1F
	2130		
15 W		0031	6.4E
	0428	0655	4.1F
	1007	1242	5.3E
	1536	1859	4.9F
	2211		
16 Th		0118	6.5E
	0506	0746	4.6F
	1110	1339	5.4E
	1641	1949	4.8F
	2249		
17 F		0203	6.8E
	0542	0835	5.2F
	1206	1434	5.6E
	1743	2038	4.6F
	2326		
18 Sa		0247	7.0E
	0617	0922	5.6F
	1258	1526	5.9E
	1842	2126	4.5F
19 Su	0003	0331	7.2E
	0652	1008	6.1F
	1347	1618	6.1E
	1938	2214	4.4F
20 M	0041	0415	7.3E
	0729	1054	6.5F
	1436	1708	6.4E
	2034	2302	4.3F
21 Tu ○	0122	0500	7.4E
	0810	1140	6.7F
	1524	1758	6.5E
	2128	2351	4.2F
22 W	0206	0546	7.5E
	0853	1228	6.9F
	1612	1849	6.7E
	2223		
23 Th		0042	4.2F
	0255	0635	7.5E
	0939	1316	7.0F
	1700	1940	6.8E
	2317		
24 F		0134	4.2F
	0349	0726	7.4E
	1028	1406	7.0F
	1748	2031	7.0E
25 Sa	0012	0229	4.3F
	0450	0820	7.2E
	1121	1458	6.8F
	1836	2123	7.1E
26 Su	0106	0326	4.4F
	0557	0917	7.0E
	1217	1550	6.6F
	1924	2215	7.1E
27 M	0159	0425	4.6F
	0710	1015	6.7E
	1317	1644	6.3F
	2012	2307	7.2E
28 Tu ◑	0252	0524	4.9F
	0826	1116	6.4E
	1421	1738	5.9F
	2059	2359	7.2E
29 W	0344	0624	5.2F
	0942	1218	6.2E
	1529	1833	5.5F
	2146		
30 Th		0051	7.2E
	0435	0723	5.5F
	1054	1321	6.0E
	1639	1929	5.1F
	2232		

Time meridian 120° W. 0000 is midnight. 1200 is noon. Times are not adjusted for Daylight Saving Time.

Deception Pass (Narrows), Washington, 2005

F–Flood, Dir. 090° True E–Ebb, Dir. 270° True

July

Day	Slack (h m)	Maximum (h m)	knots
1 F		0142	7.2E
	0524	0821	5.7F
	1201	1425	5.8E
	1749	2024	4.7F
	2319		
2 Sa		0232	7.2E
	0613	0916	6.0F
	1302	1527	5.8E
	1856	2119	4.3F
3 Su	0005	0322	7.1E
	0700	1009	6.1F
	1359	1626	5.8E
	1959	2212	4.0F
4 M	0052	0411	7.1E
	0746	1059	6.3F
	1450	1720	5.9E
	2057	2304	3.8F
5 Tu	0139	0458	7.0E
	0830	1145	6.3F
	1537	1809	6.0E
	2150	2354	3.8F
6 W ●	0226	0545	6.9E
	0911	1229	6.3F
	1620	1853	6.2E
	2238		
7 Th		0041	3.8F
	0312	0630	6.8E
	0951	1311	6.3F
	1700	1934	6.3E
	2322		
8 F		0127	3.8F
	0359	0715	6.7E
	1029	1352	6.3F
	1737	2014	6.4E
9 Sa	0003	0212	3.9F
	0445	0759	6.6E
	1106	1433	6.2F
	1812	2053	6.5E
10 Su	0041	0256	4.1F
	0532	0843	6.4E
	1144	1513	6.0F
	1846	2133	6.6E
11 M	0118	0340	4.2F
	0619	0928	6.3E
	1222	1554	5.8F
	1918	2213	6.7E
12 Tu	0154	0425	4.3F
	0710	1014	6.1E
	1302	1636	5.5F
	1950	2254	6.7E
13 W	0229	0512	4.5F
	0804	1103	5.8E
	1346	1720	5.2F
	2022	2337	6.7E
14 Th ◐	0306	0600	4.6F
	0904	1155	5.6E
	1436	1806	4.8F
	2055		
15 F		0021	6.6E
	0345	0651	4.8F
	1010	1250	5.4E
	1534	1856	4.3F
	2132		
16 Sa		0108	6.6E
	0429	0745	5.1F
	1120	1350	5.3E
	1646	1950	3.9F
	2213		
17 Su		0158	6.6E
	0517	0841	5.4F
	1228	1451	5.3E
	1810	2047	3.6F
	2302		
18 M		0251	6.7E
	0609	0937	5.8F
	1330	1554	5.6E
	1929	2147	3.6F
	2358		
19 Tu		0346	6.9E
	0702	1033	6.2F
	1426	1653	6.0E
	2034	2245	3.7F
20 W	0059	0441	7.1E
	0755	1126	6.6F
	1516	1749	6.4E
	2129	2342	4.1F
21 Th ○	0202	0535	7.3E
	0848	1217	6.9F
	1602	1840	6.9E
	2218		
22 F		0035	4.5F
	0304	0628	7.5E
	0939	1306	7.1F
	1645	1927	7.2E
	2303		
23 Sa		0127	4.9F
	0404	0720	7.6E
	1030	1353	7.1F
	1727	2013	7.5E
	2346		
24 Su		0217	5.3F
	0502	0811	7.6E
	1120	1440	7.0F
	1808	2058	7.6E
25 M	0029	0307	5.6F
	0559	0902	7.5E
	1209	1526	6.8F
	1847	2143	7.7E
26 Tu	0113	0357	5.7F
	0658	0954	7.2E
	1300	1613	6.4F
	1927	2228	7.6E
27 W ◑	0159	0448	5.7F
	0800	1047	6.7E
	1354	1701	5.8F
	2007	2315	7.4E
28 Th	0248	0542	5.7F
	0907	1143	6.2E
	1453	1752	5.1F
	2049		
29 F		0004	7.1E
	0341	0640	5.5F
	1020	1244	5.7E
	1605	1848	4.4F
	2135		
30 Sa		0057	6.8E
	0438	0742	5.4F
	1135	1352	5.2E
	1729	1949	3.7F
	2229		
31 Su		0154	6.5E
	0539	0848	5.5F
	1247	1507	5.1E
	1852	2055	3.3F
	2333		

August

Day	Slack (h m)	Maximum (h m)	knots
1 M		0254	6.4E
	0639	0953	5.6F
	1349	1629	5.3E
	2002	2202	3.2F
2 Tu	0041	0354	6.3E
	0734	1051	5.8F
	1440	1732	5.7E
	2058	2303	3.5F
3 W	0143	0450	6.4E
	0823	1139	6.0F
	1524	1813	6.1E
	2143	2353	3.9F
4 Th ●	0237	0539	6.6E
	0907	1220	6.2F
	1601	1845	6.4E
	2221		
5 F		0036	4.3F
	0324	0623	6.8E
	0946	1257	6.3F
	1635	1916	6.7E
	2255		
6 Sa		0115	4.6F
	0405	0704	6.9E
	1021	1333	6.4F
	1706	1948	6.9E
	2325		
7 Su		0152	4.9F
	0443	0743	7.0E
	1054	1408	6.4F
	1734	2021	7.1E
	2353		
8 M		0228	5.1F
	0520	0821	7.0E
	1126	1443	6.2F
	1801	2055	7.1E
9 Tu	0019	0305	5.2F
	0557	0900	6.8E
	1157	1518	5.9F
	1825	2129	7.1E
10 W	0045	0343	5.2F
	0636	0940	6.6E
	1231	1554	5.6F
	1850	2205	7.0E
11 Th	0114	0424	5.2F
	0720	1024	6.2E
	1307	1633	5.0F
	1916	2244	6.8E
12 F ◐	0147	0509	5.1F
	0812	1112	5.7E
	1350	1716	4.4F
	1946	2326	6.5E
13 Sa	0228	0601	5.0F
	0919	1209	5.2E
	1446	1808	3.7F
	2022		
14 Su		0017	6.3E
	0319	0702	5.0F
	1044	1315	4.9E
	1613	1912	3.0F
	2110		
15 M		0117	6.1E
	0425	0810	5.2F
	1210	1429	4.9E
	1820	2025	2.7F
	2219		
16 Tu		0224	6.1E
	0541	0918	5.5F
	1319	1543	5.4E
	1940	2137	3.1F
	2349		
17 W		0332	6.4E
	0652	1021	6.0F
	1413	1646	6.0E
	2033	2241	3.8F
18 Th	0113	0434	6.9E
	0754	1115	6.5F
	1458	1738	6.7E
	2116	2336	4.6F
19 F ○	0219	0530	7.3E
	0848	1204	6.9F
	1539	1823	7.3E
	2155		
20 Sa		0025	5.4F
	0316	0621	7.7E
	0936	1249	7.1F
	1617	1904	7.7E
	2232		
21 Su		0111	5.9F
	0407	0709	7.9E
	1023	1332	7.1F
	1653	1944	8.0E
	2308		
22 M		0155	6.3F
	0456	0755	8.0E
	1107	1414	7.0F
	1728	2024	8.1E
	2345		
23 Tu		0239	6.5F
	0545	0840	7.8E
	1151	1455	6.6F
	1803	2105	8.0E
24 W	0024	0323	6.4F
	0635	0927	7.4E
	1235	1538	6.0F
	1837	2146	7.7E
25 Th	0104	0410	6.2F
	0729	1015	6.8E
	1323	1623	5.3F
	1911	2230	7.3E
26 F ◑	0149	0501	5.8F
	0832	1108	6.0E
	1420	1712	4.4F
	1949	2319	6.7E
27 Sa	0242	0559	5.4F
	0947	1210	5.2E
	1544	1811	3.4F
	2035		
28 Su		0015	6.1E
	0349	0706	5.0F
	1113	1325	4.7E
	1732	1923	2.6F
	2147		
29 M		0122	5.7E
	0510	0825	4.9F
	1231	1521	4.8E
	1900	2050	2.5F
	2333		
30 Tu		0238	5.5E
	0624	0944	5.2F
	1331	1659	5.4E
	2000	2214	3.1F
31 W	0054	0350	5.8E
	0724	1043	5.5F
	1417	1732	6.0E
	2042	2307	3.9F

September

Day	Slack (h m)	Maximum (h m)	knots
1 Th	0153	0447	6.2E
	0813	1124	5.9F
	1455	1753	6.5E
	2116	2345	4.5F
2 F	0239	0530	6.6E
	0854	1159	6.2F
	1528	1816	6.9E
	2146		
3 Sa ●		0017	5.1F
	0318	0608	7.0E
	0929	1231	6.3F
	1557	1843	7.2E
	2214		
4 Su		0049	5.5F
	0353	0644	7.2E
	1002	1303	6.4F
	1624	1911	7.4E
	2238		
5 M		0121	5.8F
	0425	0719	7.4E
	1032	1335	6.3F
	1648	1942	7.5E
	2300		
6 Tu		0154	6.0F
	0457	0754	7.3E
	1101	1407	6.1F
	1709	2013	7.5E
	2322		
7 W		0228	6.0F
	0529	0830	7.2E
	1130	1440	5.8F
	1730	2045	7.3E
	2345		
8 Th		0303	5.9F
	0605	0908	6.8E
	1201	1514	5.2F
	1752	2118	7.1E
9 F	0013	0341	5.8F
	0646	0950	6.3E
	1236	1550	4.6F
	1817	2154	6.7E
10 Sa	0047	0426	5.5F
	0738	1039	5.7E
	1320	1633	3.7F
	1846	2237	6.3E
11 Su ◐	0130	0521	5.1F
	0848	1140	5.1E
	1426	1733	2.7F
	1925	2335	5.8E
12 M	0227	0631	4.9F
	1026	1257	4.7E
	1700	1856	2.0F
	2027		
13 Tu		0052	5.5E
	0351	0750	5.0F
	1157	1421	5.0E
	1842	2023	2.4F
	2226		
14 W		0213	5.7E
	0533	0905	5.4F
	1300	1537	5.7E
	1934	2138	3.5F
15 Th	0021	0327	6.2E
	0650	1008	6.0F
	1348	1633	6.6E
	2013	2236	4.6F
16 F	0131	0428	6.9E
	0749	1059	6.5F
	1429	1716	7.3E
	2048	2324	5.5F
17 Sa ○	0225	0520	7.5E
	0839	1143	6.8F
	1506	1755	7.8E
	2122		
18 Su		0007	6.3F
	0313	0607	7.9E
	0924	1224	6.9F
	1540	1833	8.2E
	2155		
19 M		0047	6.8F
	0358	0650	8.1E
	1007	1304	6.9F
	1613	1910	8.3E
	2229		
20 Tu		0128	7.0F
	0442	0733	8.0E
	1048	1343	6.6F
	1644	1947	8.3E
	2302		
21 W		0208	7.0F
	0526	0815	7.7E
	1129	1422	6.1F
	1714	2025	8.0E
	2337		
22 Th		0250	6.8F
	0612	0859	7.2E
	1211	1503	5.4F
	1744	2105	7.6E
23 F	0014	0334	6.4F
	0702	0945	6.5E
	1258	1546	4.5F
	1815	2147	6.9E
24 Sa ◑	0054	0424	5.8F
	0802	1038	5.7E
	1402	1638	3.5F
	1850	2237	6.2E
25 Su	0144	0522	5.1F
	0920	1143	4.9E
	1552	1744	2.4F
	1939	2339	5.4E
26 M	0257	0635	4.7F
	1048	1308	4.6E
	1744	1912	1.9F
	2141		
27 Tu		0058	4.9E
	0442	0801	4.6F
	1202	1553	5.1E
	1853	2108	2.6F
	2346		
28 W		0226	5.0E
	0604	0922	4.9F
	1256	1630	5.8E
	1935	2214	3.7F
29 Th	0055	0342	5.6E
	0703	1014	5.4F
	1338	1649	6.4E
	2008	2248	4.5F
30 F	0143	0430	6.2E
	0749	1051	5.8F
	1412	1709	6.9E
	2037	2317	5.2F

Time meridian 120° W. 0000 is midnight. 1200 is noon. Times are not adjusted for Daylight Saving Time.

Deception Pass (Narrows), Washington, 2005

F–Flood, Dir. 090° True E–Ebb, Dir. 270° True

October

Day	Slack h m	Maximum h m	knots
1 Sa	0223	0508	6.7E
	0828	1123	6.0F
	1443	1733	7.3E
	2103	2346	5.8F
2 Su	0257	0542	7.1E
	0902	1155	6.2F
	1510	1801	7.5E
	2127		
3 M ●		0016	6.2F
	0330	0616	7.4E
	0934	1227	6.2F
	1534	1830	7.7E
	2149		
4 Tu		0047	6.4F
	0401	0650	7.5E
	1004	1259	6.1F
	1556	1901	7.7E
	2210		
5 W		0120	6.6F
	0432	0725	7.4E
	1034	1331	5.8F
	1617	1932	7.7E
	2233		
6 Th		0153	6.5F
	0506	0802	7.2E
	1104	1404	5.3F
	1638	2003	7.4E
	2258		
7 F		0229	6.4F
	0543	0842	6.8E
	1138	1439	4.7F
	1701	2037	7.1E
	2328		
8 Sa		0309	6.1F
	0627	0926	6.2E
	1220	1518	3.9F
	1729	2115	6.6E
9 Su	0006	0357	5.7F
	0723	1020	5.6E
	1317	1610	2.9F
	1802	2204	6.0E
10 M ◐	0053	0458	5.3F
	0840	1128	5.1E
	1521	1726	2.0F
	1854	2316	5.4E
11 Tu	0159	0614	5.0F
	1015	1249	5.1E
	1726	1856	2.0F
	2040		
12 W		0043	5.2E
	0342	0734	5.1F
	1131	1412	5.6E
	1824	2023	3.1F
	2312		
13 Th		0207	5.6E
	0527	0846	5.5F
	1226	1517	6.4E
	1905	2129	4.4F
14 F	0032	0318	6.3E
	0639	0944	5.9F
	1311	1604	7.1E
	1940	2219	5.5F
15 Sa	0129	0415	7.0E
	0731	1032	6.3F
	1350	1644	7.7E
	2014	2302	6.3F
16 Su	0217	0503	7.5E
	0822	1115	6.5F
	1425	1722	8.1E
	2047	2342	6.9F
17 M ○	0301	0547	7.8E
	0906	1155	6.5F
	1458	1759	8.3E
	2119		
18 Tu		0022	7.2F
	0344	0629	7.9E
	0948	1234	6.3F
	1529	1836	8.3E
	2152		
19 W		0101	7.3F
	0426	0710	7.7E
	1029	1312	5.9F
	1559	1912	8.2E
	2225		
20 Th		0140	7.2F
	0509	0752	7.4E
	1110	1351	5.3F
	1628	1950	7.8E
	2258		
21 F		0221	6.8F
	0554	0835	6.8E
	1155	1433	4.6F
	1657	2029	7.2E
	2333		
22 Sa		0305	6.3F
	0643	0922	6.2E
	1249	1519	3.7F
	1729	2113	6.5E
23 Su	0011	0355	5.7F
	0742	1017	5.5E
	1412	1616	2.7F
	1807	2206	5.7E
24 M ◑	0059	0454	5.1F
	0855	1123	5.0E
	1600	1729	1.9F
	1913	2313	4.9E
25 Tu	0211	0605	4.6F
	1011	1246	5.0E
	1724	1901	2.1F
	2154		
26 W		0034	4.5E
	0401	0723	4.0F
	1115	1425	5.4E
	1815	2040	3.1F
	2332		
27 Th		0157	4.8E
	0525	0832	4.8F
	1205	1514	6.1E
	1852	2131	4.1F
28 F	0032	0305	5.4E
	0624	0923	5.2F
	1245	1545	6.6E
	1923	2205	4.9F
29 Sa	0116	0353	6.0E
	0711	1004	5.5F
	1320	1615	7.1E
	1951	2236	5.6F
30 Su	0155	0432	6.6E
	0752	1040	5.7F
	1350	1646	7.4E
	2017	2308	6.1F
31 M	0230	0509	6.9E
	0829	1115	5.8F
	1417	1718	7.6E
	2041	2341	6.5F

November

Day	Slack h m	Maximum h m	knots
1 Tu ●	0304	0545	7.2E
	0904	1149	5.7F
	1442	1750	7.8E
	2104		
2 W		0014	6.7F
	0338	0622	7.3E
	0938	1224	5.5F
	1505	1822	7.8E
	2128		
3 Th		0049	6.8F
	0413	0700	7.2E
	1013	1259	5.1F
	1529	1856	7.6E
	2154		
4 F		0126	6.8F
	0451	0741	7.0E
	1050	1336	4.7F
	1555	1931	7.4E
	2225		
5 Sa		0206	6.7F
	0534	0825	6.6E
	1134	1418	4.0F
	1625	2010	7.0E
	2302		
6 Su		0251	6.4F
	0623	0915	6.2E
	1231	1507	3.3F
	1702	2057	6.5E
	2346		
7 M		0344	6.0F
	0724	1014	5.8E
	1358	1610	2.7F
	1752	2157	5.9E
8 Tu ◐	0041	0447	5.6F
	0835	1121	5.7E
	1540	1727	2.5F
	1915	2312	5.5E
9 W	0156	0558	5.4F
	0948	1234	5.9E
	1652	1849	3.1F
	2130		
10 Th		0033	5.4E
	0336	0710	5.4F
	1050	1342	6.3E
	1742	2004	4.1F
	2312		
11 F		0150	5.8E
	0507	0815	5.6F
	1142	1439	6.9E
	1824	2103	5.1F
12 Sa	0020	0257	6.3E
	0615	0911	5.7F
	1227	1526	7.4E
	1902	2152	6.0F
13 Su	0114	0353	6.8E
	0712	1000	5.9F
	1306	1608	7.8E
	1938	2236	6.6F
14 M	0203	0442	7.2E
	0802	1044	5.9F
	1343	1648	8.1E
	2014	2317	7.0F
15 Tu ○	0248	0527	7.3E
	0849	1126	5.7F
	1417	1726	8.1E
	2048	2358	7.2F
16 W	0332	0610	7.3E
	0934	1206	5.4F
	1449	1805	8.0E
	2123		
17 Th		0038	7.2F
	0416	0652	7.1E
	1018	1247	5.0F
	1521	1844	7.8E
	2157		
18 F		0119	7.0F
	0500	0735	6.8E
	1105	1329	4.5F
	1553	1923	7.4E
	2232		
19 Sa		0201	6.7F
	0545	0820	6.4E
	1156	1414	3.9F
	1627	2006	6.9E
	2308		
20 Su		0246	6.2F
	0633	0908	6.0E
	1259	1504	3.2F
	1707	2053	6.2E
	2348		
21 M		0335	5.7F
	0725	1002	5.7E
	1413	1603	2.7F
	1800	2147	5.6E
22 Tu	0036	0430	5.3F
	0822	1101	5.6E
	1528	1711	2.5F
	1927	2250	5.0E
23 W ◑	0140	0531	4.9F
	0920	1205	5.6E
	1629	1823	2.9F
	2123		
24 Th		0000	4.8E
	0303	0633	4.8F
	1014	1305	5.9E
	1716	1932	3.6F
	2247		
25 F		0109	4.9E
	0424	0732	4.8F
	1101	1356	6.3E
	1755	2026	4.3F
	2348		
26 Sa		0210	5.3E
	0529	0824	4.9F
	1142	1441	6.7E
	1829	2111	5.0F
27 Su	0037	0303	5.7E
	0623	0911	5.0F
	1218	1520	7.0E
	1859	2151	5.6F
28 M	0120	0350	6.2E
	0710	0953	5.1F
	1251	1558	7.3E
	1928	2229	6.1F
29 Tu	0201	0434	6.5E
	0755	1034	5.0F
	1321	1635	7.5E
	1956	2307	6.5F
30 W	0241	0516	6.7E
	0838	1114	4.9F
	1350	1713	7.6E
	2025	2346	6.7F

December

Day	Slack h m	Maximum h m	knots
1 Th ●	0321	0559	6.8E
	0920	1154	4.7F
	1420	1751	7.6E
	2056		
2 F		0026	6.9F
	0402	0642	6.8E
	1005	1236	4.5F
	1453	1831	7.5E
	2131		
3 Sa		0108	6.9F
	0446	0728	6.7E
	1053	1321	4.1F
	1530	1913	7.3E
	2210		
4 Su		0153	6.8F
	0533	0817	6.6E
	1148	1410	3.8F
	1614	2001	7.0E
	2254		
5 M		0243	6.6F
	0623	0909	6.5E
	1250	1506	3.6F
	1708	2054	6.7E
	2345		
6 Tu		0336	6.4F
	0716	1005	6.5E
	1356	1608	3.5F
	1818	2155	6.3E
7 W	0045	0434	6.1F
	0812	1103	6.5E
	1501	1715	3.8F
	1948	2302	6.0E
8 Th ◐	0156	0535	5.8F
	0907	1203	6.7E
	1559	1822	4.3F
	2125		
9 F		0011	5.9E
	0316	0636	5.6F
	1001	1300	6.9E
	1650	1927	4.9F
	2247		
10 Sa		0120	5.9E
	0434	0736	5.4F
	1050	1354	7.2E
	1738	2027	5.6F
	2356		
11 Su		0226	6.1E
	0545	0832	5.3F
	1137	1444	7.4E
	1822	2120	6.1F
12 M	0055	0326	6.3E
	0648	0925	5.1F
	1221	1531	7.6E
	1905	2210	6.5F
13 Tu	0149	0420	6.5E
	0746	1014	4.9F
	1302	1616	7.7E
	1946	2256	6.7F
14 W	0239	0511	6.6E
	0840	1101	4.7F
	1342	1700	7.6E
	2026	2340	6.8F
15 Th ○	0327	0558	6.6E
	0931	1147	4.5F
	1421	1743	7.5E
	2106		
16 F		0024	6.8F
	0412	0643	6.5E
	1022	1233	4.2F
	1501	1826	7.3E
	2144		
17 Sa		0107	6.7F
	0456	0728	6.5E
	1111	1319	4.0F
	1542	1909	7.0E
	2222		
18 Su		0150	6.5F
	0538	0812	6.4E
	1201	1406	3.7F
	1626	1954	6.7E
	2301		
19 M		0233	6.2F
	0619	0857	6.3E
	1250	1455	3.6F
	1716	2041	6.3E
	2341		
20 Tu		0318	6.0F
	0700	0942	6.3E
	1340	1546	3.5F
	1813	2131	6.0E
21 W	0025	0404	5.7F
	0741	1029	6.3E
	1429	1640	3.6F
	1918	2224	5.6E
22 Th	0113	0453	5.4F
	0822	1117	6.3E
	1516	1734	3.8F
	2029	2319	5.4E
23 F ◑	0207	0542	5.1F
	0903	1204	6.4E
	1601	1828	4.1F
	2141		
24 Sa		0016	5.2E
	0308	0633	4.8F
	0943	1252	6.5E
	1642	1922	4.5F
	2247		
25 Su		0114	5.2E
	0414	0725	4.6F
	1023	1339	6.6E
	1722	2013	4.9F
	2348		
26 M		0211	5.3E
	0521	0816	4.4F
	1103	1425	6.7E
	1800	2103	5.4F
27 Tu	0044	0307	5.5E
	0627	0906	4.2F
	1143	1511	6.9E
	1838	2151	5.8F
28 W	0136	0401	5.8E
	0728	0956	4.1F
	1224	1557	7.0E
	1916	2238	6.2F
29 Th	0224	0453	6.1E
	0825	1046	4.0F
	1306	1643	7.2E
	1956	2325	6.5F
30 F ●	0311	0543	6.3E
	0918	1135	4.0F
	1351	1730	7.3E
	2039		
31 Sa		0011	6.8F
	0357	0632	6.6E
	1009	1225	4.1F
	1440	1817	7.4E
	2123		

Time meridian 120° W. 0000 is midnight. 1200 is noon. Times are not adjusted for Daylight Saving Time.

Rosario Strait, Washington, 2005

F–Flood, Dir. 335° True E–Ebb, Dir. 175° True

January

Day	Slack (h m)	Maximum (h m)	knots
1 Sa	0311	0621	1.9F
	0952	1338	1.5E
		1839	*
		2338	1.5E
2 Su	0345	0658	1.7F
	1007	1422	1.7E
		1933	*
3 M		0031	1.0E
	0420	0737	1.4F
	1020	1502	2.0E
◑	1932	2034	0.3F
	2201		
4 Tu		0238	0.6E
	0500	0818	1.1F
	1036	1541	2.2E
	2009	2145	0.5F
5 W	0138	0405	0.3E
	0548	0905	0.7F
	1100	1621	2.5E
	2050	2314	1.0F
6 Th	0348	0520	0.3E
	0704	1001	0.5F
	1133	1705	2.8E
	2131		
7 F		0025	1.5F
	0505	0639	0.3E
	0923	1101	0.4F
	1216	1755	3.0E
	2214		
8 Sa		0109	1.9F
	0557	0753	0.5E
	1041	1158	0.3F
	1307	1849	3.2E
	2256		
9 Su		0151	2.3F
	0636	0845	0.8E
	1137	1252	0.4F
	1403	1943	3.4E
	2338		
10 M		0233	2.5F
	0710	0927	1.0E
	1227	1344	0.5F
●	1501	2034	3.5E
11 Tu	0020	0317	2.7F
	0740	1006	1.3E
	1318	1438	0.5F
	1602	2121	3.4E
12 W	0102	0400	2.7F
	0808	1046	1.6E
	1413	1536	0.5F
	1706	2206	3.2E
13 Th	0144	0443	2.7F
	0835	1128	1.8E
	1511	1635	0.6F
	1814	2251	2.8E
14 F	0227	0524	2.5F
	0902	1214	2.1E
	1609	1735	0.6F
	1923	2340	2.2E
15 Sa	0309	0603	2.2F
	0928	1303	2.3E
	1706	1834	0.6F
	2036		
16 Su		0038	1.6E
	0352	0642	1.9F
	0953	1356	2.4E
◐	1803	1933	0.6F
	2206		
17 M		0150	1.0E
	0436	0722	1.5F
	1017	1447	2.5E
	1900	2044	0.6F
18 Tu	0022	0305	0.6E
	0525	0804	1.1F
	1039	1538	2.6E
	1958	2316	0.8F
19 W	0239	0418	0.4E
	0634	0852	0.7F
	1059	1627	2.7E
	2053		
20 Th		0025	1.2F
	0408	0543	0.3E
	0819	0949	0.3F
	1123	1718	2.7E
	2143		
21 F		0113	1.5F
	0511	0757	0.5E
		1053	*
		1810	2.7E
	2226		
22 Sa		0154	1.7F
	0556	0849	0.7E
		1153	*
		1902	2.7E
	2305		
23 Su		0230	1.8F
	0632	0925	0.9E
		1245	*
		1950	2.7E
	2341		
24 M		0300	1.9F
	0702	0949	1.1E
		1331	*
		2031	2.7E
25 Tu	0014	0322	2.0F
	0728	1007	1.3E
		1416	*
○		2108	2.7E
26 W	0046	0340	2.1F
	0750	1031	1.4E
		1503	*
		2141	2.6E
27 Th	0116	0406	2.1F
	0807	1059	1.5E
		1551	*
		2211	2.4E
28 F	0145	0438	2.1F
	0822	1130	1.7E
	1533	1639	0.3F
	1753	2237	2.1E
29 Sa	0213	0512	1.9F
	0835	1201	1.8E
	1611	1728	0.4F
	1853	2302	1.8E
30 Su	0242	0546	1.8F
	0845	1234	1.9E
	1650	1816	0.6F
	1954	2333	1.3E
31 M	0311	0621	1.5F
	0855	1309	2.0E
	1730	1906	0.7F
	2102		

February

Day	Slack (h m)	Maximum (h m)	knots
1 Tu		0017	0.9E
	0343	0657	1.2F
	0908	1349	2.2E
◑	1815	2000	0.8F
	2244		
2 W		0158	0.5E
	0418	0735	0.9F
	0928	1437	2.3E
	1906	2103	0.9F
3 Th		0340	*
		0819	0.5F
	0954	1530	2.5E
	2002	2223	1.1F
4 F		0500	*
		0919	0.3F
	1029	1627	2.6E
	2058		
5 Sa		0005	1.4F
	0453	0626	0.3E
		1032	*
		1729	2.8E
	2151		
6 Su		0059	1.8F
	0532	0752	0.6E
		1141	*
		1833	3.0E
	2239		
7 M		0140	2.1F
	0604	0834	1.0E
	1133	1242	0.4F
	1358	1934	3.1E
	2323		
8 Tu		0218	2.4F
	0631	0909	1.4E
	1216	1339	0.6F
●	1517	2028	3.2E
9 W	0006	0256	2.5F
	0656	0944	1.8E
	1300	1434	0.8F
	1634	2116	3.1E
10 Th	0047	0334	2.5F
	0720	1019	2.1E
	1346	1530	0.9F
	1748	2201	2.8E
11 F	0127	0413	2.4F
	0744	1056	2.4E
	1434	1627	1.1F
	1856	2245	2.4E
12 Sa	0207	0451	2.2F
	0807	1135	2.6E
	1525	1722	1.1F
	1959	2331	1.9E
13 Su	0247	0529	1.9F
	0829	1217	2.6E
	1617	1814	1.1F
	2103		
14 M		0023	1.4E
	0328	0607	1.5F
	0849	1303	2.6E
	1710	1906	1.1F
	2217		
15 Tu		0127	0.9E
	0412	0646	1.2F
	0906	1355	2.5E
◐	1806	2001	1.0F
16 W	0006	0240	0.5E
	0505	0727	0.8F
	0922	1450	2.4E
	1906	2115	0.8F
17 Th	0215	0356	0.3E
	0628	0813	0.4F
	0939	1547	2.3E
	2008	2355	1.0F
18 F	0337	0531	0.3E
		0911	*
		1643	2.3E
	2107		
19 Sa		0049	1.3F
	0432	0742	0.6E
		1023	*
		1741	2.2E
	2158		
20 Su		0130	1.5F
	0513	0829	0.9E
		1137	*
		1839	2.3E
	2240		
21 M		0203	1.6F
	0546	0900	1.1E
		1237	*
		1931	2.3E
	2317		
22 Tu		0228	1.7F
	0612	0917	1.3E
		1324	*
		2015	2.4E
	2350		
23 W		0242	1.8F
	0633	0932	1.6E
	1255	1408	0.4F
○	1553	2054	2.4E
24 Th	0023	0301	1.9F
	0650	0954	1.8E
	1326	1452	0.6F
	1654	2129	2.3E
25 F	0053	0328	1.8F
	0704	1019	2.0E
	1357	1536	0.8F
	1751	2202	2.1E
26 Sa	0123	0400	1.8F
	0715	1045	2.1E
	1429	1622	0.9F
	1846	2234	1.9E
27 Su	0152	0434	1.6F
	0725	1110	2.2E
	1503	1707	1.1F
	1939	2304	1.6E
28 M	0221	0510	1.4F
	0734	1132	2.3E
	1540	1753	1.2F
	2033	2337	1.2E

March

Day	Slack (h m)	Maximum (h m)	knots
1 Tu	0251	0546	1.1F
	0748	1155	2.4E
	1623	1841	1.3F
	2136		
2 W		0023	0.8E
	0326	0623	0.9F
	0808	1230	2.4E
	1712	1932	1.3F
	2306		
3 Th		0201	0.4E
	0409	0702	0.6F
	0832	1324	2.4E
◑	1811	2030	1.2F
4 F	0132	0331	0.3E
	0519	0750	0.3F
	0900	1444	2.4E
	1916	2143	1.2F
5 Sa	0309	0449	0.3E
		0857	*
		1602	2.4E
	2024	2328	1.4F
6 Su	0359	0611	0.6E
		1018	*
		1714	2.4E
	2125		
7 M		0036	1.7F
	0435	0722	0.9E
		1136	*
		1823	2.5E
	2218		
8 Tu		0115	1.9F
	0504	0803	1.4E
	1121	1243	0.5F
	1434	1926	2.6E
	2305		
9 W		0150	2.0F
	0531	0838	1.9E
	1158	1340	0.9F
	1602	2021	2.6E
	2348		
10 Th		0224	2.1F
	0556	0913	2.3E
	1236	1433	1.2F
●	1718	2110	2.6E
11 F	0029	0259	2.0F
	0620	0947	2.6E
	1316	1525	1.4F
	1824	2154	2.3E
12 Sa	0109	0336	1.8F
	0643	1021	2.8E
	1358	1616	1.6F
	1923	2237	2.0E
13 Su	0149	0414	1.6F
	0704	1056	2.9E
	1443	1705	1.6F
	2019	2321	1.6E
14 M	0231	0453	1.4F
	0724	1131	2.8E
	1529	1751	1.6F
	2114		
15 Tu		0009	1.2E
	0316	0532	1.1F
	0741	1210	2.6E
	1619	1836	1.5F
	2215		
16 W		0108	0.8E
	0410	0613	0.7F
	0757	1257	2.4E
	1712	1924	1.3F
	2339		
17 Th		0221	0.6E
	0522	0656	0.4F
	0814	1357	2.1E
◐	1810	2017	1.0F
18 F	0126	0339	0.5E
		0744	*
		1503	1.9E
	1913	2137	0.9F
19 Sa	0240	0525	0.6E
		0844	*
		1607	1.8E
	2017		
20 Su		0007	1.1F
	0329	0704	0.8E
		1001	0.3E
		1709	1.8E
	2115		
21 M		0048	1.2F
	0406	0747	1.1E
		1133	*
		1809	1.8E
	2202		
22 Tu		0116	1.4F
	0434	0809	1.4E
		1239	*
		1905	1.9E
	2243		
23 W		0132	1.5F
	0457	0823	1.7E
	1151	1320	0.5F
	1531	1954	2.0E
	2321		
24 Th		0148	1.5F
	0515	0845	1.9E
	1215	1358	0.8F
	1636	2037	2.0E
	2356		
25 F		0214	1.5F
	0530	0910	2.2E
	1241	1438	1.1F
○	1734	2115	1.9E
26 Sa	0030	0245	1.4F
	0543	0936	2.4E
	1309	1519	1.4F
	1827	2152	1.8E
27 Su	0103	0319	1.3F
	0554	1000	2.6E
	1338	1602	1.6F
	1918	2228	1.6E
28 M	0135	0355	1.1F
	0606	1022	2.7E
	1411	1646	1.8F
	2007	2304	1.3E
29 Tu	0210	0433	0.9F
	0624	1043	2.8E
	1449	1732	1.8F
	2059	2346	1.0E
30 W	0249	0513	0.7F
	0647	1109	2.8E
	1534	1820	1.8F
	2159		
31 Th		0045	0.7E
	0342	0556	0.5F
	0714	1145	2.6E
	1627	1910	1.7F
	2317		

Time meridian 120° W. 0000 is midnight. 1200 is noon. Times are not adjusted for Daylight Saving Time.
If three consecutive entries are marked (E) the middle one is not a true maximum but an intermediate value to show the current pattern.
* Current weak and variable.

Rosario Strait, Washington, 2005

F–Flood, Dir. 335° True E–Ebb, Dir. 175° True

April

Day	Slack h m	Maximum h m	knots
1 F		0208	0.5E
	0511	0643	0.3F
	0742	1238	2.4E
◑	1727	2005	1.6F
2 Sa	0056	0327	0.6E
		0739	*
		1413	2.2E
	1834	2110	1.4F
3 Su	0205	0438	0.8E
		0851	*
		1547	2.0E
	1945	2232	1.4F
4 M	0249	0542	1.1E
		1017	*
		1701	2.0E
	2051	2348	1.5F
5 Tu	0323	0637	1.5E
	1030	1149	0.3F
	1341	1811	2.0E
	2150		
6 W		0033	1.6F
	0352	0722	2.0E
	1101	1258	0.8F
	1525	1916	2.0E
	2240		
7 Th		0110	1.6F
	0420	0801	2.4E
	1135	1349	1.3F
	1644	2013	2.0E
	2326		
8 F		0144	1.6F
	0445	0838	2.8E
	1211	1434	1.6F
●	1752	2102	1.9E
9 Sa	0010	0220	1.5F
	0510	0912	3.0E
	1248	1518	1.9F
	1850	2147	1.8E
10 Su	0053	0257	1.3F
	0532	0946	3.1E
	1326	1602	2.0F
	1943	2229	1.6E
11 M	0137	0338	1.0F
	0552	1019	3.1E
	1406	1645	2.0F
	2031	2312	1.3E
12 Tu	0225	0417	0.8F
	0610	1051	2.9E
	1448	1727	1.9F
	2119	2359	1.1E
13 W	0323	0459	0.5F
	0629	1122	2.6E
	1533	1810	1.7F
	2210		
14 Th		0056	0.8E
	0433	0544	0.3F
	0648	1155	2.3E
	1621	1853	1.5F
	2313		
15 F		0208	0.7E
		0631	*
		1242	1.9E
	1714	1940	1.3F
16 Sa	0028	0327	0.7E
		0722	*
		1411	1.7E
◐	1812	2034	1.1F
17 Su	0129	0440	0.9E
		0824	0.3E
		1526	1.5E
	1912	2141	1.0F
18 M	0212	0536	1.1E
		0942	0.4E
		1631	1.4E
	2014	2301	1.1F
19 Tu	0243	0615	1.4E
		1206	*
		1732	1.4E
	2110	2348	1.1F
20 W	0307	0649	1.7E
	1055	1243	0.3F
	1438	1833	1.4E
	2200		
21 Th		0021	1.2F
	0327	0721	2.0E
	1111	1312	0.8F
	1555	1929	1.4E
	2245		
22 F		0053	1.2F
	0344	0752	2.3E
	1135	1344	1.2F
	1700	2018	1.4E
	2326		
23 Sa		0127	1.1F
	0358	0822	2.6E
	1201	1420	1.6F
	1758	2101	1.4E
24 Su	0006	0202	1.0F
	0412	0851	2.8E
	1229	1500	1.9F
○	1850	2141	1.4E
25 M	0044	0238	0.9F
	0429	0918	3.0E
	1300	1542	2.1F
	1939	2221	1.3E
26 Tu	0125	0318	0.7F
	0453	0944	3.1E
	1335	1628	2.3F
	2026	2303	1.1E
27 W	0212	0400	0.5F
	0522	1011	3.1E
	1416	1714	2.3F
	2115	2351	0.9E
28 Th	0314	0447	0.4F
	0554	1043	3.0E
	1503	1802	2.2F
	2209		
29 F		0052	0.8E
		0538	*
		1123	2.7E
	1555	1852	2.1F
	2309		
30 Sa		0206	0.8E
		0634	*
		1219	2.3E
◑	1653	1943	1.9F

May

Day	Slack h m	Maximum h m	knots
1 Su	0009	0316	1.0E
		0736	*
		1402	1.9E
	1755	2040	1.7F
2 M	0059	0415	1.4E
		0850	*
		1533	1.7E
	1900	2142	1.5F
3 Tu	0139	0506	1.8E
		1031	*
		1646	1.5E
	2008	2247	1.4F
4 W	0212	0554	2.2E
	1003	1225	0.6F
	1437	1756	1.4E
	2113	2341	1.3F
5 Th	0241	0639	2.6E
	1036	1315	1.2F
	1607	1905	1.3E
	2212		
6 F		0024	1.2F
	0308	0722	2.9E
	1111	1357	1.6F
	1720	2007	1.3E
	2305		
7 Sa		0103	1.1F
	0333	0802	3.1E
	1147	1435	1.9F
	1820	2058	1.3E
	2354		
8 Su		0141	0.9F
	0354	0839	3.2E
	1223	1512	2.1F
●	1912	2142	1.3E
9 M	0042	0220	0.7F
	0414	0915	3.2E
	1259	1550	2.2F
	1958	2223	1.2E
10 Tu	0134	0300	0.5F
	0434	0948	3.1E
	1336	1628	2.1F
	2040	2305	1.1E
11 W	0232	0344	0.3F
	0455	1019	2.9E
	1415	1707	2.1F
	2120	2351	1.0E
12 Th		0430	*
		1047	2.6E
	1455	1746	1.9F
	2202		
13 F		0046	0.9E
		0519	*
		1112	2.3E
	1537	1827	1.8F
	2246		
14 Sa		0154	0.9E
		0611	*
		1142	1.9E
	1622	1910	1.6F
	2332		
15 Su		0302	1.1E
		0705	0.3E
		1233	1.5E
	1710	1955	1.5F
16 M	0014	0352	1.3E
		0805	0.4E
		1436	1.2E
◐	1800	2043	1.3F
17 Tu	0048	0432	1.5E
		0917	0.3E
		1548	1.1E
	1857	2136	1.1F
18 W	0115	0509	1.8E
		1145	*
		1652	0.9E
	2000	2229	1.0F
19 Th	0137	0545	2.1E
	1008	1225	0.6F
	1452	1756	0.9E
	2105	2318	0.9F
20 F	0155	0622	2.4E
	1030	1253	1.1F
	1613	1900	0.9E
	2204		
21 Sa		0002	0.8F
	0211	0658	2.6E
	1057	1325	1.6F
	1721	1958	0.9E
	2256		
22 Su		0042	0.8F
	0230	0734	2.9E
	1126	1401	2.0F
	1818	2048	1.0E
	2344		
23 M		0122	0.7F
	0255	0809	3.1E
	1158	1441	2.3F
○	1908	2132	1.0E
24 Tu	0030	0203	0.6F
	0325	0844	3.3E
	1234	1525	2.5F
	1953	2214	1.0E
25 W	0120	0246	0.4F
	0359	0918	3.4E
	1312	1611	2.6F
	2035	2258	1.0E
26 Th	0220	0335	0.3F
	0437	0954	3.3E
	1355	1659	2.6F
	2116	2347	1.0E
27 F		0429	*
		1033	3.1E
	1441	1747	2.5F
	2158		
28 Sa		0044	1.1E
		0528	*
		1118	2.7E
	1531	1833	2.3F
	2238		
29 Su		0149	1.3E
		0629	*
		1219	2.2E
	1623	1920	2.1F
	2318		
30 M		0249	1.6E
		0733	*
		1353	1.7E
◑	1717	2008	1.8F
	2356		
31 Tu		0341	2.0E
		0848	*
		1516	1.3E
	1815	2059	1.5F

June

Day	Slack h m	Maximum h m	knots
1 W	0031	0428	2.4E
	0850	1114	0.4F
	1321	1628	1.0E
	1920	2153	1.2F
2 Th	0103	0514	2.7E
	0931	1226	1.0F
	1515	1739	0.8E
	2032	2249	1.0F
3 F	0132	0559	2.9E
	1010	1314	1.5F
	1639	1855	0.8E
	2144	2339	0.8F
4 Sa	0159	0645	3.1E
	1048	1356	1.8F
	1745	2008	0.8E
	2247		
5 Su		0025	0.7F
	0222	0729	3.2E
	1125	1434	2.0F
	1839	2102	0.9E
	2344		
6 M		0107	0.5F
	0244	0811	3.2E
	1202	1510	2.1F
●	1924	2144	1.0E
7 Tu	0039	0149	0.4F
	0306	0850	3.1E
	1238	1543	2.2F
	2003	2221	1.0E
8 W		0232	*
		0925	3.0E
	1313	1615	2.1F
	2038	2259	1.0E
9 Th		0318	*
		0958	2.8E
	1349	1648	2.1F
	2110	2341	1.1E
10 F		0407	*
		1026	2.5E
	1425	1724	2.1F
	2140		
11 Sa		0029	1.1E
		0459	*
		1051	2.2E
	1501	1801	2.0F
	2209		
12 Su		0122	1.2E
		0552	*
		1117	1.9E
	1538	1839	1.9F
	2236		
13 M		0214	1.3E
		0646	*
		1153	1.5E
	1615	1918	1.7F
	2302		
14 Tu		0259	1.6E
		0742	*
		1308	1.1E
◐	1656	1959	1.5F
	2324		
15 W		0338	1.8E
		0845	*
		1502	0.8E
	1740	2043	1.2F
	2343		
16 Th		0414	2.1E
		1002	*
		1613	0.6E
	1833	2131	0.9F
17 F	0000	0450	2.3E
	0917	1138	0.7F
	1503	1720	0.5E
	1946	2222	0.7F
18 Sa	0021	0528	2.6E
	0948	1227	1.2F
	1632	1831	0.5E
	2118	2314	0.6F
19 Su	0048	0608	2.8E
	1022	1305	1.7F
	1739	1940	0.5E
	2230		
20 M		0002	0.5F
	0122	0651	3.1E
	1058	1344	2.1F
	1831	2037	0.7E
	2327		
21 Tu		0049	0.4F
	0201	0737	3.3E
	1136	1425	2.4F
○	1913	2122	0.8E
22 W	0019	0136	0.4F
	0245	0822	3.4E
	1215	1510	2.6F
	1949	2204	1.0E
23 Th	0112	0225	0.4F
	0331	0905	3.4E
	1256	1556	2.7F
	2022	2246	1.2E
24 F	0211	0319	0.3F
	0423	0948	3.3E
	1339	1642	2.7F
	2054	2331	1.4E
25 Sa	0316	0419	0.3F
	0521	1032	3.0E
	1423	1727	2.6F
	2124		
26 Su		0021	1.6E
	0420	0520	0.3F
	0625	1121	2.5E
	1509	1810	2.4F
	2154		
27 M		0115	1.8E
	0520	0622	0.3F
	0735	1221	2.0E
	1555	1852	2.2F
	2223		
28 Tu		0209	2.1E
	0617	0725	0.4F
	0901	1339	1.4E
◑	1642	1934	1.8F
	2253		
29 W		0301	2.4E
	0712	0836	0.5F
	1120	1457	1.0E
	1733	2018	1.4F
	2323		
30 Th		0349	2.7E
	0806	1056	0.7F
	1351	1609	0.7E
	1834	2107	1.0F
	2352		

Time meridian 120° W. 0000 is midnight. 1200 is noon. Times are not adjusted for Daylight Saving Time.
If three consecutive entries are marked (E) the middle one is not a true maximum but an intermediate value to show the current pattern.
* Current weak and variable.

Rosario Strait, Washington, 2005

F–Flood, Dir. 335° True E–Ebb, Dir. 175° True

July

Day	Slack h m	Maximum h m	knots
1 F		0436	2.9E
	0856	1212	1.2F
	1536	1721	0.5E
	1955	2202	0.7F
2 Sa	0020	0523	3.0E
	0943	1304	1.6F
	1653	1853	0.5E
	2125	2259	0.5F
3 Su	0049	0612	3.0E
	1026	1348	1.8F
	1752	2028	0.6E
	2241	2353	0.3F
4 M	0118	0702	3.0E
	1106	1428	2.0F
	1838	2117	0.8E
5 Tu		0042	*
		0749	3.0E
	1144	1505	2.0F
	1915	2151	1.0E
6 W ●		0128	*
		0832	2.9E
	1220	1535	2.1F
	1948	2218	1.1E
7 Th		0213	*
		0910	2.8E
	1254	1559	2.1F
	2016	2246	1.2E
8 F		0300	*
		0944	2.7E
	1327	1625	2.1F
	2040	2319	1.3E
9 Sa		0350	*
		1014	2.4E
	1359	1656	2.1F
	2100	2355	1.4E
10 Su		0442	*
		1041	2.2E
	1430	1730	2.0F
	2119		
11 M		0035	1.5E
		0533	*
		1106	1.8E
	1501	1805	1.9F
	2135		
12 Tu		0116	1.6E
		0624	*
		1137	1.5E
	1533	1841	1.7F
	2150		
13 W		0158	1.8E
		0715	*
		1224	1.0E
	1607	1918	1.4F
	2201		
14 Th ◐		0239	2.0E
	0702	0811	0.3F
	0948	1413	0.6E
	1644	1957	1.1F
	2214		
15 F		0318	2.2E
	0742	0914	0.5F
	1252	1539	0.4E
	1727	2040	0.8F
	2235		

Day	Slack h m	Maximum h m	knots
16 Sa		0357	2.4E
	0824	1032	0.9F
		1651	*
		2132	0.5F
	2303		
17 Su		0439	2.6E
	0908	1155	1.3F
		1807	*
		2232	0.3F
	2342		
18 M		0526	2.8E
	0952	1247	1.7F
	1740	1927	0.4E
	2220	2331	0.3F
19 Tu	0033	0620	3.0E
	1035	1329	2.1F
	1819	2026	0.6E
	2318		
20 W		0027	0.3F
	0131	0717	3.2E
	1118	1410	2.3F
	1851	2108	0.9E
21 Th ○	0007	0120	0.4F
	0232	0810	3.3E
	1159	1452	2.5F
	1919	2146	1.2E
22 F	0055	0214	0.5F
	0336	0859	3.3E
	1241	1535	2.6F
	1945	2224	1.6E
23 Sa	0146	0311	0.6F
	0444	0945	3.1E
	1322	1617	2.6F
	2010	2303	1.9E
24 Su	0240	0411	0.7F
	0556	1031	2.8E
	1404	1659	2.5F
	2035	2346	2.1E
25 M	0336	0511	0.7F
	0710	1118	2.3E
	1446	1739	2.2F
	2100		
26 Tu		0032	2.3E
	0431	0610	0.8F
	0824	1214	1.7E
	1529	1819	1.9F
	2125		
27 W ◑		0123	2.5E
	0526	0708	0.9F
	0948	1322	1.2E
	1614	1859	1.6F
	2149		
28 Th		0215	2.6E
	0622	0811	0.9F
	1141	1437	0.7E
	1704	1941	1.2F
	2212		
29 F		0308	2.7E
	0720	1000	0.9F
	1357	1550	0.5E
	1808	2028	0.8F
	2235		
30 Sa		0359	2.7E
	0819	1151	1.2F
	1532	1707	0.4E
	1944	2122	0.4F
	2300		
31 Su		0451	2.7E
	0914	1248	1.5F
	1640	1921	0.4E
		2226	*

August

Day	Slack h m	Maximum h m	knots
1 M		0545	2.7E
	1003	1333	1.7F
	1731	2026	0.7E
		2331	*
2 Tu		0640	2.6E
	1046	1412	1.8F
	1810	2108	0.9E
3 W		0028	*
		0732	2.6E
	1125	1445	1.9F
	1842	2136	1.1E
4 Th ●		0117	*
		0817	2.6E
	1200	1510	1.9F
	1908	2155	1.3E
5 F		0202	*
		0856	2.6E
	1233	1527	2.0F
	1930	2216	1.5E
6 Sa		0248	*
		0931	2.5E
	1304	1550	2.0F
	1948	2243	1.6E
7 Su	0231	0335	0.3F
	0455	1003	2.3E
	1333	1620	1.9F
	2002	2312	1.7E
8 M	0310	0424	0.4F
	0553	1033	2.0E
	1402	1653	1.8F
	2016	2342	1.8E
9 Tu	0348	0512	0.5F
	0651	1100	1.7E
	1431	1728	1.7F
	2026		
10 W		0013	1.9E
	0426	0559	0.6F
	0750	1129	1.3E
	1501	1803	1.4F
	2034		
11 Th		0045	2.0E
	0505	0647	0.7F
	0854	1208	0.9E
	1534	1839	1.1F
	2046		
12 F ◐		0122	2.1E
	0548	0738	0.8F
	1020	1337	0.5E
	1609	1916	0.8F
	2104		
13 Sa		0207	2.2E
	0637	0835	0.9F
	1250	1517	0.3E
	1654	1957	0.5F
	2128		
14 Su		0301	2.3E
	0732	0945	1.0F
		1634	*
		2051	*
15 M		0358	2.5E
	0829	1116	1.3F
	1621	1751	0.3E
		2202	*

Day	Slack h m	Maximum h m	knots
16 Tu		0459	2.6E
	0924	1229	1.6F
	1703	1912	0.5E
		2313	*
17 W		0603	2.8E
	1014	1311	2.0F
	1734	2005	0.9E
	2311		
18 Th		0017	0.3F
	0128	0706	2.9E
	1059	1349	2.2F
	1801	2042	1.3E
	2351		
19 F ○		0115	0.6F
	0252	0802	3.0E
	1142	1427	2.3F
	1825	2118	1.7E
20 Sa	0032	0210	0.8F
	0412	0853	3.0E
	1223	1505	2.3F
	1849	2153	2.1E
21 Su	0116	0305	1.0F
	0529	0940	2.8E
	1303	1544	2.3F
	1913	2229	2.4E
22 M	0202	0402	1.2F
	0639	1025	2.5E
	1344	1624	2.1F
	1937	2307	2.7E
23 Tu	0252	0458	1.3F
	0745	1111	2.0E
	1426	1704	1.8F
	2000	2348	2.7E
24 W	0343	0552	1.4F
	0850	1202	1.5E
	1509	1745	1.5F
	2021		
25 Th		0033	2.7E
	0437	0644	1.3F
	1002	1305	1.0E
	1558	1825	1.2F
	2041		
26 F ◑		0125	2.6E
	0533	0739	1.2F
	1138	1419	0.6E
	1658	1908	0.8F
	2101		
27 Sa		0223	2.5E
	0633	0844	1.0F
	1339	1536	0.5E
	1819	1955	0.4F
	2120		
28 Su		0323	2.4E
	0736	1118	1.1F
	1502	1703	0.4E
		2051	*
29 M		0422	2.3E
	0839	1223	1.3F
	1559	1907	0.6E
		2203	*
30 Tu		0521	2.2E
	0934	1308	1.5F
	1642	2001	0.9E
		2323	*
31 W		0619	2.2E
	1020	1343	1.6F
	1716	2036	1.2E

September

Day	Slack h m	Maximum h m	knots
1 Th		0029	*
		0713	2.2E
	1059	1409	1.7F
	1743	2057	1.4E
2 F	0007	0115	0.3F
	0255	0800	2.3E
	1134	1425	1.7F
	1805	2113	1.7E
3 Sa ●	0035	0157	0.5F
	0359	0840	2.3E
	1207	1441	1.7F
	1823	2135	1.9E
4 Su	0105	0238	0.7F
	0456	0917	2.2E
	1238	1507	1.7F
	1837	2201	2.0E
5 M	0136	0320	0.8F
	0550	0951	2.0E
	1309	1538	1.6F
	1850	2226	2.2E
6 Tu	0208	0404	1.0F
	0642	1024	1.8E
	1339	1613	1.5F
	1900	2250	2.3E
7 W	0240	0448	1.2F
	0732	1055	1.5E
	1410	1649	1.2F
	1908	2311	2.3E
8 Th	0315	0534	1.3F
	0823	1128	1.2E
	1441	1725	1.0F
	1921	2330	2.4E
9 F	0356	0620	1.3F
	0921	1210	0.8E
	1517	1802	0.7F
	1940	2358	2.4E
10 Sa	0443	0709	1.3F
	1039	1339	0.5E
	1604	1842	0.5F
	2002		
11 Su ◐		0042	2.3E
	0539	0803	1.2F
	1246	1510	0.3E
		1928	*
12 M		0157	2.2E
	0642	0908	1.2F
	1433	1626	0.4E
		2029	*
13 Tu		0328	2.2E
	0749	1031	1.3F
	1524	1737	0.6E
		2148	*
14 W		0442	2.3E
	0853	1154	1.5F
	1559	1841	0.9E
		2309	*
15 Th		0551	2.3E
	0948	1240	1.8F
	1627	1928	1.4E
	2257		

Day	Slack h m	Maximum h m	knots
16 F		0018	0.5F
	0206	0656	2.4E
	1037	1317	1.9F
	1652	2007	1.9E
	2331		
17 Sa ○		0116	0.9F
	0340	0754	2.5E
	1121	1353	2.0F
	1718	2043	2.3E
18 Su	0008	0209	1.3F
	0457	0846	2.4E
	1203	1429	1.9F
	1743	2119	2.7E
19 M	0047	0300	1.6F
	0605	0932	2.3E
	1244	1507	1.8F
	1807	2154	3.0E
20 Tu	0128	0351	1.8F
	0707	1017	2.0E
	1326	1547	1.6F
	1831	2229	3.1E
21 W	0212	0441	1.9F
	0805	1102	1.7E
	1410	1627	1.3F
	1852	2305	3.0E
22 Th	0259	0530	1.8F
	0902	1151	1.3E
	1500	1710	1.0F
	1913	2344	2.8E
23 F	0350	0618	1.7F
	1004	1250	0.9E
	1601	1753	0.7F
	1932		
24 Sa ◑		0029	2.5E
	0444	0706	1.5F
	1124	1404	0.7E
	1718	1838	0.4F
	1950		
25 Su		0131	2.2E
	0543	0800	1.2F
	1301	1528	0.6E
		1928	*
26 M		0242	2.0E
	0646	0911	1.0F
	1412	1710	0.7E
		2028	*
27 Tu		0350	1.8E
	0750	1140	1.1F
	1500	1829	1.0E
		2146	0.3E
28 W		0452	1.8E
	0850	1226	1.2F
	1536	1915	1.2E
29 Th		0011	*
		0552	1.7E
	0941	1256	1.3F
	1604	1941	1.5E
30 F		0053	*
		0648	1.8E
	1024	1313	1.4F
	1626	2000	1.8E
	2334		

Time meridian 120° W. 0000 is midnight. 1200 is noon. Times are not adjusted for Daylight Saving Time.
If three consecutive entries are marked (E) the middle one is not a true maximum but an intermediate value to show the current pattern.
* Current weak and variable.

Rosario Strait, Washington, 2005

F–Flood, Dir. 335° True   E–Ebb, Dir. 175° True

October

Day	Slack (h m)	Maximum (h m)	knots
1 Sa		0120	0.5F
	0334	0738	1.8E
	1102	1328	1.4F
	1645	2024	2.1E
	2357		
2 Su		0150	0.9F
	0435	0822	1.8E
	1138	1353	1.4F
	1701	2050	2.3E
3 M ●	0023	0225	1.2F
	0530	0902	1.8E
	1213	1423	1.3F
	1714	2116	2.5E
4 Tu	0050	0302	1.4F
	0621	0939	1.7E
	1247	1456	1.2F
	1725	2141	2.6E
5 W	0118	0343	1.6F
	0709	1015	1.5E
	1322	1532	1.0F
	1736	2203	2.7E
6 Th	0149	0426	1.8F
	0756	1051	1.3E
	1357	1609	0.8F
	1752	2221	2.7E
7 F	0224	0510	1.8F
	0845	1131	1.0E
	1439	1649	0.6F
	1814	2243	2.7E
8 Sa	0305	0557	1.8F
	0940	1224	0.7E
	1538	1732	0.3F
	1839	2314	2.6E
9 Su	0355	0646	1.7F
	1049	1344	0.5E
		1819	*
		2357	2.4E
10 M ◐	0452	0738	1.6F
	1219	1506	0.6E
		1914	*
11 Tu		0108	2.1E
	0557	0837	1.5F
	1331	1614	0.8F
		2022	*
12 W		0309	1.9E
	0705	0947	1.4F
	1414	1712	1.1E
		2145	*
13 Th		0429	1.8E
	0813	1100	1.5F
	1447	1802	1.5E
		2318	*
14 F		0539	1.8E
	0915	1155	1.5F
	1515	1847	2.0E
	2236		
15 Sa		0032	0.8F
	0301	0645	1.8E
	1009	1237	1.6F
	1542	1929	2.5E
	2309		
16 Su		0124	1.3F
	0422	0746	1.8E
	1058	1315	1.6F
	1608	2008	2.9E
	2344		
17 M ○		0210	1.7F
	0532	0839	1.8E
	1144	1352	1.4F
	1633	2045	3.2E
18 Tu	0021	0254	2.0F
	0633	0926	1.7E
	1228	1430	1.3F
	1657	2121	3.3E
19 W	0059	0339	2.2F
	0728	1010	1.6E
	1314	1510	1.0F
	1719	2155	3.3E
20 Th	0140	0424	2.2F
	0819	1054	1.3E
	1405	1552	0.8F
	1741	2229	3.1E
21 F	0222	0509	2.1F
	0908	1142	1.1E
	1505	1637	0.5F
	1802	2302	2.8E
22 Sa	0308	0553	1.9F
	1001	1239	0.9E
		1724	*
		2337	2.4E
23 Su	0358	0637	1.7F
	1101	1352	0.8E
		1813	*
24 M ◑		0022	2.0E
	0452	0724	1.5F
	1209	1521	0.8E
		1907	*
25 Tu		0149	1.7E
	0548	0815	1.3F
	1306	1635	1.0E
		2008	0.4E
26 W		0309	1.4E
	0647	0916	1.1F
	1348	1721	1.3E
		2128	0.4E
27 Th		0415	1.3E
	0748	1030	1.1F
	1418	1755	1.6E
28 F		0020	*
		0516	1.3E
	0846	1125	1.1F
	1442	1828	1.9E
	2242		
29 Sa		0054	0.4F
	0242	0616	1.2E
	0938	1200	1.1F
	1502	1900	2.2E
	2257		
30 Su		0117	0.8F
	0355	0713	1.3E
	1026	1233	1.1F
	1520	1932	2.4E
	2320		
31 M		0139	1.2F
	0457	0803	1.3E
	1109	1307	1.0F
	1534	2003	2.7E
	2346		

November

Day	Slack (h m)	Maximum (h m)	knots
1 Tu ●		0209	1.6F
	0552	0847	1.3E
	1150	1341	0.9F
	1546	2033	2.8E
2 W	0013	0244	1.9F
	0642	0928	1.3E
	1230	1417	0.8F
	1602	2100	3.0E
3 Th	0042	0323	2.1F
	0728	1007	1.2E
	1311	1455	0.6F
	1624	2125	3.1E
4 F	0115	0406	2.2F
	0813	1046	1.1E
	1357	1535	0.4F
	1650	2149	3.1E
5 Sa	0152	0452	2.3F
	0859	1131	0.9E
	1500	1620	0.3F
	1720	2218	3.0E
6 Su	0234	0538	2.2F
	0947	1226	0.8E
		1711	*
		2253	2.8E
7 M	0323	0626	2.1F
	1040	1337	0.8E
		1807	*
		2337	2.4E
8 Tu ◐	0417	0716	1.9F
	1134	1449	1.0E
		1909	*
9 W		0048	1.9E
	0515	0808	1.7F
	1223	1547	1.3E
		2019	*
10 Th		0252	1.6E
	0618	0905	1.5F
	1302	1637	1.7E
		2146	*
11 F		0413	1.4E
	0726	1006	1.4F
	1336	1723	2.2E
	2139	2356	0.5F
12 Sa	0206	0524	1.2E
	0835	1104	1.3F
	1406	1807	2.6E
	2212		
13 Su		0052	1.2F
	0344	0634	1.2E
	0939	1153	1.2F
	1434	1851	3.0E
	2247		
14 M		0135	1.7F
	0500	0740	1.2E
	1036	1236	1.1F
	1501	1934	3.2E
	2323		
15 Tu ○		0214	2.0F
	0603	0836	1.3E
	1129	1317	0.9F
	1525	2014	3.4E
16 W	0000	0252	2.3F
	0657	0923	1.3E
	1219	1357	0.8F
	1549	2052	3.4E
17 Th	0037	0331	2.3F
	0744	1006	1.2E
	1311	1438	0.6F
	1612	2129	3.3E
18 F	0115	0411	2.3F
	0827	1049	1.1E
	1409	1522	0.4F
	1636	2202	3.1E
19 Sa	0155	0451	2.2F
	0908	1134	1.0E
		1609	*
		2234	2.8E
20 Su	0235	0531	2.1F
	0949	1227	1.0E
		1700	*
		2303	2.4E
21 M	0318	0611	1.9F
	1030	1332	1.0E
		1752	*
		2332	2.0E
22 Tu	0402	0652	1.7F
	1111	1442	1.1E
		1847	0.3E
23 W ◑		0014	1.6E
	0447	0735	1.5F
	1149	1535	1.3E
		1947	0.4E
24 Th		0213	1.2E
	0536	0820	1.4F
	1223	1615	1.6E
		2058	0.3E
25 F		0330	1.0E
	0629	0910	1.2F
	1251	1651	1.9E
		2357	*
26 Sa		0435	0.8E
	0732	1003	1.0F
	1314	1727	2.1E
	2157		
27 Su		0035	0.5F
	0252	0539	0.8E
	0841	1055	0.9F
	1333	1804	2.4E
	2218		
28 M		0100	1.0F
	0412	0644	0.8E
	0946	1141	0.8F
	1350	1841	2.6E
	2244		
29 Tu		0123	1.5F
	0517	0745	0.8E
	1042	1223	0.7F
	1408	1917	2.9E
	2313		
30 W		0152	1.9F
	0611	0836	0.9E
	1131	1303	0.6F
	1432	1953	3.1E
	2344		

December

Day	Slack (h m)	Maximum (h m)	knots
1 Th ●		0227	2.2F
	0658	0919	1.0E
	1218	1343	0.5F
	1501	2027	3.2E
2 F	0017	0307	2.4F
	0741	0959	1.0E
	1305	1425	0.4F
	1534	2101	3.3E
3 Sa	0053	0350	2.5F
	0819	1040	1.0E
	1400	1511	0.3F
	1611	2134	3.3E
4 Su	0132	0436	2.5F
	0856	1124	1.0E
		1603	*
		2209	3.1E
5 M	0214	0521	2.5F
	0932	1215	1.1E
		1700	*
		2248	2.8E
6 Tu	0300	0607	2.4F
	1008	1314	1.3E
		1800	*
		2336	2.3E
7 W	0348	0651	2.2F
	1043	1414	1.5E
		1903	*
8 Th ◐		0051	1.7E
	0438	0737	1.9F
	1117	1508	1.9E
		2012	*
9 F		0235	1.2E
	0532	0825	1.6F
	1151	1557	2.3E
	2020	2144	0.3F
10 Sa	0028	0354	0.9E
	0634	0917	1.3F
	1225	1643	2.6E
	2103	2359	0.9F
11 Su	0247	0507	0.7E
	0750	1014	1.0F
	1256	1730	2.9E
	2145		
12 M		0053	1.4F
	0418	0623	0.7E
	0911	1110	0.8F
	1327	1817	3.1E
	2225		
13 Tu		0137	1.9F
	0527	0744	0.8E
	1022	1201	0.6F
	1356	1904	3.2E
	2304		
14 W		0217	2.1F
	0622	0846	0.9E
	1124	1247	0.5F
	1424	1950	3.3E
	2343		
15 Th ○		0255	2.3F
	0708	0930	1.0E
	1219	1331	0.4F
	1452	2032	3.2E
16 F	0020	0331	2.3F
	0747	1008	1.1E
	1315	1415	0.3F
	1520	2111	3.1E
17 Sa	0057	0404	2.3F
	0822	1044	1.1E
		1501	*
		2146	2.9E
18 Su	0134	0435	2.2F
	0853	1123	1.2E
		1550	*
		2218	2.7E
19 M	0209	0509	2.2F
	0922	1206	1.2E
		1641	*
		2246	2.3E
20 Tu	0244	0544	2.1F
	0948	1254	1.3E
		1734	*
		2311	2.0E
21 W	0319	0620	1.9F
	1012	1345	1.4E
		1827	*
		2341	1.6E
22 Th	0354	0657	1.7F
	1035	1433	1.6E
		1922	*
23 F ◑		0031	1.1E
	0431	0736	1.5F
	1056	1515	1.8E
		2021	*
24 Sa		0236	0.7E
	0512	0818	1.2F
	1114	1554	2.0E
		2134	*
25 Su		0352	0.5E
	0602	0904	0.9F
	1131	1632	2.3E
	2103	2358	0.6F
26 M	0259	0501	0.4E
	0715	0956	0.6F
	1151	1711	2.5E
	2136		
27 Tu		0036	1.1F
	0429	0614	0.4E
	0903	1051	0.4F
	1219	1751	2.7E
	2210		
28 W		0105	1.5F
	0534	0730	0.4E
	1023	1142	0.3F
	1255	1835	2.9E
	2246		
29 Th		0136	1.9F
	0622	0829	0.6E
	1121	1231	0.3F
	1336	1921	3.1E
	2322		
30 F ●		0213	2.2F
	0701	0911	0.8E
	1209	1318	0.3F
	1422	2006	3.3E
	2359		
31 Sa		0252	2.5F
	0734	0949	1.0E
	1256	1406	0.3F
	1511	2049	3.3E

Time meridian 120° W. 0000 is midnight. 1200 is noon. Times are not adjusted for Daylight Saving Time.
If three consecutive entries are marked (E) the middle one is not a true maximum but an intermediate value to show the current pattern.
* Current weak and variable.

San Juan Channel (south entrance), Washington, 2005

F–Flood, Dir. 010° True E–Ebb, Dir. 180° True

January

Day	Slack h m	Maximum h m	knots
1 Sa	0142	0527	3.1F
	0908	1215	2.2E
	1611	1803	0.9F
	2010	2337	1.6E
2 Su	0228	0609	2.8F
	0939	1300	2.4E
	1651	1902	1.2F
	2153		
3 M		0049	1.3E
	0324	0656	2.3F
	1013	1345	2.7E
◑	1732	2008	1.7F
	2342		
4 Tu		0202	1.1E
	0433	0750	1.9F
	1049	1430	3.0E
	1814	2117	2.3F
5 W	0104	0314	1.2E
	0559	0853	1.5F
	1128	1516	3.3E
	1859	2218	3.1F
6 Th	0211	0427	1.3E
	0730	0956	1.3F
	1211	1604	3.6E
	1945	2310	3.9F
7 F	0310	0541	1.7E
	0846	1052	1.3F
	1256	1657	3.8E
	2032	2359	4.5F
8 Sa	0404	0644	2.0E
	0948	1145	1.3F
	1345	1750	4.1E
	2119		
9 Su		0047	5.0F
	0454	0736	2.4E
	1042	1236	1.3F
	1437	1843	4.3E
	2206		
10 M		0135	5.2F
	0541	0821	2.7E
	1133	1329	1.4F
●	1533	1934	4.4E
	2254		
11 Tu		0225	5.2F
	0625	0905	2.8E
	1224	1424	1.4F
	1633	2022	4.2E
	2341		
12 W		0314	5.1F
	0707	0950	2.9E
	1316	1522	1.5F
	1736	2111	3.8E
13 Th	0030	0402	4.7F
	0747	1036	2.9E
	1410	1620	1.6F
	1842	2203	3.2E
14 F	0120	0447	4.1F
	0826	1126	2.9E
	1503	1719	1.7F
	1955	2302	2.4E
15 Sa	0212	0532	3.5F
	0903	1219	2.9E
	1556	1819	1.8F
	2122		
16 Su		0012	1.7E
	0308	0617	2.7F
	0941	1311	3.0E
◐	1648	1929	1.9F
	2303		
17 M		0133	1.2E
	0411	0706	2.0F
	1019	1401	3.0E
	1740	2051	2.2F
18 Tu	0037	0256	1.0E
	0525	0804	1.4F
	1058	1449	3.0E
	1831	2205	2.6F
19 W	0153	0419	1.0E
	0651	0912	1.0F
	1140	1536	3.0E
	1920	2259	3.0F
20 Th	0253	0534	1.2E
	0812	1016	0.8F
	1223	1623	3.0E
	2005	2341	3.3F
21 F	0341	0632	1.4E
	0915	1107	0.8F
	1305	1711	3.0E
	2046		
22 Sa		0018	3.6F
	0422	0716	1.6E
	1003	1149	0.8F
	1347	1755	3.1E
	2124		
23 Su		0052	3.7F
	0458	0749	1.8E
	1042	1228	0.9F
	1428	1835	3.1E
	2159		
24 M		0124	3.8F
	0530	0817	2.0E
	1118	1306	0.9F
	1509	1912	3.2E
	2232		
25 Tu		0157	3.9F
	0600	0841	2.1E
	1153	1345	1.0F
○	1551	1946	3.2E
	2303		
26 W		0230	3.8F
	0628	0906	2.2E
	1229	1427	1.1F
	1634	2020	3.1E
	2334		
27 Th		0304	3.8F
	0653	0932	2.3E
	1305	1510	1.2F
	1720	2055	2.9E
28 F	0005	0338	3.6F
	0718	1002	2.4E
	1341	1554	1.3F
	1809	2132	2.5E
29 Sa	0038	0414	3.3F
	0741	1036	2.5E
	1419	1640	1.5F
	1904	2216	2.1E
30 Su	0117	0450	2.9F
	0805	1115	2.5E
	1458	1729	1.7F
	2011	2311	1.7E
31 M	0202	0530	2.5F
	0832	1200	2.6E
	1541	1822	1.9F
	2139		

February

Day	Slack h m	Maximum h m	knots
1 Tu		0020	1.3E
	0300	0615	1.9F
	0903	1251	2.7E
◑	1629	1925	2.2F
	2321		
2 W		0139	1.1E
	0416	0710	1.4F
	0941	1345	2.9E
	1722	2039	2.6F
3 Th	0049	0258	1.1E
	0555	0819	1.0F
	1031	1440	3.1E
	1819	2152	3.2F
4 F	0158	0420	1.3E
	0734	0934	0.9F
	1132	1539	3.3E
	1918	2253	3.9F
5 Sa	0256	0537	1.7E
	0847	1040	1.0F
	1237	1639	3.6E
	2014	2345	4.5F
6 Su	0347	0636	2.2E
	0940	1137	1.3F
	1341	1739	3.9E
	2107		
7 M		0033	4.9F
	0433	0722	2.6E
	1026	1229	1.6F
	1443	1835	4.1E
	2156		
8 Tu		0120	5.0F
	0515	0803	2.9E
	1109	1321	1.9F
●	1546	1927	4.2E
	2243		
9 W		0205	4.9F
	0554	0841	3.2E
	1153	1413	2.1F
	1647	2015	4.0E
	2329		
10 Th		0250	4.6F
	0629	0918	3.3E
	1237	1506	2.3F
	1748	2101	3.6E
11 F	0015	0333	4.1F
	0703	0956	3.3E
	1323	1559	2.4F
	1850	2149	3.0E
12 Sa	0102	0415	3.5F
	0734	1036	3.2E
	1411	1651	2.4F
	1955	2242	2.3E
13 Su	0152	0456	2.8F
	0804	1120	3.0E
	1500	1745	2.4F
	2110	2346	1.6E
14 M	0247	0537	2.1F
	0835	1209	2.8E
	1551	1843	2.3F
	2239		
15 Tu		0107	1.1E
	0352	0623	1.4F
	0907	1303	2.6E
◐	1644	1954	2.3F
16 W	0011	0237	0.9E
	0513	0719	0.8F
	0946	1400	2.5E
	1741	2121	2.4F
17 Th	0126	0403	1.0E
	0648	0835	0.5F
	1042	1457	2.4E
	1838	2229	2.7F
18 F	0223	0515	1.2E
	0812	0959	0.5F
	1150	1553	2.5E
	1932	2316	3.1F
19 Sa	0308	0609	1.5E
	0905	1057	0.7F
	1251	1647	2.6E
	2019	2353	3.3F
20 Su	0346	0649	1.8E
	0943	1139	0.9F
	1345	1737	2.8E
	2101		
21 M		0026	3.5F
	0419	0720	2.0E
	1015	1216	1.1F
	1433	1819	2.9E
	2138		
22 Tu		0056	3.6F
	0449	0744	2.2E
	1045	1251	1.4F
	1518	1857	3.0E
	2212		
23 W		0126	3.7F
	0516	0805	2.4E
	1114	1328	1.6F
○	1602	1932	3.1E
	2243		
24 Th		0157	3.6F
	0540	0827	2.6E
	1143	1406	1.8F
	1646	2006	3.0E
	2315		
25 F		0230	3.5F
	0602	0852	2.8E
	1212	1446	2.0F
	1732	2041	2.8E
	2347		
26 Sa		0304	3.2F
	0623	0919	2.9E
	1243	1528	2.2F
	1820	2119	2.5E
27 Su	0023	0339	2.9F
	0644	0950	2.9E
	1317	1613	2.4F
	1912	2203	2.2E
28 M	0105	0418	2.4F
	0707	1025	2.9E
	1356	1700	2.6F
	2015	2257	1.7E

March

Day	Slack h m	Maximum h m	knots
1 Tu	0157	0459	1.9F
	0732	1109	2.8E
	1442	1753	2.7F
	2134		
2 W		0007	1.4E
	0305	0547	1.4F
	0804	1204	2.7E
	1537	1854	2.8F
	2307		
3 Th		0131	1.2E
	0434	0645	0.9F
	0845	1308	2.7E
◑	1639	2008	3.0F
4 F	0031	0256	1.3E
	0616	0801	0.6F
	0949	1416	2.8E
	1748	2128	3.3F
5 Sa	0137	0416	1.6E
	0742	0927	0.7F
	1118	1523	3.0E
	1856	2235	3.8F
6 Su	0231	0523	2.0E
	0839	1038	1.1F
	1243	1629	3.3E
	1958	2328	4.3F
7 M	0317	0615	2.5E
	0923	1134	1.6F
	1354	1732	3.6E
	2053		
8 Tu		0014	4.5F
	0359	0657	2.9E
	1003	1223	2.1F
	1459	1829	3.7E
	2143		
9 W		0057	4.5F
	0436	0734	3.3E
	1041	1311	2.5F
	1601	1920	3.7E
	2230		
10 Th		0139	4.2F
	0510	0808	3.5E
	1119	1359	2.9F
●	1659	2007	3.6E
	2315		
11 F		0219	3.8F
	0541	0841	3.6E
	1157	1447	3.1F
	1756	2051	3.2E
12 Sa	0000	0259	3.2F
	0610	0914	3.5E
	1237	1535	3.1F
	1852	2136	2.6E
13 Su	0047	0339	2.6F
	0637	0948	3.3E
	1319	1622	3.1F
	1949	2225	2.1E
14 M	0138	0420	2.0F
	0702	1025	3.0E
	1403	1710	2.9F
	2053	2323	1.5E
15 Tu	0238	0502	1.4F
	0727	1109	2.6E
	1451	1801	2.7F
	2209		
16 W		0043	1.1E
	0350	0548	0.9F
	0753	1203	2.3E
	1545	1900	2.5F
	2331		
17 Th		0218	1.0E
	0516	0645	0.4F
	0827	1309	2.0E
◐	1644	2016	2.4F
18 F	0042	0336	1.1E
		0803	*
		1416	2.0E
	1747	2139	2.5F
19 Sa	0135	0438	1.4E
	0800	0940	0.4F
	1127	1519	2.0E
	1848	2235	2.8F
20 Su	0218	0527	1.7E
	0840	1041	0.7F
	1244	1617	2.2E
	1942	2315	3.0F
21 M	0254	0604	2.0E
	0911	1122	1.1F
	1343	1710	2.4E
	2028	2348	3.2F
22 Tu	0325	0633	2.3E
	0939	1157	1.6F
	1433	1756	2.6E
	2109		
23 W		0018	3.3F
	0353	0657	2.5E
	1006	1231	2.0F
	1521	1837	2.7E
	2146		
24 Th		0048	3.3F
	0418	0720	2.8E
	1032	1306	2.3F
	1607	1915	2.8E
	2221		
25 F		0120	3.1F
	0441	0744	3.0E
	1058	1343	2.7F
○	1653	1952	2.8E
	2257		
26 Sa		0153	2.9F
	0503	0810	3.2E
	1126	1422	3.0F
	1739	2030	2.7E
	2335		
27 Su		0229	2.6F
	0524	0839	3.3E
	1155	1504	3.2F
	1827	2111	2.5E
28 M	0018	0308	2.2F
	0547	0911	3.3E
	1230	1550	3.4F
	1920	2157	2.2E
29 Tu	0109	0351	1.7F
	0613	0947	3.2E
	1311	1639	3.4F
	2020	2253	1.8E
30 W	0214	0438	1.3F
	0642	1032	3.0E
	1401	1732	3.4F
	2133		
31 Th		0007	1.5E
	0335	0532	0.9F
	0718	1131	2.7E
	1501	1833	3.3F
	2254		

Time meridian 120° W. 0000 is midnight. 1200 is noon. Times are not adjusted for Daylight Saving Time.

* Current weak and variable.

San Juan Channel (south entrance), Washington, 2005

F–Flood, Dir. 010° True E–Ebb, Dir. 180° True

April

Day	Slack h m	Maximum h m	knots
1 F		0135	1.5E
	0506	0636	0.5F
	0809	1244	2.5E
◑	1609	1945	3.3F
2 Sa	0007	0255	1.7E
	0631	0758	0.5F
	0939	1400	2.5E
	1721	2104	3.4F
3 Su	0106	0401	2.1E
	0734	0928	0.8F
	1134	1512	2.7E
	1833	2211	3.7F
4 M	0154	0456	2.5E
	0820	1037	1.4F
	1301	1621	2.8E
	1938	2304	3.8F
5 Tu	0236	0543	2.9E
	0859	1130	2.1F
	1412	1725	3.0E
	2036	2348	3.8F
6 W	0313	0623	3.3E
	0936	1216	2.7F
	1515	1823	3.1E
	2128		
7 Th		0028	3.6F
	0346	0659	3.6E
	1011	1259	3.2F
	1613	1914	3.1E
	2216		
8 F		0107	3.3F
	0417	0732	3.7E
	1046	1343	3.5F
●	1708	2000	2.9E
	2302		
9 Sa		0145	2.8F
	0446	0803	3.7E
	1121	1426	3.7F
	1800	2043	2.6E
	2349		
10 Su		0224	2.2F
	0512	0834	3.6E
	1156	1510	3.7F
	1851	2125	2.3E
11 M	0038	0304	1.7F
	0536	0906	3.3E
	1233	1554	3.5F
	1942	2211	1.9E
12 Tu	0135	0347	1.2F
	0559	0940	3.0E
	1312	1638	3.3F
	2036	2306	1.5E
13 W	0241	0432	0.8F
	0623	1020	2.6E
	1357	1725	3.0F
	2138		
14 Th		0023	1.3E
	0356	0521	0.5F
	0649	1110	2.1E
	1448	1817	2.8F
	2245		
15 F		0151	1.3E
		0618	*
		1217	1.8E
	1546	1917	2.6F
	2347		
16 Sa		0256	1.4E
		0732	*
		1331	1.7E
◐	1648	2027	2.5F
17 Su	0037	0346	1.7E
	0727	0905	0.4F
	1102	1438	1.7E
	1751	2132	2.6F
18 M	0117	0427	1.9E
	0801	1013	0.8F
	1231	1537	1.8E
	1851	2221	2.7F
19 Tu	0151	0502	2.2E
	0830	1056	1.4F
	1334	1633	2.0E
	1944	2259	2.8F
20 W	0220	0532	2.6E
	0857	1132	2.0F
	1428	1725	2.1E
	2032	2333	2.8F
21 Th	0247	0600	2.9E
	0924	1206	2.5F
	1518	1812	2.3E
	2116		
22 F		0007	2.7F
	0312	0629	3.2E
	0950	1242	3.0F
	1607	1857	2.4E
	2159		
23 Sa		0041	2.5F
	0335	0659	3.4E
	1017	1319	3.5F
	1655	1939	2.5E
	2242		
24 Su		0118	2.2F
	0400	0731	3.6E
	1047	1400	3.8F
○	1743	2021	2.5E
	2329		
25 M		0157	1.9F
	0425	0804	3.7E
	1120	1444	4.0F
	1832	2100	2.4E
26 Tu	0022	0242	1.5F
	0454	0841	3.6E
	1158	1533	4.1F
	1925	2156	2.2E
27 W	0125	0331	1.1F
	0527	0922	3.5E
	1244	1624	4.1F
	2023	2256	1.9E
28 Th	0240	0426	0.8F
	0606	1010	3.1E
	1337	1719	4.0F
	2127		
29 F		0013	1.8E
	0359	0526	0.6F
	0654	1112	2.8E
	1439	1818	3.8F
	2234		
30 Sa		0134	2.0E
	0513	0635	0.5F
	0806	1229	2.4E
◑	1546	1923	3.6F
	2335		

May

Day	Slack h m	Maximum h m	knots
1 Su		0239	2.3E
	0617	0757	0.7F
	1003	1349	2.3E
	1657	2033	3.4F
2 M	0026	0333	2.6E
	0708	0925	1.2F
	1155	1502	2.3E
	1807	2139	3.3F
3 Tu	0109	0420	3.0E
	0751	1031	1.9F
	1317	1612	2.3E
	1915	2232	3.2F
4 W	0147	0504	3.3E
	0830	1122	2.7F
	1426	1719	2.3E
	2017	2317	3.0F
5 Th	0220	0544	3.6E
	0907	1205	3.3F
	1527	1819	2.4E
	2113	2357	2.6F
6 F	0251	0620	3.7E
	0942	1246	3.7F
	1623	1911	2.4E
	2204		
7 Sa		0035	2.2F
	0320	0655	3.8E
	1016	1326	3.9F
	1715	1957	2.3E
	2254		
8 Su		0112	1.8F
	0346	0727	3.7E
	1049	1406	4.0F
●	1802	2039	2.2E
	2343		
9 M		0151	1.4F
	0411	0759	3.5E
	1122	1447	3.9F
	1847	2119	2.0E
10 Tu	0037	0233	1.0F
	0436	0832	3.3E
	1156	1529	3.8F
	1932	2202	1.8E
11 W	0137	0318	0.7F
	0501	0906	3.0E
	1233	1612	3.6F
	2018	2252	1.6E
12 Th	0245	0407	0.4F
	0530	0944	2.6E
	1314	1656	3.3F
	2107	2357	1.5E
13 F	0354	0458	0.3F
	0605	1031	2.2E
	1400	1742	3.1F
	2159		
14 Sa		0109	1.5E
		0554	*
		1131	1.8E
	1452	1831	2.9F
	2249		
15 Su		0204	1.7E
		0658	*
		1242	1.5E
	1547	1925	2.7F
	2333		
16 M		0246	2.0E
	0636	0814	0.5F
	1029	1351	1.5E
◐	1646	2021	2.5F
17 Tu	0011	0321	2.3E
	0710	0927	1.0F
	1209	1453	1.5E
	1747	2116	2.4F
18 W	0043	0355	2.6E
	0741	1019	1.6F
	1319	1552	1.5E
	1850	2204	2.4F
19 Th	0112	0429	2.9E
	0811	1101	2.3F
	1418	1650	1.7E
	1951	2246	2.3F
20 F	0139	0504	3.2E
	0840	1139	3.0F
	1511	1747	1.8E
	2047	2326	2.1F
21 Sa	0205	0540	3.5E
	0910	1217	3.6F
	1602	1840	2.0E
	2141		
22 Su		0006	1.9F
	0233	0618	3.7E
	0942	1257	4.1F
	1653	1929	2.2E
	2234		
23 M		0047	1.6F
	0303	0657	3.9E
	1017	1341	4.5F
○	1742	2016	2.3E
	2328		
24 Tu		0132	1.4F
	0337	0738	4.0E
	1056	1429	4.7F
	1832	2104	2.3E
25 W	0028	0222	1.1F
	0415	0821	3.9E
	1140	1520	4.7F
	1923	2156	2.3E
26 Th	0134	0318	0.9F
	0500	0906	3.7E
	1229	1613	4.6F
	2016	2255	2.2E
27 F	0244	0418	0.7F
	0554	0959	3.3E
	1323	1706	4.4F
	2110		
28 Sa		0004	2.3E
	0350	0521	0.7F
	0700	1101	2.8E
	1423	1801	4.1F
	2204		
29 Su		0112	2.5E
	0449	0629	0.8F
	0830	1216	2.3E
	1526	1857	3.7F
	2254		
30 M		0208	2.8E
	0543	0747	1.1F
	1025	1335	2.0E
◑	1631	1957	3.2F
	2339		
31 Tu		0257	3.1E
	0633	0911	1.7F
	1207	1450	1.8E
	1740	2059	2.8F

June

Day	Slack h m	Maximum h m	knots
1 W	0019	0341	3.3E
	0718	1018	2.4F
	1327	1602	1.7E
	1851	2155	2.4F
2 Th	0055	0424	3.5E
	0759	1109	3.1F
	1436	1714	1.7E
	1959	2244	2.1F
3 F	0127	0504	3.6E
	0838	1153	3.6F
	1536	1819	1.8E
	2102	2327	1.7F
4 Sa	0158	0544	3.7E
	0915	1233	3.9F
	1629	1913	1.9E
	2158		
5 Su		0006	1.4F
	0226	0622	3.6E
	0950	1311	4.0F
	1716	1958	1.9E
	2250		
6 M		0045	1.1F
	0253	0658	3.5E
	1024	1349	4.0F
●	1759	2037	1.9E
	2341		
7 Tu		0125	0.9F
	0321	0733	3.4E
	1057	1428	4.0F
	1839	2114	1.8E
8 W	0034	0208	0.6F
	0350	0807	3.2E
	1131	1508	3.9F
	1917	2151	1.8E
9 Th	0130	0255	0.5F
	0424	0842	3.0E
	1206	1548	3.7F
	1955	2232	1.7E
10 F	0227	0344	0.4F
	0503	0920	2.7E
	1243	1629	3.6F
	2033	2310	1.7E
11 Sa	0322	0435	0.4F
	0550	1002	2.3E
	1323	1709	3.4F
	2112		
12 Su		0011	1.8E
	0411	0527	0.4F
	0649	1053	1.9E
	1406	1750	3.1F
	2149		
13 M		0059	2.0E
	0455	0622	0.5F
	0806	1156	1.6E
	1454	1834	2.8F
	2226		
14 Tu		0141	2.2E
	0534	0723	0.8F
	0955	1305	1.3E
◐	1546	1921	2.5F
	2300		
15 W		0219	2.5E
	0611	0831	1.2F
	1141	1411	1.2E
	1646	2013	2.2F
	2332		
16 Th		0256	2.8E
	0646	0934	1.9F
	1300	1514	1.2E
	1757	2108	1.9F
17 F	0003	0334	3.1E
	0721	1025	2.6F
	1404	1619	1.3E
	1914	2201	1.7F
18 Sa	0034	0415	3.3E
	0757	1111	3.3F
	1501	1725	1.5E
	2026	2250	1.5F
19 Su	0107	0459	3.6E
	0835	1154	4.0F
	1554	1827	1.8E
	2130	2338	1.4F
20 M	0144	0545	3.9E
	0915	1239	4.5F
	1646	1922	2.1E
	2227		
21 Tu		0025	1.3F
	0224	0633	4.0E
	0957	1325	4.8F
○	1735	2010	2.3E
	2323		
22 W		0115	1.2F
	0309	0720	4.2E
	1041	1415	5.0F
	1823	2058	2.5E
23 Th	0020	0209	1.1F
	0401	0808	4.1E
	1128	1507	5.0F
	1909	2146	2.6E
24 F	0119	0308	1.1F
	0500	0857	3.9E
	1218	1558	4.9F
	1955	2239	2.6E
25 Sa	0218	0409	1.1F
	0605	0950	3.4E
	1311	1648	4.6F
	2040	2335	2.7E
26 Su	0314	0510	1.2F
	0719	1050	2.8E
	1407	1737	4.1F
	2123		
27 M		0033	2.9E
	0409	0614	1.4F
	0848	1201	2.2E
	1505	1826	3.5F
	2206		
28 Tu		0127	3.1E
	0501	0725	1.7F
	1031	1320	1.7E
◑	1607	1919	2.8F
	2247		
29 W		0216	3.2E
	0551	0845	2.1F
	1209	1437	1.4E
	1715	2016	2.2F
	2326		
30 Th		0301	3.4E
	0640	0958	2.7F
	1330	1554	1.3E
	1831	2117	1.7F

Time meridian 120° W. 0000 is midnight. 1200 is noon. Times are not adjusted for Daylight Saving Time.

* Current weak and variable.

San Juan Channel (south entrance), Washington, 2005

F–Flood, Dir. 010° True   E–Ebb, Dir. 180° True

July

Day	Slack h m	Maximum h m	knots
1 F	0003	0345	3.4E
	0727	1054	3.2F
	1438	1711	1.3E
	1949	2215	1.3F
2 Sa	0039	0430	3.4E
	0810	1139	3.6F
	1536	1820	1.5E
	2059	2304	1.1F
3 Su	0114	0514	3.4E
	0851	1219	3.8F
	1625	1913	1.6E
	2156	2347	0.9F
4 M	0148	0557	3.3E
	0929	1256	3.9F
	1707	1954	1.8E
	2246		
5 Tu		0027	0.8F
	0222	0637	3.3E
	1004	1332	3.9F
	1745	2029	1.9E
	2330		
6 W		0108	0.7F
	0258	0715	3.2E
	1038	1408	3.9F
●	1819	2100	1.9E
7 Th	0014	0150	0.7F
	0337	0750	3.1E
	1111	1445	3.9F
	1851	2128	1.9E
8 F	0058	0234	0.7F
	0419	0825	3.0E
	1144	1521	3.8F
	1922	2159	2.0E
9 Sa	0142	0321	0.7F
	0505	0901	2.8E
	1217	1558	3.6F
	1951	2232	2.0E
10 Su	0225	0408	0.8F
	0555	0940	2.4E
	1252	1634	3.4F
	2020	2310	2.1E
11 M	0307	0456	0.9F
	0652	1024	2.0E
	1330	1711	3.1F
	2048	2352	2.2E
12 Tu	0346	0545	1.0F
	0800	1118	1.6E
	1412	1750	2.7F
	2116		
13 W		0035	2.4E
	0425	0638	1.3F
	0929	1225	1.3E
	1503	1832	2.3F
	2146		
14 Th		0119	2.6E
	0505	0739	1.6F
	1112	1336	1.1E
◐	1604	1921	1.8F
	2219		
15 F		0203	2.8E
	0547	0846	2.1F
	1238	1445	1.0E
	1722	2020	1.4F
	2256		
16 Sa		0248	3.1E
	0631	0950	2.8F
	1347	1556	1.2E
	1855	2124	1.2F
	2339		
17 Su		0336	3.3E
	0719	1045	3.5F
	1446	1710	1.4E
	2018	2225	1.1F
18 M	0026	0428	3.6E
	0807	1135	4.1F
	1540	1817	1.8E
	2123	2319	1.2F
19 Tu	0116	0522	3.9E
	0855	1222	4.7F
	1630	1911	2.2E
	2217		
20 W		0011	1.3F
	0209	0617	4.1E
	0943	1310	5.0F
	1717	1958	2.5E
	2306		
21 Th		0103	1.4F
	0307	0709	4.2E
	1030	1359	5.1F
○	1800	2041	2.8E
	2355		
22 F		0157	1.5F
	0409	0800	4.2E
	1118	1448	5.0F
	1841	2123	2.9E
23 Sa	0045	0255	1.7F
	0513	0849	3.9E
	1207	1536	4.7F
	1920	2207	3.0E
24 Su	0136	0353	1.8F
	0620	0940	3.4E
	1257	1622	4.2F
	1958	2254	3.1E
25 M	0228	0451	2.0F
	0731	1037	2.7E
	1350	1708	3.6F
	2034	2344	3.1E
26 Tu	0320	0550	2.1F
	0852	1144	2.0E
	1446	1753	2.9F
	2111		
27 W		0037	3.1E
	0413	0654	2.2F
	1026	1303	1.5E
◑	1549	1841	2.1F
	2149		
28 Th		0129	3.1E
	0506	0809	2.4F
	1201	1426	1.2E
	1701	1936	1.5F
	2229		
29 F		0220	3.1E
	0559	0929	2.7F
	1322	1548	1.1E
	1825	2043	1.0F
	2314		
30 Sa		0311	3.0E
	0652	1033	3.1F
	1427	1706	1.2E
	1950	2154	0.8F
31 Su	0001	0401	3.0E
	0742	1122	3.4F
	1519	1811	1.4E
	2058	2252	0.7F

August

Day	Slack h m	Maximum h m	knots
1 M	0049	0452	3.0E
	0828	1202	3.6F
	1603	1859	1.7E
	2148	2337	0.8F
2 Tu	0135	0540	3.0E
	0909	1238	3.7F
	1641	1936	1.9E
	2228		
3 W		0017	0.9F
	0220	0623	3.1E
	0946	1311	3.8F
	1714	2005	2.0E
	2303		
4 Th		0054	1.0F
	0303	0700	3.1E
	1020	1343	3.8F
●	1744	2030	2.1E
	2337		
5 F		0133	1.1F
	0346	0735	3.1E
	1052	1415	3.7F
	1811	2053	2.2E
6 Sa	0011	0213	1.2F
	0430	0809	3.0E
	1123	1448	3.6F
	1836	2117	2.3E
7 Su	0045	0255	1.3F
	0515	0844	2.8E
	1154	1522	3.4F
	1900	2145	2.4E
8 M	0120	0339	1.4F
	0603	0920	2.5E
	1227	1557	3.1F
	1922	2216	2.5E
9 Tu	0156	0423	1.6F
	0655	1001	2.1E
	1303	1632	2.8F
	1944	2251	2.5E
10 W	0234	0509	1.8F
	0756	1051	1.7E
	1346	1710	2.3F
	2008	2333	2.5E
11 Th	0314	0559	2.0F
	0913	1155	1.3E
	1440	1753	1.9F
	2036		
12 F		0022	2.6E
	0359	0656	2.2F
	1048	1311	1.1E
◐	1551	1843	1.4F
	2110		
13 Sa		0115	2.7E
	0450	0804	2.5F
	1217	1428	1.1E
	1722	1946	1.0F
	2157		
14 Su		0212	2.9E
	0547	0917	3.0F
	1327	1545	1.2E
	1900	2101	0.8F
	2258		
15 M		0309	3.1E
	0646	1023	3.6F
	1426	1700	1.6E
	2017	2211	0.9F
16 Tu	0006	0408	3.4E
	0744	1117	4.2F
	1517	1803	2.0E
	2113	2310	1.2F
17 W	0113	0509	3.7E
	0839	1205	4.6F
	1603	1852	2.5E
	2158		
18 Th		0003	1.6F
	0217	0607	4.0E
	0930	1252	4.8F
	1646	1934	2.9E
	2240		
19 F		0054	1.9F
	0320	0701	4.1E
	1018	1337	4.8F
○	1725	2013	3.2E
	2322		
20 Sa		0145	2.2F
	0424	0751	4.0E
	1105	1422	4.6F
	1801	2050	3.4E
21 Su	0006	0239	2.5F
	0526	0840	3.7E
	1153	1507	4.1F
	1835	2128	3.4E
22 M	0051	0333	2.7F
	0629	0929	3.2E
	1242	1551	3.5F
	1907	2208	3.4E
23 Tu	0138	0427	2.8F
	0735	1022	2.5E
	1334	1634	2.8F
	1938	2251	3.2E
24 W	0228	0521	2.8F
	0848	1126	1.8E
	1433	1719	2.1F
	2010	2341	3.0E
25 Th	0321	0619	2.7F
	1012	1247	1.3E
	1541	1806	1.4F
	2043		
26 F		0037	2.7E
	0416	0725	2.6F
	1142	1418	1.1E
◑	1701	1902	0.9F
	2125		
27 Sa		0138	2.5E
	0513	0849	2.6F
	1258	1540	1.2E
	1832	2017	0.5F
	2224		
28 Su		0238	2.5E
	0613	1004	2.8F
	1357	1650	1.4E
	1954	2145	0.5F
	2338		
29 M		0336	2.5E
	0710	1057	3.1F
	1444	1746	1.6E
	2048	2246	0.7F
30 Tu	0043	0432	2.6E
	0800	1137	3.3F
	1523	1828	1.9E
	2126	2329	1.0F
31 W	0138	0522	2.7E
	0844	1210	3.5F
	1557	1901	2.1E
	2158		

September

Day	Slack h m	Maximum h m	knots
1 Th		0005	1.2F
	0226	0606	2.8E
	0923	1240	3.5F
	1627	1927	2.3E
	2228		
2 F		0039	1.5F
	0312	0644	2.9E
	0958	1309	3.5F
	1654	1949	2.5E
	2256		
3 Sa		0114	1.7F
	0356	0719	2.9E
	1030	1339	3.4F
●	1718	2010	2.6E
	2324		
4 Su		0150	1.9F
	0439	0753	2.8E
	1102	1410	3.2F
	1740	2033	2.7E
	2352		
5 M		0229	2.1F
	0523	0828	2.7E
	1134	1443	3.0F
	1800	2059	2.8E
6 Tu	0021	0310	2.3F
	0609	0905	2.4E
	1208	1518	2.6F
	1820	2128	2.9E
7 W	0052	0352	2.5F
	0659	0946	2.1E
	1248	1556	2.2F
	1841	2200	2.8E
8 Th	0128	0438	2.6F
	0755	1035	1.7E
	1338	1637	1.7F
	1905	2240	2.7E
9 F	0211	0528	2.7F
	0906	1139	1.4E
	1445	1723	1.3F
	1934	2331	2.6E
10 Sa	0303	0624	2.7F
	1033	1300	1.2E
	1610	1818	0.8F
	2012		
11 Su		0035	2.5E
	0404	0732	2.9F
	1157	1424	1.3E
◐	1747	1928	0.6F
	2110		
12 M		0144	2.6E
	0511	0850	3.2F
	1303	1540	1.5E
	1912	2052	0.6F
	2241		
13 Tu		0251	2.8E
	0619	1000	3.6F
	1357	1645	2.0E
	2010	2207	1.0F
14 W	0012	0356	3.1E
	0724	1056	4.0F
	1443	1739	2.4E
	2055	2306	1.6F
15 Th	0127	0500	3.4E
	0822	1144	4.3F
	1525	1823	2.9E
	2134	2356	2.2F
16 F	0233	0559	3.6E
	0915	1228	4.4F
	1603	1902	3.3E
	2212		
17 Sa		0044	2.7F
	0336	0654	3.7E
	1004	1310	4.2F
○	1638	1939	3.6E
	2249		
18 Su		0132	3.1F
	0437	0743	3.6E
	1052	1351	3.8F
	1710	2014	3.7E
	2328		
19 M		0221	3.4F
	0536	0831	3.3E
	1139	1433	3.2F
	1741	2048	3.7E
20 Tu	0008	0310	3.5F
	0634	0918	2.8E
	1229	1516	2.6F
	1810	2124	3.5E
21 W	0051	0400	3.4F
	0733	1009	2.2E
	1325	1600	1.9F
	1837	2202	3.2E
22 Th	0136	0451	3.3F
	0837	1109	1.7E
	1429	1646	1.3F
	1905	2246	2.8E
23 F	0227	0543	3.0F
	0950	1233	1.3E
	1545	1735	0.8F
	1933	2341	2.4E
24 Sa	0322	0642	2.8F
	1110	1408	1.2E
	1711	1834	0.4F
◑	2011		
25 Su		0050	2.1E
	0422	0755	2.6F
	1219	1520	1.4E
		1953	*
26 M		0202	2.0E
	0526	0916	2.6F
	1313	1618	1.6E
	1946	2134	0.4F
	2322		
27 Tu		0307	2.0E
	0627	1016	2.8F
	1355	1705	1.9E
	2024	2235	0.8F
28 W	0039	0404	2.1E
	0723	1058	3.0F
	1431	1743	2.1E
	2054	2314	1.2F
29 Th	0137	0456	2.3E
	0811	1131	3.1F
	1501	1813	2.4E
	2122	2347	1.7F
30 F	0227	0542	2.4E
	0852	1201	3.2F
	1529	1838	2.6E
	2149		

Time meridian 120° W. 0000 is midnight. 1200 is noon. Times are not adjusted for Daylight Saving Time.
* Current weak and variable.

San Juan Channel (south entrance), Washington, 2005

F–Flood, Dir. 010° True   E–Ebb, Dir. 180° True

October

Day	Slack h m	Maximum h m	knots
1 Sa		0019	2.1F
	0314	0624	2.5E
	0930	1230	3.1F
	1554	1900	2.9E
	2215		
2 Su		0052	2.4F
	0359	0701	2.6E
	1006	1300	2.9F
	1616	1924	3.0E
	2240		
3 M ●		0127	2.7F
	0443	0738	2.6E
	1041	1332	2.7F
	1637	1950	3.2E
	2306		
4 Tu		0203	3.0F
	0528	0815	2.5E
	1118	1406	2.3F
	1657	2017	3.2E
	2333		
5 W		0243	3.2F
	0613	0853	2.3E
	1159	1443	1.9F
	1718	2047	3.2E
6 Th	0004	0327	3.3F
	0702	0936	2.0E
	1248	1525	1.5F
	1742	2121	3.1E
7 F	0041	0414	3.4F
	0757	1028	1.7E
	1352	1612	1.1F
	1810	2201	2.9E
8 Sa	0128	0505	3.4F
	0903	1136	1.5E
	1513	1705	0.7F
	1844	2254	2.7E
9 Su	0224	0603	3.3F
	1020	1301	1.4E
	1642	1807	0.5F
	1930		
10 M ◐		0005	2.4E
	0330	0709	3.3F
	1132	1423	1.7E
	1804	1922	0.4F
	2050		
11 Tu		0125	2.4E
	0442	0823	3.3F
	1232	1527	2.0E
	1906	2051	0.7F
	2254		
12 W		0239	2.5E
	0554	0933	3.5F
	1320	1620	2.5E
	1951	2206	1.3F
13 Th	0031	0347	2.7E
	0702	1030	3.7F
	1402	1706	2.9E
	2030	2302	2.1F
14 F	0144	0452	2.9E
	0804	1117	3.7F
	1439	1748	3.4E
	2107	2349	2.8F
15 Sa	0249	0553	3.0E
	0859	1200	3.6F
	1513	1827	3.7E
	2143		
16 Su		0034	3.4F
	0350	0648	3.1E
	0951	1240	3.3F
	1545	1903	3.9E
	2219		
17 M ○		0118	3.8F
	0447	0738	3.0E
	1040	1319	2.8F
	1615	1937	4.0E
	2255		
18 Tu		0202	4.0F
	0542	0825	2.7E
	1130	1400	2.2F
	1643	2011	3.8E
	2332		
19 W		0248	4.0F
	0635	0910	2.4E
	1222	1442	1.7F
	1710	2045	3.6E
20 Th	0010	0335	3.9F
	0727	0958	2.0E
	1322	1527	1.2F
	1735	2121	3.2E
21 F	0051	0422	3.6F
	0823	1055	1.6E
	1433	1616	0.7F
	1801	2201	2.7E
22 Sa	0137	0511	3.3F
	0923	1215	1.4E
	1552	1708	0.4F
	1829	2251	2.2E
23 Su	0229	0603	3.0F
	1028	1344	1.4E
		1808	*
		2358	1.8E
24 M ◑	0327	0701	2.7F
	1128	1446	1.6E
		1922	*
25 Tu		0115	1.6E
	0429	0808	2.6F
	1217	1533	1.8E
	1915	2100	0.4F
	2258		
26 W		0226	1.6E
	0531	0913	2.6F
	1257	1612	2.1E
	1947	2207	0.9F
27 Th	0026	0326	1.7E
	0631	1003	2.6F
	1330	1645	2.4E
	2015	2249	1.5F
28 F	0129	0421	1.8E
	0726	1042	2.7F
	1359	1714	2.7E
	2042	2323	2.0F
29 Sa	0222	0512	1.9E
	0815	1116	2.6F
	1425	1742	2.9E
	2109	2355	2.6F
30 Su	0311	0600	2.1E
	0900	1149	2.5F
	1448	1810	3.2E
	2135		
31 M		0028	3.1F
	0358	0643	2.2E
	0943	1222	2.3F
	1511	1839	3.4E
	2201		

November

Day	Slack h m	Maximum h m	knots
1 Tu ●		0103	3.5F
	0443	0724	2.3E
	1026	1256	2.1F
	1533	1910	3.5E
	2228		
2 W		0141	3.8F
	0529	0805	2.3E
	1110	1333	1.7F
	1557	1942	3.6E
	2259		
3 Th		0222	4.0F
	0615	0847	2.2E
	1200	1415	1.4F
	1624	2017	3.6E
	2333		
4 F		0308	4.1F
	0704	0933	2.1E
	1259	1502	1.0F
	1654	2055	3.4E
5 Sa	0014	0357	4.1F
	0757	1027	1.9E
	1412	1556	0.7F
	1731	2139	3.2E
6 Su	0103	0450	4.0F
	0857	1135	1.8E
	1532	1655	0.5F
	1817	2234	2.8E
7 M	0201	0546	3.8F
	1000	1256	1.9E
	1646	1801	0.4F
	1921	2347	2.4E
8 Tu ◐	0306	0647	3.6F
	1100	1405	2.2E
	1748	1918	0.6F
	2108		
9 W		0109	2.2E
	0416	0753	3.4F
	1152	1500	2.6E
	1839	2045	1.0F
	2315		
10 Th		0226	2.2E
	0528	0859	3.3F
	1236	1547	3.0E
	1922	2159	1.8F
11 F	0047	0337	2.2E
	0638	0958	3.1F
	1315	1630	3.4E
	2002	2254	2.6F
12 Sa	0159	0446	2.3E
	0745	1047	2.9F
	1350	1712	3.7E
	2040	2341	3.4F
13 Su	0303	0551	2.3E
	0846	1131	2.6F
	1422	1751	3.9E
	2117		
14 M		0023	3.9F
	0401	0648	2.4E
	0941	1211	2.3F
	1452	1829	4.0E
	2152		
15 Tu ○		0105	4.2F
	0455	0738	2.4E
	1034	1251	1.8F
	1521	1905	3.9E
	2228		
16 W		0146	4.3F
	0545	0823	2.3E
	1126	1331	1.4F
	1548	1940	3.8E
	2303		
17 Th		0229	4.3F
	0632	0906	2.1E
	1221	1414	1.0F
	1615	2015	3.5E
	2339		
18 F		0312	4.1F
	0718	0950	1.9E
	1322	1500	0.7F
	1643	2051	3.2E
19 Sa	0017	0357	3.8F
	0804	1040	1.7E
	1430	1550	0.4F
	1713	2129	2.7E
20 Su	0058	0441	3.5F
	0852	1143	1.6E
	1541	1643	0.3F
	1748	2213	2.3E
21 M	0143	0527	3.3F
	0941	1255	1.6E
		1739	*
		2310	1.8E
22 Tu	0233	0614	3.0F
	1030	1351	1.8E
		1843	*
23 W ◑		0021	1.5E
	0328	0705	2.7F
	1113	1433	2.0E
	1822	1959	0.5F
	2217		
24 Th		0134	1.3E
	0426	0759	2.5F
	1151	1508	2.3E
	1856	2117	1.0F
25 F	0002	0240	1.3E
	0527	0855	2.3F
	1223	1540	2.6E
	1927	2211	1.6F
26 Sa	0113	0340	1.3E
	0631	0916	2.2F
	1252	1613	2.9E
	1957	2252	2.3F
27 Su	0212	0438	1.4E
	0734	1028	2.1F
	1319	1647	3.1E
	2027	2328	2.9F
28 M	0304	0535	1.6E
	0832	1109	1.9F
	1345	1722	3.4E
	2056		
29 Tu		0004	3.5F
	0352	0627	1.8E
	0925	1147	1.7F
	1411	1759	3.6E
	2127		
30 W		0041	4.0F
	0439	0714	2.0E
	1016	1227	1.5F
	1439	1837	3.8E
	2200		

December

Day	Slack h m	Maximum h m	knots
1 Th ●		0122	4.3F
	0526	0758	2.2E
	1107	1309	1.3F
	1511	1916	3.9E
	2236		
2 F		0206	4.6F
	0612	0843	2.2E
	1202	1356	1.1F
	1548	1957	3.9E
	2316		
3 Sa		0254	4.6F
	0659	0929	2.2E
	1302	1448	0.9F
	1631	2041	3.7E
4 Su	0001	0344	4.6F
	0748	1022	2.2E
	1409	1546	0.7F
	1722	2128	3.4E
5 M	0051	0436	4.5F
	0838	1123	2.2E
	1514	1648	0.7F
	1824	2225	3.0E
6 Tu	0146	0528	4.2F
	0929	1230	2.4E
	1614	1752	0.8F
	1944	2334	2.4E
7 W	0247	0622	3.8F
	1018	1330	2.7E
	1709	1905	1.1F
	2133		
8 Th ◐		0054	2.0E
	0352	0719	3.3F
	1104	1422	3.0E
	1800	2027	1.6F
	2328		
9 F		0213	1.8E
	0502	0820	2.8F
	1146	1509	3.3E
	1847	2145	2.3F
10 Sa	0058	0329	1.7E
	0616	0921	2.4F
	1225	1553	3.5E
	1931	2243	3.0F
11 Su	0211	0444	1.7E
	0731	1017	2.0F
	1300	1637	3.7E
	2013	2331	3.6F
12 M	0313	0555	1.8E
	0839	1105	1.7F
	1334	1720	3.8E
	2053		
13 Tu		0014	4.0F
	0409	0654	1.9E
	0940	1149	1.4F
	1406	1802	3.8E
	2131		
14 W		0054	4.3F
	0458	0743	2.0E
	1034	1230	1.2F
	1438	1841	3.7E
	2208		
15 Th ○		0134	4.3F
	0543	0824	2.0E
	1125	1311	1.0F
	1509	1919	3.6E
	2243		
16 F		0213	4.2F
	0623	0902	2.0E
	1216	1353	0.8F
	1542	1955	3.4E
	2318		
17 Sa		0253	4.1F
	0701	0938	1.9E
	1308	1439	0.6F
	1617	2030	3.1E
	2353		
18 Su		0333	3.9F
	0738	1016	1.9E
	1402	1527	0.5F
	1657	2107	2.8E
19 M	0029	0412	3.7F
	0815	1058	1.8E
	1456	1617	0.5F
	1743	2147	2.4E
20 Tu	0107	0451	3.5F
	0851	1146	1.9E
	1546	1708	0.5F
	1838	2233	2.0E
21 W	0148	0531	3.2F
	0927	1235	2.0E
	1631	1802	0.6F
	1950	2332	1.6E
22 Th	0233	0612	2.8F
	1002	1319	2.2E
	1713	1901	0.8F
	2132		
23 F ◑		0041	1.2E
	0323	0657	2.4F
	1036	1359	2.4E
	1752	2009	1.1F
	2326		
24 Sa		0152	1.0E
	0422	0748	2.0F
	1108	1438	2.6E
	1829	2118	1.7F
25 Su	0050	0258	1.0E
	0534	0844	1.7F
	1140	1516	2.9E
	1906	2213	2.4F
26 M	0155	0404	1.1E
	0655	0941	1.5F
	1213	1557	3.1E
	1943	2258	3.1F
27 Tu	0251	0512	1.3E
	0811	1033	1.3F
	1247	1640	3.4E
	2020	2340	3.7F
28 W	0342	0614	1.6E
	0915	1120	1.3F
	1324	1726	3.6E
	2059		
29 Th		0022	4.2F
	0430	0706	1.9E
	1010	1206	1.2F
	1404	1813	3.9E
	2139		
30 F ●		0106	4.7F
	0515	0751	2.2E
	1101	1253	1.2F
	1449	1900	4.1E
	2221		
31 Sa		0152	4.9F
	0600	0835	2.4E
	1151	1343	1.2F
	1539	1946	4.1E
	2305		

Time meridian 120° W. 0000 is midnight. 1200 is noon. Times are not adjusted for Daylight Saving Time.

* Current weak and variable.

Active Pass, British Columbia, 2005

F–Flood, Dir. 045° True   E–Ebb, Dir. 225° True

January

Day	Slack (h m)	Maximum (h m)	knots
1 Sa	0135	0505	4.1F
	0835	1150	3.5E
	1535	1740	1.3F
	1950	2315	2.7E
2 Su	0220	0545	3.6F
	0905	1230	3.6E
	1615	1835	1.6F
	2110		
3 M ◑		0015	2.3E
	0310	0630	3.0F
	0935	1310	3.7E
	1655	1940	2.1F
	2235		
4 Tu		0125	1.9E
	0415	0720	2.3F
	1010	1400	3.9E
	1740	2045	2.7F
5 W	0005	0245	1.8E
	0535	0820	1.8F
	1050	1450	4.1E
	1830	2145	3.5F
6 Th	0125	0405	2.1E
	0705	0925	1.4F
	1135	1545	4.4E
	1915	2245	4.4F
7 F	0230	0515	2.5E
	0830	1035	1.2F
	1225	1640	4.7E
	2005	2340	5.2F
8 Sa	0325	0615	3.1E
	0940	1135	1.3F
	1325	1735	5.0E
	2055		
9 Su		0030	5.8F
	0415	0710	3.6E
	1035	1235	1.5F
	1425	1825	5.3E
	2145		
10 M ●		0120	6.3F
	0505	0800	4.1E
	1125	1330	1.7F
	1525	1920	5.5E
	2235		
11 Tu		0210	6.6F
	0550	0850	4.4E
	1215	1420	2.0F
	1625	2010	5.5E
	2325		
12 W		0300	6.6F
	0635	0935	4.7E
	1305	1515	2.2F
	1725	2105	5.2E
13 Th	0015	0345	6.3F
	0715	1020	4.8E
	1350	1610	2.4F
	1825	2155	4.8E
14 F	0105	0430	5.7F
	0755	1105	4.8E
	1440	1705	2.6F
	1930	2250	4.1E
15 Sa	0155	0515	4.9F
	0835	1150	4.8E
	1525	1800	2.7F
	2045	2350	3.3E
16 Su ◐	0250	0605	3.9F
	0910	1240	4.6E
	1615	1905	2.8F
	2205		
17 M		0055	2.6E
	0350	0655	2.8F
	0945	1325	4.3E
	1705	2010	3.0F
	2330		
18 Tu		0215	2.0E
	0510	0750	1.9F
	1020	1420	4.0E
	1800	2115	3.4F
19 W	0055	0340	1.9E
	0645	0900	1.1F
	1100	1515	3.8E
	1845	2215	3.8F
20 Th	0205	0455	2.2E
	0825	1010	0.7F
	1145	1605	3.7E
	1935	2310	4.1F
21 F	0300	0555	2.5E
	0940	1110	0.6F
	1235	1700	3.7E
	2020		
22 Sa		0000	4.4F
	0345	0645	2.9E
	1035	1205	0.6F
	1330	1745	3.7E
	2100		
23 Su		0040	4.7F
	0425	0725	3.1E
	1115	1250	0.8F
	1420	1825	3.8E
	2140		
24 M		0120	4.8F
	0500	0800	3.3E
	1145	1325	0.9F
	1505	1905	3.9E
	2215		
25 Tu ○		0155	4.9F
	0530	0835	3.5E
	1215	1400	1.1F
	1550	1940	4.0E
	2255		
26 W		0225	5.0F
	0600	0905	3.6E
	1240	1435	1.3F
	1630	2015	4.0E
	2325		
27 Th		0255	4.9F
	0630	0930	3.7E
	1305	1510	1.6F
	1715	2055	4.0E
28 F	0000	0330	4.8F
	0655	1000	3.9E
	1330	1545	1.8F
	1800	2130	3.8E
29 Sa	0035	0400	4.4F
	0720	1030	4.0E
	1400	1625	2.1F
	1850	2210	3.5E
30 Su	0115	0430	4.0F
	0745	1100	4.0E
	1430	1705	2.4F
	1945	2255	3.0E
31 M	0155	0505	3.4F
	0810	1135	4.1E
	1510	1755	2.6F
	2055	2350	2.5E

February

Day	Slack (h m)	Maximum (h m)	knots
1 Tu ◑	0245	0545	2.6F
	0835	1215	4.0E
	1550	1850	2.9F
	2210		
2 W		0055	2.0E
	0350	0635	1.8F
	0905	1305	4.0E
	1645	2000	3.2F
	2340		
3 Th		0220	1.8E
	0520	0740	1.2F
	0945	1405	3.9E
	1745	2115	3.7F
4 F	0105	0350	1.9E
	0705	0900	0.8F
	1040	1510	4.0E
	1845	2225	4.3F
5 Sa	0215	0505	2.4E
	0835	1020	0.8F
	1155	1620	4.3E
	1945	2325	5.0F
6 Su	0315	0605	3.1E
	0935	1130	1.2F
	1315	1725	4.7E
	2045		
7 M		0020	5.6F
	0400	0700	3.7E
	1025	1230	1.7F
	1425	1820	5.1E
	2140		
8 Tu ●		0110	6.1F
	0445	0745	4.3E
	1105	1320	2.2F
	1530	1915	5.4E
	2230		
9 W		0155	6.3F
	0525	0830	4.7E
	1150	1410	2.7F
	1635	2005	5.4E
	2315		
10 Th		0240	6.2F
	0605	0910	5.0E
	1230	1500	3.1F
	1730	2055	5.2E
11 F	0005	0320	5.8F
	0640	0950	5.2E
	1310	1550	3.4F
	1830	2145	4.7E
12 Sa	0050	0405	5.1F
	0715	1030	5.2E
	1350	1635	3.6F
	1930	2235	4.1E
13 Su	0140	0445	4.2F
	0745	1105	4.9E
	1435	1725	3.6F
	2030	2330	3.3E
14 M	0230	0525	3.1F
	0815	1145	4.5E
	1515	1820	3.5F
	2140		
15 Tu ◐		0030	2.6E
	0335	0610	2.1F
	0845	1230	4.0E
	1605	1920	3.4F
	2255		
16 W		0140	2.1E
	0455	0710	1.1F
	0915	1320	3.5E
	1700	2025	3.3F
17 Th	0015	0310	1.9E
	0645	0820	0.5F
	0950	1425	3.1E
	1755	2135	3.4F
18 F	0130	0430	2.1E
		0945	0.0
		1530	3.0E
	1855	2240	3.6F
19 Sa	0225	0535	2.5E
	0935	1055	0.4F
	1215	1635	3.1E
	1950	2330	3.9F
20 Su	0310	0620	2.9E
	1010	1150	0.7F
	1325	1725	3.3E
	2040		
21 M		0015	4.2F
	0350	0655	3.2E
	1035	1230	1.1F
	1420	1810	3.6E
	2120		
22 Tu		0055	4.5F
	0425	0730	3.5E
	1100	1305	1.4F
	1510	1850	3.8E
	2200		
23 W ○		0125	4.6F
	0455	0800	3.7E
	1125	1340	1.8F
	1550	1925	4.0E
	2235		
24 Th		0200	4.7F
	0520	0825	3.9E
	1150	1410	2.2F
	1635	2000	4.1E
	2310		
25 F		0230	4.6F
	0545	0850	4.1E
	1210	1445	2.6F
	1715	2040	4.1E
	2345		
26 Sa		0255	4.3F
	0610	0915	4.2E
	1235	1520	3.0F
	1800	2115	3.9E
27 Su	0020	0325	3.9F
	0630	0945	4.4E
	1305	1555	3.3F
	1850	2155	3.7E
28 M	0100	0400	3.4F
	0650	1015	4.4E
	1335	1635	3.6F
	1945	2240	3.2E

March

Day	Slack (h m)	Maximum (h m)	knots
1 Tu	0145	0435	2.7F
	0715	1050	4.3E
	1415	1720	3.7F
	2045	2335	2.7E
2 W	0240	0515	2.0F
	0740	1130	4.1E
	1500	1820	3.6F
	2155		
3 Th ◑		0040	2.2E
	0355	0610	1.3F
	0815	1220	3.8E
	1600	1930	3.6F
	2320		
4 F		0205	2.0E
	0535	0720	0.7F
	0900	1330	3.6E
	1710	2050	3.8F
5 Sa	0045	0340	2.1E
	0720	0855	0.5F
	1020	1455	3.6E
	1825	2205	4.2F
6 Su	0155	0455	2.7E
	0830	1020	0.9F
	1200	1615	3.9E
	1935	2310	4.7F
7 M	0250	0550	3.4E
	0920	1125	1.5F
	1330	1720	4.4E
	2035		
8 Tu		0005	5.2F
	0335	0635	4.0E
	1000	1220	2.3F
	1440	1815	4.8E
	2130		
9 W		0050	5.5F
	0415	0720	4.6E
	1035	1310	3.0F
	1540	1910	5.1E
	2220		
10 Th ●		0135	5.5F
	0450	0800	5.0E
	1115	1355	3.7F
	1640	2000	5.1E
	2305		
11 F		0215	5.2F
	0525	0835	5.3E
	1150	1440	4.2F
	1735	2045	4.9E
	2355		
12 Sa		0255	4.7F
	0555	0910	5.3E
	1225	1525	4.5F
	1830	2130	4.5E
13 Su	0040	0335	3.9F
	0625	0945	5.1E
	1305	1610	4.5F
	1920	2220	3.9E
14 M	0130	0410	3.1F
	0650	1020	4.8E
	1340	1650	4.4F
	2015	2310	3.3E
15 Tu	0225	0450	2.2F
	0720	1055	4.3E
	1420	1740	4.0F
	2110		
16 W		0005	2.7E
	0325	0540	1.3F
	0745	1140	3.6E
	1505	1830	3.6F
	2215		
17 Th ◐		0110	2.3E
	0450	0635	0.6F
	0815	1230	3.0E
	1600	1935	3.2F
	2325		
18 F		0230	2.1E
		0750	0.0
		1335	2.6E
	1705	2045	3.1F
19 Sa	0035	0350	2.2E
		0920	0.0
		1455	2.4E
	1810	2155	3.2F
20 Su	0135	0450	2.5E
	0855	1030	0.5F
	1205	1605	2.6E
	1915	2250	3.5F
21 M	0225	0540	2.9E
	0925	1120	1.0F
	1320	1700	2.9E
	2010	2340	3.7F
22 Tu	0300	0615	3.3E
	0950	1200	1.6F
	1415	1745	3.3E
	2055		
23 W		0015	4.0F
	0335	0645	3.7E
	1010	1240	2.1F
	1505	1830	3.6E
	2135		
24 Th		0050	4.1F
	0400	0715	4.0E
	1035	1310	2.7F
	1550	1905	3.9E
	2215		
25 F ○		0120	4.0F
	0425	0740	4.2E
	1055	1345	3.2F
	1630	1945	4.0E
	2255		
26 Sa		0155	3.8F
	0450	0805	4.4E
	1120	1415	3.8F
	1715	2025	4.1E
	2335		
27 Su		0225	3.5F
	0515	0835	4.6E
	1150	1450	4.2F
	1800	2105	4.0E
28 M	0015	0300	3.1F
	0535	0905	4.7E
	1220	1530	4.5F
	1850	2145	3.7E
29 Tu	0100	0335	2.5F
	0605	0935	4.6E
	1255	1610	4.6F
	1940	2235	3.4E
30 W	0150	0415	2.0F
	0630	1015	4.4E
	1335	1700	4.5F
	2035	2330	3.0E
31 Th	0255	0500	1.4F
	0705	1100	4.1E
	1430	1800	4.2F
	2145		

Time meridian 120° W. 0000 is midnight. 1200 is noon. Times are not adjusted for Daylight Saving Time.

Active Pass, British Columbia, 2005

F–Flood, Dir. 045° True E–Ebb, Dir. 225° True

April

Day	Slack h m	Maximum h m	knots
1 F		0040	2.5E
	0415	0600	0.8F
	0745	1200	3.6E
	1530	1910	4.0F
	2300		
2 Sa		0200	2.4E
	0555	0725	0.5F
	0850	1320	3.3E
	1645	2030	3.9F
3 Su	0015	0325	2.7E
	0715	0900	0.7F
	1035	1445	3.3E
	1805	2145	4.1F
4 M	0120	0430	3.2E
	0810	1015	1.4F
	1220	1605	3.6E
	1920	2245	4.4F
5 Tu	0210	0525	3.9E
	0850	1120	2.2F
	1345	1715	4.0E
	2020	2340	4.6F
6 W	0255	0605	4.5E
	0930	1210	3.1F
	1450	1810	4.3E
	2115		
7 Th		0025	4.5F
	0330	0645	4.9E
	1005	1255	3.9F
	1550	1900	4.5E
	2210		
8 F		0110	4.3F
	0405	0720	5.2E
	1040	1340	4.6F
	1645	1950	4.5E
	2300		
9 Sa		0150	3.8F
	0435	0755	5.2E
	1110	1420	5.0F
	1735	2035	4.4E
	2345		
10 Su		0225	3.2F
	0505	0830	5.1E
	1145	1500	5.2F
	1820	2120	4.1E
11 M	0035	0305	2.6F
	0530	0905	4.8E
	1220	1540	5.1F
	1910	2205	3.7E
12 Tu	0130	0345	1.9F
	0600	0940	4.4E
	1255	1620	4.8F
	1955	2255	3.3E
13 W	0225	0425	1.3F
	0625	1015	3.9E
	1330	1700	4.4F
	2045	2345	2.9E
14 Th	0330	0510	0.8F
	0655	1055	3.3E
	1415	1750	3.9F
	2135		
15 F		0040	2.6E
	0450	0610	0.4F
	0730	1145	2.7E
	1505	1845	3.4F
	2235		
16 Sa		0150	2.4E
		0720	0.0
		1250	2.3E
	1610	1955	3.1F
	2335		
17 Su		0300	2.5E
	0725	0845	0.3F
	1005	1410	2.1E
	1720	2100	3.0F
18 M	0030	0355	2.8E
	0800	0950	0.8F
	1145	1525	2.2E
	1825	2200	3.1F
19 Tu	0120	0445	3.2E
	0830	1045	1.4F
	1305	1625	2.5E
	1925	2245	3.2F
20 W	0155	0520	3.5E
	0850	1125	2.1F
	1405	1715	2.9E
	2015	2330	3.3F
21 Th	0230	0550	3.9E
	0915	1205	2.8F
	1455	1800	3.3E
	2105		
22 F		0005	3.2F
	0300	0620	4.2E
	0940	1240	3.5F
	1540	1845	3.6E
	2150		
23 Sa		0040	3.1F
	0325	0650	4.5E
	1005	1315	4.2F
	1625	1925	3.8E
	2240		
24 Su		0115	2.8F
	0350	0720	4.7E
	1035	1350	4.8F
	1710	2010	3.9E
	2325		
25 M		0155	2.5F
	0420	0755	4.8E
	1105	1430	5.2F
	1755	2055	3.9E
26 Tu	0015	0235	2.2F
	0450	0830	4.9E
	1145	1510	5.4F
	1845	2140	3.8E
27 W	0105	0315	1.8F
	0520	0910	4.7E
	1225	1555	5.4F
	1935	2235	3.6E
28 Th	0205	0405	1.4F
	0600	0955	4.4E
	1315	1650	5.1F
	2035	2330	3.3E
29 F	0310	0500	1.0F
	0645	1050	4.0E
	1410	1745	4.7F
	2135		
30 Sa		0040	3.1E
	0430	0610	0.8F
	0745	1155	3.5E
	1515	1855	4.3F
	2240		

May

Day	Slack h m	Maximum h m	knots
1 Su		0150	3.1E
	0545	0730	0.8F
	0915	1315	3.1E
	1625	2005	4.1F
	2340		
2 M		0300	3.4E
	0650	0855	1.3F
	1105	1440	3.0E
	1745	2115	3.9F
3 Tu	0035	0400	3.9E
	0735	1005	2.1F
	1240	1555	3.2E
	1900	2215	3.8F
4 W	0125	0445	4.4E
	0815	1105	3.0F
	1355	1705	3.4E
	2005	2310	3.6F
5 Th	0205	0530	4.8E
	0855	1155	3.9F
	1500	1800	3.7E
	2105	2355	3.3F
6 F	0240	0610	5.0E
	0925	1240	4.7F
	1555	1855	3.8E
	2205		
7 Sa		0040	2.8F
	0310	0645	5.1E
	1000	1320	5.2F
	1645	1940	3.9E
	2300		
8 Su		0120	2.3F
	0340	0720	5.0E
	1035	1400	5.4F
	1730	2030	3.9E
	2350		
9 M		0200	1.9F
	0410	0755	4.7E
	1105	1435	5.4F
	1815	2110	3.7E
10 Tu	0045	0240	1.4F
	0440	0825	4.4E
	1140	1515	5.3F
	1855	2155	3.6E
11 W	0135	0325	1.1F
	0510	0905	4.0E
	1215	1555	5.0F
	1935	2240	3.4E
12 Th	0230	0405	0.8F
	0540	0940	3.6E
	1255	1635	4.6F
	2015	2325	3.1E
13 F	0325	0450	0.6F
	0620	1025	3.2E
	1335	1720	4.1F
	2100		
14 Sa		0015	3.0E
	0425	0545	0.4F
	0705	1115	2.7E
	1425	1805	3.7F
	2150		
15 Su		0105	2.9E
	0525	0650	0.4F
	0810	1210	2.3E
	1520	1900	3.3F
	2235		
16 M		0200	2.9E
	0615	0755	0.6F
	0940	1320	2.0E
	1620	2000	3.0F
	2325		
17 Tu		0255	3.1E
	0655	0905	1.1F
	1120	1435	2.0E
	1725	2055	2.8F
18 W	0005	0340	3.4E
	0725	1000	1.8F
	1240	1540	2.1E
	1835	2145	2.7F
19 Th	0040	0420	3.8E
	0755	1045	2.6F
	1345	1640	2.5E
	1935	2235	2.5F
20 F	0115	0455	4.1E
	0820	1125	3.4F
	1440	1735	2.8E
	2035	2315	2.3F
21 Sa	0150	0530	4.4E
	0850	1205	4.3F
	1530	1825	3.2E
	2135		
22 Su		0000	2.1F
	0220	0605	4.7E
	0925	1245	5.0F
	1615	1910	3.5E
	2230		
23 M		0045	1.9F
	0255	0640	4.9E
	1000	1325	5.5F
	1705	2000	3.8E
	2325		
24 Tu		0130	1.7F
	0330	0720	5.0E
	1040	1410	5.9F
	1750	2050	3.9E
25 W	0020	0215	1.6F
	0415	0805	5.0E
	1125	1500	6.0F
	1840	2140	3.9E
26 Th	0115	0305	1.4F
	0500	0855	4.9E
	1210	1545	5.9F
	1930	2230	3.9E
27 F	0210	0400	1.3F
	0550	0945	4.6E
	1300	1640	5.6F
	2025	2325	3.8E
28 Sa	0310	0500	1.2F
	0650	1045	4.1E
	1400	1735	5.2F
	2115		
29 Su		0025	3.8E
	0415	0610	1.2F
	0805	1150	3.6E
	1500	1835	4.7F
	2210		
30 M		0125	3.9E
	0515	0725	1.5F
	0935	1305	3.1E
	1605	1935	4.1F
	2300		
31 Tu		0225	4.1E
	0610	0840	2.1F
	1115	1425	2.8E
	1720	2040	3.5F
	2345		

June

Day	Slack h m	Maximum h m	knots
1 W		0315	4.4E
	0655	0945	2.9F
	1245	1540	2.7E
	1835	2140	3.0F
2 Th	0030	0405	4.6E
	0735	1045	3.7F
	1400	1650	2.9E
	1950	2235	2.4F
3 F	0105	0450	4.7E
	0815	1135	4.4F
	1500	1755	3.1E
	2105	2325	2.0F
4 Sa	0140	0530	4.8E
	0855	1220	5.0F
	1555	1845	3.3E
	2210		
5 Su		0015	1.6F
	0215	0610	4.7E
	0930	1300	5.3F
	1640	1935	3.4E
	2310		
6 M		0100	1.2F
	0250	0645	4.5E
	1005	1340	5.4F
	1720	2020	3.5E
7 Tu	0000	0145	1.0F
	0320	0725	4.3E
	1040	1415	5.3F
	1800	2100	3.5E
8 W	0050	0225	0.8F
	0400	0800	4.0E
	1115	1455	5.2F
	1840	2140	3.5E
9 Th	0135	0305	0.7F
	0435	0840	3.8E
	1150	1530	5.0F
	1915	2220	3.4E
10 F	0215	0345	0.7F
	0515	0920	3.5E
	1230	1610	4.7F
	1950	2300	3.4E
11 Sa	0300	0430	0.7F
	0605	1000	3.2E
	1310	1650	4.4F
	2025	2340	3.3E
12 Su	0340	0520	0.7F
	0655	1045	2.9E
	1350	1730	4.0F
	2105		
13 M		0020	3.3E
	0425	0610	0.8F
	0800	1135	2.5E
	1435	1810	3.5F
	2140		
14 Tu		0105	3.4E
	0505	0705	1.1F
	0920	1235	2.1E
	1530	1900	3.1F
	2215		
15 W		0145	3.5E
	0540	0810	1.6F
	1045	1345	1.9E
	1630	1950	2.6F
	2250		
16 Th		0230	3.7E
	0615	0905	2.2F
	1210	1455	1.8E
	1745	2045	2.1F
	2325		
17 F		0315	3.9E
	0650	1000	3.0F
	1320	1605	2.1E
	1900	2140	1.8F
18 Sa	0000	0355	4.2E
	0730	1050	3.9F
	1425	1710	2.4E
	2015	2235	1.5F
19 Su	0040	0440	4.5E
	0810	1135	4.7F
	1515	1805	2.9E
	2125	2325	1.4F
20 M	0125	0525	4.7E
	0850	1225	5.3F
	1605	1900	3.3E
	2225		
21 Tu		0020	1.3F
	0210	0615	5.0E
	0935	1310	5.9F
	1655	1950	3.7E
	2320		
22 W		0115	1.4F
	0300	0705	5.1E
	1020	1400	6.2F
	1745	2040	4.0E
23 Th	0010	0205	1.5F
	0355	0755	5.2E
	1110	1450	6.3F
	1830	2130	4.2E
24 F	0100	0300	1.6F
	0455	0845	5.1E
	1200	1535	6.3F
	1915	2220	4.4E
25 Sa	0155	0355	1.7F
	0555	0940	4.8E
	1255	1625	6.0F
	2000	2310	4.5F
26 Su	0245	0455	1.9F
	0705	1040	4.3E
	1345	1715	5.4F
	2045		
27 M		0000	4.6E
	0335	0555	2.1F
	0820	1140	3.7E
	1440	1810	4.6F
	2130		
28 Tu		0050	4.6E
	0430	0700	2.4F
	0940	1250	3.0E
	1545	1900	3.7F
	2210		
29 W		0140	4.6E
	0520	0810	2.9F
	1110	1405	2.5E
	1655	2000	2.8F
	2250		
30 Th		0230	4.5E
	0610	0915	3.4F
	1240	1525	2.3E
	1820	2100	2.0F
	2330		

Time meridian 120° W. 0000 is midnight. 1200 is noon. Times are not adjusted for Daylight Saving Time.

Active Pass, British Columbia, 2005

F–Flood, Dir. 045° True E–Ebb, Dir. 225° True

July

Day	Slack h m	Maximum h m	knots
1 F		0325	4.4E
	0655	1015	4.0F
	1355	1640	2.4E
	1950	2205	1.4F
2 Sa	0010	0410	4.3E
	0740	1110	4.5F
	1455	1745	2.7E
	2110	2305	1.0F
3 Su	0050	0500	4.2E
	0820	1200	4.8F
	1545	1840	3.0E
	2220		
4 M		0000	0.8F
	0130	0545	4.1E
	0900	1240	5.0F
	1630	1930	3.2E
	2315		
5 Tu		0045	0.7F
	0215	0625	4.0E
	0940	1320	5.1F
	1710	2010	3.3E
6 W ●	0000	0130	0.7F
	0300	0705	3.9E
	1020	1400	5.1F
	1745	2045	3.4E
7 Th	0035	0210	0.8F
	0345	0745	3.8E
	1055	1435	5.0F
	1815	2120	3.5E
8 F	0110	0250	0.9F
	0425	0825	3.8E
	1135	1510	4.9F
	1850	2155	3.6E
9 Sa	0140	0325	1.0F
	0510	0900	3.6E
	1210	1545	4.7F
	1920	2225	3.6E
10 Su	0210	0405	1.2F
	0600	0940	3.4E
	1245	1615	4.4F
	1945	2300	3.7E
11 M	0240	0445	1.3F
	0650	1020	3.1E
	1320	1650	4.0F
	2015	2330	3.7E
12 Tu	0315	0530	1.6F
	0750	1105	2.7E
	1405	1725	3.5F
	2040		
13 W		0005	3.8E
	0350	0620	1.9F
	0855	1200	2.3E
	1450	1805	2.9F
	2110		
14 Th ◐		0045	3.8E
	0430	0715	2.2F
	1015	1305	1.9E
	1550	1850	2.2F
	2135		
15 F		0125	3.8E
	0510	0815	2.7F
	1140	1420	1.7E
	1710	1950	1.6F
	2210		
16 Sa		0215	3.9E
	0555	0915	3.3F
	1255	1535	1.8E
	1840	2050	1.1F
	2255		
17 Su		0310	4.1E
	0645	1015	4.0F
	1405	1650	2.2E
	2010	2200	0.9F
	2345		
18 M		0405	4.3E
	0735	1110	4.7F
	1500	1750	2.7E
	2120	2305	1.0F
19 Tu	0050	0505	4.6E
	0830	1205	5.4F
	1555	1845	3.3E
	2215		
20 W		0005	1.2F
	0155	0600	5.0E
	0920	1255	5.9F
	1640	1935	3.8E
	2305		
21 Th ○		0100	1.5F
	0255	0655	5.3E
	1010	1345	6.3F
	1725	2025	4.2E
	2350		
22 F		0155	1.9F
	0400	0750	5.4E
	1100	1435	6.4F
	1810	2110	4.6E
23 Sa	0035	0250	2.3F
	0500	0840	5.3E
	1150	1520	6.3F
	1850	2155	4.9E
24 Su	0120	0340	2.6F
	0605	0935	5.0E
	1240	1605	5.8F
	1930	2240	5.0E
25 M	0205	0435	2.9F
	0710	1025	4.4E
	1330	1650	5.1F
	2005	2320	5.0E
26 Tu	0250	0530	3.1F
	0820	1125	3.7E
	1425	1735	4.1F
	2040		
27 W ◑		0005	4.9E
	0340	0630	3.3F
	0935	1230	3.0E
	1525	1825	3.0F
	2115		
28 Th		0055	4.6E
	0430	0735	3.5F
	1055	1345	2.4E
	1640	1920	2.0F
	2150		
29 F		0145	4.2E
	0520	0840	3.7F
	1220	1505	2.1E
	1820	2030	1.2F
	2230		
30 Sa		0240	3.9E
	0615	0945	3.9F
	1335	1625	2.3E
	2000	2140	0.7F
	2315		
31 Su		0340	3.7E
	0705	1045	4.2F
	1435	1735	2.6E
	2125	2250	0.5F

August

Day	Slack h m	Maximum h m	knots
1 M	0010	0435	3.6E
	0755	1135	4.4F
	1525	1825	2.9E
	2220	2345	0.6F
2 Tu	0115	0525	3.6E
	0840	1220	4.6F
	1605	1910	3.2E
	2300		
3 W		0035	0.8F
	0210	0615	3.7E
	0925	1305	4.7F
	1640	1945	3.4E
	2330		
4 Th ●		0115	1.0F
	0300	0655	3.8E
	1005	1340	4.8F
	1715	2020	3.5E
	2355		
5 F		0150	1.2F
	0345	0730	3.8E
	1040	1410	4.8F
	1745	2050	3.7E
6 Sa	0025	0225	1.5F
	0430	0805	3.8E
	1115	1445	4.7F
	1810	2115	3.8E
7 Su	0050	0300	1.7F
	0510	0845	3.7E
	1150	1510	4.5F
	1835	2145	3.9E
8 M	0115	0335	2.0F
	0555	0920	3.6E
	1225	1540	4.2F
	1855	2210	4.0E
9 Tu	0140	0410	2.3F
	0645	1000	3.3E
	1300	1610	3.7F
	1920	2240	4.0E
10 W	0210	0450	2.6F
	0740	1040	2.9E
	1340	1645	3.1F
	1940	2310	4.0E
11 Th	0240	0535	2.8F
	0840	1130	2.5E
	1430	1720	2.4F
	2005	2345	4.0E
12 F ◐	0320	0625	3.0F
	0950	1235	2.0E
	1530	1810	1.7F
	2035		
13 Sa		0030	3.8E
	0410	0725	3.2F
	1110	1350	1.8E
	1655	1905	1.0F
	2110		
14 Su		0125	3.7E
	0505	0840	3.5F
	1230	1515	1.8E
	1840	2025	0.6F
	2200		
15 M		0235	3.8E
	0610	0950	4.0F
	1345	1635	2.3E
	2010	2145	0.6F
	2320		
16 Tu		0345	4.0E
	0715	1055	4.6F
	1445	1740	2.9E
	2110	2300	1.0F
17 W	0040	0455	4.5E
	0815	1150	5.3F
	1535	1830	3.5E
	2155		
18 Th		0000	1.6F
	0200	0555	4.9E
	0910	1240	5.8F
	1615	1915	4.1E
	2235		
19 F ○		0055	2.2F
	0305	0650	5.3E
	1000	1330	6.0F
	1655	2000	4.7E
	2320		
20 Sa		0145	2.8F
	0410	0740	5.4E
	1050	1415	6.0F
	1735	2040	5.1E
21 Su	0000	0235	3.4F
	0510	0830	5.3E
	1140	1455	5.7F
	1810	2120	5.3E
22 M	0040	0320	3.8F
	0610	0920	4.9E
	1230	1535	5.1F
	1845	2200	5.4E
23 Tu	0120	0410	4.1F
	0710	1015	4.3E
	1320	1620	4.2F
	1915	2240	5.2E
24 W	0200	0500	4.1F
	0810	1110	3.6E
	1415	1705	3.2F
	1950	2320	4.8E
25 Th	0245	0555	4.0F
	0915	1210	2.9E
	1515	1750	2.1F
	2020		
26 F ◑		0005	4.2E
	0335	0655	3.8F
	1030	1320	2.4E
	1640	1850	1.2F
	2050		
27 Sa		0055	3.7E
	0430	0800	3.6F
	1145	1445	2.2E
	1825	2000	0.5F
	2135		
28 Su		0200	3.2E
	0530	0910	3.6F
	1300	1605	2.3E
	2010	2125	0.3F
	2235		
29 M		0310	2.9E
	0630	1015	3.7F
	1400	1710	2.6E
	2115	2240	0.5F
30 Tu	0005	0415	3.0E
	0730	1110	3.9F
	1450	1800	3.0E
	2150	2335	0.8F
31 W	0115	0510	3.2E
	0820	1155	4.2F
	1530	1840	3.3E
	2220		

September

Day	Slack h m	Maximum h m	knots
1 Th		0015	1.2F
	0215	0555	3.4E
	0905	1235	4.3F
	1605	1910	3.6E
	2245		
2 F		0055	1.6F
	0305	0635	3.7E
	0945	1310	4.4F
	1630	1940	3.8E
	2305		
3 Sa ●		0125	2.0F
	0345	0715	3.8E
	1020	1340	4.4F
	1655	2005	4.0E
	2330		
4 Su		0200	2.4F
	0430	0750	3.8E
	1055	1410	4.2F
	1720	2030	4.1E
	2355		
5 M		0230	2.7F
	0510	0825	3.8E
	1130	1440	3.9F
	1740	2055	4.2E
6 Tu	0015	0305	3.1F
	0555	0900	3.7E
	1210	1505	3.5F
	1800	2120	4.3E
7 W	0040	0335	3.4F
	0640	0940	3.4E
	1250	1540	3.0F
	1825	2150	4.3E
8 Th	0110	0415	3.7F
	0730	1025	3.1E
	1330	1610	2.4F
	1845	2220	4.2E
9 F	0145	0500	3.7F
	0825	1115	2.7E
	1425	1650	1.7F
	1910	2300	4.0E
10 Sa	0230	0550	3.7F
	0930	1215	2.2E
	1535	1740	1.1F
	1940	2350	3.7E
11 Su ◐	0325	0655	3.6F
	1045	1335	2.0E
	1710	1850	0.6F
	2020		
12 M		0055	3.4E
	0430	0810	3.6F
	1210	1505	2.1E
	1855	2020	0.4F
	2140		
13 Tu		0215	3.4E
	0545	0930	4.0F
	1320	1620	2.6E
	2005	2145	0.8F
	2325		
14 W		0340	3.7E
	0700	1035	4.5F
	1415	1720	3.3E
	2050	2255	1.5F
15 Th	0055	0445	4.2E
	0800	1130	5.0F
	1500	1805	4.0E
	2125	2350	2.3F
16 F	0210	0545	4.7E
	0900	1220	5.3F
	1540	1845	4.6E
	2205		
17 Sa ○		0040	3.1F
	0315	0640	5.0E
	0950	1305	5.3F
	1615	1925	5.1E
	2245		
18 Su		0130	3.9F
	0415	0730	5.1E
	1040	1345	5.1F
	1650	2005	5.4E
	2320		
19 M		0215	4.5F
	0510	0820	4.9E
	1130	1425	4.6F
	1725	2040	5.5E
	2355		
20 Tu		0300	4.9F
	0605	0910	4.6E
	1220	1510	3.8F
	1755	2120	5.4E
21 W	0035	0345	5.0F
	0700	1000	4.1E
	1310	1550	3.0F
	1825	2155	5.0E
22 Th	0115	0430	4.9F
	0755	1050	3.6E
	1410	1635	2.1F
	1855	2235	4.4E
23 F	0155	0520	4.5F
	0855	1150	3.0E
	1520	1720	1.3F
	1925	2315	3.8E
24 Sa ◑	0240	0610	4.0F
	0955	1255	2.6E
	1645	1820	0.6F
	1955		
25 Su		0010	3.1E
	0335	0715	3.6F
	1105	1415	2.4E
		1940	0.0
26 M		0120	2.6E
	0440	0825	3.3F
	1215	1530	2.5E
		2105	0.0
27 Tu		0240	2.4E
	0550	0935	3.3F
	1310	1630	2.8E
	2035	2220	0.7F
28 W	0000	0350	2.5E
	0655	1030	3.5F
	1400	1720	3.2E
	2105	2310	1.2F
29 Th	0115	0450	2.8E
	0750	1120	3.7F
	1440	1755	3.5E
	2130	2350	1.8F
30 F	0210	0535	3.1E
	0835	1155	3.8F
	1510	1825	3.8E
	2155		

Time meridian 120° W. 0000 is midnight. 1200 is noon. Times are not adjusted for Daylight Saving Time.

Active Pass, British Columbia, 2005

F–Flood, Dir. 045° True E–Ebb, Dir. 225° True

October

Day	Slack h m	Maximum h m	knots
1 Sa		0025	2.3F
	0300	0615	3.4E
	0920	1230	3.8F
	1535	1850	4.1E
	2215		
2 Su		0100	2.8F
	0345	0655	3.6E
	1000	1300	3.6F
	1600	1920	4.3E
	2235		
3 M ●		0130	3.4F
	0425	0730	3.7E
	1040	1330	3.4F
	1625	1945	4.4E
	2300		
4 Tu		0200	3.9F
	0505	0810	3.7E
	1120	1405	3.0F
	1645	2010	4.5E
	2325		
5 W		0235	4.3F
	0550	0850	3.7E
	1200	1435	2.6F
	1705	2035	4.5E
	2355		
6 Th		0310	4.5F
	0635	0930	3.5E
	1245	1510	2.1F
	1730	2110	4.5E
7 F	0025	0350	4.6F
	0720	1015	3.2E
	1335	1550	1.6F
	1755	2145	4.3E
8 Sa	0105	0435	4.5F
	0815	1110	2.9E
	1435	1635	1.1F
	1825	2230	4.0E
9 Su	0155	0530	4.3F
	0915	1210	2.5E
	1550	1730	0.7F
	1905	2325	3.6E
10 M ◐	0255	0635	4.0F
	1030	1330	2.4E
	1730	1850	0.4F
	2005		
11 Tu		0040	3.2E
	0405	0750	3.9F
	1140	1450	2.6E
	1845	2020	0.6F
	2155		
12 W		0205	3.1E
	0525	0905	4.0F
	1245	1555	3.2E
	1940	2145	1.2F
	2345		
13 Th		0330	3.4E
	0640	1010	4.2F
	1335	1650	3.9E
	2020	2245	2.2F
14 F	0115	0440	3.8E
	0745	1105	4.4F
	1420	1735	4.5E
	2055	2340	3.2F
15 Sa	0225	0540	4.2E
	0845	1155	4.4F
	1455	1815	5.1E
	2130		
16 Su		0025	4.1F
	0325	0635	4.4E
	0940	1235	4.2F
	1530	1850	5.4E
	2205		
17 M ○		0110	4.9F
	0420	0725	4.5E
	1030	1320	3.7F
	1600	1925	5.5E
	2240		
18 Tu		0155	5.4F
	0515	0810	4.4E
	1125	1400	3.1F
	1630	2005	5.4E
	2320		
19 W		0235	5.6F
	0605	0900	4.2E
	1220	1440	2.5F
	1700	2040	5.1E
	2355		
20 Th		0320	5.5F
	0650	0950	3.9E
	1315	1525	1.9F
	1730	2115	4.6E
21 F	0030	0400	5.2F
	0740	1040	3.5E
	1415	1610	1.3F
	1805	2155	4.1E
22 Sa	0110	0445	4.7F
	0830	1130	3.1E
	1520	1700	0.8F
	1835	2240	3.4E
23 Su	0155	0535	4.2F
	0920	1230	2.8E
	1640	1755	0.4F
	1915	2330	2.8E
24 M ◑	0245	0630	3.6F
	1020	1335	2.6E
		1910	0.0
25 Tu		0035	2.3E
	0350	0735	3.2F
	1115	1440	2.7E
	1910	2030	0.4F
	2200		
26 W		0155	2.0E
	0455	0840	3.1F
	1210	1540	3.0E
	1945	2140	0.9F
	2340		
27 Th		0310	2.1E
	0605	0940	3.0F
	1255	1625	3.4E
	2015	2235	1.5F
28 F	0100	0410	2.3E
	0705	1025	3.1F
	1335	1700	3.7E
	2040	2315	2.2F
29 Sa	0200	0505	2.7E
	0800	1110	3.0F
	1405	1735	4.0E
	2100	2350	2.9F
30 Su	0250	0550	3.0E
	0850	1145	2.9F
	1430	1800	4.3E
	2125		
31 M		0025	3.6F
	0335	0635	3.2E
	0940	1220	2.7F
	1500	1830	4.5E
	2150		

November

Day	Slack h m	Maximum h m	knots
1 Tu ●		0100	4.3F
	0415	0715	3.5E
	1025	1255	2.4F
	1520	1900	4.6E
	2215		
2 W		0135	4.8F
	0500	0755	3.6E
	1110	1335	2.1F
	1550	1930	4.7E
	2245		
3 Th		0210	5.2F
	0540	0840	3.6E
	1200	1410	1.8F
	1615	2005	4.7E
	2320		
4 F		0250	5.4F
	0625	0925	3.6E
	1250	1450	1.5F
	1650	2040	4.6E
5 Sa	0000	0330	5.4F
	0715	1010	3.4E
	1345	1535	1.2F
	1725	2125	4.4E
6 Su	0045	0420	5.2F
	0810	1105	3.2E
	1445	1630	0.9F
	1810	2215	4.1E
7 M	0135	0515	4.8F
	0905	1205	3.1E
	1600	1735	0.7F
	1905	2315	3.6E
8 Tu ◐	0235	0620	4.5F
	1005	1315	3.1E
	1710	1850	0.7F
	2030		
9 W		0030	3.2E
	0345	0725	4.1F
	1105	1420	3.4E
	1815	2015	1.1F
	2215		
10 Th		0155	2.9E
	0500	0835	3.9F
	1200	1520	3.8E
	1905	2130	2.0F
11 F	0000	0320	3.0E
	0615	0935	3.7F
	1245	1615	4.4E
	1945	2230	3.0F
12 Sa	0125	0430	3.2E
	0730	1035	3.5F
	1325	1655	4.9E
	2025	2325	4.0F
13 Su	0235	0530	3.5E
	0835	1125	3.2F
	1405	1740	5.2E
	2100		
14 M		0010	4.8F
	0330	0625	3.8E
	0935	1210	2.8F
	1440	1815	5.3E
	2135		
15 Tu ○		0055	5.5F
	0425	0720	3.9E
	1035	1255	2.3F
	1510	1855	5.2E
	2210		
16 W		0135	5.8F
	0510	0805	3.9E
	1130	1340	1.9F
	1545	1930	5.0E
	2245		
17 Th		0215	5.8F
	0555	0855	3.9E
	1225	1420	1.5F
	1615	2010	4.7E
	2320		
18 F		0255	5.6F
	0640	0940	3.7E
	1320	1505	1.1F
	1650	2045	4.3E
19 Sa	0000	0335	5.3F
	0720	1025	3.5E
	1415	1550	0.8F
	1725	2125	3.8E
20 Su	0040	0420	4.8F
	0805	1110	3.3E
	1510	1635	0.6F
	1805	2210	3.3E
21 M	0120	0500	4.4F
	0845	1155	3.1E
	1610	1730	0.5F
	1855	2255	2.8E
22 Tu	0205	0550	3.9F
	0930	1250	3.0E
	1710	1835	0.5F
	2000	2355	2.3E
23 W ◑	0255	0640	3.4F
	1015	1345	3.1E
	1800	1940	0.7F
	2130		
24 Th		0100	1.9E
	0355	0735	3.0F
	1100	1435	3.3E
	1840	2050	1.1F
	2310		
25 F		0215	1.8E
	0500	0830	2.7F
	1140	1520	3.5E
	1910	2145	1.8F
26 Sa	0035	0325	1.8E
	0610	0925	2.4F
	1215	1600	3.8E
	1940	2235	2.6F
27 Su	0140	0430	2.1E
	0720	1015	2.2F
	1250	1635	4.1E
	2005	2315	3.4F
28 M	0235	0525	2.5E
	0825	1100	2.0F
	1320	1710	4.3E
	2035	2355	4.2F
29 Tu	0320	0615	2.9E
	0925	1145	1.7F
	1355	1745	4.5E
	2105		
30 W		0030	4.8F
	0405	0700	3.2E
	1020	1225	1.6F
	1425	1820	4.7E
	2140		

December

Day	Slack h m	Maximum h m	knots
1 Th ●		0110	5.4F
	0450	0745	3.5E
	1110	1310	1.4F
	1505	1900	4.9E
	2220		
2 F		0150	5.7F
	0535	0830	3.7E
	1200	1355	1.3F
	1545	1945	4.9E
	2300		
3 Sa		0235	5.9F
	0620	0915	3.7E
	1250	1440	1.3F
	1630	2030	4.9E
	2345		
4 Su		0325	5.9F
	0710	1005	3.8E
	1340	1530	1.2F
	1720	2120	4.7E
5 M	0035	0410	5.7F
	0755	1055	3.8E
	1435	1625	1.2F
	1815	2210	4.3E
6 Tu	0125	0505	5.4F
	0845	1150	3.8E
	1535	1730	1.3F
	1925	2310	3.8E
7 W	0225	0600	4.9F
	0935	1245	3.9E
	1635	1840	1.5F
	2050		
8 Th ◐		0020	3.2E
	0325	0655	4.2F
	1020	1345	4.1E
	1730	1955	2.0F
	2230		
9 F		0140	2.8E
	0435	0755	3.6F
	1105	1440	4.4E
	1820	2105	2.7F
10 Sa	0005	0300	2.6E
	0555	0900	2.9F
	1150	1530	4.6E
	1905	2210	3.6F
11 Su	0130	0420	2.7E
	0715	1000	2.4F
	1230	1620	4.8E
	1950	2305	4.4F
12 M	0240	0525	2.9E
	0835	1055	1.9F
	1310	1705	4.9E
	2030	2355	5.1F
13 Tu	0335	0625	3.3E
	0945	1150	1.5F
	1350	1745	4.8E
	2105		
14 W		0040	5.5F
	0420	0715	3.5E
	1050	1240	1.3F
	1430	1830	4.7E
	2145		
15 Th ○		0120	5.7F
	0505	0805	3.6E
	1140	1325	1.1F
	1505	1910	4.5E
	2225		
16 F		0200	5.6F
	0545	0845	3.7E
	1230	1410	1.0F
	1545	1950	4.3E
	2300		
17 Sa		0240	5.5F
	0625	0925	3.6E
	1315	1450	0.9F
	1625	2025	4.0E
	2340		
18 Su		0320	5.2F
	0700	1005	3.6E
	1355	1530	0.9F
	1710	2105	3.8E
19 M	0015	0355	4.9F
	0735	1040	3.5E
	1435	1615	0.9F
	1755	2145	3.4E
20 Tu	0055	0430	4.5F
	0810	1120	3.5E
	1515	1700	0.9F
	1845	2230	3.0E
21 W	0135	0510	4.1F
	0840	1200	3.5E
	1600	1750	1.0F
	1945	2315	2.5E
22 Th	0215	0550	3.6F
	0915	1240	3.5E
	1640	1845	1.2F
	2100		
23 F ◑		0010	2.1E
	0305	0635	3.0F
	0945	1325	3.5E
	1720	1945	1.5F
	2230		
24 Sa		0120	1.7E
	0405	0720	2.4F
	1020	1405	3.6E
	1800	2045	2.1F
	2355		
25 Su		0235	1.5E
	0520	0815	1.8F
	1055	1450	3.7E
	1835	2145	2.8F
26 M	0115	0350	1.7E
	0645	0915	1.4F
	1130	1535	3.9E
	1915	2235	3.6F
27 Tu	0215	0500	2.1E
	0805	1015	1.1F
	1210	1625	4.1E
	1950	2320	4.3F
28 W	0305	0555	2.6E
	0915	1110	1.0F
	1255	1710	4.4E
	2035		
29 Th		0010	5.0F
	0355	0645	3.0E
	1015	1200	1.1F
	1345	1755	4.7E
	2115		
30 F ●		0055	5.6F
	0440	0735	3.4E
	1105	1255	1.2F
	1440	1845	5.0E
	2205		
31 Sa		0140	6.0F
	0525	0820	3.8E
	1150	1345	1.4F
	1535	1935	5.2E
	2250		

Time meridian 120° W. 0000 is midnight. 1200 is noon. Times are not adjusted for Daylight Saving Time.

Burrard Inlet (First Narrows), British Columbia, 2005

F–Flood, Dir. 135° True E–Ebb, Dir. 315° True

January

Day	Slack h m	Maximum h m	knots
1 Sa	0315	0650	3.9F
	1045	1335	2.3E
	1715	1900	0.6F
	2040		
2 Su		0035	2.7E
	0350	0725	3.5F
	1105	1415	2.6E
	1805	2005	0.9F
	2205		
3 M ◑		0135	2.0E
	0430	0805	3.0F
	1130	1500	3.0E
	1855	2120	1.3F
4 Tu	0000	0240	1.3E
	0515	0850	2.5F
	1155	1545	3.5E
	1940	2235	2.0F
5 W	0200	0405	0.8E
	0610	0935	2.0F
	1225	1635	4.0E
	2030	2340	2.8F
6 Th	0345	0535	0.7E
	0730	1035	1.6F
	1300	1725	4.5E
	2115		
7 F		0040	3.7F
	0500	0650	0.9E
	0855	1135	1.3F
	1345	1815	5.0E
	2205		
8 Sa		0135	4.5F
	0555	0755	1.2E
	1015	1235	1.2F
	1435	1910	5.4E
	2250		
9 Su		0225	5.1F
	0645	0850	1.5E
	1125	1335	1.2F
	1530	2000	5.6E
	2340		
10 M ●		0315	5.6F
	0725	0945	1.9E
	1225	1435	1.3F
	1625	2050	5.8E
11 Tu	0030	0405	5.8F
	0805	1030	2.3E
	1320	1530	1.5F
	1725	2140	5.6E
12 W	0115	0450	5.8F
	0845	1120	2.6E
	1415	1625	1.6F
	1825	2235	5.3E
13 Th	0200	0530	5.6F
	0925	1205	2.9E
	1515	1725	1.6F
	1930	2325	4.6E
14 F	0245	0615	5.1F
	0955	1250	3.2E
	1610	1825	1.7F
	2035		
15 Sa		0015	3.8E
	0325	0655	4.5F
	1030	1335	3.5E
	1710	1930	1.7F
	2155		
16 Su ◐		0110	2.8E
	0410	0740	3.7F
	1100	1425	3.7E
	1810	2040	1.9F
	2330		
17 M		0215	1.8E
	0455	0820	2.8F
	1130	1510	3.8E
	1905	2200	2.2F
18 Tu	0125	0335	1.0E
	0550	0910	2.0F
	1155	1600	3.9E
	2000	2310	2.7F
19 W	0320	0510	0.6E
	0705	1005	1.3F
	1225	1655	3.9E
	2045		
20 Th		0015	3.2F
	0445	0635	0.7E
	0850	1105	0.7F
	1255	1745	4.0E
	2130		
21 F		0110	3.7F
	0535	0745	1.0E
	1025	1205	0.5F
	1335	1830	4.0E
	2215		
22 Sa		0155	4.1F
	0615	0835	1.3E
	1135	1300	0.4F
	1415	1915	4.1E
	2255		
23 Su		0235	4.3F
	0650	0915	1.5E
	1220	1350	0.4F
	1505	1955	4.1E
	2330		
24 M		0310	4.5F
	0725	0950	1.7E
	1255	1430	0.5F
	1555	2035	4.2E
25 Tu ○	0005	0345	4.5F
	0750	1020	1.8E
	1325	1505	0.7F
	1640	2110	4.2E
26 W	0040	0415	4.5F
	0815	1050	2.0E
	1355	1545	0.9F
	1725	2145	4.2E
27 Th	0110	0445	4.5F
	0840	1115	2.1E
	1420	1620	1.0F
	1815	2220	4.0E
28 F	0145	0515	4.3F
	0900	1145	2.4E
	1455	1700	1.2F
	1905	2255	3.7E
29 Sa	0215	0540	4.0F
	0920	1210	2.6E
	1530	1745	1.4F
	2000	2335	3.2E
30 Su	0245	0610	3.6F
	0935	1240	3.0E
	1610	1835	1.6F
	2105		
31 M		0020	2.5E
	0315	0640	3.1F
	0955	1315	3.3E
	1655	1930	1.8F
	2220		

February

Day	Slack h m	Maximum h m	knots
1 Tu ◑		0110	1.8E
	0350	0710	2.6F
	1015	1355	3.5E
	1745	2040	2.1F
2 W	0005	0220	1.0E
	0435	0755	2.0F
	1040	1445	3.8E
	1845	2155	2.5F
3 Th	0205	0345	0.5E
	0535	0845	1.4F
	1115	1545	4.0E
	1945	2315	3.1F
4 F	0350	0525	0.5E
	0710	0955	1.0F
	1205	1650	4.3E
	2045		
5 Sa		0025	3.8F
	0455	0650	0.8E
	0900	1115	0.9F
	1310	1755	4.7E
	2145		
6 Su		0125	4.4F
	0545	0750	1.3E
	1020	1230	1.0F
	1420	1855	5.1E
	2240		
7 M		0215	5.0F
	0620	0840	1.8E
	1120	1335	1.3F
	1530	1955	5.3E
	2330		
8 Tu ●		0300	5.3F
	0700	0925	2.3E
	1215	1435	1.7F
	1640	2045	5.4E
9 W	0015	0345	5.5F
	0730	1005	2.8E
	1305	1530	2.1F
	1740	2135	5.3E
10 Th	0100	0425	5.4F
	0805	1050	3.3E
	1350	1620	2.3F
	1845	2225	4.8E
11 F	0140	0505	5.0F
	0835	1130	3.6E
	1440	1715	2.5F
	1945	2310	4.1E
12 Sa	0220	0540	4.5F
	0900	1205	3.9E
	1530	1810	2.6F
	2055		
13 Su		0000	3.3E
	0300	0615	3.7F
	0925	1245	4.0E
	1615	1905	2.7F
	2205		
14 M		0055	2.3E
	0340	0650	2.9F
	0950	1325	4.0E
	1705	2005	2.7F
	2330		
15 Tu ◐		0155	1.5E
	0425	0730	2.0F
	1015	1410	3.8E
	1800	2115	2.7F
16 W	0110	0315	0.8E
	0525	0815	1.2F
	1040	1500	3.5E
	1855	2225	2.8F
17 Th	0255	0450	0.6E
	0705	0915	0.6F
	1105	1600	3.3E
	1955	2340	3.0F
18 F	0410	0620	0.8E
		1035	0.0
		1705	3.3E
	2050		
19 Sa		0040	3.4F
	0500	0725	1.1E
		1150	0.0
		1805	3.4E
	2140		
20 Su		0125	3.7F
	0540	0810	1.5E
	1125	1250	0.4F
	1410	1855	3.5E
	2225		
21 M		0205	3.9F
	0610	0845	1.7E
	1155	1335	0.7F
	1515	1940	3.7E
	2305		
22 Tu		0240	4.1F
	0635	0915	2.0E
	1220	1415	1.0F
	1605	2015	3.9E
	2345		
23 W ○		0315	4.2F
	0700	0940	2.2E
	1245	1455	1.3F
	1655	2055	3.9E
24 Th	0015	0340	4.2F
	0720	1005	2.5E
	1310	1530	1.6F
	1745	2130	3.9E
25 F	0050	0410	4.0F
	0740	1030	2.8E
	1335	1605	2.0F
	1835	2205	3.7E
26 Sa	0120	0435	3.8F
	0755	1055	3.1E
	1405	1645	2.3F
	1925	2245	3.3E
27 Su	0150	0500	3.4F
	0810	1120	3.4E
	1440	1725	2.6F
	2020	2325	2.8E
28 M	0225	0530	3.0F
	0825	1150	3.7E
	1515	1815	2.8F
	2125		

March

Day	Slack h m	Maximum h m	knots
1 Tu		0010	2.2E
	0255	0600	2.5F
	0845	1225	3.9E
	1600	1905	2.9F
	2240		
2 W		0105	1.5E
	0335	0630	1.9F
	0910	1310	3.9E
	1655	2010	2.9F
3 Th ◑	0010	0210	0.9E
	0425	0715	1.3F
	0940	1400	3.9E
	1800	2130	3.0F
4 F	0200	0345	0.5E
	0545	0820	0.8F
	1020	1510	3.8E
	1910	2250	3.3F
5 Sa	0330	0525	0.7E
	0745	0945	0.6F
	1130	1630	3.9E
	2020		
6 Su		0005	3.8F
	0425	0640	1.1E
	0920	1120	0.7F
	1300	1745	4.2E
	2125		
7 M		0105	4.3F
	0505	0735	1.7E
	1025	1235	1.2F
	1435	1850	4.5E
	2220		
8 Tu		0155	4.6F
	0540	0815	2.4E
	1115	1335	1.8F
	1550	1945	4.7E
	2310		
9 W		0235	4.8F
	0610	0855	3.0E
	1155	1430	2.3F
	1700	2040	4.7E
	2355		
10 Th ●		0315	4.8F
	0640	0935	3.5E
	1240	1525	2.9F
	1800	2125	4.4E
11 F	0040	0350	4.5F
	0705	1010	4.0E
	1320	1610	3.3F
	1905	2215	3.9E
12 Sa	0120	0425	4.0F
	0730	1045	4.3E
	1405	1700	3.5F
	2005	2300	3.3E
13 Su	0200	0500	3.4F
	0755	1120	4.4E
	1445	1750	3.6F
	2105	2350	2.6E
14 M	0240	0535	2.6F
	0815	1155	4.3E
	1525	1835	3.6F
	2210		
15 Tu		0040	1.9E
	0325	0610	1.9F
	0840	1230	4.1E
	1610	1930	3.4F
	2320		
16 W		0140	1.4E
	0420	0645	1.2F
	0900	1315	3.6E
	1700	2025	3.1F
17 Th ◐	0040	0255	0.9E
	0535	0735	0.6F
	0920	1405	3.2E
	1755	2135	2.9F
18 F	0205	0425	0.8E
		0840	0.0
		1505	2.8E
	1855	2245	2.9F
19 Sa	0310	0550	1.1E
		1010	0.0
		1620	2.7E
	2000	2350	3.1F
20 Su	0400	0645	1.4E
		1135	0.0
		1730	2.8E
	2100		
21 M		0040	3.3F
	0440	0725	1.8E
	1050	1235	0.6F
	1410	1825	3.0E
	2150		
22 Tu		0125	3.5F
	0510	0755	2.1E
	1115	1320	1.0F
	1520	1915	3.2E
	2230		
23 W		0200	3.6F
	0535	0825	2.4E
	1135	1400	1.5F
	1615	1955	3.3E
	2310		
24 Th		0230	3.6F
	0555	0850	2.8E
	1200	1435	2.0F
	1710	2035	3.4E
	2345		
25 F ○		0300	3.5F
	0610	0915	3.2E
	1225	1515	2.6F
	1800	2115	3.3E
26 Sa	0020	0325	3.3F
	0625	0940	3.6E
	1255	1550	3.1F
	1850	2155	3.1E
27 Su	0055	0355	2.9F
	0645	1005	3.9E
	1325	1630	3.5F
	1945	2235	2.7E
28 M	0130	0420	2.6F
	0700	1035	4.2E
	1400	1710	3.8F
	2040	2320	2.3E
29 Tu	0210	0450	2.1F
	0720	1110	4.4E
	1440	1800	3.9F
	2140		
30 W		0010	1.9E
	0255	0525	1.7F
	0745	1150	4.4E
	1525	1850	3.8F
	2250		
31 Th		0110	1.4E
	0345	0610	1.2F
	0815	1235	4.2E
	1620	1955	3.6F

Time meridian 120° W. 0000 is midnight. 1200 is noon. Times are not adjusted for Daylight Saving Time.

Burrard Inlet (First Narrows), British Columbia, 2005

F–Flood, Dir. 135° True E–Ebb, Dir. 315° True

April

Day	Slack h m	Maximum h m	knots
1 F	0015	0220	1.0E
	0455	0700	0.8F
	0855	1335	3.9E
	1730	2110	3.5F
2 Sa	0140	0350	0.9E
	0640	0820	0.4F
	0950	1450	3.6E
	1840	2225	3.6F
3 Su	0250	0515	1.3E
	0825	1000	0.4F
	1130	1615	3.5E
	1955	2335	3.8F
4 M	0335	0620	1.9E
	0930	1130	0.9F
	1325	1735	3.6E
	2100		
5 Tu		0035	4.0F
	0415	0705	2.5E
	1015	1240	1.6F
	1455	1840	3.7E
	2155		
6 W		0120	4.1F
	0445	0745	3.2E
	1055	1335	2.4F
	1615	1940	3.7E
	2245		
7 Th		0200	4.0F
	0515	0820	3.8E
	1135	1430	3.1F
	1720	2030	3.6E
	2335		
8 F		0240	3.7F
	0540	0855	4.3E
	1215	1515	3.7F
	1825	2120	3.3E
9 Sa	0020	0315	3.2F
	0600	0930	4.6E
	1250	1600	4.1F
	1925	2210	2.9E
10 Su	0105	0350	2.7F
	0625	1005	4.7E
	1330	1645	4.4F
	2020	2255	2.5E
11 M	0150	0420	2.1F
	0045	1035	4.6E
	1405	1725	4.4F
	2115	2345	2.1E
12 Tu	0235	0455	1.5F
	0705	1110	4.4E
	1440	1810	4.2F
	2210		
13 W		0035	1.7E
	0330	0535	1.0F
	0725	1145	4.0E
	1520	1855	3.9F
	2305		
14 Th		0130	1.4E
	0435	0615	0.5F
	0745	1225	3.5E
	1605	1945	3.5F
15 F	0005	0235	1.2E
		0705	0.0
		1315	3.0E
	1655	2045	3.2F
16 Sa	0105	0350	1.2E
		0820	0.0
		1415	2.6E
	1755	2150	3.0F
17 Su	0205	0455	1.4E
		0945	0.0
		1530	2.3E
	1900	2250	3.0F
18 M	0250	0550	1.8E
		1105	0.0
		1645	2.3E
	2000	2345	3.0F
19 Tu	0325	0630	2.2E
	1010	1205	0.8F
	1405	1745	2.4E
	2055		
20 W		0025	3.1F
	0355	0700	2.6E
	1030	1255	1.4F
	1520	1840	2.5E
	2140		
21 Th		0100	3.0F
	0415	0730	3.1E
	1055	1335	2.1F
	1625	1930	2.6E
	2225		
22 F		0135	2.9F
	0435	0755	3.5E
	1120	1415	2.8F
	1720	2015	2.6E
	2310		
23 Sa		0205	2.7F
	0455	0825	4.0E
	1145	1455	3.5F
	1815	2100	2.5E
	2355		
24 Su		0240	2.4F
	0510	0855	4.4E
	1215	1535	4.1F
	1910	2145	2.4E
25 M	0035	0310	2.0F
	0535	0925	4.7E
	1250	1615	4.5F
	2000	2230	2.2E
26 Tu	0120	0345	1.7F
	0600	1000	4.9E
	1330	1700	4.7F
	2055	2320	2.0E
27 W	0210	0425	1.4F
	0625	1040	4.9E
	1415	1750	4.7F
	2155		
28 Th		0015	1.7E
	0305	0510	1.1F
	0700	1125	4.7E
	1505	1840	4.6F
	2255		
29 F		0115	1.5E
	0410	0600	0.7F
	0740	1220	4.3E
	1600	1945	4.3F
30 Sa	0000	0230	1.5E
	0540	0710	0.5F
	0835	1325	3.8E
	1705	2050	4.0F

May

Day	Slack h m	Maximum h m	knots
1 Su	0105	0345	1.7E
	0715	0835	0.4F
	0955	1440	3.3E
	1810	2155	3.8F
2 M	0155	0450	2.1E
	0825	1010	0.7F
	1155	1600	3.0E
	1920	2300	3.7F
3 Tu	0240	0545	2.8E
	0915	1135	1.4F
	1350	1720	2.8E
	2025	2350	3.5F
4 W	0315	0630	3.4E
	0955	1240	2.2F
	1525	1830	2.7E
	2125		
5 Th		0040	3.3F
	0340	0710	4.0E
	1035	1335	3.1F
	1640	1930	2.5E
	2220		
6 F		0120	2.9F
	0410	0745	4.5E
	1110	1420	3.8F
	1745	2025	2.4E
	2310		
7 Sa		0200	2.4F
	0430	0820	4.8E
	1145	1505	4.4F
	1845	2115	2.2E
8 Su	0005	0235	1.9F
	0455	0850	4.9E
	1220	1545	4.7F
	1940	2205	2.1E
9 M	0055	0310	1.4F
	0515	0925	4.8E
	1255	1625	4.8F
	2030	2255	1.9E
10 Tu	0150	0350	1.0F
	0535	0955	4.6E
	1330	1705	4.7F
	2115	2340	1.8E
11 W	0245	0425	0.7F
	0600	1030	4.3E
	1405	1745	4.5F
	2200		
12 Th		0025	1.7E
	0345	0505	0.4F
	0620	1110	3.9E
	1445	1825	4.2F
	2245		
13 F		0115	1.6E
		0550	0.0
		1150	3.5E
	1525	1910	3.9F
	2330		
14 Sa		0210	1.6E
		0650	0.0
		1235	3.0E
	1610	2000	3.5F
15 Su	0015	0305	1.6E
		0750	0.0
		1330	2.5E
	1700	2050	3.3F
16 M	0055	0400	1.9E
		0910	0.0
		1435	2.1E
	1750	2145	3.1F
17 Tu	0135	0445	2.2E
	0855	1025	0.4F
	1200	1550	1.8E
	1850	2235	2.9F
18 W	0205	0525	2.7E
	0915	1130	1.1F
	1350	1700	1.7E
	1945	2315	2.7F
19 Th	0230	0600	3.2E
	0940	1225	1.9F
	1520	1805	1.7E
	2045		
20 F		0000	2.4F
	0250	0630	3.7E
	1010	1310	2.7F
	1630	1900	1.7E
	2140		
21 Sa		0035	2.2F
	0315	0705	4.2E
	1040	1355	3.5F
	1730	1955	1.8E
	2235		
22 Su		0115	1.9F
	0340	0740	4.6E
	1110	1435	4.3F
	1825	2050	1.8E
	2330		
23 M		0155	1.6F
	0405	0815	5.0E
	1150	1520	4.8F
	1920	2140	1.9E
24 Tu	0025	0235	1.4F
	0435	0855	5.2E
	1230	1605	5.2F
	2010	2230	1.9E
25 W	0120	0320	1.2F
	0510	0940	5.3E
	1315	1650	5.3F
	2100	2325	1.9E
26 Th	0215	0410	1.0F
	0555	1025	5.2E
	1400	1740	5.3F
	2155		
27 F		0015	1.9E
	0315	0505	0.8F
	0645	1115	4.9E
	1455	1835	5.1F
	2245		
28 Sa		0115	2.0E
	0425	0605	0.7F
	0740	1210	4.4E
	1545	1930	4.8F
	2335		
29 Su		0215	2.2E
	0545	0720	0.6F
	0850	1315	3.7E
	1640	2025	4.4F
30 M	0020	0315	2.5E
	0655	0840	0.7F
	1025	1425	3.0E
	1740	2120	3.9F
31 Tu	0100	0410	3.0E
	0755	1005	1.2F
	1220	1545	2.3E
	1840	2215	3.4F

June

Day	Slack h m	Maximum h m	knots
1 W	0135	0500	3.6E
	0845	1125	2.0F
	1415	1705	1.9E
	1945	2305	2.9F
2 Th	0205	0545	4.1E
	0930	1230	2.8F
	1550	1820	1.6E
	2050	2355	2.3F
3 F	0235	0630	4.5E
	1005	1320	3.6F
	1710	1925	1.6E
	2155		
4 Sa		0040	1.8F
	0300	0705	4.7E
	1045	1410	4.3F
	1810	2025	1.6E
	2300		
5 Su		0120	1.3F
	0325	0740	4.8E
	1120	1455	4.7F
	1900	2120	1.6E
6 M	0005	0200	0.9F
	0345	0815	4.8E
	1155	1535	4.9F
	1945	2205	1.6E
7 Tu	0105	0245	0.6F
	0410	0855	4.6E
	1230	1610	4.9F
	2030	2250	1.7E
8 W	0205	0325	0.4F
	0440	0930	4.4E
	1305	1650	4.8F
	2105	2330	1.7E
9 Th	0255	0405	0.3F
	0510	1005	4.2E
	1340	1725	4.6F
	2140		
10 F		0010	1.7E
		0445	0.0
		1045	3.9E
	1415	1800	4.4F
	2215		
11 Sa		0050	1.8E
		0530	0.0
		1125	3.5E
	1455	1840	4.1F
	2250		
12 Su		0130	1.9E
		0620	0.0
		1205	3.1E
	1530	1920	3.8F
	2320		
13 M		0215	2.1E
		0720	0.0
		1255	2.6E
	1610	2000	3.5F
	2350		
14 Tu		0255	2.3E
	0705	0830	0.4F
	0955	1350	2.0E
	1655	2040	3.1F
15 W	0015	0340	2.7E
	0740	0940	0.8F
	1150	1500	1.5E
	1740	2120	2.7F
16 Th	0040	0420	3.1E
	0815	1050	1.5F
	1345	1615	1.1E
	1835	2205	2.3F
17 F	0105	0500	3.6E
	0850	1150	2.3F
	1520	1730	1.0E
	1940	2255	1.9F
18 Sa	0130	0535	4.1E
	0925	1240	3.2F
	1640	1840	1.0E
	2050	2340	1.5F
19 Su	0200	0620	4.6E
	1005	1330	4.0F
	1740	1945	1.2E
	2205		
20 M		0030	1.3F
	0235	0700	5.0E
	1045	1420	4.7F
	1830	2040	1.4E
	2310		
21 Tu		0120	1.1F
	0315	0750	5.3E
	1130	1505	5.2F
	1920	2135	1.6E
22 W	0010	0215	1.1F
	0405	0835	5.5E
	1215	1555	5.5F
	2005	2225	1.8E
23 Th	0110	0310	1.1F
	0455	0925	5.5E
	1305	1640	5.7F
	2050	2315	2.1E
24 F	0210	0405	1.1F
	0550	1015	5.4E
	1350	1730	5.6F
	2135		
25 Sa		0005	2.3E
	0310	0505	1.1F
	0650	1110	5.0E
	1440	1815	5.4F
	2215		
26 Su		0055	2.6E
	0410	0605	1.2F
	0800	1205	4.3E
	1525	1905	5.0F
	2250		
27 M		0145	3.0E
	0515	0715	1.2F
	0915	1300	3.5E
	1615	1950	4.4F
	2325		
28 Tu		0235	3.3E
	0615	0830	1.5F
	1050	1410	2.5E
	1705	2035	3.6F
29 W	0000	0325	3.7E
	0715	0950	1.9F
	1240	1525	1.7E
	1800	2125	2.8F
30 Th	0030	0415	4.0E
	0805	1105	2.5F
	1435	1650	1.2E
	1905	2215	2.1F

Time meridian 120° W. 0000 is midnight. 1200 is noon. Times are not adjusted for Daylight Saving Time.

Burrard Inlet (First Narrows), British Columbia, 2005

F–Flood, Dir. 135° True E–Ebb, Dir. 315° True

July

Day	Slack (h m)	Maximum (h m)	knots
1 F	0100	0505	4.3E
	0855	1210	3.2F
	1610	1810	1.0E
	2025	2310	1.4F
2 Sa	0130	0550	4.4E
	0935	1305	3.9F
	1725	1925	1.1E
	2150		
3 Su		0005	0.9F
	0155	0635	4.5E
	1015	1355	4.3F
	1815	2025	1.2E
	2310		
4 M		0055	0.6F
	0225	0715	4.5E
	1055	1435	4.6F
	1900	2115	1.4E
5 Tu	0020	0145	0.4F
	0300	0755	4.4E
	1135	1515	4.7F
	1935	2200	1.6E
6 W ●	0110	0225	0.3F
	0335	0835	4.3E
	1210	1555	4.7F
	2010	2235	1.7E
7 Th	0155	0310	0.4F
	0420	0910	4.2E
	1245	1630	4.6F
	2040	2310	1.8E
8 F	0230	0350	0.4F
	0505	0950	4.1E
	1320	1700	4.5F
	2110	2340	1.9E
9 Sa	0305	0430	0.5F
	0550	1025	3.9E
	1355	1735	4.4F
	2135		
10 Su		0015	2.1E
	0340	0515	0.6F
	0640	1105	3.5E
	1425	1805	4.1F
	2155		
11 M		0045	2.3E
	0415	0600	0.7F
	0735	1140	3.1E
	1455	1835	3.8F
	2220		
12 Tu		0120	2.5E
	0455	0650	0.8F
	0845	1225	2.5E
	1530	1905	3.4F
	2240		
13 W		0150	2.8E
	0540	0750	1.1F
	1005	1315	1.9E
	1605	1940	2.9F
	2300		
14 Th ◐		0230	3.1E
	0625	0855	1.4F
	1145	1420	1.2E
	1645	2015	2.3F
	2320		
15 F		0310	3.5E
	0710	1005	2.0F
	1340	1540	0.7E
	1735	2100	1.8F
	2345		
16 Sa		0400	3.8E
	0755	1115	2.7F
	1525	1705	0.5E
	1850	2155	1.3F
17 Su	0020	0450	4.2E
	0845	1215	3.4F
	1645	1825	0.7E
	2025	2300	1.1F
18 M	0105	0545	4.6E
	0935	1310	4.1F
	1735	1935	1.0E
	2150		
19 Tu		0005	1.0F
	0155	0640	5.0E
	1025	1405	4.8F
	1825	2030	1.3E
	2300		
20 W		0105	1.0F
	0255	0735	5.3E
	1115	1450	5.2F
	1905	2120	1.7E
21 Th ○	0000	0205	1.2F
	0400	0825	5.5E
	1205	1540	5.5F
	1945	2205	2.2E
22 F	0055	0305	1.5F
	0500	0920	5.5E
	1250	1625	5.7F
	2020	2250	2.6E
23 Sa	0150	0400	1.7F
	0605	1010	5.3E
	1335	1705	5.5F
	2055	2335	3.0E
24 Su	0240	0500	1.9F
	0710	1100	4.8E
	1420	1750	5.2F
	2125		
25 M		0020	3.4E
	0335	0600	2.0F
	0820	1155	4.0E
	1505	1830	4.6F
	2155		
26 Tu		0105	3.7E
	0430	0700	2.1F
	0935	1250	3.0E
	1545	1910	3.8F
	2225		
27 W ◑		0150	4.0E
	0525	0810	2.3F
	1105	1350	2.0E
	1635	1950	2.9F
	2255		
28 Th		0235	4.1E
	0625	0920	2.6F
	1250	1510	1.2E
	1730	2040	2.0F
	2325		
29 F		0325	4.1E
	0720	1035	2.9F
	1445	1640	0.8E
	1845	2135	1.2F
	2350		
30 Sa		0420	4.0E
	0810	1145	3.4F
	1610	1810	0.8E
	2030	2235	0.7F
31 Su	0025	0515	4.0E
	0900	1245	3.8F
	1710	1925	1.0E
	2215	2345	0.4F

August

Day	Slack (h m)	Maximum (h m)	knots
1 M	0105	0605	3.9E
	0950	1335	4.1F
	1755	2015	1.3E
	2325		
2 Tu		0045	0.3F
	0155	0655	4.0E
	1035	1415	4.3F
	1830	2100	1.6E
3 W	0010	0135	0.4F
	0250	0740	4.0E
	1115	1455	4.4F
	1905	2135	1.8E
4 Th ●	0045	0220	0.6F
	0345	0820	4.0E
	1150	1530	4.4F
	1930	2205	1.9E
5 F	0115	0300	0.7F
	0435	0900	4.0E
	1225	1600	4.4F
	1955	2230	2.1E
6 Sa	0145	0335	0.9F
	0525	0935	3.9E
	1255	1630	4.3F
	2015	2300	2.3E
7 Su	0210	0415	1.1F
	0610	1010	3.7E
	1325	1655	4.1F
	2035	2325	2.6E
8 M	0240	0455	1.3F
	0700	1045	3.3E
	1355	1720	3.8F
	2050	2350	2.8E
9 Tu	0315	0535	1.6F
	0800	1120	2.9E
	1425	1750	3.4F
	2110		
10 W		0020	3.1E
	0350	0620	1.8F
	0900	1205	2.3E
	1500	1815	2.9F
	2125		
11 Th		0050	3.3E
	0430	0715	2.0F
	1015	1255	1.6E
	1530	1845	2.3F
	2145		
12 F ◐		0125	3.5E
	0515	0815	2.2F
	1145	1355	1.0E
	1610	1925	1.8F
	2205		
13 Sa		0215	3.7E
	0610	0925	2.5F
	1335	1520	0.5E
	1705	2010	1.2F
	2240		
14 Su		0310	3.8E
	0710	1040	3.0F
	1520	1655	0.4E
	1840	2120	0.9F
	2325		
15 M		0415	4.0E
	0815	1150	3.5F
	1630	1820	0.7E
	2030	2240	0.7F
16 Tu	0030	0525	4.4E
	0915	1250	4.1F
	1715	1920	1.1E
	2150		
17 W		0000	0.9F
	0145	0625	4.7E
	1010	1345	4.7F
	1755	2010	1.7E
	2250		
18 Th		0105	1.3F
	0305	0725	5.1E
	1100	1430	5.1F
	1830	2055	2.2E
	2345		
19 F ○		0205	1.7F
	0415	0820	5.2E
	1150	1515	5.3F
	1900	2135	2.8E
20 Sa	0035	0300	2.2F
	0520	0910	5.1E
	1235	1555	5.2F
	1930	2220	3.4E
21 Su	0120	0355	2.6F
	0625	1000	4.8E
	1315	1635	5.0F
	2000	2255	3.8E
22 M	0210	0450	2.9F
	0730	1050	4.1E
	1400	1715	4.4F
	2025	2335	4.2E
23 Tu	0255	0545	3.1F
	0840	1140	3.3E
	1440	1750	3.7F
	2055		
24 W		0015	4.3E
	0345	0640	3.2F
	0950	1235	2.5E
	1525	1830	2.9F
	2120		
25 Th		0055	4.3E
	0435	0740	3.2F
	1115	1340	1.6E
	1615	1910	2.0F
	2145		
26 F ◑		0140	4.0E
	0525	0845	3.1F
	1245	1455	1.0E
	1720	1955	1.2F
	2210		
27 Sa		0235	3.7E
	0625	0955	3.1F
	1425	1630	0.8E
	1905	2100	0.5F
	2240		
28 Su		0335	3.4E
	0725	1110	3.3F
	1540	1800	1.0E
		2220	0.0
29 M		0440	3.3E
	0825	1210	3.5F
	1630	1905	1.3E
		2340	0.0
30 Tu		0545	3.3E
	0920	1305	3.7F
	1710	1950	1.6E
	2315		
31 W		0040	0.4F
	0200	0640	3.4E
	1005	1345	3.9F
	1745	2025	1.9E
	2340		

September

Day	Slack (h m)	Maximum (h m)	knots
1 Th		0125	0.7F
	0310	0725	3.5E
	1045	1420	4.0F
	1810	2055	2.2E
2 F	0005	0210	1.1F
	0405	0805	3.6E
	1125	1455	4.0F
	1835	2120	2.4E
3 Sa ●	0030	0245	1.4F
	0455	0840	3.6E
	1155	1520	3.9F
	1855	2145	2.7E
4 Su	0055	0320	1.7F
	0540	0915	3.5E
	1230	1550	3.7F
	1910	2210	2.9E
5 M	0120	0355	2.1F
	0630	0950	3.2E
	1300	1615	3.4F
	1925	2230	3.2E
6 Tu	0150	0435	2.4F
	0720	1030	2.9E
	1330	1640	3.1F
	1940	2255	3.5E
7 W	0220	0515	2.7F
	0815	1110	2.4E
	1405	1705	2.6F
	1955	2325	3.7E
8 Th	0255	0555	2.9F
	0915	1155	1.9E
	1440	1730	2.1F
	2010	2355	3.9E
9 F	0335	0645	3.0F
	1025	1245	1.4E
	1515	1805	1.6F
	2030		
10 Sa		0035	3.9E
	0425	0745	3.0F
	1150	1350	0.9E
	1605	1845	1.1F
	2100		
11 Su ◐		0130	3.8E
	0525	0855	3.0F
	1330	1515	0.6E
	1720	1945	0.7F
	2140		
12 M		0235	3.7E
	0635	1015	3.2F
	1455	1655	0.7E
	1920	2110	0.5F
	2245		
13 Tu		0355	3.7E
	0745	1130	3.6F
	1550	1810	1.1E
	2055	2245	0.6F
14 W	0025	0510	4.0E
	0850	1230	4.1F
	1630	1900	1.7E
	2155		
15 Th		0005	1.1F
	0200	0620	4.3E
	0950	1320	4.4F
	1705	1945	2.4E
	2245		
16 F		0110	1.7F
	0325	0715	4.5E
	1040	1405	4.7F
	1735	2025	3.1E
	2325		
17 Sa ○		0205	2.4F
	0435	0810	4.5E
	1125	1445	4.6F
	1805	2105	3.7E
18 Su	0010	0255	3.1F
	0540	0900	4.3E
	1210	1520	4.4F
	1830	2140	4.3E
19 M	0055	0350	3.6F
	0645	0950	3.8E
	1255	1600	3.9F
	1855	2215	4.6E
20 Tu	0135	0435	3.9F
	0750	1040	3.3E
	1340	1635	3.3F
	1920	2250	4.7E
21 W	0215	0525	4.1F
	0850	1130	2.6E
	1425	1710	2.5F
	1945	2330	4.6E
22 Th	0300	0615	4.1F
	1000	1225	2.0E
	1515	1750	1.8F
	2010		
23 F		0005	4.3E
	0345	0710	3.8F
	1110	1330	1.5E
	1615	1830	1.1F
	2030		
24 Sa ◑		0050	3.9E
	0435	0805	3.5F
	1225	1445	1.1E
	1740	1920	0.4F
	2050		
25 Su		0140	3.3E
	0530	0915	3.3F
	1340	1610	1.1E
		2030	0.0
26 M		0245	2.9E
	0630	1025	3.2F
	1445	1730	1.3E
		2205	0.0
27 Tu		0400	2.6E
	0735	1130	3.2F
	1535	1825	1.7E
		2325	0.0
28 W		0515	2.6E
	0835	1220	3.3F
	1610	1905	2.0E
	2240		
29 Th		0025	0.6F
	0205	0610	2.8E
	0925	1300	3.5F
	1640	1940	2.4E
	2305		
30 F		0110	1.1F
	0315	0700	2.9E
	1010	1335	3.5F
	1705	2005	2.7E
	2325		

Time meridian 120° W. 0000 is midnight. 1200 is noon. Times are not adjusted for Daylight Saving Time.

Burrard Inlet (First Narrows), British Columbia, 2005

F–Flood, Dir. 135° True E–Ebb, Dir. 315° True

October

Day	Slack h m	Maximum h m	knots
1 Sa		0150	1.6F
	0415	0740	3.0E
	1045	1405	3.4F
	1725	2030	3.1E
	2350		
2 Su		0225	2.2F
	0505	0820	2.9E
	1125	1435	3.2F
	1740	2055	3.4E
3 M	0010	0305	2.7F
	0600	0900	2.8E
	1200	1500	2.9F
●	1755	2115	3.7E
4 Tu	0035	0340	3.1F
	0650	0940	2.6E
	1235	1525	2.6F
	1810	2140	4.0E
5 W	0105	0415	3.6F
	0740	1020	2.3E
	1310	1555	2.2F
	1825	2210	4.2E
6 Th	0140	0455	3.8F
	0835	1105	2.0E
	1350	1625	1.8F
	1845	2240	4.3E
7 F	0215	0535	4.0F
	0930	1150	1.6E
	1435	1655	1.4F
	1905	2320	4.3E
8 Sa	0300	0625	3.9F
	1035	1245	1.3E
	1525	1735	1.0F
	1935		
9 Su		0005	4.1E
	0350	0725	3.7F
	1150	1355	1.0E
	1630	1830	0.6F
	2010		
10 M		0100	3.8E
	0450	0835	3.6F
	1310	1520	0.9E
◐	1815	1940	0.3F
	2100		
11 Tu		0215	3.5E
	0605	0950	3.0F
	1415	1640	1.3E
	2000	2120	0.3F
	2235		
12 W		0335	3.4E
	0715	1100	3.7F
	1500	1745	1.8E
	2105	2255	0.8F
13 Th	0040	0455	3.4E
	0820	1155	3.9F
	1540	1830	2.6E
	2150		
14 F		0010	1.5F
	0225	0605	3.5E
	0920	1245	4.0F
	1610	1915	3.3E
	2230		
15 Sa		0110	2.4F
	0350	0705	3.5E
	1010	1325	3.9F
	1635	1950	4.0E
	2310		
16 Su		0200	3.2F
	0500	0800	3.4E
	1100	1405	3.6F
	1700	2025	4.6E
	2345		
17 M		0250	4.0F
	0605	0855	3.1E
	1150	1445	3.2F
○	1725	2100	5.0E
18 Tu	0025	0335	4.5F
	0705	0945	2.8E
	1235	1520	2.6F
	1750	2135	5.1E
19 W	0105	0420	4.8F
	0805	1035	2.4E
	1325	1555	2.0F
	1815	2210	5.0E
20 Th	0140	0505	4.8F
	0900	1130	2.1E
	1420	1635	1.4F
	1835	2245	4.7E
21 F	0220	0550	4.6F
	0955	1220	1.8E
	1520	1715	0.9F
	1900	2325	4.3E
22 Sa	0300	0640	4.3F
	1055	1320	1.5E
	1630	1800	0.4F
	1920		
23 Su		0005	3.7E
	0345	0730	3.8F
	1150	1425	1.4E
		1855	0.0
24 M		0055	3.1E
	0435	0830	3.5F
	1250	1535	1.4E
◑		2010	0.0
25 Tu		0200	2.6E
	0535	0930	3.2F
	1340	1645	1.6E
		2140	0.0
26 W		0310	2.2E
	0635	1030	3.1F
	1425	1735	2.0E
		2300	0.0
27 Th		0425	2.1E
	0735	1120	3.1F
	1500	1815	2.5E
	2200		
28 F		0000	0.8F
	0200	0530	2.1E
	0830	1205	3.0F
	1525	1845	2.9E
	2220		
29 Sa		0050	1.5F
	0320	0625	2.1E
	0915	1240	2.9F
	1550	1910	3.3E
	2245		
30 Su		0130	2.2F
	0425	0715	2.1E
	1005	1310	2.7F
	1605	1940	3.7E
	2305		
31 M		0205	2.9F
	0520	0800	2.1E
	1045	1345	2.4F
	1625	2005	4.1E
	2330		

November

Day	Slack h m	Maximum h m	knots
1 Tu		0245	3.5F
	0610	0845	2.1E
	1130	1415	2.1F
●	1640	2030	4.4E
2 W	0000	0320	4.1F
	0705	0930	2.0E
	1215	1445	1.7F
	1700	2100	4.7E
3 Th	0035	0400	4.5F
	0755	1015	1.9E
	1300	1520	1.4F
	1725	2135	4.8E
4 F	0110	0440	4.7F
	0845	1105	1.7E
	1350	1555	1.2F
	1750	2215	4.8E
5 Sa	0150	0525	4.7F
	0940	1155	1.6E
	1440	1640	0.9F
	1825	2255	4.7E
6 Su	0240	0615	4.6F
	1035	1250	1.4E
	1540	1730	0.7F
	1905	2345	4.4E
7 M	0330	0715	4.3F
	1135	1355	1.4E
	1705	1830	0.4F
	1950		
8 Tu		0045	3.9E
	0430	0815	4.1F
	1230	1505	1.6E
◐	1840	1955	0.3F
	2105		
9 W		0200	3.4E
	0535	0920	3.9F
	1320	1615	2.1E
	1955	2130	0.5F
	2300		
10 Th		0320	2.9E
	0640	1020	3.7F
	1400	1710	2.7E
	2045	2255	1.2F
11 F	0110	0440	2.6E
	0745	1115	3.5F
	1435	1755	3.5E
	2130		
12 Sa		0005	2.1F
	0255	0555	2.5E
	0845	1205	3.3F
	1505	1835	4.2E
	2210		
13 Su		0105	3.1F
	0415	0700	2.3E
	0945	1245	2.9F
	1535	1915	4.7E
	2245		
14 M		0155	4.0F
	0525	0800	2.2E
	1040	1330	2.4F
	1600	1950	5.1E
	2320		
15 Tu		0245	4.6F
	0630	0855	2.1E
	1135	1405	1.9F
○	1625	2025	5.2E
16 W	0000	0325	5.0F
	0725	0945	2.0E
	1230	1445	1.5F
	1650	2100	5.2E
17 Th	0035	0410	5.1F
	0815	1035	1.9E
	1330	1525	1.0F
	1715	2135	5.0E
18 F	0110	0450	5.1F
	0900	1125	1.8E
	1425	1605	0.7F
	1740	2215	4.6E
19 Sa	0150	0530	4.8F
	0945	1210	1.7E
	1530	1650	0.4F
	1805	2255	4.2E
20 Su	0230	0615	4.5F
	1030	1300	1.6E
		1735	0.0
		2335	3.7E
21 M	0310	0655	4.1F
	1115	1355	1.6E
		1830	0.0
22 Tu		0020	3.1E
	0350	0745	3.7F
	1155	1450	1.8E
		1935	0.0
23 W		0110	2.6E
	0435	0830	3.4F
	1235	1540	2.0E
◑		2100	0.0
24 Th		0215	2.0E
	0525	0920	3.1F
	1310	1630	2.4E
	2050	2215	0.4F
	2345		
25 F		0325	1.6E
	0620	1010	2.8F
	1340	1710	2.8E
	2110	2325	1.0F
26 Sa	0150	0440	1.4E
	0715	1055	2.5F
	1405	1745	3.3E
	2135		
27 Su		0015	1.8F
	0320	0550	1.3E
	0815	1135	2.2F
	1425	1815	3.7E
	2200		
28 M		0100	2.7F
	0430	0650	1.3E
	0915	1215	1.9F
	1450	1845	4.2E
	2225		
29 Tu		0145	3.4F
	0530	0745	1.4E
	1015	1255	1.6F
	1510	1920	4.6E
	2255		
30 W		0225	4.1F
	0620	0835	1.5E
	1110	1330	1.4F
	1535	1955	4.9E
	2330		

December

Day	Slack h m	Maximum h m	knots
1 Th		0305	4.7F
	0710	0925	1.6E
	1205	1415	1.2F
●	1605	2035	5.1E
2 F	0010	0345	5.1F
	0800	1015	1.7E
	1255	1455	1.1F
	1645	2115	5.3E
3 Sa	0055	0430	5.3F
	0845	1100	1.7E
	1345	1545	1.0F
	1725	2200	5.2E
4 Su	0140	0515	5.3F
	0930	1150	1.8E
	1445	1635	0.9F
	1815	2250	5.0E
5 M	0225	0605	5.1F
	1020	1245	1.9E
	1545	1730	0.8F
	1910	2340	4.6E
6 Tu	0315	0655	4.9F
	1105	1340	2.1E
	1700	1840	0.7F
	2015		
7 W		0040	4.0E
	0405	0750	4.5F
	1145	1435	2.4E
	1810	1955	0.7F
	2140		
8 Th		0145	3.2E
	0500	0840	4.0F
	1225	1535	2.9E
◐	1920	2125	1.1F
	2330		
9 F		0300	2.4E
	0555	0935	3.5F
	1300	1625	3.5E
	2015	2245	1.8F
10 Sa	0135	0420	1.8E
	0700	1025	2.9F
	1330	1715	4.1E
	2100		
11 Su		0000	2.7F
	0320	0545	1.5E
	0810	1120	2.3F
	1400	1800	4.6E
	2140		
12 M		0055	3.6F
	0445	0700	1.4E
	0920	1210	1.8F
	1430	1840	4.9E
	2220		
13 Tu		0150	4.3F
	0550	0805	1.5E
	1035	1255	1.3F
	1500	1920	5.0E
	2300		
14 W		0235	4.8F
	0645	0900	1.6E
	1140	1340	1.0F
	1525	2000	5.0E
	2340		
15 Th		0315	5.1F
	0730	0950	1.7E
	1245	1425	0.7F
○	1600	2040	4.9E
16 F	0015	0355	5.1F
	0810	1035	1.7E
	1340	1510	0.6F
	1630	2115	4.7E
17 Sa	0050	0435	5.0F
	0850	1115	1.8E
	1430	1550	0.5F
	1710	2155	4.4E
18 Su	0130	0510	4.8F
	0925	1155	1.8E
	1515	1635	0.4F
	1745	2230	4.1E
19 M	0205	0545	4.6F
	0955	1230	1.9E
	1605	1720	0.3F
	1830	2310	3.7E
20 Tu	0240	0625	4.3F
	1025	1310	2.0E
		1805	0.0
		2350	3.2E
21 W	0315	0700	3.9F
	1055	1350	2.2E
		1905	0.0
22 Th		0035	2.6E
	0350	0735	3.5F
	1125	1435	2.4E
	1840	2010	0.4F
	2135		
23 F		0125	2.0E
	0425	0810	3.1F
	1150	1515	2.7E
◑	1925	2120	0.7F
	2330		
24 Sa		0230	1.3E
	0505	0855	2.6F
	1215	1555	3.1E
	2000	2235	1.3F
25 Su	0135	0350	0.8E
	0600	0940	2.1F
	1240	1635	3.5E
	2035	2335	2.1F
26 M	0325	0515	0.6E
	0705	1025	1.6F
	1305	1720	3.9E
	2110		
27 Tu		0030	2.9F
	0440	0630	0.7E
	0830	1115	1.3F
	1335	1800	4.3E
	2150		
28 W		0120	3.7F
	0535	0735	0.9E
	0945	1210	1.1F
	1410	1845	4.7E
	2230		
29 Th		0205	4.4F
	0625	0830	1.2E
	1055	1300	1.0F
	1450	1930	5.1E
	2310		
30 F		0250	4.9F
	0705	0915	1.4E
	1150	1355	1.0F
●	1540	2015	5.4E
	2355		
31 Sa		0335	5.3F
	0750	1005	1.7E
	1245	1445	1.1F
	1635	2105	5.5E

Time meridian 120° W. 0000 is midnight. 1200 is noon. Times are not adjusted for Daylight Saving Time.

Seymour Narrows, British Columbia, 2005

F–Flood, Dir. 180° True E–Ebb, Dir. 000° True

January

Day	Slack h m	Maximum h m	knots
1 Sa	0225	0545	8.8F
	0910	1230	8.6E
	1555	1830	5.4F
	2125		
2 Su		0030	6.0E
	0320	0635	7.8F
	0950	1315	8.5E
	1640	1930	6.0F
	2235		
3 M ◑		0130	5.5E
	0420	0730	6.7F
	1030	1400	8.4E
	1725	2025	7.0F
	2345		
4 Tu		0240	5.4E
	0530	0825	5.9F
	1115	1455	8.6E
	1815	2125	8.3F
5 W	0055	0350	5.8E
	0645	0925	5.3F
	1205	1550	9.0E
	1905	2220	9.8F
6 Th	0200	0500	6.7E
	0800	1025	5.2F
	1255	1645	9.7E
	1955	2315	11.5F
7 F	0255	0600	8.0E
	0900	1125	5.3F
	1350	1735	10.4E
	2040		
8 Sa		0005	13.0F
	0350	0650	9.3E
	1000	1220	5.8F
	1445	1830	11.2E
	2130		
9 Su		0055	14.2F
	0435	0740	10.5E
	1050	1310	6.4F
	1540	1920	11.9E
	2220		
10 M ●		0140	15.1F
	0520	0825	11.5E
	1135	1400	7.1F
	1635	2005	12.2E
	2305		
11 Tu		0230	15.4F
	0605	0910	12.2E
	1220	1450	7.7F
	1730	2055	12.2E
	2355		
12 W		0315	15.1F
	0645	0955	12.5E
	1305	1540	8.2F
	1825	2145	11.8E
13 Th	0040	0400	14.2F
	0730	1035	12.5E
	1350	1630	8.5F
	1920	2235	11.0E
14 F	0130	0445	12.9F
	0810	1120	12.1E
	1435	1720	8.7F
	2020	2325	9.9E
15 Sa	0220	0535	11.2F
	0850	1205	11.5E
	1520	1815	8.7F
	2120		
16 Su ◐		0025	8.6E
	0315	0625	9.3F
	0930	1255	10.8E
	1615	1915	8.7F
	2230		
17 M		0125	7.3E
	0420	0720	7.5F
	1015	1345	9.9E
	1705	2015	8.8F
	2340		
18 Tu		0235	6.4E
	0530	0820	5.9F
	1105	1445	9.2E
	1800	2120	9.1F
19 W	0055	0350	6.2E
	0650	0925	4.7F
	1200	1545	8.7E
	1900	2220	9.6F
20 Th	0205	0500	6.5E
	0805	1030	4.1F
	1255	1645	8.6E
	1950	2315	10.2F
21 F	0305	0605	7.3E
	0915	1130	4.0F
	1355	1740	8.7E
	2040		
22 Sa		0010	10.9F
	0355	0655	8.2E
	1010	1225	4.3F
	1450	1830	8.9E
	2125		
23 Su		0055	11.4F
	0435	0740	9.0E
	1055	1310	4.7F
	1540	1915	9.1E
	2210		
24 M		0135	11.8F
	0515	0820	9.6E
	1130	1355	5.2F
	1625	1955	9.3E
	2250		
25 Tu ○		0210	12.0F
	0550	0855	10.0E
	1205	1430	5.7F
	1710	2035	9.4E
	2325		
26 W		0245	12.0F
	0620	0925	10.2E
	1240	1505	6.2F
	1750	2110	9.3E
27 Th	0005	0320	11.7F
	0650	1000	10.3E
	1310	1545	6.6F
	1830	2150	9.1E
28 F	0040	0355	11.2F
	0720	1030	10.2E
	1345	1620	7.0F
	1915	2225	8.7E
29 Sa	0120	0430	10.3F
	0750	1105	10.0E
	1420	1700	7.3F
	2000	2310	8.1E
30 Su	0205	0510	9.3F
	0825	1140	9.7E
	1455	1750	7.6F
	2055	2355	7.3E
31 M	0250	0555	7.9F
	0900	1220	9.3E
	1540	1840	7.9F
	2155		

February

Day	Slack h m	Maximum h m	knots
1 Tu ◑		0050	6.4E
	0350	0645	6.5F
	0935	1310	8.9E
	1630	1940	8.2F
	2305		
2 W		0200	5.7E
	0500	0745	5.2F
	1020	1405	8.6E
	1725	2045	8.8F
3 Th	0025	0320	5.7E
	0620	0850	4.3F
	1120	1505	8.5E
	1825	2150	9.9F
4 F	0135	0435	6.4E
	0745	1005	4.1F
	1225	1615	8.9E
	1925	2255	11.2F
5 Sa	0240	0540	7.8E
	0850	1110	4.7F
	1330	1720	9.8E
	2025	2350	12.6F
6 Su	0335	0635	9.4E
	0945	1210	5.8F
	1440	1820	10.8E
	2115		
7 M		0040	13.8F
	0420	0725	10.9E
	1035	1300	7.1F
	1540	1910	11.9E
	2210		
8 Tu ●		0130	14.6F
	0500	0810	12.1E
	1115	1350	8.4F
	1635	2000	12.6E
	2255		
9 W		0215	14.9F
	0540	0850	13.0E
	1155	1435	9.6F
	1725	2045	12.9E
	2345		
10 Th		0255	14.5F
	0620	0930	13.3E
	1235	1520	10.4F
	1820	2130	12.6E
11 F	0030	0340	13.6F
	0655	1005	13.2E
	1315	1605	10.7F
	1910	2220	11.8E
12 Sa	0115	0420	12.2F
	0735	1045	12.7E
	1355	1650	10.7F
	2000	2305	10.5E
13 Su	0205	0505	10.3F
	0810	1130	11.7E
	1440	1740	10.2F
	2100	2355	8.9E
14 M	0255	0550	8.3F
	0850	1210	10.4E
	1525	1835	9.4F
	2200		
15 Tu ◐		0055	7.3E
	0355	0640	6.2F
	0930	1300	9.1E
	1620	1935	8.7F
	2310		
16 W		0200	5.9E
	0505	0740	4.4F
	1015	1400	7.8E
	1715	2040	8.2F
17 Th	0025	0320	5.4E
	0635	0855	3.2F
	1115	1505	6.9E
	1820	2150	8.3F
18 F	0140	0440	5.7E
	0800	1010	2.8F
	1230	1620	6.8E
	1925	2255	8.8F
19 Sa	0240	0545	6.7E
	0905	1120	3.3F
	1340	1725	7.2E
	2020	2350	9.5F
20 Su	0330	0635	7.8E
	0950	1215	4.1F
	1445	1815	7.9E
	2110		
21 M		0035	10.3F
	0410	0715	8.9E
	1030	1255	5.2F
	1535	1900	8.7E
	2150		
22 Tu		0110	10.9F
	0445	0750	9.7E
	1100	1335	6.2F
	1620	1940	9.4E
	2230		
23 W ○		0145	11.3F
	0515	0825	10.3E
	1130	1405	7.2F
	1655	2015	9.9E
	2310		
24 Th		0220	11.5F
	0545	0855	10.7E
	1200	1440	8.1F
	1735	2050	10.2E
	2345		
25 F		0250	11.4F
	0610	0920	11.0E
	1225	1510	8.9F
	1810	2125	10.3E
26 Sa	0020	0325	11.0F
	0640	0950	11.0E
	1255	1550	9.5F
	1855	2205	10.0E
27 Su	0100	0400	10.1F
	0705	1025	10.8E
	1330	1625	9.8F
	1940	2245	9.3E
28 M	0145	0435	9.0F
	0740	1100	10.4E
	1405	1710	9.9F
	2030	2330	8.3E

March

Day	Slack h m	Maximum h m	knots
1 Tu	0230	0520	7.5F
	0810	1140	9.8E
	1450	1800	9.6F
	2130		
2 W		0025	7.2E
	0330	0610	5.9F
	0850	1225	9.0E
	1540	1900	9.3F
	2235		
3 Th ◑		0130	6.2E
	0440	0715	4.4F
	0940	1325	8.2E
	1645	2010	9.2F
	2355		
4 F		0255	6.0E
	0610	0830	3.6F
	1050	1440	7.8E
	1755	2125	9.7F
5 Sa	0110	0415	6.8E
	0730	0950	3.9F
	1210	1600	8.1E
	1905	2235	10.7F
6 Su	0215	0520	8.3E
	0835	1100	5.1F
	1330	1710	9.3E
	2010	2330	11.9F
7 M	0305	0615	10.0E
	0925	1200	6.9F
	1440	1810	10.7E
	2105		
8 Tu		0025	13.0F
	0350	0700	11.5E
	1005	1250	8.8F
	1540	1900	12.0E
	2200		
9 W		0110	13.6F
	0435	0740	12.8E
	1045	1335	10.5F
	1630	1950	12.9E
	2245		
10 Th ●		0155	13.7F
	0510	0820	13.5E
	1125	1415	11.8F
	1720	2035	13.2E
	2330		
11 F		0235	13.2F
	0545	0900	13.7E
	1200	1500	12.5F
	1810	2115	12.9E
12 Sa	0015	0315	12.2F
	0620	0935	13.4E
	1240	1540	12.6F
	1855	2200	12.0E
13 Su	0100	0355	10.7F
	0655	1015	12.5E
	1315	1620	12.1F
	1945	2245	10.6E
14 M	0145	0435	8.8F
	0730	1050	11.2E
	1355	1705	11.1F
	2035	2335	8.9E
15 Tu	0235	0520	6.8F
	0805	1135	9.6E
	1440	1755	9.8F
	2130		
16 W		0025	7.2E
	0335	0610	4.8F
	0845	1220	7.9E
	1530	1855	8.5F
	2235		
17 Th ◐		0135	5.8E
	0450	0710	3.1F
	0935	1320	6.3E
	1630	2000	7.5F
	2350		
18 F		0255	5.2E
	0625	0835	2.2F
	1045	1435	5.3E
	1740	2120	7.2F
19 Sa	0105	0415	5.6E
	0750	1000	2.4F
	1220	1600	5.3E
	1855	2225	7.6F
20 Su	0205	0520	6.7E
	0845	1105	3.4F
	1340	1705	6.1E
	1955	2320	8.3F
21 M	0250	0605	7.8E
	0920	1155	4.8F
	1435	1755	7.2E
	2045		
22 Tu		0005	9.1F
	0330	0640	8.9E
	0950	1230	6.2F
	1525	1840	8.3E
	2130		
23 W		0040	9.8F
	0405	0715	9.8E
	1020	1305	7.6F
	1605	1915	9.4E
	2210		
24 Th		0115	10.3F
	0430	0745	10.6E
	1045	1335	9.0F
	1640	1955	10.2E
	2250		
25 F ○		0150	10.6F
	0500	0815	11.1E
	1115	1410	10.2F
	1715	2030	10.8E
	2325		
26 Sa		0220	10.5F
	0525	0845	11.5E
	1145	1440	11.2F
	1755	2105	10.9E
27 Su	0005	0255	10.1F
	0555	0915	11.6E
	1215	1515	11.8F
	1835	2145	10.7E
28 M	0045	0330	9.2F
	0625	0945	11.4E
	1250	1555	12.0F
	1920	2225	10.0E
29 Tu	0130	0410	8.0F
	0655	1020	10.8E
	1325	1640	11.7F
	2010	2310	9.0E
30 W	0220	0455	6.6F
	0735	1105	9.9E
	1410	1730	11.1F
	2110		
31 Th		0005	7.9E
	0320	0545	5.1F
	0820	1155	8.8E
	1505	1830	10.2F
	2215		

Time meridian 120° W. 0000 is midnight. 1200 is noon. Times are not adjusted for Daylight Saving Time.

Seymour Narrows, British Columbia, 2005

F–Flood, Dir. 180° True E–Ebb, Dir. 000° True

April

Day	Slack h m	Maximum h m	knots
1 F ◑		0115	7.0E
	0435	0655	3.9F
	0920	1300	7.7E
	1610	1945	9.6F
	2325		
2 Sa		0235	6.9E
	0600	0820	3.7F
	1040	1420	7.2E
	1725	2100	9.6F
3 Su	0040	0350	7.7E
	0715	0940	4.6F
	1215	1545	7.7E
	1845	2210	10.2F
4 M	0140	0455	9.1E
	0810	1045	6.5F
	1335	1655	9.0E
	1955	2310	11.0F
5 Tu	0230	0545	10.7E
	0855	1140	8.6F
	1440	1755	10.5E
	2050		
6 W		0000	11.7F
	0315	0630	12.1E
	0935	1230	10.7F
	1535	1845	11.8E
	2145		
7 Th		0045	12.0F
	0355	0710	13.0E
	1015	1310	12.3F
	1625	1935	12.7E
	2235		
8 F ●		0130	11.9F
	0435	0750	13.5E
	1050	1355	13.4F
	1710	2015	12.9E
	2320		
9 Sa		0210	11.2F
	0510	0825	13.5E
	1125	1435	13.9F
	1755	2100	12.5E
10 Su	0005	0250	10.2F
	0545	0905	12.9E
	1205	1515	13.7F
	1840	2145	11.6E
11 M	0045	0325	8.7F
	0620	0940	11.8E
	1240	1555	12.8F
	1925	2225	10.3E
12 Tu	0135	0410	7.1F
	0655	1020	10.4E
	1320	1635	11.6F
	2015	2315	8.8E
13 W	0225	0450	5.3F
	0730	1100	8.6E
	1400	1725	10.0F
	2105		
14 Th		0005	7.4E
	0325	0545	3.7F
	0810	1145	6.9E
	1450	1820	8.5F
	2205		
15 F		0110	6.2E
	0445	0650	2.4F
	0910	1245	5.3E
	1545	1925	7.2F
	2310		
16 Sa ◐		0225	5.7E
	0610	0815	2.1F
	1030	1405	4.4E
	1655	2035	6.7F
17 Su	0015	0335	6.1E
	0715	0935	2.7F
	1210	1525	4.4E
	1815	2145	6.8F
18 M	0110	0435	6.9E
	0800	1035	4.0F
	1325	1635	5.3E
	1920	2240	7.3F
19 Tu	0200	0520	7.9E
	0835	1120	5.6F
	1420	1730	6.6E
	2015	2325	7.9F
20 W	0235	0600	8.9E
	0905	1155	7.3F
	1505	1810	7.9E
	2105		
21 Th		0005	8.5F
	0310	0630	9.8E
	0935	1230	9.0F
	1540	1850	9.1E
	2145		
22 F		0040	8.9F
	0340	0700	10.6E
	1005	1305	10.6F
	1620	1930	10.1E
	2225		
23 Sa		0115	9.1F
	0410	0735	11.3E
	1035	1340	12.0F
	1700	2005	10.7E
	2305		
24 Su ○		0150	9.0F
	0440	0805	11.7E
	1105	1415	13.0F
	1740	2045	11.0E
	2350		
25 M		0225	8.6F
	0515	0840	11.8E
	1140	1450	13.6F
	1820	2125	10.9E
26 Tu	0030	0305	7.9F
	0545	0915	11.5E
	1215	1535	13.6F
	1905	2210	10.4E
27 W	0120	0350	6.9F
	0625	0955	10.9E
	1300	1620	13.1F
	1955	2300	9.6E
28 Th	0215	0435	5.8F
	0710	1040	9.9E
	1345	1710	12.2F
	2050	2355	8.8E
29 F	0315	0535	4.8F
	0805	1135	8.7E
	1440	1810	11.0F
	2155		
30 Sa ◑		0100	8.3E
	0425	0650	4.3F
	0915	1245	7.6E
	1545	1920	10.1F
	2255		

May

Day	Slack h m	Maximum h m	knots
1 Su		0210	8.3E
	0540	0810	4.7F
	1045	1405	7.1E
	1705	2030	9.6F
2 M	0000	0315	9.0E
	0640	0920	6.1F
	1210	1525	7.5E
	1820	2140	9.6F
3 Tu	0055	0415	10.0E
	0735	1025	8.0F
	1325	1635	8.6E
	1930	2240	9.8F
4 W	0145	0510	11.1E
	0820	1120	10.1F
	1430	1735	9.9E
	2035	2330	9.9F
5 Th	0235	0555	12.1E
	0900	1205	11.9F
	1525	1830	11.0E
	2130		
6 F		0020	9.9F
	0315	0635	12.7E
	0940	1250	13.3F
	1610	1915	11.7E
	2220		
7 Sa		0100	9.5F
	0355	0715	12.9E
	1020	1330	14.1F
	1700	2000	11.9E
	2305		
8 Su ●		0145	8.9F
	0430	0755	12.7E
	1055	1410	14.3F
	1745	2045	11.6E
	2350		
9 M		0225	8.0F
	0510	0835	12.0E
	1135	1450	13.9F
	1825	2130	11.0E
10 Tu	0040	0305	6.8F
	0545	0915	10.8E
	1210	1530	13.0F
	1910	2210	10.0E
11 W	0125	0350	5.6F
	0620	0950	9.4E
	1250	1610	11.7F
	1955	2300	8.9E
12 Th	0220	0435	4.3F
	0705	1035	7.9E
	1330	1655	10.2F
	2040	2350	7.9E
13 F	0315	0530	3.3F
	0750	1120	6.4E
	1415	1745	8.8F
	2130		
14 Sa		0045	7.1E
	0420	0630	2.7F
	0855	1220	5.1E
	1510	1845	7.6F
	2225		
15 Su		0145	6.7E
	0530	0745	2.7F
	1015	1325	4.2E
	1610	1950	6.7F
	2315		
16 M ◐		0245	6.8E
	0625	0855	3.5F
	1140	1440	4.2E
	1725	2050	6.4F
17 Tu	0010	0340	7.2E
	0705	0950	4.8F
	1250	1550	4.8E
	1835	2145	6.5F
18 W	0055	0425	8.0E
	0740	1035	6.4F
	1350	1650	5.9E
	1935	2235	6.7F
19 Th	0135	0505	8.8E
	0815	1115	8.2F
	1435	1740	7.2E
	2030	2320	7.0F
20 F	0210	0545	9.7E
	0850	1155	10.0F
	1520	1825	8.4E
	2120		
21 Sa		0000	7.3F
	0250	0620	10.5E
	0920	1235	11.6F
	1600	1905	9.5E
	2205		
22 Su		0040	7.4F
	0325	0655	11.2E
	0955	1310	13.1F
	1645	1950	10.3E
	2250		
23 M ○		0120	7.4F
	0400	0735	11.6E
	1035	1350	14.1F
	1725	2030	10.7E
	2335		
24 Tu		0205	7.2F
	0440	0810	11.8E
	1110	1435	14.6F
	1810	2115	10.9E
25 W	0025	0245	6.8F
	0520	0855	11.6E
	1155	1515	14.5F
	1855	2200	10.8E
26 Th	0115	0335	6.3F
	0605	0940	11.0E
	1240	1605	14.0F
	1945	2250	10.5E
27 F	0205	0430	5.8F
	0700	1030	10.1E
	1330	1655	13.0F
	2035	2340	10.1E
28 Sa	0305	0530	5.5F
	0800	1125	9.0E
	1425	1755	11.7F
	2125		
29 Su		0040	9.8E
	0405	0635	5.6F
	0915	1235	8.0E
	1530	1855	10.5F
	2220		
30 M ◑		0140	9.8E
	0505	0715	6.3F
	1040	1345	7.4E
	1640	2000	9.5F
	2315		
31 Tu		0240	10.0E
	0600	0855	7.6F
	1200	1500	7.4E
	1755	2105	8.7F

June

Day	Slack h m	Maximum h m	knots
1 W	0010	0335	10.5E
	0650	0955	9.2F
	1310	1610	8.0E
	1905	2205	8.2F
2 Th	0100	0430	11.0E
	0740	1050	10.8F
	1415	1715	8.9E
	2015	2300	7.9F
3 F	0145	0520	11.5E
	0825	1140	12.2F
	1510	1810	9.7E
	2110	2350	7.6F
4 Sa	0235	0605	11.8E
	0910	1225	13.2F
	1600	1900	10.3E
	2205		
5 Su		0035	7.2F
	0315	0650	11.8E
	0950	1310	13.8F
	1645	1950	10.7E
	2255		
6 M ●		0125	6.7F
	0400	0730	11.5E
	1030	1350	13.8F
	1730	2035	10.7E
	2345		
7 Tu		0205	6.2F
	0440	0815	10.9E
	1110	1430	13.5F
	1815	2115	10.4E
8 W	0030	0250	5.6F
	0520	0855	10.0E
	1150	1510	12.7F
	1855	2200	9.9E
9 Th	0115	0335	4.9F
	0605	0935	8.9E
	1225	1555	11.8F
	1935	2240	9.4E
10 F	0205	0420	4.4F
	0650	1015	7.8E
	1305	1635	10.6F
	2015	2325	8.7E
11 Sa	0250	0510	3.9F
	0740	1100	6.7E
	1350	1720	9.4F
	2055		
12 Su		0010	8.2E
	0340	0600	3.8F
	0835	1150	5.7E
	1435	1805	8.3F
	2135		
13 M		0055	7.8E
	0430	0700	4.0F
	0940	1245	4.9E
	1530	1855	7.2F
	2220		
14 Tu ◐		0145	7.6E
	0515	0755	4.6F
	1055	1350	4.6E
	1635	1950	6.4F
	2300		
15 W		0235	7.7E
	0600	0855	5.6F
	1205	1500	4.7E
	1745	2050	5.9F
	2345		
16 Th		0320	8.0E
	0645	0945	7.0F
	1310	1605	5.4E
	1855	2145	5.6F
17 F	0030	0410	8.6E
	0725	1035	8.7F
	1405	1705	6.4E
	2000	2235	5.6F
18 Sa	0115	0455	9.3E
	0805	1120	10.4F
	1455	1800	7.5E
	2100	2325	5.7F
19 Su	0200	0540	10.1E
	0845	1205	12.0F
	1545	1845	8.7E
	2150		
20 M		0015	5.9F
	0245	0625	10.8E
	0930	1250	13.4F
	1630	1935	9.8E
	2240		
21 Tu ○		0100	6.1F
	0330	0710	11.4E
	1010	1335	14.5F
	1715	2020	10.6E
	2330		
22 W		0150	6.4F
	0415	0755	11.7E
	1055	1420	15.0F
	1800	2105	11.2E
23 Th	0015	0235	6.6F
	0510	0840	11.7E
	1140	1505	15.0F
	1845	2150	11.6E
24 F	0100	0325	6.8F
	0600	0930	11.4E
	1230	1550	14.5F
	1930	2235	11.6E
25 Sa	0150	0420	7.1F
	0700	1020	10.7E
	1320	1640	13.5F
	2010	2320	11.5E
26 Su	0240	0515	7.3F
	0800	1115	9.8E
	1410	1730	12.1F
	2055		
27 M		0010	11.2E
	0330	0615	7.6F
	0910	1215	8.8E
	1510	1825	10.5F
	2145		
28 Tu ◑		0100	10.9E
	0420	0715	8.1F
	1020	1320	7.8E
	1615	1925	8.9F
	2230		
29 W		0155	10.6E
	0515	0820	8.7F
	1135	1430	7.2E
	1725	2025	7.6F
	2320		
30 Th		0250	10.3E
	0610	0920	9.6F
	1250	1545	7.1E
	1840	2125	6.5F

Time meridian 120° W. 0000 is midnight. 1200 is noon. Times are not adjusted for Daylight Saving Time.

Seymour Narrows, British Columbia, 2005

F–Flood, Dir. 180° True E–Ebb, Dir. 000° True

July

Day	Slack h m	Maximum h m	knots
1 F	0010	0350	10.2E
	0700	1020	10.6F
	1355	1650	7.6E
	1955	2225	5.8F
2 Sa	0105	0445	10.2E
	0755	1115	11.5F
	1455	1755	8.3E
	2100	2325	5.4F
3 Su	0155	0540	10.3E
	0840	1205	12.2F
	1550	1850	9.0E
	2155		
4 M		0020	5.3F
	0250	0630	10.3E
	0925	1255	12.7F
	1635	1935	9.6E
	2250		
5 Tu		0110	5.3F
	0340	0715	10.2E
	1010	1335	12.9F
	1720	2020	10.1E
	2335		
6 W		0155	5.4F
	0425	0800	10.0E
	1050	1415	12.8F
●	1800	2100	10.2E
7 Th	0015	0240	5.4F
	0510	0840	9.6E
	1130	1455	12.4F
	1835	2140	10.2E
8 F	0055	0320	5.4F
	0555	0920	9.1E
	1210	1535	11.8F
	1910	2215	10.0E
9 Sa	0135	0400	5.4F
	0640	1000	8.4E
	1250	1610	11.0F
	1940	2255	9.7E
10 Su	0210	0440	5.5F
	0725	1040	7.7E
	1330	1645	10.0F
	2015	2330	9.3E
11 M	0250	0525	5.6F
	0810	1120	6.9E
	1410	1725	8.9F
	2050		
12 Tu		0005	8.9E
	0330	0610	5.8F
	0905	1210	6.1E
	1455	1810	7.7F
	2125		
13 W		0045	8.5E
	0410	0700	6.1F
	1010	1305	5.4E
	1555	1900	6.5F
	2200		
14 Th		0130	8.2E
	0455	0755	6.7F
	1115	1410	5.0E
◐	1700	1955	5.4F
	2240		
15 F		0220	8.1E
	0545	0855	7.6F
	1230	1520	5.1E
	1815	2055	4.6F
	2330		
16 Sa		0315	8.2E
	0635	0955	8.9F
	1335	1635	5.7E
	1935	2200	4.3F
17 Su	0025	0415	8.7E
	0730	1050	10.4F
	1435	1735	6.9E
	2040	2300	4.4F
18 M	0120	0510	9.4E
	0820	1145	11.9F
	1530	1830	8.4E
	2140	2355	5.0F
19 Tu	0220	0605	10.3E
	0910	1235	13.3F
	1620	1920	9.8E
	2230		
20 W		0050	5.8F
	0315	0700	11.2E
	0955	1320	14.4F
	1700	2005	11.0E
	2315		
21 Th		0140	6.8F
	0410	0745	11.9E
	1045	1405	15.0F
○	1745	2050	12.0E
22 F	0000	0225	7.8F
	0505	0835	12.3E
	1130	1450	15.1F
	1825	2130	12.6E
23 Sa	0040	0315	8.6F
	0600	0920	12.2E
	1220	1535	14.5F
	1905	2210	12.8E
24 Su	0120	0405	9.2F
	0655	1010	11.7E
	1305	1620	13.4F
	1940	2255	12.6E
25 M	0205	0450	9.5F
	0750	1100	10.8E
	1355	1705	11.8F
	2020	2335	12.1E
26 Tu	0250	0545	9.6F
	0850	1155	9.5E
	1450	1755	10.0F
	2100		
27 W		0020	11.4E
	0340	0640	9.5F
	0955	1250	8.1E
◑	1550	1845	8.0F
	2145		
28 Th		0110	10.4E
	0430	0740	9.3F
	1105	1400	6.9E
	1700	1945	6.2F
	2230		
29 F		0205	9.5E
	0525	0845	9.3F
	1220	1515	6.3E
	1820	2050	4.8F
	2325		
30 Sa		0310	8.8E
	0625	0950	9.6F
	1335	1630	6.4E
	1940	2200	4.0F
31 Su	0025	0415	8.5E
	0725	1050	10.1F
	1440	1740	7.2E
	2050	2310	4.0F

August

Day	Slack h m	Maximum h m	knots
1 M	0130	0520	8.5E
	0820	1150	10.8F
	1535	1835	8.2E
	2150		
2 Tu		0010	4.4F
	0235	0615	8.9E
	0910	1235	11.4F
	1620	1920	9.2E
	2235		
3 W		0100	5.0F
	0330	0700	9.3E
	0955	1320	11.8F
	1700	2000	9.9E
	2315		
4 Th		0140	5.7F
	0420	0745	9.6E
	1035	1400	12.0F
●	1735	2040	10.4E
	2350		
5 F		0220	6.3F
	0505	0825	9.7E
	1115	1435	11.9F
	1805	2110	10.6E
6 Sa	0020	0255	6.8F
	0545	0900	9.7E
	1155	1505	11.6F
	1835	2145	10.6E
7 Su	0055	0330	7.3F
	0620	0935	9.4E
	1230	1540	11.0F
	1905	2215	10.4E
8 M	0125	0405	7.6F
	0700	1015	8.9E
	1305	1615	10.2F
	1930	2245	10.1E
9 Tu	0155	0440	7.8F
	0745	1050	8.3E
	1345	1650	9.1F
	2000	2320	9.7E
10 W	0230	0525	7.9F
	0835	1135	7.4E
	1430	1730	7.7F
	2030	2355	9.2E
11 Th	0310	0610	7.9F
	0930	1225	6.4E
	1520	1815	6.3F
	2105		
12 F		0035	8.6E
	0355	0705	8.0F
	1035	1330	5.5E
◐	1625	1910	4.8F
	2150		
13 Sa		0130	8.1E
	0450	0810	8.2F
	1150	1445	5.1E
	1750	2020	3.7F
	2240		
14 Su		0230	7.7E
	0550	0920	8.9F
	1305	1605	5.6E
	1915	2135	3.4F
	2350		
15 M		0345	7.9E
	0655	1025	10.1F
	1415	1720	6.9E
	2030	2245	4.0F
16 Tu	0105	0455	8.8E
	0800	1125	11.5F
	1510	1815	8.6E
	2125	2345	5.2F
17 W	0215	0555	10.0E
	0855	1220	12.9F
	1555	1900	10.3E
	2210		
18 Th		0040	6.8F
	0315	0650	11.3E
	0945	1305	13.9F
	1640	1945	11.7E
	2250		
19 F		0130	8.4F
	0415	0740	12.3E
	1035	1350	14.5F
○	1720	2025	12.8E
	2330		
20 Sa		0215	9.9F
	0505	0825	13.0E
	1120	1435	14.4F
	1755	2105	13.4E
21 Su	0010	0255	11.0F
	0555	0910	13.0E
	1210	1515	13.8F
	1830	2140	13.6E
22 M	0045	0340	11.6F
	0645	0955	12.5E
	1255	1555	12.5F
	1905	2220	13.2E
23 Tu	0125	0425	11.7F
	0735	1040	11.4E
	1340	1640	10.8F
	1945	2300	12.3E
24 W	0210	0510	11.2F
	0830	1130	9.9E
	1430	1725	8.8F
	2020	2345	11.1E
25 Th	0255	0605	10.4F
	0930	1225	8.2E
	1530	1815	6.7F
	2100		
26 F		0030	9.6E
	0345	0700	9.5F
	1035	1330	6.6E
◑	1635	1910	4.7F
	2150		
27 Sa		0125	8.2E
	0445	0810	8.7F
	1150	1450	5.8E
	1805	2025	3.3F
	2250		
28 Su		0235	7.1E
	0550	0920	8.4F
	1310	1610	5.9E
	1935	2145	2.9F
29 M	0005	0355	6.7E
	0655	1030	8.7F
	1415	1720	6.8E
	2040	2300	3.4F
30 Tu	0125	0505	7.1E
	0800	1130	9.4F
	1505	1815	8.0E
	2130		
31 W		0000	4.5F
	0235	0600	7.8E
	0855	1215	10.1F
	1550	1900	9.1E
	2210		

September

Day	Slack h m	Maximum h m	knots
1 Th		0045	5.6F
	0325	0650	8.7E
	0940	1300	10.7F
	1625	1935	9.9E
	2245		
2 F		0120	6.8F
	0410	0730	9.4E
	1020	1335	11.0F
	1700	2005	10.5E
	2310		
3 Sa		0155	7.8F
	0450	0805	10.0E
	1055	1405	11.1F
●	1725	2035	10.9E
	2340		
4 Su		0225	8.6F
	0525	0840	10.2E
	1135	1435	10.9F
	1755	2105	11.0E
5 M	0010	0255	9.3F
	0600	0915	10.2E
	1210	1510	10.5F
	1820	2135	10.9E
6 Tu	0035	0330	9.9F
	0640	0950	9.9E
	1245	1540	9.7F
	1845	2205	10.7E
7 W	0105	0405	10.1F
	0720	1025	9.3E
	1325	1615	8.6F
	1915	2235	10.3E
8 Th	0140	0445	10.0F
	0805	1110	8.3E
	1410	1655	7.2F
	1945	2310	9.6E
9 F	0220	0530	9.7F
	0900	1200	7.2E
	1505	1740	5.6F
	2020	2355	8.7E
10 Sa	0310	0630	9.2F
	1005	1300	6.2E
	1610	1840	4.2F
	2110		
11 Su		0050	7.8E
	0405	0735	8.9F
	1120	1420	5.7E
◐	1740	1955	3.2F
	2210		
12 M		0200	7.2E
	0515	0850	9.0F
	1235	1545	6.2E
	1905	2120	3.4F
	2335		
13 Tu		0325	7.3E
	0630	1005	9.9F
	1345	1655	7.6E
	2010	2235	4.6F
14 W	0105	0440	8.4E
	0740	1105	11.0F
	1440	1750	9.4E
	2100	2335	6.5F
15 Th	0215	0545	9.9E
	0840	1200	12.2F
	1525	1835	11.0E
	2140		
16 F		0025	8.6F
	0315	0640	11.4E
	0935	1245	13.0F
	1605	1915	12.4E
	2220		
17 Sa		0110	10.6F
	0410	0725	12.6E
	1025	1330	13.3F
○	1645	1955	13.4E
	2255		
18 Su		0155	12.1F
	0500	0810	13.3E
	1110	1410	13.0F
	1720	2035	13.8E
	2335		
19 M		0235	13.1F
	0545	0855	13.3E
	1155	1450	12.2F
	1755	2110	13.7E
20 Tu	0010	0315	13.5F
	0635	0940	12.6E
	1240	1530	10.9F
	1830	2150	13.0E
21 W	0050	0355	13.2F
	0720	1025	11.4E
	1325	1610	9.2F
	1905	2225	11.9E
22 Th	0130	0440	12.3F
	0810	1110	9.8E
	1415	1655	7.2F
	1940	2310	10.3E
23 F	0215	0530	10.9F
	0905	1205	8.1E
	1515	1745	5.2F
	2025	2355	8.5E
24 Sa	0305	0625	9.4F
	1010	1305	6.7E
	1625	1850	3.5F
◑	2115		
25 Su		0050	6.8E
	0400	0735	8.1F
	1120	1425	5.8E
	1755	2010	2.6F
	2225		
26 M		0205	5.6E
	0510	0850	7.5F
	1230	1545	6.0E
	1920	2135	2.7F
27 Tu	0000	0335	5.3E
	0630	1000	7.6F
	1335	1655	6.9E
	2020	2245	3.8F
28 W	0125	0445	6.0E
	0735	1100	8.1F
	1425	1745	8.0E
	2100	2340	5.2F
29 Th	0225	0540	7.0E
	0830	1145	8.7F
	1510	1825	9.0E
	2135		
30 F		0015	6.6F
	0315	0625	8.2E
	0915	1225	9.3F
	1540	1855	9.9E
	2200		

Time meridian 120° W. 0000 is midnight. 1200 is noon. Times are not adjusted for Daylight Saving Time.

Seymour Narrows, British Columbia, 2005

F–Flood, Dir. 180° True E–Ebb, Dir. 000° True

October

Day	Slack h m	Maximum h m	knots
1 Sa		0050	8.0F
	0355	0705	9.1E
	1000	1300	9.7F
	1610	1925	10.5E
	2230		
2 Su		0120	9.3F
	0430	0740	9.9E
	1035	1330	9.8F
	1640	1955	10.9E
	2255		
3 M ●		0155	10.4F
	0505	0815	10.4E
	1110	1405	9.7F
	1705	2025	11.2E
	2325		
4 Tu		0225	11.3F
	0540	0850	10.6E
	1150	1435	9.3F
	1735	2055	11.2E
	2355		
5 W		0300	11.9F
	0620	0925	10.4E
	1230	1510	8.5F
	1800	2125	11.0E
6 Th	0025	0335	12.1F
	0700	1005	9.8E
	1310	1545	7.5F
	1830	2200	10.5E
7 F	0105	0415	11.8F
	0750	1050	9.0E
	1400	1630	6.3F
	1905	2240	9.7E
8 Sa	0145	0505	11.2F
	0840	1145	8.0E
	1455	1720	4.9F
	1950	2325	8.7E
9 Su	0235	0600	10.3F
	0945	1245	7.2E
	1605	1825	3.8F
	2045		
10 M ◐		0025	7.6E
	0335	0710	9.6F
	1050	1400	6.9E
	1725	1945	3.5F
	2206		
11 Tu		0145	6.9E
	0450	0825	9.3F
	1200	1515	7.5E
	1840	2110	4.3F
	2340		
12 W		0310	7.1E
	0610	0935	9.6F
	1305	1620	8.7E
	1940	2215	6.1F
13 Th	0105	0425	8.2E
	0720	1040	10.3F
	1400	1715	10.2E
	2025	2315	8.3F
14 F	0215	0530	9.8E
	0825	1130	11.0F
	1445	1800	11.6E
	2105		
15 Sa		0005	10.5F
	0310	0620	11.3E
	0920	1220	11.4F
	1525	1845	12.8E
	2145		
16 Su		0045	12.4F
	0400	0710	12.3E
	1010	1305	11.4F
	1605	1925	13.5E
	2225		
17 M ○		0130	13.8F
	0450	0755	12.8E
	1055	1345	11.0F
	1640	2000	13.6E
	2300		
18 Tu		0210	14.5F
	0535	0840	12.8E
	1145	1425	10.1F
	1720	2040	13.3E
	2340		
19 W		0250	14.5F
	0620	0925	12.1E
	1230	1505	8.8F
	1755	2120	12.4E
20 Th	0020	0335	13.8F
	0705	1010	11.0E
	1315	1550	7.3F
	1830	2155	11.0E
21 F	0100	0415	12.6F
	0755	1055	9.7E
	1410	1635	5.7F
	1910	2240	9.3E
22 Sa	0140	0505	11.1F
	0845	1145	8.3E
	1510	1725	4.2F
	1955	2330	7.5E
23 Su	0225	0555	9.4F
	0940	1245	7.1E
	1620	1835	3.0F
	2055		
24 M ◑		0025	5.9E
	0325	0700	8.0F
	1040	1355	6.5E
	1740	1955	2.6F
	2215		
25 Tu		0140	4.7E
	0430	0810	7.1F
	1145	1505	6.6E
	1845	2115	3.2F
	2350		
26 W		0305	4.5E
	0550	0920	6.8F
	1245	1610	7.2E
	1935	2215	4.4F
27 Th	0110	0415	5.1E
	0700	1015	7.0F
	1330	1655	8.0E
	2015	2305	5.9F
28 F	0205	0510	6.2E
	0800	1105	7.3F
	1415	1735	8.8E
	2045	2340	7.4F
29 Sa	0250	0555	7.4E
	0850	1145	7.7F
	1450	1810	9.6E
	2115		
30 Su		0015	8.9F
	0330	0635	8.5E
	0930	1220	8.0F
	1520	1845	10.2E
	2145		
31 M		0050	10.4F
	0410	0715	9.4E
	1010	1255	8.1F
	1550	1915	10.8E
	2215		

November

Day	Slack h m	Maximum h m	knots
1 Tu ●		0120	11.7F
	0445	0750	10.0E
	1055	1330	8.0F
	1620	1945	11.1E
	2245		
2 W		0155	12.7F
	0525	0830	10.4E
	1135	1405	7.7F
	1650	2020	11.3E
	2320		
3 Th		0235	13.3F
	0605	0910	10.4E
	1215	1445	7.2F
	1725	2055	11.1E
	2355		
4 F		0310	13.4F
	0650	0955	10.1E
	1300	1525	6.4F
	1800	2135	10.6E
5 Sa	0035	0355	13.1F
	0735	1040	9.6E
	1355	1615	5.6F
	1845	2215	9.8E
6 Su	0120	0445	12.4F
	0825	1130	9.0E
	1450	1710	4.8F
	1935	2310	8.8E
7 M	0210	0540	11.3F
	0925	1230	8.6E
	1555	1815	4.4F
	2045		
8 Tu ◐		0015	7.8E
	0315	0645	10.3F
	1020	1335	8.5E
	1705	1930	4.7F
	2210		
9 W		0130	7.1E
	0425	0755	9.6F
	1120	1440	8.9E
	1805	2045	5.9F
	2335		
10 Th		0250	7.2E
	0545	0905	9.3F
	1220	1540	9.8E
	1900	2150	7.7F
11 F	0055	0405	8.1E
	0700	1005	9.3F
	1310	1635	10.8E
	1945	2250	9.8F
12 Sa	0200	0505	9.3E
	0805	1100	9.4F
	1400	1725	11.8E
	2030	2340	11.8F
13 Su	0300	0600	10.5E
	0900	1150	9.3F
	1445	1810	12.5E
	2110		
14 M		0025	13.3F
	0350	0655	11.3E
	0955	1235	9.1F
	1525	1850	12.9E
	2155		
15 Tu ○		0105	14.4F
	0435	0740	11.8E
	1045	1320	8.6F
	1605	1930	12.8E
	2230		
16 W		0150	14.8F
	0525	0825	11.8E
	1135	1405	7.9F
	1645	2015	12.3E
	2310		
17 Th		0230	14.6F
	0610	0910	11.4E
	1220	1445	7.0F
	1725	2055	11.4E
	2350		
18 F		0315	13.8F
	0655	0955	10.6E
	1310	1530	5.9F
	1805	2135	10.1E
19 Sa	0030	0355	12.6F
	0740	1040	9.8E
	1400	1620	4.9F
	1850	2220	8.6E
20 Su	0115	0440	11.2F
	0825	1130	8.8E
	1455	1715	4.0F
	1940	2310	7.1E
21 M	0200	0530	9.6F
	0910	1225	8.0E
	1555	1815	3.4F
	2045		
22 Tu		0005	5.7E
	0250	0625	8.2F
	1000	1320	7.5E
	1700	1925	3.4F
	2200		
23 W ◑		0105	4.7E
	0350	0725	7.1F
	1050	1415	7.3E
	1755	2030	3.9F
	2320		
24 Th		0220	4.3E
	0500	0825	6.4F
	1140	1510	7.5E
	1840	2125	5.0F
25 F	0035	0330	4.6E
	0610	0920	6.1F
	1225	1600	7.9E
	1920	2215	6.3F
26 Sa	0130	0430	5.4E
	0715	1010	6.0F
	1305	1645	8.4E
	1955	2300	7.8F
27 Su	0220	0520	6.4E
	0815	1100	6.1F
	1345	1720	9.1E
	2030	2340	9.4F
28 M	0305	0610	7.5E
	0905	1140	6.2F
	1425	1800	9.8E
	2105		
29 Tu		0015	11.0F
	0350	0650	8.5E
	0950	1220	6.4F
	1500	1835	10.4E
	2135		
30 W		0055	12.3F
	0430	0735	9.4E
	1035	1305	6.4F
	1535	1915	10.9E
	2215		

December

Day	Slack h m	Maximum h m	knots
1 Th ●		0135	13.4F
	0510	0815	10.0E
	1120	1345	6.4F
	1615	1950	11.2E
	2250		
2 F		0215	14.1F
	0555	0900	10.4E
	1210	1430	6.2F
	1655	2035	11.2E
	2335		
3 Sa		0255	14.3F
	0640	0940	10.6E
	1255	1515	6.0F
	1745	2120	10.9E
4 Su	0015	0340	14.0F
	0725	1030	10.5E
	1345	1605	5.9F
	1835	2205	10.2E
5 M	0105	0430	13.3F
	0810	1115	10.4E
	1435	1700	5.8F
	1935	2300	9.4E
6 Tu	0200	0525	12.2F
	0900	1210	10.2E
	1530	1805	6.0F
	2045		
7 W		0000	8.5E
	0255	0620	10.9F
	0950	1305	10.1E
	1630	1910	6.6F
	2200		
8 Th ◐		0110	7.7E
	0405	0725	9.7F
	1040	1400	10.2E
	1725	2015	7.6F
	2320		
9 F		0220	7.4E
	0515	0825	8.7F
	1130	1455	10.4E
	1815	2120	9.0F
10 Sa	0035	0335	7.7E
	0630	0930	8.0F
	1220	1555	10.8E
	1910	2220	10.6F
11 Su	0145	0445	8.4E
	0740	1025	7.5F
	1315	1645	11.2E
	1955	2310	12.0F
12 M	0245	0545	9.2E
	0845	1120	7.1F
	1400	1735	11.6E
	2040		
13 Tu		0000	13.2F
	0340	0640	10.0E
	0945	1215	6.8F
	1450	1825	11.7E
	2125		
14 W		0050	13.9F
	0425	0730	10.5E
	1035	1300	6.6F
	1535	1910	11.6E
	2210		
15 Th ○		0135	14.2F
	0515	0815	10.8E
	1125	1350	6.3F
	1620	1955	11.2E
	2250		
16 F		0215	14.0F
	0555	0900	10.8E
	1215	1435	5.9F
	1705	2040	10.5E
	2335		
17 Sa		0300	13.4F
	0640	0945	10.6E
	1300	1520	5.5F
	1755	2120	9.6E
18 Su	0015	0340	12.5F
	0720	1025	10.2E
	1345	1605	5.1F
	1840	2205	8.6E
19 M	0055	0420	11.3F
	0800	1105	9.6E
	1430	1655	4.8F
	1930	2250	7.4E
20 Tu	0140	0505	10.0F
	0835	1150	9.0E
	1515	1745	4.6F
	2025	2335	6.3E
21 W	0225	0545	8.7F
	0915	1235	8.5E
	1605	1835	4.6F
	2125		
22 Th		0030	5.4E
	0315	0635	7.5F
	0955	1320	8.0E
	1650	1935	5.0F
	2235		
23 F ◑		0125	4.7E
	0410	0725	6.4F
	1035	1405	7.8E
	1735	2030	5.6F
	2340		
24 Sa		0230	4.5E
	0520	0820	5.5F
	1115	1455	7.8E
	1820	2120	6.7F
25 Su	0050	0340	4.7E
	0630	0915	4.9F
	1200	1545	8.0E
	1900	2215	8.0F
26 M	0150	0445	5.5E
	0740	1010	4.7F
	1245	1630	8.5E
	1945	2300	9.5F
27 Tu	0240	0540	6.5E
	0840	1105	4.7F
	1330	1720	9.1E
	2025	2345	11.0F
28 W	0330	0630	7.7E
	0935	1155	4.9F
	1420	1805	9.8E
	2110		
29 Th		0030	12.4F
	0415	0720	8.9E
	1025	1240	5.3F
	1505	1850	10.5E
	2150		
30 F ●		0115	13.6F
	0500	0805	10.0E
	1110	1330	5.8F
	1555	1935	11.1E
	2235		
31 Sa		0200	14.4F
	0540	0845	10.8E
	1155	1415	6.3F
	1645	2025	11.5E
	2320		

Time meridian 120° W. 0000 is midnight. 1200 is noon. Times are not adjusted for Daylight Saving Time.

Snow Passage, Alaska, 2005

F–Flood, Dir. 120° True E–Ebb, Dir. 294° True

January

Day	Slack h m	Maximum h m	knots
1 Sa	0215	0627	1.9E
	1059	1307	0.7F
	1523	1846	1.0E
	2025		
2 Su		0029	1.1F
	0313	0718	1.7E
	1208	1418	0.5F
	1600	1942	1.0E
	2126		
3 M		0144	1.1F
	0418	0817	1.6E
	1309	1504	0.6F
◑	1641	2043	1.2E
	2300		
4 Tu		0256	1.3F
	0529	0915	1.6E
	1346	1538	0.7F
	1732	2140	1.7E
5 W	0108	0347	1.6F
	0710	1009	1.7E
	1409	1608	0.9F
	1846	2235	2.2E
6 Th	0208	0439	1.6F
	0830	1101	1.9E
	1425	1643	1.2F
	2009	2330	2.7E
7 F	0259	0551	1.6F
	0917	1151	2.3E
	1441	1737	1.5F
	2116		
8 Sa		0022	3.1E
	0350	0702	1.6F
	0955	1236	2.7E
	1507	1852	1.8F
	2217		
9 Su		0108	3.3E
	0446	0753	1.5F
	1029	1319	3.0E
	1543	1939	2.1F
	2314		
10 M		0153	3.3E
	0543	0841	1.3F
	1102	1402	3.1E
●	1628	2020	2.2F
11 Tu	0005	0239	3.0E
	0634	0934	1.0F
	1136	1448	3.1E
	1718	2108	2.1F
12 W	0052	0328	2.6E
	0717	1023	0.9F
	1211	1539	3.0E
	1809	2200	2.0F
13 Th	0137	0419	2.2E
	0746	1100	0.8F
	1250	1633	2.8E
	1900	2236	1.7F
14 F	0218	0508	2.0E
	0751	1122	0.8F
	1337	1726	2.6E
	1958	2310	1.3F
15 Sa	0252	0554	1.7E
	0811	1145	0.8F
	1443	1819	2.3E
	2117	2349	0.9F

Day	Slack h m	Maximum h m	knots
16 Su	0315	0643	1.4E
	0857	1234	0.8F
	1613	1918	2.0E
◐	2259		
17 M		0043	0.5F
	0339	0740	1.3E
	1027	1451	0.9F
	1802	2024	1.7E
18 Tu	0034	0218	0.4F
	0415	0842	1.4E
	1222	1542	1.2F
	1934	2129	1.7E
19 W	0142	0312	0.4F
		0604	*
		0700	*
		0942	1.7E
	1332	1634†	1.3F
20 Th	0224	0351	0.6F
	0634	1039	2.1E
	1426	1932	1.5F
	2136	2328	1.9E
21 F	0259	0434	0.8F
	0836	1134	2.6E
	1515	1938	1.7F
	2226		
22 Sa		0018	2.0E
	0334	0534	0.9F
	0921	1224	3.0E
	1605	1958	1.8F
	2310		
23 Su		0059	2.1E
	0412	0654	1.2F
	1008	1307	3.2E
	1655	2026	1.7F
	2348		
24 M		0137	2.1E
	0455	0746	1.4F
	1056	1348	3.1E
	1743	2056	1.5F
25 Tu	0017	0216	2.1E
	0543	0836	1.5F
	1142	1430	2.7E
○	1821	2123	1.3F
26 W	0029	0256	2.1E
	0630	0934	1.5F
	1223	1514	2.3E
	1848	2143	1.1F
27 Th	0020	0340	2.1E
	0717	1026	1.4F
	1300	1602	1.9E
	1846	2205	1.1F
28 F	0033	0426	2.2E
	0806	1106	1.3F
	1334	1648	1.6E
	1850	2237	1.2F
29 Sa	0102	0509	2.2E
	0859	1140	1.1F
	1406	1731	1.4E
	1919	2316	1.3F
30 Su	0147	0552	2.1E
	0953	1215	0.9F
	1441	1815	1.4E
	2004		
31 M		0002	1.3F
	0246	0639	1.9E
	1047	1307	0.7F
	1522	1906	1.4E
	2107		

February

Day	Slack h m	Maximum h m	knots
1 Tu		0109	1.2F
	0353	0736	1.7E
	1146	1420	0.6F
◑	1609	2008	1.5E
	2240		
2 W		0234	1.3F
	0501	0838	1.6E
	1243	1508	0.8F
	1707	2110	1.8E
3 Th	0047	0331	1.5F
	0621	0936	1.7E
	1324	1545	1.0F
	1826	2209	2.2E
4 F	0155	0422	1.5F
	0751	1031	2.0E
	1354	1622	1.3F
	2008	2306	2.5E
5 Sa	0249	0529	1.4F
	0847	1124	2.3E
	1423	1712	1.6F
	2117		
6 Su		0001	2.9E
	0341	0647	1.4F
	0931	1213	2.7E
	1453	1841	1.8F
	2215		
7 M		0049	3.1E
	0435	0742	1.3F
	1012	1259	3.0E
	1530	1929	2.0F
	2306		
8 Tu		0133	3.1E
	0529	0831	1.2F
	1052	1342	3.1E
●	1612	1959	2.0F
	2351		
9 W		0216	2.9E
	0615	0922	1.0F
	1131	1428	3.1E
	1701	2029	1.9F
10 Th	0030	0302	2.6E
	0649	1010	1.0F
	1210	1518	2.9E
	1753	2111	1.7F
11 F	0103	0351	2.3E
	0709	1048	1.0F
	1250	1612	2.7E
	1846	2200	1.5F
12 Sa	0129	0440	2.1E
	0718	1112	1.0F
	1336	1705	2.5E
	1944	2244	1.2F
13 Su	0153	0527	1.9E
	0745	1127	1.0F
	1439	1757	2.3E
	2102	2327	0.9F
14 M	0220	0615	1.6E
	0832	1205	1.0F
	1606	1853	1.9E
	2232		
15 Tu		0016	0.6F
	0257	0708	1.5E
	1000	1417	0.9F
◐	1742	1957	1.6E
	2357		

Day	Slack h m	Maximum h m	knots
16 W		0132	0.4F
	0345	0811	1.5E
	1153	1520	1.1F
	1911	2103	1.5E
17 Th	0113	0254	0.5F
	0440	0914	1.7E
	1312	1610	1.3F
	2021	2206	1.5E
18 F	0205	0339	0.6F
	0550	1012	2.0E
	1409	1718	1.4F
		1813	1.4F
		1927†	1.5F
19 Sa	0245	0422	0.7F
	0816	1109	2.4E
	1457	2000	1.6F
	2206		
20 Su		0000	1.8E
	0324	0523	0.9F
	0911	1201	2.7E
	1542	1929	1.6F
	2246		
21 M		0040	2.0E
	0403	0649	1.1F
	1002	1245	2.8E
	1625	1951	1.5F
	2319		
22 Tu		0115	2.2E
	0445	0739	1.3F
	1049	1325	2.8E
	1706	2012	1.4F
	2341		
23 W		0150	2.3E
	0528	0824	1.4F
	1132	1404	2.5E
○	1740	2029	1.3F
	2342		
24 Th		0227	2.4E
	0608	0912	1.5F
	1207	1446	2.2E
	1800	2050	1.2F
	2342		
25 F		0306	2.4E
	0643	0959	1.4F
	1235	1530	2.0E
	1806	2127	1.3F
26 Sa	0006	0350	2.4E
	0708	1037	1.4F
	1301	1617	1.9E
	1829	2213	1.4F
27 Su	0043	0436	2.4E
	0725	1107	1.3F
	1331	1703	1.9E
	1906	2257	1.4F
28 M	0128	0521	2.3E
	0753	1136	1.1F
	1408	1748	1.9E
	1954	2344	1.4F

March

Day	Slack h m	Maximum h m	knots
1 Tu	0224	0607	2.1E
	0836	1213	1.0F
	1455	1838	1.8E
	2055		
2 W		0041	1.3F
	0327	0700	1.9E
	0933	1317	0.9F
	1552	1938	1.7E
	2227		
3 Th		0209	1.2F
	0430	0803	1.8E
	1050	1441	1.0F
◑	1657	2043	1.8E
4 F	0029	0314	1.3F
	0536	0905	1.9E
	1231	1529	1.2F
	1829	2144	2.0E
5 Sa	0144	0406	1.2F
	0700	1002	2.1E
	1327	1608	1.4F
	2013	2243	2.2E
6 Su	0239	0508	1.2F
	0819	1057	2.3E
	1408	1650	1.5F
	2113	2339	2.5E
7 M	0328	0632	1.2F
	0915	1151	2.6E
	1443	1819	1.6F
	2203		
8 Tu		0028	2.8E
	0416	0731	1.2F
	1006	1239	2.9E
	1519	1913	1.7F
	2248		
9 W		0111	2.9E
	0502	0819	1.2F
	1053	1324	3.0E
	1600	1931	1.7F
	2326		
10 Th		0153	2.9E
	0542	0907	1.2F
	1136	1409	3.0E
●	1648	1956	1.6F
	2359		
11 F		0236	2.7E
	0613	0953	1.2F
	1215	1457	2.8E
	1740	2035	1.5F
12 Sa	0027	0323	2.5E
	0636	1032	1.2F
	1253	1550	2.6E
	1834	2128	1.4F
13 Su	0051	0413	2.2E
	0700	1059	1.2F
	1335	1644	2.5E
	1930	2223	1.2F
14 M	0118	0502	2.1E
	0734	1117	1.1F
	1433	1735	2.2E
	2038	2309	1.0F
15 Tu	0152	0550	1.9E
	0828	1150	1.1F
	1556	1828	1.9E
	2200	2356	0.8F

Day	Slack h m	Maximum h m	knots
16 W	0234	0641	1.7E
	1002	1309	1.0F
	1720	1928	1.5E
	2320		
17 Th		0100	0.6F
	0326	0741	1.6E
	1130	1455	1.1F
◐	1845	2035	1.3E
18 F	0043	0234	0.5F
	0421	0845	1.7E
	1248	1544	1.2F
	1957	2141	1.3E
19 Sa	0148	0328	0.6F
	0526	0944	1.8E
	1347	1629	1.3F
		1806	1.1F
		1912†	1.2F
20 Su	0235	0415	0.7F
	0809	1041	2.0E
	1433	1727	1.3F
		1847	1.3F
		1948†	1.3F
21 M	0315	0515	0.8F
	0907	1134	2.2E
	1513	1831	1.3F
	2214		
22 Tu		0016	1.9E
	0353	0637	1.0F
	0955	1221	2.4E
	1550	1904	1.3F
	2243		
23 W		0050	2.2E
	0430	0725	1.2F
	1037	1301	2.4E
	1625	1928	1.3F
	2259		
24 Th		0124	2.4E
	0504	0803	1.4F
	1111	1339	2.4E
	1655	1950	1.3F
	2300		
25 F		0158	2.5E
	0533	0839	1.4F
	1137	1418	2.4E
○	1720	2020	1.4F
	2317		
26 Sa		0235	2.5E
	0551	0917	1.5F
	1201	1501	2.3E
	1746	2103	1.4F
	2349		
27 Su		0318	2.5E
	0605	0955	1.5F
	1230	1548	2.3E
	1820	2155	1.5F
28 M	0028	0405	2.5E
	0634	1031	1.5F
	1305	1637	2.3E
	1900	2243	1.5F
29 Tu	0112	0453	2.4E
	0713	1106	1.4F
	1348	1725	2.2E
	1946	2328	1.4F
30 W	0200	0541	2.4E
	0759	1144	1.4F
	1442	1814	2.1E
	2044		
31 Th		0017	1.2F
	0256	0631	2.2E
	0856	1236	1.2F
	1547	1911	1.8E
	2217		

Time meridian 135° W. 0000 is midnight. 1200 is noon. Times are not adjusted for Daylight Saving Time.
If three or more consecutive entries are marked (F) or (E) the middle ones are not true maximums but intermediate values to show the current pattern.
* Current weak and variable.
† See page 164 for the remaining currents on this day.

Snow Passage, Alaska, 2005

F–Flood, Dir. 120° True   E–Ebb, Dir. 294° True

April

Day	Slack h m	Maximum h m	knots
1 F		0137	1.0F
	0354	0731	2.1E
	1006	1414	1.2F
	1657	2016	1.7E
2 Sa	0013	0256	1.0F
	0455	0835	2.0E
	1140	1515	1.3F
	1841	2119	1.7E
3 Su	0132	0351	1.0F
	0614	0935	2.1E
	1304	1554	1.3F
	2005	2217	1.9E
4 M	0226	0449	1.0F
	0800	1033	2.3E
	1356	1627	1.4F
	2058	2313	2.2E
5 Tu	0309	0616	1.1F
	0908	1129	2.5E
	1437	1703	1.3F
	2142		
6 W		0004	2.5E
	0348	0719	1.3F
	1003	1221	2.7E
	1515	1800	1.3F
	2221		
7 Th		0048	2.8E
	0425	0805	1.4F
	1052	1307	2.8E
	1554	1853	1.3F
	2256		
8 F		0130	2.9E
	0503	0849	1.4F
	1136	1351	2.8E
	1639	1929	1.4F
	2326		
9 Sa		0211	2.8E
	0540	0933	1.4F
	1215	1437	2.7E
	1729	2011	1.4F
	2355		
10 Su		0256	2.7E
	0617	1014	1.3F
	1253	1527	2.5E
	1820	2102	1.4F
11 M	0024	0346	2.5E
	0655	1047	1.3F
	1333	1621	2.3E
	1911	2202	1.3F
12 Tu	0057	0438	2.3E
	0740	1113	1.2F
	1427	1713	2.1E
	2012	2254	1.2F
13 W	0134	0527	2.2E
	0841	1142	1.1F
	1541	1803	1.8E
	2130	2340	1.0F
14 Th	0217	0616	2.0E
	0953	1231	1.0F
	1654	1859	1.5E
	2250		
15 F		0038	0.7F
	0309	0711	1.7E
	1103	1416	1.0F
	1809	2003	1.3E
16 Sa	0015	0216	0.5F
	0407	0813	1.5E
	1214	1510	1.1F
	1920	2110	1.2E
17 Su	0133	0319	0.6F
	0513	0914	1.5E
	1316	1547	1.1F
		1759	0.7F
		1852†	0.8F
18 M		0007	1.4E
	0224	0407	0.6F
	0804	1010	1.5E
	1403	1620	1.2F
	2059	2301	1.6E
19 Tu	0304	0459	0.7F
	0859	1104	1.7E
	1441	1657	1.2F
	2134	2344	1.9E
20 W	0337	0610	0.9F
	0940	1153	1.9E
	1515	1752	1.2F
	2202		
21 Th		0021	2.1E
	0406	0700	1.1F
	1011	1235	2.2E
	1547	1846	1.3F
	2219		
22 F		0056	2.4E
	0429	0733	1.3F
	1033	1314	2.5E
	1621	1925	1.5F
	2233		
23 Sa		0131	2.5E
	0441	0800	1.4F
	1056	1353	2.7E
	1658	2003	1.5F
	2301		
24 Su		0208	2.6E
	0450	0827	1.5F
	1128	1435	2.8E
	1738	2047	1.6F
	2336		
25 M		0249	2.6E
	0520	0905	1.7F
	1206	1521	2.7E
	1818	2140	1.5F
26 Tu	0013	0336	2.6E
	0600	0956	1.7F
	1249	1613	2.6E
	1858	2230	1.5F
27 W	0051	0427	2.7E
	0646	1042	1.8F
	1339	1704	2.4E
	1941	2313	1.3F
28 Th	0133	0516	2.7E
	0735	1124	1.7F
	1438	1753	2.1E
	2033	2356	1.1F
29 F	0221	0606	2.6E
	0832	1211	1.5F
	1546	1847	1.8E
	2207		
30 Sa		0058	0.8F
	0318	0702	2.3E
	0940	1330	1.3F
	1656	1947	1.5E
	2350		

May

Day	Slack h m	Maximum h m	knots
1 Su		0238	0.7F
	0424	0806	2.1E
	1108	1456	1.2F
	1825	2050	1.5E
2 M	0109	0336	0.8F
	0547	0910	2.1E
	1248	1536	1.1F
	1942	2148	1.6E
3 Tu	0201	0431	1.0F
	0751	1010	2.1E
	1351	1602	1.0F
	2034	2243	1.9E
4 W	0240	0552	1.1F
	0901	1108	2.2E
	1435	1631	1.0F
	2114	2336	2.3E
5 Th	0315	0704	1.4F
	0956	1202	2.5E
	1513	1713	1.0F
	2150		
6 F		0023	2.6E
	0351	0750	1.5F
	1045	1250	2.6E
	1550	1813	1.1F
	2222		
7 Sa		0107	2.9E
	0431	0830	1.6F
	1129	1333	2.7E
	1630	1907	1.2F
	2254		
8 Su		0149	3.0E
	0518	0912	1.5F
	1210	1417	2.6E
	1715	1952	1.4F
	2328		
9 M		0233	2.9E
	0606	0955	1.4F
	1249	1504	2.4E
	1802	2042	1.4F
10 Tu	0004	0322	2.8E
	0653	1033	1.4F
	1329	1556	2.2E
	1852	2144	1.4F
11 W	0040	0414	2.6E
	0742	1104	1.3F
	1416	1649	2.0E
	1950	2241	1.3F
12 Th	0119	0504	2.4E
	0834	1131	1.2F
	1514	1739	1.8E
	2106	2329	1.1F
13 F	0204	0552	2.0E
	0929	1202	1.1F
	1611	1829	1.6E
	2227		
14 Sa		0023	0.8F
	0258	0642	1.6E
	1025	1252	0.9F
	1706	1926	1.4E
	2349		
15 Su		0156	0.6F
	0400	0739	1.2E
	1125	1416	0.9F
	1811	2029	1.3E
		2208†	1.0E
16 M	0112	0305	0.6F
	0513	0841	1.0E
	1230	1504	1.0F
	1917	2127	1.4E
		2249†	1.2E
17 Tu	0208	0351	0.6F
		0601	*
		0701	*
		0937	1.1E
	1324	1539†	1.1F
18 W	0246	0433	0.6F
	0836	1030	1.3E
	1406	1616	1.2F
	2050	2305	1.7E
19 Th	0314	0519	0.7F
	0906	1120	1.7E
	1443	1701	1.3F
	2122	2348	2.0E
20 F	0334	0612	0.9F
	0922	1207	2.2E
	1519	1808	1.5F
	2149		
21 Sa		0028	2.2E
	0346	0651	1.2F
	0942	1249	2.7E
	1558	1907	1.6F
	2216		
22 Su		0106	2.5E
	0348	0720	1.4F
	1018	1330	3.0E
	1643	1951	1.7F
	2247		
23 M		0144	2.6E
	0409	0751	1.7F
	1104	1412	3.1E
	1732	2035	1.6F
	2319		
24 Tu		0224	2.8E
	0448	0832	1.9F
	1152	1458	3.0E
	1818	2127	1.5F
	2353		
25 W		0310	2.9E
	0535	0927	2.0F
	1242	1550	2.7E
	1901	2218	1.3F
26 Th	0028	0401	2.9E
	0624	1024	2.1F
	1335	1642	2.4E
	1941	2300	1.1F
27 F	0106	0453	2.9E
	0715	1108	2.0F
	1435	1732	2.1E
	2024	2338	0.9F
28 Sa	0150	0544	2.8E
	0811	1148	1.7F
	1538	1822	1.8E
	2133		
29 Su		0025	0.7F
	0249	0638	2.5E
	0918	1237	1.3F
	1636	1917	1.5E
	2302		
30 M		0214	0.6F
	0403	0740	2.1E
	1050	1425	0.9F
	1732	2018	1.3E
31 Tu	0021	0319	0.8F
	0539	0846	2.0E
	1238	1514	0.7F
	1858	2117	1.4E

June

Day	Slack h m	Maximum h m	knots
1 W	0121	0410	1.0F
	0740	0948	1.9E
	1348	1539	0.7F
	1958	2212	1.7E
2 Th	0205	0516	1.2F
	0849	1047	2.0E
	1433	1607	0.7F
	2040	2307	2.1E
3 F	0244	0646	1.4F
	0944	1143	2.2E
	1507	1646	0.8F
	2114	2358	2.5E
4 Sa	0325	0735	1.6F
	1033	1232	2.4E
	1540	1741	1.0F
	2148		
5 Su		0045	2.9E
	0411	0814	1.7F
	1118	1315	2.5E
	1616	1847	1.2F
	2226		
6 M		0128	3.2E
	0503	0853	1.6F
	1200	1358	2.4E
	1657	1937	1.4F
	2306		
7 Tu		0211	3.2E
	0556	0936	1.5F
	1240	1442	2.3E
	1745	2027	1.5F
	2347		
8 W		0258	3.0E
	0644	1015	1.4F
	1317	1531	2.1E
	1836	2131	1.5F
9 Th	0028	0349	2.7E
	0729	1045	1.3F
	1351	1623	2.0E
	1935	2234	1.4F
10 F	0110	0439	2.3E
	0811	1108	1.2F
	1414	1712	1.9E
	2048	2323	1.2F
11 Sa	0158	0527	1.9E
	0849	1131	1.1F
	1430	1759	1.8E
	2204		
12 Su		0013	0.9F
	0255	0613	1.4E
	0916	1203	1.0F
	1501	1847	1.6E
	2319		
13 M		0128	0.6F
	0358	0704	1.0E
	0920	1253	0.9F
	1545	1943	1.5E
14 Tu	0039	0241	0.5F
	0447	0802	0.8E
	1019	1411	0.9F
	1637	2042	1.4E
15 W	0142	0326	0.5F
		0545	*
		0639	*
		0900	0.9E
	1229	1506†	1.1F
16 Th		0014	1.1E
	0220	0400	0.6F
	0739	0954	1.2E
	1329	1550	1.3F
	1955	2224	1.6E
17 F	0243	0428	0.7F
	0732	1046	1.8E
	1415	1636	1.5F
	2048	2313	1.8E
18 Sa	0258	0458	0.9F
	0814	1136	2.3E
	1458	1739	1.6F
	2125	2358	2.1E
19 Su	0304	0544	1.2F
	0902	1224	2.9E
	1543	1850	1.6F
	2157		
20 M		0040	2.4E
	0312	0641	1.6F
	0956	1308	3.2E
	1633	1939	1.6F
	2227		
21 Tu		0121	2.8E
	0341	0726	1.9F
	1052	1351	3.3E
	1726	2024	1.5F
	2258		
22 W		0202	3.0E
	0424	0811	2.2F
	1146	1437	3.1E
	1817	2115	1.3F
	2332		
23 Th		0246	3.1E
	0513	0904	2.3F
	1237	1527	2.8E
	1902	2207	1.1F
24 F	0007	0337	3.1E
	0605	1005	2.2F
	1327	1619	2.4E
	1939	2251	1.0F
25 Sa	0045	0430	3.0E
	0657	1049	2.0F
	1419	1709	2.1E
	2000	2326	0.9F
26 Su	0131	0523	2.8E
	0752	1124	1.6F
	1510	1756	1.8E
	2022	2359	0.8F
27 M	0233	0616	2.5E
	0901	1201	1.1F
	1550	1846	1.5E
	2105		
28 Tu		0130	0.7F
	0353	0715	2.1E
	1041	1255	0.7F
	1615	1942	1.3E
	2223		
29 W		0256	0.9F
	0535	0821	1.9E
	1227	1441	0.4F
	1638	2044	1.3E
30 Th	0023	0347	1.1F
	0723	0925	1.8E
	1342	1516	0.4F
		1808	*
		1900†	*

Time meridian 135° W. 0000 is midnight. 1200 is noon. Times are not adjusted for Daylight Saving Time.
If three or more consecutive entries are marked (F) or (E) the middle ones are not true maximums but intermediate values to show the current pattern.
* Current weak and variable.
† See page 164 for the remaining currents on this day.

Snow Passage, Alaska, 2005

F–Flood, Dir. 120° True E–Ebb, Dir. 294° True

July

Day	Slack h m	Maximum h m	knots
1 F	0130	0440	1.3F
	0833	1024	1.8E
	1423	1547	0.5F
	1834	2238	2.0E
2 Sa	0220	0623	1.4F
	0929	1122	2.0E
	1455	1625	0.7F
	2034	2333	2.5E
3 Su	0307	0728	1.6F
	1020	1213	2.1E
	1526	1716	0.9F
	2116		
4 M		0023	3.0E
	0356	0801	1.7F
	1106	1257	2.3E
	1601	1828	1.1F
	2202		
5 Tu		0108	3.3E
	0448	0835	1.7F
	1148	1339	2.3E
	1643	1928	1.4F
	2249		
6 W ●		0151	3.3E
	0540	0911	1.6F
	1225	1420	2.2E
	1732	2020	1.5F
	2336		
7 Th		0235	3.1E
	0626	0947	1.4F
	1255	1505	2.1E
	1826	2126	1.5F
8 F	0022	0323	2.6E
	0706	1015	1.3F
	1310	1554	2.1E
	1923	2227	1.5F
9 Sa	0107	0413	2.1E
	0738	1036	1.2F
	1310	1642	2.1E
	2027	2314	1.3F
10 Su	0155	0501	1.7E
	0745	1058	1.1F
	1327	1727	2.1E
	2135	2357	1.1F
11 M	0246	0545	1.3E
	0743	1129	1.1F
	1401	1810	1.9E
	2241		
12 Tu		0047	0.7F
	0325	0630	1.0E
	0812	1211	1.0F
	1452	1858	1.7E
	2349		
13 W		0158	0.5F
	0350	0721	0.9E
	0904	1316	1.0F
	1555	1955	1.5E
		2219†	0.6E
14 Th ◐	0057	0250	0.4F
	0420	0821	1.0E
	1029	1436	1.2F
	1702	2053	1.4E
		2308†	0.6E
15 F	0141	0324	0.5F
	0501	0918	1.4E
	1249	1529	1.4F
	1830	2147	1.5E
16 Sa	0204	0352	0.7F
	0602	1013	1.9E
	1351	1617	1.5F
	2009	2238	1.7E
17 Su	0218	0422	1.0F
	0728	1107	2.4E
	1442	1715	1.6F
	2056	2328	2.0E
18 M	0228	0503	1.4F
	0844	1200	2.9E
	1531	1831	1.5F
	2131		
19 Tu		0015	2.5E
	0248	0607	1.7F
	0947	1248	3.2E
	1624	1927	1.5F
	2204		
20 W		0058	2.9E
	0321	0709	2.1F
	1046	1332	3.3E
	1719	2015	1.3F
	2238		
21 Th ○		0141	3.1E
	0404	0755	2.3F
	1140	1416	3.1E
	1811	2105	1.1F
	2315		
22 F		0225	3.2E
	0454	0841	2.3F
	1227	1503	2.8E
	1852	2158	1.0F
	2354		
23 Sa		0315	3.1E
	0548	0938	2.1F
	1310	1554	2.4E
	1918	2242	1.0F
24 Su	0035	0409	3.0E
	0641	1026	1.8F
	1349	1643	2.1E
	1923	2315	1.0F
25 M	0123	0503	2.8E
	0738	1059	1.4F
	1422	1730	1.8E
	1942	2336	0.9F
26 Tu	0224	0555	2.5E
	0851	1133	1.0F
	1447	1816	1.6E
	2020		
27 W ◑		0009	0.9F
	0346	0651	2.1E
	1028	1217	0.6F
	1510	1909	1.4E
	2119		
28 Th		0224	0.9F
	0522	0755	1.8E
	1203	1328	0.3F
	1541	2010	1.4E
	2314		
29 F		0323	1.1F
	0702	0900	1.6E
	1321	1450	0.3F
	1626	2112	1.6E
30 Sa	0059	0412	1.3F
	0815	1001	1.6E
	1405	1530	0.5F
	1729	2210	2.0E
31 Su	0201	0557	1.4F
	0914	1100	1.7E
	1439	1609	0.7F
	1931	2307	2.5E

August

Day	Slack h m	Maximum h m	knots
1 M	0251	0751	1.6F
	1005	1155	1.9E
	1513	1658	0.9F
	2053		
2 Tu		0001	2.9E
	0339	0749	1.7F
	1050	1240	2.1E
	1550	1819	1.1F
	2147		
3 W		0047	3.2E
	0428	0811	1.7F
	1129	1320	2.2E
	1633	1927	1.3F
	2241		
4 Th ●		0130	3.2E
	0517	0839	1.5F
	1201	1358	2.3E
	1723	2019	1.5F
	2333		
5 F		0212	2.9E
	0602	0906	1.3F
	1222	1439	2.3E
	1814	2118	1.5F
6 Sa	0020	0257	2.4E
	0637	0929	1.2F
	1221	1522	2.3E
	1904	2214	1.5F
7 Su	0103	0345	2.0E
	0658	0953	1.1F
	1225	1608	2.3E
	1957	2257	1.4F
8 M	0141	0432	1.6E
	0652	1025	1.1F
	1249	1653	2.2E
	2053	2333	1.2F
9 Tu	0207	0516	1.4E
	0708	1102	1.2F
	1328	1736	2.1E
	2149		
10 W		0005	0.9F
	0230	0558	1.3E
	0744	1144	1.2F
	1422	1820	1.9E
	2244		
11 Th		0043	0.6F
	0258	0645	1.2E
	0837	1240	1.2F
	1527	1912	1.6E
	2340		
12 F ◐		0150	0.5F
	0337	0741	1.3E
	0955	1404	1.3F
	1634	2013	1.5E
13 Sa	0032	0245	0.6F
	0428	0844	1.6E
	1212	1509	1.4F
	1744	2111	1.5E
14 Su	0108	0323	0.8F
	0534	0943	1.9E
	1332	1559	1.5F
	1913	2205	1.7E
15 M	0134	0359	1.2F
	0711	1040	2.3E
	1430	1654	1.4F
	2019	2258	2.1E
16 Tu	0159	0440	1.6F
	0840	1137	2.7E
	1521	1813	1.3F
	2105	2349	2.5E
17 W	0229	0537	1.8F
	0943	1227	3.0E
	1614	1916	1.3F
	2146		
18 Th		0037	2.9E
	0306	0654	2.1F
	1038	1312	3.2E
	1706	2006	1.2F
	2227		
19 F ○		0121	3.1E
	0349	0739	2.2F
	1126	1355	3.0E
	1752	2055	1.1F
	2309		
20 Sa		0206	3.2E
	0439	0817	2.1F
	1208	1439	2.8E
	1826	2146	1.1F
	2351		
21 Su		0254	3.1E
	0534	0900	1.8F
	1244	1526	2.4E
	1843	2230	1.1F
22 M	0033	0348	2.9E
	0630	0951	1.5F
	1313	1615	2.2E
	1849	2302	1.1F
23 Tu	0119	0442	2.7E
	0727	1033	1.2F
	1336	1703	1.9E
	1914	2316	1.1F
24 W	0215	0534	2.5E
	0837	1110	0.9F
	1358	1750	1.8E
	1955	2341	1.1F
25 Th	0335	0627	2.1E
	1003	1151	0.6F
	1424	1840	1.6E
	2054		
26 F ◑		0035	1.0F
	0505	0727	1.7E
	1124	1247	0.4F
	1504	1939	1.5E
	2244		
27 Sa		0258	1.1F
	0638	0833	1.5E
	1246	1418	0.3F
	1558	2043	1.7E
28 Su	0031	0349	1.3F
	0755	0937	1.4E
	1344	1514	0.5F
	1700	2143	2.0E
29 M	0140	0442	1.4F
		0604	1.3F
		0708	1.4F
	0856	1038	1.5E
	1426	1557†	0.6F
30 Tu	0232	0748	1.6F
	0945	1136	1.7E
	1503	1647	0.8F
	2044	2337	2.7E
31 W	0319	0729	1.6F
	1028	1222	1.9E
	1542	1821	1.0F
	2144		

September

Day	Slack h m	Maximum h m	knots
1 Th		0026	2.9E
	0405	0739	1.5F
	1103	1259	2.2E
	1624	1924	1.3F
	2239		
2 F		0110	2.9E
	0450	0802	1.4F
	1130	1335	2.4E
	1709	2012	1.5F
	2328		
3 Sa ●		0150	2.6E
	0531	0822	1.2F
	1142	1411	2.4E
	1754	2100	1.6F
4 Su	0011	0231	2.3E
	0604	0840	1.2F
	1137	1450	2.4E
	1836	2149	1.5F
5 M	0044	0315	2.0E
	0619	0912	1.2F
	1153	1533	2.4E
	1914	2230	1.4F
6 Tu	0106	0401	1.7E
	0622	0956	1.3F
	1226	1619	2.3E
	1946	2301	1.2F
7 W	0123	0446	1.7E
	0648	1041	1.4F
	1309	1704	2.3E
	2002	2326	1.0F
8 Th	0148	0530	1.7E
	0728	1125	1.4F
	1400	1748	2.1E
	2025	2355	0.9F
9 F	0223	0614	1.7E
	0820	1215	1.4F
	1501	1837	1.9E
	2109		
10 Sa		0039	0.8F
	0313	0708	1.6E
	0933	1330	1.3F
	1603	1934	1.7E
	2207		
11 Su ◐		0158	0.8F
	0414	0812	1.7E
	1144	1448	1.3F
	1705	2037	1.7E
	2328		
12 M		0300	1.1F
	0525	0916	1.9E
	1317	1543	1.3F
	1816	2135	1.9E
13 Tu	0048	0341	1.4F
	0711	1015	2.2E
	1419	1637	1.2F
	1943	2231	2.1E
14 W	0137	0421	1.6F
	0836	1112	2.5E
	1510	1755	1.1F
	2047	2325	2.5E
15 Th	0217	0509	1.8F
	0932	1205	2.8E
	1557	1905	1.2F
	2139		
16 F		0017	2.8E
	0257	0632	1.8F
	1021	1251	3.0E
	1641	1956	1.3F
	2228		
17 Sa ○		0103	3.0E
	0340	0722	1.8F
	1105	1333	3.0E
	1720	2042	1.3F
	2313		
18 Su		0148	3.1E
	0429	0752	1.7F
	1143	1414	2.8E
	1749	2130	1.3F
	2354		
19 M		0235	3.0E
	0524	0826	1.6F
	1214	1459	2.5E
	1808	2213	1.2F
20 Tu	0034	0326	2.8E
	0619	0913	1.4F
	1240	1547	2.3E
	1827	2246	1.2F
21 W	0115	0420	2.6E
	0713	1005	1.2F
	1303	1637	2.1E
	1859	2302	1.2F
22 Th	0206	0512	2.4E
	0813	1049	1.0F
	1329	1726	2.0E
	1943	2325	1.2F
23 F	0322	0604	2.1E
	0929	1132	0.8F
	1400	1814	1.9E
	2047		
24 Sa ◑		0011	1.1F
	0447	0659	1.7E
	1047	1222	0.5F
	1443	1910	1.8E
	2230		
25 Su		0224	1.1F
	0612	0804	1.3E
	1210	1345	0.4F
	1539	2013	1.7E
26 M	0002	0322	1.2F
	0730	0911	1.2E
	1325	1503	0.4F
	1642	2116	1.9E
27 Tu	0115	0407	1.3F
		0553	1.1F
		0650	1.2F
	0830	1014	1.4E
	1415	1552†	0.6F
28 W	0210	0455	1.3F
		0624	1.3F
		0733	1.3F
	0918	1113	1.6E
	1455	1645†	0.8F
29 Th	0256	0610	1.3F
	0958	1159	1.9E
	1532	1813	1.0F
	2142		
30 F		0004	2.4E
	0339	0654	1.3F
	1030	1235	2.2E
	1610	1913	1.3F
	2231		

Time meridian 135° W. 0000 is midnight. 1200 is noon. Times are not adjusted for Daylight Saving Time.
If three or more consecutive entries are marked (F) or (E) the middle ones are not true maximums but intermediate values to show the current pattern.
† See page 164 for the remaining currents on this day.

Snow Passage, Alaska, 2005

F–Flood, Dir. 120° True E–Ebb, Dir. 294° True

October

Day	Slack h m	Maximum h m	knots
1 Sa		0048	2.5E
	0419	0722	1.2F
	1054	1310	2.4E
	1648	1955	1.4F
	2314		
2 Su		0127	2.4E
	0457	0745	1.2F
	1100	1344	2.5E
	1725	2033	1.5F
	2347		
3 M ●		0205	2.3E
	0528	0809	1.2F
	1105	1420	2.5E
	1756	2111	1.5F
4 Tu	0008	0245	2.2E
	0547	0845	1.3F
	1133	1500	2.5E
	1817	2149	1.4F
5 W	0026	0330	2.1E
	0605	0933	1.4F
	1210	1546	2.4E
	1833	2222	1.3F
6 Th	0050	0417	2.1E
	0636	1023	1.5F
	1252	1634	2.4E
	1902	2251	1.3F
7 F	0122	0503	2.1E
	0717	1108	1.5F
	1338	1720	2.4E
	1941	2324	1.3F
8 Sa	0206	0550	2.0E
	0807	1153	1.4F
	1431	1808	2.3E
	2029		
9 Su		0006	1.2F
	0304	0642	1.9E
	0916	1255	1.1F
	1529	1902	2.1E
	2127		
10 M ◐		0110	1.1F
	0409	0743	1.7E
	1125	1426	1.0F
	1629	2004	1.9E
	2238		
11 Tu		0237	1.2F
	0519	0848	1.8E
	1301	1528	1.0F
	1736	2107	2.0E
12 W	0011	0326	1.4F
	0704	0948	1.9E
	1404	1622	1.1F
	1918	2205	2.1E
13 Th	0123	0403	1.5F
	0820	1045	2.2E
	1451	1735	1.1F
	2039	2303	2.3E
14 F	0213	0442	1.5F
	0912	1139	2.5E
	1530	1852	1.3F
	2137	2357	2.6E
15 Sa	0255	0535	1.4F
	0957	1227	2.7E
	1607	1943	1.4F
	2227		
16 Su		0046	2.8E
	0337	0700	1.4F
	1038	1309	2.9E
	1641	2027	1.4F
	2313		
17 M ○		0131	2.9E
	0422	0726	1.4F
	1114	1351	2.8E
	1713	2110	1.4F
	2353		
18 Tu		0216	2.9E
	0512	0759	1.4F
	1145	1434	2.7E
	1745	2154	1.3F
19 W	0031	0304	2.7E
	0603	0842	1.3F
	1212	1521	2.5E
	1819	2230	1.3F
20 Th	0111	0356	2.5E
	0652	0938	1.2F
	1240	1612	2.4E
	1857	2255	1.3F
21 F	0159	0449	2.3E
	0746	1031	1.1F
	1310	1703	2.3E
	1944	2317	1.3F
22 Sa	0309	0540	2.0E
	0858	1116	1.0F
	1345	1751	2.1E
	2048	2355	1.2F
23 Su	0424	0633	1.6E
	1019	1205	0.7F
	1429	1843	1.9E
	2209		
24 M ◑		0107	1.1F
	0537	0733	1.3E
	1142	1322	0.4F
	1527	1943	1.7E
	2328		
25 Tu		0246	1.1F
	0651	0841	1.2E
	1305	1455	0.5F
	1635	2047	1.6E
26 W	0043	0330	1.2F
	0753	0944	1.3E
		1056	1.2E
		1143	1.3E
	1403	1547†	0.6F
27 Th	0143	0406	1.2F
		0619	0.8F
		0715	0.9F
	0842	1039	1.5E
	1443	1637†	0.7F
28 F	0231	0444	1.1F
	0921	1127	1.8E
	1518	1748	0.9F
	2130	2338	1.9E
29 Sa	0311	0538	1.1F
	0954	1207	2.0E
	1550	1849	1.1F
	2211		
30 Su		0023	2.2E
	0348	0638	1.2F
	1017	1242	2.3E
	1619	1927	1.3F
	2244		
31 M		0102	2.4E
	0423	0715	1.3F
	1027	1317	2.4E
	1644	1957	1.4F
	2307		

November

Day	Slack h m	Maximum h m	knots
1 Tu ●		0139	2.5E
	0455	0749	1.4F
	1045	1353	2.5E
	1659	2023	1.4F
	2326		
2 W		0218	2.5E
	0526	0827	1.5F
	1117	1432	2.6E
	1716	2053	1.5F
	2353		
3 Th		0301	2.5E
	0556	0914	1.5F
	1153	1515	2.6E
	1749	2136	1.6F
4 F	0028	0349	2.4E
	0630	1006	1.5F
	1232	1604	2.6E
	1829	2221	1.7F
5 Sa	0110	0440	2.3E
	0709	1052	1.4F
	1312	1654	2.7E
	1913	2302	1.7F
6 Su	0201	0529	2.2E
	0756	1135	1.3F
	1359	1742	2.6E
	2003	2344	1.6F
7 M	0300	0619	2.0E
	0900	1226	1.0F
	1455	1835	2.4E
	2102		
8 Tu ◐		0036	1.4F
	0402	0715	1.7E
	1059	1402	0.8F
	1559	1936	2.1E
	2213		
9 W		0208	1.2F
	0504	0819	1.6E
	1232	1512	0.9F
	1713	2041	2.0E
	2354		
10 Th		0309	1.2F
	0629	0919	1.7E
	1337	1607	1.0F
	1909	2143	2.0E
11 F	0120	0346	1.1F
	0754	1015	1.9E
	1421	1711	1.2F
	2033	2241	2.2E
12 Sa	0214	0418	1.1F
	0846	1110	2.2E
	1458	1834	1.3F
	2131	2338	2.4E
13 Su	0255	0457	1.0F
	0928	1201	2.5E
	1533	1929	1.5F
	2221		
14 M		0028	2.6E
	0333	0555	1.1F
	1007	1246	2.8E
	1609	2011	1.5F
	2306		
15 Tu ○		0113	2.7E
	0412	0656	1.2F
	1043	1328	2.9E
	1650	2052	1.5F
	2348		
16 W		0156	2.7E
	0455	0736	1.3F
	1116	1411	2.9E
	1734	2134	1.4F
17 Th	0027	0242	2.6E
	0542	0820	1.4F
	1148	1457	2.8E
	1818	2213	1.4F
18 F	0106	0332	2.3E
	0631	0915	1.3F
	1221	1548	2.6E
	1900	2244	1.4F
19 Sa	0150	0425	2.1E
	0726	1016	1.2F
	1256	1639	2.5E
	1945	2308	1.3F
20 Su	0245	0517	2.0E
	0838	1107	1.1F
	1336	1729	2.2E
	2036	2337	1.3F
21 M	0347	0607	1.8E
	0958	1156	0.8F
	1425	1818	1.9E
	2139		
22 Tu		0018	1.1F
	0445	0701	1.5E
	1117	1310	0.6F
	1529	1913	1.5E
	2247		
23 W ◑		0135	1.0F
	0547	0802	1.4E
	1240	1443	0.5F
	1646	2016	1.2E
24 Th	0002	0247	1.0F
	0656	0905	1.4E
	1344	1535	0.6F
	1924	2118	1.2E
25 F	0110	0329	1.0F
		0611	0.3F
		0654	0.4F
	0755	0959	1.5E
		1116†	1.3E
26 Sa	0201	0406	1.1F
	0841	1047	1.6E
	1459	1704	0.8F
	2105	2306	1.6E
27 Su	0242	0448	1.1F
	0918	1132	1.8E
	1525	1801	0.9F
	2135	2353	2.0E
28 M	0318	0548	1.2F
	0945	1213	2.1E
	1545	1845	1.1F
	2156		
29 Tu		0035	2.4E
	0353	0650	1.4F
	1005	1250	2.3E
	1554	1915	1.3F
	2217		
30 W		0114	2.7E
	0431	0733	1.5F
	1028	1327	2.6E
	1603	1940	1.5F
	2251		

December

Day	Slack h m	Maximum h m	knots
1 Th ●		0154	2.9E
	0511	0812	1.5F
	1058	1406	2.7E
	1634	2014	1.8F
	2333		
2 F		0237	2.8E
	0553	0858	1.5F
	1132	1448	2.9E
	1716	2059	1.9F
3 Sa	0018	0325	2.7E
	0632	0950	1.4F
	1209	1537	2.9E
	1803	2156	2.0F
4 Su	0105	0417	2.5E
	0709	1038	1.3F
	1247	1629	2.9E
	1852	2244	2.0F
5 M	0155	0507	2.3E
	0747	1120	1.1F
	1332	1720	2.9E
	1943	2325	1.8F
6 Tu	0250	0556	2.0E
	0834	1204	0.9F
	1428	1812	2.6E
	2043		
7 W		0009	1.4F
	0344	0647	1.7E
	0953	1328	0.8F
	1540	1911	2.3E
	2201		
8 Th ◐		0114	1.0F
	0432	0745	1.5E
	1132	1453	0.9F
	1702	2016	2.0E
	2352		
9 F		0249	0.8F
	0518	0847	1.5E
	1249	1548	1.1F
	1901	2120	1.9E
10 Sa	0121	0328	0.7F
	0637	0944	1.7E
	1343	1644	1.2F
	2021	2220	2.0E
11 Su	0214	0356	0.7F
	0809	1039	2.0E
	1426	1811	1.3F
	2120	2317	2.1E
12 M	0250	0429	0.8F
	0854	1134	2.4E
	1506	1918	1.5F
	2210		
13 Tu		0009	2.3E
	0323	0516	0.9F
	0932	1223	2.8E
	1548	1958	1.6F
	2256		
14 W		0054	2.5E
	0356	0621	1.1F
	1010	1307	3.1E
	1635	2035	1.6F
	2339		
15 Th ○		0137	2.5E
	0435	0716	1.3F
	1050	1350	3.2E
	1724	2114	1.5F
16 F	0018	0220	2.4E
	0522	0803	1.4F
	1130	1435	3.1E
	1810	2152	1.5F
17 Sa	0055	0308	2.3E
	0616	0901	1.4F
	1209	1524	2.8E
	1852	2224	1.4F
18 Su	0130	0400	2.1E
	0714	1011	1.3F
	1250	1616	2.5E
	1932	2248	1.4F
19 M	0201	0451	2.1E
	0823	1104	1.2F
	1336	1706	2.1E
	2011	2312	1.3F
20 Tu	0227	0539	2.0E
	0938	1152	1.0F
	1433	1754	1.7E
	2053	2345	1.1F
21 W	0257	0627	1.8E
	1050	1254	0.7F
	1548	1843	1.3E
	2148		
22 Th		0030	1.0F
	0338	0720	1.6E
	1205	1419	0.6F
	1719	1941	0.9E
	2304		
23 F ◑		0146	0.9F
	0426	0820	1.4E
		1012	1.0E
		1057	1.0E
	1318	1512†	0.6F
24 Sa	0023	0253	1.0F
	0527	0916	1.4E
		1056	1.0E
		1154	1.1E
	1406	1549†	0.6F
25 Su	0124	0338	1.1F
	0753	1006	1.4E
	1435	1619	0.6F
	2017	2229	1.5E
26 M	0210	0421	1.3F
	0842	1055	1.6E
	1454	1647	0.8F
	2032	2319	2.0E
27 Tu	0250	0514	1.4F
	0916	1140	1.9E
	1505	1724	1.0F
	2056		
28 W		0007	2.5E
	0331	0624	1.5F
	0940	1222	2.2E
	1509	1819	1.4F
	2139		
29 Th		0050	2.9E
	0414	0717	1.5F
	1005	1302	2.6E
	1528	1906	1.8F
	2230		
30 F ●		0132	3.1E
	0503	0800	1.5F
	1035	1342	2.9E
	1604	1948	2.1F
	2322		
31 Sa		0215	3.0E
	0552	0845	1.3F
	1110	1424	3.1E
	1651	2035	2.3F

Time meridian 135° W. 0000 is midnight. 1200 is noon. Times are not adjusted for Daylight Saving Time.
If three or more consecutive entries are marked (F) or (E) the middle ones are not true maximums but intermediate values to show the current pattern.
† See page 164 for the remaining currents on this day.

Wrangell Narrows (off Petersburg), Alaska, 2005

F–Flood, Dir. 246° True E–Ebb, Dir. 062° True

January

Day	Slack h m	Maximum h m	knots
1 Sa		0154	3.0F
	0502	0654	1.6E
	1033	1410	2.2F
	1659	1903	2.0E
	2245		
2 Su		0239	2.7F
	0548	0737	1.5E
	1123	1504	2.0F
	1754	1950	1.7E
	2327		
3 M		0333	2.5F
	0637	0826	1.5E
	1222	1608	2.0F
◑	1855	2041	1.4E
4 Tu	0019	0433	2.4F
	0729	0923	1.5E
	1338	1716	2.1F
	2001	2141	1.2E
5 W	0126	0534	2.4F
	0825	1041	1.5E
	1502	1822	2.4F
	2111		
6 Th		0017	1.0E
	0254	0635	2.5F
	0922	1254	1.8E
	1609	1925	2.8F
	2220		
7 F		0135	1.2E
	0415	0735	2.7F
	1018	1357	2.3E
	1706	2022	3.3F
	2324		
8 Sa		0235	1.6E
	0518	0831	3.0F
	1113	1449	2.7E
	1757	2115	3.8F
9 Su	0021	0326	2.0E
	0613	0923	3.3F
	1205	1537	3.1E
	1847	2204	4.2F
10 M	0111	0413	2.3E
	0704	1012	3.6F
	1256	1623	3.4E
●	1934	2252	4.5F
11 Tu	0157	0459	2.6E
	0753	1101	3.7F
	1345	1709	3.5E
	2020	2338	4.6F
12 W	0242	0546	2.7E
	0840	1148	3.8F
	1432	1755	3.5E
	2105		
13 Th		0023	4.6F
	0325	0632	2.7E
	0927	1235	3.7F
	1520	1842	3.2E
	2149		
14 F		0107	4.3F
	0410	0719	2.6E
	1016	1322	3.5F
	1610	1929	2.8E
	2233		
15 Sa		0151	4.0F
	0455	0807	2.4E
	1107	1412	3.2F
	1702	2018	2.3E
	2317		
16 Su		0239	3.5F
	0542	0858	2.1E
	1202	1509	2.8F
◐	1758	2112	1.8E
17 M	0004	0334	3.0F
	0631	0954	1.8E
	1303	1615	2.5F
	1859	2216	1.3E
18 Tu	0058	0435	2.6F
	0723	1055	1.6E
	1413	1722	2.3F
	2009	2324	0.9E
19 W	0210	0538	2.3F
	0819	1200	1.5E
	1522	1827	2.3F
	2140		
20 Th		0039	0.8E
	0328	0640	2.2F
	0919	1309	1.6E
	1624	1927	2.5F
	2326		
21 F		0200	0.9E
	0434	0738	2.3F
	1018	1411	1.8E
	1716	2021	2.7F
22 Sa	0022	0256	1.1E
	0529	0831	2.4F
	1111	1458	2.0E
	1803	2110	3.0F
23 Su	0057	0333	1.4E
	0616	0919	2.6F
	1158	1536	2.2E
	1844	2154	3.3F
24 M	0121	0407	1.6E
	0658	1001	2.8F
	1240	1612	2.4E
	1922	2234	3.5F
25 Tu	0144	0441	1.8E
	0735	1040	2.9F
	1318	1646	2.5E
○	1955	2310	3.6F
26 W	0211	0515	1.9E
	0807	1115	3.0F
	1355	1717	2.6E
	2024	2344	3.7F
27 Th	0241	0547	2.0E
	0837	1150	3.1F
	1432	1741	2.5E
	2051		
28 F		0016	3.6F
	0314	0613	2.0E
	0905	1224	3.0F
	1509	1741	2.5E
	2116		
29 Sa		0048	3.5F
	0349	0613	2.0E
	0934	1300	2.9F
	1549	1804	2.5E
	2143		
30 Su		0123	3.3F
	0426	0633	2.1E
	1008	1340	2.8F
	1633	1841	2.3E
	2214		
31 M		0201	3.1F
	0506	0709	2.1E
	1047	1425	2.5F
	1722	1923	2.1E
	2250		

February

Day	Slack h m	Maximum h m	knots
1 Tu		0246	2.7F
	0551	0752	2.1E
	1134	1521	2.3F
◑	1819	2010	1.7E
	2333		
2 W		0343	2.3F
	0642	0841	1.9E
	1235	1630	2.2F
	1925	2103	1.3E
3 Th	0030	0450	2.1F
	0740	0939	1.7E
	1403	1747	2.3F
	2041	2210	0.8E
4 F	0200	0602	2.1F
	0846	1215	1.6E
	1539	1902	2.6F
	2201		
5 Sa		0117	1.0E
	0358	0715	2.3F
	0953	1339	2.0E
	1650	2007	3.1F
	2317		
6 Su		0226	1.4E
	0511	0819	2.8F
	1057	1439	2.6E
	1747	2103	3.7F
7 M	0016	0318	2.0E
	0608	0914	3.3F
	1155	1529	3.1E
	1838	2152	4.2F
8 Tu	0103	0404	2.5E
	0658	1003	3.7F
	1247	1614	3.5E
●	1924	2238	4.6F
9 W	0144	0447	2.8E
	0744	1050	4.0F
	1335	1658	3.6E
	2007	2321	4.7F
10 Th	0223	0529	3.0E
	0827	1134	4.2F
	1420	1741	3.6E
	2048		
11 F		0002	4.7F
	0301	0610	3.1E
	0909	1217	4.1F
	1504	1822	3.3E
	2126		
12 Sa		0041	4.5F
	0339	0651	3.0E
	0951	1259	3.9F
	1549	1903	2.9E
	2203		
13 Su		0119	4.1F
	0417	0730	2.7E
	1032	1341	3.5F
	1635	1941	2.4E
	2238		
14 M		0159	3.5F
	0458	0809	2.3E
	1116	1428	3.0F
	1724	2017	1.7E
	2313		
15 Tu		0244	2.9F
	0542	0850	1.9E
	1205	1523	2.5F
◐	1818	2012	1.1E
	2351		
16 W		0339	2.3F
	0631	0950	1.5E
	1308	1635	2.1F
	1923	2044	0.6E
17 Th	0043	0449	1.9F
	0728	1108	1.1E
	1433	1750	2.0F
	2049		
18 F		0000	0.4E
	0245	0606	1.7F
	0834	1231	1.1E
	1552	1859	2.1F
	2333		
19 Sa		0236	0.6E
	0413	0715	1.8F
	0948	1355	1.3E
	1653	1958	2.4F
20 Su	0021	0313	1.0E
	0514	0813	2.1F
	1055	1448	1.7E
	1743	2049	2.8F
21 M	0049	0330	1.3E
	0603	0902	2.5F
	1147	1523	2.1E
	1825	2133	3.2F
22 Tu	0107	0354	1.7E
	0643	0944	2.8F
	1227	1556	2.4E
	1902	2212	3.5F
23 W	0123	0422	2.0E
	0718	1022	3.1F
	1304	1628	2.6E
○	1934	2247	3.7F
24 Th	0146	0452	2.2E
	0749	1056	3.3F
	1339	1658	2.7E
	2002	2318	3.8F
25 F	0213	0521	2.4E
	0817	1128	3.4F
	1414	1724	2.7E
	2027	2348	3.8F
26 Sa	0243	0544	2.5E
	0843	1201	3.5F
	1450	1733	2.7E
	2051		
27 Su		0019	3.7F
	0316	0547	2.5E
	0910	1235	3.4F
	1528	1746	2.6E
	2117		
28 M		0052	3.5F
	0350	0606	2.6E
	0941	1313	3.2F
	1610	1819	2.5E
	2147		

March

Day	Slack h m	Maximum h m	knots
1 Tu		0129	3.2F
	0428	0641	2.6E
	1017	1356	3.0F
	1658	1859	2.2E
	2221		
2 W		0212	2.8F
	0512	0722	2.4E
	1102	1448	2.6F
	1754	1944	1.7E
	2303		
3 Th		0306	2.3F
	0604	0810	2.1E
	1159	1557	2.3F
◑	1901	2036	1.2E
	2359		
4 F		0418	1.9F
	0707	0906	1.7E
	1326	1723	2.3F
	2021	2143	0.7E
5 Sa	0143	0544	1.9F
	0820	1158	1.4E
	1520	1845	2.6F
	2152		
6 Su		0106	0.9E
	0357	0704	2.2F
	0937	1326	1.9E
	1636	1951	3.1F
	2313		
7 M		0217	1.5E
	0505	0809	2.8F
	1048	1429	2.5E
	1733	2046	3.7F
8 Tu	0005	0307	2.2E
	0559	0902	3.4F
	1148	1517	3.0E
	1822	2133	4.2F
9 W	0045	0348	2.7E
	0645	0949	3.9F
	1238	1600	3.4E
	1906	2216	4.5F
10 Th	0121	0427	3.1E
	0728	1033	4.2F
	1323	1641	3.5E
●	1946	2257	4.7F
11 F	0156	0505	3.3E
	0808	1115	4.4F
	1405	1720	3.5E
	2023	2335	4.6F
12 Sa	0230	0542	3.3E
	0845	1154	4.3F
	1445	1758	3.2E
	2058		
13 Su		0011	4.3F
	0304	0618	3.2E
	0921	1233	4.1F
	1526	1834	2.7E
	2130		
14 M		0046	3.9F
	0340	0650	2.8E
	0957	1311	3.7F
	1607	1904	2.2E
	2159		
15 Tu		0123	3.3F
	0417	0709	2.4E
	1032	1352	3.2F
	1652	1904	1.7E
	2228		
16 W		0203	2.7F
	0459	0711	2.0E
	1111	1440	2.6F
	1743	1923	1.2E
	2259		
17 Th		0252	2.1F
	0546	0742	1.6E
	1159	1545	2.1F
◐	1844	2001	0.7E
	2339		
18 F		0402	1.6F
	0644	0825	1.1E
	1326	1712	1.8F
		2050	*
		2157†	*
19 Sa		0533	1.4F
	0754	0921	0.7E
		1014	0.7E
		1150	0.8E
	1512	1827†	2.0F
20 Su		0225	0.5E
	0350	0648	1.6F
	0916	1322	1.1E
	1621	1929	2.3F
	2350		
21 M		0257	1.0E
	0452	0748	2.0F
	1032	1424	1.5E
	1713	2020	2.7F
22 Tu	0012	0306	1.4E
	0539	0837	2.5F
	1126	1459	1.9E
	1756	2104	3.1F
23 W	0026	0326	1.8E
	0618	0919	2.9F
	1207	1531	2.3E
	1832	2142	3.5F
24 Th	0045	0354	2.2E
	0652	0956	3.3F
	1243	1602	2.5E
	1904	2215	3.7F
25 F	0110	0422	2.5E
	0722	1030	3.6F
	1318	1632	2.7E
○	1932	2246	3.8F
26 Sa	0139	0450	2.7E
	0750	1103	3.7F
	1354	1659	2.7E
	1958	2317	3.8F
27 Su	0210	0511	2.8E
	0816	1136	3.8F
	1430	1715	2.7E
	2024	2349	3.7F
28 M	0243	0518	2.9E
	0845	1212	3.8F
	1509	1727	2.6E
	2052		
29 Tu		0025	3.5F
	0318	0540	3.0E
	0918	1251	3.6F
	1552	1759	2.4E
	2125		
30 W		0104	3.2F
	0358	0616	2.8E
	0956	1335	3.3F
	1641	1839	2.0E
	2202		
31 Th		0148	2.7F
	0443	0658	2.6E
	1043	1428	2.8F
	1739	1925	1.6E
	2250		

Time meridian 135° W. 0000 is midnight. 1200 is noon. Times are not adjusted for Daylight Saving Time.

* Current weak and variable.

† See page 164 for the remaining currents on this day.

Wrangell Narrows (off Petersburg), Alaska, 2005

F–Flood, Dir. 246° True E–Ebb, Dir. 062° True

April

Day	Slack h m	Maximum h m	knots
1 F ◑		0245	2.2F
	0539	0747	2.1E
	1144	1539	2.5F
	1847	2020	1.0E
	2358		
2 Sa		0402	1.9F
	0647	0846	1.6E
	1316	1709	2.4F
	2009	2324	0.8E
3 Su	0212	0535	1.9F
	0805	1148	1.5E
	1503	1827	2.8F
	2140		
4 M		0052	1.2E
	0350	0651	2.3F
	0926	1309	1.9E
	1616	1929	3.2F
	2251		
5 Tu		0200	1.8E
	0452	0752	2.9F
	1039	1412	2.4E
	1712	2022	3.7F
	2338		
6 W		0247	2.4E
	0542	0843	3.5F
	1137	1459	2.8E
	1759	2108	4.1F
7 Th	0015	0326	2.9E
	0626	0929	4.0F
	1226	1541	3.1E
	1841	2150	4.3F
8 F ●	0050	0402	3.3E
	0706	1012	4.3F
	1308	1619	3.2E
	1920	2229	4.4F
9 Sa	0123	0438	3.4E
	0744	1052	4.4F
	1347	1656	3.1E
	1956	2305	4.2F
10 Su	0156	0512	3.3E
	0819	1130	4.3F
	1425	1732	2.8E
	2028	2341	3.9F
11 M	0230	0545	3.1E
	0852	1207	4.0F
	1503	1806	2.4E
	2058		
12 Tu		0016	3.5F
	0304	0611	2.8E
	0924	1244	3.6F
	1543	1831	1.9E
	2126		
13 W		0052	3.1F
	0341	0610	2.4E
	0956	1323	3.2F
	1626	1825	1.5E
	2154		
14 Th		0131	2.5F
	0422	0631	2.1E
	1031	1408	2.6F
	1715	1853	1.2E
	2226		
15 F		0219	2.0F
	0510	0708	1.7E
	1114	1507	2.2F
	1814	1934	0.8E
	2312		
16 Sa ◐		0324	1.5F
	0608	0753	1.3E
	1220	1631	1.9F
	1925	2025	0.4E
		2136†	0.3E
17 Su	0102	0457	1.4F
	0718	0848	0.8E
		1004	0.7E
		1104	0.8E
	1414	1749†	2.0F
18 M		0015	0.5E
	0315	0614	1.6F
	0836	1227	0.9E
	1535	1850	2.3F
	2224		
19 Tu		0137	0.9E
	0418	0714	2.0F
	0951	1334	1.3E
	1630	1941	2.7F
	2257		
20 W		0215	1.4E
	0505	0804	2.5F
	1049	1420	1.7E
	1714	2025	3.0F
	2326		
21 Th		0247	1.9E
	0544	0847	2.9F
	1135	1457	2.0E
	1752	2104	3.3F
	2357		
22 F		0318	2.3E
	0618	0925	3.3F
	1215	1531	2.3E
	1826	2138	3.6F
23 Sa	0029	0348	2.7E
	0650	1001	3.7F
	1254	1604	2.5E
	1857	2212	3.7F
24 Su ○	0102	0417	2.9E
	0720	1036	3.9F
	1332	1634	2.5E
	1928	2246	3.7F
25 M	0137	0440	3.0E
	0751	1113	4.0F
	1412	1659	2.5E
	1959	2323	3.6F
26 Tu	0213	0455	3.1E
	0825	1153	4.0F
	1454	1718	2.4E
	2033		
27 W		0003	3.4F
	0252	0521	3.1E
	0903	1235	3.8F
	1540	1749	2.1E
	2112		
28 Th		0046	3.1F
	0336	0559	2.9E
	0946	1323	3.5F
	1631	1831	1.8E
	2158		
29 F		0135	2.7F
	0425	0644	2.5E
	1038	1418	3.1F
	1730	1923	1.4E
	2257		
30 Sa ◑		0235	2.3F
	0525	0737	2.1E
	1144	1529	2.8F
	1838	2145	1.1E

May

Day	Slack h m	Maximum h m	knots
1 Su	0025	0355	2.0F
	0634	0959	1.6E
	1310	1652	2.8F
	1954	2311	1.2E
2 M	0212	0521	2.2F
	0751	1129	1.6E
	1438	1801	3.0F
	2110		
3 Tu		0025	1.5E
	0330	0630	2.5F
	0911	1243	1.8E
	1547	1900	3.3F
	2212		
4 W		0129	2.0E
	0429	0728	3.0F
	1024	1346	2.2E
	1643	1952	3.6F
	2259		
5 Th		0219	2.5E
	0519	0820	3.5F
	1122	1436	2.4E
	1730	2039	3.8F
	2338		
6 F		0259	2.9E
	0602	0906	3.8F
	1210	1518	2.6E
	1813	2121	3.9F
7 Sa ●	0015	0336	3.1E
	0643	0949	4.0F
	1252	1557	2.6E
	1853	2200	3.8F
8 Su	0049	0411	3.2E
	0720	1029	4.1F
	1330	1634	2.5E
	1929	2237	3.7F
9 M	0124	0445	3.1E
	0755	1108	4.0F
	1407	1710	2.3E
	2002	2313	3.4F
10 Tu	0159	0517	2.9E
	0828	1145	3.8F
	1444	1745	2.0E
	2032	2350	3.1F
11 W	0235	0540	2.6E
	0859	1223	3.5F
	1523	1815	1.7E
	2101		
12 Th		0028	2.8F
	0313	0537	2.3E
	0931	1302	3.2F
	1605	1804	1.4E
	2133		
13 F		0109	2.4F
	0354	0604	2.1E
	1006	1346	2.8F
	1653	1833	1.1E
	2211		
14 Sa		0156	2.0F
	0442	0644	1.8E
	1048	1439	2.4F
	1748	1916	0.9E
	2304		
15 Su ◐		0255	1.6F
	0539	0730	1.4E
	1141	1547	2.2F
	1848	2008	0.6E
		2110†	0.6E
16 M	0030	0413	1.5F
	0643	0824	1.1E
	1255	1700	2.2F
	1952	2317	0.7E
17 Tu	0221	0529	1.7F
	0753	0957	0.8E
	1423	1800	2.3F
	2053		
18 W		0021	1.0E
	0330	0630	2.0F
	0902	1234	1.1E
	1531	1852	2.6F
	2144		
19 Th		0115	1.4E
	0421	0723	2.4F
	1004	1331	1.4E
	1622	1939	2.8F
	2228		
20 F		0200	1.9E
	0503	0809	2.8F
	1057	1418	1.7E
	1706	2021	3.1F
	2308		
21 Sa		0239	2.3E
	0541	0852	3.3F
	1145	1459	2.0E
	1745	2101	3.3F
	2348		
22 Su		0314	2.7E
	0618	0932	3.6F
	1229	1537	2.2E
	1824	2140	3.4F
23 M ○	0028	0347	2.9E
	0654	1012	3.9F
	1313	1614	2.3E
	1902	2220	3.5F
24 Tu	0108	0419	3.1E
	0732	1055	4.0F
	1357	1651	2.3E
	1942	2302	3.4F
25 W	0150	0449	3.2E
	0813	1139	4.1F
	1443	1732	2.2E
	2025	2348	3.3F
26 Th	0234	0522	3.1E
	0857	1226	4.0F
	1531	1819	2.0F
	2112		
27 F		0036	3.1F
	0321	0604	2.9E
	0945	1315	3.8F
	1623	1917	1.8E
	2207		
28 Sa		0128	2.8F
	0414	0659	2.5E
	1040	1410	3.5F
	1721	2024	1.6E
	2312		
29 Su		0227	2.6F
	0514	0825	2.1E
	1142	1514	3.2F
	1822	2135	1.5E
30 M ◑	0029	0340	2.4F
	0621	0949	1.9E
	1261	1624	3.1F
	1925	2244	1.6E
31 Tu	0149	0456	2.5F
	0732	1102	1.7E
	1403	1728	3.1F
	2028	2349	1.8E

June

Day	Slack h m	Maximum h m	knots
1 W	0301	0602	2.7F
	0848	1210	1.7E
	1510	1826	3.2F
	2126		
2 Th		0051	2.1E
	0401	0700	3.0F
	1001	1314	1.8E
	1609	1919	3.3F
	2215		
3 F		0145	2.4E
	0452	0754	3.3F
	1104	1410	1.9E
	1700	2008	3.3F
	2300		
4 Sa		0231	2.7E
	0538	0842	3.5F
	1156	1457	2.0E
	1746	2053	3.3F
	2340		
5 Su		0311	2.8E
	0620	0927	3.7F
	1240	1537	2.0E
	1828	2135	3.3F
6 M ●	0019	0349	2.8E
	0659	1010	3.7F
	1318	1616	2.0E
	1907	2214	3.2F
7 Tu	0057	0424	2.8E
	0736	1050	3.7F
	1353	1653	1.8E
	1943	2253	3.0F
8 W	0134	0458	2.6E
	0811	1129	3.6F
	1429	1731	1.7E
	2017	2332	2.9F
9 Th	0212	0529	2.4E
	0844	1207	3.4F
	1507	1808	1.5E
	2050		
10 F		0011	2.7F
	0252	0535	2.2E
	0916	1246	3.2F
	1547	1843	1.4E
	2124		
11 Sa		0051	2.4F
	0334	0549	2.1E
	0950	1326	3.0F
	1631	1832	1.2E
	2204		
12 Su		0135	2.2F
	0420	0627	1.9E
	1027	1411	2.8F
	1719	1905	1.2E
	2252		
13 M		0225	2.0F
	0511	0711	1.7E
	1108	1502	2.5F
	1809	1951	1.1E
	2350		
14 Tu ◐		0325	1.8F
	0608	0759	1.5E
	1156	1601	2.4F
	1901	2046	1.1E
15 W	0103	0434	1.8F
	0710	0854	1.2E
	1253	1701	2.4F
	1953	2313	1.2E
16 Th	0222	0538	2.0F
	0814	0959	1.0E
	1403	1756	2.4F
	2045		
17 F		0012	1.4E
	0326	0636	2.3F
	0919	1234	1.1E
	1516	1848	2.6F
	2136		
18 Sa		0108	1.7E
	0419	0730	2.7F
	1020	1337	1.3E
	1616	1939	2.7F
	2224		
19 Su		0159	2.1E
	0506	0820	3.1F
	1116	1430	1.6E
	1709	2027	2.9F
	2312		
20 M		0244	2.5E
	0551	0908	3.5F
	1209	1516	1.8E
	1758	2114	3.1F
	2359		
21 Tu ○		0327	2.9E
	0635	0955	3.8F
	1258	1601	2.1E
	1847	2201	3.3F
22 W	0046	0408	3.1E
	0720	1041	4.1F
	1346	1646	2.2E
	1934	2249	3.4F
23 Th	0133	0452	3.2E
	0806	1129	4.2F
	1433	1734	2.3E
	2023	2337	3.4F
24 F	0221	0538	3.2E
	0853	1217	4.3F
	1520	1824	2.3E
	2114		
25 Sa		0027	3.4F
	0311	0629	3.1E
	0941	1304	4.1F
	1610	1916	2.2E
	2208		
26 Su		0118	3.2F
	0403	0722	2.8E
	1032	1354	3.9F
	1701	2010	2.1E
	2306		
27 M		0212	3.0F
	0500	0819	2.4E
	1124	1448	3.6F
	1754	2108	2.0E
28 Tu ◑	0008	0315	2.8F
	0600	0922	2.0E
	1220	1548	3.3F
	1847	2208	2.0E
29 W	0115	0423	2.7F
	0706	1027	1.7E
	1321	1649	3.1F
	1942	2309	1.9E
30 Th	0223	0529	2.7F
	0816	1133	1.4E
	1428	1748	2.9F
	2038		

Time meridian 135° W. 0000 is midnight. 1200 is noon. Times are not adjusted for Daylight Saving Time.

† See page 164 for the remaining currents on this day.

Wrangell Narrows (off Petersburg), Alaska, 2005

F–Flood, Dir. 246° True   E–Ebb, Dir. 062° True

July

Day	Slack h m	Maximum h m	knots
1 F		0010	2.0E
	0328	0630	2.8F
	0933	1240	1.3E
	1533	1845	2.8F
	2132		
2 Sa		0111	2.1E
	0424	0727	2.9F
	1049	1346	1.3E
	1631	1939	2.8F
	2224		
3 Su		0206	2.2E
	0515	0820	3.1F
	1152	1441	1.4E
	1723	2029	2.8F
	2311		
4 M		0252	2.4E
	0601	0908	3.3F
	1239	1525	1.5E
	1811	2115	2.8F
	2356		
5 Tu		0333	2.5E
	0643	0953	3.4F
	1314	1604	1.6E
	1854	2159	2.8F
6 W ●	0037	0410	2.5E
	0722	1035	3.5F
	1345	1641	1.7E
	1933	2239	2.8F
7 Th	0117	0446	2.5E
	0758	1115	3.5F
	1417	1719	1.7E
	2009	2318	2.8F
8 F	0156	0521	2.4E
	0831	1152	3.5F
	1450	1755	1.7E
	2042	2356	2.8F
9 Sa	0235	0550	2.3E
	0901	1227	3.4F
	1526	1829	1.6E
	2114		
10 Su		0033	2.7F
	0315	0550	2.2E
	0930	1302	3.3F
	1604	1854	1.6E
	2148		
11 M		0112	2.6F
	0357	0611	2.1E
	1000	1339	3.1F
	1644	1848	1.6E
	2225		
12 Tu		0154	2.4F
	0443	0648	2.0E
	1033	1420	2.9F
	1727	1922	1.6E
	2308		
13 W		0242	2.2F
	0533	0731	1.8E
	1109	1508	2.6F
	1813	2005	1.6E
	2357		
14 Th ◐		0339	2.1F
	0629	0818	1.5E
	1153	1603	2.4F
	1902	2055	1.6E
15 F	0059	0444	2.0F
	0731	0912	1.2E
	1247	1703	2.3F
	1954	2152	1.5E
16 Sa	0217	0550	2.2F
	0839	1014	1.0E
	1401	1803	2.3F
	2051		
17 Su		0007	1.6E
	0334	0655	2.5F
	0948	1300	1.0E
	1533	1905	2.4F
	2148		
18 M		0125	1.9E
	0437	0756	2.9F
	1055	1409	1.3E
	1647	2004	2.7F
	2245		
19 Tu		0224	2.4E
	0532	0850	3.4F
	1155	1503	1.7E
	1746	2058	3.0F
	2339		
20 W		0314	2.8E
	0623	0941	3.9F
	1248	1551	2.1E
	1840	2149	3.3F
21 Th ○	0032	0401	3.2E
	0711	1029	4.2F
	1335	1637	2.4E
	1929	2238	3.6F
22 F	0122	0446	3.4E
	0758	1116	4.5F
	1419	1723	2.6E
	2017	2326	3.8F
23 Sa	0210	0533	3.5E
	0843	1201	4.6F
	1502	1809	2.7E
	2104		
24 Su		0013	3.8F
	0258	0619	3.3E
	0927	1245	4.5F
	1546	1855	2.7E
	2152		
25 M		0100	3.7F
	0348	0706	3.0E
	1011	1329	4.2F
	1630	1942	2.6E
	2242		
26 Tu		0149	3.5F
	0439	0754	2.6E
	1055	1414	3.8F
	1716	2031	2.4E
	2335		
27 W ◑		0242	3.1F
	0534	0847	2.1E
	1141	1506	3.3F
	1805	2125	2.1E
28 Th	0033	0344	2.8F
	0634	0947	1.5E
	1233	1605	2.8F
	1856	2225	1.8E
29 F	0139	0452	2.5F
	0741	1054	1.1E
	1338	1709	2.5F
	1952	2329	1.7E
30 Sa	0251	0559	2.5F
	0902	1207	0.9E
	1458	1813	2.3F
	2052		
31 Su		0038	1.6E
	0357	0701	2.5F
	1052	1329	0.9E
	1609	1914	2.3F
	2154		

August

Day	Slack h m	Maximum h m	knots
1 M		0147	1.8E
	0454	0759	2.7F
	1202	1440	1.1E
	1709	2011	2.4F
	2252		
2 Tu		0240	2.0E
	0544	0850	3.0F
	1244	1521	1.4E
	1759	2100	2.6F
	2342		
3 W		0322	2.2E
	0628	0936	3.2F
	1312	1554	1.6E
	1843	2145	2.8F
4 Th ●	0025	0358	2.4E
	0707	1017	3.4F
	1334	1627	1.8E
	1921	2225	2.9F
5 F	0104	0432	2.5E
	0741	1055	3.6F
	1358	1701	1.9E
	1955	2302	3.0F
6 Sa	0141	0505	2.5E
	0812	1129	3.6F
	1426	1733	2.0E
	2026	2336	3.1F
7 Su	0218	0534	2.5E
	0839	1201	3.6F
	1457	1802	2.0E
	2054		
8 M		0010	3.1F
	0255	0546	2.4E
	0904	1233	3.5F
	1531	1815	2.0E
	2121		
9 Tu		0045	3.0F
	0333	0550	2.4E
	0929	1306	3.3F
	1606	1816	2.1E
	2152		
10 W		0122	2.8F
	0415	0622	2.3E
	0957	1342	3.1F
	1644	1848	2.1E
	2227		
11 Th		0204	2.6F
	0501	0702	2.1E
	1029	1423	2.7F
	1727	1929	2.1E
	2309		
12 F ◐		0255	2.3F
	0554	0746	1.7E
	1107	1514	2.3F
	1815	2015	1.9E
13 Sa	0002	0358	2.1F
	0657	0836	1.3E
	1157	1618	2.0F
	1911	2108	1.7E
14 Su	0115	0514	2.1F
	0808	0936	0.9E
	1310	1730	2.0F
	2014	2214	1.5E
15 M	0258	0630	2.4F
	0926	1240	0.8E
	1514	1844	2.1F
	2122		
16 Tu		0105	1.7E
	0418	0738	2.9F
	1043	1357	1.2E
	1641	1951	2.5F
	2227		
17 W		0211	2.3E
	0519	0836	3.4F
	1146	1453	1.8E
	1741	2048	3.0F
	2327		
18 Th		0304	2.8E
	0611	0926	4.0F
	1235	1539	2.3E
	1832	2138	3.5F
19 F ○	0021	0350	3.3E
	0658	1012	4.4F
	1318	1621	2.8E
	1918	2225	3.9F
20 Sa	0111	0433	3.6E
	0742	1056	4.7F
	1357	1703	3.1E
	2002	2310	4.2F
21 Su	0157	0516	3.6E
	0823	1138	4.7F
	1436	1745	3.2E
	2045	2354	4.3F
22 M	0243	0559	3.4E
	0903	1218	4.6F
	1515	1826	3.1E
	2127		
23 Tu		0037	4.1F
	0328	0641	3.1E
	0942	1258	4.2F
	1554	1907	2.9E
	2210		
24 W		0120	3.8F
	0415	0723	2.5E
	1020	1338	3.7F
	1636	1948	2.6E
	2255		
25 Th		0207	3.3F
	0505	0806	1.9E
	1059	1423	3.1F
	1721	2033	2.1E
	2345		
26 F ◑		0302	2.8F
	0600	0859	1.3E
	1141	1517	2.5F
	1810	2132	1.6E
27 Sa	0047	0412	2.3F
	0704	1015	0.8E
	1240	1628	2.0F
	1907	2247	1.3E
28 Su	0208	0528	2.2F
	0830	1138	0.5E
	1427	1745	1.8F
	2014		
29 M		0007	1.2E
	0328	0637	2.2F
	1102	1407	0.7E
	1554	1854	1.9F
	2129		
30 Tu		0132	1.4E
	0432	0737	2.5F
	1156	1456	1.0E
	1656	1953	2.2F
	2240		
31 W		0232	1.7E
	0524	0828	2.8F
	1229	1517	1.4E
	1745	2043	2.5F
	2332		

September

Day	Slack h m	Maximum h m	knots
1 Th		0308	2.1E
	0607	0913	3.2F
	1250	1538	1.8E
	1826	2126	2.9F
2 F	0013	0340	2.4E
	0644	0953	3.5F
	1307	1606	2.0E
	1901	2204	3.1F
3 Sa ●	0048	0412	2.6E
	0717	1028	3.7F
	1328	1635	2.3E
	1933	2239	3.3F
4 Su	0123	0442	2.6E
	0745	1100	3.8F
	1354	1703	2.4E
	2001	2311	3.4F
5 M	0157	0509	2.6E
	0810	1129	3.7F
	1424	1727	2.4E
	2026	2343	3.5F
6 Tu	0233	0521	2.5E
	0833	1200	3.6F
	1455	1729	2.5E
	2051		
7 W		0017	3.4F
	0309	0525	2.5E
	0856	1232	3.4F
	1528	1742	2.6E
	2119		
8 Th		0053	3.2F
	0349	0555	2.4E
	0924	1307	3.1F
	1604	1816	2.6E
	2152		
9 F		0133	2.9F
	0435	0634	2.1E
	0956	1348	2.7F
	1646	1856	2.4E
	2233		
10 Sa		0221	2.6F
	0528	0718	1.7E
	1034	1438	2.2F
	1736	1942	2.1E
	2325		
11 Su ◐		0325	2.2F
	0632	0808	1.3E
	1125	1546	1.8F
	1837	2036	1.7E
12 M	0040	0449	2.2F
	0748	0909	0.8E
	1251	1711	1.8F
	1949	2145	1.4E
13 Tu	0238	0614	2.4F
	0914	1231	0.8E
	1522	1834	2.1F
	2106		
14 W		0051	1.7E
	0403	0722	3.0F
	1034	1346	1.4E
	1637	1940	2.6F
	2217		
15 Th		0159	2.3E
	0503	0818	3.5F
	1130	1439	2.1E
	1731	2035	3.2F
	2318		
16 F		0250	2.9E
	0553	0906	4.1F
	1214	1521	2.7E
	1818	2123	3.8F
17 Sa ○	0011	0334	3.3E
	0638	0949	4.5F
	1252	1601	3.1E
	1901	2207	4.3F
18 Su	0058	0415	3.5E
	0720	1030	4.7F
	1329	1639	3.4E
	1942	2249	4.5F
19 M	0142	0455	3.5E
	0758	1110	4.6F
	1404	1717	3.5E
	2021	2331	4.5F
20 Tu	0224	0535	3.3E
	0835	1148	4.4F
	1440	1754	3.3E
	2059		
21 W		0011	4.3F
	0306	0613	2.9E
	0910	1225	4.0F
	1517	1829	3.0E
	2137		
22 Th		0052	3.9F
	0349	0649	2.3E
	0943	1304	3.4F
	1556	1859	2.6E
	2216		
23 F		0134	3.4F
	0436	0718	1.7E
	1017	1345	2.8F
	1639	1903	2.1E
	2258		
24 Sa ◑		0223	2.8F
	0528	0715	1.2E
	1053	1436	2.2F
	1728	1928	1.6E
	2352		
25 Su		0329	2.2F
	0630	0749	0.6E
	1145	1546	1.7F
	1827	2010	1.1E
26 M	0116	0455	2.0F
	0753	1110	0.4E
	1359	1718	1.5F
	1938	2334	0.9E
27 Tu	0252	0608	2.1F
	1041	1358	0.6E
	1534	1830	1.7F
	2102		
28 W		0103	1.1E
	0402	0708	2.4F
	1125	1437	1.1E
	1635	1929	2.1F
	2221		
29 Th		0208	1.5E
	0454	0759	2.8F
	1150	1452	1.5E
	1722	2018	2.5F
	2313		
30 F		0243	1.9E
	0536	0843	3.2F
	1207	1510	1.9E
	1800	2100	3.0F
	2352		

Time meridian 135° W. 0000 is midnight. 1200 is noon. Times are not adjusted for Daylight Saving Time.

Wrangell Narrows (off Petersburg), Alaska, 2005

F–Flood, Dir. 246° True E–Ebb, Dir. 062° True

October

Day	Slack h m	Maximum h m	knots
1 Sa		0314	2.2E
	0613	0921	3.5F
	1225	1536	2.3E
	1834	2138	3.3F
2 Su	0027	0345	2.5E
	0645	0955	3.7F
	1250	1604	2.5E
	1905	2212	3.6F
3 M ●	0101	0415	2.6E
	0713	1026	3.7F
	1318	1631	2.7E
	1932	2244	3.7F
4 Tu	0135	0441	2.6E
	0738	1056	3.7F
	1348	1651	2.8E
	1957	2316	3.7F
5 W	0211	0456	2.5E
	0802	1128	3.6F
	1420	1653	2.8E
	2023	2351	3.7F
6 Th	0248	0502	2.4E
	0828	1202	3.4F
	1454	1713	2.9E
	2053		
7 F		0029	3.5F
	0329	0532	2.3E
	0858	1240	3.0F
	1532	1748	2.8E
	2128		
8 Sa		0111	3.2F
	0416	0611	2.0E
	0933	1322	2.6F
	1615	1830	2.6E
	2211		
9 Su		0200	2.8F
	0511	0657	1.6E
	1017	1415	2.1F
	1708	1918	2.2E
	2307		
10 M ◐		0306	2.4F
	0617	0749	1.1E
	1119	1527	1.8F
	1815	2014	1.7E
11 Tu	0027	0434	2.3F
	0735	0901	0.7E
		0939	0.6E
		1048	0.7E
	1318	1701†	1.8F
12 W	0221	0556	2.6F
	0900	1216	1.1E
	1518	1821	2.2F
	2053		
13 Th		0034	1.7E
	0342	0700	3.1F
	1012	1326	1.7E
	1623	1923	2.8F
	2206		
14 F		0140	2.2E
	0441	0753	3.6F
	1103	1417	2.3E
	1714	2016	3.4F
	2307		
15 Sa		0231	2.7E
	0530	0841	4.1F
	1144	1450	2.9E
	1759	2103	4.0F
	2358		
16 Su		0315	3.1E
	0614	0923	4.3F
	1221	1537	3.3E
	1840	2147	4.3F
17 M ○	0044	0355	3.2E
	0654	1003	4.4F
	1257	1613	3.5E
	1919	2228	4.5F
18 Tu	0125	0433	3.1E
	0732	1041	4.3F
	1332	1649	3.5E
	1957	2308	4.4F
19 W	0205	0511	2.9E
	0807	1119	4.0F
	1408	1724	3.3E
	2032	2347	4.2F
20 Th	0245	0548	2.5E
	0840	1156	3.6F
	1444	1754	3.0E
	2107		
21 F		0027	3.8F
	0326	0622	2.0E
	0912	1234	3.1F
	1522	1807	2.5E
	2142		
22 Sa		0108	3.3F
	0410	0635	1.5E
	0943	1315	2.6F
	1604	1818	2.1E
	2220		
23 Su		0153	2.8F
	0500	0641	1.1E
	1020	1403	2.0F
	1652	1852	1.7E
	2307		
24 M ◑		0252	2.3F
	0600	0720	0.7E
	1113	1508	1.6F
	1751	1936	1.2E
25 Tu	0016	0414	2.0F
	0712	0811	0.3E
		0858	0.3E
		1034	0.4F
	1319	1642†	1.4F
26 W	0157	0531	2.1F
	0852	1200	0.6E
	1500	1757	1.6F
	2021		
27 Th		0009	1.0E
	0315	0630	2.3F
	1012	1325	1.0E
	1602	1856	2.0F
	2138		
28 F		0116	1.3E
	0411	0721	2.7F
	1042	1400	1.5E
	1649	1946	2.5F
	2236		
29 Sa		0204	1.6E
	0456	0806	3.0F
	1109	1430	1.9E
	1728	2029	2.9F
	2320		
30 Su		0241	2.0E
	0534	0845	3.3F
	1138	1501	2.3E
	1802	2108	3.3F
	2359		
31 M		0315	2.2E
	0608	0920	3.5F
	1209	1531	2.6E
	1834	2143	3.6F

November

Day	Slack h m	Maximum h m	knots
1 Tu ●	0037	0347	2.3E
	0639	0952	3.6F
	1242	1559	2.8E
	1903	2217	3.8F
2 W	0114	0416	2.4E
	0707	1025	3.5F
	1315	1621	3.0E
	1932	2253	3.9F
3 Th	0152	0439	2.3E
	0736	1100	3.4F
	1351	1631	3.0E
	2002	2331	3.8F
4 F	0232	0451	2.2E
	0808	1139	3.3F
	1428	1653	3.0E
	2037		
5 Sa		0012	3.7F
	0316	0519	2.1E
	0844	1221	3.0F
	1509	1730	2.9E
	2117		
6 Su		0057	3.4F
	0404	0600	1.8E
	0926	1308	2.6F
	1556	1814	2.6E
	2205		
7 M		0149	3.1F
	0501	0648	1.5E
	1020	1403	2.2F
	1653	1904	2.2E
	2305		
8 Tu ◐		0253	2.8F
	0605	0750	1.1E
	1137	1516	2.0F
	1800	2005	1.7E
9 W	0023	0414	2.7F
	0717	1034	1.1E
	1328	1645	2.0F
	1916	2251	1.5E
10 Th	0155	0529	2.9F
	0831	1149	1.4E
	1457	1759	2.4F
	2035		
11 F		0007	1.7E
	0312	0630	3.2F
	0935	1255	1.9E
	1600	1900	2.9F
	2150		
12 Sa		0113	2.0E
	0412	0724	3.5F
	1026	1349	2.5E
	1652	1954	3.4F
	2253		
13 Su		0209	2.4E
	0503	0812	3.8F
	1110	1434	2.9E
	1737	2042	3.9F
	2345		
14 M		0254	2.6E
	0548	0856	3.9F
	1149	1513	3.2E
	1820	2126	4.1F
15 Tu ○	0030	0335	2.7E
	0630	0937	3.9F
	1227	1550	3.4E
	1859	2208	4.2F
16 W	0112	0414	2.6E
	0709	1016	3.8F
	1303	1626	3.3E
	1936	2249	4.2F
17 Th	0150	0452	2.4E
	0745	1055	3.6F
	1340	1700	3.1E
	2012	2328	4.0F
18 F	0228	0530	2.1E
	0819	1133	3.3F
	1417	1732	2.8E
	2046		
19 Sa		0007	3.7F
	0308	0606	1.8E
	0851	1212	2.9F
	1456	1745	2.4E
	2120		
20 Su		0048	3.3F
	0350	0639	1.4E
	0925	1253	2.5F
	1537	1753	2.1E
	2156		
21 M		0131	2.9F
	0437	0627	1.1E
	1004	1339	2.1F
	1624	1828	1.8E
	2237		
22 Tu		0221	2.6F
	0530	0703	0.9E
	1056	1435	1.7F
	1719	1912	1.5E
	2328		
23 W ◑		0324	2.3F
	0629	0753	0.7E
		0837	0.6E
		0945	0.7E
	1220	1550†	1.5F
24 Th	0035	0437	2.2F
	0731	1057	0.8E
	1400	1709	1.6F
	1932	2307	0.9E
25 F	0200	0540	2.3F
	0832	1200	1.0E
	1512	1811	1.9F
	2042		
26 Sa		0014	1.0E
	0310	0633	2.5F
	0924	1256	1.4E
	1605	1905	2.3F
	2147		
27 Su		0113	1.2E
	0404	0721	2.7F
	1009	1343	1.8E
	1649	1953	2.7F
	2242		
28 M		0202	1.5E
	0450	0804	2.9F
	1050	1423	2.2E
	1728	2036	3.1F
	2329		
29 Tu		0244	1.8E
	0530	0844	3.1F
	1129	1459	2.5E
	1803	2116	3.5F
30 W	0013	0322	2.0E
	0607	0922	3.2F
	1208	1532	2.8E
	1838	2155	3.7F

December

Day	Slack h m	Maximum h m	knots
1 Th ●	0055	0358	2.1E
	0643	1000	3.3F
	1248	1603	3.0E
	1914	2235	3.9F
2 F	0138	0433	2.1E
	0721	1041	3.3F
	1328	1629	3.1E
	1951	2317	4.0F
3 Sa	0221	0508	2.1E
	0801	1124	3.2F
	1411	1655	3.1E
	2032		
4 Su		0002	3.9F
	0306	0547	2.0E
	0844	1210	3.1F
	1456	1731	2.9E
	2117		
5 M		0049	3.8F
	0355	0636	1.9E
	0934	1300	2.9F
	1546	1818	2.7E
	2207		
6 Tu		0140	3.6F
	0449	0740	1.7E
	1033	1354	2.6F
	1642	1914	2.3E
	2303		
7 W		0237	3.3F
	0547	0854	1.6E
	1144	1500	2.4F
	1746	2057	1.9E
8 Th ◐	0007	0344	3.1F
	0648	1006	1.6E
	1305	1617	2.4F
	1855	2222	1.7E
9 F	0119	0453	3.0F
	0750	1113	1.7E
	1423	1729	2.6F
	2010	2334	1.6E
10 Sa	0233	0555	3.1F
	0850	1216	2.0E
	1530	1832	2.9F
	2126		
11 Su		0041	1.7E
	0338	0651	3.2F
	0945	1316	2.3E
	1626	1928	3.2F
	2237		
12 M		0144	1.8E
	0435	0743	3.3F
	1034	1408	2.6E
	1716	2020	3.5F
	2336		
13 Tu		0237	2.0E
	0525	0832	3.3F
	1119	1452	2.9E
	1801	2108	3.7F
14 W	0024	0321	2.1E
	0610	0916	3.3F
	1201	1532	3.0E
	1843	2152	3.8F
15 Th ○	0105	0401	2.1E
	0653	0958	3.3F
	1241	1610	2.9E
	1922	2234	3.8F
16 F	0142	0440	2.0E
	0732	1039	3.2F
	1320	1647	2.8E
	1959	2315	3.8F
17 Sa	0218	0518	1.9E
	0808	1118	3.0F
	1358	1722	2.7E
	2034	2354	3.6F
18 Su	0254	0557	1.7E
	0843	1157	2.9F
	1437	1752	2.4E
	2107		
19 M		0031	3.4F
	0332	0634	1.6E
	0917	1236	2.6F
	1518	1753	2.2E
	2139		
20 Tu		0110	3.2F
	0413	0708	1.4E
	0953	1318	2.4F
	1602	1814	2.0E
	2213		
21 W		0151	2.9F
	0457	0658	1.3E
	1036	1404	2.1F
	1650	1853	1.8E
	2249		
22 Th		0237	2.7F
	0545	0732	1.2E
	1127	1458	1.9F
	1744	1937	1.5E
	2331		
23 F ◑		0332	2.4F
	0635	0819	1.1E
	1231	1604	1.8F
	1844	2027	1.2E
24 Sa	0021	0433	2.3F
	0727	1043	1.1E
	1350	1713	1.9F
	1948	2124	1.0E
25 Su	0124	0532	2.3F
	0820	1148	1.2E
	1504	1815	2.1F
	2055		
26 M		0010	0.8E
	0246	0628	2.3F
	0913	1248	1.5E
	1603	1912	2.4F
	2201		
27 Tu		0119	1.0E
	0357	0721	2.5F
	1004	1343	1.8E
	1652	2005	2.8F
	2300		
28 W		0215	1.3E
	0454	0811	2.7F
	1053	1431	2.2E
	1737	2053	3.2F
	2354		
29 Th		0303	1.6E
	0544	0858	2.9F
	1141	1513	2.6E
	1820	2139	3.6F
30 F ●	0042	0346	1.9E
	0631	0944	3.1F
	1227	1553	2.0E
	1903	2223	3.9F
31 Sa	0127	0428	2.1E
	0716	1029	3.3F
	1314	1634	3.1E
	1946	2308	4.2F

Time meridian 135° W. 0000 is midnight. 1200 is noon. Times are not adjusted for Daylight Saving Time.

† See page 164 for the remaining currents on this day.

Sergius Narrows, Peril Strait, Alaska, 2005

F–Flood, Dir. 059° True E–Ebb, Dir. 241° True

January

Day	Slack h m	Maximum h m	knots
1 Sa	0220	0444	4.1E
	0819	1113	5.2F
	1434	1703	4.1E
	2037	2338	5.4F
2 Su	0306	0530	3.9E
	0910	1203	4.9F
	1526	1752	3.7E
	2128		
3 M ◑		0029	5.1F
	0357	0623	3.6E
	1008	1259	4.7F
	1626	1850	3.4E
	2228		
4 Tu		0125	4.9F
	0456	0726	3.6E
	1113	1358	4.8F
	1734	1956	3.2E
	2335		
5 W		0225	4.9F
	0600	0832	3.7E
	1219	1500	5.0F
	1842	2105	3.4E
6 Th	0043	0327	5.2F
	0703	0941	4.2E
	1321	1607	5.5F
	1944	2219	3.8E
7 F	0145	0432	5.7F
	0801	1053	4.8E
	1420	1712	6.3F
	2041	2333	4.5E
8 Sa	0243	0534	6.4F
	0856	1158	5.6E
	1514	1810	7.2F
	2135		
9 Su		0032	5.3E
	0337	0628	7.2F
	0948	1252	6.4E
	1606	1900	7.9F
	2226		
10 M ●		0123	6.0E
	0428	0718	7.8F
	1039	1341	6.9E
	1655	1948	8.4F
	2315		
11 Tu		0211	6.5E
	0517	0807	8.2F
	1128	1430	7.2E
	1743	2037	8.6F
12 W	0002	0258	6.7E
	0606	0856	8.2F
	1217	1516	7.2E
	1830	2124	8.5F
13 Th	0050	0344	6.6E
	0654	0945	8.0F
	1306	1601	6.8E
	1918	2211	8.1F
14 F	0138	0428	6.3E
	0744	1033	7.5F
	1357	1645	6.1E
	2007	2258	7.5F
15 Sa	0227	0512	5.6E
	0836	1122	6.8F
	1449	1731	5.2E
	2058	2346	6.6F
16 Su ◐	0318	0603	4.8E
	0930	1214	5.9F
	1545	1827	4.2E
	2152		
17 M		0038	5.8F
	0413	0713	4.0E
	1028	1312	5.1F
	1646	1955	3.4E
	2250		
18 Tu		0137	5.0F
	0513	0850	3.6E
	1129	1417	4.5F
	1754	2133	3.0E
	2352		
19 W		0242	4.4F
	0618	1009	3.5E
	1231	1537	4.3F
	1904	2246	3.0E
20 Th	0055	0403	4.2F
	0721	1114	3.7E
	1331	1658	4.4F
	2006	2345	3.3E
21 F	0153	0517	4.4F
	0817	1207	4.0E
	1425	1754	4.8F
	2058		
22 Sa		0034	3.7E
	0246	0606	4.7F
	0905	1251	4.3E
	1514	1835	5.3F
	2142		
23 Su		0116	4.0E
	0333	0643	5.1F
	0948	1329	4.6E
	1557	1906	5.7F
	2221		
24 M		0151	4.2E
	0415	0712	5.5F
	1027	1357	4.7E
	1636	1933	6.0F
	2257		
25 Tu ○		0215	4.4E
	0453	0740	5.9F
	1103	1412	4.9E
	1713	2001	6.3F
	2330		
26 W		0225	4.6E
	0529	0812	6.2F
	1137	1426	5.1E
	1746	2032	6.6F
27 Th	0003	0242	4.8E
	0603	0846	6.4F
	1211	1451	5.3E
	1819	2107	6.7F
28 F	0036	0310	5.0E
	0636	0924	6.5F
	1246	1522	5.3E
	1851	2144	6.7F
29 Sa	0110	0342	5.2E
	0709	1003	6.4F
	1322	1557	5.3E
	1923	2224	6.6F
30 Su	0145	0418	5.1E
	0745	1044	6.2F
	1402	1636	5.0E
	1959	2306	6.2F
31 M	0225	0459	4.9E
	0827	1130	5.8F
	1447	1719	4.5E
	2042	2353	5.8F

February

Day	Slack h m	Maximum h m	knots
1 Tu ◑	0311	0545	4.5E
	0918	1221	5.4F
	1542	1810	3.9E
	2137		
2 W		0046	5.2F
	0407	0642	4.0E
	1024	1320	5.0F
	1650	1913	3.3E
	2249		
3 Th		0147	4.9F
	0517	0750	3.7E
	1140	1425	4.8F
	1809	2028	3.1E
4 F	0010	0255	4.8F
	0633	0909	3.7E
	1256	1539	5.0F
	1923	2200	3.3E
5 Sa	0124	0410	5.1F
	0743	1048	4.3E
	1402	1700	5.8F
	2028	2343	4.2E
6 Su	0229	0526	6.0F
	0844	1206	5.4E
	1501	1805	6.9F
	2124		
7 M		0041	5.3E
	0326	0624	7.0F
	0939	1259	6.4E
	1554	1855	7.9F
	2214		
8 Tu ●		0127	6.3E
	0417	0713	7.9F
	1029	1344	7.1E
	1642	1941	8.6F
	2301		
9 W		0210	6.9E
	0504	0759	8.5F
	1116	1426	7.5E
	1727	2024	8.9F
	2345		
10 Th		0250	7.3E
	0549	0843	8.7F
	1201	1506	7.5E
	1811	2106	8.9F
11 F	0028	0328	7.2E
	0633	0926	8.5F
	1246	1543	7.1E
	1854	2147	8.5F
12 Sa	0110	0403	6.8E
	0717	1007	8.0F
	1330	1618	6.4E
	1937	2227	7.8F
13 Su	0153	0438	6.1E
	0803	1049	7.2F
	1416	1654	5.4E
	2021	2308	6.8F
14 M	0238	0515	5.1E
	0850	1133	6.2F
	1505	1733	4.3E
	2109	2353	5.8F
15 Tu ◐	0326	0557	4.1E
	0942	1221	5.1F
	1600	1820	3.2E
	2203		
16 W		0043	4.7F
	0422	0652	3.1E
	1042	1318	4.2F
	1706	1931	2.2E
	2307		
17 Th		0143	3.8F
	0531	0929	2.5E
	1150	1430	3.6F
	1825	2221	2.1E
18 F	0018	0300	3.3F
	0647	1052	2.8E
	1259	1628	3.6F
	1941	2328	2.6E
19 Sa	0126	0457	3.6F
	0754	1150	3.3E
	1400	1737	4.2F
	2040		
20 Su		0018	3.3E
	0224	0554	4.2F
	0848	1236	4.0E
	1452	1821	4.9F
	2125		
21 M		0100	3.9E
	0314	0633	4.9F
	0931	1313	4.5E
	1537	1853	5.6F
	2202		
22 Tu		0134	4.4E
	0356	0701	5.6F
	1009	1342	4.9E
	1615	1918	6.2F
	2235		
23 W ○		0158	4.7E
	0433	0725	6.2F
	1043	1359	5.1E
	1651	1942	6.7F
	2306		
24 Th		0208	5.1E
	0507	0753	6.6F
	1116	1409	5.4E
	1723	2010	7.1F
	2337		
25 F		0221	5.4E
	0540	0824	7.0F
	1149	1430	5.7E
	1754	2042	7.3F
26 Sa	0008	0246	5.8E
	0611	0859	7.2F
	1221	1459	5.9E
	1824	2117	7.4F
27 Su	0039	0317	6.0E
	0642	0937	7.2F
	1256	1533	5.9E
	1855	2156	7.2F
28 M	0113	0352	5.9E
	0716	1017	6.9F
	1334	1610	5.5E
	1929	2237	6.8F

March

Day	Slack h m	Maximum h m	knots
1 Tu	0152	0431	5.6E
	0755	1102	6.4F
	1419	1652	5.0E
	2010	2323	6.1F
2 W	0237	0516	5.0E
	0845	1152	5.7F
	1513	1741	4.1E
	2105		
3 Th ◑		0016	5.3F
	0335	0611	4.2E
	0952	1251	5.1F
	1623	1844	3.3E
	2223		
4 F		0120	4.7F
	0449	0722	3.5E
	1116	1401	4.7F
	1748	2007	2.8E
	2352		
5 Sa		0234	4.5F
	0615	0856	3.4E
	1238	1523	4.8F
	1910	2231	3.2E
6 Su	0111	0401	4.9F
	0731	1108	4.2E
	1347	1657	5.7F
	2016	2348	4.4E
7 M	0216	0525	5.9F
	0834	1208	5.4E
	1446	1800	6.9F
	2110		
8 Tu		0037	5.7E
	0312	0620	7.1F
	0928	1254	6.5E
	1538	1846	7.9F
	2158		
9 W		0118	6.6E
	0401	0704	8.1F
	1015	1335	7.2E
	1624	1926	8.6F
	2241		
10 Th ●		0156	7.3E
	0446	0744	8.7F
	1059	1412	7.5E
	1707	2004	8.9F
	2322		
11 F		0231	7.5E
	0528	0823	8.8F
	1141	1446	7.4E
	1747	2041	8.8F
12 Sa	0001	0303	7.3E
	0608	0900	8.6F
	1221	1517	7.0E
	1826	2117	8.4F
13 Su	0040	0332	6.9E
	0648	0938	8.0F
	1302	1547	6.3E
	1905	2154	7.6F
14 M	0118	0402	6.1E
	0728	1016	7.2F
	1343	1617	5.3E
	1945	2232	6.7F
15 Tu	0159	0434	5.2E
	0811	1056	6.2F
	1427	1651	4.3E
	2028	2314	5.6F
16 W	0243	0510	4.1E
	0858	1140	5.1F
	1518	1731	3.2E
	2119		
17 Th ◐		0001	4.5F
	0336	0554	3.1E
	0956	1233	4.1F
	1621	1823	2.1E
	2224		
18 F		0058	3.5F
	0444	0658	2.1E
	1107	1339	3.3F
	1743	2152	1.6E
	2342		
19 Sa		0211	3.0F
	0608	1024	2.2E
	1223	1524	3.2F
	1907	2302	2.3E
20 Su	0056	0419	3.2F
	0725	1124	2.9E
	1328	1706	3.8F
	2009	2352	3.1E
21 M	0156	0527	4.0F
	0821	1210	3.7E
	1422	1752	4.7F
	2053		
22 Tu		0032	3.9E
	0246	0607	4.9F
	0905	1246	4.3E
	1507	1824	5.5F
	2130		
23 W		0103	4.5E
	0328	0635	5.7F
	0942	1312	4.8E
	1546	1848	6.2F
	2203		
24 Th		0124	5.0E
	0405	0659	6.4F
	1016	1327	5.2E
	1621	1912	6.9F
	2235		
25 F ○		0134	5.5E
	0439	0726	7.0F
	1050	1340	5.7E
	1655	1941	7.4F
	2306		
26 Sa		0151	5.9E
	0512	0758	7.5F
	1123	1403	6.0E
	1726	2014	7.6F
	2337		
27 Su		0218	6.3E
	0544	0833	7.7F
	1157	1434	6.2E
	1758	2050	7.7F
28 M	0011	0251	6.4E
	0617	0912	7.7F
	1233	1510	6.1E
	1831	2130	7.4F
29 Tu	0047	0328	6.3E
	0653	0954	7.3F
	1314	1549	5.7E
	1908	2213	6.9F
30 W	0128	0409	5.9E
	0736	1040	6.7F
	1401	1633	5.0E
	1954	2301	6.2F
31 Th	0217	0456	5.1E
	0830	1132	5.9F
	1459	1723	4.1E
	2056	2357	5.3F

Time meridian 135° W. 0000 is midnight. 1200 is noon. Times are not adjusted for Daylight Saving Time.

Sergius Narrows, Peril Strait, Alaska, 2005

F–Flood, Dir. 059° True E–Ebb, Dir. 241° True

April

Day	Slack h m	Maximum h m	knots
1 F	0319	0552	4.2E
	0940	1233	5.2F
	1611	1829	3.2E
◑	2217		
2 Sa		0104	4.6F
	0437	0709	3.4E
	1103	1347	4.7F
	1735	2012	2.8E
	2343		
3 Su		0223	4.5F
	0604	0926	3.5E
	1222	1516	5.0F
	1855	2232	3.7E
4 M	0058	0359	5.0F
	0719	1100	4.5E
	1329	1646	5.8F
	1958	2333	4.9E
5 Tu	0200	0515	6.1F
	0820	1154	5.6E
	1426	1744	6.9F
	2050		
6 W		0019	6.0E
	0253	0606	7.2F
	0911	1238	6.4E
	1516	1827	7.8F
	2135		
7 Th		0100	6.8E
	0340	0647	8.0F
	0956	1317	6.9E
	1601	1905	8.3F
	2217		
8 F		0135	7.2E
	0423	0724	8.4F
	1038	1352	7.0E
●	1642	1939	8.4F
	2255		
9 Sa		0207	7.2E
	0504	0758	8.5F
	1118	1422	6.8E
	1721	2012	8.3F
	2333		
10 Su		0234	6.9E
	0542	0832	8.2F
	1156	1449	6.3E
	1758	2046	7.8F
11 M	0009	0300	6.4E
	0619	0907	7.6F
	1234	1516	5.7E
	1835	2122	7.1F
12 Tu	0046	0328	5.8E
	0657	0944	6.9F
	1313	1545	5.0E
	1913	2200	6.3F
13 W	0125	0400	5.0E
	0736	1024	6.0F
	1356	1619	4.1E
	1954	2241	5.4F
14 Th	0208	0436	4.1E
	0821	1108	5.1F
	1445	1658	3.2E
	2043	2328	4.4F
15 F	0259	0519	3.2E
	0917	1159	4.2F
	1544	1747	2.3E
	2148		
16 Sa		0024	3.6F
	0404	0615	2.3E
	1026	1301	3.5F
◐	1659	1900	1.6E
	2304		
17 Su		0132	3.1F
	0524	0937	1.9E
	1140	1418	3.3F
	1818	2221	2.1E
18 M	0017	0256	3.2F
	0641	1043	2.5E
	1246	1554	3.7F
	1922	2313	2.9E
19 Tu	0118	0431	3.9F
	0740	1130	3.3E
	1341	1659	4.5F
	2009	2352	3.7E
20 W	0208	0520	4.8F
	0826	1205	3.9E
	1428	1737	5.4F
	2048		
21 Th		0021	4.4E
	0251	0554	5.7F
	0906	1229	4.5E
	1509	1806	6.2F
	2124		
22 F		0038	5.0E
	0331	0624	6.5F
	0944	1244	5.1E
	1547	1837	6.8F
	2158		
23 Sa		0053	5.6E
	0407	0655	7.2F
	1020	1306	5.6E
	1623	1910	7.4F
	2233		
24 Su		0119	6.1E
	0443	0730	7.7F
	1057	1336	6.0E
○	1659	1946	7.7F
	2308		
25 M		0151	6.5E
	0519	0808	7.9F
	1135	1411	6.1E
	1735	2026	7.7F
	2346		
26 Tu		0229	6.6E
	0557	0850	7.8F
	1215	1451	6.0E
	1814	2109	7.4F
27 W	0027	0310	6.4E
	0639	0936	7.5F
	1301	1534	5.6E
	1858	2156	6.9F
28 Th	0114	0355	5.9E
	0727	1025	6.9F
	1353	1622	5.0E
	1951	2247	6.2F
29 F	0209	0445	5.1E
	0826	1119	6.2F
	1453	1716	4.2E
	2058	2345	5.5F
30 Sa	0314	0545	4.3E
	0935	1222	5.5F
	1603	1829	3.5E
◑	2213		

May

Day	Slack h m	Maximum h m	knots
1 Su		0054	4.9F
	0429	0712	3.6E
	1049	1338	5.2F
	1720	2043	3.5E
	2329		
2 M		0216	4.9F
	0549	0923	3.9E
	1201	1502	5.4F
	1832	2209	4.3E
3 Tu	0037	0343	5.4F
	0700	1036	4.7E
	1305	1620	6.0F
	1933	2307	5.2E
4 W	0137	0453	6.2F
	0759	1131	5.4E
	1400	1718	6.7F
	2023	2355	5.9E
5 Th	0229	0545	7.0F
	0849	1217	5.9E
	1450	1803	7.3F
	2108		
6 F		0037	6.4E
	0316	0626	7.5F
	0935	1257	6.2E
	1535	1840	7.5F
	2150		
7 Sa		0113	6.6E
	0359	0702	7.7F
	1016	1331	6.1E
●	1617	1913	7.6F
	2229		
8 Su		0143	6.5E
	0439	0734	7.7F
	1055	1400	5.8E
	1656	1944	7.3F
	2306		
9 M		0208	6.2E
	0517	0806	7.4F
	1133	1424	5.4E
	1733	2018	7.0F
	2342		
10 Tu		0232	5.7E
	0554	0840	7.0F
	1211	1449	5.0E
	1810	2054	6.5F
11 W	0019	0300	5.3E
	0631	0918	6.5F
	1250	1519	4.5E
	1847	2133	5.9F
12 Th	0058	0333	4.7E
	0710	0958	5.8F
	1332	1554	3.9E
	1928	2215	5.2F
13 F	0141	0411	4.1E
	0754	1042	5.2F
	1419	1634	3.3E
	2017	2302	4.6F
14 Sa	0231	0453	3.4E
	0845	1131	4.6F
	1513	1722	2.7E
	2116	2355	4.0F
15 Su	0329	0545	2.8E
	0946	1228	4.1F
	1615	1822	2.2E
◐	2223		
16 M		0056	3.6F
	0437	0652	2.3E
	1052	1331	3.9F
	1722	1948	2.2E
	2330		
17 Tu		0202	3.6F
	0547	0823	2.4E
	1155	1435	4.1F
	1824	2159	2.7E
18 W	0031	0308	4.0F
	0649	1009	2.8E
	1252	1537	4.5F
	1916	2243	3.3E
19 Th	0123	0410	4.7F
	0741	1054	3.4E
	1342	1632	5.2F
	2001	2312	4.1E
20 F	0210	0501	5.5F
	0827	1126	4.1E
	1428	1719	5.9F
	2042	2340	4.8E
21 Sa	0254	0545	6.3F
	0909	1158	4.7E
	1511	1800	6.6F
	2122		
22 Su		0013	5.5E
	0336	0625	7.1F
	0951	1233	5.3E
	1553	1840	7.1F
	2203		
23 M		0050	6.1E
	0417	0705	7.6F
	1033	1312	5.7E
○	1635	1922	7.5F
	2244		
24 Tu		0129	6.5E
	0459	0748	7.9F
	1117	1354	5.9E
	1718	2006	7.6F
	2328		
25 W		0213	6.6E
	0543	0834	7.9F
	1203	1439	5.9E
	1803	2053	7.4F
26 Th	0015	0300	6.4E
	0630	0923	7.6F
	1252	1528	5.6E
	1853	2144	7.0F
27 F	0106	0349	6.0E
	0722	1015	7.2F
	1346	1619	5.2E
	1950	2238	6.5F
28 Sa	0203	0442	5.4E
	0820	1110	6.7F
	1445	1716	4.6E
	2053	2336	6.0F
29 Su	0306	0543	4.7E
	0923	1211	6.1F
	1548	1831	4.2E
	2159		
30 M		0043	5.5F
	0414	0711	4.2E
	1028	1321	5.8F
◑	1655	2020	4.2E
	2306		
31 Tu		0158	5.4F
	0525	0854	4.2E
	1133	1433	5.7F
	1801	2136	4.6E

June

Day	Slack h m	Maximum h m	knots
1 W	0010	0313	5.6F
	0633	1005	4.5E
	1235	1543	5.9F
	1901	2237	5.1E
2 Th	0109	0422	5.9F
	0733	1104	4.9E
	1331	1645	6.1F
	1954	2329	5.5E
3 F	0202	0519	6.3F
	0825	1154	5.1E
	1422	1736	6.4F
	2041		
4 Sa		0015	5.7E
	0250	0604	6.6F
	0913	1238	5.2E
	1509	1816	6.5F
	2124		
5 Su		0055	5.7E
	0335	0642	6.8F
	0956	1317	5.1E
	1553	1851	6.5F
	2205		
6 M		0128	5.6E
	0417	0714	6.7F
	1036	1349	4.9E
●	1634	1922	6.4F
	2244		
7 Tu		0154	5.3E
	0456	0746	6.6F
	1115	1412	4.6E
	1713	1955	6.2F
	2321		
8 W		0215	5.1E
	0534	0819	6.4F
	1153	1434	4.4E
	1751	2032	6.0F
	2359		
9 Th		0242	4.8E
	0612	0857	6.1F
	1231	1503	4.2E
	1829	2112	5.7F
10 F	0038	0315	4.6E
	0650	0937	5.9F
	1312	1537	3.9E
	1909	2154	5.3F
11 Sa	0120	0351	4.3E
	0731	1020	5.5F
	1355	1615	3.7E
	1954	2239	5.0F
12 Su	0205	0432	3.9E
	0815	1105	5.2F
	1441	1658	3.4E
	2044	2327	4.6F
13 M	0255	0517	3.5E
	0905	1154	4.9F
	1532	1747	3.1E
	2139		
14 Tu		0019	4.4F
	0350	0610	3.1E
	1000	1248	4.6F
◐	1626	1845	3.0E
	2237		
15 W		0116	4.3F
	0451	0712	2.9E
	1058	1344	4.6F
	1724	1950	3.1E
	2337		
16 Th		0214	4.4F
	0554	0816	3.0E
	1157	1440	4.7F
	1820	2051	3.4E
17 F	0034	0313	4.7F
	0653	0919	3.2E
	1254	1537	5.1F
	1913	2151	3.9E
18 Sa	0128	0412	5.3F
	0747	1020	3.7E
	1347	1634	5.6F
	2003	2248	4.6E
19 Su	0219	0508	6.0F
	0838	1118	4.3E
	1438	1727	6.2F
	2051	2341	5.3E
20 M	0308	0559	6.7F
	0927	1209	4.9E
	1528	1817	6.8F
	2139		
21 Tu		0029	5.9E
	0356	0646	7.4F
	1015	1257	5.4E
○	1617	1904	7.3F
	2227		
22 W		0117	6.4E
	0444	0734	7.8F
	1104	1346	5.8E
	1705	1952	7.6F
	2316		
23 Th		0206	6.6E
	0532	0822	8.0F
	1152	1436	6.0E
	1754	2043	7.7F
24 F	0006	0256	6.7E
	0621	0913	8.0F
	1242	1526	6.0E
	1845	2135	7.5F
25 Sa	0057	0347	6.5E
	0712	1005	7.8F
	1334	1617	5.8E
	1939	2227	7.2F
26 Su	0152	0437	6.0E
	0806	1057	7.4F
	1428	1710	5.4E
	2036	2322	6.7F
27 M	0249	0533	5.3E
	0902	1151	6.8F
	1524	1812	4.9E
	2135		
28 Tu		0021	6.2F
	0349	0642	4.6E
	1000	1251	6.2F
◑	1622	1936	4.6E
	2236		
29 W		0125	5.7F
	0453	0812	4.2E
	1100	1354	5.8F
	1724	2056	4.5E
	2337		
30 Th		0234	5.4F
	0600	0930	4.0E
	1201	1500	5.4F
	1825	2204	4.5E

Time meridian 135° W. 0000 is midnight. 1200 is noon. Times are not adjusted for Daylight Saving Time.

Sergius Narrows, Peril Strait, Alaska, 2005

F–Flood, Dir. 059° True E–Ebb, Dir. 241° True

July

Day	Slack h m	Maximum h m	knots
1 F	0037	0345	5.3F
	0704	1037	4.0E
	1259	1608	5.3F
	1923	2305	4.6E
2 Sa	0134	0452	5.4F
	0802	1136	4.2E
	1355	1710	5.3F
	2016	2358	4.8E
3 Su	0226	0546	5.6F
	0853	1226	4.3E
	1446	1759	5.5F
	2103		
4 M		0044	4.9E
	0314	0629	5.8F
	0939	1309	4.3E
	1533	1838	5.6F
	2147		
5 Tu		0123	4.9E
	0358	0704	6.0F
	1021	1347	4.3E
	1616	1910	5.7F
	2227		
6 W ●		0155	4.8E
	0439	0734	6.1F
	1100	1417	4.3E
	1656	1942	5.8F
	2306		
7 Th		0215	4.8E
	0517	0805	6.1F
	1137	1433	4.2E
	1734	2016	5.8F
	2343		
8 F		0233	4.7E
	0554	0839	6.2F
	1213	1452	4.3E
	1811	2053	5.8F
9 Sa	0020	0300	4.7E
	0630	0917	6.2F
	1249	1520	4.3E
	1848	2133	5.8F
10 Su	0057	0332	4.7E
	0706	0956	6.1F
	1327	1554	4.3E
	1926	2214	5.6F
11 M	0137	0408	4.5E
	0743	1036	5.9F
	1406	1631	4.2E
	2007	2257	5.4F
12 Tu	0219	0448	4.3E
	0822	1120	5.6F
	1448	1712	4.1E
	2052	2343	5.1F
13 W	0306	0532	3.9E
	0907	1207	5.3F
	1534	1800	3.8E
	2142		
14 Th ◐		0034	4.9F
	0359	0623	3.5E
	0959	1259	5.0F
	1627	1855	3.6E
	2241		
15 F		0130	4.7F
	0501	0723	3.2E
	1101	1355	4.8F
	1727	1957	3.6E
	2346		
16 Sa		0229	4.7F
	0609	0828	3.1E
	1208	1454	4.9F
	1831	2102	3.8E
17 Su	0050	0332	5.0F
	0714	0937	3.4E
	1314	1558	5.2F
	1932	2212	4.2E
18 M	0151	0439	5.6F
	0814	1052	3.9E
	1414	1703	5.8F
	2029	2323	4.9E
19 Tu	0248	0541	6.4F
	0909	1201	4.7E
	1511	1801	6.6F
	2123		
20 W		0023	5.7E
	0341	0635	7.3F
	1001	1256	5.5E
	1603	1853	7.3F
	2214		
21 Th ○		0115	6.5E
	0431	0724	8.0F
	1051	1345	6.1E
	1654	1943	7.9F
	2304		
22 F		0204	7.0E
	0519	0812	8.4F
	1138	1433	6.6E
	1742	2032	8.2F
	2353		
23 Sa		0252	7.2E
	0606	0900	8.6F
	1225	1520	6.8E
	1830	2121	8.2F
24 Su	0042	0338	7.0E
	0654	0948	8.4F
	1313	1604	6.6E
	1919	2210	7.9F
25 M	0132	0422	6.6E
	0742	1035	8.0F
	1401	1649	6.2E
	2011	2258	7.4F
26 Tu	0224	0508	5.8E
	0833	1122	7.2F
	1452	1737	5.5E
	2104	2349	6.6F
27 W ◑	0318	0559	4.8E
	0926	1213	6.4F
	1545	1835	4.6E
	2200		
28 Th		0045	5.7F
	0417	0709	3.9E
	1023	1310	5.5F
	1643	2002	4.0E
	2301		
29 F		0148	5.0F
	0523	0852	3.3E
	1125	1413	4.8F
	1748	2133	3.7E
30 Sa	0004	0302	4.5F
	0634	1015	3.1E
	1229	1528	4.4F
	1854	2246	3.7E
31 Su	0106	0428	4.5F
	0741	1121	3.4E
	1330	1652	4.4F
	1955	2345	4.0E

August

Day	Slack h m	Maximum h m	knots
1 M	0204	0534	4.8F
	0838	1214	3.7E
	1426	1750	4.7F
	2047		
2 Tu		0034	4.4E
	0255	0620	5.2F
	0926	1259	4.0E
	1515	1832	5.1F
	2133		
3 W		0115	4.6E
	0341	0656	5.6F
	1007	1338	4.3E
	1559	1904	5.5F
	2213		
4 Th ●		0149	4.8E
	0421	0725	6.0F
	1043	1409	4.5E
	1639	1931	5.8F
	2249		
5 F		0211	4.9E
	0458	0750	6.3F
	1117	1426	4.6E
	1715	2000	6.1F
	2324		
6 Sa		0222	5.0E
	0533	0819	6.5F
	1149	1435	4.8E
	1749	2032	6.3F
	2358		
7 Su		0240	5.1E
	0605	0852	6.7F
	1221	1457	4.9E
	1822	2108	6.4F
8 M	0032	0308	5.2E
	0637	0927	6.7F
	1254	1527	5.1E
	1855	2145	6.4F
9 Tu	0106	0341	5.2E
	0708	1005	6.5F
	1328	1601	5.1E
	1929	2225	6.2F
10 W	0144	0417	4.9E
	0741	1046	6.2F
	1405	1638	4.9E
	2006	2308	5.8F
11 Th	0226	0457	4.5E
	0819	1130	5.7F
	1447	1721	4.5E
	2052	2356	5.3F
12 F ◐	0316	0544	3.9E
	0907	1220	5.2F
	1538	1812	4.0E
	2150		
13 Sa		0052	4.9F
	0418	0641	3.3E
	1012	1318	4.7F
	1643	1916	3.6E
	2304		
14 Su		0154	4.6F
	0533	0752	2.9E
	1134	1423	4.6F
	1758	2030	3.5E
15 M	0022	0304	4.7F
	0651	0913	3.0E
	1252	1534	4.8F
	1911	2155	3.9E
16 Tu	0131	0421	5.3F
	0758	1057	3.7E
	1359	1650	5.5F
	2015	2329	4.8E
17 W	0232	0532	6.4F
	0855	1209	4.8E
	1457	1754	6.6F
	2111		
18 Th		0027	5.9E
	0326	0627	7.5F
	0947	1257	5.9E
	1550	1846	7.6F
	2202		
19 F ○		0114	6.8E
	0416	0713	8.3F
	1034	1340	6.7E
	1638	1932	8.4F
	2250		
20 Sa		0157	7.4E
	0502	0757	8.9F
	1119	1422	7.3E
	1724	2017	8.8F
	2336		
21 Su		0239	7.6E
	0546	0840	9.0F
	1202	1502	7.4E
	1809	2101	8.8F
22 M	0022	0319	7.4E
	0630	0923	8.8F
	1246	1540	7.2E
	1854	2145	8.4F
23 Tu	0107	0357	6.8E
	0714	1005	8.2F
	1330	1618	6.5E
	1940	2228	7.6F
24 W	0154	0436	5.9E
	0800	1048	7.3F
	1416	1657	5.6E
	2029	2314	6.7F
25 Th	0244	0516	4.7E
	0849	1134	6.2F
	1506	1740	4.5E
	2122		
26 F ◑		0003	5.6F
	0340	0605	3.5E
	0944	1225	5.0F
	1602	1837	3.4E
	2222		
27 Sa		0101	4.5F
	0445	0731	2.5E
	1048	1327	4.1F
	1710	2105	2.8E
	2329		
28 Su		0216	3.9F
	0604	0956	2.4E
	1159	1448	3.6F
	1827	2228	3.0E
29 M	0038	0405	3.8F
	0721	1105	2.9E
	1307	1637	3.7F
	1937	2329	3.5E
30 Tu	0141	0518	4.3F
	0822	1158	3.5E
	1406	1737	4.4F
	2031		
31 W		0017	4.1E
	0233	0605	5.0F
	0907	1241	4.1E
	1455	1819	5.0F
	2115		

September

Day	Slack h m	Maximum h m	knots
1 Th		0056	4.6E
	0318	0639	5.6F
	0944	1317	4.5E
	1538	1849	5.6F
	2153		
2 F		0128	4.9E
	0358	0705	6.1F
	1018	1344	4.8E
	1616	1913	6.1F
	2227		
3 Sa ●		0149	5.1E
	0433	0727	6.6F
	1049	1358	5.1E
	1650	1937	6.6F
	2300		
4 Su		0157	5.3E
	0506	0753	6.9F
	1119	1407	5.3E
	1723	2006	6.9F
	2332		
5 M		0214	5.5E
	0537	0823	7.1F
	1150	1428	5.6E
	1753	2040	7.0F
6 Tu	0004	0240	5.6E
	0606	0857	7.1F
	1220	1457	5.7E
	1823	2116	7.0F
7 W	0037	0312	5.6E
	0635	0934	6.9F
	1252	1530	5.7E
	1854	2155	6.7F
8 Th	0113	0348	5.3E
	0706	1014	6.5F
	1328	1608	5.4E
	1929	2238	6.3F
9 F	0154	0428	4.8E
	0742	1058	5.9F
	1410	1650	4.9E
	2014	2326	5.6F
10 Sa	0245	0514	4.1E
	0831	1149	5.2F
	1504	1740	4.1E
	2116		
11 Su ◐		0022	5.0F
	0350	0611	3.3E
	0944	1250	4.6F
	1614	1846	3.4E
	2239		
12 M		0129	4.5F
	0512	0728	2.7E
	1116	1401	4.3F
	1739	2012	3.2E
13 Tu	0004	0246	4.6F
	0636	0915	2.9E
	1239	1521	4.6F
	1859	2215	3.8E
14 W	0116	0413	5.3F
	0744	1112	4.0E
	1346	1646	5.6F
	2004	2334	5.0E
15 Th	0216	0525	6.5F
	0840	1205	5.3E
	1443	1748	6.9F
	2058		
16 F		0023	6.2E
	0309	0615	7.7F
	0928	1247	6.4E
	1533	1835	8.0F
	2147		
17 Sa ○		0104	7.0E
	0356	0657	8.6F
	1013	1325	7.2E
	1619	1916	8.7F
	2232		
18 Su		0142	7.5E
	0440	0737	9.1F
	1055	1401	7.6E
	1702	1956	9.1F
	2316		
19 M		0218	7.6E
	0522	0815	9.1F
	1136	1436	7.6E
	1744	2036	8.9F
	2358		
20 Tu		0254	7.2E
	0603	0854	8.7F
	1216	1510	7.2E
	1826	2116	8.4F
21 W	0040	0328	6.6E
	0644	0934	8.0F
	1257	1544	6.5E
	1908	2157	7.6F
22 Th	0124	0402	5.6E
	0726	1014	7.0F
	1340	1619	5.5E
	1953	2239	6.5F
23 F	0211	0438	4.5E
	0812	1057	5.8F
	1427	1657	4.4E
	2043	2324	5.3F
24 Sa ◑	0303	0519	3.3E
	0906	1145	4.7F
	1522	1743	3.2E
	2142		
25 Su		0019	4.3F
	0408	0615	2.2E
	1012	1245	3.7F
	1631	1855	2.2E
	2252		
26 M		0129	3.5F
	0528	0933	2.0E
	1127	1405	3.1F
	1754	2204	2.5E
27 Tu	0006	0328	3.4F
	0652	1040	2.6E
	1239	1608	3.4F
	1910	2303	3.1E
28 W	0110	0448	4.0F
	0753	1132	3.4E
	1339	1711	4.2F
	2006	2350	3.9E
29 Th	0204	0536	4.8F
	0836	1213	4.1E
	1428	1752	5.0F
	2048		
30 F		0028	4.5E
	0248	0609	5.6F
	0912	1246	4.6E
	1510	1822	5.7F
	2125		

Time meridian 135° W. 0000 is midnight. 1200 is noon. Times are not adjusted for Daylight Saving Time.

Sergius Narrows, Peril Strait, Alaska, 2005

F–Flood, Dir. 059° True E–Ebb, Dir. 241° True

October

Day	Slack h m	Maximum h m	knots
1 Sa		0057	4.9E
	0327	0634	6.2F
	0945	1310	5.1E
	1547	1845	6.4F
	2159		
2 Su		0115	5.2E
	0403	0656	6.8F
	1016	1321	5.4E
	1621	1909	6.9F
	2232		
3 M ●		0125	5.5E
	0436	0722	7.2F
	1047	1333	5.7E
	1653	1938	7.3F
	2304		
4 Tu		0144	5.7E
	0507	0753	7.4F
	1117	1357	6.0E
	1724	2012	7.5F
	2337		
5 W		0212	5.8E
	0537	0828	7.4F
	1149	1428	6.2E
	1755	2049	7.4F
6 Th	0011	0246	5.8E
	0608	0906	7.1F
	1222	1504	6.0E
	1828	2129	7.1F
7 F	0049	0324	5.4E
	0641	0948	6.6F
	1301	1543	5.7E
	1906	2214	6.5F
8 Sa	0133	0406	4.9E
	0722	1034	6.0F
	1347	1628	5.0E
	1955	2304	5.8F
9 Su	0228	0454	4.1E
	0818	1128	5.2F
	1446	1720	4.2E
	2102		
10 M ◐		0002	5.1F
	0337	0554	3.2E
	0939	1231	4.5F
	1601	1829	3.4E
	2226		
11 Tu		0113	4.7F
	0459	0720	2.7E
	1107	1348	4.4F
	1727	2013	3.2E
	2348		
12 W		0235	4.8F
	0620	0949	3.3E
	1225	1514	4.8F
	1846	2220	4.1E
13 Th	0057	0402	5.6F
	0726	1059	4.5E
	1329	1637	5.9F
	1949	2321	5.2E
14 F	0156	0509	6.7F
	0819	1148	5.8E
	1424	1734	7.1F
	2041		
15 Sa		0007	6.2E
	0247	0557	7.7F
	0906	1229	6.7E
	1513	1819	8.0F
	2129		

Day	Slack h m	Maximum h m	knots
16 Su		0047	6.9E
	0334	0637	8.4F
	0949	1306	7.3E
	1557	1858	8.6F
	2212		
17 M ○		0123	7.2E
	0417	0713	8.7F
	1030	1339	7.5E
	1639	1934	8.8F
	2254		
18 Tu		0157	7.1E
	0458	0749	8.6F
	1109	1411	7.3E
	1720	2011	8.6F
	2335		
19 W		0228	6.6E
	0537	0825	8.1F
	1148	1441	6.8E
	1759	2048	8.0F
20 Th	0015	0259	6.0E
	0616	0903	7.4F
	1227	1513	6.1E
	1839	2127	7.2F
21 F	0056	0331	5.2E
	0656	0943	6.5F
	1308	1546	5.2E
	1921	2208	6.2F
22 Sa	0141	0406	4.2E
	0740	1025	5.5F
	1353	1623	4.2E
	2008	2252	5.2F
23 Su	0231	0446	3.3E
	0832	1113	4.5F
	1446	1706	3.2E
	2105	2344	4.3F
24 M ◑	0332	0535	2.3E
	0937	1209	3.7F
	1552	1803	2.3E
	2212		
25 Tu		0047	3.6F
	0445	0854	1.8E
	1050	1320	3.2F
	1710	2123	2.1E
	2324		
26 W		0210	3.4F
	0603	1002	2.4E
	1201	1455	3.3F
	1827	2225	2.7E
27 Th	0029	0347	3.8F
	0706	1054	3.1E
	1301	1621	3.9F
	1926	2313	3.4E
28 F	0123	0446	4.6F
	0753	1135	3.8E
	1351	1709	4.8F
	2011	2351	4.0E
29 Sa	0210	0524	5.3F
	0831	1208	4.4E
	1434	1742	5.6F
	2051		
30 Su		0019	4.5E
	0251	0553	6.0F
	0906	1228	4.9E
	1513	1810	6.3F
	2127		
31 M		0035	4.9E
	0329	0620	6.6F
	0940	1240	5.4E
	1550	1839	6.9F
	2202		

November

Day	Slack h m	Maximum h m	knots
1 Tu ●		0050	5.3E
	0404	0651	7.1F
	1014	1300	5.9E
	1624	1911	7.4F
	2238		
2 W		0116	5.6E
	0439	0725	7.3F
	1048	1329	6.2E
	1659	1947	7.6F
	2314		
3 Th		0148	5.8E
	0513	0802	7.4F
	1123	1404	6.3E
	1734	2026	7.6F
	2352		
4 F		0226	5.7E
	0549	0844	7.2F
	1202	1444	6.2E
	1812	2110	7.3F
5 Sa	0034	0307	5.4E
	0630	0929	6.7F
	1245	1527	5.8E
	1856	2157	6.8F
6 Su	0123	0353	4.9E
	0718	1019	6.1F
	1337	1616	5.2E
	1951	2250	6.1F
7 M	0220	0445	4.2E
	0820	1114	5.5F
	1439	1711	4.4E
	2058	2349	5.5F
8 Tu ◐	0328	0548	3.6E
	0936	1219	4.9F
	1551	1822	3.7E
	2213		
9 W		0059	5.2F
	0442	0722	3.3E
	1053	1335	4.8F
	1710	2019	3.6E
	2326		
10 Th		0218	5.3F
	0556	0928	3.9E
	1204	1459	5.2F
	1825	2158	4.3E
11 F	0033	0337	5.8F
	0700	1033	4.9E
	1307	1615	6.0F
	1928	2258	5.1E
12 Sa	0131	0443	6.6F
	0754	1125	5.8E
	1401	1714	6.9F
	2021	2348	5.8E
13 Su	0222	0534	7.3F
	0841	1209	6.5E
	1450	1800	7.6F
	2109		
14 M		0030	6.2E
	0310	0615	7.7F
	0925	1247	6.8E
	1535	1839	8.0F
	2153		
15 Tu ○		0108	6.3E
	0354	0652	7.8F
	1006	1322	6.8E
	1618	1915	8.0F
	2235		

Day	Slack h m	Maximum h m	knots
16 W		0141	6.2E
	0435	0726	7.7F
	1046	1352	6.6E
	1658	1949	7.8F
	2315		
17 Th		0211	5.8E
	0515	0801	7.3F
	1125	1420	6.1E
	1737	2025	7.3F
	2354		
18 F		0239	5.3E
	0554	0838	6.8F
	1203	1449	5.6E
	1816	2103	6.8F
19 Sa	0035	0309	4.7E
	0633	0917	6.1F
	1243	1522	4.9E
	1856	2143	6.1F
20 Su	0117	0343	4.1E
	0715	1000	5.4F
	1327	1558	4.3E
	1940	2227	5.4F
21 M	0204	0422	3.5E
	0804	1046	4.7F
	1416	1640	3.6E
	2030	2314	4.7F
22 Tu	0257	0507	2.8E
	0901	1137	4.1F
	1513	1729	2.9E
	2129		
23 W ◑		0009	4.2F
	0357	0605	2.3E
	1005	1237	3.7F
	1618	1833	2.3E
	2232		
24 Th		0111	3.9F
	0502	0737	2.2E
	1111	1343	3.6F
	1728	2054	2.3E
	2335		
25 F		0216	4.0F
	0605	0953	2.6E
	1212	1451	3.9F
	1832	2210	2.7E
26 Sa	0033	0319	4.3F
	0659	1039	3.2E
	1300	1554	4.4F
	1926	2255	3.2E
27 Su	0124	0415	4.9F
	0745	1112	3.8E
	1354	1647	5.1F
	2012	2325	3.8E
28 M	0211	0502	5.5F
	0826	1135	4.5E
	1437	1730	5.9F
	2054	2349	4.3E
29 Tu	0254	0543	6.1F
	0905	1200	5.1E
	1519	1808	6.6F
	2134		
30 W		0019	4.8E
	0335	0622	6.7F
	0944	1232	5.7E
	1559	1847	7.2F
	2215		

December

Day	Slack h m	Maximum h m	knots
1 Th ●		0053	5.3E
	0416	0702	7.1F
	1024	1309	6.1E
	1639	1927	7.5F
	2256		
2 F		0132	5.6E
	0457	0743	7.3F
	1106	1349	6.3E
	1721	2010	7.7F
	2340		
3 Sa		0214	5.7E
	0539	0828	7.3F
	1150	1433	6.3E
	1804	2057	7.6F
4 Su	0026	0300	5.6E
	0625	0917	7.0F
	1238	1521	6.1E
	1852	2147	7.3F
5 M	0116	0349	5.3E
	0717	1009	6.7F
	1331	1611	5.6E
	1947	2239	6.8F
6 Tu	0212	0442	4.9E
	0817	1104	6.2F
	1430	1706	5.0E
	2047	2336	6.3F
7 W	0312	0542	4.4E
	0922	1205	5.7F
	1536	1812	4.4E
	2151		
8 Th ◐		0040	5.9F
	0416	0706	4.1E
	1029	1314	5.4F
	1645	1950	4.0E
	2257		
9 F		0150	5.7F
	0523	0850	4.3E
	1136	1429	5.5F
	1757	2124	4.2E
10 Sa	0002	0301	5.8F
	0627	1001	4.8E
	1238	1543	5.8F
	1902	2232	4.6E
11 Su	0102	0410	6.0F
	0725	1101	5.3E
	1335	1650	6.2F
	2000	2329	5.0E
12 M	0157	0509	6.3F
	0817	1151	5.7E
	1427	1743	6.7F
	2051		
13 Tu		0017	5.3E
	0247	0557	6.6F
	0903	1235	5.9E
	1515	1826	6.9F
	2137		
14 W		0100	5.3E
	0334	0637	6.7F
	0947	1314	5.9E
	1559	1902	7.0F
	2220		
15 Th ○		0138	5.2E
	0417	0712	6.7F
	1028	1347	5.7E
	1641	1936	6.9F
	2300		

Day	Slack h m	Maximum h m	knots
16 F		0209	5.0E
	0458	0745	6.5F
	1108	1414	5.5E
	1720	2009	6.7F
	2339		
17 Sa		0234	4.7E
	0537	0821	6.3F
	1146	1438	5.2E
	1759	2045	6.5F
18 Su	0018	0258	4.5E
	0616	0859	6.0F
	1225	1507	4.9E
	1837	2123	6.2F
19 M	0057	0328	4.2E
	0656	0939	5.7F
	1305	1540	4.6E
	1916	2204	5.8F
20 Tu	0138	0403	4.0E
	0738	1022	5.3F
	1348	1617	4.2E
	1958	2247	5.5F
21 W	0222	0442	3.7E
	0824	1107	4.9F
	1435	1658	3.7E
	2044	2333	5.1F
22 Th	0309	0526	3.3E
	0916	1156	4.5F
	1527	1746	3.2E
	2135		
23 F ◑		0023	4.7F
	0401	0618	3.0E
	1012	1251	4.2F
	1625	1843	2.8E
	2232		
24 Sa		0118	4.4F
	0458	0720	2.9E
	1113	1349	4.1F
	1729	1948	2.6E
	2333		
25 Su		0215	4.4F
	0557	0825	3.0E
	1213	1448	4.3F
	1833	2053	2.8E
26 M	0033	0313	4.5F
	0654	0928	3.4E
	1309	1549	4.7F
	1930	2159	3.1E
27 Tu	0129	0412	4.9F
	0746	1030	3.9E
	1401	1649	5.3F
	2022	2303	3.6E
28 W	0221	0508	5.5F
	0834	1126	4.6E
	1451	1742	6.1F
	2111	2355	4.3E
29 Th	0311	0559	6.2F
	0921	1214	5.3E
	1539	1829	6.9F
	2157		
30 F ●		0041	5.0E
	0358	0645	6.9F
	1008	1259	5.9E
	1625	1914	7.5F
	2244		
31 Sa		0126	5.5E
	0445	0731	7.4F
	1054	1344	6.4E
	1711	2000	7.9F
	2330		

Time meridian 135° W. 0000 is midnight. 1200 is noon. Times are not adjusted for Daylight Saving Time.

North Inian Pass, Cross Sound, Alaska, 2005

F–Flood, Dir. 075° True E–Ebb, Dir. 260° True

January

Day	Slack h m	Maximum h m	knots
1 Sa		0137	1.9F
	0401	0719	4.3E
	1138	1338	0.8F
	1529	1930	4.8E
	2341		
2 Su		0221	1.5F
	0440	0808	4.2E
	1254	1443	0.5F
	1617	2021	4.3E
3 M ◑	0039	0316	1.2F
	0524	0903	4.2E
	1423	1608	0.5F
	1728	2120	3.8E
4 Tu	0149	0422	1.0F
	0615	1004	4.3E
	1531	1727	0.7F
	1857	2227	3.5E
5 W	0302	0530	1.0F
	0714	1107	4.7E
	1618	1832	1.3F
	2021	2337	3.5E
6 Th	0404	0630	1.3F
	0816	1209	5.3E
	1658	1927	2.1F
	2133		
7 F		0042	3.8E
	0452	0722	1.7F
	0917	1306	6.0E
	1736	2016	2.8F
	2234		
8 Sa		0140	4.3E
	0535	0811	2.3F
	1013	1358	6.8E
	1815	2102	3.5F
	2328		
9 Su		0231	4.9E
	0617	0857	2.8F
	1106	1447	7.4E
	1856	2146	4.1F
10 M ●	0016	0319	5.4E
	0700	0943	3.3F
	1157	1534	7.9E
	1937	2229	4.4F
11 Tu	0103	0406	5.8E
	0745	1029	3.5F
	1246	1621	8.0E
	2020	2313	4.6F
12 W	0148	0452	6.1E
	0832	1116	3.6F
	1335	1707	7.8E
	2104	2357	4.4F
13 Th	0232	0538	6.1E
	0923	1205	3.3F
	1423	1753	7.3E
	2151		
14 F		0044	4.0F
	0316	0626	6.0E
	1020	1258	2.9F
	1513	1841	6.6E
	2241		
15 Sa		0133	3.4F
	0401	0716	5.7E
	1125	1357	2.3F
	1605	1931	5.6E
	2337		
16 Su ◐		0226	2.7F
	0447	0809	5.3E
	1242	1504	1.8F
	1701	2026	4.6E
17 M	0044	0327	2.1F
	0535	0908	4.9E
	1411	1619	1.4F
	1806	2128	3.8E
18 Tu	0206	0435	1.6F
	0628	1015	4.6E
	1535	1733	1.4F
	1920	2243	3.2E
19 W	0335	0543	1.3F
	0726	1129	4.5E
	1641	1841	1.6F
	2036		
20 Th		0008	3.0E
	0448	0646	1.3F
	0826	1238	4.7E
	1732	1938	2.0F
	2144		
21 F		0121	3.1E
	0542	0741	1.5F
	0924	1334	5.0E
	1813	2027	2.3F
	2240		
22 Sa		0212	3.4E
	0624	0828	1.7F
	1015	1417	5.3E
	1849	2110	2.7F
	2326		
23 Su		0250	3.8E
	0658	0910	2.0F
	1101	1453	5.7E
	1920	2148	3.0F
24 M	0006	0323	4.2E
	0728	0947	2.2F
	1143	1527	6.0E
	1948	2224	3.1F
25 Tu ○	0043	0354	4.5E
	0757	1022	2.3F
	1221	1559	6.2E
	2015	2257	3.2F
26 W	0118	0426	4.9E
	0825	1055	2.3F
	1257	1633	6.3E
	2042	2329	3.2F
27 Th	0151	0459	5.1E
	0855	1127	2.3F
	1333	1707	6.3E
	2111	2359	3.0F
28 F	0224	0534	5.2E
	0929	1159	2.1F
	1409	1743	6.1E
	2142		
29 Sa		0029	2.8F
	0256	0611	5.3E
	1007	1233	1.9F
	1445	1821	5.7E
	2216		
30 Su		0058	2.4F
	0328	0650	5.2E
	1051	1311	1.6F
	1524	1902	5.2E
	2254		
31 M		0129	2.0F
	0400	0733	5.1E
	1145	1359	1.2F
	1609	1948	4.6E
	2340		

February

Day	Slack h m	Maximum h m	knots
1 Tu ◑		0207	1.5F
	0435	0821	5.0E
	1253	1506	0.9F
	1705	2042	3.9E
2 W	0040	0300	1.1F
	0517	0918	4.8E
	1417	1634	0.9F
	1820	2145	3.4E
3 Th	0200	0422	0.8F
	0615	1022	4.9E
	1536	1756	1.2F
	1946	2258	3.2E
4 F	0330	0550	0.9F
	0729	1133	5.2E
	1635	1903	1.8F
	2106		
5 Sa		0014	3.4E
	0437	0659	1.4F
	0844	1241	5.7E
	1722	1959	2.6F
	2214		
6 Su		0122	4.0E
	0528	0757	2.0F
	0953	1341	6.4E
	1805	2047	3.4F
	2310		
7 M		0219	4.7E
	0612	0848	2.7F
	1053	1435	7.0E
	1846	2132	4.0F
8 Tu ●	0000	0308	5.4E
	0655	0935	3.3F
	1147	1523	7.5E
	1925	2215	4.4F
9 W	0045	0354	6.0E
	0738	1021	3.7F
	1238	1608	7.7E
	2005	2256	4.6F
10 Th	0128	0438	6.5E
	0822	1106	3.9F
	1326	1652	7.6E
	2046	2337	4.5F
11 F	0208	0520	6.7E
	0909	1152	3.8F
	1413	1735	7.2E
	2127		
12 Sa		0018	4.1F
	0248	0603	6.6E
	0958	1239	3.3F
	1459	1819	6.5E
	2211		
13 Su		0101	3.5F
	0327	0647	6.3E
	1053	1330	2.7F
	1546	1903	5.6E
	2259		
14 M		0146	2.7F
	0406	0733	5.8E
	1158	1428	2.0F
	1634	1950	4.6E
	2357		
15 Tu ◐		0238	1.9F
	0446	0822	5.2E
	1319	1536	1.4F
	1729	2044	3.6E
16 W	0118	0344	1.1F
	0531	0920	4.6E
	1455	1653	1.1F
	1834	2149	2.9E
17 Th	0312	0501	0.7F
	0625	1030	4.2E
	1618	1808	1.1F
	1951	2317	2.5E
18 F	0445	0616	0.7F
	0733	1153	4.1E
	1716	1913	1.5F
	2108		
19 Sa		0055	2.7E
	0540	0718	0.9F
	0845	1307	4.4E
	1759	2005	1.9F
	2211		
20 Su		0157	3.1E
	0619	0810	1.4F
	0948	1358	4.8E
	1833	2049	2.4F
	2302		
21 M		0236	3.7E
	0649	0853	1.8F
	1041	1437	5.3E
	1902	2127	2.8F
	2343		
22 Tu		0307	4.2E
	0716	0931	2.2F
	1126	1510	5.7E
	1928	2202	3.1F
23 W ○	0019	0336	4.8E
	0741	1006	2.5F
	1207	1542	6.1E
	1953	2233	3.3F
24 Th	0052	0406	5.2E
	0806	1038	2.7F
	1245	1614	6.3E
	2017	2302	3.4F
25 F	0124	0437	5.7E
	0833	1110	2.8F
	1322	1647	6.4E
	2043	2330	3.3F
26 Sa	0154	0509	6.0E
	0903	1141	2.8F
	1358	1722	6.2E
	2111	2357	3.1F
27 Su	0224	0544	6.1E
	0937	1213	2.6F
	1436	1758	5.9E
	2143		
28 M		0023	2.7F
	0253	0621	6.1E
	1016	1249	2.4F
	1515	1838	5.4E
	2219		

March

Day	Slack h m	Maximum h m	knots
1 Tu		0052	2.3F
	0324	0702	6.0E
	1104	1331	2.0F
	1558	1922	4.8E
	2302		
2 W		0126	1.8F
	0357	0748	5.7E
	1204	1429	1.5F
	1649	2013	4.1E
	2359		
3 Th ◑		0213	1.2F
	0436	0843	5.3E
	1324	1553	1.1F
	1756	2114	3.5E
4 F	0122	0333	0.7F
	0532	0948	5.0E
	1458	1726	1.2F
	1919	2230	3.1E
5 Sa	0316	0524	0.7F
	0654	1105	5.0E
	1614	1841	1.8F
	2043	2355	3.3E
6 Su	0435	0645	1.2F
	0823	1222	5.3E
	1708	1940	2.5F
	2154		
7 M		0110	3.9E
	0526	0747	2.0F
	0939	1328	5.9E
	1752	2030	3.2F
	2251		
8 Tu		0208	4.8E
	0608	0839	2.8F
	1043	1423	6.5E
	1832	2114	3.9F
	2339		
9 W		0256	5.6E
	0649	0926	3.4F
	1139	1511	7.0E
	1910	2156	4.3F
10 Th ●	0022	0339	6.4E
	0729	1010	3.9F
	1229	1554	7.2E
	1947	2235	4.5F
11 F	0102	0419	6.9E
	0809	1053	4.1F
	1315	1635	7.1E
	2025	2313	4.3F
12 Sa	0139	0458	7.1E
	0851	1135	4.0F
	1359	1715	6.7E
	2103	2350	3.9F
13 Su	0216	0537	7.0E
	0935	1218	3.6F
	1442	1754	6.1E
	2142		
14 M		0028	3.3F
	0251	0616	6.7E
	1022	1303	3.0F
	1524	1835	5.3E
	2226		
15 Tu		0108	2.5F
	0325	0656	6.1E
	1116	1353	2.2F
	1606	1917	4.5E
	2318		
16 W		0152	1.7F
	0359	0741	5.4E
	1226	1453	1.5F
	1653	2005	3.6E
17 Th ◐	0032	0251	0.9F
	0435	0832	4.6E
	1404	1609	0.9F
	1749	2103	2.9E
18 F	0255	0416	0.3F
	0517	0936	4.0E
	1545	1730	0.8F
	1903	2223	2.4E
19 Sa		0542	*
		1058	3.7E
	1651	1840	1.1F
	2024		
20 Su		0012	2.5E
	0534	0652	0.6F
	0802	1226	3.9E
	1734	1936	1.6F
	2134		
21 M		0128	3.1E
	0605	0746	1.1F
	0917	1328	4.3E
	1807	2020	2.1F
	2227		
22 Tu		0209	3.7E
	0631	0830	1.7F
	1017	1410	4.9E
	1835	2058	2.6F
	2309		
23 W		0240	4.4E
	0654	0909	2.2F
	1105	1445	5.4E
	1900	2132	3.0F
	2346		
24 Th		0309	5.1E
	0717	0944	2.7F
	1148	1518	5.8E
	1924	2203	3.2F
25 F ○	0018	0338	5.7E
	0740	1016	3.0F
	1228	1550	6.1E
	1948	2231	3.3F
26 Sa	0049	0409	6.3E
	0806	1048	3.3F
	1307	1624	6.2E
	2013	2258	3.3F
27 Su	0120	0442	6.7E
	0835	1120	3.3F
	1345	1659	6.1E
	2042	2325	3.2F
28 M	0150	0517	6.9E
	0909	1153	3.2F
	1424	1736	5.9E
	2115	2353	2.9F
29 Tu	0221	0554	6.9E
	0949	1230	3.0F
	1504	1817	5.5E
	2152		
30 W		0024	2.5F
	0254	0636	6.6E
	1036	1314	2.5F
	1548	1901	4.9E
	2239		
31 Th		0103	1.9F
	0330	0723	6.1E
	1134	1411	2.0F
	1640	1953	4.2E
	2339		

Time meridian 135° W. 0000 is midnight. 1200 is noon. Times are not adjusted for Daylight Saving Time.
* Current weak and variable.

North Inian Pass, Cross Sound, Alaska, 2005

F–Flood, Dir. 075° True E–Ebb, Dir. 260° True

April

Day	Slack h m	Maximum h m	knots
1 F		0155	1.3F
	0413	0818	5.6E
	1252	1531	1.5F
	1744	2056	3.6E
2 Sa	0111	0322	0.7F
	0512	0925	5.0E
	1426	1702	1.5F
	1902	2214	3.3E
3 Su	0314	0512	0.7F
	0639	1044	4.8E
	1548	1817	1.9F
	2021	2342	3.5E
4 M	0430	0633	1.3F
	0811	1206	5.0E
	1647	1918	2.5F
	2130		
5 Tu		0057	4.2E
	0519	0735	2.1F
	0929	1314	5.4E
	1733	2008	3.1F
	2225		
6 W		0154	5.1E
	0601	0827	2.9F
	1034	1409	5.9E
	1814	2052	3.6F
	2312		
7 Th		0239	5.9E
	0639	0913	3.5F
	1129	1455	6.2E
	1851	2133	3.9F
	2354		
8 F		0319	6.6E
	0717	0956	3.9F
	1217	1537	6.4E
	1928	2211	4.0F
9 Sa	0031	0357	7.0E
	0754	1037	4.1F
	1301	1615	6.3E
	2003	2247	3.8F
10 Su	0107	0433	7.2E
	0832	1117	4.0F
	1343	1652	6.0E
	2039	2323	3.5F
11 M	0141	0509	7.1E
	0911	1157	3.8F
	1422	1729	5.6E
	2117	2358	2.9F
12 Tu	0215	0545	6.7E
	0954	1237	3.0F
	1501	1807	5.0E
	2158		
13 W		0034	2.2F
	0247	0624	6.2E
	1041	1322	2.3F
	1540	1848	4.4E
	2247		
14 Th		0113	1.5F
	0318	0706	5.5E
	1141	1416	1.6F
	1623	1933	3.7E
	2357		
15 F		0205	0.7F
	0348	0753	4.7E
	1304	1526	1.0F
	1714	2028	3.0E
16 Sa		0329	*
		0851	4.1E
	1449	1646	0.8F
	1820	2140	2.6E
17 Su		0503	*
		1005	3.6E
	1606	1758	1.0F
	1937	2313	2.7E
18 M	0516	0617	0.3F
	0713	1129	3.6E
	1654	1856	1.4F
	2047		
19 Tu		0035	3.2E
	0540	0714	0.9F
	0839	1241	4.0E
	1730	1942	1.9F
	2142		
20 W		0125	3.9E
	0603	0801	1.5F
	0946	1331	4.4E
	1759	2022	2.3F
	2226		
21 Th		0201	4.6E
	0625	0841	2.2F
	1039	1411	4.9E
	1825	2056	2.7F
	2303		
22 F		0233	5.4E
	0647	0917	2.8F
	1126	1447	5.3E
	1850	2127	2.9F
	2338		
23 Sa		0305	6.1E
	0711	0951	3.2F
	1208	1523	5.7E
	1915	2157	3.1F
24 Su	0011	0338	6.7E
	0738	1025	3.6F
	1249	1559	5.9E
	1943	2226	3.2F
25 M	0044	0414	7.2E
	0810	1059	3.7F
	1329	1636	5.9E
	2015	2256	3.1F
26 Tu	0118	0451	7.4E
	0840	1135	3.7F
	1410	1716	5.8E
	2051	2329	2.9F
27 W	0154	0532	7.3E
	0928	1216	3.4F
	1453	1759	5.4E
	2134		
28 Th		0007	2.5F
	0233	0616	7.0E
	1017	1304	3.0F
	1540	1846	4.9E
	2226		
29 F		0053	2.0F
	0315	0705	6.5E
	1116	1402	2.4F
	1633	1940	4.4E
	2334		
30 Sa		0155	1.4F
	0404	0801	5.8E
	1229	1516	2.0F
	1734	2044	3.9E

May

Day	Slack h m	Maximum h m	knots
1 Su	0111	0325	0.9F
	0509	0908	5.1E
	1355	1637	1.9F
	1845	2201	3.7E
2 M	0300	0500	1.0F
	0632	1025	4.7E
	1515	1749	2.1F
	1956	2325	4.0E
3 Tu	0413	0617	1.5F
	0800	1146	4.7E
	1619	1850	2.5F
	2100		
4 W		0037	4.7E
	0505	0719	2.2F
	0917	1256	4.9E
	1709	1941	2.9F
	2154		
5 Th		0133	5.4E
	0548	0811	2.9F
	1021	1351	5.2E
	1753	2027	3.2F
	2240		
6 F		0217	6.1E
	0627	0858	3.4F
	1115	1437	5.4E
	1832	2108	3.4F
	2322		
7 Sa		0256	6.6E
	0704	0940	3.8F
	1203	1518	5.5E
	1909	2146	3.4F
	2359		
8 Su		0333	6.9E
	0740	1020	3.9F
	1245	1555	5.5E
	1944	2223	3.2F
9 M	0035	0407	7.0E
	0815	1059	3.7F
	1324	1630	5.3E
	2020	2258	2.9F
10 Tu	0109	0442	6.9E
	0852	1137	3.4F
	1402	1706	5.1E
	2056	2333	2.5F
11 W	0142	0518	6.6E
	0931	1216	2.9F
	1439	1744	4.7E
	2137		
12 Th		0008	1.9F
	0215	0556	6.1E
	1014	1258	2.4F
	1518	1824	4.3E
	2225		
13 F		0046	1.3F
	0246	0637	5.6E
	1105	1346	1.8F
	1559	1908	3.8E
	2331		
14 Sa		0134	0.7F
	0317	0723	4.9E
	1209	1446	1.3F
	1647	2001	3.4E
15 Su		0245	*
		0816	4.3E
	1331	1557	1.0F
	1744	2103	3.1E
16 M		0416	*
		0919	3.8E
	1455	1706	1.0F
	1848	2217	3.1E
17 Tu		0534	*
		1031	3.6E
	1557	1806	1.2F
	1951	2330	3.4E
18 W	0504	0635	0.7F
	0755	1142	3.7E
	1641	1856	1.5F
	2047		
19 Th		0028	4.1E
	0527	0726	1.4F
	0908	1242	4.0E
	1715	1938	1.9F
	2134		
20 F		0113	4.8E
	0550	0809	2.1F
	1008	1331	4.4E
	1745	2015	2.2F
	2216		
21 Sa		0152	5.6E
	0615	0848	2.7F
	1059	1414	4.8E
	1814	2049	2.5F
	2255		
22 Su		0230	6.4E
	0642	0925	3.3F
	1146	1454	5.2E
	1843	2123	2.8F
	2334		
23 M		0308	7.0E
	0713	1002	3.7F
	1229	1534	5.5E
	1916	2157	3.0F
24 Tu	0013	0348	7.5E
	0749	1040	3.9F
	1313	1615	5.6E
	1952	2233	3.1F
25 W	0053	0430	7.7E
	0828	1121	4.0F
	1357	1659	5.6E
	2034	2313	3.0F
26 Th	0135	0513	7.6E
	0913	1205	3.8F
	1442	1744	5.4E
	2121	2358	2.7F
27 F	0219	0600	7.3E
	1003	1254	3.4F
	1530	1834	5.1E
	2218		
28 Sa		0051	2.2F
	0307	0651	6.7E
	1100	1351	3.0F
	1622	1929	4.8E
	2329		
29 Su		0157	1.7F
	0402	0747	5.9E
	1206	1457	2.5F
	1720	2032	4.5E
30 M	0059	0317	1.3F
	0506	0850	5.2E
	1321	1608	2.3F
	1821	2143	4.4E
31 Tu	0234	0441	1.3F
	0623	1003	4.6E
	1438	1716	2.2F
	1924	2259	4.5E

June

Day	Slack h m	Maximum h m	knots
1 W	0349	0555	1.7F
	0745	1120	4.3E
	1547	1818	2.4F
	2024		
2 Th		0009	5.0E
	0445	0658	2.2F
	0900	1231	4.3E
	1644	1912	2.5F
	2118		
3 F		0106	5.5E
	0532	0753	2.7F
	1005	1331	4.4E
	1734	2001	2.7F
	2206		
4 Sa		0153	6.0E
	0614	0841	3.1F
	1059	1419	4.5E
	1817	2044	2.7F
	2249		
5 Su		0234	6.3E
	0652	0924	3.4F
	1146	1459	4.6E
	1855	2125	2.7F
	2329		
6 M		0310	6.5E
	0728	1004	3.5F
	1227	1536	4.7E
	1931	2203	2.6F
7 Tu	0006	0345	6.6E
	0802	1043	3.4F
	1306	1611	4.7E
	2007	2239	2.4F
8 W	0042	0420	6.5E
	0837	1120	3.2F
	1343	1647	4.7E
	2043	2314	2.2F
9 Th	0116	0456	6.4E
	0913	1158	2.9F
	1420	1723	4.6E
	2122	2349	1.8F
10 F	0151	0533	6.1E
	0951	1237	2.5F
	1458	1803	4.4E
	2207		
11 Sa		0027	1.4F
	0224	0613	5.7E
	1034	1319	2.1F
	1538	1846	4.1E
	2303		
12 Su		0111	0.9F
	0259	0656	5.2E
	1123	1407	1.7F
	1620	1934	3.8E
13 M	0019	0208	0.5F
	0336	0744	4.6E
	1220	1503	1.4F
	1706	2027	3.7E
14 Tu	0203	0322	0.3F
	0425	0839	4.1E
	1328	1605	1.1F
	1757	2127	3.7E
15 W	0327	0441	0.3F
	0541	0940	3.7E
	1439	1706	1.1F
	1851	2229	3.9E
16 Th	0412	0549	0.7F
	0709	1046	3.5E
	1540	1801	1.2F
	1945	2329	4.3E
17 F	0444	0646	1.2F
	0827	1151	3.6E
	1627	1850	1.4F
	2038		
18 Sa		0023	4.9E
	0514	0735	1.9F
	0934	1249	3.9E
	1705	1933	1.7F
	2127		
19 Su		0112	5.7E
	0544	0819	2.5F
	1031	1340	4.3E
	1740	2014	2.1F
	2215		
20 M		0158	6.4E
	0617	0901	3.2F
	1123	1427	4.7E
	1816	2054	2.5F
	2301		
21 Tu		0242	7.1E
	0653	0942	3.7F
	1210	1512	5.2E
	1854	2135	2.9F
	2348		
22 W		0327	7.5E
	0732	1024	4.0F
	1256	1557	5.5E
	1935	2217	3.1F
23 Th	0034	0412	7.8E
	0813	1107	4.2F
	1342	1643	5.7E
	2021	2302	3.1F
24 F	0122	0458	7.8E
	0858	1152	4.1F
	1428	1731	5.7E
	2111	2351	3.0F
25 Sa	0210	0546	7.5E
	0947	1240	3.9F
	1515	1821	5.6E
	2209		
26 Su		0046	2.7F
	0301	0636	6.9E
	1039	1333	3.4F
	1604	1914	5.4E
	2316		
27 M		0148	2.2F
	0356	0730	6.1E
	1138	1430	3.0F
	1655	2012	5.2E
28 Tu	0035	0259	1.8F
	0457	0829	5.2E
	1245	1533	2.5F
	1749	2115	5.0E
29 W	0201	0415	1.6F
	0607	0934	4.4E
	1400	1639	2.2F
	1845	2224	4.9E
30 Th	0320	0529	1.7F
	0722	1047	3.9E
	1516	1744	2.0F
	1943	2333	5.0E

Time meridian 135° W. 0000 is midnight. 1200 is noon. Times are not adjusted for Daylight Saving Time.

* Current weak and variable.

North Inian Pass, Cross Sound, Alaska, 2005

F–Flood, Dir. 075° True E–Ebb, Dir. 260° True

July

Day	Slack h m	Maximum h m	knots
1 F	0424	0635	2.0F
	0837	1203	3.7E
	1624	1843	2.0F
	2039		
2 Sa		0036	5.3E
	0517	0733	2.4F
	0943	1309	3.7E
	1720	1936	2.0F
	2131		
3 Su		0129	5.5E
	0602	0823	2.7F
	1040	1402	3.8E
	1808	2024	2.1F
	2219		
4 M		0214	5.8E
	0642	0908	3.0F
	1127	1445	4.0E
	1848	2107	2.2F
	2303		
5 Tu		0252	6.0E
	0718	0949	3.1F
	1209	1521	4.2E
	1924	2147	2.2F
	2344		
6 W ●		0328	6.1E
	0751	1027	3.1F
	1247	1556	4.4E
	1958	2224	2.2F
7 Th	0022	0402	6.2E
	0823	1103	3.1F
	1324	1630	4.6E
	2031	2259	2.1F
8 F	0058	0437	6.2E
	0854	1139	3.0F
	1400	1705	4.6E
	2107	2334	1.9F
9 Sa	0134	0513	6.1E
	0927	1214	2.8F
	1435	1743	4.7E
	2146		
10 Su		0010	1.7F
	0210	0551	5.9E
	1002	1250	2.5F
	1511	1822	4.6E
	2230		
11 M		0048	1.4F
	0246	0631	5.5E
	1041	1327	2.1F
	1548	1904	4.5E
	2324		
12 Tu		0133	1.0F
	0325	0714	5.0E
	1125	1407	1.7F
	1625	1950	4.4E
13 W	0029	0228	0.7F
	0410	0802	4.4E
	1216	1454	1.3F
	1705	2041	4.3E
14 Th ◐	0147	0340	0.6F
	0509	0856	3.9E
	1319	1552	1.0F
	1749	2137	4.4E
15 F	0301	0457	0.7F
	0625	0958	3.5E
	1431	1657	0.9F
	1842	2237	4.6E
16 Sa	0356	0605	1.1F
	0747	1105	3.3E
	1539	1800	1.0F
	1941	2338	5.0E
17 Su	0440	0703	1.7F
	0901	1211	3.5E
	1632	1856	1.3F
	2043		
18 M		0037	5.6E
	0519	0754	2.3F
	1005	1312	3.9E
	1716	1947	1.8F
	2142		
19 Tu		0131	6.3E
	0558	0840	3.0F
	1101	1405	4.5E
	1757	2034	2.4F
	2238		
20 W		0222	6.9E
	0637	0924	3.6F
	1151	1455	5.0E
	1839	2120	2.9F
	2331		
21 Th ○		0310	7.5E
	0717	1007	4.1F
	1238	1542	5.6E
	1923	2206	3.3F
22 F	0022	0357	7.8E
	0758	1050	4.4F
	1324	1629	6.0E
	2009	2253	3.5F
23 Sa	0112	0444	7.8E
	0841	1134	4.4F
	1408	1715	6.2E
	2059	2342	3.5F
24 Su	0202	0531	7.5E
	0926	1219	4.2F
	1452	1803	6.3E
	2153		
25 M		0033	3.2F
	0252	0618	6.9E
	1014	1307	3.7F
	1537	1852	6.1E
	2253		
26 Tu		0130	2.7F
	0344	0708	6.0E
	1107	1358	3.1F
	1622	1944	5.8E
27 W ◑	0002	0233	2.2F
	0440	0801	5.1E
	1208	1455	2.5F
	1710	2040	5.4E
28 Th	0123	0344	1.8F
	0542	0901	4.2E
	1322	1600	1.9F
	1801	2142	5.0E
29 F	0248	0458	1.6F
	0652	1010	3.5E
	1450	1709	1.5F
	1857	2252	4.8E
30 Sa	0403	0609	1.7F
	0807	1130	3.1E
	1612	1816	1.4F
	1958		
31 Su		0005	4.8E
	0503	0712	2.0F
	0917	1250	3.1E
	1715	1916	1.5F
	2058		

August

Day	Slack h m	Maximum h m	knots
1 M		0108	5.0E
	0551	0805	2.3F
	1017	1350	3.4E
	1804	2007	1.7F
	2154		
2 Tu		0158	5.2E
	0631	0851	2.6F
	1107	1434	3.7E
	1843	2052	1.9F
	2243		
3 W		0238	5.5E
	0706	0931	2.9F
	1149	1509	4.1E
	1916	2132	2.1F
	2326		
4 Th ●		0313	5.8E
	0736	1008	3.1F
	1226	1541	4.4E
	1946	2209	2.3F
5 F	0006	0346	6.0E
	0804	1042	3.2F
	1301	1612	4.8E
	2015	2243	2.3F
6 Sa	0044	0419	6.2E
	0831	1115	3.1F
	1335	1645	5.0E
	2046	2316	2.3F
7 Su	0120	0453	6.2E
	0859	1146	3.0F
	1408	1719	5.2E
	2118	2349	2.2F
8 M	0156	0528	6.0E
	0929	1216	2.8F
	1439	1754	5.3E
	2155		
9 Tu		0023	1.9F
	0233	0605	5.7E
	1002	1245	2.4F
	1511	1832	5.3E
	2237		
10 W		0100	1.7F
	0311	0645	5.2E
	1039	1315	2.0F
	1541	1913	5.2E
	2326		
11 Th		0144	1.3F
	0353	0729	4.6E
	1122	1349	1.5F
	1614	1959	5.0E
12 F ◐	0028	0242	1.0F
	0443	0819	4.0E
	1217	1434	1.0F
	1651	2052	4.8E
13 Sa	0147	0402	0.8F
	0550	0918	3.4E
	1331	1545	0.7F
	1740	2153	4.8E
14 Su	0308	0526	1.0F
	0711	1027	3.2E
	1502	1716	0.7F
	1851	2301	4.9E
15 M	0411	0636	1.5F
	0832	1141	3.3E
	1614	1830	1.1F
	2009		
16 Tu		0009	5.4E
	0459	0733	2.2F
	0942	1251	3.7E
	1705	1930	1.7F
	2121		
17 W		0112	6.0E
	0542	0822	3.0F
	1041	1350	4.4E
	1748	2022	2.4F
	2224		
18 Th		0207	6.7E
	0621	0907	3.7F
	1132	1441	5.2E
	1830	2110	3.1F
	2320		
19 F ○		0256	7.2E
	0701	0949	4.2F
	1218	1527	5.9E
	1912	2156	3.6F
20 Sa	0013	0343	7.6E
	0740	1031	4.5F
	1301	1612	6.5E
	1956	2241	3.9F
21 Su	0103	0428	7.6E
	0820	1112	4.5F
	1343	1656	6.8E
	2042	2328	3.9F
22 M	0151	0512	7.3E
	0902	1153	4.3F
	1423	1739	6.9E
	2132		
23 Tu		0015	3.7F
	0239	0557	6.7E
	0946	1236	3.7F
	1503	1824	6.7E
	2225		
24 W		0106	3.1F
	0327	0642	5.9E
	1034	1322	3.0F
	1544	1910	6.2E
	2327		
25 Th		0203	2.5F
	0417	0730	4.9E
	1131	1414	2.2F
	1626	2000	5.6E
26 F ◑	0041	0309	1.8F
	0512	0824	3.9E
	1246	1517	1.4F
	1711	2057	4.9E
27 Sa	0213	0425	1.4F
	0616	0929	3.2E
	1432	1634	0.9F
	1806	2206	4.4E
28 Su	0341	0541	1.3F
	0730	1053	2.7E
	1611	1750	0.8F
	1913	2329	4.2E
29 M	0447	0648	1.6F
	0846	1231	2.8E
	1715	1856	1.0F
	2025		
30 Tu		0046	4.4E
	0536	0743	2.0F
	0951	1339	3.2E
	1758	1950	1.4F
	2130		
31 W		0141	4.8E
	0614	0829	2.4F
	1042	1421	3.7E
	1832	2035	1.8F
	2224		

September

Day	Slack h m	Maximum h m	knots
1 Th		0222	5.2E
	0646	0909	2.8F
	1124	1452	4.2E
	1901	2115	2.2F
	2310		
2 F		0255	5.5E
	0713	0944	3.1F
	1200	1521	4.7E
	1928	2150	2.5F
	2351		
3 Sa ●		0327	5.8E
	0738	1016	3.2F
	1233	1550	5.2E
	1953	2224	2.7F
4 Su	0029	0358	6.1E
	0803	1046	3.3F
	1304	1620	5.6E
	2020	2255	2.8F
5 M	0106	0430	6.1E
	0828	1114	3.1F
	1334	1651	5.9E
	2048	2326	2.7F
6 Tu	0142	0504	6.0E
	0856	1140	2.9F
	1403	1725	6.1E
	2120	2358	2.6F
7 W	0218	0539	5.7E
	0926	1205	2.6F
	1432	1800	6.0E
	2158		
8 Th		0031	2.3F
	0256	0618	5.3E
	1001	1232	2.2F
	1500	1840	5.9E
	2242		
9 F		0110	1.9F
	0337	0700	4.7E
	1042	1302	1.7F
	1530	1924	5.6E
	2338		
10 Sa		0202	1.5F
	0424	0749	4.0E
	1136	1344	1.1F
	1605	2016	5.2E
11 Su ◐	0053	0319	1.1F
	0525	0847	3.4E
	1254	1452	0.6F
	1653	2118	4.9E
12 M	0226	0454	1.1F
	0645	0959	3.1E
	1449	1648	0.5F
	1813	2232	4.8E
13 Tu	0345	0612	1.5F
	0809	1121	3.2E
	1611	1815	0.9F
	1947	2348	5.1E
14 W	0440	0712	2.2F
	0921	1237	3.8E
	1700	1919	1.7F
	2108		
15 Th		0057	5.7E
	0524	0802	3.0F
	1020	1337	4.7E
	1742	2012	2.6F
	2214		
16 F		0153	6.3E
	0604	0847	3.7F
	1109	1427	5.6E
	1822	2059	3.3F
	2312		
17 Sa ○		0242	6.8E
	0642	0928	4.2F
	1153	1511	6.4E
	1901	2144	3.9F
18 Su	0003	0327	7.1E
	0720	1008	4.4F
	1234	1552	7.0E
	1942	2227	4.2F
19 M	0051	0410	7.1E
	0758	1046	4.4F
	1313	1633	7.3E
	2024	2311	4.2F
20 Tu	0137	0451	6.8E
	0837	1125	4.1F
	1351	1713	7.3E
	2109	2355	3.9F
21 W	0222	0533	6.3E
	0918	1204	3.5F
	1428	1753	7.0E
	2156		
22 Th		0041	3.3F
	0306	0615	5.5E
	1003	1245	2.7F
	1504	1836	6.4E
	2251		
23 F		0132	2.6F
	0351	0659	4.7E
	1057	1332	1.9F
	1541	1921	5.7E
	2359		
24 Sa ◑		0233	1.8F
	0439	0748	3.8E
	1211	1433	1.0F
	1620	2014	4.8E
25 Su	0131	0348	1.2F
	0537	0848	3.0E
	1421	1556	0.4F
	1708	2118	4.1E
26 M	0312	0508	1.0F
	0649	1011	2.6E
	1615	1722	0.3F
	1821	2243	3.8E
27 Tu	0423	0618	1.3F
	0808	1201	2.7E
	1709	1832	0.7F
	1948		
28 W		0012	3.9E
	0511	0715	1.7F
	0916	1315	3.2E
	1745	1928	1.2F
	2103		
29 Th		0115	4.3E
	0548	0801	2.2F
	1009	1356	3.9E
	1814	2013	1.8F
	2202		
30 F		0157	4.8E
	0618	0839	2.6F
	1051	1426	4.5E
	1839	2053	2.3F
	2251		

Time meridian 135° W. 0000 is midnight. 1200 is noon. Times are not adjusted for Daylight Saving Time.

North Inian Pass, Cross Sound, Alaska, 2005

F–Flood, Dir. 075° True E–Ebb, Dir. 260° True

October

Day	Slack h m	Maximum h m	knots
1 Sa		0230	5.2E
	0644	0914	3.0F
	1126	1454	5.1E
	1903	2128	2.7F
	2333		
2 Su		0302	5.6E
	0708	0945	3.1F
	1159	1522	5.7E
	1927	2201	3.0F
3 M ●	0012	0333	5.8E
	0732	1013	3.2F
	1229	1551	6.2E
	1951	2232	3.2F
4 Tu	0049	0405	6.0E
	0757	1040	3.1F
	1258	1622	6.6E
	2019	2303	3.2F
5 W	0126	0439	5.9E
	0824	1105	3.0F
	1327	1655	6.7E
	2050	2334	3.1F
6 Th	0203	0515	5.7E
	0855	1131	2.7F
	1357	1732	6.7E
	2127		
7 F		0008	2.9F
	0242	0554	5.3E
	0931	1159	2.3F
	1427	1811	6.5E
	2211		
8 Sa		0048	2.4F
	0323	0637	4.7E
	1014	1233	1.8F
	1500	1856	6.0E
	2306		
9 Su		0140	1.9F
	0411	0726	4.1E
	1111	1319	1.2F
	1538	1949	5.5E
10 M ◐	0017	0255	1.4F
	0511	0826	3.6E
	1237	1435	0.6F
	1630	2052	5.0E
11 Tu	0149	0428	1.3F
	0627	0940	3.3E
	1445	1635	0.5F
	1756	2209	4.7E
12 W	0315	0546	1.7F
	0747	1106	3.5E
	1604	1802	1.0F
	1934	2330	4.8E
13 Th	0416	0648	2.3F
	0857	1223	4.2E
	1652	1906	1.8F
	2058		
14 F		0041	5.3E
	0503	0739	3.0F
	0954	1322	5.1E
	1733	2000	2.7F
	2205		
15 Sa		0139	5.8E
	0544	0824	3.5F
	1043	1410	6.0E
	1812	2047	3.5F
	2302		
16 Su		0227	6.2E
	0622	0905	3.9F
	1126	1452	6.8E
	1850	2131	4.0F
	2353		
17 M ○		0311	6.4E
	0659	0944	4.1F
	1205	1531	7.3E
	1928	2212	4.3F
18 Tu	0039	0351	6.4E
	0736	1021	4.0F
	1242	1609	7.5E
	2007	2253	4.3F
19 W	0122	0430	6.2E
	0814	1058	3.6F
	1318	1646	7.4E
	2047	2335	3.9F
20 Th	0203	0509	5.8E
	0853	1136	3.1F
	1353	1725	7.1E
	2131		
21 F		0017	3.4F
	0244	0549	5.2E
	0936	1214	2.4F
	1428	1804	6.5E
	2220		
22 Sa		0103	2.6F
	0325	0630	4.5E
	1027	1256	1.6F
	1502	1847	5.7E
	2319		
23 Su		0158	1.9F
	0409	0717	3.8E
	1139	1351	0.8F
	1535	1935	4.9E
24 M ◑	0040	0307	1.2F
	0501	0813	3.1E
		1513	*
		2034	4.1E
25 Tu	0222	0426	1.0F
	0606	0926	2.7E
		1645	*
		2148	3.6E
26 W	0342	0538	1.1F
	0721	1103	2.7E
	1655	1800	0.4F
	1900	2316	3.6E
27 Th	0434	0637	1.5F
	0829	1227	3.3E
	1724	1858	1.0F
	2026		
28 F		0029	3.9E
	0512	0724	1.9F
	0924	1315	3.9E
	1749	1946	1.6F
	2132		
29 Sa		0119	4.3E
	0544	0804	2.3F
	1008	1349	4.7E
	1813	2026	2.2F
	2225		
30 Su		0158	4.7E
	0611	0839	2.6F
	1046	1419	5.4E
	1836	2102	2.7F
	2311		
31 M		0233	5.1E
	0636	0911	2.8F
	1120	1449	6.0E
	1859	2136	3.2F
	2352		

November

Day	Slack h m	Maximum h m	knots
1 Tu ●		0306	5.4E
	0701	0940	3.0F
	1152	1521	6.6E
	1924	2209	3.4F
2 W	0031	0341	5.6E
	0727	1008	3.0F
	1224	1554	7.0E
	1953	2241	3.6F
3 Th	0110	0416	5.7E
	0756	1036	2.9F
	1256	1630	7.2E
	2027	2315	3.5F
4 F	0149	0454	5.6E
	0830	1106	2.7F
	1329	1708	7.2E
	2105	2352	3.3F
5 Sa	0229	0535	5.3E
	0909	1140	2.4F
	1405	1750	7.0E
	2150		
6 Su		0036	2.9F
	0313	0619	4.9E
	0957	1221	2.0F
	1444	1837	6.5E
	2244		
7 M		0129	2.4F
	0402	0711	4.4E
	1059	1315	1.4F
	1529	1930	5.8E
	2351		
8 Tu ◐		0238	2.0F
	0501	0811	3.9E
	1228	1438	0.9F
	1628	2033	5.2E
9 W	0113	0359	1.8F
	0609	0924	3.7E
	1423	1621	0.8F
	1750	2147	4.7E
10 Th	0237	0515	1.9F
	0721	1046	4.0E
	1544	1744	1.2F
	1923	2308	4.6E
11 F	0345	0619	2.3F
	0827	1202	4.6E
	1637	1850	2.0F
	2045		
12 Sa		0021	4.8E
	0438	0712	2.8F
	0923	1301	5.4E
	1721	1945	2.8F
	2154		
13 Su		0122	5.1E
	0524	0759	3.2F
	1012	1350	6.2E
	1801	2033	3.4F
	2251		
14 M		0211	5.4E
	0605	0842	3.4F
	1056	1432	6.8E
	1839	2117	3.9F
	2341		
15 Tu ○		0254	5.5E
	0643	0922	3.5F
	1136	1510	7.2E
	1916	2158	4.1F
16 W	0025	0334	5.6E
	0720	1000	3.4F
	1214	1547	7.3E
	1953	2238	4.0F
17 Th	0106	0411	5.5E
	0757	1037	3.1F
	1250	1623	7.2E
	2031	2318	3.7F
18 F	0145	0449	5.2E
	0836	1114	2.7F
	1325	1700	6.9E
	2111	2358	3.3F
19 Sa	0224	0526	4.9E
	0917	1151	2.2F
	1359	1738	6.4E
	2155		
20 Su		0040	2.7F
	0303	0606	4.4E
	1005	1230	1.5F
	1432	1819	5.8E
	2244		
21 M		0128	2.1F
	0344	0651	3.9E
	1109	1318	0.9F
	1505	1904	5.1E
	2346		
22 Tu		0225	1.5F
	0430	0742	3.5E
	1255	1427	0.3F
	1538	1956	4.4E
23 W ◑	0105	0334	1.1F
	0524	0843	3.1E
		1555	*
		2057	3.8E
24 Th	0232	0445	1.0F
	0627	0957	3.1E
		1715	*
		2210	3.5E
25 F	0340	0547	1.2F
	0731	1114	3.4E
	1654	1820	0.7F
	1937	2326	3.5E
26 Sa	0428	0639	1.5F
	0828	1216	4.0E
	1720	1912	1.3F
	2053		
27 Su		0029	3.7E
	0506	0723	1.8F
	0917	1301	4.7E
	1744	1956	2.0F
	2153		
28 M		0118	4.1E
	0537	0802	2.1F
	0959	1339	5.4E
	1808	2035	2.6F
	2244		
29 Tu		0200	4.5E
	0605	0836	2.4F
	1038	1416	6.1E
	1833	2112	3.1F
	2330		
30 W		0239	4.9E
	0633	0909	2.6F
	1116	1452	6.7E
	1901	2147	3.5F

December

Day	Slack h m	Maximum h m	knots
1 Th ●	0012	0317	5.2E
	0702	0941	2.8F
	1153	1529	7.2E
	1933	2223	3.7F
2 F	0053	0356	5.4E
	0735	1015	2.9F
	1231	1609	7.5E
	2009	2300	3.8F
3 Sa	0135	0437	5.5E
	0812	1051	2.9F
	1311	1650	7.6E
	2049	2341	3.7F
4 Su	0218	0520	5.4E
	0855	1131	2.7F
	1353	1735	7.3E
	2135		
5 M		0026	3.4F
	0303	0607	5.2E
	0946	1219	2.3F
	1438	1822	6.9E
	2226		
6 Tu		0117	3.0F
	0351	0658	4.9E
	1050	1317	1.8F
	1529	1915	6.2E
	2326		
7 W		0217	2.6F
	0445	0756	4.6E
	1211	1432	1.4F
	1628	2014	5.4E
8 Th ◐	0036	0327	2.2F
	0544	0903	4.4E
	1349	1559	1.2F
	1742	2122	4.7E
9 F	0154	0438	2.1F
	0646	1017	4.5E
	1514	1720	1.5F
	1906	2239	4.3E
10 Sa	0309	0544	2.2F
	0749	1131	4.9E
	1616	1829	2.0F
	2027	2356	4.2E
11 Su	0414	0643	2.4F
	0847	1235	5.4E
	1707	1927	2.6F
	2138		
12 M		0102	4.3E
	0507	0735	2.6F
	0940	1328	6.0E
	1750	2018	3.1F
	2237		
13 Tu		0156	4.5E
	0553	0822	2.8F
	1027	1413	6.4E
	1831	2104	3.5F
	2327		
14 W		0241	4.7E
	0635	0905	2.8F
	1110	1453	6.7E
	1908	2146	3.7F
15 Th ○	0011	0321	4.8E
	0713	0945	2.8F
	1150	1530	6.8E
	1945	2225	3.7F
16 F	0051	0357	4.9E
	0749	1022	2.7F
	1228	1606	6.8E
	2020	2304	3.5F
17 Sa	0129	0433	4.9E
	0826	1059	2.4F
	1304	1642	6.6E
	2056	2341	3.2F
18 Su	0205	0509	4.8E
	0904	1135	2.1F
	1339	1718	6.3E
	2133		
19 M		0020	2.8F
	0242	0547	4.6E
	0948	1213	1.7F
	1413	1757	5.9E
	2214		
20 Tu		0100	2.4F
	0320	0628	4.3E
	1039	1255	1.2F
	1448	1838	5.4E
	2259		
21 W		0144	1.9F
	0400	0713	4.1E
	1147	1346	0.7F
	1524	1923	4.8E
	2352		
22 Th		0236	1.5F
	0443	0804	3.8E
	1324	1454	0.3F
	1606	2014	4.2E
23 F ◑	0058	0336	1.1F
	0531	0901	3.7E
		1615	*
		2113	3.6E
24 Sa	0217	0441	0.9F
	0623	1003	3.7E
	1608	1729	0.5F
	1838	2220	3.3E
25 Su	0330	0542	1.0F
	0719	1107	4.1E
	1643	1831	1.0F
	2004	2330	3.3F
26 M	0425	0636	1.1F
	0815	1206	4.6E
	1713	1923	1.6F
	2115		
27 Tu		0033	3.5E
	0505	0722	1.4F
	0907	1257	5.2E
	1741	2008	2.2F
	2215		
28 W		0126	3.9E
	0538	0803	1.8F
	0956	1343	5.9E
	1811	2049	2.9F
	2306		
29 Th		0213	4.4E
	0609	0842	2.2F
	1043	1426	6.6E
	1842	2128	3.4F
	2353		
30 F ●		0256	4.9E
	0643	0921	2.6F
	1129	1509	7.2E
	1917	2207	3.8F
31 Sa	0037	0339	5.3E
	0719	1000	3.0F
	1215	1552	7.6E
	1955	2247	4.1F

Time meridian 135° W. 0000 is midnight. 1200 is noon. Times are not adjusted for Daylight Saving Time.
* Current weak and variable.

Isanotski Strait (False Pass Cannery), Alaska, 2005

F–Flood, Dir. 000° True   E–Ebb, Dir. 185° True

January

Day	Slack h m	Maximum h m	knots
1 Sa	0259	0621	3.5F
	1015	1300	2.2E
	1614	1831	2.5F
	2105		
2 Su		0047	2.1E
	0333	0701	3.5F
	1042	1345	2.4E
	1709	1929	2.6F
	2228		
3 M		0142	1.9E
	0414	0746	3.4F
	1114	1434	2.7E
	1804	2032	2.8F
4 Tu	0008	0244	1.7E
	0505	0836	3.4F
	1151	1525	3.0E
	1859	2138	3.1F
5 W	0139	0350	1.7E
	0605	0930	3.4F
	1235	1617	3.2E
	1953	2242	3.5F
6 Th	0252	0456	1.7E
	0713	1026	3.4F
	1324	1710	3.5E
	2044	2343	3.8F
7 F	0352	0558	1.9E
	0821	1123	3.5F
	1415	1802	3.7E
	2134		
8 Sa		0039	4.2F
	0444	0656	2.2E
	0926	1220	3.6F
	1509	1854	3.9E
	2223		
9 Su		0131	4.4F
	0533	0751	2.5E
	1027	1315	3.7F
	1603	1945	4.0E
	2311		
10 M		0220	4.6F
	0619	0843	2.7E
	1126	1410	3.7F
	1658	2035	4.1E
	2358		
11 Tu		0308	4.7F
	0703	0934	2.9E
	1223	1505	3.8F
	1754	2126	4.0E
12 W	0045	0356	4.7F
	0748	1024	3.1E
	1321	1600	3.8F
	1852	2217	3.9E
13 Th	0132	0442	4.7F
	0832	1115	3.2E
	1419	1656	3.7F
	1952	2309	3.6E
14 F	0219	0530	4.5F
	0917	1207	3.2E
	1518	1754	3.6F
	2057		
15 Sa		0003	3.3E
	0308	0618	4.3F
	1002	1301	3.2E
	1620	1856	3.5F
	2208		
16 Su		0101	2.9E
	0357	0707	4.0F
	1049	1358	3.2E
	1725	2005	3.4F
	2327		
17 M		0203	2.5E
	0450	0759	3.7F
	1138	1457	3.2E
	1830	2124	3.3F
18 Tu	0052	0311	2.1E
	0547	0854	3.4F
	1228	1558	3.2E
	1932	2248	3.4F
19 W	0215	0425	1.8E
	0651	0952	3.2F
	1319	1657	3.2E
	2031		
20 Th		0001	3.6F
	0330	0539	1.7E
	0757	1052	3.0F
	1409	1752	3.2E
	2123		
21 F		0100	3.7F
	0432	0643	1.7E
	0901	1149	2.8F
	1457	1841	3.2E
	2210		
22 Sa		0148	3.8F
	0522	0737	1.8E
	0959	1239	2.8F
	1541	1925	3.2E
	2251		
23 Su		0227	3.9F
	0605	0821	1.9E
	1049	1322	2.7F
	1621	2003	3.2E
	2328		
24 M		0259	3.9F
	0641	0858	2.0E
	1133	1400	2.7F
	1658	2038	3.1E
25 Tu	0002	0323	3.8F
	0713	0930	2.0E
	1212	1436	2.7F
	1731	2111	3.0E
26 W	0033	0344	3.8F
	0741	1000	2.1E
	1249	1512	2.8F
	1803	2143	2.9E
27 Th	0101	0408	3.8F
	0806	1030	2.2E
	1324	1550	2.8F
	1837	2216	2.8E
28 F	0128	0435	3.7F
	0828	1102	2.3E
	1401	1630	2.9F
	1915	2252	2.6E
29 Sa	0154	0506	3.7F
	0849	1137	2.5E
	1442	1714	2.9F
	2002	2331	2.4E
30 Su	0222	0541	3.6F
	0911	1216	2.6E
	1528	1803	3.0F
	2100		
31 M		0016	2.1E
	0255	0620	3.5F
	0938	1301	2.8E
	1622	1858	3.0F
	2217		

February

Day	Slack h m	Maximum h m	knots
1 Tu		0110	1.9E
	0335	0706	3.4F
	1014	1351	2.9E
	1720	2001	3.1F
	2350		
2 W		0213	1.6E
	0426	0758	3.3F
	1059	1446	3.1E
	1821	2108	3.3F
3 Th	0121	0324	1.6E
	0534	0857	3.3F
	1153	1545	3.3E
	1921	2217	3.6F
4 F	0235	0435	1.7E
	0652	1000	3.3F
	1254	1644	3.5E
	2018	2322	3.9F
5 Sa	0334	0542	2.0E
	0809	1104	3.4F
	1356	1742	3.7E
	2112		
6 Su		0020	4.2F
	0423	0641	2.3E
	0918	1206	3.5F
	1458	1837	3.9E
	2203		
7 M		0113	4.4F
	0508	0735	2.7E
	1019	1304	3.7F
	1558	1929	4.0E
	2252		
8 Tu		0201	4.6F
	0551	0824	3.0E
	1115	1359	3.9F
	1656	2020	4.0E
	2339		
9 W		0247	4.7F
	0632	0912	3.2E
	1209	1453	4.0F
	1752	2110	3.9E
10 Th	0024	0331	4.6F
	0713	0959	3.3E
	1302	1545	4.0F
	1849	2159	3.7E
11 F	0109	0415	4.5F
	0754	1046	3.4E
	1355	1638	3.9F
	1947	2249	3.4E
12 Sa	0154	0458	4.4F
	0835	1134	3.4E
	1450	1733	3.8F
	2048	2340	3.1E
13 Su	0238	0542	4.1F
	0917	1225	3.3E
	1548	1830	3.6F
	2156		
14 M		0035	2.6E
	0325	0628	3.8F
	1002	1318	3.2E
	1649	1935	3.4F
	2312		
15 Tu		0135	2.2E
	0415	0718	3.4F
	1050	1417	3.1E
	1754	2053	3.3F
16 W	0036	0244	1.8E
	0514	0812	3.1F
	1144	1521	3.0E
	1859	2224	3.3F
17 Th	0200	0403	1.5E
	0623	0914	2.8F
	1242	1626	3.0E
	2000	2341	3.4F
18 F	0312	0522	1.5E
	0739	1024	2.6F
	1341	1726	3.0E
	2054		
19 Sa		0038	3.6F
	0408	0626	1.7E
	0847	1132	2.6F
	1437	1819	3.0E
	2141		
20 Su		0122	3.7F
	0451	0715	1.9E
	0944	1227	2.7F
	1527	1903	3.0E
	2223		
21 M		0157	3.7F
	0526	0753	2.1E
	1031	1309	2.8F
	1611	1941	3.0E
	2259		
22 Tu		0223	3.8F
	0557	0825	2.2E
	1110	1344	2.9F
	1651	2015	3.0E
	2332		
23 W		0244	3.8F
	0624	0853	2.3E
	1146	1419	3.0F
	1728	2047	2.9E
24 Th	0002	0305	3.7F
	0649	0921	2.5E
	1219	1453	3.1F
	1803	2120	2.9E
25 F	0030	0330	3.7F
	0710	0950	2.6E
	1252	1530	3.2F
	1840	2153	2.7E
26 Sa	0057	0359	3.7F
	0729	1023	2.7E
	1327	1609	3.3F
	1921	2230	2.6E
27 Su	0125	0431	3.6F
	0750	1058	2.8E
	1407	1653	3.4F
	2009	2310	2.4E
28 M	0155	0507	3.6F
	0815	1138	3.0E
	1453	1741	3.4F
	2108	2357	2.1E

March

Day	Slack h m	Maximum h m	knots
1 Tu	0231	0548	3.5F
	0848	1224	3.0E
	1545	1836	3.4F
	2220		
2 W		0053	1.9E
	0316	0636	3.3F
	0930	1317	3.1E
	1644	1937	3.5F
	2344		
3 Th		0158	1.7E
	0416	0733	3.2F
	1023	1416	3.2E
	1748	2045	3.6F
4 F	0104	0310	1.7E
	0533	0837	3.1F
	1127	1520	3.3E
	1852	2154	3.7F
5 Sa	0211	0422	1.9E
	0655	0946	3.2F
	1239	1623	3.4E
	1952	2300	4.0F
6 Su	0305	0528	2.3E
	0809	1054	3.4F
	1351	1724	3.6E
	2049	2358	4.2F
7 M	0352	0625	2.6E
	0913	1158	3.6F
	1457	1821	3.7E
	2141		
8 Tu		0050	4.4F
	0435	0715	3.0E
	1010	1256	3.8F
	1559	1915	3.7E
	2230		
9 W		0137	4.5F
	0516	0802	3.2E
	1102	1350	4.0F
	1656	2005	3.7E
	2317		
10 Th		0221	4.5F
	0555	0847	3.4E
	1152	1441	4.1F
	1752	2054	3.6E
11 F	0001	0303	4.4F
	0634	0931	3.5E
	1241	1531	4.1F
	1847	2142	3.4E
12 Sa	0045	0344	4.3F
	0712	1016	3.5E
	1330	1620	4.0F
	1943	2230	3.1E
13 Su	0127	0425	4.0F
	0750	1101	3.4E
	1421	1711	3.9F
	2042	2319	2.7E
14 M	0210	0507	3.8F
	0829	1148	3.3E
	1514	1805	3.7F
	2146		
15 Tu		0012	2.3E
	0256	0550	3.4F
	0911	1239	3.1E
	1611	1904	3.5F
	2258		
16 W		0110	1.9E
	0347	0638	3.1F
	0958	1335	2.9E
	1712	2015	3.3F
17 Th	0017	0219	1.6E
	0450	0734	2.7F
	1055	1439	2.7E
	1816	2143	3.2F
18 F	0132	0338	1.5E
	0607	0839	2.5F
	1202	1547	2.7E
	1917	2301	3.3F
19 Sa	0235	0454	1.6E
	0725	0955	2.4F
	1313	1651	2.7E
	2012	2356	3.4F
20 Su	0323	0552	1.8E
	0830	1109	2.5F
	1416	1745	2.7E
	2100		
21 M		0037	3.5F
	0401	0636	2.1E
	0921	1205	2.7F
	1510	1831	2.7E
	2142		
22 Tu		0107	3.5F
	0433	0710	2.3E
	1003	1245	2.9F
	1557	1910	2.8E
	2220		
23 W		0130	3.6F
	0501	0740	2.5E
	1039	1321	3.1F
	1640	1945	2.8E
	2254		
24 Th		0154	3.6F
	0526	0809	2.7E
	1113	1356	3.3F
	1720	2020	2.8E
	2325		
25 F		0220	3.6F
	0548	0839	2.8E
	1146	1432	3.5F
	1800	2055	2.7E
	2356		
26 Sa		0250	3.6F
	0609	0911	3.0E
	1220	1510	3.6F
	1841	2132	2.6E
27 Su	0026	0322	3.6F
	0631	0946	3.1E
	1258	1551	3.7F
	1927	2212	2.5E
28 M	0059	0358	3.6F
	0657	1025	3.2E
	1339	1636	3.8F
	2018	2256	2.3E
29 Tu	0136	0439	3.5F
	0730	1108	3.3E
	1426	1725	3.8F
	2117	2347	2.2E
30 W	0221	0525	3.4F
	0810	1157	3.3E
	1518	1820	3.8F
	2224		
31 Th		0045	2.0E
	0317	0618	3.2F
	0900	1252	3.3E
	1617	1920	3.8F
	2335		

Time meridian 135° W. 0000 is midnight. 1200 is noon. Times are not adjusted for Daylight Saving Time.

Isanotski Strait (False Pass Cannery), Alaska, 2005

F–Flood, Dir. 000° True E–Ebb, Dir. 185° True

April

Day	Slack h m	Maximum h m	knots
1 F		0151	2.0E
	0427	0719	3.1F
	1002	1354	3.2E
	1719	2025	3.9F
2 Sa	0041	0301	2.1E
	0546	0828	3.1F
	1117	1459	3.3E
	1823	2132	3.9F
3 Su	0139	0409	2.3E
	0702	0939	3.2F
	1236	1605	3.3E
	1924	2235	4.0F
4 M	0230	0510	2.7E
	0808	1049	3.4F
	1352	1708	3.3E
	2022	2332	4.2F
5 Tu	0315	0604	3.0E
	0906	1152	3.7F
	1500	1806	3.4E
	2116		
6 W		0022	4.2F
	0357	0653	3.3E
	0959	1249	3.9F
	1601	1900	3.4E
	2205		
7 Th		0108	4.2F
	0437	0738	3.4E
	1048	1342	4.1F
	1659	1950	3.3E
	2252		
8 F		0151	4.2F
	0515	0821	3.6E
	1135	1431	4.1F
	1754	2039	3.2E
	2337		
9 Sa		0232	4.0F
	0552	0904	3.6E
	1221	1518	4.1F
	1849	2126	2.9E
10 Su	0020	0312	3.9F
	0628	0946	3.5E
	1307	1605	4.0F
	1943	2213	2.7E
11 M	0103	0352	3.6F
	0704	1029	3.4E
	1354	1653	3.9F
	2040	2302	2.4E
12 Tu	0146	0432	3.3F
	0740	1114	3.2E
	1442	1742	3.7F
	2140	2353	2.0E
13 W	0233	0516	3.0F
	0818	1201	3.0E
	1533	1835	3.5F
	2244		
14 Th		0050	1.8E
	0329	0604	2.7F
	0901	1254	2.7E
	1627	1934	3.3F
	2350		
15 F		0155	1.6E
	0437	0659	2.4F
	0958	1350	2.5E
	1724	2041	3.2F

Day	Slack h m	Maximum h m	knots
16 Sa	0051	0305	1.6E
	0553	0804	2.2F
	1114	1457	2.4E
	1821	2147	3.2F
17 Su	0142	0411	1.8E
	0705	0917	2.2F
	1237	1601	2.3E
	1916	2241	3.2F
18 M	0224	0502	2.0E
	0803	1030	2.4F
	1348	1658	2.3E
	2005	2322	3.3F
19 Tu	0259	0543	2.3E
	0849	1126	2.7F
	1447	1747	2.4E
	2049	2355	3.4F
20 W	0329	0618	2.5E
	0928	1211	3.0F
	1538	1830	2.4E
	2129		
21 Th		0027	3.4F
	0356	0651	2.8E
	1004	1250	3.3F
	1625	1910	2.5E
	2206		
22 F		0100	3.5F
	0421	0724	3.0E
	1039	1329	3.5F
	1710	1950	2.5E
	2242		
23 Sa		0134	3.5F
	0445	0759	3.2E
	1115	1409	3.8F
	1754	2030	2.5E
	2319		
24 Su		0210	3.5F
	0511	0836	3.3E
	1153	1450	3.9F
	1840	2112	2.5E
	2357		
25 M		0249	3.5F
	0541	0915	3.5E
	1233	1535	4.1F
	1929	2157	2.4E
26 Tu	0039	0331	3.5F
	0616	0958	3.5E
	1317	1622	4.1F
	2022	2246	2.3E
27 W	0126	0417	3.4F
	0658	1045	3.5E
	1405	1712	4.2F
	2118	2341	2.3E
28 Th	0222	0509	3.3F
	0747	1136	3.5E
	1457	1807	4.2F
	2216		
29 F		0040	2.3E
	0326	0607	3.2F
	0845	1233	3.4E
	1553	1904	4.1F
	2315		
30 Sa		0143	2.4E
	0438	0711	3.1F
	0955	1335	3.3E
	1653	2005	4.1F

May

Day	Slack h m	Maximum h m	knots
1 Su	0011	0248	2.5E
	0551	0821	3.1F
	1115	1440	3.2E
	1754	2106	4.1F
2 M	0103	0351	2.8E
	0658	0933	3.3F
	1237	1546	3.1E
	1854	2205	4.1F
3 Tu	0151	0448	3.0E
	0759	1043	3.5F
	1353	1649	3.0E
	1952	2301	4.1F
4 W	0236	0540	3.3E
	0855	1146	3.7F
	1502	1749	3.0E
	2047	2351	4.0F
5 Th	0318	0629	3.5E
	0946	1243	3.9F
	1605	1845	2.9E
	2138		
6 F		0037	3.9F
	0358	0714	3.6E
	1034	1336	4.1F
	1704	1937	2.8E
	2227		
7 Sa		0121	3.8F
	0436	0757	3.6E
	1120	1424	4.1F
	1759	2027	2.6E
	2313		
8 Su		0202	3.6F
	0513	0839	3.6E
	1204	1510	4.1F
	1854	2115	2.4E
	2358		
9 M		0242	3.4F
	0547	0920	3.5E
	1248	1555	4.0F
	1947	2202	2.2E
10 Tu	0042	0321	3.2F
	0621	1002	3.3E
	1331	1638	3.9F
	2039	2250	2.0E
11 W	0129	0402	2.9F
	0654	1044	3.1E
	1414	1722	3.7F
	2132	2339	1.8E
12 Th	0219	0446	2.7F
	0729	1128	2.9E
	1457	1807	3.6F
	2224		
13 F		0031	1.7E
	0317	0534	2.4F
	0809	1215	2.6E
	1542	1854	3.5F
	2314		
14 Sa		0126	1.7E
	0422	0628	2.2F
	0900	1307	2.4E
	1629	1942	3.3F
15 Su	0000	0222	1.8E
	0530	0728	2.1F
	1015	1404	2.2F
	1717	2031	3.3F

Day	Slack h m	Maximum h m	knots
16 M	0041	0315	2.0E
	0632	0834	2.1F
	1147	1503	2.1E
	1807	2118	3.2F
17 Tu	0117	0403	2.2E
	0724	0939	2.4F
	1309	1601	2.0E
	1855	2203	3.2F
18 W	0150	0445	2.5E
	0809	1038	2.7F
	1417	1655	2.0E
	1943	2246	3.3F
19 Th	0219	0524	2.7E
	0849	1130	3.1F
	1516	1746	2.0E
	2028	2328	3.3F
20 F	0247	0603	3.0E
	0928	1217	3.4F
	1609	1833	2.1E
	2113		
21 Sa		0009	3.4F
	0317	0642	3.2E
	1007	1302	3.7F
	1658	1920	2.2E
	2158		
22 Su		0051	3.4F
	0349	0723	3.5E
	1047	1347	4.0F
	1746	2006	2.3E
	2243		
23 M		0135	3.5F
	0424	0805	3.6E
	1129	1432	4.2F
	1835	2054	2.3E
	2332		
24 Tu		0221	3.5F
	0504	0849	3.7E
	1212	1519	4.3F
	1924	2144	2.4E
25 W	0023	0309	3.5F
	0549	0936	3.8E
	1258	1608	4.4F
	2013	2236	2.5E
26 Th	0119	0401	3.4F
	0639	1026	3.7E
	1347	1658	4.5F
	2104	2330	2.6E
27 F	0220	0457	3.3F
	0735	1119	3.6E
	1437	1751	4.4F
	2155		
28 Sa		0028	2.7E
	0326	0557	3.3F
	0839	1216	3.5E
	1531	1845	4.4F
	2246		
29 Su		0127	2.8E
	0433	0702	3.2F
	0952	1316	3.3E
	1626	1941	4.3F
	2336		
30 M		0227	3.0E
	0540	0811	3.3F
	1112	1420	3.1E
	1724	2037	4.2F
31 Tu	0025	0326	3.1E
	0644	0923	3.4F
	1233	1525	2.9E
	1822	2133	4.0F

June

Day	Slack h m	Maximum h m	knots
1 W	0112	0422	3.3E
	0744	1034	3.6F
	1351	1630	2.7E
	1920	2227	3.9F
2 Th	0158	0515	3.4E
	0840	1140	3.8F
	1503	1733	2.5E
	2017	2319	3.7F
3 F	0241	0605	3.5E
	0931	1240	3.9F
	1609	1832	2.4E
	2111		
4 Sa		0008	3.6F
	0323	0652	3.6E
	1020	1334	4.0F
	1709	1928	2.3E
	2204		
5 Su		0053	3.4F
	0402	0737	3.6E
	1106	1423	4.1F
	1805	2020	2.2E
	2253		
6 M		0136	3.2F
	0440	0819	3.5E
	1149	1507	4.0F
	1857	2108	2.1E
	2342		
7 Tu		0217	3.0F
	0515	0900	3.4E
	1230	1548	4.0F
	1945	2154	2.0E
8 W	0029	0257	2.8F
	0548	0939	3.2E
	1309	1625	3.9F
	2031	2239	1.9E
9 Th	0117	0338	2.7F
	0621	1018	3.0E
	1347	1701	3.8F
	2113	2322	1.8E
10 F	0206	0421	2.5F
	0655	1058	2.8E
	1424	1736	3.7F
	2153		
11 Sa		0005	1.8E
	0259	0507	2.3F
	0733	1140	2.6E
	1459	1813	3.5F
	2230		
12 Su		0049	1.9E
	0354	0557	2.2F
	0821	1225	2.4E
	1536	1852	3.5F
	2304		
13 M		0134	2.0E
	0450	0651	2.2F
	0924	1314	2.1E
	1613	1933	3.4F
	2336		
14 Tu		0219	2.2E
	0545	0749	2.3F
	1050	1408	1.9E
	1654	2016	3.3F
15 W	0006	0305	2.4E
	0636	0851	2.5F
	1224	1505	1.7E
	1740	2101	3.3F

Day	Slack h m	Maximum h m	knots
16 Th	0037	0350	2.7E
	0724	0952	2.8F
	1345	1605	1.7E
	1831	2149	3.3F
17 F	0109	0435	2.9E
	0809	1050	3.1F
	1453	1703	1.7E
	1926	2237	3.3F
18 Sa	0144	0520	3.2E
	0853	1145	3.5F
	1552	1759	1.8E
	2022	2327	3.3F
19 Su	0223	0606	3.4E
	0937	1236	3.8F
	1644	1853	2.0E
	2119		
20 M		0017	3.4F
	0306	0653	3.7E
	1022	1326	4.1F
	1733	1945	2.2E
	2216		
21 Tu		0108	3.5F
	0352	0740	3.8E
	1107	1414	4.4F
	1820	2037	2.4E
	2313		
22 W		0159	3.5F
	0442	0828	3.9E
	1153	1502	4.5F
	1906	2128	2.6E
23 Th	0010	0253	3.6F
	0534	0918	3.9E
	1240	1551	4.6F
	1952	2220	2.7E
24 F	0109	0347	3.6F
	0630	1009	3.9E
	1328	1640	4.6F
	2038	2312	2.9E
25 Sa	0209	0444	3.6F
	0730	1102	3.7E
	1416	1729	4.6F
	2124		
26 Su		0006	3.0E
	0311	0544	3.5F
	0835	1157	3.5E
	1507	1820	4.5F
	2211		
27 M		0102	3.1E
	0415	0647	3.5F
	0947	1256	3.2E
	1559	1912	4.3F
	2258		
28 Tu		0158	3.2E
	0519	0754	3.5F
	1105	1357	2.9E
	1653	2005	4.1F
	2346		
29 W		0256	3.3E
	0623	0907	3.5F
	1226	1502	2.6E
	1750	2059	3.9F
30 Th	0034	0354	3.4E
	0724	1022	3.6F
	1347	1610	2.3E
	1849	2154	3.6F

Time meridian 135° W. 0000 is midnight. 1200 is noon. Times are not adjusted for Daylight Saving Time.

Isanotski Strait (False Pass Cannery), Alaska, 2005

F–Flood, Dir. 000° True E–Ebb, Dir. 185° True

July

Day	Slack (h m)	Maximum (h m)	knots
1 F	0122	0450	3.4E
	0822	1134	3.7F
	1502	1718	2.1E
	1949	2249	3.4F
2 Sa	0209	0544	3.5E
	0915	1238	3.9F
	1610	1823	2.0E
	2049	2342	3.2F
3 Su	0254	0634	3.5E
	1005	1333	3.9F
	1710	1922	1.9E
	2147		
4 M		0032	3.1F
	0337	0720	3.5E
	1051	1421	4.0F
	1803	2015	1.9E
	2241		
5 Tu		0118	2.9F
	0418	0803	3.4E
	1133	1502	4.0F
	1848	2101	1.9E
	2331		
6 W ●		0200	2.8F
	0457	0843	3.3E
	1211	1537	3.9F
	1929	2142	1.9E
7 Th	0017	0240	2.7F
	0532	0920	3.1E
	1247	1606	3.9F
	2006	2219	2.0E
8 F	0102	0320	2.6F
	0606	0956	3.0E
	1320	1633	3.8F
	2039	2254	2.0E
9 Sa	0145	0400	2.6F
	0640	1032	2.8E
	1351	1701	3.7F
	2109	2329	2.0E
10 Su	0228	0442	2.5F
	0717	1109	2.6E
	1420	1732	3.6F
	2136		
11 M		0006	2.1E
	0312	0526	2.5F
	0801	1148	2.3E
	1449	1806	3.5F
	2202		
12 Tu		0045	2.3E
	0400	0615	2.5F
	0857	1232	2.1E
	1519	1843	3.4F
	2226		
13 W		0127	2.4E
	0450	0709	2.6F
	1011	1322	1.8E
	1554	1925	3.3F
	2254		
14 Th ◐		0212	2.6E
	0543	0808	2.7F
	1145	1420	1.6E
	1637	2011	3.3F
	2327		
15 F		0301	2.8E
	0637	0911	2.9F
	1317	1523	1.5E
	1732	2102	3.2F
16 Sa	0007	0352	3.0E
	0729	1015	3.3F
	1432	1629	1.5E
	1838	2158	3.2F
17 Su	0054	0445	3.3E
	0820	1116	3.6F
	1533	1732	1.7E
	1948	2255	3.3F
18 M	0146	0537	3.5E
	0910	1212	3.9F
	1624	1831	2.0E
	2056	2352	3.4F
19 Tu	0240	0628	3.7E
	0958	1305	4.2F
	1711	1926	2.3E
	2159		
20 W		0049	3.5F
	0336	0719	3.9E
	1046	1354	4.5F
	1755	2017	2.6E
	2259		
21 Th ○		0144	3.7F
	0432	0810	4.0E
	1133	1442	4.6F
	1837	2107	2.9E
	2356		
22 F		0239	3.8F
	0529	0901	4.0E
	1220	1529	4.7F
	1920	2157	3.1E
23 Sa	0053	0334	3.8F
	0626	0952	3.9E
	1306	1616	4.7F
	2003	2247	3.2E
24 Su	0150	0429	3.8F
	0727	1044	3.7E
	1353	1702	4.6F
	2046	2338	3.3E
25 M	0248	0527	3.8F
	0830	1137	3.4E
	1441	1750	4.4F
	2131		
26 Tu		0030	3.3E
	0348	0627	3.7F
	0939	1234	3.1E
	1531	1839	4.2F
	2217		
27 W ◑		0126	3.3E
	0451	0733	3.6F
	1055	1334	2.7E
	1623	1930	3.9F
	2305		
28 Th		0224	3.3E
	0555	0846	3.5F
	1217	1440	2.3E
	1720	2025	3.6F
	2356		
29 F		0324	3.3E
	0659	1006	3.5F
	1340	1552	2.0E
	1822	2123	3.3F
30 Sa	0049	0425	3.3E
	0800	1124	3.6F
	1457	1707	1.8E
	1930	2224	3.1F
31 Su	0142	0524	3.3E
	0856	1229	3.8F
	1603	1815	1.8E
	2036	2325	2.9F

August

Day	Slack (h m)	Maximum (h m)	knots
1 M	0234	0617	3.3E
	0946	1322	3.9F
	1657	1913	1.9E
	2137		
2 Tu		0021	2.9F
	0323	0705	3.3E
	1031	1407	3.9F
	1742	2001	2.0E
	2230		
3 W		0108	2.8F
	0408	0747	3.2E
	1112	1443	3.9F
	1820	2041	2.1E
	2317		
4 Th ●		0149	2.8F
	0450	0825	3.2E
	1148	1511	3.9F
	1854	2116	2.1E
	2359		
5 F		0226	2.8F
	0527	0900	3.0E
	1221	1534	3.8F
	1924	2146	2.2E
6 Sa	0037	0302	2.8F
	0603	0933	2.9E
	1251	1557	3.7F
	1951	2216	2.2E
7 Su	0113	0338	2.8F
	0637	1006	2.7E
	1318	1622	3.6F
	2014	2247	2.3E
8 M	0149	0417	2.8F
	0713	1040	2.5E
	1344	1651	3.6F
	2035	2321	2.4E
9 Tu	0227	0458	2.9F
	0755	1117	2.3E
	1409	1724	3.5F
	2054	2358	2.5E
10 W	0310	0544	2.9F
	0848	1159	2.0E
	1437	1800	3.4F
	2118		
11 Th		0039	2.6E
	0359	0636	2.9F
	0957	1249	1.8E
	1512	1842	3.3F
	2149		
12 F ◐		0127	2.8E
	0454	0734	3.0F
	1126	1348	1.5E
	1558	1932	3.2F
	2230		
13 Sa		0220	2.9E
	0553	0839	3.2F
	1256	1456	1.4E
	1701	2029	3.1F
	2322		
14 Su		0317	3.1E
	0652	0946	3.4F
	1409	1606	1.5E
	1819	2131	3.1F
15 M	0021	0416	3.3E
	0750	1050	3.7F
	1507	1713	1.8E
	1938	2235	3.3F
16 Tu	0125	0513	3.5E
	0843	1149	4.0F
	1555	1812	2.2E
	2048	2337	3.4F
17 W	0229	0609	3.7E
	0935	1242	4.3F
	1639	1906	2.6E
	2150		
18 Th		0036	3.7F
	0330	0702	3.8E
	1023	1331	4.5F
	1720	1955	2.9E
	2246		
19 F ○		0132	3.9F
	0429	0753	3.9E
	1111	1417	4.6F
	1801	2043	3.2E
	2340		
20 Sa		0226	4.0F
	0527	0844	3.9E
	1157	1502	4.6F
	1842	2130	3.4E
21 Su	0033	0319	4.1F
	0624	0934	3.7E
	1243	1547	4.6F
	1922	2217	3.5E
22 M	0126	0412	4.1F
	0723	1024	3.5E
	1328	1632	4.4F
	2004	2306	3.5E
23 Tu	0221	0507	4.0F
	0824	1116	3.2E
	1415	1717	4.2F
	2047	2356	3.4E
24 W	0318	0605	3.8F
	0931	1212	2.8E
	1503	1805	3.9F
	2133		
25 Th		0050	3.3E
	0419	0708	3.7F
	1045	1312	2.4E
	1555	1855	3.6F
	2222		
26 F ◑		0149	3.2E
	0522	0821	3.5F
	1206	1420	2.0E
	1655	1951	3.2F
	2317		
27 Sa		0252	3.1E
	0628	0946	3.5F
	1327	1536	1.7E
	1805	2054	2.9F
28 Su	0017	0358	3.1E
	0731	1106	3.6F
	1439	1654	1.7E
	1919	2205	2.8F
29 M	0119	0501	3.1E
	0828	1208	3.7F
	1537	1800	1.9E
	2028	2316	2.7F
30 Tu	0219	0557	3.1E
	0919	1257	3.7F
	1624	1852	2.0E
	2126		
31 W		0015	2.8F
	0313	0645	3.1E
	1003	1337	3.8F
	1702	1933	2.2E
	2214		

September

Day	Slack (h m)	Maximum (h m)	knots
1 Th		0100	2.9F
	0400	0726	3.0E
	1042	1407	3.8F
	1735	2008	2.3E
	2256		
2 F		0136	3.0F
	0443	0803	3.0E
	1117	1431	3.7F
	1805	2038	2.4E
	2332		
3 Sa ●		0209	3.0F
	0523	0836	2.9E
	1149	1452	3.7F
	1830	2106	2.5E
4 Su	0006	0242	3.1F
	0559	0908	2.8E
	1218	1515	3.6F
	1853	2134	2.6E
5 M	0038	0316	3.2F
	0636	0941	2.6E
	1244	1542	3.5F
	1912	2205	2.7E
6 Tu	0111	0353	3.2F
	0714	1015	2.4E
	1310	1612	3.5F
	1930	2239	2.7E
7 W	0148	0434	3.3F
	0758	1053	2.2E
	1337	1646	3.4F
	1950	2316	2.8E
8 Th	0230	0519	3.3F
	0852	1137	2.0E
	1409	1724	3.3F
	2019	2359	2.9E
9 F	0318	0611	3.3F
	1000	1230	1.7E
	1450	1810	3.2F
	2058		
10 Sa		0050	2.9E
	0414	0709	3.4F
	1120	1332	1.6E
	1547	1904	3.0F
	2148		
11 Su ◐		0147	3.0E
	0516	0814	3.5F
	1237	1442	1.6E
	1702	2008	3.0F
	2252		
12 M		0249	3.1E
	0619	0921	3.6F
	1341	1552	1.8E
	1825	2116	3.1F
13 Tu	0005	0353	3.3E
	0719	1025	3.9F
	1434	1657	2.2E
	1940	2224	3.3F
14 W	0119	0454	3.4E
	0816	1124	4.1F
	1519	1753	2.6E
	2044	2328	3.5F
15 Th	0228	0551	3.6E
	0909	1216	4.3F
	1602	1844	3.0E
	2140		
16 F		0027	3.8F
	0331	0646	3.6E
	0959	1304	4.4F
	1642	1931	3.3E
	2233		
17 Sa ○		0122	4.0F
	0430	0737	3.7E
	1047	1349	4.5F
	1721	2017	3.5E
	2323		
18 Su		0214	4.2F
	0527	0827	3.6E
	1133	1433	4.4F
	1801	2102	3.6E
19 M	0013	0305	4.2F
	0623	0916	3.4E
	1218	1516	4.3F
	1840	2147	3.6E
20 Tu	0103	0356	4.2F
	0721	1006	3.2E
	1303	1559	4.1F
	1920	2234	3.6E
21 W	0154	0448	4.1F
	0820	1057	2.9E
	1349	1643	3.8F
	2001	2322	3.5E
22 Th	0247	0542	3.9F
	0925	1152	2.5E
	1438	1730	3.5F
	2045		
23 F		0014	3.3E
	0344	0642	3.7F
	1036	1252	2.1E
	1533	1820	3.2F
	2135		
24 Sa ◑		0112	3.1E
	0445	0752	3.5F
	1151	1400	1.8E
	1638	1918	2.8F
	2234		
25 Su		0215	2.9E
	0549	0913	3.4F
	1303	1518	1.7E
	1755	2026	2.6F
	2344		
26 M		0324	2.8E
	0651	1031	3.5F
	1406	1632	1.8E
	1911	2146	2.5F
27 Tu	0057	0430	2.7E
	0749	1130	3.5F
	1456	1731	2.0E
	2016	2305	2.6F
28 W	0202	0528	2.7E
	0839	1215	3.6F
	1536	1817	2.2E
	2108		
29 Th		0002	2.8F
	0259	0616	2.7E
	0924	1249	3.6F
	1611	1854	2.4E
	2150		
30 F		0042	3.0F
	0349	0657	2.7E
	1003	1316	3.6F
	1641	1925	2.6E
	2227		

Time meridian 135° W. 0000 is midnight. 1200 is noon. Times are not adjusted for Daylight Saving Time.

Isanotski Strait (False Pass Cannery), Alaska, 2005

F–Flood, Dir. 000° True E–Ebb, Dir. 185° True

October

Day	Slack h m	Maximum h m	knots
1 Sa		0115	3.1F
	0433	0734	2.7E
	1039	1339	3.6F
	1707	1954	2.7E
	2301		
2 Su		0146	3.3F
	0515	0808	2.6E
	1111	1404	3.5F
	1730	2023	2.8E
	2333		
3 M ●		0219	3.4F
	0554	0842	2.6E
	1141	1432	3.5F
	1750	2054	2.9E
4 Tu	0006	0255	3.5F
	0634	0917	2.4E
	1210	1502	3.4F
	1809	2127	3.0E
5 W	0040	0333	3.6F
	0717	0955	2.3E
	1240	1536	3.4F
	1831	2203	3.1E
6 Th	0118	0415	3.7F
	0805	1037	2.1E
	1314	1614	3.3F
	1900	2243	3.1E
7 F	0200	0501	3.7F
	0900	1125	2.0E
	1355	1657	3.2F
	1936	2329	3.1E
8 Sa	0249	0553	3.7F
	1002	1220	1.9E
	1449	1748	3.0F
	2023		
9 Su		0022	3.1E
	0344	0650	3.7F
	1108	1323	1.8E
	1558	1848	2.9F
	2123		
10 M ◐		0122	3.1E
	0444	0752	3.8F
	1211	1431	2.0E
	1717	1956	2.9F
	2238		
11 Tu		0226	3.1E
	0547	0850	3.9F
	1307	1538	2.3E
	1833	2107	3.1F
12 W	0000	0332	3.2E
	0649	0958	4.0F
	1356	1638	2.6E
	1939	2217	3.3F
13 Th	0119	0435	3.2E
	0747	1055	4.1F
	1441	1731	3.0E
	2037	2321	3.6F
14 F	0230	0534	3.3E
	0842	1147	4.2F
	1523	1821	3.3E
	2130		
15 Sa		0019	3.9F
	0334	0630	3.3E
	0933	1235	4.2F
	1603	1907	3.5E
	2219		
16 Su		0113	4.1F
	0433	0722	3.3E
	1021	1320	4.2F
	1642	1952	3.7E
	2307		
17 M ○		0204	4.3F
	0530	0812	3.2E
	1108	1403	4.1F
	1720	2036	3.7E
	2355		
18 Tu		0253	4.3F
	0625	0902	3.0E
	1154	1446	4.0F
	1758	2120	3.7E
19 W	0042	0342	4.2F
	0721	0951	2.8E
	1239	1528	3.7F
	1836	2204	3.6E
20 Th	0130	0431	4.1F
	0819	1041	2.5E
	1326	1611	3.4F
	1915	2251	3.4E
21 F	0219	0522	3.9F
	0919	1135	2.2E
	1417	1657	3.1F
	1957	2340	3.2E
22 Sa	0310	0617	3.7F
	1023	1233	1.9E
	1516	1747	2.8F
	2044		
23 Su		0034	2.9E
	0405	0716	3.6F
	1127	1339	1.8E
	1626	1845	2.5F
	2144		
24 M ◑		0134	2.7E
	0502	0822	3.4F
	1227	1449	1.8E
	1742	1953	2.2F
	2302		
25 Tu		0239	2.5E
	0600	0929	3.4F
	1319	1555	1.9E
	1854	2112	2.2F
26 W	0026	0345	2.4E
	0655	1025	3.3F
	1402	1648	2.2E
	1952	2232	2.4F
27 Th	0138	0445	2.3E
	0747	1109	3.4F
	1440	1730	2.4E
	2040	2330	2.7F
28 F	0240	0536	2.3E
	0833	1143	3.4F
	1512	1806	2.6E
	2119		
29 Sa		0011	3.0F
	0333	0620	2.3E
	0914	1214	3.4F
	1540	1839	2.8E
	2155		
30 Su		0046	3.2F
	0420	0700	2.3E
	0952	1245	3.4F
	1605	1911	3.0E
	2229		
31 M		0120	3.5F
	0505	0739	2.3E
	1028	1317	3.4F
	1628	1943	3.1E
	2302		

November

Day	Slack h m	Maximum h m	knots
1 Tu ●		0156	3.7F
	0548	0817	2.3E
	1102	1351	3.4F
	1651	2018	3.3E
	2337		
2 W		0235	3.8F
	0632	0856	2.3E
	1138	1427	3.3F
	1717	2054	3.4E
3 Th	0014	0315	3.9F
	0717	0939	2.2E
	1217	1507	3.3F
	1748	2134	3.4E
4 F	0054	0359	4.0F
	0805	1025	2.2E
	1301	1551	3.2F
	1826	2218	3.4E
5 Sa	0138	0447	4.1F
	0857	1116	2.1E
	1353	1640	3.1F
	1911	2307	3.4E
6 Su	0227	0538	4.1F
	0951	1212	2.2E
	1456	1735	3.0F
	2006		
7 M		0001	3.3E
	0320	0632	4.1F
	1045	1312	2.3E
	1606	1838	3.0F
	2114		
8 Tu ◐		0101	3.2E
	0417	0730	4.1F
	1139	1415	2.5E
	1719	1947	3.0F
	2234		
9 W		0205	3.1E
	0516	0830	4.1F
	1229	1517	2.7E
	1827	2058	3.2F
10 Th	0000	0311	3.0E
	0617	0928	4.1F
	1317	1614	3.0E
	1920	2209	3.4F
11 F	0120	0416	3.0E
	0716	1024	4.1F
	1402	1708	3.3E
	2025	2314	3.7F
12 Sa	0232	0517	2.9E
	0812	1117	4.0F
	1445	1757	3.5E
	2117		
13 Su		0013	4.0F
	0337	0615	2.9E
	0906	1206	4.0F
	1526	1845	3.7E
	2207		
14 M		0108	4.1F
	0438	0710	2.8E
	0957	1252	3.9F
	1606	1930	3.8E
	2254		
15 Tu ○		0158	4.2F
	0535	0801	2.7E
	1046	1336	3.7F
	1644	2014	3.8E
	2340		
16 W		0247	4.3F
	0630	0851	2.5E
	1133	1419	3.6F
	1722	2057	3.7E
17 Th	0025	0333	4.2F
	0724	0941	2.4E
	1221	1501	3.3F
	1759	2140	3.5E
18 F	0109	0419	4.1F
	0817	1030	2.2E
	1310	1544	3.1F
	1836	2224	3.3E
19 Sa	0153	0504	4.0F
	0910	1120	2.0E
	1402	1629	2.8F
	1915	2309	3.1E
20 Su	0238	0550	3.8F
	1002	1213	1.9E
	1501	1718	2.5F
	1958	2357	2.8E
21 M	0323	0637	3.6F
	1052	1309	1.9E
	1607	1813	2.3F
	2052		
22 Tu		0050	2.5E
	0410	0725	3.5F
	1139	1406	1.9E
	1715	1914	2.1F
	2207		
23 W ◑		0147	2.3E
	0459	0815	3.3F
	1222	1501	2.1E
	1820	2021	2.1F
	2339		
24 Th		0247	2.0E
	0549	0903	3.3F
	1301	1550	2.3E
	1915	2132	2.3F
25 F	0103	0348	1.9E
	0639	0949	3.2F
	1335	1634	2.5E
	2002	2236	2.6F
26 Sa	0213	0445	1.9E
	0728	1032	3.2F
	1407	1714	2.7E
	2042	2328	2.9F
27 Su	0313	0536	1.9E
	0814	1113	3.2F
	1435	1751	2.9E
	2120		
28 M		0012	3.3F
	0406	0624	1.9E
	0859	1153	3.2F
	1503	1829	3.1E
	2157		
29 Tu		0053	3.6F
	0454	0709	2.0E
	0942	1234	3.3F
	1532	1907	3.3E
	2234		
30 W		0134	3.8F
	0539	0753	2.1E
	1026	1315	3.3F
	1604	1947	3.5E
	2312		

December

Day	Slack h m	Maximum h m	knots
1 Th ●		0216	4.0F
	0624	0837	2.2E
	1112	1359	3.3F
	1641	2028	3.6E
	2353		
2 F		0259	4.2F
	0708	0924	2.2E
	1200	1445	3.3F
	1722	2112	3.7E
3 Sa	0035	0344	4.3F
	0753	1012	2.3E
	1253	1534	3.3F
	1809	2159	3.7E
4 Su	0120	0431	4.4F
	0839	1103	2.5E
	1350	1627	3.3F
	1902	2250	3.6E
5 M	0207	0521	4.4F
	0926	1157	2.6E
	1452	1725	3.2F
	2003	2344	3.5E
6 Tu	0258	0612	4.4F
	1013	1253	2.7E
	1558	1827	3.2F
	2112		
7 W		0042	3.3E
	0351	0705	4.3F
	1102	1351	2.9E
	1705	1934	3.2F
	2232		
8 Th ◐		0144	3.1E
	0446	0800	4.2F
	1150	1450	3.1E
	1811	2045	3.4F
	2356		
9 F		0249	2.8E
	0545	0856	4.0F
	1237	1548	3.3E
	1912	2157	3.5F
10 Sa	0118	0356	2.7E
	0644	0952	3.9F
	1324	1643	3.5E
	2010	2307	3.7F
11 Su	0233	0501	2.5E
	0743	1047	3.8F
	1410	1736	3.6E
	2104		
12 M		0010	3.9F
	0342	0604	2.4E
	0841	1139	3.7F
	1454	1825	3.7E
	2154		
13 Tu		0107	4.1F
	0444	0702	2.3E
	0937	1228	3.5F
	1536	1912	3.7E
	2242		
14 W		0159	4.2F
	0541	0756	2.3E
	1030	1315	3.4F
	1618	1957	3.7E
	2327		
15 Th ○		0246	4.2F
	0633	0847	2.2E
	1121	1359	3.2F
	1657	2040	3.6E
16 F	0010	0328	4.2F
	0722	0934	2.1E
	1210	1442	3.0F
	1735	2121	3.4E
17 Sa	0051	0408	4.1F
	0807	1019	2.1E
	1259	1524	2.9F
	1812	2202	3.2E
18 Su	0130	0444	3.9F
	0849	1103	2.0E
	1348	1607	2.7F
	1849	2242	3.0E
19 M	0207	0520	3.8F
	0929	1146	2.0E
	1440	1652	2.5F
	1929	2324	2.7E
20 Tu	0243	0556	3.7F
	1007	1230	2.0E
	1534	1740	2.3F
	2015		
21 W		0008	2.4E
	0319	0634	3.5F
	1043	1315	2.1E
	1631	1833	2.2F
	2114		
22 Th		0056	2.1E
	0356	0714	3.4F
	1118	1401	2.2E
	1729	1930	2.2F
	2236		
23 F ◑		0149	1.8E
	0435	0756	3.3F
	1150	1448	2.3E
	1823	2033	2.3F
24 Sa	0014	0247	1.6E
	0519	0842	3.2F
	1222	1534	2.5E
	1913	2137	2.6F
25 Su	0140	0349	1.4E
	0609	0929	3.1F
	1254	1621	2.8E
	1959	2239	2.9F
26 M	0252	0450	1.4E
	0706	1018	3.1F
	1329	1700	3.0E
	2043	2335	3.3F
27 Tu	0350	0547	1.5E
	0804	1108	3.1F
	1406	1751	3.3E
	2125		
28 W		0024	3.6F
	0439	0641	1.7E
	0902	1158	3.2F
	1448	1836	3.5E
	2207		
29 Th		0111	3.9F
	0523	0730	2.0E
	0958	1248	3.3F
	1532	1921	3.7E
	2249		
30 F ●		0156	4.2F
	0605	0818	2.2E
	1052	1338	3.4F
	1620	2007	3.8E
	2332		
31 Sa		0241	4.4F
	0646	0906	2.5E
	1147	1429	3.5F
	1710	2054	3.9E

Time meridian 135° W. 0000 is midnight. 1200 is noon. Times are not adjusted for Daylight Saving Time.

EXPLANATION OF PREDICTIONS FOR UNIMAK PASS

The predictions for Unimak Pass contain the predicted times of slack water, times and speeds of maximum flood and ebb, and times and speeds of minimum flood. The currents are identified by an "F" accompanying a flood speed and "E" with an ebb speed. The middle one of three consecutive floods or a 0.0 speed is called "minimum flood".

The currents in this waterway are dominated primarily by the declination of the Moon. When the Moon is near the Equator there are two flood and two ebb currents each day, but as the Moon's declination increases it gives rise to a diminishing speed in one ebb and an increasing speed in the other. As the Moon approaches its extreme declination, north or south of the Equator, the diurnal inequality in the ebb current may become so pronounced that one ebb entirely disappears and instead of two floods and two ebbs there are but one flood and one ebb in a day.

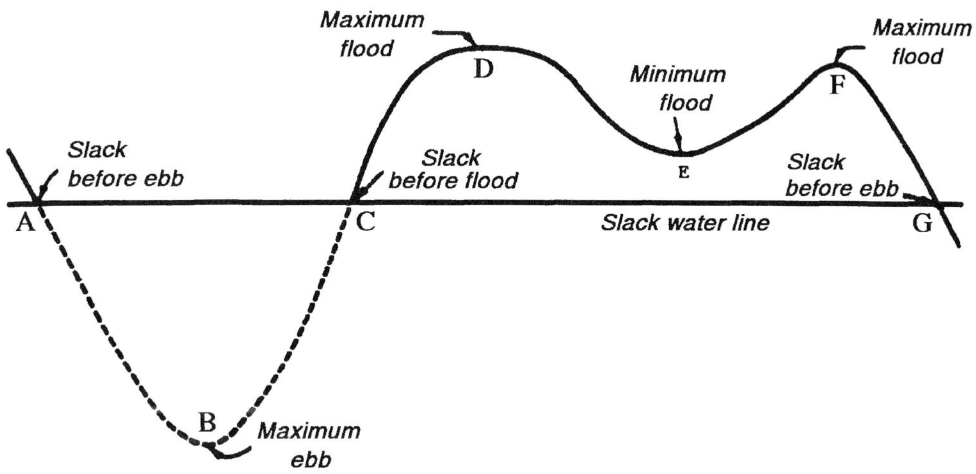

The current on days when there is but one flood and one ebb is represented by the figure above, which is characteristic of the current on a day when the Moon is near its maximum declination. The speed at any moment is given by the distance of the curve from the horizontal line, while the occurrence of slack water is denoted by the intersection of the curve with the horizontal or slack water line. The flood current is represented by the curve above the line and the ebb current by the curve below the line. It will be noted that when there are but one ebb and one flood in a day the flood lasts for the greater part of the day and is of varying strength. Starting with the slack before ebb, *A*, the figure shows that the current gradually increases to a maximum ebb at *B* after which it begins to decrease to a slack before flood at *C*. From this slack the current gradually increases to a maximum flood at *D* and then begins to decrease gradually to a smaller speed at *E*, called the minimum flood. From this point the current again increases to a second maximum flood, *F*, after which it gradually decreases to a slack before ebb at *G*.

Unimak Pass (off Scotch Cap), Aleutian Islands, 2005

F–Flood, Dir. 295° True   E–Ebb, Dir. 105° True

January

Day	Slack h m	Maximum h m	knots
1 Sa		0115	3.3E
	0437	0754	3.9F
	1240	1426	0.8E
	1631	1936	2.2F
	2229		
2 Su		0156	2.5E
	0501	0827	3.8F
	1250	1509	1.4E
	1752	2038	2.1F
	2349		
3 M ◑		0244	1.6E
	0520	0903	3.6F
	1304	1556	2.1E
	1910	2149	2.1F
4 Tu	0152	0342	0.7E
	0529	0943	3.4F
	1324	1648	2.9E
	2022	2309	2.4F
5 W		0453	0.1F
		1027	3.2F
	1353	1744	3.7E
	2129		
6 Th		0041	2.9F
		0618	0.7F
		1117	3.1F
	1433	1841	4.6E
	2230		
7 F		0208	3.7F
		0742	0.9F
		1212	3.2F
	1522	1938	5.4E
	2328		
8 Sa		0311	4.4F
		0851	0.9F
		1310	3.3F
	1617	2033	6.1E
9 Su	0022	0402	5.0F
		0946	0.6F
		1409	3.6F
	1715	2127	6.6E
10 M ●	0113	0447	5.4F
		1035	0.3F
		1507	3.8F
	1814	2218	6.7E
11 Tu	0200	0529	5.5F
	1034	1122	0.2E
	1210	1605	3.9F
	1913	2308	6.5E
12 W	0245	0607	5.5F
	1041	1208	0.8E
	1337	1703	3.9F
	2013	2357	5.9E
13 Th	0326	0644	5.3F
	1101	1255	1.4E
	1453	1801	3.7F
	2114		
14 F		0045	4.9E
	0404	0719	5.1F
	1126	1344	2.1E
	1607	1901	3.3F
	2220		
15 Sa		0133	3.7E
	0437	0753	4.7F
	1155	1435	2.6E
	1720	2006	2.9F
	2337		
16 Su ◐		0223	2.4E
	0506	0828	4.3F
	1227	1528	3.0E
	1833	2121	2.5F
17 M	0112	0318	1.2E
	0527	0904	3.8F
	1301	1625	3.4E
	1946	2301	2.4F
18 Tu	0336	0426	0.1E
	0518	0943	3.3F
	1340	1724	3.6E
	2055		
19 W		0049	2.6F
		0554	0.6F
		1028	2.9F
	1422	1824	3.8E
	2159		
20 Th		0207	3.1F
		0722	0.8F
		1118	2.5F
	1509	1921	4.0E
	2256		
21 F		0305	3.5F
		0831	0.8F
		1214	2.3F
	1558	2012	4.2E
	2345		
22 Sa		0353	3.8F
		0925	0.6F
		1311	2.3F
	1647	2058	4.4E
23 Su	0029	0432	4.0F
		1009	0.4F
		1405	2.4F
	1732	2139	4.6E
24 M	0108	0505	4.1F
		1046	0.2F
		1453	2.6F
	1814	2215	4.6E
25 Tu ○	0143	0532	4.2F
		1118	0.0
		1537	2.7F
	1854	2247	4.6E
26 W	0215	0550	4.2F
	1045	1146	0.3E
	1246	1620	2.9F
	1932	2317	4.4E
27 Th	0244	0603	4.2F
	1044	1211	0.6E
	1343	1703	3.0F
	2012	2348	4.1E
28 F	0310	0620	4.2F
	1049	1237	1.1E
	1437	1747	3.0F
	2055		
29 Sa		0020	3.5E
	0334	0643	4.2F
	1057	1306	1.6E
	1531	1833	3.0F
	2145		
30 Su		0054	2.9E
	0355	0710	4.1F
	1106	1339	2.1E
	1627	1924	2.9F
	2246		
31 M		0132	2.1E
	0410	0740	3.9F
	1119	1417	2.7E
	1728	2020	2.9F

February

Day	Slack h m	Maximum h m	knots
1 Tu ◑	0008	0215	1.1E
	0417	0814	3.7F
	1139	1502	3.2E
	1834	2124	2.8F
2 W	0218	0307	0.2E
	0357	0853	3.4F
	1209	1556	3.7E
	1946	2240	2.8F
3 Th		0415	0.7F
		0939	3.2F
	1252	1659	4.2E
	2059		
4 F		0018	3.0F
		0546	1.2F
		1035	3.0F
	1348	1809	4.7E
	2209		
5 Sa		0203	3.6F
		0729	1.3F
		1142	3.0F
	1456	1919	5.2E
	2312		
6 Su		0306	4.2F
		0842	0.9F
		1253	3.2F
	1609	2024	5.7E
7 M	0008	0353	4.7F
		0935	0.3F
		1402	3.5F
	1719	2121	6.0E
8 Tu ●	0058	0432	5.0F
	0912	1021	0.5E
	1128	1508	3.8F
	1825	2213	6.0E
9 W	0142	0506	5.2F
	0916	1104	1.3E
	1250	1609	4.1F
	1927	2301	5.6E
10 Th	0221	0537	5.2F
	0932	1145	2.1E
	1358	1707	4.2F
	2028	2346	4.9E
11 F	0256	0607	5.0F
	0955	1227	2.8E
	1500	1802	4.1F
	2129		
12 Sa		0030	3.9E
	0328	0636	4.8F
	1020	1308	3.4E
	1600	1857	3.9F
	2234		
13 Su		0114	2.8E
	0355	0706	4.4F
	1048	1351	3.7E
	1659	1953	3.5F
	2347		
14 M		0158	1.6E
	0415	0738	4.0F
	1118	1436	3.7E
	1758	2054	3.0F
15 Tu ◐	0120	0247	0.6E
	0420	0812	3.5F
	1153	1526	3.6E
	1901	2211	2.7F
16 W		0346	0.3F
		0850	3.0F
	1233	1625	3.4E
	2007		
17 Th		0010	2.5F
		0516	0.9F
		0936	2.5F
	1323	1734	3.3E
	2116		
18 F		0142	2.8F
		0700	1.1F
		1031	2.2F
	1424	1846	3.4E
	2221		
19 Sa		0245	3.1F
		0813	0.9F
		1138	2.0F
	1531	1949	3.6E
	2316		
20 Su		0331	3.4F
		0905	0.5F
		1250	2.1F
	1633	2040	3.8E
21 M	0002	0407	3.7F
		0945	0.1F
		1355	2.3F
	1727	2122	4.0E
22 Tu	0041	0435	3.8F
	0917	1019	0.3E
	1117	1448	2.6F
	1815	2158	4.1E
23 W ○	0114	0454	3.9F
	0909	1047	0.8E
	1218	1534	2.9F
	1859	2230	4.0E
24 Th	0143	0505	3.9F
	0910	1110	1.3E
	1308	1617	3.2F
	1942	2300	3.7E
25 F	0209	0517	4.0F
	0916	1133	1.8E
	1354	1658	3.5F
	2027	2331	3.3E
26 Sa	0233	0535	4.0F
	0924	1157	2.4E
	1438	1741	3.7F
	2115		
27 Su		0002	2.8E
	0254	0559	4.0F
	0934	1225	3.0E
	1523	1825	3.8F
	2209		
28 M		0036	2.1E
	0310	0627	3.9F
	0947	1258	3.5E
	1611	1912	3.8F
	2314		

March

Day	Slack h m	Maximum h m	knots
1 Tu		0114	1.3E
	0319	0658	3.7F
	1007	1336	3.9E
	1705	2005	3.7F
2 W	0041	0157	0.5E
	0315	0733	3.5F
	1037	1422	4.1E
	1806	2105	3.4F
3 Th ◑		0249	0.3F
		0814	3.3F
	1118	1519	4.2E
	1916	2219	3.2F
4 F		0358	1.0F
		0905	3.0F
	1213	1628	4.3E
	2032		
5 Sa		0002	3.2F
		0541	1.3F
		1011	2.8F
	1325	1749	4.4E
	2145		
6 Su		0147	3.5F
		0727	0.9F
		1130	2.7F
	1452	1908	4.6E
	2250		
7 M		0244	4.0F
		0830	0.2F
		1253	2.9F
	1619	2016	4.8E
	2344		
8 Tu		0325	4.4F
	0746	0918	0.7E
	1044	1411	3.4F
	1735	2114	4.9E
9 W	0031	0358	4.6F
	0750	1000	1.8E
	1202	1519	3.9F
	1843	2204	4.7E
10 Th ●	0111	0428	4.7F
	0806	1039	2.7E
	1305	1618	4.3F
	1945	2250	4.2E
11 F	0146	0455	4.6F
	0827	1117	3.6E
	1401	1711	4.6F
	2046	2333	3.5E
12 Sa	0218	0523	4.5F
	0851	1153	4.1E
	1452	1800	4.6F
	2146		
13 Su		0015	2.6E
	0245	0551	4.2F
	0917	1230	4.4E
	1541	1847	4.4F
	2248		
14 M		0055	1.7E
	0308	0620	3.9F
	0944	1308	4.4E
	1629	1934	4.0F
	2358		
15 Tu		0136	0.9E
	0322	0652	3.5F
	1014	1347	4.1E
	1719	2024	3.5F
16 W	0142	0220	0.1E
	0302	0727	3.1F
	1047	1432	3.7E
	1814	2123	3.0F
17 Th ◐		0314	0.5F
		0806	2.6F
	1128	1527	3.3E
	1915	2302	2.6F
18 F		0442	1.0F
		0853	2.2F
	1223	1637	2.9E
	2023		
19 Sa		0100	2.6F
		0634	0.9F
		0954	1.9F
	1338	1801	2.8E
	2131		
20 Su		0205	2.8F
		0744	0.6F
		1111	1.7F
	1504	1914	2.9E
	2230		
21 M		0249	3.1F
		0832	0.1F
		1236	1.8F
	1620	2010	3.1E
	2317		
22 Tu		0320	3.3F
	0743	0909	0.5E
	1027	1350	2.2F
	1722	2055	3.2E
	2356		
23 W		0341	3.4F
	0735	0939	1.1E
	1130	1446	2.7F
	1816	2133	3.2E
24 Th	0029	0352	3.5F
	0738	1004	1.8E
	1220	1531	3.3F
	1907	2207	3.0E
25 F ○	0058	0404	3.6F
	0745	1026	2.5E
	1304	1612	3.8F
	1956	2240	2.7E
26 Sa	0124	0423	3.7F
	0755	1050	3.2E
	1345	1652	4.2F
	2046	2313	2.3E
27 Su	0147	0447	3.7F
	0806	1117	3.9E
	1426	1733	4.5F
	2139	2347	1.8E
28 M	0205	0515	3.7F
	0822	1149	4.4E
	1509	1816	4.6F
	2239		
29 Tu		0023	1.2E
	0218	0547	3.6F
	0845	1225	4.8E
	1556	1903	4.6F
	2352		
30 W		0103	0.5E
	0220	0622	3.5F
	0916	1307	4.9E
	1648	1954	4.3F
31 Th		0149	0.1F
		0701	3.3F
	0956	1356	4.8E
	1747	2054	3.9F

Time meridian 135° W. 0000 is midnight. 1200 is noon. Times are not adjusted for Daylight Saving Time.
NOTE–See detailed explanation on page 97.

Unimak Pass (off Scotch Cap), Aleutian Islands, 2005

F–Flood, Dir. 295° True E–Ebb, Dir. 105° True

April

Day	Slack h m	Maximum h m	knots
1 F ◑		0246	0.7F
		0749	3.0F
	1046	1456	4.5E
	1854	2207	3.5F
2 Sa		0406	1.0F
		0849	2.7F
	1152	1610	4.1E
	2005	2342	3.4F
3 Su		0555	0.7F
		1006	2.4F
	1322	1734	3.8E
	2114		
4 M		0106	3.6F
	0644	0714	0.1E
	0744	1136	2.4F
	1506	1856	3.7E
	2214		
5 Tu		0159	3.8F
	0610	0808	1.1E
	0957	1312	2.8F
	1639	2004	3.6E
	2305		
6 W		0238	4.0F
	0620	0853	2.2E
	1113	1432	3.5F
	1757	2102	3.4E
	2349		
7 Th		0310	4.1F
	0640	0933	3.3E
	1214	1535	4.1F
	1906	2152	3.0E
8 F ●	0027	0339	4.1F
	0702	1010	4.1E
	1306	1626	4.6F
	2009	2238	2.5E
9 Sa	0101	0407	4.0F
	0727	1045	4.7E
	1352	1712	4.9F
	2108	2320	1.9E
10 Su	0131	0436	3.8F
	0752	1120	5.0E
	1436	1754	4.9F
	2207		
11 M		0000	1.4E
	0150	0505	3.6F
	0819	1154	5.0E
	1518	1834	4.6F
	2309		
12 Tu		0039	0.8E
	0215	0537	3.3F
	0846	1229	4.7E
	1601	1914	4.3F
13 W	0025	0119	0.2E
	0217	0611	3.0F
	0916	1306	4.3E
	1645	1957	3.8F
14 Th		0201	0.3F
		0648	2.7F
	0951	1349	3.8E
	1734	2047	3.3F
15 F		0255	0.6F
		0730	2.3F
	1033	1439	3.2E
	1828	2151	2.8F

Day	Slack h m	Maximum h m	knots
16 Sa ◐		0418	0.8F
		0821	1.9F
	1128	1542	2.7E
	1927	2334	2.6F
17 Su		0557	0.6F
		0926	1.6F
	1248	1659	2.4E
	2028		
18 M		0051	2.7F
		0701	0.1F
		1047	1.5F
	1431	1818	2.2E
	2123		
19 Tu		0132	2.8F
	0603	0746	0.6E
	0916	1218	1.7F
	1603	1923	2.1E
	2211		
20 W		0156	2.9F
	0555	0820	1.3E
	1028	1339	2.3F
	1717	2015	2.1E
	2251		
21 Th		0213	3.0F
	0601	0848	2.1E
	1122	1436	2.9F
	1821	2059	1.9E
	2326		
22 F		0234	3.2F
	0611	0914	3.0E
	1207	1522	3.6F
	1920	2139	1.7E
	2357		
23 Sa		0259	3.3F
	0623	0941	3.8E
	1249	1603	4.3F
	2016	2217	1.4E
24 Su ○	0024	0329	3.4F
	0639	1010	4.6E
	1330	1644	4.8F
	2112	2255	1.1E
25 M	0046	0401	3.4F
	0700	1043	5.2E
	1412	1726	5.1F
	2212	2333	0.7E
26 Tu	0103	0435	3.5F
	0728	1120	5.6E
	1457	1810	5.2F
	2322		
27 W		0014	0.3E
	0110	0513	3.5F
	0803	1202	5.7E
	1545	1858	5.0F
28 Th		0058	0.1F
		0555	3.4F
	0844	1249	5.6E
	1637	1950	4.7F
29 F		0151	0.4F
		0642	3.1F
	0934	1342	5.1E
	1733	2049	4.3F
30 Sa ◑		0257	0.5F
		0739	2.8F
	1034	1444	4.5E
	1832	2155	4.0F

May

Day	Slack h m	Maximum h m	knots
1 Su		0422	0.3F
		0849	2.3F
	1152	1556	3.7E
	1932	2305	3.8F
2 M	0418	0545	0.4E
	0707	1015	2.1F
	1338	1717	3.0E
	2030		
3 Tu		0008	3.7F
	0426	0649	1.4E
	0859	1159	2.3F
	1529	1838	2.5E
	2123		
4 W		0058	3.7F
	0447	0740	2.6E
	1018	1340	2.9F
	1706	1947	2.1E
	2210		
5 Th		0138	3.7F
	0511	0824	3.6E
	1120	1450	3.7F
	1828	2046	1.7E
	2252		
6 F		0213	3.6F
	0538	0904	4.4E
	1212	1545	4.3F
	1939	2139	1.4E
	2330		
7 Sa ●		0246	3.5F
	0605	0941	5.0E
	1257	1631	4.7F
	2043	2226	1.0E
8 Su	0004	0318	3.4F
	0633	1016	5.3E
	1339	1713	4.9F
	2144	2309	0.7E
9 M	0033	0351	3.3F
	0701	1050	5.3E
	1419	1750	4.8F
	2245	2350	0.4E
10 Tu	0054	0425	3.1F
	0730	1124	5.1E
	1458	1825	4.6F
11 W		0028	0.0
		0500	2.9F
	0801	1159	4.8E
	1537	1901	4.3F
12 Th		0107	0.2F
		0537	2.7F
	0834	1236	4.4E
	1618	1938	3.9F
13 F		0150	0.4F
		0617	2.4F
	0911	1317	3.9E
	1701	2020	3.6F
14 Sa		0242	0.5F
		0703	2.1F
	0954	1403	3.3E
	1746	2107	3.3F
15 Su ◐		0349	0.4F
		0758	1.8F
	1048	1456	2.8E
	1832	2156	3.1F

Day	Slack h m	Maximum h m	knots
16 M		0502	0.1F
		0904	1.5F
	1205	1558	2.2E
	1918	2245	3.0F
17 Tu	0408	0600	0.5E
	0742	1023	1.4F
	1355	1707	1.6E
	2003	2329	2.9F
18 W	0408	0644	1.3E
	0907	1150	1.7F
	1546	1819	1.2E
	2045		
19 Th		0009	2.9F
	0418	0719	2.2E
	1009	1314	2.3F
	1720	1923	0.9E
	2124		
20 F		0046	3.0F
	0432	0752	3.1E
	1100	1418	3.1F
	1841	2020	0.7E
	2159		
21 Sa		0124	3.0F
	0449	0825	4.0E
	1146	1508	3.9F
	1952	2109	0.5E
	2229		
22 Su		0201	3.1F
	0511	0900	4.8E
	1230	1553	4.6F
	2100	2155	0.3E
	2253		
23 M ○		0240	3.3F
	0539	0937	5.6E
	1314	1637	5.1F
	2214	2238	0.1E
	2304		
24 Tu		0321	3.4F
	0614	1018	6.1E
	1400	1721	5.4F
		2322	0.1F
25 W		0404	3.5F
	0655	1102	6.3E
	1448	1807	5.4F
26 Th		0007	0.2F
		0450	3.5F
	0741	1149	6.2E
	1536	1855	5.3F
27 F		0057	0.3F
		0540	3.4F
	0833	1239	5.8E
	1626	1945	5.0F
28 Sa		0154	0.1F
		0636	3.1F
	0931	1334	5.1E
	1715	2036	4.7F
29 Su	0159	0300	0.3E
	0405	0740	2.6F
	1040	1433	4.2E
	1804	2127	4.4F
30 M ◑	0217	0409	0.9E
	0604	0856	2.2F
	1209	1540	3.1E
	1851	2218	4.1F
31 Tu	0242	0516	1.8E
	0745	1029	2.1F
	1400	1654	2.1E
	1935	2307	3.8F

June

Day	Slack h m	Maximum h m	knots
1 W	0310	0614	2.8E
	0909	1224	2.4F
	1557	1813	1.3E
	2017	2353	3.6F
2 Th	0341	0706	3.7E
	1017	1352	3.1F
	1747	1926	0.7E
	2056		
3 F		0036	3.4F
	0413	0753	4.4E
	1113	1456	3.8F
	1924	2031	0.4E
	2133		
4 Sa		0117	3.2F
	0446	0835	5.0E
	1201	1547	4.3F
	2047	2127	0.1E
	2206		
5 Su		0157	3.1F
	0519	0915	5.2E
	1244	1632	4.6F
		2217	0.0
6 M ●		0236	2.9F
	0552	0953	5.3E
	1325	1713	4.7F
		2301	0.1F
7 Tu		0314	2.8F
	0625	1028	5.2E
	1403	1749	4.6F
		2341	0.2F
8 W		0353	2.8F
	0659	1103	5.0E
	1441	1821	4.5F
9 Th		0019	0.3F
		0432	2.7F
	0733	1138	4.8E
	1518	1851	4.3F
10 F		0056	0.4F
		0513	2.6F
	0809	1215	4.4E
	1555	1920	4.1F
11 Sa		0135	0.3F
		0556	2.4F
	0848	1252	4.0E
	1631	1951	3.9F
12 Su		0218	0.1F
		0645	2.1F
	0933	1333	3.4E
	1706	2024	3.7F
13 M	0208	0305	0.2E
	0408	0740	1.9F
	1027	1417	2.7E
	1738	2100	3.5F
14 Tu ◐	0207	0354	0.7E
	0555	0843	1.7F
	1142	1507	2.0E
	1809	2137	3.4F
15 W	0216	0443	1.3E
	0723	0955	1.7F
	1331	1606	1.2E
	1835	2216	3.3F

Day	Slack h m	Maximum h m	knots
16 Th	0230	0529	2.1E
	0836	1115	1.9F
	1540	1713	0.5E
	1853	2258	3.1F
17 F	0247	0613	2.9E
	0937	1239	2.5F
		1827	0.0
		2341	3.1F
18 Sa	0310	0657	3.8E
	1032	1355	3.3F
		1938	0.4F
19 Su		0026	3.1F
	0340	0741	4.7E
	1123	1454	4.0F
		2040	0.5F
20 M		0113	3.2F
	0418	0827	5.5E
	1212	1545	4.7F
		2135	0.6F
21 Tu ○		0202	3.3F
	0502	0914	6.1E
	1301	1633	5.1F
		2224	0.5F
22 W		0253	3.5F
	0551	1002	6.5E
	1350	1718	5.4F
		2311	0.4F
23 Th		0345	3.7F
	0643	1051	6.6E
	1438	1802	5.4F
		2359	0.1F
24 F		0439	3.7F
	0739	1141	6.3E
	1524	1845	5.4F
	2353		
25 Sa		0049	0.3E
	0148	0536	3.5F
	0838	1232	5.7E
	1608	1926	5.2F
26 Su	0008	0142	0.9E
	0322	0637	3.2F
	0943	1324	4.7E
	1649	2007	4.9F
27 M	0032	0239	1.5E
	0452	0744	2.8F
	1058	1419	3.5E
	1727	2048	4.5F
28 Tu ◑	0100	0337	2.3E
	0618	0900	2.5F
	1229	1519	2.3E
	1802	2128	4.2F
29 W	0132	0437	3.0E
	0740	1035	2.4F
	1421	1627	1.1E
	1831	2210	3.8F
30 Th	0206	0535	3.7E
	0854	1224	2.7F
	1644	1746	0.2E
	1846	2254	3.4F

Time meridian 135° W. 0000 is midnight. 1200 is noon. Times are not adjusted for Daylight Saving Time.

NOTE–See detailed explanation on page 97.

Unimak Pass (off Scotch Cap), Aleutian Islands, 2005

F–Flood, Dir. 295° True E–Ebb, Dir. 105° True

July

Day	Slack h m	Maximum h m	knots
1 F	0244	0630	4.2E
	0959	1348	3.2F
		1907	0.3F
		2341	3.1F
2 Sa	0324	0722	4.6E
	1056	1451	3.7F
		2017	0.5F
3 Su		0029	2.8F
	0405	0811	4.8E
	1145	1543	4.1F
		2117	0.5F
4 M		0118	2.7F
	0447	0856	4.9E
	1230	1628	4.3F
		2207	0.5F
5 Tu		0205	2.6F
	0528	0937	5.0E
	1311	1708	4.4F
		2250	0.4F
6 W		0250	2.6F
	0608	1015	4.9E
	1349	1742	4.4F
●		2329	0.3F
7 Th		0333	2.7F
	0646	1050	4.8E
	1424	1810	4.3F
8 F		0003	0.2F
		0416	2.7F
	0724	1124	4.6E
	1458	1831	4.2F
9 Sa		0034	0.0
		0459	2.6F
	0803	1157	4.3E
	1529	1850	4.1F
10 Su	0006	0105	0.2E
	0208	0544	2.5F
	0843	1230	3.8E
	1558	1912	4.0F
11 M	0008	0137	0.6E
	0318	0632	2.4F
	0930	1306	3.2E
	1624	1939	3.9F
12 Tu	0016	0212	1.0E
	0427	0724	2.3F
	1026	1344	2.5E
	1646	2009	3.8F
13 W	0027	0251	1.6E
	0536	0821	2.2F
	1140	1428	1.6E
	1703	2042	3.6F
14 Th	0040	0333	2.2E
	0646	0926	2.2F
	1326	1519	0.8E
◐	1709	2119	3.4F
15 F	0059	0421	2.8E
	0754	1038	2.3F
		1621	0.1F
		2200	3.2F
16 Sa	0126	0514	3.5E
	0900	1202	2.7F
		1738	0.7F
		2247	3.1F
17 Su	0204	0610	4.2E
	1002	1333	3.3F
		1903	1.0F
		2341	3.1F
18 M	0252	0708	4.9E
	1101	1445	3.9F
		2019	1.1F
19 Tu		0039	3.2F
	0347	0805	5.6E
	1157	1539	4.5F
		2118	0.9F
20 W		0139	3.4F
	0447	0901	6.1E
	1248	1624	5.0F
		2209	0.5F
21 Th		0239	3.7F
	0547	0953	6.4E
	1337	1705	5.2F
○	2226	2255	0.1E
	2326		
22 F		0339	3.9F
	0649	1044	6.3E
	1421	1742	5.3F
	2216	2341	0.8E
23 Sa	0108	0438	4.0F
	0750	1133	5.8E
	1503	1818	5.3F
	2232		
24 Su		0027	1.5E
	0227	0538	3.9F
	0853	1222	5.0E
	1540	1852	5.1F
	2254		
25 M		0115	2.3E
	0340	0639	3.7F
	1000	1310	3.9E
	1614	1927	4.8F
	2321		
26 Tu		0204	2.9E
	0452	0743	3.3F
	1116	1400	2.7E
	1644	2002	4.4F
	2352		
27 W		0255	3.4E
	0603	0854	3.0F
	1246	1455	1.4E
◑	1707	2039	4.0F
28 Th	0027	0350	3.7E
	0714	1021	2.8F
	1449	1559	0.3E
	1712	2119	3.5F
29 F	0107	0450	3.9E
	0823	1207	2.8F
		1720	0.5F
		2203	3.1F
30 Sa	0152	0552	4.0E
	0930	1334	3.1F
		1850	0.8F
		2254	2.7F
31 Su	0243	0654	4.1E
	1031	1439	3.5F
		2005	0.8F
		2352	2.4F

August

Day	Slack h m	Maximum h m	knots
1 M	0337	0751	4.3E
	1125	1532	3.8F
		2104	0.6F
2 Tu		0052	2.3F
	0430	0841	4.4E
	1212	1615	4.0F
		2151	0.4F
3 W		0149	2.4F
	0520	0925	4.5E
	1252	1650	4.1F
		2230	0.1F
4 Th		0240	2.5F
	0605	1003	4.5E
	1328	1719	4.1F
●	2223	2304	0.1E
	2344		
5 F		0326	2.7F
	0648	1037	4.4E
	1400	1739	4.1F
	2217	2333	0.4E
6 Sa	0046	0409	2.8F
	0728	1108	4.2E
	1429	1751	4.0F
	2219	2359	0.8E
7 Su	0139	0452	2.9F
	0809	1138	3.8E
	1456	1805	4.0F
	2226		
8 M		0024	1.2E
	0230	0535	3.0F
	0852	1209	3.3E
	1519	1826	4.0F
	2235		
9 Tu		0050	1.7E
	0320	0620	3.0F
	0941	1241	2.7E
	1539	1851	3.9F
	2244		
10 W		0121	2.2E
	0411	0708	3.0F
	1039	1317	1.9E
	1553	1920	3.7F
	2256		
11 Th		0156	2.7E
	0506	0800	2.9F
	1154	1357	1.1E
	1600	1953	3.5F
	2315		
12 F		0237	3.1E
	0606	0858	2.8F
	1347	1445	0.2E
◐	1544	2029	3.3F
	2343		
13 Sa		0327	3.5E
	0713	1007	2.8F
		1545	0.6F
		2113	3.1F
14 Su	0023	0426	3.9E
	0825	1132	2.9F
		1705	1.1F
		2206	3.0F
15 M	0117	0535	4.3E
	0936	1319	3.3F
		1846	1.3F
		2310	2.9F
16 Tu	0223	0646	4.8E
	1041	1434	3.8F
		2008	1.0F
17 W		0021	3.1F
	0337	0753	5.3E
	1139	1522	4.3F
		2104	0.4F
18 Th		0132	3.4F
	0450	0852	5.6E
	1229	1601	4.7F
	2046	2150	0.4E
	2254		
19 F		0240	3.8F
	0559	0945	5.7E
	1314	1635	4.9F
○	2044	2233	1.3E
20 Sa	0022	0343	4.2F
	0704	1035	5.4E
	1355	1707	5.0F
	2059	2315	2.2E
21 Su	0131	0443	4.4F
	0807	1122	4.8E
	1431	1738	4.9F
	2121	2357	3.1E
22 M	0235	0540	4.4F
	0911	1208	3.9E
	1503	1809	4.8F
	2146		
23 Tu		0040	3.7E
	0335	0636	4.3F
	1017	1253	2.8E
	1532	1841	4.5F
	2215		
24 W		0124	4.1E
	0434	0734	3.9F
	1131	1340	1.7E
	1555	1915	4.0F
	2247		
25 Th		0210	4.2E
	0534	0835	3.5F
	1302	1431	0.7E
	1607	1951	3.6F
	2324		
26 F		0301	4.0E
	0636	0950	3.1F
		1533	0.2F
◑		2031	3.1F
27 Sa	0007	0400	3.8E
	0742	1136	2.8F
		1659	0.8F
		2118	2.6F
28 Su	0101	0509	3.5E
	0850	1311	2.9F
		1837	0.9F
		2215	2.2F
29 M	0206	0624	3.5E
	0957	1417	3.2F
		1950	0.7F
		2324	2.0F
30 Tu	0316	0729	3.6E
	1055	1507	3.5F
		2043	0.3F
31 W		0040	2.0F
	0422	0823	3.7E
	1142	1545	3.6F
	2052	2125	0.1E
	2159		

September

Day	Slack h m	Maximum h m	knots
1 Th		0148	2.3F
	0518	0908	3.8E
	1222	1615	3.8F
	2038	2200	0.5E
	2316		
2 F		0242	2.6F
	0608	0945	3.9E
	1255	1635	3.8F
	2037	2229	1.0E
3 Sa	0012	0326	2.9F
	0653	1018	3.7E
	1324	1647	3.8F
●	2043	2254	1.5E
4 Su	0059	0407	3.2F
	0737	1048	3.4E
	1350	1657	3.8F
	2051	2316	2.0E
5 M	0143	0446	3.5F
	0822	1118	3.0E
	1414	1715	3.8F
	2100	2339	2.5E
6 Tu	0224	0526	3.7F
	0909	1148	2.5E
	1434	1738	3.7F
	2109		
7 W		0005	3.0E
	0305	0608	3.8F
	1001	1220	1.9E
	1449	1805	3.6F
	2122		
8 Th		0035	3.5E
	0350	0652	3.8F
	1102	1256	1.2E
	1457	1835	3.5F
	2140		
9 F		0111	3.8E
	0439	0741	3.7F
	1224	1336	0.4E
	1451	1908	3.3F
	2208		
10 Sa		0154	4.0E
	0535	0837	3.4F
		1424	0.3F
		1948	3.1F
	2247		
11 Su		0247	4.0E
	0641	0944	3.2F
		1527	0.9F
◐		2037	2.9F
	2338		
12 M		0352	4.0E
	0755	1111	3.1F
		1659	1.2F
		2140	2.7F
13 Tu	0047	0510	4.1E
	0909	1259	3.3F
		1849	1.0F
		2257	2.6F
14 W	0214	0631	4.3E
	1015	1405	3.8F
		1957	0.2F
15 Th		0021	2.8F
	0345	0742	4.5E
	1111	1448	4.1F
	1911	2045	0.7E
	2214		
16 F		0141	3.3F
	0506	0842	4.6E
	1159	1522	4.4F
	1914	2128	1.8E
	2335		
17 Sa		0251	3.9F
	0618	0935	4.4E
	1240	1553	4.5F
○	1930	2207	2.9E
18 Su	0039	0352	4.5F
	0724	1024	4.0E
	1317	1623	4.6F
	1952	2246	3.9E
19 M	0136	0447	4.8F
	0827	1109	3.3E
	1350	1653	4.5F
	2016	2325	4.5E
20 Tu	0228	0539	4.9F
	0930	1153	2.5E
	1419	1724	4.3F
	2044		
21 W		0004	4.9E
	0319	0629	4.8F
	1035	1237	1.6E
	1444	1756	3.9F
	2114		
22 Th		0044	4.9E
	0409	0718	4.4F
	1148	1321	0.8E
	1501	1830	3.5F
	2146		
23 F		0126	4.6E
	0500	0811	3.9F
	1337	1410	0.1E
	1445	1907	3.1F
	2223		
24 Sa		0213	4.1E
	0555	0913	3.3F
		1510	0.5F
◑		1948	2.6F
	2307		
25 Su		0309	3.5E
	0656	1050	2.9F
		1643	0.8F
		2038	2.1F
26 M	0006	0420	3.0E
	0802	1233	2.8F
		1819	0.7F
		2141	1.8F
27 Tu	0125	0544	2.8E
	0908	1339	3.0F
		1925	0.3F
		2302	1.6F
28 W	0255	0657	2.8E
	1007	1424	3.2F
	1917	2013	0.2E
	2107		
29 Th		0036	1.8F
	0412	0754	2.9E
	1054	1457	3.3F
	1903	2050	0.8E
	2228		
30 F		0153	2.2F
	0516	0840	2.9E
	1133	1519	3.4F
	1903	2121	1.5E
	2325		

Time meridian 135° W. 0000 is midnight. 1200 is noon. Times are not adjusted for Daylight Saving Time.
NOTE–See detailed explanation on page 97.

Unimak Pass (off Scotch Cap), Aleutian Islands, 2005

F–Flood, Dir. 295° True E–Ebb, Dir. 105° True

October

Day	Slack h m	Maximum h m	knots
1 Sa		0244	2.7F
	0611	0919	2.9E
	1206	1531	3.4F
	1910	2147	2.1E
2 Su	0011	0325	3.2F
	0702	0954	2.7E
	1236	1542	3.4F
	1919	2210	2.7E
3 M	0053	0402	3.7F
	0750	1026	2.3E
	1301	1601	3.5F
●	1930	2232	3.4E
4 Tu	0131	0439	4.1F
	0839	1058	1.9E
	1323	1624	3.5F
	1942	2258	3.9E
5 W	0209	0517	4.4F
	0931	1130	1.5E
	1340	1652	3.5F
	1956	2327	4.4E
6 Th	0249	0557	4.5F
	1028	1204	0.9E
	1350	1722	3.4F
	2017		
7 F		0000	4.7E
	0332	0640	4.5F
	1139	1241	0.4E
	1349	1756	3.3F
	2046		
8 Sa		0039	4.8E
	0420	0728	4.3F
		1324	0.2F
		1834	3.2F
	2124		
9 Su		0125	4.7E
	0515	0824	3.9F
		1417	0.7F
		1919	2.9F
	2211		
10 M		0221	4.4E
	0618	0930	3.6F
		1530	1.0F
◐		2017	2.6F
	2312		
11 Tu		0330	4.0E
	0727	1051	3.4F
		1714	0.8F
		2132	2.3F
12 W	0037	0452	3.7E
	0835	1216	3.5F
		1840	0.0
		2301	2.3F
13 Th	0225	0616	3.5E
	0936	1315	3.7F
	1735	1936	1.1E
	2127		
14 F		0036	2.7F
	0406	0729	3.3E
	1029	1357	3.9F
	1744	2021	2.3E
	2246		
15 Sa		0201	3.4F
	0531	0830	3.1E
	1115	1432	4.0F
	1803	2102	3.4E
	2348		

Day	Slack h m	Maximum h m	knots
16 Su		0307	4.2F
	0644	0923	2.8E
	1154	1504	4.1F
	1827	2140	4.4E
17 M	0041	0402	4.8F
	0750	1012	2.3E
	1230	1536	4.1F
○	1853	2218	5.1E
18 Tu	0130	0451	5.2F
	0853	1057	1.8E
	1301	1608	3.9F
	1921	2255	5.5E
19 W	0216	0536	5.2F
	0954	1140	1.2E
	1329	1641	3.7F
	1950	2332	5.5E
20 Th	0300	0619	5.0F
	1100	1223	0.7E
	1350	1715	3.4F
	2021		
21 F		0009	5.2E
	0344	0702	4.6F
	1221	1306	0.2E
	1353	1751	3.1F
	2055		
22 Sa		0049	4.7E
	0429	0748	4.1F
		1353	0.3F
		1829	2.7F
	2132		
23 Su		0133	4.1E
	0518	0840	3.5F
		1453	0.6F
		1913	2.3F
	2215		
24 M		0224	3.4E
	0611	0947	3.1F
		1619	0.7F
◑		2005	1.8F
	2311		
25 Tu		0326	2.8E
	0708	1121	2.8F
		1745	0.4F
		2112	1.5F
26 W	0035	0443	2.3E
	0805	1229	2.0F
	1753	1846	0.2E
	1936	2238	1.3F
27 Th	0223	0603	2.1E
	0859	1311	2.9F
	1728	1931	0.8E
	2118		
28 F		0022	1.6F
	0357	0708	1.9E
	0946	1336	2.9F
	1729	2005	1.6E
	2224		
29 Sa		0145	2.2F
	0514	0801	1.8E
	1027	1353	3.0F
	1737	2035	2.4E
	2314		
30 Su		0237	2.9F
	0618	0846	1.6E
	1102	1413	3.1F
	1749	2100	3.2E
	2357		
31 M		0317	3.5F
	0716	0926	1.4E
	1132	1438	3.2F
	1803	2126	3.9E

November

Day	Slack h m	Maximum h m	knots
1 Tu	0036	0354	4.1F
	0811	1003	1.1E
	1158	1507	3.2F
●	1818	2153	4.5E
2 W	0115	0431	4.6F
	0906	1039	0.8E
	1219	1538	3.3F
	1838	2224	5.1E
3 Th	0154	0508	4.9F
	1004	1114	0.5E
	1231	1612	3.4F
	1904	2258	5.4E
4 F	0236	0549	5.0F
	1117	1152	0.1E
	1230	1648	3.4F
	1936	2337	5.6E
5 Sa	0320	0633	4.9F
		1233	0.2F
		1728	3.3F
	2015		
6 Su		0021	5.5E
	0409	0722	4.7F
		1321	0.5F
		1813	3.1F
	2101		
7 M		0110	5.1E
	0501	0816	4.3F
		1422	0.6F
		1907	2.8F
	2156		
8 Tu		0207	4.5E
	0557	0915	4.0F
		1540	0.4F
◐		2014	2.4F
	2307		
9 W		0314	3.8E
	0654	1019	3.8F
	1551	1705	0.3E
	1820	2136	2.0F
10 Th	0048	0432	3.0E
	0751	1120	3.7F
	1553	1813	1.3E
	2025	2315	2.1F
11 F	0249	0555	2.4E
	0844	1213	3.7F
	1612	1907	2.5E
	2149		
12 Sa		0103	2.7F
	0436	0711	1.9E
	0932	1258	3.7F
	1637	1953	3.7E
	2254		
13 Su		0222	3.6F
	0606	0816	1.5E
	1016	1338	3.7F
	1704	2035	4.7E
	2348		
14 M		0321	4.4F
	0722	0912	1.2E
	1056	1415	3.6F
	1734	2115	5.4E
15 Tu	0036	0411	4.9F
	0830	1003	0.9E
	1132	1452	3.5F
○	1805	2153	5.7E

Day	Slack h m	Maximum h m	knots
16 W	0120	0455	5.2F
	0932	1049	0.6E
	1204	1528	3.4F
	1838	2230	5.8E
17 Th	0202	0535	5.2F
	1035	1132	0.3E
	1229	1605	3.3F
	1910	2307	5.6E
18 F	0243	0614	4.9F
		1213	0.0
		1642	3.1F
	1944	2344	5.3E
19 Sa	0323	0651	4.6F
		1255	0.2F
		1720	2.8F
	2019		
20 Su		0022	4.7E
	0404	0729	4.2F
		1339	0.4F
		1801	2.5F
	2057		
21 M		0103	4.2E
	0446	0810	3.8F
		1432	0.4F
		1847	2.2F
	2140		
22 Tu		0147	3.5E
	0529	0853	3.4F
		1537	0.3F
		1941	1.8F
	2233		
23 W		0238	2.8E
	0611	0939	3.2F
	1603	1647	0.1E
◑	1729	2047	1.4F
	2349		
24 Th		0337	2.1E
	0654	1024	3.0F
	1540	1745	0.7E
	1936	2207	1.3F
25 F	0143	0447	1.5E
	0736	1107	2.9F
	1546	1830	1.4E
	2059	2342	1.5F
26 Sa	0341	0601	1.0E
	0815	1147	2.9F
	1558	1907	2.2E
	2200		
27 Su		0118	2.2F
	0522	0710	0.6E
	0853	1225	2.9F
	1614	1940	3.1E
	2250		
28 M		0219	2.9F
	0648	0808	0.4E
	0926	1303	2.9F
	1632	2012	3.9E
	2334		
29 Tu		0305	3.7F
	0802	0858	0.2E
	0954	1341	3.0F
	1654	2046	4.7E
30 W	0016	0345	4.3F
		0942	0.0
		1419	3.1F
	1722	2121	5.3E

December

Day	Slack h m	Maximum h m	knots
1 Th	0058	0424	4.8F
		1023	0.1F
		1459	3.3F
●	1755	2159	5.8E
2 F	0142	0504	5.1F
		1103	0.2F
		1541	3.4F
	1833	2241	6.1E
3 Sa	0226	0546	5.2F
		1144	0.3F
		1626	3.5F
	1917	2325	6.2E
4 Su	0312	0629	5.2F
		1229	0.3F
		1714	3.4F
	2005		
5 M		0012	5.9E
	0358	0715	5.0F
		1321	0.1F
		1807	3.2F
	2100		
6 Tu		0102	5.3E
	0445	0801	4.7F
	1326	1419	0.2E
	1518	1908	2.8F
	2203		
7 W		0157	4.4E
	0530	0848	4.5F
	1341	1525	0.8E
	1718	2019	2.4F
	2324		
8 Th		0258	3.3E
	0614	0936	4.2F
	1405	1633	1.7E
◐	1904	2144	2.2F
9 F	0113	0409	2.1E
	0656	1024	3.9F
	1433	1736	2.7E
	2033	2332	2.3F
10 Sa	0318	0529	1.2E
	0735	1112	3.7F
	1505	1832	3.7E
	2147		
11 Su		0119	3.0F
	0524	0651	0.5E
	0812	1159	3.5F
	1539	1923	4.5E
	2247		
12 M		0231	3.8F
	0719	0804	0.2E
	0846	1246	3.3F
	1616	2010	5.1E
	2339		
13 Tu		0327	4.4F
		0905	0.0
		1332	3.2F
	1654	2054	5.5E
14 W	0026	0415	4.8F
		0957	0.1F
		1416	3.1F
	1732	2136	5.6E
15 Th	0109	0458	4.9F
		1044	0.1F
		1458	3.0F
○	1810	2215	5.6E

Day	Slack h m	Maximum h m	knots
16 F	0150	0536	4.9F
		1126	0.2F
		1539	2.9F
	1848	2252	5.4E
17 Sa	0228	0611	4.7F
		1205	0.2F
		1620	2.8F
	1925	2328	5.1E
18 Su	0306	0642	4.5F
		1243	0.2F
		1701	2.7F
	2002		
19 M		0004	4.7E
	0341	0709	4.2F
		1320	0.1F
		1744	2.5F
	2041		
20 Tu		0040	4.1E
	0416	0736	4.0F
	1327	1359	0.1E
	1434	1831	2.3F
	2124		
21 W		0118	3.5E
	0448	0805	3.8F
	1325	1443	0.4E
	1609	1923	2.0F
	2216		
22 Th		0159	2.8E
	0517	0837	3.6F
	1335	1529	0.8E
	1739	2023	1.7F
	2326		
23 F		0245	1.9E
	0543	0913	3.4F
	1349	1618	1.3E
◑	1903	2132	1.6F
24 Sa	0111	0340	1.1E
	0603	0950	3.2F
	1406	1707	2.0E
	2017	2252	1.8F
25 Su	0332	0446	0.3E
	0606	1031	3.0F
	1426	1754	2.7E
	2121		
26 M		0026	2.2F
		0604	0.3F
		1115	2.9F
	1450	1840	3.5E
	2217		
27 Tu		0153	2.9F
		0723	0.6F
		1202	2.8F
	1522	1926	4.3E
	2308		
28 W		0252	3.6F
		0830	0.7F
		1251	2.9F
	1600	2012	5.0E
	2357		
29 Th		0338	4.3F
		0923	0.7F
		1341	3.1F
	1644	2058	5.7E
30 F	0044	0421	4.8F
		1009	0.6F
		1432	3.3F
●	1732	2144	6.2E
31 Sa	0131	0501	5.1F
		1051	0.4F
		1523	3.6F
	1823	2231	6.4E

Time meridian 135° W. 0000 is midnight. 1200 is noon. Times are not adjusted for Daylight Saving Time.
NOTE–See detailed explanation on page 97.

Akutan Pass, Aleutian Islands, 2005

F–Flood, Dir. 295° True   E–Ebb, Dir. 115° True

January

Day	Slack h m	Maximum h m	knots
1 Sa		0024	6.6E
	0351	0707	6.8F
	1104	1318	3.8E
	1626	1910	3.8F
	2155		
2 Su		0112	5.5E
	0426	0745	6.4F
	1128	1401	4.7E
	1724	2009	4.1F
	2316		
3 M ◑		0203	4.4E
	0503	0825	5.8F
	1152	1447	5.7E
	1823	2112	4.6F
4 Tu	0048	0301	3.2E
	0542	0908	5.3F
	1217	1535	6.8E
	1921	2217	5.4F
5 W	0219	0403	2.2E
	0627	0954	4.8F
	1248	1627	7.6E
	2020	2324	6.4F
6 Th	0343	0512	1.5E
	0718	1045	4.5F
	1328	1721	8.1E
	2117		
7 F		0030	7.4F
	0459	0624	1.2E
	0817	1139	4.4F
	1416	1817	8.4E
	2214		
8 Sa		0132	7.8F
	0603	0734	1.2E
	0920	1236	4.6F
	1511	1914	8.8E
	2310		
9 Su		0230	8.1F
	0655	0834	1.6E
	1024	1333	4.9F
	1611	2010	9.0E
10 M ●	0003	0322	8.3F
	0739	0928	2.1E
	1127	1430	5.3F
	1713	2104	9.0E
11 Tu	0054	0411	8.4F
	0818	1018	2.8E
	1229	1528	5.6F
	1815	2157	8.9E
12 W	0143	0456	8.3F
	0855	1106	3.6E
	1332	1625	5.8F
	1919	2250	8.6E
13 Th	0229	0539	8.2F
	0930	1153	4.5E
	1435	1724	5.8F
	2024	2341	8.0E
14 F	0312	0620	7.9F
	1005	1241	5.3E
	1539	1824	5.6F
	2133		
15 Sa		0033	7.1E
	0354	0701	7.6F
	1039	1329	6.1E
	1643	1927	5.4F
	2247		

Day	Slack h m	Maximum h m	knots
16 Su ◐		0126	5.3E
	0434	0742	6.9F
	1115	1418	6.6E
	1747	2033	5.2F
17 M	0005	0224	3.5E
	0513	0824	5.9F
	1151	1508	7.0E
	1849	2148	5.1F
18 Tu	0129	0331	2.1E
	0554	0909	5.0F
	1230	1600	7.2E
	1949	2316	5.3F
19 W	0258	0452	1.2E
	0639	0957	4.2F
	1311	1655	7.3E
	2046		
20 Th		0033	5.7F
	0431	0610	0.7E
	0732	1048	3.6F
	1356	1749	7.4E
	2139		
21 F		0132	6.1F
	0544	0715	0.7E
	0833	1141	3.4F
	1444	1841	7.5E
	2229		
22 Sa		0222	6.5F
	0632	0809	0.9E
	0934	1234	3.3F
	1533	1928	7.6E
	2315		
23 Su		0305	6.8F
	0706	0854	1.1E
	1029	1324	3.5F
	1620	2010	7.6E
	2357		
24 M		0338	7.0F
	0734	0933	1.5E
	1119	1411	3.7F
	1705	2049	7.7E
25 Tu ○	0035	0402	7.1F
	0800	1005	1.9E
	1205	1457	4.0F
	1749	2125	7.7E
26 W	0111	0425	7.2F
	0826	1032	2.3E
	1251	1541	4.2F
	1832	2202	7.7E
27 Th	0144	0452	7.3F
	0851	1058	3.0E
	1336	1626	4.5F
	1917	2240	7.5E
28 F	0216	0522	7.2F
	0916	1127	3.7E
	1421	1712	4.7F
	2005	2320	6.9E
29 Sa	0247	0554	7.0F
	0939	1201	4.6E
	1507	1759	5.0F
	2101		
30 Su		0002	6.1E
	0317	0628	6.6F
	0959	1239	5.5E
	1556	1850	5.2F
	2205		
31 M		0048	5.0E
	0349	0703	6.1F
	1019	1320	6.4E
	1648	1945	5.5F
	2319		

February

Day	Slack h m	Maximum h m	knots
1 Tu ◑		0138	3.8E
	0421	0741	5.6F
	1042	1406	7.2E
	1744	2044	5.8F
2 W	0042	0233	2.6E
	0455	0824	5.0F
	1113	1456	7.6E
	1845	2149	6.2F
3 Th	0209	0335	1.6E
	0534	0913	4.5F
	1155	1552	7.9E
	1950	2258	6.6F
4 F	0335	0445	0.9E
	0624	1010	4.3F
	1248	1652	8.1E
	2055		
5 Sa		0009	7.2F
	0452	0603	0.7E
	0735	1113	4.3F
	1352	1756	8.4E
	2157		
6 Su		0118	7.6F
	0547	0719	1.1E
	0858	1218	4.6F
	1502	1859	8.6E
	2256		
7 M		0217	7.9F
	0629	0820	1.9E
	1015	1322	5.1F
	1613	2000	8.6E
	2349		
8 Tu ●		0307	8.0F
	0704	0911	2.9E
	1123	1424	5.8F
	1721	2057	8.6E
9 W	0038	0350	8.1F
	0737	0958	4.0E
	1226	1523	6.3F
	1827	2150	8.4E
10 Th	0123	0430	8.0F
	0809	1041	5.2E
	1326	1620	6.7F
	1930	2240	8.0E
11 F	0205	0508	7.8F
	0841	1123	6.2E
	1423	1716	6.8F
	2033	2329	7.4E
12 Sa	0244	0545	7.6F
	0913	1204	6.9E
	1519	1811	6.7F
	2137		
13 Su		0016	5.8E
	0322	0622	6.9F
	0945	1246	7.4E
	1613	1907	6.4F
	2243		
14 M		0105	4.2E
	0358	0701	6.0F
	1019	1329	7.5E
	1709	2004	6.0F
	2353		
15 Tu ◐		0157	2.8E
	0434	0741	5.1F
	1054	1415	7.3E
	1806	2106	5.6F

Day	Slack h m	Maximum h m	knots
16 W	0109	0255	1.6E
	0511	0826	4.3F
	1134	1506	7.0E
	1905	2220	5.3F
17 Th	0232	0409	0.8E
	0553	0915	3.6F
	1221	1603	6.7E
	2005	2353	5.2F
18 F	0404	0535	0.5E
	0649	1011	3.2F
	1317	1706	6.6E
	2105		
19 Sa		0102	5.4F
	0513	0645	0.6E
	0759	1110	3.1F
	1417	1810	6.7E
	2200		
20 Su		0155	5.8F
	0552	0740	0.9E
	0909	1209	3.2F
	1515	1905	6.9E
	2248		
21 M		0236	6.1F
	0619	0824	1.4E
	1010	1305	3.6F
	1610	1952	7.2E
	2331		
22 Tu		0306	6.3F
	0643	0900	2.1E
	1104	1356	4.0F
	1701	2032	7.3E
23 W ○	0009	0325	6.5F
	0707	0929	2.8E
	1152	1444	4.6F
	1750	2108	7.3E
24 Th	0043	0347	6.7F
	0731	0953	3.6E
	1236	1529	5.1F
	1838	2145	7.1E
25 F	0115	0414	6.7F
	0754	1019	4.5E
	1317	1613	5.7F
	1927	2222	6.7E
26 Sa	0146	0444	6.6F
	0814	1049	5.5E
	1359	1657	6.2F
	2018	2302	6.0E
27 Su	0216	0516	6.4F
	0833	1123	6.5E
	1441	1743	6.6F
	2114	2344	5.2E
28 M	0247	0549	6.0F
	0852	1202	7.3E
	1526	1832	6.9F
	2214		

March

Day	Slack h m	Maximum h m	knots
1 Tu		0030	4.2E
	0317	0625	5.5F
	0915	1244	7.6E
	1617	1924	7.0F
	2322		
2 W		0119	3.1E
	0349	0704	5.0F
	0946	1332	7.8E
	1713	2022	6.9F
3 Th ◑	0037	0214	2.1E
	0423	0750	4.6F
	1028	1425	7.9E
	1817	2126	6.8F
4 F	0158	0315	1.2E
	0505	0845	4.2F
	1121	1525	8.0E
	1925	2236	6.8F
5 Sa	0317	0428	0.8E
	0610	0949	4.0F
	1229	1631	8.0E
	2034	2349	7.0F
6 Su	0420	0552	1.0E
	0740	1100	4.2F
	1349	1741	8.0E
	2138		
7 M		0058	7.3F
	0504	0705	1.9E
	0907	1211	4.7F
	1511	1851	8.0E
	2236		
8 Tu		0153	7.6F
	0540	0801	3.1E
	1021	1319	5.5F
	1626	1954	8.0E
	2328		
9 W		0238	7.6F
	0613	0848	4.5E
	1125	1423	6.3F
	1735	2051	7.9E
10 Th ●	0014	0317	7.6F
	0645	0930	5.8E
	1222	1521	7.1F
	1839	2142	7.6E
11 F	0057	0354	7.5F
	0716	1009	7.0E
	1314	1614	7.5F
	1940	2230	6.8E
12 Sa	0136	0430	7.1F
	0747	1047	7.6E
	1403	1705	7.6F
	2038	2316	5.6E
13 Su	0214	0506	6.5F
	0818	1124	7.8E
	1451	1754	7.6F
	2136		
14 M		0000	4.5E
	0250	0543	5.8F
	0849	1203	7.8E
	1539	1842	7.3F
	2235		
15 Tu		0044	3.3E
	0325	0621	5.0F
	0921	1243	7.6E
	1628	1932	6.7F
	2336		

Day	Slack h m	Maximum h m	knots
16 W		0130	2.3E
	0401	0702	4.3F
	0957	1328	7.4E
	1721	2025	6.0F
17 Th ◐	0042	0221	1.5E
	0439	0748	3.7F
	1040	1418	6.8E
	1818	2125	5.4F
18 F	0153	0325	0.9E
	0525	0839	3.2F
	1133	1515	6.2E
	1920	2239	5.0F
19 Sa	0305	0451	0.7E
	0628	0939	2.9F
	1239	1620	5.9E
	2021		
20 Su		0007	5.0F
	0402	0605	0.9E
	0744	1043	2.9F
	1351	1729	5.8E
	2119		
21 M		0104	5.2F
	0441	0659	1.5E
	0856	1147	3.2F
	1459	1832	6.0E
	2209		
22 Tu		0140	5.4F
	0511	0741	2.3E
	0957	1247	3.8F
	1601	1923	6.1E
	2252		
23 W		0204	5.7F
	0537	0814	3.2E
	1048	1341	4.6F
	1658	2007	6.1E
	2331		
24 Th		0230	5.8F
	0603	0841	4.2E
	1133	1429	5.5F
	1751	2047	6.0E
25 F ○	0006	0259	5.9F
	0626	0907	5.4E
	1214	1514	6.3F
	1843	2126	5.8E
26 Sa	0040	0330	5.9F
	0647	0937	6.5E
	1254	1558	7.2F
	1935	2205	5.3E
27 Su	0113	0403	5.7F
	0706	1011	7.5E
	1334	1642	7.6F
	2027	2246	4.8E
28 M	0145	0437	5.5F
	0727	1048	7.8E
	1417	1728	7.8F
	2122	2329	4.1E
29 Tu	0218	0513	5.2F
	0753	1130	8.0E
	1503	1816	7.8F
	2220		
30 W		0015	3.3E
	0251	0553	4.9F
	0826	1215	8.1E
	1554	1908	7.7F
	2324		
31 Th		0105	2.5E
	0326	0637	4.5F
	0908	1306	8.1E
	1652	2005	7.6F

Time meridian 135° W. 0000 is midnight. 1200 is noon. Times are not adjusted for Daylight Saving Time.

Akutan Pass, Aleutian Islands, 2005

F–Flood, Dir. 295° True   E–Ebb, Dir. 115° True

April

Day	Slack h m	Maximum h m	knots
1 F	0031	0201	1.8E
	0410	0729	4.2F
	1000	1402	8.0E
	1755	2108	7.2F
2 Sa	0140	0306	1.4E
	0512	0831	3.9F
	1106	1505	7.8E
	1902	2215	6.9F
3 Su	0241	0424	1.5E
	0637	0942	3.8F
	1230	1615	7.6E
	2009	2323	6.8F
4 M	0329	0543	2.3E
	0805	1058	4.2F
	1402	1730	7.3E
	2111		
5 Tu		0024	6.8F
	0409	0645	3.6E
	0921	1213	4.9F
	1527	1843	7.0E
	2207		
6 W		0115	6.8F
	0445	0735	5.0E
	1025	1323	5.9F
	1641	1948	6.6E
	2257		
7 Th		0158	6.7F
	0518	0819	6.4E
	1121	1425	7.0F
	1748	2044	6.1E
	2343		
8 F		0237	6.5F
	0550	0858	7.5E
	1211	1519	7.6F
	1849	2134	5.5E
9 Sa	0025	0314	6.1F
	0621	0935	7.8E
	1256	1607	7.8F
	1946	2220	4.7E
10 Su	0104	0350	5.7F
	0652	1010	8.0E
	1340	1651	7.8F
	2040	2303	4.0E
11 M	0142	0427	5.2F
	0723	1046	8.0E
	1423	1734	7.8F
	2132	2344	3.3E
12 Tu	0219	0505	4.6F
	0754	1124	7.9E
	1506	1816	7.6F
	2224		
13 W		0024	2.6E
	0255	0545	4.1F
	0827	1205	7.7E
	1551	1901	7.1F
	2318		
14 Th		0106	2.0E
	0333	0627	3.7F
	0906	1249	7.4E
	1640	1950	6.4F
15 F	0014	0153	1.5E
	0415	0714	3.2F
	0951	1339	6.7E
	1733	2043	5.7F
16 Sa	0112	0250	1.2E
	0510	0809	2.9F
	1049	1434	6.0E
	1830	2141	5.3F
17 Su	0206	0402	1.3E
	0620	0910	2.7F
	1202	1534	5.5E
	1928	2240	5.0F
18 M	0251	0512	1.7E
	0735	1017	2.8F
	1325	1640	5.1E
	2023	2333	5.0F
19 Tu	0328	0604	2.5E
	0842	1124	3.4F
	1442	1745	4.8E
	2113		
20 W		0017	5.1F
	0359	0644	3.6E
	0938	1226	4.2F
	1551	1843	4.7E
	2159		
21 Th		0056	5.2F
	0427	0716	4.8E
	1025	1322	5.3F
	1653	1935	4.6E
	2242		
22 F		0133	5.2F
	0452	0748	6.0E
	1108	1411	6.5F
	1751	2022	4.4E
	2321		
23 Sa		0209	5.2F
	0515	0822	7.2E
	1149	1457	7.5F
	1846	2106	4.2E
	2359		
24 Su		0245	5.1F
	0537	0858	7.8E
	1230	1541	7.9F
	1939	2149	3.9E
25 M	0036	0323	5.0F
	0602	0937	8.2E
	1312	1627	8.1F
	2032	2232	3.5E
26 Tu	0112	0401	4.9F
	0633	1019	8.4E
	1358	1714	8.2F
	2126	2317	3.0E
27 W	0149	0443	4.7F
	0710	1105	8.5E
	1447	1803	8.2F
	2222		
28 Th		0005	2.6E
	0229	0528	4.5F
	0755	1154	8.5E
	1539	1855	8.0F
	2319		
29 F		0058	2.2E
	0317	0620	4.3F
	0847	1247	8.2E
	1636	1951	7.8F
30 Sa	0015	0157	2.0E
	0419	0719	4.0F
	0951	1345	7.9E
	1735	2050	7.5F

May

Day	Slack h m	Maximum h m	knots
1 Su	0108	0305	2.3E
	0537	0827	3.8F
	1111	1450	7.5E
	1836	2149	7.1F
2 M	0155	0418	3.1E
	0701	0943	3.9F
	1246	1601	6.4E
	1936	2248	6.7F
3 Tu	0236	0523	4.2E
	0818	1102	4.4F
	1418	1719	5.5E
	2033	2341	6.3F
4 W	0313	0618	5.6E
	0923	1221	5.4F
	1541	1834	4.8E
	2127		
5 Th		0029	6.0F
	0349	0705	6.9E
	1020	1332	6.5F
	1655	1938	4.3E
	2218		
6 F		0112	5.6F
	0423	0747	7.6E
	1109	1429	7.5F
	1801	2035	3.8E
	2305		
7 Sa		0153	5.2F
	0457	0825	8.0E
	1154	1516	7.8F
	1859	2125	3.4E
	2349		
8 Su		0233	4.9F
	0529	0901	8.1E
	1237	1558	7.9F
	1951	2210	3.1E
9 M	0030	0312	4.5F
	0601	0937	8.1E
	1317	1636	7.9F
	2040	2251	2.7E
10 Tu	0110	0351	4.2F
	0634	1013	8.0E
	1358	1714	7.8F
	2126	2329	2.3E
11 W	0148	0431	3.9F
	0707	1052	7.9E
	1439	1754	7.6F
	2212		
12 Th		0006	2.0E
	0226	0513	3.6F
	0744	1133	7.7E
	1522	1835	7.3F
	2258		
13 F		0045	1.8E
	0307	0557	3.3F
	0826	1218	7.4E
	1606	1920	6.7F
	2345		
14 Sa		0129	1.7E
	0357	0646	3.0F
	0914	1305	6.7E
	1653	2006	6.2F
15 Su	0029	0219	1.8E
	0459	0742	2.8F
	1014	1357	5.9E
	1741	2054	5.8F
16 M	0110	0314	2.2E
	0608	0844	2.7F
	1130	1453	5.1E
	1830	2142	5.5F
17 Tu	0146	0408	2.9E
	0716	0950	3.0F
	1259	1552	4.4E
	1919	2229	5.3F
18 W	0218	0456	3.9E
	0816	1056	3.7F
	1424	1655	3.8E
	2008	2313	5.0F
19 Th	0247	0539	5.1E
	0908	1159	4.8F
	1539	1758	3.4E
	2056	2356	4.9F
20 F	0313	0620	6.4E
	0955	1256	6.0F
	1647	1858	3.1E
	2143		
21 Sa		0039	4.7F
	0338	0700	7.6E
	1039	1348	7.3F
	1749	1953	2.9E
	2228		
22 Su		0120	4.6F
	0405	0742	8.0E
	1123	1437	7.8F
	1846	2044	2.8E
	2312		
23 M		0203	4.6F
	0437	0825	8.4E
	1208	1525	8.2F
	1940	2132	2.6E
	2354		
24 Tu		0246	4.6F
	0515	0910	8.7E
	1255	1613	8.4F
	2033	2219	2.5E
25 W	0037	0332	4.7F
	0559	0958	8.8E
	1344	1702	8.4F
	2124	2307	2.4E
26 Th	0123	0420	4.7F
	0648	1047	8.8E
	1435	1752	8.3F
	2213	2358	2.4E
27 F	0216	0513	4.6F
	0743	1139	8.6E
	1527	1842	8.1F
	2300		
28 Sa		0053	2.6E
	0319	0611	4.4F
	0845	1234	8.2E
	1619	1934	7.9F
	2344		
29 Su		0151	3.1E
	0431	0715	4.2F
	1000	1332	7.7E
	1711	2026	7.6F
30 M	0026	0254	3.9E
	0549	0826	4.1F
	1127	1435	6.4E
	1804	2117	7.1F
31 Tu	0105	0355	4.9E
	0705	0943	4.3F
	1300	1546	4.9E
	1856	2207	6.4F

June

Day	Slack h m	Maximum h m	knots
1 W	0143	0453	6.1E
	0813	1106	5.0F
	1429	1704	3.7E
	1948	2256	5.8F
2 Th	0220	0545	7.2E
	0912	1229	5.9F
	1553	1820	2.9E
	2040	2343	5.2F
3 F	0257	0632	7.7E
	1005	1335	6.9F
	1709	1926	2.4E
	2132		
4 Sa		0029	4.7F
	0334	0715	7.9E
	1052	1428	7.5F
	1813	2024	2.1E
	2223		
5 Su		0113	4.3F
	0410	0755	8.0E
	1136	1513	7.7F
	1908	2114	2.0E
	2310		
6 M		0157	4.0F
	0447	0833	8.1E
	1218	1551	7.7F
	1954	2158	1.9E
	2355		
7 Tu		0239	3.8F
	0523	0910	8.0E
	1258	1625	7.7F
	2036	2238	1.8E
8 W	0036	0321	3.7F
	0559	0948	8.0E
	1338	1659	7.6F
	2116	2314	1.7E
9 Th	0117	0403	3.6F
	0637	1027	7.9E
	1417	1734	7.5F
	2154	2348	1.8E
10 F	0159	0447	3.5F
	0717	1108	7.7E
	1456	1811	7.3F
	2231		
11 Sa		0022	1.9E
	0247	0533	3.3F
	0801	1151	7.5E
	1536	1850	7.0F
	2307		
12 Su		0101	2.2E
	0341	0623	3.2F
	0852	1236	6.7E
	1615	1929	6.7F
	2341		
13 M		0143	2.7E
	0441	0718	3.1F
	0953	1324	5.8E
	1654	2010	6.3F
14 Tu	0012	0228	3.4E
	0543	0817	3.2F
	1111	1416	4.8E
	1734	2052	5.9F
15 W	0041	0313	4.3E
	0643	0920	3.6F
	1240	1512	3.8E
	1816	2134	5.4F
16 Th	0107	0359	5.3E
	0739	1024	4.4F
	1407	1612	2.9E
	1900	2218	5.0F
17 F	0133	0445	6.5E
	0831	1128	5.4F
	1527	1716	2.3E
	1948	2303	4.6F
18 Sa	0200	0533	7.6E
	0922	1228	6.6F
	1640	1821	1.9E
	2039	2350	4.4F
19 Su	0233	0621	8.0E
	1011	1325	7.6F
	1745	1924	1.7E
	2130		
20 M		0038	4.4F
	0313	0710	8.4E
	1101	1418	8.0F
	1844	2021	1.7E
	2222		
21 Tu		0128	4.5F
	0359	0800	8.7E
	1152	1510	8.2F
	1936	2115	1.8E
	2314		
22 W		0219	4.7F
	0450	0851	9.0E
	1242	1600	8.4F
	2023	2205	2.1E
23 Th	0009	0312	5.0F
	0545	0942	9.0E
	1333	1649	8.4F
	2106	2256	2.5E
24 F	0108	0407	5.1F
	0644	1035	8.8E
	1422	1737	8.3F
	2146	2346	3.1E
25 Sa	0211	0505	5.2F
	0748	1128	8.5E
	1510	1823	8.1F
	2224		
26 Su		0038	3.8E
	0319	0606	5.1F
	0857	1222	8.0E
	1557	1909	7.9F
	2301		
27 M		0131	4.7E
	0430	0711	4.9F
	1015	1319	7.0E
	1643	1954	7.6F
	2337		
28 Tu		0226	5.7E
	0541	0821	4.9F
	1139	1419	5.1E
	1728	2039	6.9F
29 W	0014	0320	6.6E
	0649	0936	5.1F
	1306	1527	3.5E
	1813	2125	6.0F
30 Th	0052	0415	7.3E
	0752	1102	5.5F
	1434	1646	2.2E
	1900	2212	5.2F

Time meridian 135° W. 0000 is midnight. 1200 is noon. Times are not adjusted for Daylight Saving Time.

Akutan Pass, Aleutian Islands, 2005

F–Flood, Dir. 295° True E–Ebb, Dir. 115° True

July

Day	Slack h m	Maximum h m	knots
1 F	0131	0507	7.6E
	0850	1223	6.2F
	1602	1803	1.5E
	1952	2301	4.5F
2 Sa	0211	0558	7.8E
	0943	1326	6.8F
	1721	1911	1.2E
	2047	2351	4.0F
3 Su	0253	0646	7.8E
	1032	1419	7.2F
	1823	2009	1.2E
	2142		
4 M		0040	3.7F
	0337	0731	7.9E
	1118	1505	7.4F
	1910	2058	1.3E
	2235		
5 Tu		0128	3.6F
	0420	0813	7.9E
	1200	1544	7.5F
	1948	2142	1.4E
	2323		
6 W ●		0214	3.6F
	0502	0852	7.9E
	1241	1615	7.5F
	2021	2220	1.5E
7 Th	0009	0258	3.6F
	0544	0930	7.8E
	1319	1642	7.4F
	2052	2253	1.8E
8 F	0054	0343	3.7F
	0625	1008	7.8E
	1356	1710	7.4F
	2121	2322	2.1E
9 Sa	0140	0428	3.8F
	0708	1047	7.6E
	1431	1742	7.3F
	2150	2351	2.6E
10 Su	0229	0514	3.8F
	0754	1127	7.2E
	1505	1815	7.1F
	2218		
11 M		0024	3.2E
	0319	0603	3.8F
	0847	1210	6.4E
	1538	1850	6.7F
	2244		
12 Tu		0101	3.9E
	0412	0655	4.0F
	0949	1255	5.4E
	1611	1926	6.3F
	2308		
13 W		0142	4.8E
	0506	0750	4.2F
	1103	1344	4.3E
	1645	2004	5.8F
	2330		
14 Th ◐		0225	5.7E
	0600	0849	4.6F
	1226	1438	3.2E
	1720	2045	5.2F
	2354		
15 F		0311	6.6E
	0656	0951	5.2F
	1351	1537	2.2E
	1759	2129	4.8F

Day	Slack h m	Maximum h m	knots
16 Sa	0023	0401	7.5E
	0753	1055	5.9F
	1514	1640	1.5E
	1844	2217	4.4F
17 Su	0100	0454	7.9E
	0850	1200	6.8F
	1631	1748	1.0E
	1937	2310	4.3F
18 M	0147	0549	8.2E
	0948	1302	7.6F
	1738	1857	1.0E
	2039		
19 Tu		0006	4.4F
	0242	0645	8.6E
	1043	1401	7.9F
	1832	2001	1.2E
	2146		
20 W		0104	4.7F
	0342	0742	8.8E
	1137	1455	8.1F
	1915	2057	1.8E
	2253		
21 Th ○		0202	5.2F
	0445	0837	8.9E
	1229	1544	8.2F
	1953	2148	2.6E
	2358		
22 F		0301	5.6F
	0549	0932	8.8E
	1318	1630	8.2F
	2028	2236	3.5E
23 Sa	0102	0400	6.0F
	0654	1025	8.6E
	1404	1713	8.1F
	2102	2323	4.6E
24 Su	0206	0459	6.2F
	0801	1117	8.1E
	1448	1754	8.0F
	2136		
25 M		0011	5.6E
	0310	0559	6.2F
	0911	1210	7.5E
	1530	1835	7.7F
	2210		
26 Tu		0058	6.5E
	0414	0701	6.1F
	1024	1304	5.7E
	1611	1917	7.1F
	2245		
27 W ◑		0147	7.2E
	0517	0806	5.9F
	1141	1401	3.9E
	1651	1959	6.2F
	2322		
28 Th		0237	7.5E
	0619	0917	5.8F
	1302	1506	2.4E
	1732	2044	5.2F
29 F	0001	0331	7.6E
	0720	1038	5.8F
	1429	1623	1.4E
	1817	2133	4.4F
30 Sa	0045	0427	7.6E
	0819	1202	6.0F
	1559	1743	0.8E
	1909	2225	3.8F
31 Su	0133	0525	7.5E
	0916	1308	6.3F
	1718	1851	0.8E
	2010	2320	3.5F

August

Day	Slack h m	Maximum h m	knots
1 M	0224	0621	7.5E
	1009	1402	6.5F
	1811	1948	0.9E
	2113		
2 Tu		0015	3.4F
	0317	0713	7.6E
	1057	1448	6.7F
	1847	2036	1.2E
	2211		
3 W		0107	3.5F
	0407	0758	7.6E
	1141	1526	6.9F
	1915	2117	1.5E
	2304		
4 Th ●		0156	3.7F
	0455	0838	7.6E
	1221	1553	6.9F
	1941	2152	2.0E
	2352		
5 F		0243	4.0F
	0541	0915	7.6E
	1257	1613	7.0F
	2006	2220	2.5E
6 Sa	0038	0328	4.3F
	0626	0950	7.5E
	1330	1637	7.0F
	2031	2245	3.1E
7 Su	0123	0413	4.5F
	0712	1027	7.2E
	1402	1705	6.9F
	2055	2311	3.8E
8 M	0207	0458	4.8F
	0800	1105	6.6E
	1432	1736	6.7F
	2118	2343	4.7E
9 Tu	0251	0544	5.1F
	0853	1146	5.8E
	1502	1809	6.3F
	2138		
10 W		0019	5.5E
	0337	0632	5.3F
	0953	1230	4.8E
	1532	1843	5.8F
	2157		
11 Th		0058	6.2E
	0425	0724	5.6F
	1100	1318	3.7E
	1603	1920	5.3F
	2219		
12 F ◐		0142	6.9E
	0518	0820	5.8F
	1216	1410	2.7E
	1636	2000	4.8F
	2248		
13 Sa		0230	7.5E
	0616	0920	6.0F
	1338	1508	1.7E
	1711	2047	4.4F
	2328		
14 Su		0324	7.7E
	0719	1026	6.3F
	1500	1612	1.0E
	1755	2142	4.2F
15 M	0019	0423	8.0E
	0823	1135	6.8F
	1616	1724	0.7E
	1859	2243	4.2F

Day	Slack h m	Maximum h m	knots
16 Tu	0121	0525	8.2E
	0927	1242	7.4F
	1715	1839	1.0E
	2021	2348	4.5F
17 W	0231	0628	8.4E
	1026	1342	7.7F
	1757	1943	1.7E
	2141		
18 Th		0053	5.1F
	0344	0730	8.5E
	1120	1434	7.9F
	1833	2037	2.8E
	2252		
19 F ○		0156	5.8F
	0454	0828	8.5E
	1211	1519	8.0F
	1906	2124	4.0E
	2356		
20 Sa		0256	6.5F
	0602	0923	8.3E
	1257	1600	7.9F
	1938	2209	5.4E
21 Su	0057	0355	7.1F
	0708	1015	8.0E
	1340	1640	7.8F
	2009	2252	6.5E
22 M	0155	0452	7.4F
	0813	1106	7.5E
	1421	1718	7.6F
	2042	2335	7.5E
23 Tu	0251	0548	7.4F
	0918	1156	6.0E
	1500	1757	7.0F
	2115		
24 W		0019	7.7E
	0347	0645	7.2F
	1025	1247	4.4E
	1538	1838	6.1F
	2150		
25 Th		0104	7.7E
	0443	0743	6.8F
	1135	1341	3.0E
	1617	1920	5.2F
	2228		
26 F ◑		0152	7.6E
	0541	0845	6.2F
	1249	1441	1.8E
	1657	2006	4.4F
	2311		
27 Sa		0244	7.4E
	0641	0958	5.8F
	1409	1556	1.0E
	1744	2057	3.7F
28 Su	0001	0343	7.0E
	0743	1128	5.6F
	1533	1717	0.7E
	1843	2153	3.3F
29 M	0100	0449	6.7E
	0843	1239	5.6F
	1640	1825	0.8E
	1952	2254	3.1F
30 Tu	0202	0555	6.7E
	0940	1334	5.8F
	1723	1921	1.2E
	2100	2355	3.2F
31 W	0304	0652	6.8E
	1030	1417	6.1F
	1752	2006	1.7E
	2200		

September

Day	Slack h m	Maximum h m	knots
1 Th		0052	3.6F
	0400	0740	6.9E
	1113	1449	6.2F
	1818	2043	2.3E
	2253		
2 F		0144	4.1F
	0452	0821	7.0E
	1152	1509	6.3F
	1842	2113	3.1E
	2339		
3 Sa ●		0231	4.6F
	0542	0857	6.9E
	1226	1529	6.3F
	1907	2137	3.8E
4 Su	0022	0315	5.2F
	0630	0932	6.6E
	1258	1555	6.3F
	1930	2201	4.7E
5 M	0102	0358	5.7F
	0718	1008	6.2E
	1329	1624	6.2F
	1951	2229	5.6E
6 Tu	0142	0441	6.2F
	0808	1046	5.6E
	1359	1655	5.9F
	2010	2302	6.4E
7 W	0222	0525	6.6F
	0900	1126	4.8E
	1429	1728	5.5F
	2027	2339	7.1E
8 Th	0304	0611	6.9F
	0957	1210	4.0E
	1458	1802	5.1F
	2049		
9 F		0020	7.6E
	0351	0700	6.9F
	1100	1257	3.1E
	1529	1840	4.7F
	2119		
10 Sa		0106	7.7E
	0444	0755	6.8F
	1210	1349	2.1E
	1601	1924	4.3F
	2159		
11 Su ◐		0157	7.8E
	0544	0855	6.7F
	1326	1447	1.4E
	1641	2017	4.0F
	2250		
12 M		0255	7.8E
	0651	1002	6.6F
	1441	1554	1.0E
	1742	2120	3.9F
	2355		
13 Tu		0359	7.8E
	0759	1111	6.8F
	1543	1711	1.1E
	1908	2230	4.0F
14 W	0114	0507	7.9E
	0904	1218	7.1F
	1629	1825	1.9E
	2036	2341	4.6F
15 Th	0239	0616	7.9E
	1003	1315	7.4F
	1706	1924	3.2E
	2150		

Day	Slack h m	Maximum h m	knots
16 F		0050	5.5F
	0358	0722	7.9E
	1057	1402	7.5F
	1739	2013	4.6E
	2255		
17 Sa ○		0155	6.5F
	0510	0821	7.8E
	1145	1444	7.5F
	1811	2056	6.1E
	2353		
18 Su		0254	7.5F
	0616	0915	7.5E
	1229	1524	7.4F
	1843	2138	7.4E
19 M	0047	0350	7.8F
	0719	1006	6.7E
	1310	1602	7.0F
	1914	2218	7.8E
20 Tu	0138	0443	7.9F
	0820	1055	5.6E
	1350	1640	6.5F
	1946	2258	8.0E
21 W	0227	0533	7.8F
	0920	1142	4.5E
	1429	1719	5.8F
	2020	2339	8.0E
22 Th	0317	0624	7.7F
	1020	1229	3.4E
	1507	1759	5.0F
	2055		
23 F		0022	7.9E
	0407	0715	7.3F
	1122	1319	2.4E
	1546	1843	4.3F
	2134		
24 Sa ◑		0109	7.6E
	0501	0810	6.5F
	1227	1415	1.6E
	1629	1930	3.6F
	2220		
25 Su		0200	7.0E
	0558	0912	5.8F
	1336	1525	1.0E
	1722	2024	3.1F
	2317		
26 M		0258	6.3E
	0700	1029	5.3F
	1444	1645	1.0E
	1829	2125	2.8F
27 Tu	0027	0406	5.9E
	0801	1151	5.2F
	1537	1752	1.3E
	1944	2231	2.8F
28 W	0141	0518	5.7E
	0858	1245	5.3F
	1614	1844	1.9E
	2052	2337	3.2F
29 Th	0250	0621	5.7E
	0948	1322	5.4F
	1645	1925	2.7E
	2149		
30 F		0038	3.8F
	0352	0713	5.7E
	1032	1346	5.5F
	1712	1959	3.6E
	2238		

Time meridian 135° W. 0000 is midnight. 1200 is noon. Times are not adjusted for Daylight Saving Time.

Akutan Pass, Aleutian Islands, 2005

F–Flood, Dir. 295° True   E–Ebb, Dir. 115° True

October

Day	Slack h m	Maximum h m	knots
1 Sa		0131	4.6F
	0449	0756	5.6E
	1111	1410	5.6F
	1738	2025	4.6E
	2321		
2 Su		0217	5.5F
	0542	0835	5.5E
	1147	1439	5.6F
	1802	2050	5.6E
3 M ●	0000	0300	6.3F
	0633	0912	5.2E
	1221	1509	5.5F
	1823	2117	6.6E
4 Tu	0038	0342	7.1F
	0723	0949	4.8E
	1253	1541	5.4F
	1842	2150	7.5E
5 W	0116	0424	7.6F
	0813	1028	4.4E
	1325	1615	5.1F
	1901	2226	7.8E
6 Th	0155	0507	7.7F
	0905	1109	3.8E
	1356	1650	4.8F
	1925	2305	8.0E
7 F	0238	0552	7.8F
	1000	1153	3.1E
	1427	1727	4.5F
	1957	2349	8.0E
8 Sa	0326	0642	7.7F
	1059	1240	2.4E
	1501	1810	4.2F
	2037		
9 Su		0038	8.0E
	0421	0736	7.6F
	1204	1334	1.8E
	1542	1859	3.9F
	2126		
10 M ◐		0132	7.9E
	0521	0836	7.3F
	1309	1434	1.5E
	1612	2000	3.7F
	2229		
11 Tu		0232	7.8E
	0626	0939	6.9F
	1408	1545	1.6E
	1805	2111	3.7F
	2350		
12 W		0339	7.5E
	0732	1044	6.8F
	1455	1702	2.4E
	1933	2226	4.0F
13 Th	0126	0452	7.1E
	0834	1144	6.8F
	1535	1807	3.7E
	2051	2342	4.9F
14 F	0256	0606	6.7E
	0932	1237	6.8F
	1610	1900	5.2E
	2156		
15 Sa		0053	6.0F
	0414	0714	6.3E
	1024	1323	6.7F
	1644	1945	6.8E
	2253		
16 Su		0157	7.2F
	0525	0815	5.9E
	1112	1404	6.5F
	1716	2027	7.7E
	2344		
17 M ○		0254	7.8F
	0629	0908	5.3E
	1156	1444	6.2F
	1749	2106	8.1E
18 Tu	0032	0344	8.0F
	0728	0958	4.6E
	1238	1524	5.7F
	1821	2145	8.2E
19 W	0118	0431	8.1F
	0824	1044	3.9E
	1318	1603	5.3F
	1854	2224	8.3E
20 Th	0203	0516	8.0F
	0918	1128	3.2E
	1358	1643	4.7F
	1929	2304	8.1E
21 F	0248	0601	7.8F
	1011	1212	2.5E
	1437	1725	4.2F
	2005	2346	7.9E
22 Sa	0334	0647	7.5F
	1105	1257	2.0E
	1518	1809	3.6F
	2046		
23 Su		0032	7.6E
	0423	0736	6.7F
	1200	1348	1.5E
	1606	1857	3.2F
	2134		
24 M ◑		0121	6.9E
	0516	0829	6.0F
	1256	1450	1.3E
	1705	1953	2.8F
	2234		
25 Tu		0216	6.0E
	0611	0926	5.5F
	1347	1602	1.5E
	1817	2056	2.6F
	2351		
26 W		0318	5.3E
	0708	1024	5.2F
	1430	1705	2.1E
	1931	2205	2.7F
27 Th	0115	0425	4.8E
	0802	1115	5.1F
	1505	1754	2.9E
	2036	2314	3.3F
28 F	0232	0533	4.4E
	0852	1158	5.0F
	1537	1832	3.9E
	2129		
29 Sa		0018	4.1F
	0342	0633	4.2E
	0938	1236	5.0F
	1605	1904	5.0E
	2215		
30 Su		0112	5.2F
	0444	0724	4.1E
	1021	1312	5.0F
	1630	1933	6.2E
	2256		
31 M		0159	6.3F
	0541	0809	3.9E
	1100	1348	4.9F
	1654	2004	7.3E
	2335		

November

Day	Slack h m	Maximum h m	knots
1 Tu ●		0242	7.3F
	0634	0851	3.7E
	1138	1424	4.8F
	1715	2039	7.8E
2 W	0013	0324	7.8F
	0725	0931	3.5E
	1214	1500	4.7F
	1739	2116	8.1E
3 Th	0053	0407	8.0F
	0815	1012	3.2E
	1248	1538	4.6F
	1808	2156	8.3E
4 F	0135	0451	8.1F
	0906	1054	2.8E
	1323	1618	4.5F
	1843	2240	8.4E
5 Sa	0221	0538	8.1F
	0959	1140	2.4E
	1401	1701	4.3F
	1925	2327	8.4E
6 Su	0311	0628	8.0F
	1053	1229	2.1E
	1446	1750	4.1F
	2014		
7 M		0018	8.2E
	0404	0721	7.8F
	1147	1324	2.0E
	1545	1847	3.9F
	2114		
8 Tu ◐		0114	8.0E
	0501	0816	7.6F
	1237	1427	2.3E
	1702	1954	3.7F
	2228		
9 W		0215	7.5E
	0559	0914	7.2F
	1322	1536	3.0E
	1826	2108	3.8F
10 Th	0004	0322	6.5E
	0658	1010	6.8F
	1403	1643	4.2E
	1945	2227	4.3F
11 F	0143	0438	5.4E
	0756	1104	6.4F
	1440	1741	5.7E
	2053	2347	5.3F
12 Sa	0312	0556	4.6E
	0852	1154	6.0F
	1516	1831	7.1E
	2152		
13 Su		0101	6.6F
	0430	0707	4.0E
	0944	1240	5.7F
	1551	1916	7.8E
	2244		
14 M		0203	7.6F
	0540	0808	3.6E
	1034	1325	5.3F
	1626	1958	8.2E
	2331		
15 Tu ○		0254	7.9F
	0641	0902	3.3E
	1121	1407	5.0F
	1701	2038	8.3E
16 W	0016	0340	8.1F
	0735	0950	3.0E
	1205	1449	4.7F
	1737	2117	8.4E
17 Th	0059	0421	8.1F
	0825	1035	2.6E
	1247	1530	4.4F
	1813	2156	8.3E
18 F	0141	0501	8.0F
	0912	1116	2.3E
	1328	1612	4.0F
	1850	2236	8.1E
19 Sa	0224	0541	7.8F
	0958	1155	2.0E
	1409	1655	3.7F
	1929	2317	7.9E
20 Su	0307	0622	7.5F
	1043	1236	1.8E
	1453	1740	3.4F
	2011		
21 M		0001	7.6E
	0350	0705	7.0F
	1128	1320	1.8E
	1545	1830	3.0F
	2100		
22 Tu		0048	6.9E
	0436	0749	6.4F
	1210	1409	1.9E
	1648	1925	2.8F
	2200		
23 W ◑		0138	5.9E
	0522	0836	5.9F
	1249	1503	2.4E
	1757	2027	2.7F
	2317		
24 Th		0233	5.0E
	0609	0922	5.5F
	1325	1556	3.1E
	1905	2134	2.9F
25 F	0047	0333	4.1E
	0656	1008	5.2F
	1357	1643	4.0E
	2005	2242	3.5F
26 Sa	0212	0437	3.3E
	0745	1052	4.9F
	1427	1725	5.1E
	2057	2347	4.5F
27 Su	0328	0542	2.9E
	0833	1136	4.6F
	1454	1805	6.3E
	2143		
28 M		0045	5.7F
	0436	0644	2.6E
	0920	1218	4.5F
	1520	1844	7.5E
	2226		
29 Tu		0135	6.8F
	0537	0739	2.4E
	1006	1300	4.4F
	1547	1924	7.9E
	2308		
30 W		0222	7.6F
	0632	0828	2.4E
	1049	1342	4.3F
	1617	2006	8.2E
	2351		

December

Day	Slack h m	Maximum h m	knots
1 Th ●		0307	8.0F
	0724	0913	2.3E
	1130	1424	4.4F
	1653	2049	8.6E
2 F	0036	0352	8.2F
	0814	0957	2.2E
	1211	1508	4.5F
	1735	2135	8.7E
3 Sa	0122	0439	8.3F
	0901	1042	2.2E
	1255	1555	4.6F
	1822	2222	8.7E
4 Su	0210	0526	8.2F
	0948	1129	2.3E
	1345	1646	4.6F
	1914	2312	8.6E
5 M	0258	0614	8.1F
	1032	1220	2.6E
	1445	1741	4.5F
	2013		
6 Tu		0004	8.3E
	0348	0703	8.0F
	1114	1314	3.1E
	1555	1843	4.3F
	2122		
7 W		0100	7.8E
	0438	0752	7.7F
	1153	1412	3.9E
	1711	1951	4.2F
	2247		
8 Th ◐		0159	6.7E
	0528	0842	7.3F
	1232	1513	4.9E
	1828	2106	4.4F
9 F	0022	0306	5.1E
	0619	0931	6.6F
	1309	1612	6.1E
	1938	2226	5.0F
10 Sa	0156	0422	3.6E
	0711	1021	5.9F
	1347	1708	7.3E
	2042	2352	5.9F
11 Su	0325	0544	2.7E
	0805	1111	5.3F
	1425	1800	7.8E
	2138		
12 M		0107	6.9F
	0446	0658	2.2E
	0900	1200	4.8F
	1505	1848	8.1E
	2229		
13 Tu		0206	7.6F
	0556	0801	2.0E
	0955	1249	4.4F
	1545	1934	8.2E
	2316		
14 W		0255	7.8F
	0653	0854	1.9E
	1047	1335	4.1F
	1626	2016	8.2E
15 Th ○	0001	0338	7.9F
	0740	0942	1.9E
	1136	1421	4.0F
	1707	2056	8.2E
16 F	0043	0415	7.8F
	0822	1024	1.9E
	1219	1505	3.9F
	1747	2136	8.1E
17 Sa	0124	0449	7.8F
	0901	1102	1.9E
	1302	1548	3.8F
	1828	2215	8.0E
18 Su	0204	0522	7.6F
	0937	1137	1.9E
	1346	1632	3.7F
	1909	2254	7.8E
19 M	0242	0557	7.5F
	1012	1211	2.1E
	1434	1718	3.5F
	1953	2336	7.6E
20 Tu	0320	0633	7.2F
	1046	1246	2.4E
	1526	1807	3.4F
	2043		
21 W		0019	6.8E
	0357	0710	6.8F
	1118	1325	2.9E
	1624	1900	3.3F
	2142		
22 Th		0105	5.8E
	0434	0749	6.3F
	1149	1408	3.6E
	1724	1958	3.3F
	2256		
23 F ◑		0155	4.7E
	0512	0829	5.8F
	1218	1452	4.3E
	1824	2100	3.5F
24 Sa	0022	0250	3.6E
	0551	0911	5.3F
	1245	1538	5.2E
	1921	2204	4.1F
25 Su	0149	0349	2.6E
	0633	0954	4.7F
	1311	1624	6.2E
	2015	2308	4.9F
26 M	0311	0453	1.8E
	0720	1040	4.3F
	1339	1712	7.2E
	2105		
27 Tu		0010	5.9F
	0426	0600	1.4E
	0811	1127	4.1F
	1413	1801	7.8E
	2155		
28 W		0108	7.0F
	0532	0705	1.3E
	0904	1216	4.1F
	1452	1850	8.2E
	2244		
29 Th		0201	7.7F
	0629	0803	1.3E
	0957	1306	4.2F
	1538	1939	8.5E
	2333		
30 F ●		0251	8.0F
	0718	0854	1.6E
	1050	1357	4.5F
	1629	2029	8.8E
31 Sa	0022	0339	8.2F
	0802	0942	1.9E
	1144	1449	4.9F
	1723	2119	8.9E

Time meridian 135° W. 0000 is midnight. 1200 is noon. Times are not adjusted for Daylight Saving Time.

Kvichak Bay (off Naknek River ent.), Alaska, 2005

F–Flood, Dir. 055° True E–Ebb, Dir. 240° True

January

Day	Slack h m	Maximum h m	knots
1 Sa	0234	0416	1.5F
	0720	1029	1.6E
	1416	1626	2.4F
	1944	2319	2.4E
2 Su	0315	0503	1.7F
	0814	1117	1.6E
	1507	1712	2.2F
	2019	2350	2.5E
3 M	0356	0550	2.0F
	0908	1207	1.6E
	1559	1759	2.0F
	2055		
4 Tu		0024	2.5E
	0435	0637	2.2F
	1002	1258	1.6E
	1653	1847	1.9F
	2130		
5 W		0059	2.6E
	0515	0725	2.5F
	1055	1350	1.7E
	1748	1935	1.7F
	2206		
6 Th		0136	2.7E
	0554	0813	2.8F
	1147	1443	1.7E
	1842	2024	1.6F
	2245		
7 F		0217	2.8E
	0635	0902	3.1F
	1238	1535	1.9E
	1936	2114	1.5F
	2329		
8 Sa		0302	2.8E
	0717	0951	3.3F
	1328	1628	2.0E
	2029	2206	1.5F
9 Su	0019	0351	2.9E
	0803	1041	3.5F
	1418	1721	2.2E
	2121	2258	1.6F
10 M	0116	0445	2.9E
	0851	1132	3.6F
	1507	1814	2.4E
	2212	2352	1.7F
11 Tu	0218	0543	2.9E
	0944	1223	3.6F
	1556	1907	2.7E
	2303		
12 W		0048	1.9F
	0324	0644	2.8E
	1039	1315	3.6F
	1645	2001	2.9E
	2354		
13 Th		0144	2.1F
	0432	0748	2.7E
	1136	1408	3.5F
	1734	2055	3.1E
14 F	0045	0241	2.3F
	0540	0853	2.7E
	1235	1502	3.4F
	1823	2148	3.3E
15 Sa	0136	0339	2.5F
	0648	0959	2.6E
	1335	1556	3.2F
	1913	2242	3.4E
16 Su	0227	0437	2.7F
	0755	1106	2.5E
	1435	1650	2.9F
	2002	2336	3.4E
17 M	0318	0535	2.9F
	0900	1215	2.5E
	1536	1744	2.6F
	2051		
18 Tu		0030	3.4E
	0409	0633	3.0F
	1003	1329	2.4E
	1636	1838	2.3F
	2141		
19 W		0123	3.3E
	0500	0729	3.1F
	1104	1454	2.4E
	1735	1932	2.0F
	2230		
20 Th		0216	3.1E
	0550	0824	3.1F
	1202	1606	2.5E
	1833	2025	1.8F
	2319		
21 F		0307	2.9E
	0640	0915	3.1F
	1257	1708	2.5E
	1931	2118	1.5F
22 Sa	0008	0357	2.7E
	0729	1004	3.0F
	1348	1804	2.5E
	2027	2210	1.4F
23 Su	0057	0446	2.5E
	0817	1049	2.9F
	1435	1857	2.5E
	2121	2300	1.2F
24 M	0146	0532	2.3E
	0905	1133	2.8F
	1519	1948	2.5E
	2212	2350	1.2F
25 Tu	0237	0618	2.1E
	0952	1216	2.7F
	1601	2036	2.5E
	2302		
26 W		0038	1.2F
	0328	0703	1.9E
	1040	1259	2.6F
	1641	2121	2.4E
	2348		
27 Th		0126	1.3F
	0420	0749	1.8E
	1128	1343	2.5F
	1719	2201	2.4E
28 F	0033	0214	1.4F
	0513	0835	1.7E
	1217	1428	2.4F
	1756	2152	2.3E
29 Sa	0116	0301	1.5F
	0607	0922	1.7E
	1307	1513	2.3F
	1833	2210	2.3E
30 Su	0158	0348	1.7F
	0700	1010	1.7E
	1357	1559	2.1F
	1909	2239	2.3E
31 M	0239	0434	2.0F
	0753	1059	1.7E
	1449	1645	2.0F
	1943	2312	2.4E

February

Day	Slack h m	Maximum h m	knots
1 Tu	0319	0521	2.2F
	0845	1147	1.7E
	1541	1732	1.8F
	2018	2346	2.5E
2 W	0359	0608	2.5F
	0937	1236	1.7E
	1633	1820	1.7F
	2054		
3 Th		0022	2.6E
	0438	0656	2.7F
	1027	1325	1.8E
	1725	1908	1.6F
	2133		
4 F		0102	2.7E
	0519	0744	3.0F
	1117	1415	1.9E
	1816	1957	1.6F
	2216		
5 Sa		0146	2.8E
	0602	0833	3.2F
	1207	1505	2.0E
	1908	2048	1.6F
	2306		
6 Su		0236	2.8E
	0649	0923	3.3F
	1256	1557	2.2E
	1958	2140	1.7F
7 M	0003	0330	2.9E
	0738	1013	3.4F
	1345	1650	2.4E
	2049	2234	1.8F
8 Tu	0105	0428	2.8E
	0831	1105	3.4F
	1435	1743	2.6E
	2140	2329	2.0F
9 W	0210	0530	2.8E
	0927	1158	3.4F
	1524	1838	2.8E
	2231		
10 Th		0025	2.2F
	0317	0634	2.7E
	1025	1251	3.3F
	1614	1933	2.9E
	2322		
11 F		0122	2.4F
	0425	0739	2.7E
	1124	1345	3.2F
	1705	2028	3.1E
12 Sa	0014	0220	2.6F
	0531	0845	2.6E
	1223	1439	3.0F
	1755	2122	3.2E
13 Su	0106	0317	2.8F
	0636	0951	2.6E
	1322	1534	2.8F
	1845	2217	3.3E
14 M	0157	0415	3.0F
	0740	1058	2.6E
	1421	1628	2.6F
	1935	2311	3.2E
15 Tu	0249	0511	3.0F
	0841	1207	2.5E
	1519	1721	2.3F
	2025		
16 W		0004	3.2E
	0340	0606	3.1F
	0939	1322	2.5E
	1616	1814	2.1F
	2115		
17 Th		0056	3.0E
	0430	0659	3.0F
	1035	1436	2.5E
	1713	1907	1.9F
	2204		
18 F		0147	2.9E
	0520	0749	3.0F
	1128	1539	2.5E
	1808	1958	1.7F
	2253		
19 Sa		0237	2.7E
	0610	0838	2.9F
	1218	1636	2.5E
	1902	2049	1.6F
	2343		
20 Su		0326	2.5E
	0659	0924	2.8F
	1306	1730	2.4E
	1955	2139	1.5F
21 M	0033	0414	2.3E
	0748	1010	2.7F
	1351	1821	2.4E
	2046	2229	1.4F
22 Tu	0123	0502	2.1E
	0837	1055	2.6F
	1435	1909	2.4E
	2135	2318	1.4F
23 W	0215	0550	2.0E
	0927	1141	2.5F
	1517	1954	2.3E
	2223		
24 Th		0007	1.5F
	0308	0639	1.9E
	1017	1227	2.4F
	1558	2028	2.3E
	2308		
25 F		0055	1.6F
	0401	0728	1.9E
	1107	1313	2.3F
	1638	2023	2.2E
	2353		
26 Sa		0143	1.7F
	0455	0818	1.8E
	1158	1400	2.1F
	1717	2051	2.2E
27 Su	0035	0231	1.9F
	0548	0907	1.8E
	1250	1446	2.0F
	1755	2123	2.2E
28 M	0117	0318	2.1F
	0640	0955	1.8E
	1341	1533	1.9F
	1831	2157	2.3E

March

Day	Slack h m	Maximum h m	knots
1 Tu	0159	0405	2.4F
	0731	1042	1.9E
	1433	1620	1.7F
	1908	2232	2.4E
2 W	0239	0452	2.6F
	0820	1128	1.9E
	1523	1707	1.7F
	1945	2310	2.4E
3 Th	0320	0539	2.8F
	0909	1214	1.9E
	1613	1755	1.6F
	2026	2350	2.5E
4 F	0403	0627	2.9F
	0958	1301	2.0E
	1702	1844	1.6F
	2111		
5 Sa		0035	2.6E
	0447	0715	3.1F
	1046	1349	2.1E
	1751	1934	1.7F
	2202		
6 Su		0125	2.7E
	0535	0805	3.2F
	1134	1438	2.3E
	1840	2026	1.8F
	2259		
7 M		0220	2.7E
	0627	0856	3.2F
	1222	1529	2.4E
	1929	2119	2.0F
8 Tu	0000	0319	2.7E
	0721	0948	3.2F
	1312	1622	2.6E
	2019	2213	2.1F
9 W	0104	0421	2.7E
	0818	1041	3.1F
	1402	1716	2.7E
	2110	2309	2.4F
10 Th	0210	0525	2.6E
	0917	1135	3.0F
	1452	1811	2.8E
	2201		
11 F		0006	2.6F
	0315	0631	2.6E
	1016	1229	2.9F
	1543	1906	2.9E
	2252		
12 Sa		0102	2.7F
	0420	0736	2.6E
	1115	1324	2.7F
	1634	2001	3.0E
	2344		
13 Su		0159	2.9F
	0523	0842	2.6E
	1213	1418	2.5F
	1726	2056	3.0E
14 M	0036	0255	3.0F
	0623	0948	2.6E
	1311	1512	2.3F
	1817	2150	3.0E
15 Tu	0127	0350	3.1F
	0721	1055	2.6E
	1407	1606	2.2F
	1908	2243	2.9E
16 W	0218	0443	3.1F
	0816	1203	2.6E
	1502	1658	2.0F
	1958	2334	2.8E
17 Th	0309	0533	3.0F
	0909	1308	2.6E
	1556	1750	1.9F
	2048		
18 F		0024	2.6E
	0359	0622	2.9F
	0959	1408	2.5E
	1649	1840	1.8F
	2138		
19 Sa		0114	2.5E
	0448	0710	2.8F
	1046	1502	2.5E
	1741	1930	1.7F
	2228		
20 Su		0202	2.3E
	0538	0756	2.7F
	1132	1553	2.4E
	1831	2019	1.6F
	2319		
21 M		0251	2.2E
	0628	0843	2.6F
	1217	1638	2.3E
	1920	2108	1.6F
22 Tu	0010	0340	2.1E
	0718	0930	2.5F
	1301	1709	2.3E
	2008	2157	1.7F
23 W	0103	0431	2.0E
	0810	1018	2.3F
	1345	1731	2.3E
	2055	2246	1.8F
24 Th	0156	0523	2.0E
	0902	1106	2.2F
	1428	1805	2.2E
	2141	2335	1.9F
25 F	0250	0615	1.9E
	0954	1154	2.1F
	1510	1842	2.2E
	2225		
26 Sa		0024	2.0F
	0343	0708	2.0E
	1047	1242	1.9F
	1551	1920	2.2E
	2309		
27 Su		0112	2.2F
	0436	0759	2.0E
	1140	1330	1.8F
	1632	1957	2.2E
	2351		
28 M		0200	2.4F
	0527	0849	2.0E
	1233	1418	1.7F
	1711	2035	2.2E
29 Tu	0034	0247	2.6F
	0617	0936	2.1E
	1324	1506	1.6F
	1751	2114	2.3E
30 W	0116	0334	2.8F
	0706	1022	2.1E
	1414	1554	1.6F
	1833	2154	2.3E
31 Th	0159	0422	2.9F
	0753	1107	2.2E
	1503	1642	1.6F
	1917	2238	2.4E

Time meridian 135° W. 0000 is midnight. 1200 is noon. Times are not adjusted for Daylight Saving Time.

Kvichak Bay (off Naknek River ent.), Alaska, 2005

F–Flood, Dir. 055° True E–Ebb, Dir. 240° True

April

Day	Slack h m	Maximum h m	knots
1 F ◑	0243	0510	3.0F
	0840	1152	2.3E
	1551	1732	1.6F
	2006	2325	2.5E
2 Sa	0330	0559	3.1F
	0927	1238	2.4E
	1638	1822	1.7F
	2100		
3 Su		0017	2.5E
	0420	0649	3.1F
	1014	1325	2.5E
	1726	1914	1.9F
	2158		
4 M		0113	2.5E
	0514	0739	3.1F
	1101	1414	2.6E
	1813	2007	2.1F
	2300		
5 Tu		0213	2.5E
	0610	0831	3.0F
	1149	1506	2.7E
	1902	2101	2.3F
6 W	0004	0316	2.5E
	0709	0924	2.9F
	1239	1558	2.8E
	1952	2156	2.5F
7 Th	0108	0421	2.5E
	0808	1018	2.7F
	1329	1652	2.9E
	2042	2252	2.7F
8 F ●	0212	0527	2.5E
	0908	1113	2.5F
	1420	1747	2.9E
	2133	2349	2.9F
9 Sa	0315	0634	2.5E
	1008	1208	2.3F
	1512	1842	2.9E
	2224		
10 Su		0044	3.0F
	0415	0743	2.6E
	1106	1303	2.2F
	1804	1936	2.9E
	2315		
11 M		0139	3.1F
	0513	0853	2.6E
	1203	1358	2.0F
	1656	2030	2.8E
12 Tu	0006	0232	3.1F
	0608	1003	2.6E
	1259	1451	1.9F
	1748	2122	2.7E
13 W	0057	0323	3.1F
	0700	1105	2.7E
	1353	1543	1.8F
	1839	2213	2.5E
14 Th	0147	0411	3.0F
	0749	1200	2.6E
	1445	1634	1.7F
	1930	2302	2.4E
15 F	0236	0458	2.9F
	0835	1249	2.6E
	1535	1723	1.6F
	2021	2350	2.2E
16 Sa ◐	0325	0544	2.8F
	0919	1331	2.5E
	1624	1812	1.6F
	2112		
17 Su		0037	2.1E
	0415	0631	2.7F
	1002	1358	2.5E
	1712	1900	1.7F
	2203		
18 M		0126	2.0E
	0505	0717	2.5F
	1044	1428	2.4E
	1759	1948	1.7F
	2256		
19 Tu		0215	1.9E
	0556	0804	2.4F
	1127	1504	2.4E
	1844	2037	1.8F
	2349		
20 W		0307	1.9E
	0648	0852	2.3F
	1209	1542	2.4E
	1930	2126	2.0F
21 Th	0043	0400	1.9E
	0742	0941	2.1F
	1252	1623	2.3E
	2014	2215	2.1F
22 F	0137	0455	1.9E
	0837	1031	1.9F
	1335	1703	2.3E
	2058	2303	2.3F
23 Sa	0230	0550	1.9E
	0932	1120	1.8F
	1417	1744	2.3E
	2141	2352	2.5F
24 Su ○	0323	0644	2.0E
	1026	1210	1.6F
	1459	1825	2.3E
	2224		
25 M		0040	2.7F
	0414	0736	2.1E
	1120	1259	1.5F
	1542	1906	2.3E
	2307		
26 Tu		0128	2.9F
	0503	0825	2.2E
	1212	1349	1.5F
	1625	1948	2.3E
	2350		
27 W		0216	3.1F
	0552	0912	2.3E
	1303	1438	1.5F
	1712	2033	2.4E
28 Th	0034	0304	3.2F
	0638	0957	2.4E
	1352	1528	1.5F
	1801	2121	2.4E
29 F	0120	0353	3.3F
	0725	1042	2.5E
	1440	1619	1.6F
	1856	2212	2.4E
30 Sa ◑	0210	0442	3.3F
	0810	1128	2.6E
	1526	1710	1.8F
	1954	2307	2.4E

May

Day	Slack h m	Maximum h m	knots
1 Su	0302	0532	3.2F
	0856	1215	2.7E
	1613	1803	2.0F
	2055		
2 M		0005	2.4E
	0358	0623	3.1F
	0943	1303	2.8E
	1700	1856	2.2F
	2158		
3 Tu		0107	2.4E
	0456	0716	3.0F
	1030	1353	2.9E
	1748	1951	2.5F
	2302		
4 W		0211	2.3E
	0556	0809	2.8F
	1119	1445	3.0E
	1837	2046	2.7F
5 Th	0007	0318	2.3E
	0657	0903	2.5F
	1208	1538	3.0E
	1927	2142	2.9F
6 F	0110	0426	2.4E
	0758	0957	2.3F
	1259	1632	3.0E
	2017	2238	3.0F
7 Sa ●	0212	0539	2.4E
	0858	1053	2.1F
	1350	1726	2.9E
	2108	2333	3.1F
8 Su	0312	0701	2.5E
	0958	1148	1.9F
	1442	1820	2.8E
	2158		
9 M		0027	3.2F
	0408	0816	2.6E
	1055	1243	1.7F
	1535	1913	2.7E
	2249		
10 Tu		0119	3.2F
	0501	0916	2.7E
	1151	1337	1.6F
	1627	2004	2.5E
	2338		
11 W		0208	3.1F
	0550	1010	2.7E
	1245	1430	1.5F
	1719	2054	2.3E
12 Th	0027	0254	3.0F
	0636	1100	2.7E
	1336	1520	1.4F
	1811	2142	2.2E
13 F	0115	0339	2.9F
	0719	1146	2.7E
	1425	1608	1.4F
	1902	2228	2.0E
14 Sa	0204	0424	2.8F
	0800	1225	2.6E
	1512	1656	1.5F
	1954	2314	1.9E
15 Su ◐	0252	0509	2.7F
	0839	1238	2.5E
	1558	1743	1.6F
	2046		
16 M		0002	1.8E
	0342	0555	2.5F
	0919	1301	2.5E
	1642	1831	1.7F
	2139		
17 Tu		0050	1.7E
	0433	0641	2.4F
	0959	1334	2.5E
	1726	1918	1.9F
	2233		
18 W		0141	1.7E
	0525	0729	2.2F
	1039	1411	2.5E
	1809	2006	2.1F
	2327		
19 Th		0234	1.7E
	0619	0817	2.1F
	1119	1450	2.5E
	1851	2055	2.3F
20 F	0021	0329	1.8E
	0715	0907	1.9F
	1200	1529	2.5E
	1934	2143	2.5F
21 Sa	0114	0425	1.8E
	0811	0956	1.7F
	1241	1610	2.5E
	2016	2232	2.7F
22 Su	0207	0520	1.9E
	0907	1046	1.5F
	1323	1651	2.5E
	2058	2320	2.9F
23 M ○	0258	0613	2.0E
	1002	1137	1.4F
	1406	1734	2.5E
	2140		
24 Tu		0009	3.1F
	0348	0704	2.2E
	1055	1227	1.4F
	1452	1819	2.5E
	2224		
25 W		0057	3.3F
	0436	0753	2.3E
	1147	1319	1.4F
	1543	1908	2.5E
	2309		
26 Th		0146	3.4F
	0523	0840	2.5E
	1236	1410	1.5F
	1638	1959	2.5E
	2357		
27 F		0236	3.5F
	0609	0927	2.6E
	1324	1502	1.6F
	1737	2054	2.5E
28 Sa	0049	0325	3.5F
	0655	1014	2.8E
	1411	1555	1.8F
	1840	2152	2.4E
29 Su	0143	0416	3.4F
	0741	1102	2.9E
	1459	1649	2.1F
	1944	2253	2.4E
30 M ◑	0240	0508	3.2F
	0827	1151	3.1E
	1546	1744	2.3F
	2050	2357	2.4E
31 Tu	0339	0600	3.0F
	0914	1242	3.2E
	1634	1840	2.6F
	2156		

June

Day	Slack h m	Maximum h m	knots
1 W		0102	2.3E
	0440	0653	2.8F
	1002	1333	3.2E
	1723	1936	2.8F
	2301		
2 Th		0210	2.3E
	0542	0747	2.5F
	1051	1426	3.2E
	1813	2032	3.0F
3 F	0005	0322	2.3E
	0643	0842	2.2F
	1141	1519	3.2E
	1903	2129	3.1F
4 Sa	0107	0445	2.4E
	0745	0937	2.0F
	1232	1613	3.0E
	1954	2224	3.2F
5 Su	0207	0610	2.5E
	0845	1033	1.7F
	1323	1706	2.9E
	2044	2318	3.2F
6 M ●	0303	0714	2.6E
	0944	1128	1.5F
	1415	1759	2.7E
	2134		
7 Tu		0009	3.2F
	0355	0812	2.7E
	1040	1223	1.4F
	1507	1850	2.5E
	2223		
8 W		0057	3.1F
	0443	0905	2.7E
	1135	1316	1.3F
	1559	1939	2.3E
	2311		
9 Th		0141	3.0F
	0527	0955	2.7E
	1226	1406	1.2F
	1651	2027	2.1E
	2359		
10 F		0225	2.9F
	0609	1042	2.7E
	1315	1454	1.2F
	1743	2112	1.9E
11 Sa	0047	0308	2.8F
	0647	1125	2.6E
	1401	1541	1.3F
	1836	2157	1.8E
12 Su	0134	0352	2.6F
	0725	1156	2.5E
	1445	1628	1.4F
	1928	2243	1.7E
13 M	0223	0437	2.5F
	0802	1150	2.5E
	1528	1714	1.6F
	2021	2330	1.6E
14 Tu ◐	0313	0522	2.4F
	0839	1217	2.5E
	1610	1801	1.8F
	2115		
15 W		0019	1.6E
	0404	0609	2.2F
	0917	1250	2.5E
	1652	1848	2.0F
	2209		
16 Th		0111	1.6E
	0458	0657	2.0F
	0954	1326	2.6E
	1733	1936	2.3F
	2303		
17 F		0204	1.6E
	0553	0745	1.9F
	1032	1403	2.6E
	1814	2024	2.5F
	2356		
18 Sa		0258	1.7E
	0648	0834	1.7F
	1111	1442	2.6E
	1855	2112	2.8F
19 Su	0048	0352	1.8E
	0744	0924	1.5F
	1150	1522	2.6E
	1936	2200	3.0F
20 M	0139	0445	1.9E
	0839	1014	1.4F
	1233	1605	2.7E
	2018	2249	3.2F
21 Tu ○	0229	0537	2.0E
	0933	1105	1.4F
	1320	1652	2.7E
	2102	2338	3.4F
22 W	0317	0627	2.2E
	1025	1157	1.4F
	1413	1742	2.7E
	2148		
23 Th		0027	3.5F
	0405	0717	2.4E
	1115	1249	1.5F
	1512	1836	2.7E
	2237		
24 F		0117	3.6F
	0451	0806	2.6E
	1204	1343	1.7F
	1615	1934	2.6E
	2330		
25 Sa		0208	3.6F
	0538	0855	2.8E
	1252	1437	1.9F
	1721	2035	2.6E
26 Su	0025	0259	3.5F
	0625	0945	3.0E
	1340	1532	2.1F
	1828	2138	2.5E
27 M	0123	0352	3.3F
	0712	1036	3.2E
	1429	1628	2.4F
	1935	2243	2.5E
28 Tu ◑	0223	0444	3.2F
	0759	1127	3.3E
	1518	1725	2.7F
	2042	2349	2.4E
29 W	0323	0538	2.9F
	0848	1220	3.4E
	1608	1822	2.9F
	2147		
30 Th		0057	2.4E
	0425	0632	2.6F
	0937	1313	3.4E
	1658	1920	3.1F
	2252		

Time meridian 135° W. 0000 is midnight. 1200 is noon. Times are not adjusted for Daylight Saving Time.

Kvichak Bay (off Naknek River ent.), Alaska, 2005

F–Flood, Dir. 055° True E–Ebb, Dir. 240° True

July

Day	Slack h m	Maximum h m	knots
1 F		0209	2.4E
	0526	0727	2.3F
	1027	1406	3.3E
	1749	2017	3.2F
	2354		
2 Sa		0333	2.4E
	0627	0822	2.0F
	1117	1500	3.2E
	1840	2113	3.3F
3 Su	0054	0455	2.5E
	0727	0917	1.8F
	1208	1554	3.0E
	1930	2207	3.2F
4 M	0151	0600	2.6E
	0826	1012	1.5F
	1259	1647	2.8E
	2020	2258	3.2F
5 Tu	0244	0658	2.6E
	0923	1107	1.4F
	1351	1739	2.6E
	2110	2346	3.1F
6 W ●	0332	0752	2.7E
	1019	1200	1.2F
	1443	1829	2.3E
	2158		
7 Th		0031	2.9F
	0417	0843	2.7E
	1111	1251	1.2F
	1535	1917	2.1E
	2247		
8 F		0113	2.8F
	0459	0932	2.6E
	1200	1340	1.2F
	1627	2002	1.9E
	2334		
9 Sa		0156	2.7F
	0538	1017	2.6E
	1247	1427	1.2F
	1720	2047	1.8E
10 Su	0022	0238	2.6F
	0615	1057	2.5E
	1331	1513	1.3F
	1812	2132	1.7E
11 M	0110	0322	2.4F
	0651	1057	2.4E
	1414	1559	1.5F
	1905	2218	1.6E
12 Tu	0159	0407	2.3F
	0727	1107	2.4E
	1455	1645	1.7F
	1958	2305	1.6E
13 W	0250	0453	2.2F
	0803	1137	2.4E
	1536	1731	2.0F
	2051	2354	1.6E
14 Th ◐	0342	0540	2.0F
	0838	1210	2.5E
	1617	1818	2.2F
	2144		
15 F		0045	1.6E
	0435	0627	1.9F
	0914	1244	2.5E
	1657	1905	2.5F
	2236		
16 Sa		0135	1.7E
	0529	0715	1.7F
	0950	1321	2.6E
	1737	1953	2.7F
	2327		
17 Su		0226	1.7E
	0623	0804	1.6F
	1028	1400	2.7E
	1817	2041	2.9F
18 M	0017	0317	1.8E
	0716	0853	1.5F
	1110	1443	2.7E
	1859	2129	3.1F
19 Tu	0107	0408	1.9E
	0809	0944	1.4F
	1158	1529	2.8E
	1943	2218	3.3F
20 W	0155	0458	2.1E
	0900	1035	1.5F
	1251	1620	2.8E
	2030	2308	3.4F
21 Th ○	0243	0549	2.3E
	0950	1128	1.6F
	1351	1716	2.8E
	2120	2359	3.5F
22 F	0330	0640	2.5E
	1039	1222	1.8F
	1455	1815	2.7E
	2214		
23 Sa		0050	3.5F
	0418	0731	2.7E
	1128	1317	2.0F
	1602	1917	2.7E
	2310		
24 Su		0142	3.4F
	0506	0823	3.0E
	1218	1413	2.3F
	1710	2022	2.6E
25 M	0009	0235	3.3F
	0554	0916	3.1E
	1308	1510	2.5F
	1817	2127	2.6E
26 Tu	0108	0328	3.2F
	0643	1009	3.3E
	1358	1607	2.8F
	1923	2233	2.5E
27 W ◑	0208	0422	2.9F
	0732	1103	3.4E
	1449	1704	3.0F
	2029	2340	2.5E
28 Th	0308	0517	2.7F
	0822	1157	3.4E
	1540	1802	3.1F
	2132		
29 F		0049	2.5E
	0408	0611	2.4F
	0913	1251	3.3E
	1631	1859	3.2F
	2234		
30 Sa		0206	2.5E
	0508	0706	2.2F
	1004	1345	3.2E
	1723	1955	3.2F
	2333		
31 Su		0328	2.5E
	0607	0801	1.9F
	1055	1440	3.1E
	1814	2049	3.2F

August

Day	Slack h m	Maximum h m	knots
1 M	0029	0436	2.6E
	0705	0855	1.7F
	1147	1534	2.8E
	1905	2141	3.1F
2 Tu	0123	0536	2.6E
	0802	0949	1.5F
	1239	1627	2.6E
	1955	2230	3.0F
3 W	0213	0631	2.6E
	0856	1042	1.4F
	1331	1718	2.4E
	2045	2316	2.8F
4 Th ●	0259	0724	2.6E
	0949	1133	1.3F
	1423	1808	2.2E
	2134	2359	2.7F
5 F	0342	0814	2.6E
	1039	1222	1.3F
	1515	1856	2.0E
	2223		
6 Sa		0042	2.6F
	0423	0901	2.5E
	1126	1310	1.3F
	1607	1942	1.9E
	2312		
7 Su		0126	2.4F
	0501	0943	2.4E
	1212	1357	1.4F
	1700	2028	1.8E
8 M	0001	0210	2.3F
	0539	1011	2.3E
	1255	1443	1.6F
	1752	2114	1.7E
9 Tu	0051	0255	2.2F
	0615	0958	2.3E
	1337	1529	1.8F
	1844	2200	1.7E
10 W	0141	0340	2.1F
	0651	1025	2.3E
	1419	1615	2.0F
	1936	2247	1.7E
11 Th	0232	0426	1.9F
	0727	1057	2.3E
	1459	1701	2.2F
	2027	2334	1.7E
12 F ◐	0323	0513	1.8F
	0802	1130	2.4E
	1539	1748	2.4F
	2117		
13 Sa		0021	1.7E
	0415	0600	1.7F
	0838	1205	2.5E
	1619	1835	2.7F
	2206		
14 Su		0108	1.8E
	0506	0648	1.6F
	0916	1243	2.5E
	1700	1922	2.9F
	2255		
15 M		0155	1.8E
	0557	0736	1.5F
	0958	1326	2.6E
	1743	2010	3.0F
	2343		
16 Tu		0243	2.0E
	0647	0826	1.5F
	1045	1412	2.7E
	1827	2059	3.1F
17 W	0031	0332	2.1E
	0736	0917	1.6F
	1139	1504	2.7E
	1915	2149	3.2F
18 Th	0118	0421	2.3E
	0825	1009	1.8F
	1238	1600	2.7E
	2007	2240	3.3F
19 F ○	0206	0513	2.4E
	0914	1103	1.9F
	1342	1700	2.7E
	2102	2332	3.3F
20 Sa	0254	0605	2.6E
	1003	1158	2.2F
	1448	1803	2.7E
	2159		
21 Su		0024	3.2F
	0343	0658	2.8E
	1053	1254	2.4F
	1555	1908	2.7E
	2258		
22 M		0118	3.1F
	0433	0753	3.0E
	1144	1351	2.7F
	1701	2013	2.7E
	2357		
23 Tu		0212	3.0F
	0524	0847	3.2E
	1235	1448	2.9F
	1806	2119	2.7E
24 W	0056	0306	2.8F
	0615	0942	3.2E
	1327	1545	3.1F
	1910	2224	2.6E
25 Th	0155	0401	2.6F
	0706	1037	3.3E
	1419	1641	3.2F
	2011	2331	2.6E
26 F ◑	0253	0456	2.4F
	0758	1132	3.2E
	1511	1737	3.2F
	2110		
27 Sa		0041	2.6E
	0351	0550	2.2F
	0850	1227	3.1E
	1603	1832	3.2F
	2208		
28 Su		0155	2.6E
	0448	0644	2.0F
	0942	1321	3.0E
	1655	1925	3.1F
	2302		
29 M		0305	2.6E
	0544	0738	1.9F
	1034	1416	2.8E
	1747	2016	3.0F
	2354		
30 Tu		0407	2.6E
	0639	0831	1.7F
	1127	1509	2.6E
	1838	2105	2.9F
31 W	0044	0504	2.6E
	0732	0923	1.6F
	1219	1602	2.4E
	1929	2152	2.7F

September

Day	Slack h m	Maximum h m	knots
1 Th	0131	0557	2.5E
	0824	1013	1.5F
	1312	1655	2.2E
	2019	2238	2.5F
2 F	0215	0647	2.5E
	0913	1103	1.5F
	1404	1746	2.1E
	2110	2324	2.4F
3 Sa ●	0258	0734	2.4E
	1001	1151	1.6F
	1457	1836	2.0E
	2201		
4 Su		0009	2.3F
	0339	0815	2.3E
	1047	1239	1.7F
	1549	1925	1.9E
	2251		
5 M		0055	2.1F
	0419	0820	2.2E
	1131	1325	1.8F
	1641	2013	1.9E
	2343		
6 Tu		0141	2.0F
	0458	0837	2.2E
	1214	1412	2.0F
	1733	2100	1.9E
7 W	0034	0228	1.9F
	0536	0908	2.2E
	1257	1458	2.1F
	1823	2146	1.9E
8 Th	0125	0314	1.8F
	0614	0941	2.2E
	1338	1544	2.3F
	1913	2231	1.9E
9 F	0216	0400	1.7F
	0651	1016	2.2E
	1419	1631	2.5F
	2001	2315	1.9E
10 Sa	0306	0447	1.6F
	0728	1052	2.3E
	1500	1717	2.7F
	2048	2358	1.9E
11 Su ◐	0356	0534	1.5F
	0808	1130	2.4E
	1542	1804	2.8F
	2135		
12 M		0042	2.0E
	0444	0623	1.5F
	0851	1213	2.4E
	1626	1852	2.9F
	2221		
13 Tu		0126	2.1E
	0531	0712	1.6F
	0940	1300	2.5E
	1712	1941	3.0F
	2307		
14 W		0212	2.2E
	0618	0802	1.7F
	1034	1352	2.5E
	1802	2030	3.0F
	2353		
15 Th		0259	2.3E
	0705	0854	1.9F
	1134	1449	2.6E
	1855	2121	3.0F
16 F	0041	0349	2.5E
	0752	0947	2.1F
	1236	1549	2.6E
	1951	2213	3.0F
17 Sa ○	0129	0441	2.6E
	0841	1042	2.4F
	1341	1653	2.6E
	2049	2306	2.9F
18 Su	0218	0534	2.8E
	0930	1137	2.6F
	1446	1758	2.6E
	2149		
19 M		0000	2.8F
	0309	0629	2.9E
	1021	1233	2.9F
	1550	1903	2.6E
	2248		
20 Tu		0055	2.7F
	0401	0724	3.0E
	1113	1329	3.1F
	1653	2009	2.7E
	2347		
21 W		0150	2.5F
	0453	0820	3.1E
	1205	1426	3.2F
	1754	2114	2.7E
22 Th	0045	0245	2.4F
	0546	0916	3.1E
	1257	1521	3.3F
	1853	2219	2.7E
23 F	0142	0340	2.2F
	0640	1011	3.0E
	1349	1615	3.3F
	1950	2325	2.8E
24 Sa ◑	0238	0435	2.1F
	0733	1106	2.9E
	1441	1708	3.2F
	2043		
25 Su		0031	2.7E
	0333	0528	2.0F
	0827	1200	2.8E
	1533	1800	3.1F
	2135		
26 M		0135	2.7E
	0427	0621	1.9F
	0920	1254	2.6E
	1625	1849	3.0F
	2224		
27 Tu		0234	2.7E
	0519	0713	1.8F
	1013	1347	2.4E
	1717	1937	2.8F
	2311		
28 W		0328	2.6E
	0610	0803	1.7F
	1106	1440	2.2E
	1809	2025	2.6F
	2357		
29 Th		0418	2.5E
	0659	0853	1.7F
	1200	1533	2.1E
	1900	2112	2.4F
30 F	0041	0502	2.4E
	0747	0942	1.8F
	1253	1627	2.0E
	1953	2200	2.3F

Time meridian 135° W. 0000 is midnight. 1200 is noon. Times are not adjusted for Daylight Saving Time.

Kvichak Bay (off Naknek River ent.), Alaska, 2005

F–Flood, Dir. 055° True E–Ebb, Dir. 240° True

October

Day	Slack h m	Maximum h m	knots
1 Sa	0125	0529	2.3E
	0834	1031	1.8F
	1346	1720	2.0E
	2045	2247	2.1F
2 Su	0207	0554	2.3E
	0920	1119	1.9F
	1439	1814	1.9E
	2138	2335	2.0F
3 M ●	0250	0628	2.2E
	1005	1206	2.1F
	1531	1907	2.0E
	2231		
4 Tu		0023	1.8F
	0331	0704	2.2E
	1048	1254	2.2F
	1622	1958	2.0E
	2324		
5 W		0111	1.7F
	0412	0741	2.2E
	1131	1341	2.4F
	1712	2045	2.0E
6 Th	0017	0159	1.6F
	0453	0819	2.2E
	1214	1427	2.6F
	1801	2129	2.1E
7 F	0108	0247	1.5F
	0533	0857	2.2E
	1256	1514	2.7F
	1848	2211	2.1E
8 Sa	0158	0334	1.4F
	0614	0935	2.2E
	1338	1601	2.9F
	1933	2253	2.2E
9 Su	0247	0422	1.4F
	0657	1016	2.2E
	1422	1648	2.9F
	2018	2334	2.2E
10 M ◐	0334	0510	1.5F
	0744	1101	2.3E
	1507	1735	3.0F
	2102		
11 Tu		0016	2.3E
	0419	0600	1.6F
	0836	1150	2.3E
	1555	1824	3.0F
	2147		
12 W		0059	2.4E
	0504	0650	1.8F
	0932	1243	2.3E
	1647	1913	3.0F
	2232		
13 Th		0144	2.5E
	0550	0742	2.0F
	1033	1341	2.4E
	1742	2004	2.9F
	2318		
14 F		0232	2.6E
	0636	0835	2.3F
	1135	1443	2.4E
	1840	2056	2.8F
15 Sa	0005	0323	2.8E
	0723	0929	2.5F
	1239	1547	2.4E
	1939	2149	2.6F
16 Su	0054	0415	2.9E
	0812	1023	2.8F
	1342	1652	2.4E
	2040	2243	2.5F
17 M ○	0145	0509	2.9E
	0902	1119	3.0F
	1445	1758	2.5E
	2139	2338	2.3F
18 Tu	0237	0604	3.0E
	0953	1214	3.2F
	1546	1905	2.6E
	2239		
19 W		0034	2.2F
	0330	0659	3.0E
	1044	1310	3.3F
	1645	2012	2.7E
	2337		
20 Th		0129	2.0F
	0424	0755	2.9E
	1136	1404	3.4F
	1741	2119	2.7E
21 F	0033	0225	1.9F
	0519	0850	2.8E
	1228	1457	3.3F
	1834	2225	2.8E
22 Sa	0128	0319	1.8F
	0614	0944	2.7E
	1320	1548	3.2F
	1925	2325	2.8E
23 Su	0222	0413	1.8F
	0708	1038	2.5E
	1411	1637	3.1F
	2013		
24 M ◑		0019	2.8E
	0313	0505	1.7F
	0803	1131	2.3E
	1502	1725	2.9F
	2059		
25 Tu		0108	2.7E
	0403	0556	1.7F
	0857	1223	2.2E
	1554	1812	2.8F
	2143		
26 W		0149	2.7E
	0452	0645	1.8F
	0951	1315	2.0E
	1645	1859	2.6F
	2226		
27 Th		0221	2.6E
	0539	0734	1.8F
	1044	1407	1.9E
	1738	1946	2.4F
	2308		
28 F		0254	2.5E
	0625	0822	1.9F
	1138	1500	1.9E
	1831	2034	2.2F
	2350		
29 Sa		0330	2.4E
	0710	0910	2.0F
	1232	1555	1.9E
	1925	2122	2.0F
30 Su	0032	0409	2.4E
	0755	0958	2.2F
	1325	1650	1.9E
	2019	2211	1.9F
31 M	0115	0448	2.3E
	0839	1046	2.4F
	1418	1747	1.9E
	2114	2301	1.7F

November

Day	Slack h m	Maximum h m	knots
1 Tu ●	0157	0529	2.3E
	0922	1134	2.5F
	1509	1842	2.0E
	2209	2350	1.6F
2 W	0240	0609	2.3E
	1005	1222	2.7F
	1559	1934	2.1E
	2303		
3 Th		0040	1.4F
	0322	0650	2.2E
	1048	1309	2.9F
	1648	2020	2.2E
	2356		
4 F		0129	1.4F
	0406	0731	2.2E
	1131	1356	3.0F
	1735	2103	2.2E
5 Sa	0047	0218	1.3F
	0451	0813	2.2E
	1214	1443	3.1F
	1820	2144	2.3E
6 Su	0136	0307	1.4F
	0539	0858	2.2E
	1259	1531	3.1F
	1904	2225	2.4E
7 M	0222	0357	1.5F
	0631	0947	2.2E
	1346	1619	3.1F
	1947	2307	2.5E
8 Tu ◐	0308	0447	1.6F
	0728	1039	2.2E
	1437	1708	3.1F
	2031	2349	2.6E
9 W	0352	0538	1.8F
	0827	1135	2.2E
	1531	1757	3.0F
	2115		
10 Th		0034	2.8E
	0437	0630	2.1F
	0929	1234	2.2E
	1627	1848	2.9F
	2159		
11 F		0121	2.9E
	0523	0724	2.4F
	1033	1337	2.2E
	1727	1940	2.7F
	2246		
12 Sa		0210	3.0E
	0609	0818	2.7F
	1137	1442	2.2E
	1827	2033	2.5F
	2334		
13 Su		0301	3.0E
	0658	0912	2.9F
	1240	1548	2.3E
	1928	2127	2.3F
14 M	0023	0354	3.0E
	0747	1008	3.1F
	1342	1657	2.4E
	2029	2222	2.1F
15 Tu ○	0115	0448	3.0E
	0837	1103	3.3F
	1442	1808	2.5E
	2129	2318	1.9F
16 W	0208	0543	2.9E
	0928	1157	3.4F
	1539	1926	2.6E
	2228		
17 Th		0014	1.8F
	0302	0638	2.8E
	1019	1250	3.4F
	1634	2037	2.7E
	2324		
18 F		0109	1.6F
	0357	0732	2.7E
	1110	1342	3.3F
	1725	2137	2.8E
19 Sa	0019	0204	1.6F
	0453	0826	2.5E
	1201	1431	3.2F
	1813	2230	2.8E
20 Su	0112	0258	1.5F
	0548	0919	2.3E
	1252	1519	3.1F
	1858	2319	2.8E
21 M	0203	0350	1.5F
	0643	1010	2.1E
	1342	1605	2.9F
	1941		
22 Tu		0003	2.8E
	0251	0440	1.6F
	0738	1100	2.0E
	1432	1651	2.7F
	2022		
23 W ◑		0034	2.7E
	0338	0528	1.6F
	0832	1150	1.9E
	1523	1736	2.6F
	2102		
24 Th		0053	2.6E
	0423	0615	1.8F
	0926	1241	1.8E
	1614	1823	2.4F
	2141		
25 F		0124	2.6E
	0507	0703	1.9F
	1020	1332	1.7E
	1707	1910	2.2F
	2221		
26 Sa		0159	2.6E
	0550	0750	2.1F
	1114	1426	1.7E
	1801	1958	2.0F
	2301		
27 Su		0236	2.5E
	0633	0838	2.3F
	1208	1520	1.7E
	1857	2047	1.8F
	2342		
28 M		0315	2.5E
	0716	0926	2.5F
	1301	1616	1.8E
	1953	2136	1.7F
29 Tu	0023	0355	2.5E
	0758	1014	2.7F
	1352	1712	1.9E
	2049	2226	1.5F
30 W	0104	0436	2.4E
	0840	1102	2.9F
	1443	1805	2.0E
	2144	2317	1.4F

December

Day	Slack h m	Maximum h m	knots
1 Th ●	0147	0518	2.4E
	0923	1149	3.1F
	1532	1856	2.1E
	2238		
2 F		0007	1.3F
	0233	0602	2.4E
	1006	1237	3.2F
	1619	1942	2.2E
	2329		
3 Sa		0058	1.3F
	0321	0648	2.4E
	1051	1325	3.3F
	1704	2026	2.4E
4 Su	0019	0149	1.3F
	0414	0736	2.4E
	1137	1414	3.3F
	1749	2110	2.5E
5 M	0106	0240	1.5F
	0512	0828	2.3E
	1226	1502	3.3F
	1833	2153	2.6E
6 Tu	0152	0332	1.6F
	0613	0924	2.3E
	1318	1552	3.3F
	1916	2237	2.8E
7 W	0237	0424	1.9F
	0716	1023	2.3E
	1413	1642	3.1F
	2000	2323	2.9E
8 Th ◐	0323	0518	2.2F
	0820	1125	2.2E
	1511	1733	3.0F
	2045		
9 F		0010	3.1E
	0409	0612	2.5F
	0925	1228	2.2E
	1611	1825	2.8F
	2131		
10 Sa		0100	3.2E
	0456	0706	2.8F
	1030	1334	2.2E
	1712	1918	2.5F
	2219		
11 Su		0151	3.2E
	0545	0802	3.0F
	1133	1442	2.2E
	1814	2012	2.3F
	2308		
12 M		0243	3.2E
	0634	0857	3.2F
	1236	1553	2.3E
	1915	2107	2.0F
	2358		
13 Tu		0337	3.1E
	0724	0953	3.3F
	1336	1716	2.4E
	2015	2203	1.8F
14 W	0051	0431	3.0E
	0815	1047	3.4F
	1433	1836	2.5E
	2115	2259	1.6F
15 Th ○	0144	0525	2.9E
	0906	1140	3.4F
	1527	1939	2.7E
	2212	2355	1.5F
16 F	0239	0620	2.7E
	0957	1231	3.3F
	1618	2035	2.7E
	2308		
17 Sa		0050	1.4F
	0334	0713	2.4E
	1047	1319	3.2F
	1705	2128	2.8E
18 Su	0001	0144	1.3F
	0430	0805	2.2E
	1137	1405	3.0F
	1748	2217	2.8E
19 M	0051	0235	1.3F
	0525	0855	2.0E
	1226	1450	2.9F
	1829	2302	2.7E
20 Tu	0139	0324	1.4F
	0619	0944	1.9E
	1315	1534	2.7F
	1908	2340	2.7E
21 W	0225	0412	1.5F
	0714	1032	1.7E
	1405	1619	2.5F
	1946	2347	2.6E
22 Th	0309	0458	1.6F
	0807	1120	1.7E
	1455	1704	2.4F
	2023		
23 F ◑		0006	2.6E
	0351	0544	1.8F
	0901	1209	1.6E
	1547	1750	2.2F
	2100		
24 Sa		0038	2.6E
	0433	0631	2.0F
	0954	1300	1.6E
	1640	1837	2.0F
	2138		
25 Su		0113	2.6E
	0514	0718	2.3F
	1047	1352	1.6E
	1734	1925	1.9F
	2216		
26 M		0150	2.6E
	0556	0805	2.5F
	1140	1445	1.7E
	1829	2014	1.7F
	2255		
27 Tu		0228	2.6E
	0637	0853	2.7F
	1231	1539	1.8E
	1925	2104	1.5F
	2335		
28 W		0308	2.6E
	0719	0941	2.9F
	1322	1632	1.9E
	2020	2154	1.4F
29 Th	0017	0350	2.6E
	0801	1029	3.1F
	1411	1723	2.0E
	2114	2244	1.3F
30 F ●	0102	0434	2.6E
	0845	1117	3.3F
	1459	1812	2.1E
	2206	2335	1.3F
31 Sa	0152	0522	2.5E
	0930	1206	3.3F
	1545	1859	2.3E
	2256		

Time meridian 135° W. 0000 is midnight. 1200 is noon. Times are not adjusted for Daylight Saving Time.

Tokyo Wan Entrance (N. of Kannon Saki), Japan, 2005

F–Flood, Dir. 313° True E–Ebb, Dir. 133° True

January

Day	Slack (h m)	Maximum (h m)	knots
1 Sa	0222	0617	1.4F
	0939	1201	0.7E
	1552	1821	0.5F
	2015	2323	1.1E
2 Su	0259	0653	1.3F
	1008	1246	0.7E
	1650	1913	0.5F
	2114		
3 M		0007	0.8E
	0346	0733	1.2F
	1042	1351	0.8E
	1749	2009	0.6F
	2229		
4 Tu		0107	0.5E
	0445	0817	1.1F
	1120	1524	0.9E
◑	1848	2112	0.7F
5 W	0019	0344	0.3E
	0557	0906	0.9F
	1205	1635	1.1E
	1946	2224	0.8F
6 Th	0303	0524	0.4E
	0715	1001	0.7F
	1257	1732	1.3E
	2042	2351	1.0F
7 F	0426	0629	0.5E
	0830	1102	0.6F
	1357	1822	1.5E
	2135		
8 Sa		0120	1.3F
	0523	0725	0.6E
	0942	1213	0.5F
	1459	1909	1.7E
	2227		
9 Su		0223	1.6F
	0610	0816	0.7E
	1050	1331	0.5F
	1556	1953	1.9E
	2317		
10 M		0312	1.8F
	0653	0902	0.8E
	1151	1437	0.6F
●	1649	2035	1.9E
11 Tu	0003	0355	1.9F
	0733	0945	0.8E
	1246	1531	0.6F
	1737	2115	1.9E
12 W	0048	0435	1.9F
	0809	1026	0.8E
	1335	1618	0.7F
	1824	2153	1.7E
13 Th	0130	0511	1.8F
	0841	1104	0.8E
	1421	1702	0.7F
	1909	2230	1.5E
14 F	0209	0546	1.7F
	0910	1141	0.8E
	1507	1745	0.7F
	1955	2306	1.3E
15 Sa	0246	0618	1.5F
	0935	1218	0.8E
	1556	1829	0.7F
	2043	2342	1.0E
16 Su	0321	0651	1.3F
	0959	1302	0.8E
	1648	1916	0.7F
	2138		
17 M		0021	0.7E
	0356	0724	1.1F
	1024	1402	0.9E
◐	1744	2006	0.7F
	2248		
18 Tu		0115	0.3E
	0434	0759	0.9F
	1053	1520	0.9E
	1842	2104	0.6F
19 W	0134	0401	0.1E
	0523	0840	0.7F
	1128	1626	1.0E
	1940	2215	0.7F
20 Th	0406	0524	0.1E
	0634	0926	0.6F
	1211	1721	1.1E
	2036		
21 F		0010	0.8F
	0458	0624	0.2E
	0754	1021	0.4F
	1310	1809	1.3E
	2127		
22 Sa		0141	1.0F
	0535	0716	0.3E
	0913	1130	0.3F
	1423	1853	1.4E
	2213		
23 Su		0224	1.3F
	0607	0800	0.5E
	1026	1300	0.3F
	1528	1932	1.5E
	2255		
24 M		0259	1.4F
	0637	0840	0.6E
	1127	1412	0.4F
	1620	2008	1.5E
	2332		
25 Tu		0330	1.6F
	0705	0914	0.7E
	1215	1500	0.4F
○	1704	2041	1.5E
26 W	0005	0359	1.6F
	0729	0945	0.8E
	1253	1537	0.5F
	1742	2111	1.5E
27 Th	0036	0426	1.6F
	0752	1011	0.8E
	1324	1611	0.6F
	1819	2141	1.5E
28 F	0105	0452	1.6F
	0813	1036	0.9E
	1353	1645	0.7F
	1856	2209	1.4E
29 Sa	0133	0518	1.5F
	0832	1100	0.9E
	1425	1720	0.7F
	1937	2240	1.3E
30 Su	0202	0547	1.4F
	0853	1128	1.0E
	1503	1800	0.8F
	2022	2313	1.1E
31 M	0233	0617	1.3F
	0916	1159	1.0E
	1550	1846	0.8F
	2114	2350	0.9E

February

Day	Slack (h m)	Maximum (h m)	knots
1 Tu	0308	0653	1.1F
	0943	1238	1.1E
	1647	1937	0.9F
	2218		
2 W		0035	0.5E
	0351	0734	0.9F
	1015	1332	1.0E
◑	1753	2037	0.9F
	2358		
3 Th		0159	0.2E
	0455	0821	0.7F
	1053	1520	1.0E
	1903	2148	0.9F
4 F	0330	0512	0.2E
	0635	0918	0.5F
	1142	1658	1.2E
	2011	2323	1.0F
5 Sa	0438	0625	0.3E
	0816	1027	0.4F
	1255	1803	1.4E
	2115		
6 Su		0114	1.3F
	0521	0721	0.5E
	0942	1159	0.4F
	1435	1857	1.6E
	2213		
7 M		0215	1.6F
	0559	0808	0.7E
	1053	1337	0.5F
	1556	1945	1.8E
	2305		
8 Tu		0300	1.8F
	0633	0850	0.9E
	1147	1441	0.7F
	1657	2029	1.8E
	2353		
9 W		0339	1.9F
	0705	0928	1.0E
	1232	1529	0.9F
●	1749	2110	1.8E
10 Th	0036	0414	1.9F
	0734	1002	1.1E
	1312	1611	1.0F
	1835	2147	1.7E
11 F	0115	0447	1.8F
	0759	1034	1.2E
	1350	1650	1.1F
	1918	2222	1.5E
12 Sa	0149	0517	1.6F
	0821	1102	1.2E
	1428	1728	1.1F
	2001	2254	1.2E
13 Su	0218	0544	1.4F
	0841	1129	1.2E
	1508	1806	1.0F
	2043	2323	0.9E
14 M	0243	0611	1.2F
	0900	1156	1.1E
	1553	1846	0.9F
	2129	2351	0.6E
15 Tu	0303	0640	1.0F
	0921	1226	1.1E
	1644	1930	0.8F
	2225		
16 W		0019	0.3E
	0315	0711	0.8F
	0944	1303	1.0E
◐	1742	2021	0.7F
17 Th		0056	0.0
		0748	0.6F
	1012	1431	0.8E
	1847	2123	0.7F
18 F		0511	0.0
		0836	0.4F
	1048	1641	0.9E
	1952	2254	0.7F
19 Sa	0501	0617	0.1E
	0738	0939	0.2F
	1141	1745	1.0E
	2051		
20 Su		0123	0.9F
	0517	0705	0.3E
	0921	1106	0.2F
	1336	1835	1.2E
	2144		
21 M		0202	1.2F
	0539	0746	0.6E
	1035	1310	0.3F
	1526	1918	1.3E
	2229		
22 Tu		0234	1.4F
	0603	0820	0.8E
	1123	1413	0.5F
	1626	1956	1.4E
	2308		
23 W		0303	1.5F
	0626	0851	0.9E
	1158	1453	0.6F
	1711	2030	1.5E
	2344		
24 Th		0331	1.6F
	0649	0918	1.1E
	1228	1527	0.8F
○	1751	2101	1.5E
25 F	0016	0357	1.6F
	0709	0943	1.1E
	1254	1558	0.9F
	1828	2131	1.5E
26 Sa	0046	0422	1.5F
	0728	1005	1.2E
	1320	1629	1.0F
	1904	2200	1.4E
27 Su	0113	0447	1.4F
	0746	1028	1.3E
	1348	1702	1.1F
	1942	2229	1.2E
28 M	0141	0512	1.3F
	0804	1053	1.3E
	1421	1739	1.1F
	2024	2300	1.0E

March

Day	Slack (h m)	Maximum (h m)	knots
1 Tu	0207	0541	1.1F
	0824	1120	1.3E
	1502	1821	1.1F
	2113	2333	0.8E
2 W	0235	0613	0.9F
	0848	1152	1.3E
	1555	1911	1.1F
	2215		
3 Th		0012	0.4E
	0303	0653	0.7F
	0917	1231	1.2E
	1704	2010	1.0F
4 F	0001	0113	0.1E
	0337	0744	0.5F
	0953	1333	1.0E
◑	1823	2122	0.9F
5 Sa	0357	0518	0.1E
	0633	0849	0.3F
	1043	1633	1.0E
	1943	2300	1.0F
6 Su	0428	0623	0.4E
	0830	1015	0.2F
	1217	1751	1.2E
	2054		
7 M		0058	1.2F
	0458	0710	0.7E
	0950	1212	0.3F
	1447	1848	1.4E
	2156		
8 Tu		0155	1.5F
	0528	0750	0.9E
	1044	1341	0.7F
	1609	1936	1.6E
	2249		
9 W		0236	1.6F
	0556	0826	1.1E
	1127	1434	1.0F
	1706	2019	1.7E
	2335		
10 Th		0312	1.7F
	0623	0859	1.3E
	1205	1517	1.2F
●	1755	2059	1.7E
11 F	0015	0344	1.7F
	0648	0930	1.4E
	1241	1555	1.3F
	1839	2134	1.6E
12 Sa	0051	0414	1.6F
	0710	0958	1.5E
	1316	1631	1.4F
	1920	2206	1.4E
13 Su	0121	0441	1.4F
	0729	1023	1.5E
	1350	1706	1.4F
	1959	2235	1.1E
14 M	0146	0506	1.2F
	0747	1046	1.4E
	1424	1741	1.3F
	2038	2301	0.8E
15 Tu	0204	0530	1.0F
	0805	1109	1.4E
	1501	1817	1.1F
	2119	2324	0.6E
16 W	0214	0555	0.8F
	0823	1131	1.2E
	1544	1857	1.0F
	2208	2348	0.3E
17 Th	0211	0625	0.6F
	0844	1157	1.1E
	1637	1943	0.8F
18 F		0017	0.0
		0705	0.4F
	0909	1229	0.9E
◐	1744	2041	0.7F
19 Sa		0508	0.0
		0802	0.2F
	0941	1320	0.7E
	1857	2154	0.7F
20 Su	0421	0605	0.2E
	0803	0918	0.1F
	1033	1718	0.7E
	2005	2348	0.8F
21 M	0432	0644	0.5E
	0933	1101	0.1F
	1321	1814	0.9E
	2103		
22 Tu		0109	1.0F
	0452	0718	0.8E
	1020	1304	0.3F
	1530	1859	1.1E
	2153		
23 W		0148	1.2F
	0514	0749	1.0E
	1055	1356	0.6F
	1628	1937	1.3E
	2236		
24 Th		0221	1.4F
	0537	0817	1.2E
	1125	1433	0.9F
	1712	2013	1.4E
	2315		
25 F		0251	1.4F
	0559	0844	1.3E
	1153	1506	1.1F
	1752	2047	1.4E
	2350		
26 Sa		0319	1.4F
	0620	0909	1.4E
	1220	1539	1.3F
○	1830	2118	1.4E
27 Su	0023	0346	1.3F
	0639	0933	1.5E
	1247	1611	1.4F
	1908	2148	1.3E
28 M	0052	0411	1.2F
	0657	0957	1.6E
	1317	1645	1.4F
	1948	2218	1.1E
29 Tu	0119	0438	1.0F
	0715	1021	1.6E
	1350	1722	1.4F
	2030	2249	0.8E
30 W	0144	0506	0.9F
	0735	1048	1.5E
	1429	1805	1.3F
	2120	2323	0.6E
31 Th	0209	0539	0.7F
	0759	1118	1.4E
	1519	1854	1.2F
	2224		

Time meridian 135° E. 0000 is midnight. 1200 is noon. Times are not adjusted for Daylight Saving Time.

Tokyo Wan Entrance (N. of Kannon Saki), Japan, 2005

F–Flood, Dir. 313° True E–Ebb, Dir. 133° True

April

Day	Slack h m	Maximum h m	knots
1 F		0002	0.3E
	0233	0623	0.5F
	0828	1154	1.2E
	1626	1953	1.1F
2 Sa ◑		0343	0.0
		0723	0.3F
	0905	1244	0.9E
	1751	2103	1.0F
3 Su	0322	0518	0.2E
	0713	0842	0.2F
	1005	1617	0.8E
	1915	2231	1.0F
4 M	0348	0607	0.5E
	0844	1020	0.2F
	1229	1739	1.0E
	2028		
5 Tu		0012	1.1F
	0415	0646	0.8E
	0939	1219	0.5F
	1506	1835	1.2E
	2129		
6 W		0115	1.3F
	0442	0721	1.1E
	1022	1331	0.8F
	1617	1921	1.4E
	2222		
7 Th		0158	1.4F
	0508	0754	1.4E
	1100	1418	1.2F
	1709	2003	1.5E
	2307		
8 F		0234	1.4F
	0533	0826	1.6E
	1136	1459	1.4F
	1756	2041	1.4E
	2347		
9 Sa ●		0306	1.4F
	0557	0855	1.7E
	1211	1536	1.6F
	1838	2117	1.3E
10 Su	0021	0336	1.2F
	0619	0923	1.7E
	1244	1612	1.6F
	1918	2148	1.1E
11 M	0049	0402	1.1F
	0638	0947	1.7F
	1317	1646	1.5F
	1957	2216	0.9E
12 Tu	0112	0427	0.9F
	0656	1010	1.6E
	1347	1719	1.4F
	2035	2241	0.7E
13 W	0128	0450	0.7F
	0713	1031	1.5E
	1419	1753	1.3F
	2114	2306	0.4E
14 Th	0139	0515	0.6F
	0730	1053	1.4E
	1453	1831	1.1F
	2200	2332	0.2E
15 F	0145	0547	0.4F
	0751	1118	1.2E
	1537	1914	1.0F
	2309		
16 Sa ◐		0008	0.1E
	0135	0637	0.2F
	0817	1147	1.0E
	1636	2006	0.9F
17 Su	0301	0449	0.1E
	0624	0745	0.1F
	0852	1226	0.7E
	1753	2108	0.8F
18 M	0304	0534	0.4E
	0813	0907	0.1F
	1002	1638	0.5E
	1907	2219	0.9F
19 Tu	0324	0608	0.7E
	0907	1041	0.2F
	1316	1743	0.7E
	2011	2334	1.0F
20 W	0349	0640	0.9E
	0943	1222	0.5F
	1523	1831	0.9E
	2107		
21 Th		0036	1.1F
	0415	0711	1.2E
	1016	1322	0.8F
	1621	1913	1.1E
	2157		
22 F		0123	1.2F
	0441	0740	1.4E
	1047	1405	1.1F
	1707	1952	1.2E
	2241		
23 Sa		0202	1.2F
	0505	0809	1.6E
	1119	1443	1.3F
	1750	2029	1.2E
	2321		
24 Su ○		0236	1.1F
	0529	0836	1.7E
	1150	1520	1.5F
	1832	2104	1.2E
	2358		
25 M		0308	1.0F
	0550	0903	1.7E
	1221	1556	1.6F
	1914	2138	1.0E
26 Tu	0032	0338	0.9F
	0611	0930	1.8F
	1254	1634	1.6F
	1957	2212	0.9E
27 W	0104	0408	0.8F
	0631	0957	1.7E
	1330	1714	1.6F
	2043	2246	0.6E
28 Th	0134	0441	0.6F
	0654	1026	1.6E
	1411	1758	1.5F
	2135	2324	0.4E
29 F	0210	0522	0.4F
	0721	1058	1.4E
	1500	1847	1.4F
	2243		
30 Sa		0017	0.2E
	0324	0616	0.3F
	0755	1135	1.2E
	1603	1941	1.2F

May

Day	Slack h m	Maximum h m	knots
1 Su ◑	0038	0352	0.2E
	0548	0725	0.2F
	0844	1226	0.9E
	1722	2043	1.1F
2 M	0206	0453	0.4E
	0723	0844	0.2F
	1008	1556	0.7E
	1841	2151	1.0F
3 Tu	0244	0535	0.7E
	0823	1016	0.3F
	1306	1718	0.8E
	1951	2305	1.0F
4 W	0315	0612	1.0E
	0910	1159	0.6F
	1513	1814	1.0E
	2052		
5 Th		0012	1.0F
	0344	0647	1.3E
	0951	1310	1.0F
	1618	1901	1.1E
	2144		
6 F		0104	1.1F
	0412	0720	1.5E
	1030	1359	1.3F
	1710	1944	1.1E
	2231		
7 Sa		0147	1.1F
	0440	0753	1.7E
	1108	1441	1.5F
	1756	2023	1.1E
	2312		
8 Su ●		0223	1.0F
	0505	0823	1.8E
	1144	1520	1.6F
	1839	2059	1.0E
	2348		
9 M		0256	0.9F
	0529	0851	1.8E
	1218	1556	1.6F
	1919	2133	0.8E
10 Tu	0019	0325	0.8F
	0551	0917	1.8E
	1250	1630	1.6F
	1958	2203	0.7E
11 W	0046	0352	0.6F
	0610	0941	1.7E
	1320	1704	1.5F
	2036	2231	0.5E
12 Th	0109	0419	0.5F
	0629	1004	1.5E
	1348	1737	1.4F
	2113	2259	0.4E
13 F	0134	0449	0.4F
	0649	1028	1.4E
	1419	1812	1.3F
	2153	2331	0.3E
14 Sa	0217	0531	0.3F
	0714	1054	1.2E
	1454	1851	1.2F
	2240		
15 Su		0017	0.2E
	0427	0629	0.2F
	0749	1126	1.0E
	1541	1936	1.1F
	2339		
16 M ◐		0352	0.3E
	0620	0735	0.1F
	0841	1208	0.7E
	1647	2027	1.0F
17 Tu	0047	0441	0.5E
	0728	0847	0.2F
	1009	1339	0.4E
	1803	2122	1.0F
18 W	0142	0519	0.8E
	0816	1003	0.3F
	1241	1700	0.5E
	1913	2220	1.0F
19 Th	0225	0554	1.0E
	0856	1124	0.6F
	1503	1758	0.7E
	2016	2318	1.0F
20 F	0303	0629	1.3E
	0934	1237	0.9F
	1610	1847	0.8E
	2113		
21 Sa		0015	1.0F
	0336	0702	1.5E
	1011	1334	1.2F
	1702	1931	0.9E
	2205		
22 Su		0106	0.9F
	0408	0735	1.7E
	1048	1422	1.5F
	1750	2013	1.0E
	2253		
23 M		0152	0.9F
	0438	0808	1.8E
	1125	1506	1.7F
	1836	2054	0.9E
	2338		
24 Tu ○		0234	0.8F
	0506	0840	1.9E
	1203	1548	1.8F
	1922	2134	0.8E
25 W	0022	0314	0.7F
	0533	0911	1.9E
	1241	1630	1.8F
	2008	2213	0.7E
26 Th	0105	0355	0.5F
	0602	0943	1.8E
	1321	1712	1.7F
	2055	2254	0.5E
27 F	0153	0438	0.4F
	0633	1017	1.6E
	1404	1755	1.6F
	2143	2341	0.4E
28 Sa	0257	0528	0.3F
	0711	1052	1.4E
	1452	1840	1.5F
	2235		
29 Su		0051	0.3E
	0423	0624	0.3F
	0758	1133	1.1E
	1547	1926	1.3F
	2330		
30 M ◑		0302	0.4E
	0543	0727	0.3F
	0900	1228	0.8E
	1651	2015	1.2F
31 Tu	0024	0405	0.6E
	0649	0835	0.3F
	1031	1514	0.6E
	1758	2106	1.0F

June

Day	Slack h m	Maximum h m	knots
1 W	0112	0451	0.8E
	0744	0950	0.5F
	1310	1648	0.6E
	1903	2201	0.9F
2 Th	0154	0531	1.1E
	0833	1117	0.7F
	1513	1749	0.6E
	2003	2257	0.9F
3 F	0233	0610	1.4E
	0919	1242	1.0F
	1620	1839	0.7E
	2059	2354	0.8F
4 Sa	0309	0647	1.6E
	1002	1341	1.3F
	1713	1925	0.7E
	2150		
5 Su		0048	0.7F
	0343	0722	1.7E
	1043	1428	1.5F
	1759	2007	0.7E
	2237		
6 M		0137	0.7F
	0416	0755	1.8E
	1121	1509	1.6F
	1842	2047	0.7E
	2321		
7 Tu ●		0220	0.6F
	0446	0827	1.8E
	1157	1546	1.6F
	1922	2123	0.6E
8 W	0002	0259	0.5F
	0513	0856	1.7E
	1230	1620	1.6F
	1959	2157	0.5E
9 Th	0041	0334	0.5F
	0539	0923	1.6E
	1300	1652	1.6F
	2033	2228	0.5E
10 F	0120	0409	0.4F
	0604	0948	1.5E
	1328	1723	1.5F
	2103	2259	0.4E
11 Sa	0206	0447	0.3F
	0633	1014	1.4E
	1356	1754	1.4F
	2132	2331	0.4E
12 Su	0306	0531	0.3F
	0708	1043	1.2E
	1427	1828	1.3F
	2202		
13 M		0011	0.4E
	0417	0621	0.3F
	0754	1117	1.0E
	1505	1905	1.3F
	2235		
14 Tu		0113	0.5E
	0525	0717	0.3F
	0853	1159	0.8E
	1555	1947	1.2F
	2313		
15 W ◐		0306	0.6E
	0623	0817	0.4F
	1009	1300	0.5E
	1658	2033	1.1F
	2357		
16 Th		0411	0.8E
	0716	0921	0.5F
	1200	1556	0.4E
	1811	2123	1.0F
17 F	0045	0500	1.0E
	0805	1031	0.7F
	1438	1722	0.5E
	1922	2216	0.9F
18 Sa	0135	0545	1.3E
	0852	1149	0.9F
	1605	1823	0.6E
	2029	2313	0.8F
19 Su	0224	0627	1.5E
	0938	1307	1.2F
	1704	1915	0.7E
	2132		
20 M		0013	0.7F
	0310	0707	1.7E
	1023	1408	1.5F
	1754	2003	0.7E
	2233		
21 Tu		0116	0.6F
	0354	0747	1.8E
	1107	1459	1.7F
	1841	2049	0.8E
	2330		
22 W ○		0215	0.6F
	0435	0825	1.9E
	1151	1545	1.9F
	1926	2133	0.7E
23 Th	0025	0309	0.5F
	0517	0903	1.9E
	1235	1627	1.9F
	2008	2216	0.7E
24 F	0118	0359	0.5F
	0558	0940	1.8E
	1317	1707	1.9F
	2047	2259	0.7E
25 Sa	0211	0446	0.5F
	0641	1017	1.6E
	1400	1746	1.7F
	2124	2342	0.6E
26 Su	0305	0534	0.5F
	0727	1054	1.4E
	1443	1823	1.6F
	2158		
27 M		0031	0.6E
	0402	0623	0.5F
	0819	1135	1.1E
	1527	1901	1.4F
	2229		
28 Tu		0135	0.6E
	0501	0715	0.5F
	0918	1223	0.8E
	1614	1940	1.2F
	2301		
29 W ◑		0250	0.7E
	0559	0811	0.6F
	1035	1348	0.5E
	1706	2020	1.0F
	2335		
30 Th ◐		0352	0.9E
	0657	0913	0.6F
	1247	1608	0.3E
	1803	2104	0.9F

Time meridian 135° E. 0000 is midnight. 1200 is noon. Times are not adjusted for Daylight Saving Time.

Tokyo Wan Entrance (N. of Kannon Saki), Japan, 2005

F–Flood, Dir. 313° True E–Ebb, Dir. 133° True

July

Day	Slack h m	Maximum h m	knots
1 F	0015	0445	1.1E
	0752	1028	0.7F
	1520	1722	0.3E
	1905	2152	0.7F
2 Sa	0101	0532	1.3E
	0845	1207	0.9F
	1631	1820	0.4E
	2009	2245	0.6F
3 Su	0153	0616	1.4E
	0934	1330	1.1F
	1722	1911	0.4E
	2112	2347	0.5F
4 M	0246	0657	1.6E
	1020	1421	1.3F
	1805	1957	0.5E
	2215		
5 Tu		0058	0.4F
	0334	0735	1.6E
	1102	1501	1.5F
	1842	2039	0.6E
	2313		
6 W ●		0203	0.4F
	0418	0811	1.7E
	1140	1537	1.6F
	1916	2117	0.6E
7 Th	0006	0253	0.4F
	0457	0843	1.6E
	1214	1609	1.6F
	1946	2151	0.6E
8 F	0053	0335	0.4F
	0532	0912	1.6E
	1244	1638	1.6F
	2012	2221	0.6E
9 Sa	0133	0411	0.4F
	0606	0940	1.5E
	1311	1705	1.6F
	2035	2248	0.7E
10 Su	0209	0446	0.4F
	0641	1007	1.4E
	1338	1731	1.5F
	2056	2314	0.7E
11 M	0246	0523	0.5F
	0720	1036	1.3E
	1406	1759	1.4F
	2117	2343	0.8E
12 Tu	0328	0604	0.5F
	0805	1109	1.1E
	1437	1831	1.4F
	2142		
13 W		0017	0.8E
	0417	0650	0.6F
	0857	1147	0.9E
	1514	1906	1.2F
	2210		
14 Th		0102	0.8E
	0513	0742	0.6F
	1000	1233	0.6E
	1601	1947	1.1F
	2243		
15 F ◐		0211	0.9E
	0613	0840	0.7F
	1129	1350	0.3E
	1705	2033	0.9F
	2323		
16 Sa		0348	1.0E
	0713	0947	0.8F
	1424	1649	0.2E
	1827	2125	0.7F
17 Su	0010	0500	1.2E
	0812	1108	0.9F
	1612	1807	0.3E
	1953	2224	0.6F
18 M	0108	0557	1.4E
	0909	1247	1.2F
	1709	1906	0.5E
	2114	2335	0.5F
19 Tu	0219	0647	1.6E
	1004	1401	1.5F
	1754	1958	0.6E
	2228		
20 W		0100	0.5F
	0329	0734	1.8E
	1055	1453	1.7F
	1835	2044	0.8E
	2332		
21 Th ○		0216	0.5F
	0429	0817	1.9E
	1143	1536	1.9F
	1913	2127	0.8E
22 F	0026	0313	0.6F
	0523	0859	1.9E
	1228	1615	1.9F
	1947	2205	0.9E
23 Sa	0112	0400	0.7F
	0611	0938	1.8E
	1311	1651	1.9F
	2017	2242	0.9E
24 Su	0155	0444	0.8F
	0657	1016	1.6E
	1350	1725	1.8F
	2045	2316	0.9E
25 M	0237	0526	0.8F
	0743	1052	1.4E
	1426	1757	1.6F
	2109	2349	0.9E
26 Tu	0322	0608	0.8F
	0830	1127	1.1E
	1500	1828	1.4F
	2132		
27 W		0023	0.9E
	0411	0652	0.8F
	0921	1203	0.8E
	1532	1900	1.1F
	2154		
28 Th ◑		0105	0.9E
	0505	0740	0.8F
	1023	1245	0.4E
	1605	1934	0.9F
	2220		
29 F		0210	0.9E
	0604	0834	0.7F
	1210	1455	0.1E
	1644	2012	0.7F
	2250		
30 Sa		0342	1.0E
	0706	0939	0.7F
	1601	1700	0.1E
	1753	2056	0.6F
	2329		
31 Su		0453	1.1E
	0808	1117	0.7F
	1654	1809	0.2E
	1926	2151	0.4F

August

Day	Slack h m	Maximum h m	knots
1 M	0022	0549	1.2E
	0905	1328	1.0F
	1727	1904	0.3E
	2057	2303	0.3F
2 Tu	0147	0638	1.3E
	0957	1414	1.2F
	1757	1950	0.5E
	2218		
3 W		0048	0.3F
	0313	0721	1.4E
	1041	1449	1.4F
	1824	2029	0.6E
	2321		
4 Th		0209	0.4F
	0414	0759	1.5E
	1121	1520	1.5F
	1850	2104	0.7E
5 F ●	0008	0256	0.5F
	0501	0833	1.5E
	1155	1548	1.6F
	1914	2133	0.8E
6 Sa	0044	0332	0.6F
	0541	0904	1.5E
	1225	1614	1.6F
	1935	2159	0.9E
7 Su	0113	0404	0.7F
	0617	0932	1.5E
	1253	1638	1.6F
	1954	2222	1.0E
8 M	0140	0435	0.7F
	0652	0959	1.4E
	1318	1702	1.5F
	2011	2244	1.0E
9 Tu	0206	0506	0.8F
	0728	1027	1.3E
	1344	1726	1.4F
	2029	2307	1.1E
10 W	0238	0541	0.9F
	0809	1057	1.1E
	1411	1753	1.3F
	2050	2334	1.1E
11 Th	0318	0622	0.9F
	0855	1130	0.9E
	1440	1825	1.1F
	2114		
12 F		0007	1.1E
	0409	0710	0.9F
	0951	1208	0.6E
	1512	1903	1.0F
	2142		
13 Sa ◐		0047	1.1E
	0511	0805	0.9F
	1110	1259	0.3E
	1555	1948	0.8F
	2217		
14 Su		0151	1.0E
	0623	0911	0.9F
	1509	1635	0.1E
	1737	2043	0.5F
	2300		
15 M		0412	1.0E
	0736	1035	0.9F
	1624	1803	0.2E
	1941	2152	0.4F
16 Tu	0004	0535	1.2E
	0844	1233	1.1F
	1703	1901	0.5E
	2117	2321	0.3F
17 W	0153	0635	1.5E
	0946	1350	1.4F
	1738	1948	0.7E
	2230		
18 Th		0108	0.4F
	0333	0725	1.7E
	1041	1437	1.7F
	1811	2029	0.9E
	2324		
19 F		0219	0.7F
	0440	0811	1.8E
	1131	1517	1.8F
	1841	2106	1.1E
20 Sa ○	0008	0309	0.9F
	0534	0853	1.8E
	1215	1553	1.9F
	1909	2140	1.2E
21 Su	0047	0351	1.1F
	0621	0931	1.7E
	1255	1625	1.8F
	1935	2211	1.3E
22 M	0124	0430	1.2F
	0705	1007	1.6E
	1330	1656	1.6F
	1957	2240	1.3E
23 Tu	0201	0508	1.2F
	0748	1040	1.3E
	1401	1724	1.4F
	2017	2306	1.3E
24 W	0239	0546	1.2F
	0831	1110	1.0E
	1427	1751	1.2F
	2036	2332	1.2E
25 Th	0321	0625	1.1F
	0916	1138	0.7E
	1448	1818	1.0F
	2055	2359	1.2E
26 F	0410	0708	1.0F
	1009	1206	0.4E
	1459	1848	0.8F
	2117		
27 Sa ◑		0030	1.0E
	0507	0757	0.8F
	1137	1238	0.1E
	1411	1924	0.6F
	2142		
28 Su		0114	0.9E
	0614	0856	0.7F
		1654	0.1F
		2011	0.4F
	2214		
29 M		0406	0.8E
	0723	1018	0.7F
	1654	1805	0.1E
	1923	2115	0.2F
	2300		
30 Tu		0527	0.9E
	0829	1309	0.8F
	1705	1854	0.3E
	2116	2245	0.1F
31 W	0053	0622	1.1E
	0926	1349	1.1F
	1724	1933	0.6E
	2228		

September

Day	Slack h m	Maximum h m	knots
1 Th		0107	0.3F
	0319	0706	1.2E
	1013	1420	1.3F
	1746	2006	0.8E
	2312		
2 F		0207	0.5F
	0422	0745	1.3E
	1054	1448	1.4F
	1808	2036	1.0E
	2346		
3 Sa		0245	0.7F
	0507	0819	1.4E
	1129	1515	1.5F
	1829	2103	1.1E
4 Su ●	0014	0317	0.9F
	0546	0851	1.4E
	1201	1540	1.5F
	1849	2127	1.2E
5 M	0039	0347	1.0F
	0622	0920	1.4E
	1229	1604	1.5F
	1906	2148	1.3E
6 Tu	0103	0416	1.1F
	0656	0947	1.3E
	1256	1627	1.4F
	1923	2210	1.3E
7 W	0129	0446	1.2F
	0732	1014	1.2E
	1320	1650	1.2F
	1940	2232	1.4E
8 Th	0158	0519	1.2F
	0811	1043	1.0E
	1345	1716	1.1F
	1959	2257	1.4E
9 F	0233	0558	1.2F
	0855	1114	0.8E
	1409	1746	0.9F
	2021	2326	1.4E
10 Sa	0319	0644	1.1F
	0949	1150	0.5E
	1434	1823	0.7F
	2047		
11 Su ◐		0001	1.3E
	0420	0740	1.0F
	1110	1236	0.2E
	1454	1912	0.5F
	2121		
12 M		0046	1.1E
	0539	0846	1.0F
		1649	0.0
		2017	0.3F
	2205		
13 Tu		0316	0.9E
	0703	1011	0.9F
	1606	1800	0.3E
	2001	2140	0.2F
	2324		
14 W		0522	1.0E
	0820	1204	1.1F
	1633	1847	0.6E
	2125	2328	0.3F
15 Th	0207	0624	1.3E
	0925	1322	1.3F
	1702	1927	0.9E
	2219		
16 F		0112	0.6F
	0346	0715	1.5E
	1021	1408	1.5F
	1730	2003	1.2E
	2302		
17 Sa		0210	0.9F
	0448	0759	1.7E
	1110	1446	1.6F
	1757	2036	1.4E
	2341		
18 Su ○		0255	1.2F
	0539	0840	1.7E
	1153	1520	1.6F
	1822	2107	1.5E
19 M	0017	0335	1.4F
	0624	0917	1.6E
	1230	1552	1.5F
	1845	2136	1.6E
20 Tu	0052	0412	1.5F
	0707	0951	1.4E
	1303	1620	1.4F
	1906	2203	1.6E
21 W	0127	0448	1.5F
	0748	1023	1.2E
	1331	1647	1.2F
	1925	2228	1.5E
22 Th	0202	0524	1.4F
	0829	1050	0.9E
	1352	1712	1.0F
	1943	2251	1.4E
23 F	0238	0601	1.3F
	0911	1116	0.6E
	1406	1737	0.8F
	2001	2314	1.3E
24 Sa	0319	0641	1.1F
	1000	1141	0.3E
	1407	1806	0.6F
	2021	2339	1.2E
25 Su ◑	0410	0726	0.9F
	1119	1209	0.1E
	1315	1845	0.4F
	2044		
26 M		0007	0.9E
	0514	0820	0.8F
		1656	0.0
		1942	0.2F
	2113		
27 Tu		0043	0.7E
	0628	0929	0.7F
	1604	1752	0.2E
	2000	2059	0.1F
	2157		
28 W		0459	0.6E
	0740	1104	0.8F
	1613	1830	0.5E
	2128	2242	0.1F
29 Th	0039	0559	0.8E
	0841	1242	1.0F
	1631	1902	0.8E
	2209		
30 F		0055	0.3F
	0324	0645	1.0E
	0932	1326	1.1F
	1653	1932	1.0E
	2241		

Time meridian 135° E. 0000 is midnight. 1200 is noon. Times are not adjusted for Daylight Saving Time.

Tokyo Wan Entrance (N. of Kannon Saki), Japan, 2005

F–Flood, Dir. 313° True E–Ebb, Dir. 133° True

October

Day	Slack h m	Maximum h m	knots
1 Sa		0146	0.6F
	0421	0724	1.2E
	1016	1400	1.3F
	1715	2000	1.2E
	2310		
2 Su		0221	0.9F
	0505	0759	1.3E
	1055	1430	1.3F
	1737	2027	1.4E
	2337		
3 M ●		0253	1.1F
	0544	0833	1.3E
	1131	1458	1.3F
	1757	2052	1.5E
4 Tu	0004	0324	1.3F
	0621	0904	1.3E
	1203	1524	1.2F
	1816	2115	1.6E
5 W	0030	0356	1.4F
	0658	0934	1.2E
	1232	1549	1.1F
	1833	2138	1.6E
6 Th	0057	0428	1.4F
	0736	1003	1.0E
	1258	1614	1.0F
	1851	2202	1.6E
7 F	0128	0504	1.4F
	0817	1033	0.9E
	1323	1641	0.8F
	1910	2227	1.6E
8 Sa	0204	0543	1.4F
	0903	1105	0.6E
	1347	1712	0.7F
	1933	2256	1.5E
9 Su	0248	0630	1.3F
	1000	1143	0.4E
	1414	1755	0.5F
	2001	2330	1.3E
10 M	0347	0725	1.2F
	1129	1240	0.1E
	1514	1855	0.3F
	2037		
11 Tu ◐		0014	1.1E
	0505	0829	1.1F
	1440	1648	0.2E
	1837	2011	0.2F
	2133		
12 W		0155	0.7E
	0632	0944	1.0F
	1514	1740	0.5E
	2011	2141	0.2F
	2326		
13 Th		0506	0.9E
	0749	1112	1.1F
	1543	1820	0.8E
	2109	2329	0.4F
14 F	0229	0608	1.1E
	0855	1230	1.2F
	1611	1856	1.1E
	2153		
15 Sa		0058	0.8F
	0353	0658	1.3E
	0951	1323	1.3F
	1639	1930	1.4E
	2234		
16 Su		0153	1.1F
	0450	0742	1.4E
	1040	1405	1.3F
	1705	2003	1.6E
	2311		
17 M ○		0237	1.4F
	0539	0823	1.4E
	1123	1441	1.3F
	1731	2034	1.7E
	2348		
18 Tu		0317	1.6F
	0624	0900	1.3E
	1200	1513	1.2F
	1754	2103	1.8E
19 W	0023	0354	1.7F
	0707	0935	1.1E
	1233	1542	1.1F
	1816	2130	1.8E
20 Th	0058	0431	1.6F
	0748	1006	0.9E
	1301	1610	0.9F
	1836	2155	1.7E
21 F	0131	0506	1.5F
	0829	1035	0.7E
	1323	1635	0.7F
	1854	2218	1.6E
22 Sa	0205	0542	1.4F
	0911	1102	0.4E
	1339	1703	0.5F
	1913	2241	1.4E
23 Su	0239	0620	1.2F
	0957	1130	0.3E
	1352	1736	0.4F
	1933	2305	1.2E
24 M	0320	0702	1.1F
	1101	1207	0.1E
	1408	1825	0.2F
	1958	2333	1.0E
25 Tu ◑	0413	0750	1.0F
	1411	1631	0.1E
	1820	1931	0.1F
	2031		
26 W		0007	0.7E
	0520	0846	0.9F
	1432	1715	0.4E
	2001	2049	0.1F
	2137		
27 Th		0409	0.5E
	0636	0948	0.9F
	1455	1750	0.6E
	2051	2217	0.2F
28 F	0033	0524	0.6E
	0742	1056	0.9F
	1521	1821	0.9E
	2127	2358	0.4F
29 Sa	0312	0614	0.8E
	0840	1159	1.0F
	1547	1852	1.2E
	2159		
30 Su		0105	0.7F
	0411	0657	0.9E
	0931	1251	1.1F
	1614	1921	1.4E
	2230		
31 M		0149	1.0F
	0458	0736	1.1E
	1016	1333	1.1F
	1639	1950	1.6E
	2301		

November

Day	Slack h m	Maximum h m	knots
1 Tu		0228	1.3F
	0541	0813	1.1E
	1058	1410	1.1F
	1703	2018	1.7E
	2332		
2 W ●		0304	1.5F
	0622	0849	1.1E
	1136	1443	1.0F
	1725	2045	1.7E
3 Th	0003	0341	1.6F
	0703	0923	1.0E
	1211	1514	0.8F
	1747	2112	1.8E
4 F	0035	0418	1.6F
	0745	0957	0.8E
	1245	1545	0.7F
	1808	2139	1.7E
5 Sa	0110	0457	1.6F
	0829	1032	0.7E
	1317	1619	0.6F
	1831	2208	1.7E
6 Su	0148	0539	1.6F
	0917	1109	0.5E
	1355	1659	0.4F
	1859	2240	1.5E
7 M	0233	0625	1.4F
	1014	1156	0.3E
	1501	1753	0.3F
	1935	2317	1.3E
8 Tu	0329	0715	1.3F
	1127	1434	0.2E
	1707	1859	0.2F
	2024		
9 W ◐		0003	1.0E
	0440	0811	1.2F
	1256	1616	0.4E
	1842	2013	0.2F
	2140		
10 Th		0149	0.7E
	0557	0911	1.1F
	1354	1703	0.7E
	1947	2134	0.3F
	2353		
11 F		0441	0.7E
	0710	1016	1.0F
	1433	1743	1.0E
	2037	2305	0.5F
12 Sa	0236	0545	0.9E
	0815	1121	1.0F
	1507	1820	1.3E
	2122		
13 Su		0031	0.9F
	0354	0637	1.0E
	0912	1221	1.0F
	1539	1855	1.5E
	2204		
14 M		0132	1.2F
	0451	0723	1.0E
	1003	1312	1.0F
	1610	1930	1.7E
	2244		
15 Tu		0219	1.5F
	0541	0805	1.0E
	1048	1355	1.0F
	1639	2003	1.8E
	2323		
16 W ○		0302	1.7F
	0627	0844	1.0E
	1129	1433	0.9F
	1706	2034	1.9E
17 Th	0000	0342	1.7F
	0710	0921	0.8E
	1206	1508	0.8F
	1731	2103	1.8E
18 F	0035	0419	1.7F
	0752	0956	0.7E
	1240	1541	0.6F
	1754	2130	1.7E
19 Sa	0109	0454	1.6F
	0832	1028	0.5E
	1312	1612	0.5F
	1817	2155	1.6E
20 Su	0140	0529	1.5F
	0910	1059	0.4E
	1347	1646	0.4F
	1839	2220	1.4E
21 M	0211	0603	1.4F
	0948	1132	0.3E
	1444	1728	0.3F
	1906	2246	1.2E
22 Tu	0244	0640	1.2F
	1027	1217	0.2E
	1629	1820	0.2F
	1941	2316	1.0E
23 W	0324	0720	1.1F
	1112	1517	0.3E
	1800	1921	0.1F
	2031	2354	0.8E
24 Th ◑	0417	0805	1.1F
	1202	1614	0.5E
	1905	2026	0.2F
	2149		
25 F		0052	0.5E
	0526	0854	1.0F
	1255	1654	0.7E
	1953	2136	0.3F
	2355		
26 Sa		0430	0.4E
	0636	0946	1.0F
	1342	1731	1.0E
	2035	2252	0.5F
27 Su	0244	0535	0.5E
	0741	1041	0.9F
	1424	1807	1.2E
	2113		
28 M		0009	0.8F
	0358	0627	0.7E
	0841	1136	0.9F
	1503	1841	1.5E
	2151		
29 Tu		0113	1.1F
	0452	0713	0.8E
	0936	1230	0.8F
	1538	1916	1.6E
	2229		
30 W		0205	1.4F
	0540	0757	0.8E
	1028	1321	0.8F
	1610	1949	1.8E
	2306		

December

Day	Slack h m	Maximum h m	knots
1 Th		0250	1.6F
	0625	0838	0.8E
	1116	1408	0.7F
	1641	2022	1.8E
	2344		
2 F ●		0333	1.7F
	0710	0919	0.8E
	1203	1452	0.6F
	1711	2055	1.8E
3 Sa	0023	0414	1.8F
	0754	0959	0.7E
	1248	1535	0.5F
	1742	2128	1.8E
4 Su	0102	0455	1.8F
	0838	1039	0.6E
	1337	1620	0.4F
	1816	2201	1.7E
5 M	0143	0536	1.7F
	0921	1122	0.5E
	1435	1709	0.4F
	1856	2237	1.5E
6 Tu	0228	0618	1.6F
	1004	1214	0.4E
	1547	1803	0.3F
	1944	2317	1.3E
7 W	0318	0701	1.4F
	1047	1341	0.4E
	1700	1901	0.3F
	2043		
8 Th ◐		0005	1.0E
	0414	0746	1.3F
	1131	1515	0.6E
	1806	2003	0.4F
	2200		
9 F		0128	0.6E
	0517	0833	1.1F
	1216	1611	0.8E
	1905	2111	0.5F
	2358		
10 Sa		0404	0.5E
	0622	0922	1.0F
	1301	1659	1.0E
	1958	2227	0.7F
11 Su	0235	0518	0.6E
	0725	1015	0.9F
	1346	1741	1.3E
	2048	2356	0.9F
12 M	0359	0615	0.6E
	0825	1111	0.8F
	1429	1822	1.5E
	2135		
13 Tu		0113	1.2F
	0458	0705	0.7E
	0922	1210	0.7F
	1510	1900	1.7E
	2220		
14 W		0209	1.4F
	0547	0751	0.7E
	1016	1308	0.6F
	1549	1938	1.8E
	2303		
15 Th		0254	1.6F
	0633	0834	0.7E
	1106	1401	0.6F
	1626	2013	1.8E
	2342		
16 F ○		0335	1.7F
	0714	0914	0.6E
	1155	1448	0.5F
	1659	2045	1.8E
17 Sa	0019	0411	1.7F
	0751	0952	0.6E
	1241	1529	0.5F
	1731	2115	1.7E
18 Su	0052	0444	1.6F
	0825	1025	0.5E
	1327	1608	0.4F
	1801	2143	1.5E
19 M	0122	0515	1.6F
	0854	1057	0.5E
	1414	1646	0.4F
	1832	2209	1.4E
20 Tu	0150	0545	1.5F
	0920	1127	0.5E
	1505	1727	0.3F
	1908	2237	1.2E
21 W	0218	0615	1.4F
	0944	1201	0.5E
	1601	1811	0.3F
	1950	2308	1.1E
22 Th	0249	0647	1.3F
	1011	1245	0.5E
	1658	1859	0.3F
	2042	2344	0.9E
23 F	0327	0724	1.2F
	1042	1356	0.6E
	1753	1953	0.4F
	2148		
24 Sa ◑		0031	0.6E
	0418	0805	1.1F
	1118	1527	0.8E
	1847	2052	0.5F
	2319		
25 Su		0206	0.3E
	0524	0851	1.0F
	1201	1628	1.0E
	1938	2157	0.6F
26 M	0201	0451	0.3E
	0639	0941	0.9F
	1248	1718	1.2E
	2027	2313	0.8F
27 Tu	0352	0600	0.4E
	0752	1035	0.7F
	1340	1804	1.4E
	2115		
28 W		0038	1.1F
	0453	0656	0.5E
	0902	1135	0.6F
	1433	1847	1.6E
	2202		
29 Th		0149	1.4F
	0543	0747	0.6E
	1008	1242	0.6F
	1524	1928	1.7E
	2247		
30 F		0242	1.6F
	0628	0833	0.7E
	1110	1350	0.5F
	1613	2008	1.8E
	2332		
31 Sa ●		0328	1.8F
	0710	0917	0.7E
	1207	1450	0.5F
	1659	2047	1.9E

Time meridian 135° E. 0000 is midnight. 1200 is noon. Times are not adjusted for Daylight Saving Time.

Tomogashima Suido, Japan, 2005

F–Flood, Dir. Northward E–Ebb, Dir. Southward

January

Day	Slack h m	Maximum h m	knots
1 Sa	0034	0411	2.6E
	0723	1122	2.8F
	1453	1724	1.7E
	2048	2335	1.1F
2 Su	0152	0507	2.2E
	0822	1208	2.5F
	1528	1815	1.8E
	2142		
3 M		0037	1.4F
	0314	0617	1.8E
	0929	1255	2.1F
	1603	1912	2.0E
	2236		
4 Tu ◑		0138	1.8F
	0444	0742	1.5E
	1041	1344	1.7F
	1640	2012	2.2E
	2328		
5 W		0239	2.3F
	0619	0910	1.4E
	1155	1433	1.4F
	1717	2110	2.5E
6 Th	0018	0340	2.6F
	0744	1022	1.5E
	1305	1522	1.0F
	1752	2202	2.7E
7 F	0105	0441	3.0F
	0853	1121	1.6E
	1411	1610	0.7F
	1825	2249	2.9E
8 Sa	0150	0540	3.3F
	0950	1211	1.7E
	1513	1656	0.5F
	1856	2333	3.1E
9 Su	0234	0635	3.5F
	1040	1257	1.7E
	1609	1739	0.3F
	1926		
10 M ●		0013	3.2E
	0315	0725	3.6F
	1125	1337	1.7E
	1653	1820	0.3F
	2002		
11 Tu		0053	3.3E
	0356	0809	3.6F
	1204	1414	1.6E
	1726	1901	0.3F
	2051		
12 W		0132	3.3E
	0436	0848	3.5F
	1240	1449	1.6E
	1753	1947	0.5F
	2152		
13 Th		0211	3.2E
	0517	0925	3.3F
	1311	1524	1.7E
	1823	2042	0.7F
	2257		
14 F		0253	3.0E
	0601	1003	3.0F
	1341	1600	1.7E
	1901	2144	0.9F
15 Sa	0005	0338	2.8E
	0647	1042	2.7F
	1411	1639	1.8E
	1947	2247	1.2F
16 Su	0115	0429	2.4E
	0739	1124	2.4F
	1441	1722	1.9E
	2041	2349	1.5F
17 M ◐	0229	0527	1.9E
	0839	1210	2.0F
	1513	1812	2.0E
	2138		
18 Tu		0051	1.9F
	0352	0641	1.5E
	0948	1258	1.7F
	1547	1911	2.2E
	2236		
19 W		0152	2.2F
	0525	0813	1.3E
	1105	1348	1.3F
	1623	2014	2.3E
	2332		
20 Th		0254	2.6F
	0659	0939	1.3E
	1222	1439	0.9F
	1701	2116	2.5E
21 F	0025	0358	2.9F
	0816	1046	1.4E
	1335	1531	0.7F
	1740	2211	2.8E
22 Sa	0115	0501	3.2F
	0917	1141	1.6E
	1440	1623	0.5F
	1820	2300	3.0E
23 Su	0203	0600	3.4F
	1009	1227	1.7E
	1536	1713	0.4F
	1903	2345	3.2E
24 M	0247	0653	3.5F
	1053	1308	1.7E
	1620	1759	0.4F
	1952		
25 Tu ○		0027	3.3E
	0330	0739	3.6F
	1133	1345	1.7E
	1652	1843	0.5F
	2045		
26 W		0107	3.3E
	0411	0819	3.4F
	1207	1419	1.7E
	1717	1926	0.6F
	2142		
27 Th		0147	3.3E
	0452	0855	3.2F
	1238	1451	1.8E
	1743	2012	0.8F
	2240		
28 F		0227	3.1E
	0532	0929	3.0F
	1305	1524	1.8E
	1815	2104	1.1F
	2340		
29 Sa		0309	2.9E
	0615	1004	2.7F
	1331	1558	1.9E
	1854	2201	1.4F
30 Su	0042	0354	2.5E
	0702	1042	2.3F
	1358	1636	2.0E
	1942	2301	1.6F
31 M	0150	0446	2.1E
	0754	1125	1.9F
	1426	1720	2.1E
	2037		

February

Day	Slack h m	Maximum h m	knots
1 Tu		0003	1.9F
	0305	0548	1.6E
	0858	1211	1.6F
	1457	1811	2.2E
	2138		
2 W ◑		0105	2.2F
	0430	0711	1.3E
	1014	1302	1.2F
	1531	1912	2.2E
	2240		
3 Th		0208	2.5F
	0606	0848	1.1E
	1137	1356	0.9F
	1609	2022	2.4E
	2340		
4 F		0313	2.8F
	0733	1007	1.3E
	1256	1451	0.6F
	1652	2128	2.6E
5 Sa	0036	0418	3.1F
	0841	1107	1.5E
	1405	1547	0.5F
	1741	2225	2.8E
6 Su	0129	0521	3.3F
	0934	1156	1.7E
	1502	1643	0.4F
	1837	2315	3.0E
7 M	0217	0617	3.5F
	1019	1238	1.8E
	1546	1736	0.5F
	1937		
8 Tu		0000	3.2E
	0302	0706	3.5F
	1059	1315	1.8E
	1618	1823	0.6F
	2035		
9 W ●		0041	3.3E
	0345	0748	3.4F
	1133	1348	1.8E
	1643	1905	0.8F
	2130		
10 Th		0122	3.3E
	0426	0824	3.2F
	1203	1419	1.9E
	1707	1946	1.1F
	2224		
11 F		0201	3.2E
	0506	0857	2.9F
	1229	1450	1.9E
	1735	2030	1.3F
	2318		
12 Sa		0241	3.0E
	0546	0928	2.5F
	1253	1521	2.0E
	1809	2120	1.6F
13 Su	0015	0323	2.7E
	0629	1002	2.2F
	1316	1555	2.1E
	1851	2215	1.8F
14 M	0116	0410	2.3E
	0716	1040	1.8F
	1340	1634	2.2E
	1941	2315	2.1F
15 Tu	0223	0505	1.8E
	0812	1125	1.5F
	1408	1718	2.3E
	2039		
16 W ◐		0017	2.3F
	0341	0613	1.3E
	0923	1216	1.1F
	1440	1811	2.3E
	2143		
17 Th		0121	2.5F
	0511	0747	1.1E
	1050	1311	0.8F
	1517	1919	2.3E
	2249		
18 F		0226	2.7F
	0644	0923	1.1E
	1216	1410	0.6F
	1602	2036	2.4E
	2352		
19 Sa		0332	3.0F
	0759	1031	1.4E
	1329	1510	0.5F
	1658	2144	2.6E
20 Su	0050	0437	3.2F
	0856	1123	1.6E
	1427	1611	0.5F
	1805	2241	2.9E
21 M	0143	0537	3.3F
	0942	1206	1.8E
	1511	1708	0.6F
	1916	2330	3.1E
22 Tu	0232	0629	3.4F
	1023	1243	1.9E
	1543	1759	0.8F
	2019		
23 W		0015	3.2E
	0317	0714	3.3F
	1058	1317	2.0E
	1610	1842	1.0F
	2116		
24 Th ○		0056	3.2E
	0359	0752	3.1F
	1127	1347	2.0E
	1634	1922	1.3F
	2208		
25 F		0135	3.2E
	0440	0824	2.8F
	1153	1417	2.1E
	1700	2001	1.6F
	2259		
26 Sa		0215	3.0E
	0519	0853	2.4F
	1214	1447	2.2E
	1730	2044	1.8F
	2351		
27 Su		0255	2.7E
	0559	0923	2.0F
	1235	1518	2.3E
	1807	2134	2.1F
28 M	0047	0339	2.4E
	0642	0956	1.7F
	1256	1553	2.4E
	1851	2230	2.2F

March

Day	Slack h m	Maximum h m	knots
1 Tu	0148	0428	1.9E
	0732	1037	1.3F
	1320	1632	2.4E
	1943	2331	2.4F
2 W	0258	0526	1.5E
	0836	1127	1.0F
	1349	1719	2.4E
	2044		
3 Th		0035	2.5F
	0418	0646	1.1E
	1002	1225	0.7F
	1425	1818	2.3E
	2153		
4 F ◑		0140	2.7F
	0549	0829	1.0E
	1135	1327	0.5F
	1511	1935	2.3E
	2303		
5 Sa		0246	2.9F
	0711	0950	1.3E
	1253	1431	0.4F
	1611	2056	2.4E
6 Su	0007	0351	3.1F
	0813	1047	1.5E
	1352	1535	0.5F
	1727	2203	2.6E
7 M	0105	0453	3.2F
	0902	1132	1.8E
	1435	1635	0.7F
	1847	2259	2.9E
8 Tu	0159	0548	3.2F
	0944	1210	1.9E
	1508	1729	0.9F
	1958	2346	3.1E
9 W	0247	0636	3.1F
	1019	1244	2.1E
	1536	1816	1.2F
	2057		
10 Th ●		0029	3.2E
	0332	0716	2.9F
	1050	1316	2.1E
	1601	1857	1.5F
	2150		
11 F		0110	3.1E
	0413	0750	2.6F
	1115	1345	2.2E
	1627	1934	1.8F
	2240		
12 Sa		0149	3.0E
	0453	0819	2.2F
	1136	1413	2.3E
	1655	2013	2.1F
	2330		
13 Su		0229	2.7E
	0532	0846	1.9F
	1154	1443	2.4E
	1729	2058	2.3F
14 M	0022	0310	2.4E
	0612	0914	1.5F
	1212	1515	2.5E
	1808	2148	2.4F
15 Tu	0118	0355	2.0E
	0658	0950	1.2F
	1233	1551	2.5E
	1854	2246	2.6F
16 W	0220	0447	1.6E
	0754	1037	0.9F
	1258	1633	2.5E
	1950	2349	2.7F
17 Th	0332	0553	1.2E
	0914	1136	0.6F
	1332	1724	2.4E
	2055		
18 F ◐		0053	2.7F
	0454	0726	1.0E
	1054	1243	0.4F
	1417	1831	2.2E
	2208		
19 Sa		0159	2.9F
	0617	0903	1.2E
	1217	1350	0.4F
	1521	1958	2.2E
	2319		
20 Su		0303	3.0F
	0725	1007	1.4E
	1316	1457	0.5F
	1643	2119	2.4E
21 M	0023	0405	3.1F
	0818	1055	1.7E
	1359	1559	0.7F
	1813	2222	2.7E
22 Tu	0121	0503	3.1F
	0901	1135	2.0E
	1433	1657	1.0F
	1931	2315	2.9E
23 W	0214	0554	3.0F
	0938	1211	2.1E
	1502	1746	1.4F
	2035		
24 Th		0001	3.0E
	0301	0637	2.8F
	1010	1242	2.2E
	1529	1829	1.7F
	2130		
25 F		0043	3.0E
	0345	0714	2.5F
	1036	1312	2.3E
	1555	1908	2.0F
	2220		
26 Sa ○		0124	2.9E
	0426	0744	2.1F
	1057	1341	2.4E
	1623	1946	2.3F
	2309		
27 Su		0203	2.7E
	0505	0809	1.7F
	1114	1410	2.5E
	1654	2026	2.5F
	2359		
28 M		0243	2.4E
	0545	0834	1.4F
	1129	1440	2.6E
	1730	2112	2.7F
29 Tu	0051	0326	2.1E
	0627	0904	1.0F
	1146	1514	2.7E
	1812	2205	2.8F
30 W	0148	0413	1.7E
	0718	0946	0.7F
	1207	1553	2.6E
	1902	2305	2.8F
31 Th	0253	0511	1.3E
	0830	1044	0.5F
	1238	1639	2.5E
	2001		

Time meridian 135° E. 0000 is midnight. 1200 is noon. Times are not adjusted for Daylight Saving Time.

Tomogashima Suido, Japan, 2005

F–Flood, Dir. Northward E–Ebb, Dir. Southward

April

Day	Slack (h m)	Maximum (h m)	knots
1 F		0009	2.8F
	0405	0627	1.1E
	1011	1155	0.3F
	1322	1736	2.3E
	2111		
2 Sa ◑		0113	2.9F
	0523	0806	1.1E
	1141	1308	0.3F
	1428	1853	2.2E
	2225		
3 Su		0216	2.9F
	0633	0923	1.4E
	1239	1417	0.4F
	1555	2025	2.2E
	2336		
4 M		0317	3.0F
	0730	1015	1.7E
	1321	1521	0.7F
	1731	2141	2.4E
5 Tu	0040	0415	3.0F
	0816	1058	2.0E
	1356	1620	1.1F
	1858	2240	2.7E
6 W	0136	0508	2.8F
	0855	1135	2.2E
	1428	1713	1.5F
	2008	2330	2.8E
7 Th	0228	0554	2.6F
	0928	1208	2.3E
	1457	1759	1.9F
	2107		
8 F		0016	2.9E
	0315	0635	2.3F
	0955	1239	2.4E
	1524	1841	2.2F
	2159		
9 Sa ●		0058	2.8E
	0359	0707	1.9F
	1017	1309	2.5E
	1553	1919	2.5F
	2249		
10 Su		0138	2.6E
	0439	0733	1.5F
	1034	1338	2.6E
	1623	1959	2.7F
	2338		
11 M		0217	2.4E
	0519	0756	1.2F
	1047	1408	2.7E
	1657	2041	2.9F
12 Tu	0028	0259	2.1E
	0600	0822	0.9F
	1100	1440	2.8E
	1736	2130	2.9F
13 W	0121	0343	1.7E
	0647	0857	0.6F
	1117	1516	2.8E
	1821	2226	3.0F
14 Th	0219	0435	1.4E
	0751	0949	0.4F
	1142	1559	2.7E
	1913	2326	3.0F
15 F	0322	0538	1.2E
	0928	1105	0.2F
	1224	1649	2.5E
	2016		
16 Sa ◐		0028	3.0F
	0431	0702	1.1E
	1103	1223	0.2F
	1333	1755	2.2E
	2129		
17 Su		0129	2.9F
	0538	0830	1.3E
	1201	1334	0.4F
	1504	1922	2.1E
	2244		
18 M		0229	2.9F
	0637	0931	1.6E
	1242	1440	0.7F
	1643	2051	2.2E
	2353		
19 Tu		0327	2.9F
	0726	1017	1.9E
	1318	1541	1.1F
	1817	2200	2.4E
20 W	0055	0420	2.7F
	0807	1057	2.2E
	1352	1636	1.6F
	1936	2257	2.6E
21 Th	0151	0509	2.5F
	0843	1133	2.4E
	1423	1726	2.0F
	2040	2345	2.7E
22 F	0242	0552	2.2F
	0912	1205	2.5E
	1453	1811	2.4F
	2136		
23 Sa		0030	2.7E
	0329	0628	1.8F
	0936	1237	2.7E
	1523	1853	2.7F
	2228		
24 Su ○		0112	2.5E
	0412	0657	1.4F
	0953	1306	2.8E
	1553	1933	2.9F
	2317		
25 M		0153	2.3E
	0454	0720	1.0F
	1005	1337	2.8E
	1627	2014	3.0F
26 Tu	0006	0234	2.0E
	0535	0744	0.7F
	1014	1408	2.9E
	1703	2059	3.1F
27 W	0056	0316	1.8E
	0621	0813	0.5F
	1026	1443	2.9E
	1745	2151	3.1F
28 Th	0149	0403	1.5E
	0717	0858	0.3F
	1047	1523	2.8E
	1832	2247	3.1F
29 F	0245	0457	1.3E
	0844	1009	0.2F
	1126	1610	2.6E
	1929	2346	3.1F
30 Sa	0345	0605	1.2E
	1022	1135	0.1F
	1237	1706	2.4E
	2035		

May

Day	Slack (h m)	Maximum (h m)	knots
1 Su ◑		0045	3.0F
	0445	0728	1.2E
	1119	1250	0.3F
	1412	1820	2.1E
	2149		
2 M		0142	2.9F
	0542	0839	1.5E
	1201	1357	0.7F
	1552	1951	2.0E
	2302		
3 Tu		0238	2.8F
	0633	0932	1.8E
	1239	1459	1.1F
	1730	2114	2.1E
4 W	0009	0331	2.6F
	0717	1016	2.1E
	1314	1556	1.6F
	1858	2218	2.3E
5 Th	0110	0421	2.4F
	0754	1055	2.4E
	1348	1649	2.0F
	2009	2312	2.4E
6 F	0205	0507	2.0F
	0826	1130	2.6E
	1421	1738	2.5F
	2110		
7 Sa		0001	2.5E
	0256	0547	1.7F
	0853	1204	2.7E
	1453	1824	2.8F
	2205		
8 Su ●		0045	2.4E
	0344	0619	1.3F
	0912	1235	2.8E
	1525	1907	3.0F
	2255		
9 M		0128	2.2E
	0429	0646	0.9F
	0924	1306	2.9E
	1558	1949	3.2F
	2345		
10 Tu		0209	2.0E
	0512	0709	0.6F
	0931	1339	3.0E
	1634	2033	3.3F
11 W	0033	0250	1.8E
	0557	0736	0.4F
	0938	1413	3.0E
	1713	2120	3.3F
12 Th	0122	0334	1.5E
	0649	0814	0.3F
	0955	1451	3.0E
	1757	2212	3.3F
13 F	0212	0423	1.4E
	0801	0915	0.1F
	1032	1534	2.8E
	1847	2306	3.2F
14 Sa	0304	0519	1.3E
	0932	1044	0.1F
	1145	1625	2.6E
	1946		
15 Su		0002	3.1F
	0357	0626	1.3E
	1032	1204	0.3F
	1323	1728	2.2E
	2054		
16 M ◐		0057	2.9F
	0449	0739	1.4E
	1116	1312	0.6F
	1501	1848	2.0E
	2208		
17 Tu		0151	2.8F
	0538	0841	1.7E
	1155	1415	1.1F
	1639	2018	1.9E
	2319		
18 W		0243	2.6F
	0623	0930	2.0E
	1234	1514	1.5F
	1813	2135	2.0E
19 Th	0025	0334	2.3F
	0703	1014	2.3E
	1311	1610	2.0F
	1934	2236	2.2E
20 F	0125	0420	1.9F
	0737	1053	2.6E
	1347	1702	2.5F
	2041	2329	2.2E
21 Sa	0221	0503	1.6F
	0806	1129	2.8E
	1422	1752	2.8F
	2139		
22 Su		0017	2.2E
	0313	0541	1.2F
	0829	1204	2.9E
	1456	1840	3.1F
	2232		
23 M		0102	2.1E
	0402	0611	0.8F
	0843	1237	3.0E
	1530	1924	3.3F
	2323		
24 Tu ○		0144	1.9E
	0449	0638	0.5F
	0850	1310	3.1E
	1606	2009	3.4F
25 W	0010	0226	1.7E
	0535	0705	0.3F
	0855	1344	3.1E
	1644	2054	3.4F
26 Th	0057	0307	1.6E
	0624	0730	0.2F
	0910	1422	3.1E
	1726	2141	3.4F
27 F	0142	0351	1.4E
	0722	0830	0.1F
	0948	1503	3.0E
	1812	2230	3.3F
28 Sa	0228	0439	1.3E
	0832	0951	0.1F
	1102	1550	2.7E
	1905	2322	3.2F
29 Su	0314	0534	1.3E
	0934	1116	0.3F
	1238	1645	2.4E
	2005		
30 M ◑		0013	3.0F
	0359	0636	1.4E
	1023	1226	0.6F
	1412	1752	2.1E
	2114		
31 Tu		0105	2.8F
	0445	0741	1.6E
	1107	1329	1.0F
	1547	1915	1.9E
	2226		

June

Day	Slack (h m)	Maximum (h m)	knots
1 W		0156	2.5F
	0528	0838	1.9E
	1149	1429	1.5F
	1722	2042	1.8E
	2336		
2 Th		0246	2.2F
	0609	0928	2.2E
	1230	1527	2.0F
	1851	2155	1.9E
3 F	0041	0334	1.9F
	0646	1012	2.5E
	1310	1623	2.4F
	2007	2255	2.0E
4 Sa	0142	0419	1.5F
	0717	1053	2.7E
	1348	1717	2.8F
	2111	2347	2.0E
5 Su	0239	0500	1.1F
	0743	1130	2.9E
	1426	1809	3.1F
	2207		
6 M		0035	2.0E
	0334	0537	0.8F
	0802	1207	3.1E
	1503	1858	3.4F
	2259		
7 Tu ●		0119	1.9E
	0425	0608	0.5F
	0812	1242	3.1E
	1540	1944	3.5F
	2347		
8 W		0201	1.7E
	0514	0637	0.3F
	0819	1317	3.2E
	1618	2029	3.5F
9 Th	0031	0241	1.6E
	0600	0711	0.2F
	0836	1354	3.2E
	1658	2113	3.5F
10 F	0114	0322	1.5E
	0646	0756	0.2F
	0918	1434	3.1E
	1741	2158	3.4F
11 Sa	0155	0405	1.4E
	0736	0904	0.2F
	1032	1518	2.9E
	1829	2244	3.2F
12 Su	0235	0450	1.4E
	0829	1026	0.3F
	1201	1607	2.6E
	1923	2332	3.0F
13 M	0315	0541	1.5E
	0922	1138	0.6F
	1329	1705	2.3E
	2023		
14 Tu		0021	2.8F
	0354	0638	1.6E
	1012	1243	1.0F
	1457	1816	1.9E
	2132		
15 W ◐		0110	2.5F
	0435	0738	1.9E
	1100	1344	1.5F
	1629	1941	1.7E
	2244		
16 Th		0159	2.2F
	0514	0835	2.1E
	1146	1443	1.9F
	1802	2106	1.7E
	2354		
17 F		0247	1.8F
	0552	0926	2.4E
	1230	1541	2.4F
	1928	2216	1.7E
18 Sa	0100	0335	1.4F
	0626	1013	2.7E
	1313	1639	2.8F
	2039	2315	1.8E
19 Su	0203	0419	1.1F
	0656	1055	2.9E
	1354	1735	3.1F
	2140		
20 M		0006	1.8E
	0302	0501	0.7F
	0720	1135	3.0E
	1434	1829	3.4F
	2233		
21 Tu		0053	1.8E
	0359	0538	0.5F
	0738	1214	3.2E
	1513	1918	3.5F
	2321		
22 W ○		0135	1.7E
	0450	0612	0.3F
	0752	1251	3.2E
	1552	2004	3.6F
23 Th	0005	0216	1.6E
	0535	0647	0.2F
	0815	1329	3.2E
	1632	2047	3.5F
24 F	0046	0254	1.5E
	0612	0730	0.2F
	0902	1407	3.2E
	1713	2128	3.4F
25 Sa	0123	0332	1.5E
	0648	0827	0.3F
	1012	1449	3.1E
	1757	2210	3.3F
26 Su	0159	0412	1.5E
	0729	0939	0.4F
	1131	1535	2.8E
	1846	2253	3.0F
27 M	0233	0455	1.6E
	0818	1050	0.7F
	1251	1626	2.5E
	1940	2338	2.8F
28 Tu	0308	0542	1.7E
	0911	1155	1.1F
	1411	1726	2.1E
	2041		
29 W ◑		0024	2.5F
	0344	0636	1.8E
	1004	1256	1.4F
	1536	1840	1.7E
	2150		
30 Th		0112	2.1F
	0420	0735	2.0E
	1056	1357	1.9F
	1709	2009	1.5E
	2303		

Time meridian 135° E. 0000 is midnight. 1200 is noon. Times are not adjusted for Daylight Saving Time.

Tomogashima Suido, Japan, 2005

F–Flood, Dir. Northward E–Ebb, Dir. Southward

July

Day	Slack h m	Maximum h m	knots
1 F		0201	1.7F
	0457	0834	2.3E
	1146	1457	2.3F
	1842	2132	1.5E
2 Sa	0015	0250	1.4F
	0533	0928	2.5E
	1234	1557	2.7F
	2003	2239	1.6E
3 Su	0123	0338	1.0F
	0607	1017	2.8E
	1319	1658	3.1F
	2109	2335	1.7E
4 M	0228	0424	0.7F
	0638	1102	3.0E
	1403	1756	3.3F
	2205		
5 Tu		0024	1.7E
	0329	0508	0.5F
	0705	1144	3.1E
	1445	1849	3.5F
	2253		
6 W ●		0109	1.7E
	0423	0549	0.3F
	0730	1224	3.2E
	1526	1937	3.6F
	2337		
7 Th		0149	1.7E
	0506	0628	0.2F
	0805	1303	3.3E
	1607	2020	3.6F
8 F	0017	0226	1.6E
	0538	0709	0.3F
	0855	1342	3.3E
	1647	2059	3.5F
9 Sa	0052	0301	1.6E
	0607	0758	0.4F
	0959	1422	3.2E
	1729	2138	3.3F
10 Su	0124	0337	1.6E
	0640	0858	0.6F
	1108	1505	3.0E
	1813	2216	3.0F
11 M	0155	0414	1.7E
	0720	1003	0.8F
	1219	1552	2.7E
	1901	2257	2.7F
12 Tu	0225	0455	1.8E
	0808	1106	1.2F
	1331	1644	2.3E
	1955	2341	2.4F
13 W	0257	0541	1.9E
	0903	1208	1.5F
	1449	1746	1.9E
	2058		
14 Th		0027	2.0F
	0329	0633	2.0E
	0959	1310	1.9F
	1615	1906	1.5E
	2210		
15 F ◐		0115	1.7F
	0404	0733	2.2E
	1056	1411	2.2F
	1750	2039	1.3E
	2326		
16 Sa		0205	1.3F
	0440	0835	2.4E
	1150	1513	2.6F
	1920	2159	1.4E
17 Su	0041	0255	1.0F
	0517	0934	2.6E
	1241	1616	2.9F
	2034	2301	1.5E
18 M	0152	0346	0.7F
	0553	1026	2.8E
	1329	1718	3.2F
	2133	2354	1.6E
19 Tu	0256	0436	0.5F
	0630	1113	3.0E
	1416	1816	3.5F
	2223		
20 W		0040	1.7E
	0352	0524	0.4F
	0710	1157	3.2E
	1459	1907	3.6F
	2307		
21 Th ○		0121	1.7E
	0434	0609	0.3F
	0756	1238	3.3E
	1541	1952	3.6F
	2346		
22 F		0157	1.7E
	0505	0651	0.4F
	0850	1317	3.3E
	1622	2031	3.4F
23 Sa	0020	0231	1.7E
	0530	0735	0.6F
	0949	1357	3.2E
	1702	2107	3.2F
24 Su	0050	0303	1.7E
	0557	0824	0.8F
	1049	1437	3.1E
	1744	2142	3.0F
25 M	0118	0337	1.8E
	0631	0920	1.0F
	1152	1521	2.8E
	1828	2218	2.7F
26 Tu	0144	0412	1.9E
	0712	1019	1.3F
	1257	1608	2.5E
	1916	2258	2.3F
27 W	0212	0452	2.0E
	0803	1120	1.6F
	1407	1702	2.0E
	2011	2341	2.0F
28 Th ◑	0241	0538	2.1E
	0859	1222	1.9F
	1525	1809	1.6E
	2117		
29 F		0029	1.6F
	0313	0631	2.2E
	1000	1324	2.2F
	1654	1937	1.2E
	2235		
30 Sa		0119	1.2F
	0347	0735	2.3E
	1100	1427	2.5F
	1831	2111	1.2E
	2357		
31 Su		0212	0.9F
	0425	0843	2.4E
	1158	1532	2.8F
	1954	2225	1.3E

August

Day	Slack h m	Maximum h m	knots
1 M	0114	0307	0.6F
	0507	0945	2.7E
	1252	1636	3.1F
	2058	2322	1.5E
2 Tu	0222	0402	0.5F
	0554	1039	2.9E
	1343	1738	3.4F
	2149		
3 W		0010	1.7E
	0318	0456	0.4F
	0647	1127	3.1E
	1430	1833	3.5F
	2234		
4 Th		0051	1.8E
	0400	0547	0.4F
	0745	1211	3.2E
	1514	1920	3.5F
	2313		
5 F ●		0127	1.8E
	0430	0632	0.6F
	0842	1252	3.3E
	1556	2001	3.4F
	2347		
6 Sa		0200	1.8E
	0455	0714	0.7F
	0938	1332	3.3E
	1637	2036	3.2F
7 Su	0016	0231	1.8E
	0520	0757	1.0F
	1033	1411	3.1E
	1717	2109	2.9F
8 M	0041	0302	1.9E
	0549	0843	1.2F
	1129	1452	2.9E
	1758	2141	2.6F
9 Tu	0105	0334	2.0E
	0625	0936	1.5F
	1228	1536	2.6E
	1841	2216	2.2F
10 W	0129	0410	2.1E
	0709	1033	1.8F
	1331	1624	2.2E
	1931	2256	1.9F
11 Th	0154	0449	2.2E
	0801	1134	2.0F
	1441	1722	1.7E
	2029	2341	1.5F
12 F	0223	0536	2.2E
	0901	1236	2.3F
	1603	1836	1.3E
	2144		
13 Sa ◐		0033	1.1F
	0256	0633	2.2E
	1005	1341	2.5F
	1736	2015	1.1E
	2311		
14 Su		0128	0.8F
	0334	0743	2.3E
	1110	1446	2.8F
	1907	2143	1.2E
15 M	0035	0226	0.6F
	0418	0857	2.5E
	1210	1552	3.0F
	2018	2247	1.4E
16 Tu	0146	0325	0.5F
	0513	1001	2.7E
	1306	1656	3.2F
	2112	2337	1.6E
17 W	0242	0424	0.5F
	0618	1055	2.9E
	1358	1754	3.4F
	2158		
18 Th		0019	1.8E
	0325	0520	0.6F
	0726	1143	3.1E
	1445	1845	3.4F
	2237		
19 F		0056	1.9E
	0356	0610	0.7F
	0828	1226	3.2E
	1529	1928	3.3F
	2311		
20 Sa ○		0129	1.9E
	0421	0652	0.9F
	0924	1307	3.3E
	1611	2004	3.1F
	2340		
21 Su		0159	2.0E
	0445	0732	1.2F
	1017	1346	3.2E
	1651	2036	2.8F
22 M	0005	0229	2.0E
	0512	0812	1.5F
	1109	1426	3.0E
	1730	2105	2.4F
23 Tu	0027	0259	2.1E
	0544	0858	1.7F
	1203	1507	2.7E
	1811	2136	2.1F
24 W	0047	0331	2.2E
	0622	0950	2.0F
	1300	1551	2.3E
	1855	2211	1.7F
25 Th	0110	0407	2.3E
	0708	1047	2.2F
	1404	1642	1.9E
	1947	2254	1.4F
26 F	0135	0448	2.3E
	0803	1149	2.4F
	1517	1745	1.4E
	2055	2345	1.0F
27 Sa ◑	0205	0537	2.3E
	0907	1253	2.5F
	1641	1911	1.1E
	2223		
28 Su		0042	0.7F
	0242	0640	2.3E
	1016	1359	2.7F
	1814	2054	1.1E
	2354		
29 M		0144	0.5F
	0328	0759	2.3E
	1123	1505	2.9F
	1932	2208	1.3E
30 Tu	0110	0247	0.5F
	0428	0917	2.5E
	1225	1611	3.1F
	2031	2302	1.6E
31 W	0207	0350	0.5F
	0543	1019	2.7E
	1322	1711	3.2F
	2119	2346	1.8E

September

Day	Slack h m	Maximum h m	knots
1 Th	0249	0449	0.7F
	0701	1112	3.0E
	1413	1805	3.3F
	2159		
2 F		0023	1.9E
	0321	0542	0.9F
	0809	1158	3.1E
	1500	1851	3.2F
	2234		
3 Sa		0057	2.0E
	0348	0627	1.2F
	0907	1241	3.2E
	1544	1930	3.0F
	2303		
4 Su ●		0127	2.1E
	0412	0707	1.4F
	0959	1321	3.1E
	1624	2003	2.7F
	2328		
5 M		0156	2.1E
	0438	0745	1.7F
	1049	1400	3.0E
	1704	2031	2.3F
	2348		
6 Tu		0225	2.2E
	0507	0825	2.0F
	1140	1440	2.7E
	1742	2058	1.9F
7 W	0007	0255	2.3E
	0542	0911	2.2F
	1233	1522	2.4E
	1824	2128	1.6F
8 Th	0026	0328	2.4E
	0623	1005	2.4F
	1331	1608	2.0E
	1911	2206	1.2F
9 F	0047	0405	2.5E
	0712	1104	2.5F
	1437	1703	1.5E
	2011	2255	0.9F
10 Sa	0114	0449	2.4E
	0810	1207	2.6F
	1553	1814	1.2E
	2135	2355	0.6F
11 Su ◐	0149	0543	2.3E
	0918	1313	2.7F
	1718	1954	1.0E
	2314		
12 M		0100	0.4F
	0235	0655	2.2E
	1030	1418	2.9F
	1841	2124	1.2E
13 Tu	0034	0207	0.4F
	0340	0823	2.3E
	1139	1523	3.0F
	1946	2224	1.5E
14 W	0130	0313	0.5F
	0502	0938	2.5E
	1241	1625	3.1F
	2036	2310	1.8E
15 Th	0212	0415	0.7F
	0629	1038	2.8E
	1337	1721	3.1F
	2118	2349	2.0E
16 F	0246	0511	1.0F
	0744	1128	3.0E
	1428	1811	3.0F
	2154		
17 Sa		0023	2.1E
	0314	0559	1.3F
	0846	1213	3.1E
	1514	1852	2.8F
	2224		
18 Su ○		0054	2.2E
	0340	0641	1.7F
	0940	1255	3.1E
	1557	1927	2.5F
	2249		
19 M		0124	2.3E
	0406	0718	1.9F
	1030	1335	2.9E
	1637	1956	2.1F
	2310		
20 Tu		0152	2.4E
	0435	0757	2.2F
	1119	1414	2.7E
	1716	2021	1.8F
	2327		
21 W		0221	2.5E
	0506	0838	2.4F
	1210	1454	2.4E
	1755	2047	1.4F
	2342		
22 Th		0253	2.6E
	0544	0926	2.6F
	1304	1538	2.1E
	1839	2119	1.1F
23 F	0000	0327	2.6E
	0628	1022	2.7F
	1403	1627	1.7E
	1932	2204	0.8F
24 Sa	0023	0407	2.6E
	0720	1123	2.7F
	1510	1728	1.3E
	2049	2304	0.5F
25 Su ◑	0055	0455	2.5E
	0822	1227	2.8F
	1626	1851	1.0E
	2232		
26 M		0014	0.3F
	0141	0557	2.3E
	0934	1332	2.9F
	1746	2032	1.1E
	2357		
27 Tu		0125	0.3F
	0248	0719	2.2E
	1048	1435	2.9F
	1855	2142	1.4E
28 W	0053	0234	0.5F
	0416	0848	2.3E
	1156	1537	3.0F
	1949	2231	1.7E
29 Th	0135	0337	0.7F
	0551	0959	2.5E
	1257	1634	3.0F
	2033	2312	2.0E
30 F	0209	0435	1.1F
	0714	1055	2.8E
	1352	1726	2.9F
	2111	2348	2.2E

Time meridian 135° E. 0000 is midnight. 1200 is noon. Times are not adjusted for Daylight Saving Time.

Tomogashima Suido, Japan, 2005

F–Flood, Dir. Northward   E–Ebb, Dir. Southward

October

Day	Slack h m	Maximum h m	knots
1 Sa	0240	0527	1.5F
	0821	1143	2.9E
	1442	1811	2.7F
	2143		
2 Su		0021	2.3E
	0308	0611	1.8F
	0918	1227	2.9E
	1528	1849	2.4F
	2209		
3 M ●		0051	2.4E
	0335	0652	2.2F
	1010	1309	2.9E
	1610	1920	2.0F
	2230		
4 Tu		0120	2.5E
	0404	0730	2.4F
	1059	1348	2.7E
	1650	1945	1.6F
	2247		
5 W		0149	2.6E
	0434	0810	2.6F
	1148	1428	2.4E
	1729	2008	1.2F
	2259		
6 Th		0219	2.7E
	0509	0853	2.8F
	1239	1510	2.1E
	1811	2035	0.9F
	2313		
7 F		0252	2.7E
	0549	0944	2.9F
	1334	1556	1.7E
	1900	2113	0.7F
	2332		
8 Sa		0329	2.7E
	0636	1042	2.9F
	1434	1649	1.4E
	2007	2210	0.4F
9 Su	0000	0413	2.6E
	0731	1144	2.9F
	1541	1757	1.1E
	2149	2325	0.2F
10 M	0045	0507	2.4E
	0838	1247	2.9F
	1652	1929	1.1E
	2320		
11 Tu ◐		0041	0.2F
	0154	0617	2.2E
	0952	1348	2.9F
	1800	2053	1.3E
12 W	0016	0152	0.4F
	0326	0748	2.1E
	1106	1448	2.9F
	1858	2149	1.6E
13 Th	0057	0257	0.7F
	0505	0912	2.3E
	1213	1546	2.9F
	1945	2233	1.9E
14 F	0132	0357	1.1F
	0636	1017	2.5E
	1313	1639	2.8F
	2025	2311	2.2E
15 Sa	0205	0451	1.6F
	0751	1111	2.7E
	1406	1726	2.5F
	2059	2346	2.4E
16 Su	0235	0539	2.0F
	0853	1158	2.8E
	1456	1808	2.2F
	2127		
17 M ○		0018	2.5E
	0304	0623	2.3F
	0947	1241	2.7E
	1541	1842	1.8F
	2149		
18 Tu		0048	2.6E
	0334	0704	2.6F
	1038	1323	2.6E
	1624	1909	1.4F
	2206		
19 W		0118	2.7E
	0404	0743	2.8F
	1127	1403	2.3E
	1705	1932	1.1F
	2217		
20 Th		0148	2.8E
	0438	0825	3.0F
	1217	1444	2.1E
	1746	1956	0.8F
	2227		
21 F		0220	2.9E
	0516	0912	3.1F
	1308	1528	1.8E
	1832	2028	0.5F
	2241		
22 Sa		0255	2.9E
	0559	1005	3.1F
	1402	1616	1.5E
	1932	2116	0.3F
	2305		
23 Su		0336	2.8E
	0648	1103	3.1F
	1501	1714	1.2E
	2105	2232	0.2F
	2347		
24 M		0425	2.6E
	0747	1203	3.0F
	1603	1827	1.1E
	2240	2355	0.2F
25 Tu ◑	0100	0525	2.3E
	0857	1303	3.0F
	1706	1953	1.2E
	2335		
26 W		0109	0.4F
	0235	0645	2.1E
	1012	1401	2.9F
	1803	2100	1.5E
27 Th	0016	0215	0.7F
	0415	0817	2.1E
	1124	1457	2.8F
	1853	2150	1.9E
28 F	0053	0316	1.1F
	0553	0934	2.2E
	1229	1550	2.6F
	1935	2231	2.2E
29 Sa	0127	0412	1.6F
	0716	1035	2.4E
	1327	1639	2.4F
	2011	2309	2.4E
30 Su	0200	0504	2.0F
	0824	1127	2.5E
	1421	1723	2.1F
	2042	2343	2.6E
31 M	0232	0552	2.4F
	0923	1213	2.5E
	1510	1801	1.7F
	2106		

November

Day	Slack h m	Maximum h m	knots
1 Tu		0016	2.7E
	0304	0636	2.8F
	1016	1257	2.4E
	1556	1832	1.3F
	2124		
2 W ●		0047	2.8E
	0335	0718	3.0F
	1106	1338	2.3E
	1640	1857	0.9F
	2135		
3 Th		0117	2.9E
	0409	0800	3.1F
	1155	1419	2.0E
	1722	1920	0.7F
	2142		
4 F		0149	3.0E
	0445	0844	3.2F
	1244	1501	1.8E
	1807	1948	0.4F
	2151		
5 Sa		0224	3.0E
	0525	0933	3.2F
	1334	1546	1.5E
	1902	2029	0.3F
	2211		
6 Su		0303	2.9E
	0611	1026	3.2F
	1426	1637	1.3E
	2020	2136	0.1F
	2250		
7 M		0348	2.7E
	0704	1122	3.1F
	1520	1737	1.2E
	2153	2305	0.1F
8 Tu	0007	0441	2.5E
	0805	1219	3.0F
	1615	1849	1.3E
	2249		
9 W ◐		0023	0.3F
	0145	0548	2.2E
	0917	1315	2.9F
	1709	2002	1.5E
	2332		
10 Th		0131	0.6F
	0324	0714	2.0E
	1030	1409	2.8F
	1758	2100	1.8E
11 F	0011	0233	1.1F
	0503	0842	2.0E
	1140	1502	2.6F
	1842	2147	2.1E
12 Sa	0048	0331	1.6F
	0634	0953	2.1E
	1244	1551	2.3F
	1921	2229	2.4E
13 Su	0124	0426	2.1F
	0750	1052	2.3E
	1342	1637	2.0F
	1954	2307	2.6E
14 M	0159	0517	2.5F
	0855	1143	2.3E
	1436	1719	1.6F
	2021	2342	2.8E
15 Tu	0233	0605	2.8F
	0952	1229	2.3E
	1527	1754	1.2F
	2041		
16 W ○		0015	2.9E
	0307	0651	3.1F
	1044	1313	2.2E
	1614	1823	0.8F
	2054		
17 Th		0048	3.0E
	0341	0735	3.3F
	1133	1355	2.0E
	1700	1848	0.6F
	2100		
18 F		0121	3.1E
	0417	0819	3.3F
	1221	1436	1.7E
	1746	1915	0.4F
	2105		
19 Sa		0155	3.1E
	0456	0905	3.4F
	1308	1519	1.6E
	1835	1951	0.2F
	2123		
20 Su		0233	3.0E
	0538	0954	3.3F
	1355	1604	1.4E
	1939	2047	0.1F
	2203		
21 M		0315	2.9E
	0626	1045	3.2F
	1442	1655	1.3E
	2055	2213	0.1F
	2321		
22 Tu		0404	2.7E
	0721	1137	3.1F
	1530	1753	1.3E
	2155	2336	0.3F
23 W	0059	0502	2.4E
	0825	1230	3.0F
	1617	1859	1.4E
	2241		
24 Th ◑		0045	0.6F
	0234	0614	2.0E
	0935	1323	2.8F
	1703	2003	1.7E
	2324		
25 F		0148	1.0F
	0410	0741	1.9E
	1048	1414	2.5F
	1747	2058	2.0E
26 Sa	0005	0247	1.5F
	0546	0905	1.9E
	1156	1504	2.2F
	1827	2145	2.3E
27 Su	0045	0344	2.0F
	0711	1013	2.0E
	1300	1551	1.9F
	1903	2227	2.6E
28 M	0123	0439	2.5F
	0823	1110	2.1E
	1359	1635	1.5F
	1933	2306	2.8E
29 Tu	0201	0532	2.9F
	0925	1200	2.1E
	1454	1715	1.1F
	1957	2343	2.9E
30 W	0237	0622	3.2F
	1020	1247	2.0E
	1547	1749	0.8F
	2013		

December

Day	Slack h m	Maximum h m	knots
1 Th		0018	3.0E
	0314	0710	3.3F
	1110	1330	1.9E
	1637	1818	0.5F
	2021		
2 F ●		0053	3.1E
	0350	0755	3.4F
	1158	1411	1.7E
	1725	1847	0.3F
	2027		
3 Sa		0128	3.2E
	0429	0840	3.5F
	1243	1453	1.6E
	1811	1921	0.2F
	2045		
4 Su		0205	3.1E
	0509	0925	3.4F
	1326	1534	1.5E
	1901	2010	0.1F
	2129		
5 M		0246	3.0E
	0554	1011	3.3F
	1408	1618	1.4E
	1955	2123	0.2F
	2247		
6 Tu		0331	2.8E
	0643	1059	3.2F
	1449	1706	1.4E
	2052	2247	0.3F
7 W	0019	0422	2.6E
	0739	1148	3.0F
	1530	1801	1.5E
	2143	2358	0.6F
8 Th ◐	0148	0523	2.2E
	0843	1237	2.8F
	1612	1900	1.6E
	2232		
9 F		0101	1.0F
	0319	0639	1.9E
	0953	1327	2.5F
	1653	2000	1.9E
	2318		
10 Sa		0202	1.5F
	0453	0807	1.7E
	1105	1416	2.2F
	1732	2055	2.2E
11 Su	0002	0300	2.0F
	0625	0928	1.7E
	1214	1505	1.8F
	1809	2144	2.5E
12 M	0045	0358	2.4F
	0747	1033	1.8E
	1318	1551	1.4F
	1842	2228	2.7E
13 Tu	0126	0455	2.8F
	0855	1129	1.9E
	1419	1635	1.1F
	1911	2309	2.9E
14 W	0206	0550	3.1F
	0953	1219	1.9E
	1517	1714	0.7F
	1932	2347	3.1E
15 Th	0246	0642	3.4F
	1045	1304	1.8E
	1611	1749	0.4F
	1946		
16 F ○		0024	3.2E
	0324	0730	3.5F
	1133	1347	1.7E
	1702	1821	0.3F
	1958		
17 Sa		0101	3.2E
	0403	0815	3.6F
	1217	1426	1.6E
	1747	1856	0.2F
	2020		
18 Su		0139	3.2E
	0443	0858	3.5F
	1258	1505	1.5E
	1825	1941	0.2F
	2110		
19 M		0218	3.2E
	0525	0940	3.4F
	1335	1545	1.5E
	1904	2042	0.2F
	2224		
20 Tu		0301	3.0E
	0610	1023	3.2F
	1412	1626	1.5E
	1949	2158	0.4F
	2347		
21 W		0348	2.7E
	0700	1108	3.0F
	1448	1711	1.6E
	2040	2309	0.7F
22 Th	0108	0441	2.4E
	0757	1154	2.8F
	1524	1802	1.7E
	2132		
23 F		0014	1.0F
	0231	0545	2.0E
	0901	1241	2.5F
	1600	1858	1.8E
	2224		
24 Sa ◑		0115	1.5F
	0359	0705	1.7E
	1011	1329	2.1F
	1638	1957	2.1E
	2314		
25 Su		0215	1.9F
	0533	0834	1.5E
	1124	1418	1.7F
	1715	2053	2.3E
26 M	0002	0315	2.3F
	0704	0951	1.6E
	1234	1506	1.4F
	1750	2145	2.6E
27 Tu	0049	0415	2.7F
	0821	1055	1.7E
	1341	1553	1.0F
	1822	2232	2.8E
28 W	0133	0514	3.1F
	0923	1149	1.7E
	1444	1638	0.7F
	1850	2315	3.0E
29 Th	0216	0610	3.4F
	1018	1237	1.8E
	1543	1720	0.4F
	1914	2356	3.1E
30 F	0257	0702	3.5F
	1106	1320	1.7E
	1636	1758	0.3F
	1936		
31 Sa ●		0035	3.2E
	0337	0749	3.6F
	1149	1400	1.7E
	1719	1836	0.2F
	2008		

Time meridian 135° E. 0000 is midnight. 1200 is noon. Times are not adjusted for Daylight Saving Time.

Naruto, Japan, 2005

F–Flood, Dir. Northward E–Ebb, Dir. Southward

January

Day	Slack h m	Maximum h m	knots
1 Sa		0307	7.2E
	0607	0934	7.3F
	1303	1609	7.3E
	1926	2203	5.3F
2 Su	0046	0359	6.9E
	0657	1019	6.8F
	1344	1653	7.1E
	2010	2255	5.4F
3 M	0150	0458	6.4E
	0754	1110	6.2F
	1427	1741	6.9E
	2056	2354	5.6F
4 Tu	0305	0605	6.0E
	0902	1207	5.5F
	1512	1833	6.8E
◑	2147		
5 W		0059	5.9F
	0425	0720	5.8E
	1020	1312	4.8F
	1601	1929	6.8E
	2239		
6 Th		0206	6.3F
	0542	0836	6.0E
	1144	1421	4.4F
	1653	2025	6.9E
	2332		
7 F		0309	6.9F
	0650	0944	6.4E
	1301	1526	4.2F
	1746	2119	7.2E
8 Sa	0024	0406	7.4F
	0747	1044	6.9E
	1405	1624	4.4F
	1839	2212	7.6E
9 Su	0116	0458	7.9F
	0839	1135	7.4E
	1458	1716	4.7F
	1931	2302	8.0E
10 M	0206	0546	8.3F
	0925	1223	7.8E
	1543	1804	5.1F
●	2022	2351	8.3E
11 Tu	0255	0632	8.5F
	1010	1307	8.1E
	1625	1850	5.5F
	2114		
12 W		0041	8.5E
	0344	0717	8.5F
	1053	1351	8.2E
	1707	1937	5.8F
	2206		
13 Th		0130	8.5E
	0434	0803	8.4F
	1136	1435	8.2E
	1749	2025	6.0F
	2302		
14 F		0222	8.3E
	0525	0850	8.1F
	1219	1521	8.1E
	1835	2116	6.1F
15 Sa	0001	0317	7.9E
	0618	0939	7.5F
	1304	1609	7.8E
	1923	2211	6.1F
16 Su	0106	0417	7.3E
	0717	1031	6.8F
	1350	1701	7.5E
	2015	2312	6.0F
17 M	0222	0524	6.7E
	0823	1130	6.0F
	1441	1759	7.1E
◐	2113		
18 Tu		0023	6.0F
	0347	0642	6.1E
	0943	1240	5.1F
	1536	1900	6.8E
	2213		
19 W		0139	6.1F
	0514	0807	5.9E
	1116	1359	4.4F
	1636	2005	6.6E
	2314		
20 Th		0252	6.4F
	0630	0925	6.0E
	1247	1514	4.0F
	1736	2105	6.5E
21 F	0009	0352	6.7F
	0730	1029	6.4E
	1357	1616	4.0F
	1830	2156	6.6E
22 Sa	0056	0440	7.1F
	0818	1119	6.8E
	1448	1703	4.1F
	1916	2240	6.8E
23 Su	0137	0519	7.3F
	0856	1157	7.1E
	1525	1740	4.3F
	1953	2316	7.0E
24 M	0212	0552	7.6F
	0928	1229	7.3E
	1553	1810	4.6F
	2026	2348	7.3E
25 Tu	0245	0621	7.7F
	0955	1255	7.5E
	1616	1837	5.0F
○	2058		
26 W		0019	7.6E
	0317	0650	7.9F
	1022	1321	7.7E
	1637	1905	5.4F
	2132		
27 Th		0051	7.8E
	0351	0719	8.0F
	1048	1347	7.9E
	1700	1935	5.8F
	2209		
28 F		0126	8.0E
	0427	0751	7.9F
	1116	1416	8.0E
	1727	2008	6.1F
	2249		
29 Sa		0203	8.1E
	0505	0825	7.8F
	1146	1447	8.0E
	1757	2044	6.4F
	2333		
30 Su		0244	7.9E
	0547	0902	7.4F
	1218	1522	7.9E
	1831	2125	6.5F
31 M	0022	0329	7.5E
	0633	0941	6.8F
	1252	1600	7.6E
	1910	2211	6.4F

February

Day	Slack h m	Maximum h m	knots
1 Tu	0119	0422	7.0E
	0725	1025	6.1F
	1328	1643	7.2E
	1954	2304	6.3F
2 W	0228	0525	6.2E
	0828	1117	5.1F
	1407	1734	6.8E
◑	2047		
3 Th		0009	6.2F
	0352	0644	5.6E
	0951	1223	4.0F
	1454	1836	6.4E
	2149		
4 F		0128	6.3F
	0524	0816	5.5E
	1138	1348	3.2F
	1557	1949	6.4E
	2258		
5 Sa		0250	6.7F
	0644	0941	6.0E
	1313	1516	3.3F
	1716	2103	6.7E
6 Su	0007	0359	7.2F
	0746	1045	6.7E
	1414	1624	4.0F
	1830	2207	7.2E
7 M	0110	0455	7.7F
	0835	1135	7.3E
	1457	1717	4.8F
	1933	2303	7.8E
8 Tu	0205	0543	8.1F
	0918	1216	7.8E
	1532	1801	5.5F
	2028	2352	8.3E
9 W	0255	0626	8.3F
	0957	1254	8.2E
	1606	1843	6.2F
●	2118		
10 Th		0039	8.6E
	0343	0707	8.4F
	1033	1331	8.4E
	1640	1924	6.6F
	2207		
11 F		0124	8.8E
	0429	0747	8.2F
	1109	1409	8.5E
	1716	2006	6.9F
	2257		
12 Sa		0210	8.6E
	0515	0828	7.9F
	1144	1447	8.4E
	1753	2049	7.0F
	2348		
13 Su		0258	8.2E
	0603	0910	7.3F
	1221	1527	8.1E
	1835	2136	6.9F
14 M	0045	0349	7.6E
	0654	0954	6.5F
	1259	1611	7.6E
	1920	2228	6.6F
15 Tu	0149	0449	6.8E
	0753	1045	5.6F
	1341	1702	7.0E
	2012	2330	6.3F
16 W	0306	0602	6.0E
	0910	1148	4.4F
	1430	1802	6.4E
◐	2112		
17 Th		0047	6.0F
	0437	0731	5.5E
	1054	1315	3.4F
	1535	1915	5.9E
	2221		
18 F		0212	6.1F
	0602	0902	5.6E
	1242	1451	3.1F
	1657	2031	5.8E
	2330		
19 Sa		0324	6.4F
	0707	1012	6.2E
	1352	1602	3.5F
	1811	2134	6.0E
20 Su	0029	0418	6.8F
	0754	1100	6.7E
	1433	1649	4.0F
	1905	2223	6.4E
21 M	0116	0458	7.1F
	0831	1135	7.1E
	1501	1723	4.6F
	1945	2301	6.9E
22 Tu	0154	0530	7.4F
	0900	1202	7.4E
	1522	1750	5.1F
	2017	2333	7.4E
23 W	0229	0559	7.6F
	0925	1225	7.7E
	1539	1814	5.6F
	2048		
24 Th		0003	7.8E
	0302	0626	7.8F
	0948	1247	7.9E
○	1556	1839	6.2F
	2120		
25 F		0034	8.2E
	0335	0654	7.9F
	1012	1311	8.2E
	1617	1907	6.7F
	2155		
26 Sa		0107	8.5E
	0411	0723	7.8F
	1037	1337	8.3E
	1641	1938	7.1F
	2233		
27 Su		0142	8.5E
	0448	0756	7.6F
	1104	1406	8.3E
	1711	2012	7.3F
	2315		
28 M		0222	8.4E
	0529	0830	7.2F
	1133	1439	8.2E
	1743	2051	7.4F

March

Day	Slack h m	Maximum h m	knots
1 Tu	0001	0305	7.9E
	0613	0907	6.6F
	1204	1514	7.9E
	1821	2135	7.2F
2 W	0056	0355	7.2E
	0704	0949	5.7F
	1236	1555	7.4E
	1905	2226	6.8F
3 Th	0203	0457	6.3E
	0809	1039	4.5F
	1311	1645	6.7E
	1959	2333	6.4F
4 F	0330	0624	5.4E
	0945	1150	3.1F
	1357	1756	6.0E
◑	2111		
5 Sa		0103	6.2F
	0514	0812	5.4E
	1200	1343	2.3F
	1524	1933	5.8E
	2239		
6 Su		0242	6.5F
	0638	0941	6.1E
	1324	1525	3.2F
	1723	2104	6.3E
7 M	0003	0356	7.0F
	0736	1039	6.9E
	1406	1628	4.4F
	1845	2211	7.1E
8 Tu	0111	0450	7.5F
	0821	1122	7.5E
	1438	1713	5.5F
	1944	2304	7.9E
9 W	0205	0534	7.8F
	0859	1158	8.0E
	1507	1752	6.3F
	2034	2350	8.4E
10 Th	0253	0613	8.0F
	0932	1231	8.3E
	1536	1829	6.9F
●	2118		
11 F		0032	8.8E
	0337	0649	8.0F
	1003	1303	8.5E
	1606	1905	7.4F
	2202		
12 Sa		0112	8.9E
	0419	0725	7.8F
	1034	1335	8.6E
	1638	1941	7.6F
	2246		
13 Su		0153	8.7E
	0501	0801	7.4F
	1104	1409	8.5E
	1712	2020	7.6F
	2332		
14 M		0236	8.3E
	0545	0838	6.8F
	1135	1445	8.1E
	1749	2101	7.5F
15 Tu	0021	0323	7.7E
	0632	0918	6.0F
	1208	1523	7.6E
	1829	2147	7.1F
16 W	0118	0417	6.8E
	0728	1004	5.0F
	1244	1608	6.9E
	1916	2242	6.6F
17 Th	0227	0525	6.0E
	0843	1102	3.8F
	1327	1705	6.0E
	2012	2353	6.0F
18 F	0353	0654	5.4E
	1035	1232	2.6F
	1434	1823	5.3E
◐	2123		
19 Sa		0123	5.8F
	0520	0829	5.6E
	1227	1425	2.6F
	1622	1953	5.1E
	2244		
20 Su		0246	6.0F
	0627	0938	6.2E
	1323	1538	3.5F
	1752	2106	5.6E
	2354		
21 M		0344	6.4F
	0715	1024	6.7E
	1356	1622	4.3F
	1849	2158	6.2E
22 Tu	0048	0426	6.8F
	0750	1056	7.1E
	1418	1653	5.1F
	1929	2237	6.8E
23 W	0131	0459	7.1F
	0819	1122	7.5E
	1435	1719	5.7F
	2001	2310	7.5E
24 Th	0208	0528	7.3F
	0843	1144	7.8E
	1452	1743	6.4F
	2032	2341	8.0E
25 F	0242	0555	7.4F
	0906	1207	8.1E
	1511	1809	7.0F
	2104		
26 Sa		0012	8.4E
	0317	0624	7.5F
	0930	1231	8.4E
○	1533	1837	7.5F
	2139		
27 Su		0046	8.7E
	0353	0654	7.4F
	0955	1258	8.5E
	1559	1909	7.9F
	2217		
28 M		0123	8.7E
	0432	0727	7.2F
	1022	1328	8.6E
	1629	1944	8.0F
	2259		
29 Tu		0202	8.5E
	0513	0802	6.7F
	1050	1401	8.4E
	1704	2023	8.0F
	2347		
30 W		0247	8.0E
	0559	0840	6.0F
	1121	1437	8.0E
	1744	2108	7.7F
31 Th	0041	0338	7.2E
	0654	0924	5.0F
	1154	1520	7.3E
	1830	2202	7.2F

Time meridian 135° E. 0000 is midnight. 1200 is noon. Times are not adjusted for Daylight Saving Time.

Naruto, Japan, 2005

F–Flood, Dir. Northward E–Ebb, Dir. Southward

April

Day	Slack h m	Maximum h m	knots
1 F	0150	0445	6.2E
	0808	1019	3.8F
	1233	1616	6.5E
	1929	2312	6.5F
2 Sa	0319	0620	5.6E
	1002	1146	2.5F
	1334	1741	5.7E
	2051		
3 Su		0051	6.2F
	0459	0807	5.7E
	1203	1355	2.7F
	1547	1935	5.6E
	2232		
4 M		0232	6.4F
	0617	0925	6.4E
	1300	1524	4.1F
	1745	2105	6.3E
5 Tu	0001	0343	6.8F
	0711	1017	7.1E
	1336	1617	5.3F
	1855	2209	7.2E
6 W	0108	0435	7.2F
	0753	1056	7.7E
	1406	1658	6.2F
	1947	2258	7.9E
7 Th	0201	0516	7.4F
	0828	1130	8.1E
	1434	1735	7.0F
	2032	2341	8.4E
8 F	0247	0553	7.4F
	0859	1201	8.3E
	1502	1809	7.5F
	2113		
9 Sa		0020	8.6E
	0328	0627	7.3F
	0927	1231	8.5E
	1531	1842	7.9F
	2153		
10 Su		0058	8.7E
	0408	0700	7.0F
	0954	1301	8.5E
	1601	1917	8.0F
	2233		
11 M		0136	8.5E
	0447	0734	6.6F
	1022	1332	8.3E
	1634	1953	8.0F
	2316		
12 Tu		0216	8.1E
	0529	0809	6.1F
	1051	1405	8.0E
	1708	2031	7.7F
13 W	0001	0300	7.5E
	0615	0847	5.4F
	1122	1441	7.5E
	1746	2113	7.3F
14 Th	0052	0351	6.9E
	0710	0931	4.5F
	1156	1523	6.8E
	1829	2203	6.8F
15 F	0152	0453	6.2E
	0823	1028	3.5F
	1238	1617	5.9E
	1920	2305	6.3F
16 Sa	0305	0613	5.7E
	1003	1152	2.6F
	1347	1733	5.1E
	2027		
17 Su		0024	5.9F
	0422	0737	5.8E
	1135	1337	2.7F
	1543	1905	4.8E
	2148		
18 M		0147	5.8F
	0528	0843	6.2E
	1227	1452	3.7F
	1719	2023	5.3E
	2307		
19 Tu		0252	6.1F
	0618	0930	6.7E
	1258	1538	4.7F
	1818	2120	6.0E
20 W	0009	0340	6.3F
	0656	1005	7.1E
	1321	1611	5.5F
	1901	2204	6.7E
21 Th	0059	0417	6.6F
	0727	1033	7.4E
	1341	1640	6.3F
	1937	2241	7.4E
22 F	0141	0450	6.8F
	0754	1059	7.8E
	1402	1708	6.9F
	2011	2316	8.0E
23 Sa	0221	0521	6.9F
	0819	1124	8.1E
	1425	1737	7.6F
	2047	2351	8.4E
24 Su	0259	0553	6.9F
	0845	1152	8.4E
	1452	1809	8.0F
	2124		
25 M		0027	8.6E
	0339	0626	6.7F
	0912	1222	8.6E
	1523	1844	8.3F
	2205		
26 Tu		0106	8.6E
	0420	0701	6.5F
	0942	1255	8.6E
	1557	1923	8.4F
	2250		
27 W		0149	8.3E
	0505	0739	6.0F
	1014	1332	8.4E
	1636	2005	8.3F
	2339		
28 Th		0237	7.9E
	0555	0822	5.4F
	1049	1413	8.0E
	1720	2053	7.9F
29 F	0035	0333	7.2E
	0656	0912	4.5F
	1130	1503	7.3E
	1812	2150	7.4F
30 Sa	0143	0444	6.5E
	0816	1018	3.6F
	1225	1609	6.4E
	1017	2304	6.7F

May

Day	Slack h m	Maximum h m	knots
1 Su	0305	0612	6.1E
	0956	1154	3.1F
	1359	1744	5.7E
	2043		
2 M		0036	6.3F
	0429	0741	6.3E
	1121	1345	3.8F
	1611	1930	5.8E
	2223		
3 Tu		0208	6.3F
	0539	0850	6.8E
	1215	1501	5.0F
	1746	2054	6.4E
	2350		
4 W		0317	6.5F
	0633	0941	7.3E
	1254	1554	6.0F
	1851	2156	7.2E
5 Th	0057	0410	6.6F
	0716	1022	7.7E
	1329	1636	6.8F
	1942	2246	7.7E
6 F	0152	0453	6.6F
	0751	1058	7.9E
	1400	1714	7.4F
	2026	2329	8.1E
7 Sa	0238	0530	6.5F
	0821	1129	8.1E
	1430	1748	7.8F
	2106		
8 Su		0008	8.2E
	0320	0605	6.4F
	0849	1200	8.2E
	1459	1822	8.0F
	2145		
9 M		0046	8.2E
	0359	0637	6.1F
	0916	1230	8.2E
	1530	1856	8.1F
	2223		
10 Tu		0123	8.0E
	0438	0711	5.7F
	0943	1301	8.0E
	1602	1930	8.0F
	2303		
11 W		0201	7.8E
	0519	0745	5.3F
	1012	1334	7.8E
	1635	2007	7.9F
	2344		
12 Th		0243	7.4E
	0604	0823	4.8F
	1045	1410	7.4E
	1712	2047	7.5F
13 F	0029	0329	7.0E
	0655	0907	4.3F
	1123	1452	6.8E
	1753	2132	7.1F
14 Sa	0119	0424	6.6E
	0757	1001	3.7F
	1211	1543	6.1E
	1841	2225	6.6F
15 Su	0216	0526	6.3E
	0907	1109	3.3F
	1319	1649	5.4E
	1941	2328	6.2F
16 M	0317	0632	6.2E
	1015	1229	3.4F
	1454	1809	5.1E
	2053		
17 Tu		0038	5.9F
	0416	0733	6.4E
	1107	1342	4.1F
	1625	1927	5.3E
	2210		
18 W		0145	5.8F
	0507	0823	6.6E
	1146	1438	5.0F
	1734	2031	5.8E
	2322		
19 Th		0242	5.8F
	0551	0904	7.0E
	1217	1521	5.8F
	1826	2124	6.5E
20 F	0022	0329	5.9F
	0627	0939	7.3E
	1245	1559	6.6F
	1910	2210	7.1E
21 Sa	0114	0409	6.0F
	0659	1011	7.7E
	1314	1634	7.3F
	1952	2252	7.6E
22 Su	0201	0447	6.0F
	0730	1044	8.0E
	1345	1710	7.8F
	2033	2332	8.0E
23 M	0245	0524	6.0F
	0801	1117	8.3E
	1418	1747	8.3F
	2115		
24 Tu		0014	8.2E
	0329	0603	5.9F
	0834	1153	8.5E
	1456	1827	8.5F
	2159		
25 W		0057	8.3E
	0415	0643	5.7F
	0910	1232	8.5E
	1536	1909	8.6F
	2245		
26 Th		0143	8.1E
	0503	0727	5.5F
	0950	1316	8.3E
	1621	1955	8.4F
	2336		
27 F		0234	7.8E
	0556	0815	5.1F
	1035	1404	7.9E
	1710	2047	8.1F
28 Sa	0031	0330	7.4E
	0655	0911	4.6F
	1129	1501	7.4E
	1806	2144	7.6F
29 Su	0131	0435	7.0E
	0803	1019	4.3F
	1241	1611	6.7E
	1912	2252	7.0F
30 M	0238	0547	6.8E
	0917	1142	4.3F
	1416	1737	6.1E
	2033		
31 Tu		0010	6.4F
	0346	0659	6.8E
	1024	1310	4.8F
	1602	1909	6.1E
	2203		

June

Day	Slack h m	Maximum h m	knots
1 W		0130	6.1F
	0450	0804	7.0E
	1122	1425	5.6F
	1729	2030	6.4E
	2329		
2 Th		0241	5.9F
	0545	0859	7.2E
	1209	1524	6.4F
	1837	2136	6.9E
3 F	0042	0340	5.8F
	0631	0945	7.5E
	1251	1612	7.0F
	1932	2231	7.3E
4 Sa	0142	0429	5.7F
	0711	1025	7.6E
	1328	1654	7.5F
	2019	2318	7.5E
5 Su	0234	0510	5.5F
	0745	1101	7.7E
	1402	1732	7.8F
	2101		
6 M		0000	7.7E
	0318	0547	5.3F
	0816	1134	7.8E
	1434	1806	7.9F
	2140		
7 Tu		0038	7.7E
	0358	0622	5.1F
	0845	1206	7.8E
	1506	1840	8.0F
	2216		
8 W		0115	7.6E
	0435	0655	5.0F
	0914	1238	7.7E
	1538	1914	8.0F
	2252		
9 Th		0150	7.6E
	0512	0729	4.8F
	0947	1312	7.6E
	1612	1948	7.9F
	2328		
10 F		0228	7.4E
	0551	0806	4.7F
	1023	1349	7.4E
	1649	2025	7.7F
11 Sa	0005	0307	7.3E
	0632	0847	4.5F
	1105	1430	7.0E
	1729	2105	7.4F
12 Su	0046	0350	7.1E
	0716	0933	4.4F
	1155	1517	6.6E
	1814	2150	7.0F
13 M	0129	0436	6.9E
	0804	1025	4.4F
	1254	1611	6.2E
	1905	2239	6.6F
14 Tu	0214	0526	6.8E
	0853	1123	4.5F
	1406	1716	5.8E
	2005	2334	6.1F
15 W	0302	0618	6.7E
	0941	1227	4.8F
	1525	1826	5.6E
	2115		
16 Th		0035	5.7F
	0350	0709	6.7E
	1026	1329	5.4F
	1642	1937	5.8E
	2230		
17 F		0137	5.3F
	0437	0758	6.8E
	1109	1426	6.0F
	1748	2042	6.1E
	2344		
18 Sa		0236	5.1F
	0521	0844	7.0E
	1151	1517	6.7F
	1846	2141	6.6E
19 Su	0050	0329	5.0F
	0603	0927	7.3E
	1231	1604	7.3F
	1937	2233	7.1E
20 M	0148	0418	5.0F
	0644	1010	7.7E
	1313	1649	7.8F
	2024	2321	7.5E
21 Tu	0240	0505	5.1F
	0725	1053	8.0E
	1356	1733	8.2F
	2111		
22 W		0007	7.8E
	0327	0549	5.2F
	0809	1136	8.3E
	1440	1817	8.5F
	2156		
23 Th		0053	8.0E
	0413	0635	5.3F
	0854	1223	8.4E
	1527	1903	8.6F
	2242		
24 F		0140	8.1E
	0459	0722	5.4F
	0943	1311	8.4E
	1616	1950	8.5F
	2329		
25 Sa		0228	8.0E
	0546	0811	5.4F
	1037	1403	8.2E
	1707	2040	8.2F
26 Su	0017	0318	7.8E
	0636	0905	5.4F
	1137	1500	7.8E
	1803	2132	7.7F
27 M	0107	0411	7.6E
	0729	1004	5.4F
	1246	1604	7.2E
	1904	2229	7.1F
28 Tu	0200	0508	7.3E
	0826	1111	5.4F
	1406	1716	6.7E
	2014	2334	6.4F
29 W	0256	0609	7.1E
	0926	1225	5.6F
	1536	1837	6.3E
	2135		
30 Th		0045	5.7F
	0353	0712	7.0E
	1026	1341	6.0F
	1704	1959	6.2E
	2303		

Time meridian 135° E. 0000 is midnight. 1200 is noon. Times are not adjusted for Daylight Saving Time.

Naruto, Japan, 2005

F–Flood, Dir. Northward E–Ebb, Dir. Southward

July

Day	Slack (h m)	Maximum (h m)	knots
1 F		0200	5.2F
	0451	0812	6.9E
	1123	1450	6.5F
	1819	2115	6.4E
2 Sa	0027	0310	4.8F
	0545	0908	7.0E
	1214	1548	6.9F
	1922	2219	6.8E
3 Su	0139	0408	4.7F
	0634	0957	7.1E
	1259	1637	7.3F
	2013	2312	7.1E
4 M	0236	0457	4.6F
	0717	1040	7.2E
	1340	1719	7.6F
	2056	2356	7.3E
5 Tu	0321	0538	4.6F
	0754	1118	7.3E
	1417	1756	7.8F
	2133		
6 W ●		0033	7.4E
	0358	0613	4.6F
	0828	1152	7.4E
	1450	1829	7.9F
	2206		
7 Th		0106	7.5E
	0429	0645	4.7F
	0901	1225	7.5E
	1523	1900	7.9F
	2236		
8 F		0136	7.6E
	0457	0716	4.9F
	0935	1258	7.6E
	1557	1931	7.9F
	2306		
9 Sa		0206	7.6E
	0525	0748	5.1F
	1012	1333	7.6E
	1632	2004	7.8F
	2336		
10 Su		0237	7.7E
	0555	0823	5.3F
	1053	1411	7.5E
	1710	2038	7.7F
11 M	0008	0311	7.6E
	0627	0900	5.5F
	1137	1452	7.3E
	1751	2116	7.3F
12 Tu	0041	0347	7.5E
	0702	0942	5.6F
	1228	1538	7.0E
	1837	2156	6.9F
13 W	0117	0426	7.3E
	0741	1029	5.6F
	1326	1631	6.6E
	1928	2241	6.3F
14 Th	0155	0509	7.0E
	0824	1122	5.7F
	1434	1733	6.1E
	2029	2333	5.5F
15 F ◐	0236	0558	6.8E
	0912	1224	5.8F
	1553	1846	5.7E
	2144		
16 Sa		0034	4.8F
	0321	0653	6.7E
	1005	1333	6.1F
	1714	2005	5.7E
	2313		
17 Su		0145	4.1F
	0413	0752	6.7E
	1101	1441	6.6F
	1828	2121	6.0E
18 M	0040	0258	3.9F
	0511	0852	6.9E
	1158	1543	7.1F
	1929	2225	6.6E
19 Tu	0149	0402	4.0F
	0611	0948	7.3E
	1253	1638	7.7F
	2021	2318	7.1E
20 W	0242	0457	4.5F
	0708	1042	7.7E
	1345	1727	8.1F
	2107		
21 Th ○		0005	7.6E
	0325	0545	5.0F
	0802	1132	8.2E
	1435	1813	8.4F
	2150		
22 F		0047	8.0E
	0404	0630	5.5F
	0855	1221	8.5E
	1525	1857	8.5F
	2231		
23 Sa		0129	8.2E
	0442	0715	6.0F
	0947	1310	8.6E
	1614	1941	8.5F
	2311		
24 Su		0210	8.3E
	0521	0800	6.3F
	1040	1359	8.6E
	1703	2025	8.2F
	2352		
25 M		0252	8.3E
	0602	0847	6.4F
	1135	1450	8.3E
	1754	2111	7.7F
26 Tu	0033	0337	8.0E
	0647	0938	6.5F
	1235	1546	7.7E
	1849	2200	7.0F
27 W	0116	0424	7.7E
	0735	1034	6.3F
	1343	1648	7.0E
	1950	2253	6.2F
28 Th ◑	0202	0517	7.3E
	0829	1138	6.2F
	1503	1801	6.3E
	2104	2357	5.2F
29 F	0253	0617	6.8E
	0929	1254	6.1F
	1633	1927	5.9E
	2237		
30 Sa		0117	4.3F
	0353	0725	6.5E
	1035	1415	6.3F
	1759	2055	5.9E
31 Su	0020	0243	3.8F
	0502	0834	6.4E
	1139	1526	6.7F
	1908	2209	6.4E

August

Day	Slack (h m)	Maximum (h m)	knots
1 M	0141	0356	3.8F
	0608	0936	6.5E
	1235	1623	7.0F
	2002	2305	6.8E
2 Tu	0237	0451	4.1F
	0703	1027	6.7E
	1323	1707	7.3F
	2044	2347	7.2E
3 W	0316	0531	4.4F
	0747	1108	7.0E
	1403	1743	7.6F
	2118		
4 Th		0020	7.4E
	0344	0604	4.8F
	0823	1142	7.3E
	1438	1814	7.7F
	2146		
5 F ●		0047	7.6E
	0406	0631	5.1F
	0855	1214	7.6E
	1511	1841	7.8F
	2211		
6 Sa		0111	7.8E
	0426	0657	5.5F
	0927	1244	7.8E
	1543	1909	7.9F
	2235		
7 Su		0135	7.9E
	0446	0723	5.9F
	1001	1316	8.0E
	1616	1938	7.9F
	2300		
8 M		0200	8.0E
	0510	0753	6.3F
	1037	1350	8.1E
	1652	2009	7.7F
	2326		
9 Tu		0228	8.1E
	0536	0826	6.5F
	1117	1427	8.0E
	1730	2042	7.4F
	2355		
10 W		0259	8.0E
	0607	0903	6.7F
	1202	1508	7.7E
	1812	2118	6.9F
11 Th	0025	0333	7.7E
	0641	0944	6.6F
	1254	1555	7.2E
	1859	2158	6.2F
12 F	0057	0411	7.4E
	0722	1033	6.5F
	1356	1653	6.4E
	1956	2244	5.3F
13 Sa ◐	0131	0457	6.9E
	0810	1132	6.3F
	1514	1806	5.7E
	2113	2342	4.1F
14 Su	0212	0555	6.5E
	0910	1248	6.2F
	1648	1940	5.3E
	2302		
15 M		0106	3.1F
	0310	0710	6.2E
	1022	1416	6.4F
	1817	2113	5.7E
16 Tu	0051	0245	2.9F
	0436	0832	6.4E
	1137	1533	6.9F
	1923	2222	6.4E
17 W	0155	0401	3.7F
	0604	0943	6.9E
	1245	1632	7.4F
	2013	2313	7.1E
18 Th	0236	0456	4.6F
	0712	1041	7.6E
	1343	1721	7.9F
	2056	2354	7.7E
19 F	0309	0541	5.5F
	0808	1132	8.2E
	1435	1805	8.2F
	2133		
20 Sa ○		0031	8.1E
	0341	0621	6.2F
	0859	1218	8.7E
	1523	1845	8.3F
	2209		
21 Su		0107	8.4E
	0413	0701	6.8F
	0947	1303	8.9E
	1609	1924	8.2F
	2243		
22 M		0142	8.6E
	0447	0741	7.2F
	1035	1347	8.9E
	1654	2004	8.0F
	2317		
23 Tu		0219	8.5E
	0523	0823	7.3F
	1124	1434	8.6E
	1741	2044	7.5F
	2352		
24 W		0258	8.3E
	0603	0907	7.3F
	1217	1523	8.0E
	1830	2127	6.7F
25 Th	0029	0340	7.9E
	0646	0956	7.0F
	1317	1618	7.2E
	1926	2215	5.8F
26 F	0108	0427	7.2E
	0735	1054	6.6F
	1429	1727	6.3E
	2037	2313	4.6F
27 Sa ◑	0153	0524	6.5E
	0834	1206	6.2F
	1559	1855	5.6E
	2218		
28 Su		0035	3.4F
	0255	0638	5.9E
	0945	1336	6.1F
	1731	2033	5.7E
29 M	0018	0223	3.0F
	0425	0803	5.7E
	1102	1500	6.3F
	1845	2152	6.2E
30 Tu	0135	0346	3.5F
	0554	0917	6.0E
	1211	1602	6.7F
	1938	2245	6.8E
31 W	0219	0439	4.2F
	0657	1012	6.4E
	1305	1647	7.1F
	2017	2323	7.2E

September

Day	Slack (h m)	Maximum (h m)	knots
1 Th	0249	0515	4.8F
	0741	1054	6.9E
	1347	1721	7.3F
	2048	2351	7.5E
2 F	0310	0543	5.3F
	0815	1127	7.3E
	1423	1750	7.5F
	2113		
3 Sa		0014	7.7E
	0327	0606	5.8F
	0845	1156	7.7E
	1454	1816	7.6F
	2135		
4 Su ●		0035	7.9E
	0343	0629	6.3F
	0914	1225	8.1E
	1526	1841	7.7F
	2156		
5 M		0056	8.1E
	0401	0654	6.8F
	0945	1255	8.4E
	1558	1908	7.6F
	2218		
6 Tu		0119	8.3E
	0423	0722	7.2F
	1020	1327	8.5E
	1633	1937	7.5F
	2243		
7 W		0146	8.3E
	0448	0753	7.4F
	1058	1403	8.4E
	1711	2009	7.2F
	2309		
8 Th		0215	8.3E
	0519	0829	7.5F
	1141	1443	8.0E
	1752	2044	6.6F
	2337		
9 F		0247	8.0E
	0554	0909	7.4F
	1231	1529	7.4E
	1840	2123	5.8F
10 Sa	0006	0325	7.6E
	0635	0957	7.1F
	1332	1626	6.6E
	1938	2209	4.7F
11 Su ◐	0039	0411	6.9E
	0724	1056	6.6F
	1451	1744	5.6E
	2105	2311	3.4F
12 M	0119	0514	6.2E
	0830	1219	6.2F
	1631	1930	5.2E
	2317		
13 Tu		0054	2.3F
	0233	0649	5.7E
	0957	1401	6.3F
	1803	2107	5.8E
14 W	0054	0250	2.8F
	0443	0829	6.0E
	1129	1524	6.7F
	1907	2210	6.6E
15 Th	0139	0400	4.2F
	0618	0944	6.8E
	1242	1623	7.2F
	1953	2255	7.3E
16 F	0211	0448	5.3F
	0721	1040	7.6E
	1341	1709	7.6F
	2032	2331	7.8E
17 Sa	0240	0528	6.2F
	0811	1127	8.3E
	1431	1749	7.8F
	2106		
18 Su ○		0005	8.2E
	0309	0605	7.0F
	0857	1210	8.7E
	1516	1826	7.9F
	2137		
19 M		0037	8.5E
	0339	0641	7.5F
	0941	1251	8.9E
	1559	1902	7.7F
	2207		
20 Tu		0110	8.6E
	0411	0718	7.8F
	1025	1332	8.9E
	1641	1938	7.4F
	2238		
21 W		0143	8.6E
	0445	0756	7.9F
	1110	1415	8.5E
	1725	2016	6.9F
	2309		
22 Th		0219	8.3E
	0522	0837	7.8F
	1159	1500	8.0E
	1812	2055	6.1F
	2342		
23 F		0257	7.8E
	0602	0923	7.4F
	1254	1553	7.2E
	1906	2140	5.2F
24 Sa	0018	0341	7.1E
	0647	1015	6.9F
	1359	1658	6.3E
	2018	2236	4.0F
25 Su ◑	0100	0436	6.3E
	0742	1122	6.3F
	1521	1823	5.7E
	2204		
26 M		0001	2.9F
	0205	0553	5.4E
	0853	1251	6.0F
	1650	2000	5.7E
27 Tu	0000	0159	2.8F
	0358	0729	5.1E
	1018	1421	6.0F
	1803	2116	6.3E
28 W	0103	0322	3.7F
	0539	0850	5.5E
	1137	1526	6.4F
	1855	2206	6.8E
29 Th	0139	0410	4.5F
	0642	0947	6.2E
	1236	1612	6.7F
	1934	2242	7.2E
30 F	0204	0444	5.2F
	0724	1029	6.8E
	1322	1647	6.9F
	2004	2309	7.5E

Time meridian 135° E. 0000 is midnight. 1200 is noon. Times are not adjusted for Daylight Saving Time.

Naruto, Japan, 2005

F–Flood, Dir. Northward E–Ebb, Dir. Southward

October

Day	Slack (h m)	Maximum (h m)	knots
1 Sa	0223	0510	5.8F
	0757	1102	7.3E
	1359	1717	7.1F
	2029	2331	7.8E
2 Su	0239	0533	6.4F
	0826	1132	7.8E
	1433	1743	7.2F
	2050	2352	8.0E
3 M ●	0255	0557	7.0F
	0856	1202	8.2E
	1506	1809	7.2F
	2111		
4 Tu		0014	8.2E
	0315	0623	7.5F
	0928	1233	8.5E
	1540	1837	7.2F
	2134		
5 W		0039	8.4E
	0339	0652	7.8F
	1003	1306	8.6E
	1616	1908	7.0F
	2159		
6 Th		0106	8.5E
	0408	0725	8.1F
	1042	1344	8.5E
	1656	1941	6.6F
	2226		
7 F		0137	8.4E
	0441	0802	8.1F
	1127	1426	8.1E
	1740	2017	6.0F
	2255		
8 Sa		0212	8.1E
	0518	0844	7.9F
	1218	1514	7.4E
	1831	2059	5.2F
	2328		
9 Su		0253	7.6E
	0602	0934	7.4F
	1319	1615	6.6E
	1937	2151	4.2F
10 M	0005	0344	6.8E
	0656	1037	6.8F
	1438	1738	5.8E
	2114	2305	3.0F
11 Tu ◐	0100	0459	6.0E
	0809	1204	6.3F
	1612	1919	5.6E
	2311		
12 W		0101	2.5F
	0253	0647	5.5E
	0946	1345	6.2F
	1735	2044	6.2E
13 Th	0020	0243	3.7F
	0503	0826	6.0E
	1121	1505	6.5F
	1835	2141	6.8E
14 F	0101	0344	5.0F
	0624	0937	6.9E
	1236	1603	6.9F
	1921	2224	7.4E
15 Sa	0134	0429	6.1F
	0721	1031	7.6E
	1334	1648	7.1F
	1950	2300	7.9E
16 Su	0205	0508	6.9F
	0808	1117	8.2E
	1423	1728	7.2F
	2030	2334	8.2E
17 M ○	0235	0544	7.5F
	0851	1158	8.6E
	1507	1804	7.1F
	2100		
18 Tu		0005	8.4E
	0305	0620	7.9F
	0933	1237	8.7E
	1549	1838	6.9F
	2129		
19 W		0037	8.5E
	0337	0655	8.1F
	1015	1317	8.6E
	1630	1913	6.6F
	2158		
20 Th		0110	8.4E
	0410	0732	8.2F
	1058	1358	8.2E
	1712	1949	6.1F
	2228		
21 F		0144	8.1E
	0446	0812	8.0F
	1143	1442	7.8E
	1759	2029	5.4F
	2301		
22 Sa		0222	7.7E
	0525	0855	7.6F
	1234	1533	7.1E
	1854	2114	4.6F
	2337		
23 Su		0305	7.0E
	0609	0944	7.1F
	1332	1634	6.5E
	2003	2210	3.7F
24 M	0023	0358	6.1E
	0700	1043	6.5F
	1440	1748	6.1E
	2134	2329	2.9F
25 Tu ◑	0132	0511	5.3E
	0805	1159	6.1F
	1555	1910	6.0E
	2305		
26 W		0111	3.0F
	0323	0643	5.0E
	0925	1322	5.9F
	1702	2019	6.3E
27 Th	0003	0232	3.9F
	0504	0805	5.3E
	1047	1431	6.0F
	1756	2110	6.7E
28 F	0039	0323	4.8F
	0609	0906	5.9E
	1154	1523	6.2F
	1836	2147	7.1E
29 Sa	0105	0359	5.5F
	0654	0953	6.5E
	1247	1603	6.3F
	1909	2217	7.4E
30 Su	0127	0429	6.2F
	0730	1031	7.1E
	1331	1636	6.4F
	1938	2243	7.6E
31 M	0147	0456	6.8F
	0804	1105	7.6E
	1410	1706	6.5F
	2000	2307	7.9E

November

Day	Slack (h m)	Maximum (h m)	knots
1 Tu	0209	0524	7.4F
	0837	1138	8.0E
	1447	1737	6.5F
	2024	2334	8.2E
2 W ●	0234	0554	7.9F
	0912	1213	8.3E
	1525	1808	6.4F
	2050		
3 Th		0002	8.4E
	0303	0627	8.2F
	0951	1250	8.4E
	1605	1842	6.3F
	2118		
4 F		0034	8.5E
	0336	0704	8.4F
	1033	1331	8.3E
	1648	1919	5.9F
	2150		
5 Sa		0110	8.4E
	0414	0745	8.4F
	1119	1417	7.9E
	1736	2001	5.4F
	2226		
6 Su		0151	8.1E
	0457	0831	8.1F
	1212	1509	7.4E
	1832	2049	4.8F
	2307		
7 M		0238	7.6E
	0546	0924	7.6F
	1313	1612	6.8E
	1941	2149	4.0F
8 Tu	0000	0338	6.8E
	0646	1029	7.0F
	1424	1729	6.3E
	2104	2309	3.4F
9 W ◐	0120	0500	6.0E
	0802	1148	6.5F
	1541	1851	6.2E
	2227		
10 Th		0048	3.7F
	0316	0640	5.7E
	0935	1317	6.2F
	1653	2004	6.6E
	2329		
11 F		0215	4.7F
	0501	0811	6.1E
	1106	1434	6.2F
	1752	2101	7.0E
12 Sa	0015	0317	5.7F
	0616	0920	6.8E
	1222	1534	6.3F
	1839	2147	7.4E
13 Su	0054	0405	6.6F
	0713	1017	7.4E
	1323	1623	6.4F
	1918	2227	7.8E
14 M	0129	0447	7.3F
	0802	1104	7.9E
	1415	1705	6.3F
	1952	2302	8.0E
15 Tu	0203	0525	7.7F
	0846	1147	8.1E
	1501	1742	6.1F
	2023	2335	8.1E
16 W ○	0235	0601	8.0F
	0927	1228	8.2E
	1543	1818	5.9F
	2053		
17 Th		0009	8.2E
	0309	0638	8.2F
	1008	1307	8.1E
	1624	1853	5.6F
	2123		
18 F		0042	8.1E
	0343	0714	8.2F
	1048	1347	7.9E
	1706	1930	5.3F
	2155		
19 Sa		0117	7.9E
	0419	0753	8.0F
	1131	1429	7.6E
	1751	2010	4.9F
	2230		
20 Su		0156	7.5E
	0458	0833	7.7F
	1215	1516	7.2E
	1841	2055	4.4F
	2311		
21 M		0239	7.0E
	0540	0918	7.3F
	1303	1607	6.9E
	1938	2147	4.0F
22 Tu	0002	0330	6.3E
	0628	1009	6.8F
	1356	1705	6.6E
	2042	2251	3.7F
23 W	0108	0432	5.7E
	0724	1107	6.4F
	1452	1807	6.5E
	2146		
24 Th ◑		0005	3.8F
	0236	0547	5.3E
	0831	1212	6.0F
	1548	1906	6.5E
	2240		
25 F		0118	4.3F
	0406	0705	5.3E
	0947	1319	5.7F
	1641	1959	6.6E
	2323		
26 Sa		0218	5.0F
	0519	0812	5.6E
	1100	1419	5.6F
	1726	2042	6.9E
	2357		
27 Su		0305	5.7F
	0615	0909	6.2E
	1205	1509	5.6F
	1804	2119	7.1E
28 M	0027	0343	6.4F
	0700	0956	6.7E
	1300	1552	5.6F
	1837	2153	7.4E
29 Tu	0056	0419	7.0F
	0741	1039	7.2E
	1348	1630	5.6F
	1908	2225	7.7E
30 W	0127	0455	7.6F
	0822	1119	7.6E
	1433	1708	5.6F
	1939	2258	8.0E

December

Day	Slack (h m)	Maximum (h m)	knots
1 Th	0200	0531	8.1F
	0902	1200	7.9E
	1517	1746	5.6F
	2012	2334	8.3E
2 F ●	0236	0610	8.4F
	0944	1242	8.1E
	1600	1825	5.5F
	2048		
3 Sa		0013	8.4E
	0317	0651	8.5F
	1029	1326	8.1E
	1646	1908	5.4F
	2129		
4 Su		0056	8.4E
	0400	0736	8.5F
	1116	1413	7.9E
	1734	1954	5.2F
	2214		
5 M		0143	8.1E
	0448	0824	8.3F
	1206	1505	7.6E
	1827	2047	4.9F
	2308		
6 Tu		0236	7.7E
	0542	0917	7.8F
	1300	1602	7.3E
	1925	2147	4.7F
7 W	0012	0339	7.1E
	0642	1017	7.3F
	1359	1705	7.0E
	2028	2257	4.6F
8 Th ◐	0133	0454	6.5E
	0753	1124	6.6F
	1500	1811	6.8E
	2132		
9 F		0016	4.8F
	0309	0619	6.1E
	0915	1240	6.1F
	1602	1916	6.8E
	2233		
10 Sa		0135	5.4F
	0442	0744	6.2E
	1043	1355	5.7F
	1700	2015	7.0E
	2327		
11 Su		0243	6.1F
	0600	0859	6.6E
	1204	1501	5.5F
	1752	2108	7.2E
12 M	0014	0340	6.8F
	0703	1002	7.0E
	1314	1558	5.4F
	1836	2154	7.4E
13 Tu	0057	0428	7.3F
	0756	1055	7.3E
	1412	1646	5.3F
	1916	2235	7.6E
14 W	0137	0511	7.7F
	0843	1142	7.6E
	1502	1728	5.1F
	1952	2314	7.7E
15 Th	0214	0549	7.9F
	0925	1224	7.7E
	1546	1800	5.0F
	2026	2350	7.8E
16 F ○	0250	0626	8.1F
	1004	1302	7.7E
	1625	1842	4.9F
	2100		
17 Sa		0025	7.7E
	0325	0702	8.1F
	1041	1340	7.7E
	1702	1918	4.9F
	2136		
18 Su		0101	7.7E
	0402	0738	8.0F
	1117	1417	7.6E
	1739	1956	4.8F
	2215		
19 M		0140	7.5E
	0439	0815	7.8F
	1153	1454	7.5E
	1817	2036	4.8F
	2258		
20 Tu		0220	7.2E
	0519	0853	7.5F
	1230	1534	7.3E
	1858	2120	4.8F
	2346		
21 W		0305	6.8E
	0602	0935	7.2F
	1310	1617	7.2E
	1941	2207	4.7F
22 Th	0042	0356	6.4E
	0650	1020	6.7F
	1351	1702	7.0E
	2026	2301	4.8F
23 F	0147	0454	6.0E
	0745	1109	6.2F
	1434	1751	6.8E
	2112	2359	5.0F
24 Sa ◑	0302	0600	5.6E
	0848	1205	5.6F
	1519	1840	6.7E
	2159		
25 Su		0102	5.3F
	0419	0711	5.6E
	1002	1306	5.1F
	1605	1930	6.7E
	2244		
26 M		0202	5.8F
	0530	0821	6.8E
	1121	1409	4.8F
	1651	2018	6.8E
	2327		
27 Tu		0257	6.4F
	0631	0924	6.2E
	1234	1507	4.6F
	1735	2105	7.0E
28 W	0010	0347	7.0F
	0724	1019	6.7E
	1336	1600	4.6F
	1819	2149	7.4E
29 Th	0053	0433	7.5F
	0811	1108	7.1E
	1429	1648	4.7F
	1903	2234	7.7E
30 F	0137	0517	8.0F
	0856	1153	7.5E
	1515	1733	4.9F
	1948	2318	8.1E
31 Sa ●	0222	0601	8.4F
	0940	1237	7.8E
	1557	1817	5.2F
	2035		

Time meridian 135° E. 0000 is midnight. 1200 is noon. Times are not adjusted for Daylight Saving Time.

Akashi Kaikyo, Japan, 2005

F–Flood, Dir. Westward E–Ebb, Dir. Eastward

January

Day	Slack h m	Maximum h m	knots
1 Sa	0127	0451	3.9E
	0814	1134	5.3F
	1513	1753	3.3E
	2147	2357	1.9F
2 Su	0228	0543	3.4E
	0901	1220	4.8F
	1552	1846	3.4E
	2235		
3 M		0059	2.0F
	0340	0646	2.8E
	0953	1311	3.9F
	1631	1941	3.5E
	2325		
4 Tu		0214	2.3F
	0508	0803	2.2E
	1055	1412	3.0F
◑	1712	2037	3.5E
5 W	0018	0329	2.9F
	0647	0927	1.9E
	1219	1523	2.1F
	1756	2135	3.6E
6 Th	0110	0436	3.9F
	0814	1043	2.1E
	1408	1636	1.6F
	1845	2230	3.8E
7 F	0159	0531	5.0F
	0920	1146	2.5E
	1539	1738	1.6F
	1935	2322	4.2E
8 Sa	0245	0618	5.9F
	1011	1234	2.9E
	1639	1828	1.8F
	2023		
9 Su		0009	4.7E
	0329	0701	6.7F
	1055	1314	3.3E
	1724	1912	2.1F
	2108		
10 M		0054	5.1E
	0415	0742	7.2F
	1137	1352	3.5E
●	1803	1954	2.4F
	2154		
11 Tu		0138	5.5E
	0502	0825	7.4F
	1219	1432	3.6E
	1841	2036	2.7F
	2242		
12 W		0223	5.7E
	0552	0908	7.3F
	1301	1513	3.6E
	1919	2120	2.8F
	2335		
13 Th		0309	5.6E
	0642	0953	7.0F
	1342	1559	3.6E
	2001	2208	2.9F
14 F	0035	0358	5.3E
	0732	1040	6.4F
	1423	1649	3.7E
	2046	2301	3.0F
15 Sa	0141	0452	4.6E
	0821	1129	5.7F
	1503	1743	3.8E
	2134		
16 Su		0000	3.0F
	0252	0553	3.8E
	0912	1219	4.7F
	1541	1841	3.9E
	2225		
17 M		0110	3.1F
	0413	0704	2.9E
	1008	1314	3.6F
◐	1621	1940	3.8E
	2320		
18 Tu		0226	3.3F
	0549	0826	2.2E
	1119	1420	2.4F
	1704	2041	3.7E
19 W	0018	0344	3.7F
	0728	0953	1.9E
	1308	1541	1.5F
	1756	2143	3.5E
20 Th	0116	0454	4.3F
	0848	1120	2.1E
	1506	1703	1.1F
	1901	2244	3.4E
21 F	0211	0550	5.0F
	0946	1232	2.5E
	1620	1808	1.2F
	2007	2338	3.5E
22 Sa	0258	0635	5.5F
	1029	1318	2.9E
	1706	1855	1.5F
	2100		
23 Su		0022	3.6E
	0338	0711	5.8F
	1104	1346	3.1E
	1739	1930	1.7F
	2141		
24 M		0059	3.8E
	0412	0742	5.9F
	1134	1407	3.1E
	1804	1959	1.9F
	2211		
25 Tu		0130	4.1E
	0444	0811	6.0F
	1200	1427	3.1E
○	1824	2024	2.0F
	2239		
26 W		0200	4.3E
	0516	0840	6.1F
	1226	1449	3.2E
	1844	2050	2.2F
	2307		
27 Th		0232	4.6E
	0550	0910	6.1F
	1253	1517	3.4E
	1908	2119	2.4F
	2343		
28 F		0306	4.7E
	0626	0943	6.1F
	1323	1549	3.7E
	1938	2155	2.7F
29 Sa	0026	0344	4.7E
	0705	1019	5.9F
	1353	1625	3.9E
	2013	2236	2.9F
30 Su	0117	0426	4.4E
	0747	1058	5.4F
	1423	1705	4.1E
	2051	2322	3.1F
31 M	0215	0514	3.8E
	0832	1139	4.6F
	1454	1747	4.1E
	2131		

February

Day	Slack h m	Maximum h m	knots
1 Tu		0016	3.3F
	0322	0609	3.1E
	0923	1224	3.5F
	1525	1834	3.9E
	2216		
2 W		0120	3.3F
	0441	0721	2.3E
	1027	1319	2.2F
◑	1559	1930	3.6E
	2309		
3 Th		0239	3.5F
	0617	0853	1.7E
	1205	1438	1.1F
	1641	2041	3.3E
4 F	0012	0400	4.0F
	0756	1027	1.8E
	1434	1615	0.7F
	1747	2159	3.3E
5 Sa	0124	0510	4.8F
	0910	1144	2.2E
	1556	1730	1.0F
	1911	2307	3.7E
6 Su	0232	0605	5.7F
	1004	1235	2.7E
	1640	1823	1.6F
	2023		
7 M		0002	4.4E
	0331	0651	6.4F
	1048	1310	3.1E
	1714	1906	2.3F
	2119		
8 Tu		0049	5.1E
	0421	0733	6.9F
	1127	1343	3.4E
	1745	1946	3.0F
	2208		
9 W		0133	5.6E
	0507	0813	7.1F
	1203	1417	3.7E
●	1816	2025	3.5F
	2255		
10 Th		0215	5.9E
	0550	0853	7.1F
	1237	1454	4.0E
	1848	2107	3.8F
	2345		
11 F		0258	5.7E
	0632	0933	6.7F
	1309	1534	4.2E
	1923	2151	4.0F
12 Sa	0039	0343	5.3E
	0714	1014	6.0F
	1340	1615	4.4E
	2002	2238	4.2F
13 Su	0138	0431	4.5E
	0758	1056	5.1F
	1410	1659	4.5E
	2042	2330	4.2F
14 M	0242	0526	3.7E
	0846	1140	4.0F
	1441	1747	4.3E
	2126		
15 Tu		0028	4.1F
	0353	0632	2.8E
	0941	1229	2.8F
	1513	1839	3.9E
	2214		
16 W		0134	3.9F
	0517	0753	2.1E
	1057	1330	1.5F
◐	1551	1942	3.3E
	2309		
17 Th		0251	3.8F
	0651	0925	1.8E
	1310	1504	0.6F
	1646	2058	2.8E
18 F	0015	0411	3.9F
	0815	1104	2.0E
	1517	1650	0.6F
	1827	2217	2.6E
19 Sa	0128	0520	4.3F
	0917	1223	2.5E
	1612	1802	1.1F
	2004	2323	2.9E
20 Su	0233	0612	4.8F
	1002	1306	2.9E
	1645	1847	1.7F
	2104		
21 M		0012	3.3E
	0322	0651	5.2F
	1037	1331	3.1E
	1710	1918	2.1F
	2142		
22 Tu		0048	3.8E
	0400	0723	5.6F
	1106	1347	3.3E
	1730	1942	2.4F
	2211		
23 W		0118	4.2E
	0432	0749	5.8F
	1130	1402	3.4E
	1747	2003	2.7F
	2236		
24 Th		0145	4.6E
	0502	0815	6.0F
	1153	1421	3.7E
○	1804	2027	3.1F
	2304		
25 F		0213	4.9E
	0532	0843	6.1F
	1215	1444	4.0E
	1825	2055	3.5F
	2338		
26 Sa		0246	5.0E
	0605	0914	6.0F
	1238	1513	4.4E
	1850	2130	4.0F
27 Su	0021	0323	4.9E
	0643	0948	5.6F
	1302	1547	4.7E
	1921	2209	4.3F
28 M	0110	0404	4.5E
	0726	1025	4.9F
	1328	1623	4.8E
	1956	2253	4.6F

March

Day	Slack h m	Maximum h m	knots
1 Tu	0206	0451	4.0E
	0814	1105	3.9F
	1355	1702	4.6E
	2035	2342	4.5F
2 W	0308	0544	3.2E
	0910	1149	2.7F
	1423	1746	4.2E
	2120		
3 Th		0041	4.3F
	0421	0652	2.3E
	1023	1242	1.5F
	1455	1840	3.5E
	2217		
4 F		0159	4.0F
	0552	0829	1.7E
	1230	1410	0.4F
◑	1538	2003	2.9E
	2333		
5 Sa		0332	4.0F
	0735	1022	1.8E
	1459	1609	0.4F
	1720	2144	2.9E
6 Su	0107	0453	4.5F
	0855	1153	2.3E
	1548	1727	1.1F
	1922	2302	3.5E
7 M	0229	0553	5.3F
	0949	1236	2.8E
	1621	1818	2.1F
	2039	2359	4.4E
8 Tu	0331	0640	6.1F
	1031	1300	3.3E
	1650	1858	3.0F
	2134		
9 W		0044	5.1E
	0419	0720	6.5F
	1105	1327	3.7E
	1718	1935	3.8F
	2220		
10 Th		0124	5.6E
	0500	0757	6.7F
	1133	1357	4.2E
●	1745	2012	4.5F
	2305		
11 F		0204	5.6E
	0539	0832	6.4F
	1159	1429	4.6E
	1812	2051	5.0F
	2351		
12 Sa		0243	5.4E
	0617	0908	5.9F
	1222	1503	4.9E
	1841	2131	5.2F
13 Su	0040	0325	4.8E
	0656	0945	5.0F
	1245	1539	5.0E
	1912	2214	5.3F
14 M	0133	0410	4.2E
	0739	1024	4.1F
	1310	1617	4.8E
	1948	2259	5.2F
15 Tu	0228	0501	3.4E
	0829	1105	3.0F
	1339	1657	4.4E
	2028	2348	4.9F
16 W	0328	0602	2.7E
	0929	1151	1.9F
	1411	1744	3.7E
	2113		
17 Th		0045	4.4F
	0436	0721	2.1E
	1053	1252	0.9F
	1448	1846	2.8E
	2207		
18 F		0155	3.8F
	0559	0854	1.9E
	1317	1436	0.3F
◐	1550	2018	2.2E
	2317		
19 Sa		0319	3.5F
	0725	1029	2.0E
	1459	1632	0.5F
	1810	2150	2.1E
20 Su	0044	0440	3.7F
	0833	1145	2.5E
	1540	1743	1.2F
	1954	2302	2.5E
21 M	0201	0540	4.2F
	0923	1230	2.9E
	1609	1825	1.9F
	2050	2352	3.2E
22 Tu	0256	0622	4.8F
	0959	1257	3.3E
	1631	1854	2.5F
	2128		
23 W		0028	3.8E
	0336	0653	5.2F
	1027	1314	3.6E
	1649	1917	3.0F
	2158		
24 Th		0057	4.2E
	0410	0720	5.5F
	1049	1328	3.9E
	1705	1938	3.6F
	2227		
25 F		0124	4.6E
	0440	0745	5.6F
	1108	1346	4.3E
	1721	2003	4.2F
	2258		
26 Sa		0153	4.8E
	0512	0813	5.6F
	1125	1410	4.7E
○	1739	2032	4.8F
	2335		
27 Su		0226	4.9E
	0547	0844	5.2F
	1144	1438	5.1E
	1802	2106	5.4F
28 M	0017	0304	4.8E
	0628	0918	4.6F
	1205	1511	5.3E
	1830	2145	5.7F
29 Tu	0105	0346	4.4E
	0716	0957	3.8F
	1230	1547	5.2E
	1906	2229	5.8F
30 W	0159	0433	3.9E
	0811	1038	2.9F
	1259	1628	4.8E
	1949	2317	5.5F
31 Th	0258	0526	3.2E
	0915	1125	1.9F
	1332	1714	4.1E
	2041		

Time meridian 135° E. 0000 is midnight. 1200 is noon. Times are not adjusted for Daylight Saving Time.
If three consecutive entries are marked (E) the middle one is not a true maximum but an intermediate value to show the current pattern.

Akashi Kaikyo, Japan, 2005

F–Flood, Dir. Westward E–Ebb, Dir. Eastward

April

Day	Slack h m	Maximum h m	knots
1 F		0015	4.9F
	0406	0635	2.4E
	1038	1224	0.9F
	1411	1813	3.3E
	2147		
2 Sa ◑		0132	4.3F
	0532	0817	1.9E
	1251	1403	0.3F
	1514	1948	2.7E
	2315		
3 Su		0309	4.0F
	0710	1015	2.0E
	1433	1603	0.6F
	1738	2136	2.8E
4 M	0056	0433	4.4F
	0827	1135	2.5E
	1516	1717	1.6F
	1934	2254	3.5E
5 Tu	0218	0535	5.0F
	0920	1212	3.1E
	1550	1806	2.7F
	2046	2350	4.3E
6 W	0319	0622	5.6F
	0958	1237	3.7E
	1620	1846	3.8F
	2140		
7 Th		0034	4.8E
	0407	0700	5.8F
	1027	1303	4.2E
	1647	1923	4.8F
	2227		
8 F		0113	5.0E
	0448	0735	5.7F
	1051	1330	4.7E
	1711	1958	5.5F
	2311		
9 Sa ●		0151	5.0E
	0527	0809	5.2F
	1110	1400	5.1E
	1733	2034	6.0F
	2354		
10 Su		0229	4.7E
	0605	0843	4.5F
	1127	1431	5.2E
	1756	2110	6.2F
11 M	0039	0309	4.2E
	0646	0918	3.7F
	1147	1503	5.1E
	1823	2148	6.2F
12 Tu	0124	0351	3.7E
	0732	0955	2.8F
	1213	1538	4.7E
	1857	2229	5.9F
13 W	0211	0438	3.2E
	0825	1036	2.0F
	1244	1617	4.1E
	1937	2313	5.4F
14 Th	0300	0534	2.7E
	0927	1123	1.3F
	1320	1701	3.4E
	2024		
15 F		0004	4.7F
	0356	0647	2.3E
	1046	1224	0.6F
	1407	1801	2.6E
	2120		
16 Sa ◐		0107	4.0F
	0503	0817	2.1E
	1238	1404	0.3F
	1523	1937	2.0E
	2230		
17 Su		0225	3.5F
	0622	0939	2.2E
	1405	1549	0.6F
	1733	2112	2.0E
	2353		
18 M		0345	3.5F
	0734	1048	2.5E
	1451	1702	1.2F
	1917	2223	2.3E
19 Tu	0111	0451	3.8F
	0828	1138	3.0E
	1522	1748	2.0F
	2020	2317	2.9E
20 W	0213	0538	4.3F
	0906	1209	3.4E
	1545	1821	2.8F
	2105	2358	3.4E
21 Th	0301	0614	4.6F
	0934	1230	3.9E
	1605	1847	3.6F
	2143		
22 F		0031	3.9E
	0341	0644	4.8F
	0956	1248	4.3E
	1621	1911	4.4F
	2218		
23 Sa		0102	4.2E
	0418	0713	4.7F
	1014	1310	4.8E
	1637	1938	5.3F
	2254		
24 Su ○		0134	4.4E
	0457	0744	4.5F
	1030	1336	5.2E
	1655	2010	6.0F
	2332		
25 M		0209	4.5E
	0539	0817	4.0F
	1048	1406	5.5E
	1719	2045	6.5F
26 Tu	0014	0248	4.5E
	0626	0854	3.4F
	1112	1441	5.5E
	1751	2125	6.7F
27 W	0101	0331	4.2E
	0720	0935	2.8F
	1141	1521	5.2E
	1832	2210	6.6F
28 Th	0153	0418	3.7E
	0819	1020	2.1F
	1218	1605	4.8E
	1923	2300	6.1F
29 F	0250	0513	3.1E
	0924	1112	1.4F
	1302	1658	4.1E
	2026	2359	5.3F
30 Sa	0356	0623	2.5E
	1042	1217	0.9F
	1401	1805	3.4E
	2138		

May

Day	Slack h m	Maximum h m	knots
1 Su ◑		0114	4.6F
	0512	0807	2.2E
	1218	1356	0.6F
	1539	1941	2.9E
	2302		
2 M		0242	4.2F
	0634	0942	2.4E
	1340	1542	1.2F
	1747	2119	3.0E
3 Tu	0032	0403	4.3F
	0743	1047	2.9E
	1432	1656	2.2F
	1931	2235	3.3E
4 W	0153	0506	4.5F
	0833	1130	3.5E
	1512	1748	3.4F
	2045	2335	3.8E
5 Th	0259	0555	4.7F
	0911	1203	4.1E
	1545	1830	4.6F
	2142		
6 F		0022	4.1E
	0353	0635	4.6F
	0939	1232	4.6E
	1612	1907	5.6F
	2231		
7 Sa		0103	4.2E
	0440	0712	4.2F
	1001	1302	5.0E
	1636	1942	6.3F
	2315		
8 Su ●		0141	4.1E
	0524	0747	3.7F
	1019	1331	5.2E
	1656	2016	6.6F
	2355		
9 M		0217	3.9E
	0607	0821	3.0F
	1037	1402	5.1E
	1717	2050	6.7F
10 Tu	0034	0255	3.6E
	0650	0856	2.4F
	1100	1434	4.8E
	1744	2125	6.5F
11 W	0111	0334	3.4E
	0736	0933	1.9F
	1120	1600	4.4E
	1819	2203	6.1F
12 Th	0151	0417	3.1E
	0824	1014	1.4F
	1205	1547	3.9E
	1901	2244	5.5F
13 F	0234	0506	2.8E
	0917	1059	1.0F
	1251	1633	3.3E
	1952	2332	4.9F
14 Sa	0322	0611	2.5E
	1016	1156	0.8F
	1347	1730	2.7E
	2048		
15 Su		0027	4.4F
	0416	0730	2.4E
	1128	1315	0.6F
	1501	1848	2.3E
	2150		
16 M ◐		0132	3.9F
	0518	0841	2.5E
	1243	1444	0.7F
	1636	2016	2.1E
	2257		
17 Tu		0241	3.7F
	0621	0942	2.7E
	1341	1600	1.2F
	1819	2129	2.2E
18 W	0009	0346	3.6F
	0715	1033	3.1E
	1421	1658	2.0F
	1941	2231	2.4E
19 Th	0118	0442	3.6F
	0758	1111	3.5E
	1450	1740	3.0F
	2041	2323	2.8E
20 F	0221	0528	3.7F
	0830	1140	4.0E
	1515	1813	4.1F
	2129		
21 Sa		0005	3.2E
	0316	0607	3.6F
	0856	1206	4.5E
	1536	1844	5.1F
	2211		
22 Su		0043	3.6E
	0407	0643	3.5F
	0917	1235	4.9E
	1557	1916	6.1F
	2250		
23 M		0119	3.9E
	0455	0719	3.2F
	0938	1306	5.3E
	1620	1950	6.8F
	2329		
24 Tu ○		0155	4.1E
	0544	0757	2.9F
	1003	1341	5.5E
	1650	2028	7.2F
25 W	0011	0235	4.1E
	0634	0837	2.6F
	1034	1421	5.4E
	1730	2110	7.2F
26 Th	0058	0318	3.9E
	0726	0921	2.2F
	1113	1505	5.2E
	1820	2156	6.9F
27 F	0148	0406	3.5E
	0820	1008	1.9F
	1200	1554	4.9E
	1919	2248	6.4F
28 Sa	0243	0501	3.1E
	0917	1102	1.6F
	1300	1650	4.4E
	2023	2346	5.7F
29 Su	0341	0610	2.8E
	1019	1207	1.4F
	1416	1758	3.8E
	2129		
30 M ◑		0053	5.0F
	0443	0736	2.7E
	1129	1335	1.4F
	1551	1922	3.2E
	2240		
31 Tu		0206	4.4F
	0546	0852	2.9E
	1239	1508	1.9F
	1739	2051	2.9E
	2356		

June

Day	Slack h m	Maximum h m	knots
1 W		0320	4.0F
	0643	0953	3.3E
	1339	1625	2.8F
	1921	2209	2.9E
2 Th	0119	0427	3.6F
	0733	1043	3.8E
	1426	1724	4.0F
	2041	2317	3.0E
3 F	0239	0524	3.4F
	0814	1125	4.2E
	1505	1811	5.1F
	2143		
4 Sa		0013	3.2E
	0347	0612	3.1F
	0847	1201	4.6E
	1537	1851	6.0F
	2233		
5 Su		0058	3.4E
	0444	0653	2.7F
	0915	1235	4.8E
	1604	1926	6.5F
	2315		
6 M		0136	3.4E
	0534	0731	2.4F
	0940	1308	4.8E
	1627	1959	6.7F
	2351		
7 Tu ●		0210	3.4E
	0617	0807	2.0F
	1005	1341	4.6E
	1652	2032	6.7F
8 W	0024	0244	3.3E
	0656	0842	1.7F
	1034	1414	4.4E
	1722	2106	6.4F
9 Th	0057	0318	3.1E
	0732	0917	1.5F
	1108	1450	4.1E
	1759	2142	6.0F
10 F	0131	0356	3.0E
	0808	0955	1.4F
	1149	1529	3.9E
	1843	2221	5.7F
11 Sa	0209	0439	2.9E
	0847	1036	1.3F
	1237	1613	3.6E
	1932	2304	5.3F
12 Su	0250	0529	2.8E
	0931	1123	1.2F
	1332	1702	3.3E
	2022	2350	4.9F
13 M	0334	0629	2.8E
	1021	1221	1.2F
	1435	1759	2.9E
	2112		
14 Tu		0041	4.5F
	0419	0731	2.9E
	1117	1332	1.2F
	1549	1909	2.4E
	2206		
15 W ◐		0136	4.0F
	0506	0827	3.1E
	1213	1447	1.0F
	1720	2026	2.1E
	2307		
16 Th		0236	3.4F
	0552	0918	3.3E
	1303	1557	2.3F
	1857	2141	2.0E
17 F	0021	0341	2.9F
	0637	1005	3.5E
	1345	1653	3.3F
	2016	2248	2.2E
18 Sa	0147	0442	2.5F
	0717	1048	3.9E
	1421	1739	4.4F
	2115	2343	2.6E
19 Su	0307	0535	2.3F
	0754	1128	4.3E
	1453	1819	5.5F
	2203		
20 M		0029	3.1E
	0413	0621	2.2F
	0828	1207	4.7E
	1526	1857	6.4F
	2245		
21 Tu		0107	3.4E
	0505	0703	2.2F
	0902	1246	5.1E
	1601	1935	7.0F
	2326		
22 W ○		0145	3.6E
	0551	0744	2.3F
	0938	1327	5.3E
	1641	2016	7.3F
23 Th	0008	0224	3.7E
	0634	0826	2.3F
	1020	1411	5.5E
	1729	2059	7.3F
24 F	0053	0306	3.6E
	0717	0910	2.4F
	1108	1457	5.5E
	1823	2145	7.1F
25 Sa	0139	0353	3.5E
	0802	0957	2.3F
	1204	1546	5.2E
	1919	2235	6.8F
26 Su	0227	0445	3.4E
	0850	1050	2.3F
	1309	1641	4.8E
	2015	2327	6.0F
27 M	0315	0544	3.3E
	0942	1150	2.3F
	1423	1742	4.1E
	2111		
28 Tu		0023	5.3F
	0401	0650	3.4E
	1038	1304	2.3F
	1547	1855	3.4E
	2210		
29 W ◑		0123	4.3F
	0447	0754	3.5E
	1138	1426	2.7F
	1724	2018	2.7E
	2318		
30 Th		0230	3.4F
	0535	0855	3.7E
	1238	1546	3.4F
	1907	2141	2.3E

Time meridian 135° E. 0000 is midnight. 1200 is noon. Times are not adjusted for Daylight Saving Time.
If three consecutive entries are marked (E) the middle one is not a true maximum but an intermediate value to show the current pattern.

Akashi Kaikyo, Japan, 2005

F–Flood, Dir. Westward   E–Ebb, Dir. Eastward

July

Day	Slack h m	Maximum h m	knots
1 F	0047	0343	2.5F
	0624	0953	3.8E
	1333	1655	4.3F
	2034	2302	2.3E
2 Sa	0232	0454	2.0F
	0715	1047	3.9E
	1422	1749	5.2F
	2138		
3 Su		0011	2.6E
	0356	0555	1.8F
	0804	1135	4.1E
	1504	1834	5.9F
	2227		
4 M		0102	2.9E
	0455	0644	1.7F
	0849	1217	4.1E
	1540	1911	6.3F
	2307		
5 Tu		0138	3.1E
	0540	0725	1.7F
	0928	1255	4.2E
	1612	1945	6.4F
	2341		
6 W		0207	3.1E
	0615	0800	1.8F
	1002	1330	4.2E
●	1643	2017	6.3F
7 Th	0011	0234	3.1E
	0643	0832	1.8F
	1034	1404	4.2E
	1716	2049	6.2F
8 F	0039	0302	3.1E
	0709	0903	1.8F
	1107	1438	4.2E
	1751	2122	6.0F
9 Sa	0109	0333	3.1E
	0735	0934	1.8F
	1144	1513	4.2E
	1830	2157	5.9F
10 Su	0141	0407	3.2E
	0806	1009	1.9F
	1227	1552	4.1E
	1911	2234	5.7F
11 M	0214	0447	3.3E
	0842	1049	2.0F
	1317	1634	3.9E
	1953	2312	5.4F
12 Tu	0248	0529	3.5E
	0922	1136	2.1F
	1413	1721	3.5E
	2037	2354	4.8F
13 W	0322	0615	3.6E
	1005	1232	2.2F
	1519	1816	2.9E
	2124		
14 Th		0039	4.0F
	0357	0705	3.6E
	1051	1338	2.4F
	1639	1927	2.2E
	2221		
15 F		0132	3.0F
	0433	0759	3.5E
	1141	1453	2.8F
◐	1815	2053	1.8E
	2339		
16 Sa		0240	2.0F
	0514	0858	3.4E
	1234	1605	3.5F
	1949	2217	1.8E
17 Su	0135	0402	1.3F
	0603	1000	3.5E
	1329	1707	4.5F
	2101	2329	2.2E
18 M	0323	0514	1.2F
	0702	1059	3.8E
	1422	1758	5.4F
	2154		
19 Tu		0021	2.7E
	0425	0609	1.5F
	0759	1150	4.3E
	1511	1843	6.3F
	2239		
20 W		0100	3.1E
	0508	0654	1.9F
	0851	1236	4.9E
	1600	1924	6.9F
	2320		
21 Th		0136	3.4E
	0543	0735	2.4F
	0939	1320	5.4E
○	1647	2006	7.2F
22 F	0000	0213	3.5E
	0617	0816	2.7F
	1026	1404	5.7E
	1735	2047	7.3F
23 Sa	0040	0252	3.7E
	0653	0858	3.0F
	1116	1448	5.8E
	1823	2131	7.1F
24 Su	0119	0334	3.8E
	0732	0943	3.2F
	1212	1535	5.6E
	1910	2215	6.7F
25 M	0157	0420	3.9E
	0814	1033	3.4F
	1314	1625	5.0E
	1958	2301	6.0F
26 Tu	0233	0509	4.0E
	0858	1127	3.4F
	1423	1721	4.2E
	2047	2349	5.0F
27 W	0309	0600	4.1E
	0946	1229	3.5F
	1538	1826	3.3E
	2141		
28 Th		0040	3.8F
	0345	0656	4.0E
	1036	1341	3.6F
◑	1705	1946	2.4E
	2248		
29 F		0140	2.5F
	0423	0757	3.7E
	1133	1500	3.8F
	1845	2115	1.9E
30 Sa	0031	0300	1.4F
	0512	0904	3.4E
	1235	1617	4.2F
	2015	2250	2.0E
31 Su	0245	0434	0.9F
	0622	1013	3.3E
	1338	1724	4.8F
	2122		

August

Day	Slack h m	Maximum h m	knots
1 M		0017	2.5E
	0407	0549	1.1F
	0744	1117	3.3E
	1437	1816	5.4F
	2211		
2 Tu		0112	2.9E
	0453	0642	1.5F
	0851	1208	3.6E
	1526	1857	5.7F
	2250		
3 W		0141	3.1E
	0527	0721	1.9F
	0938	1249	3.9E
	1605	1931	5.9F
	2322		
4 Th		0200	3.2E
	0553	0752	2.1F
	1012	1323	4.1E
	1639	2002	6.0F
	2349		
5 F		0219	3.2E
	0614	0818	2.3F
	1040	1353	4.3E
●	1709	2030	6.0F
6 Sa	0014	0240	3.3E
	0633	0843	2.4F
	1107	1423	4.5E
	1739	2059	6.0F
7 Su	0038	0304	3.5E
	0653	0910	2.6F
	1139	1454	4.6E
	1811	2128	6.0F
8 M	0103	0332	3.7E
	0719	0941	2.9F
	1218	1529	4.6E
	1847	2200	5.8F
9 Tu	0129	0405	4.0E
	0750	1018	3.1F
	1305	1607	4.3E
	1925	2235	5.3F
10 W	0156	0440	4.2E
	0824	1100	3.4F
	1359	1651	3.9E
	2008	2313	4.6F
11 Th	0224	0518	4.3E
	0901	1148	3.5F
	1459	1742	3.2E
	2056	2354	3.6F
12 F	0252	0600	4.1E
	0942	1245	3.6F
	1610	1845	2.4E
	2155		
13 Sa		0042	2.4F
	0322	0650	3.7E
	1031	1357	3.5F
◐	1738	2012	1.8E
	2323		
14 Su		0149	1.2F
	0358	0757	3.2E
	1133	1522	3.7F
	1920	2154	1.6E
15 M	0202	0335	0.5F
	0457	0924	3.0E
	1249	1641	4.3F
	2044	2324	2.0E
16 Tu	0339	0505	0.7F
	0635	1043	3.3E
	1406	1742	5.1F
	2143		
17 W		0022	2.5E
	0421	0603	1.4F
	0801	1143	4.0E
	1511	1832	6.0F
	2229		
18 Th		0055	3.0E
	0453	0647	2.2F
	0902	1231	4.9E
	1603	1914	6.6F
	2307		
19 F		0125	3.3E
	0522	0726	2.9F
	0951	1314	5.5E
	1649	1953	7.0F
	2342		
20 Sa		0158	3.7E
	0551	0805	3.6F
	1037	1355	5.9E
○	1731	2032	7.1F
21 Su	0013	0232	4.1E
	0621	0845	4.1F
	1125	1437	5.9E
	1812	2111	6.8F
22 M	0043	0309	4.4E
	0654	0927	4.4F
	1218	1520	5.5E
	1854	2150	6.2F
23 Tu	0112	0348	4.6E
	0730	1012	4.6F
	1315	1607	4.9E
	1938	2231	5.3F
24 W	0141	0429	4.7E
	0809	1101	4.7F
	1417	1659	4.0E
	2026	2314	4.2F
25 Th	0210	0513	4.6E
	0851	1155	4.6F
	1523	1759	3.1E
	2122		
26 F		0001	2.9F
	0241	0602	4.1E
	0937	1256	4.3F
	1639	1916	2.3E
	2235		
27 Sa		0058	1.6F
	0317	0702	3.5E
	1030	1410	4.0F
◑	1809	2050	1.9E
28 Su	0037	0227	0.7F
	0407	0820	2.8E
	1137	1534	3.9F
	1940	2232	2.0E
29 M	0256	0422	0.5F
	0552	0948	2.6E
	1257	1653	4.2F
	2051		
30 Tu		0003	2.5E
	0354	0545	1.1F
	0747	1103	2.8E
	1413	1753	4.7F
	2142		
31 W		0054	3.0E
	0429	0635	1.8F
	0855	1159	3.3E
	1510	1837	5.2F
	2222		

September

Day	Slack h m	Maximum h m	knots
1 Th		0121	3.3E
	0456	0709	2.3F
	0939	1239	3.8E
	1552	1911	5.5F
	2253		
2 F		0139	3.4E
	0518	0735	2.7F
	1010	1311	4.2E
	1624	1940	5.7F
	2317		
3 Sa		0154	3.5E
	0536	0758	3.0F
	1036	1337	4.5E
	1652	2005	5.8F
	2337		
4 Su		0210	3.7E
	0551	0820	3.3F
	1102	1404	4.7E
●	1719	2029	5.8F
	2355		
5 M		0230	4.1E
	0608	0845	3.7F
	1134	1433	4.7E
	1749	2057	5.7F
6 Tu	0015	0255	4.4E
	0629	0915	4.1F
	1211	1506	4.7E
	1824	2127	5.3F
7 W	0035	0325	4.7E
	0656	0950	4.5F
	1257	1545	4.4E
	1905	2202	4.7F
8 Th	0059	0358	4.9E
	0728	1030	4.8F
	1347	1629	4.0E
	1952	2239	3.8F
9 F	0124	0435	4.7E
	0804	1116	4.8F
	1444	1718	3.4E
	2046	2321	2.8F
10 Sa	0152	0516	4.3E
	0847	1209	4.6F
	1549	1818	2.6E
	2153		
11 Su		0010	1.6F
	0222	0605	3.7E
	0941	1317	4.1F
◐	1710	1944	1.9E
	2338		
12 M		0121	0.6F
	0258	0717	2.9E
	1053	1448	3.8F
	1852	2140	1.7E
13 Tu	0225	0325	0.2F
	0423	0906	2.7E
	1228	1619	4.1F
	2022	2322	2.0E
14 W	0323	0458	0.8F
	0643	1035	3.2E
	1358	1727	4.9F
	2123		
15 Th		0015	2.6E
	0357	0555	1.8F
	0812	1136	4.0E
	1505	1817	5.7F
	2206		
16 F		0041	3.1E
	0426	0636	2.8F
	0911	1223	4.9E
	1556	1858	6.3F
	2240		
17 Sa		0106	3.7E
	0453	0714	3.8F
	1000	1304	5.4E
	1640	1935	6.5F
	2309		
18 Su		0135	4.2E
	0520	0751	4.6F
	1046	1343	5.6E
○	1719	2010	6.3F
	2333		
19 M		0206	4.7E
	0545	0829	5.2F
	1132	1423	5.5E
	1758	2046	5.8F
	2355		
20 Tu		0239	5.0E
	0612	0908	5.6F
	1221	1504	5.0E
	1839	2123	5.1F
21 W	0017	0314	5.2E
	0643	0950	5.8F
	1312	1549	4.4E
	1924	2201	4.1F
22 Th	0042	0351	5.0E
	0718	1034	5.7F
	1406	1638	3.7E
	2015	2243	3.0F
23 F	0110	0431	4.6E
	0758	1122	5.4F
	1504	1735	3.0E
	2116	2329	2.0F
24 Sa	0143	0517	3.9E
	0843	1216	4.8F
	1607	1848	2.4E
	2235		
25 Su		0029	1.0F
	0223	0616	3.1E
	0937	1323	4.2F
◑	1723	2022	2.1E
26 M	0038	0203	0.4F
	0325	0744	2.3E
	1047	1446	3.7F
	1850	2157	2.1E
27 Tu	0228	0402	0.5F
	0541	0923	2.1E
	1215	1611	3.7F
	2005	2318	2.5E
28 W	0317	0523	1.3F
	0737	1041	2.5E
	1339	1718	4.1F
	2100		
29 Th		0010	3.0E
	0351	0611	2.0F
	0841	1137	3.1E
	1440	1806	4.6F
	2140		
30 F		0041	3.4E
	0417	0645	2.7F
	0923	1217	3.6E
	1523	1841	5.0F
	2209		

Time meridian 135° E. 0000 is midnight. 1200 is noon. Times are not adjusted for Daylight Saving Time.
If three consecutive entries are marked (E) the middle one is not a true maximum but an intermediate value to show the current pattern.

Akashi Kaikyo, Japan, 2005

F–Flood, Dir. Westward E–Ebb, Dir. Eastward

October

Day	Slack h m	Maximum h m	knots
1 Sa	0437 0956 1558 2231	0102 0711 1249 1908	3.7E 3.2F 4.0E 5.2F
2 Su	0453 1026 1628 2248	0117 0732 1316 1932	4.0E 3.7F 4.3E 5.3F
3 M ●	0507 1056 1658 2303	0133 0754 1343 1957	4.3E 4.3F 4.5E 5.1F
4 Tu	0522 1129 1731 2319	0153 0819 1413 2025	4.7E 4.9F 4.6E 4.8F
5 W	0541 1207 1810 2337	0219 0850 1447 2058	5.0E 5.4F 4.5E 4.3F
6 Th	0606 1250 1857	0249 0925 1527 2134	5.2E 5.8F 4.3E 3.6F
7 F	0001 0639 1339 1951	0323 1006 1611 2214	5.2E 6.0F 4.0E 2.8F
8 Sa	0029 0721 1433 2052	0402 1052 1700 2259	4.9E 5.8F 3.4E 2.0F
9 Su	0103 0811 1535 2205	0447 1146 1801 2353	4.3E 5.2F 2.7E 1.1F
10 M	0142 0915 1652 2348	0542 1254 1928	3.6E 4.5F 2.1E
11 Tu ◐	0239 1036 1824	0112 0703 1424 2125	0.5F 2.9E 4.0F 1.9E
12 W	0147 0440 1212 1948	0315 0854 1554 2252	0.4F 2.7E 4.1F 2.3E
13 Th	0243 0650 1341 2047	0442 1020 1703 2341	1.3F 3.2E 4.7F 2.9E
14 F	0321 0814 1448 2128	0539 1122 1754	2.4F 3.9E 5.2F
15 Sa	0353 0915 1541 2159	0011 0622 1211 1835	3.5E 3.6F 4.5E 5.5F
16 Su	0421 1006 1627 2223	0039 0700 1252 1912	4.2E 4.7F 4.8E 5.4F
17 M ○	0446 1052 1709 2243	0107 0736 1331 1947	4.7E 5.6F 4.9E 5.0F
18 Tu	0508 1137 1751 2301	0137 0812 1410 2022	5.2E 6.3F 4.7E 4.4F
19 W	0531 1222 1835 2322	0209 0849 1450 2058	5.4E 6.6F 4.3E 3.6F
20 Th	0559 1306 1923 2348	0242 0928 1532 2137	5.3E 6.6F 3.9E 2.8F
21 F	0632 1353 2016	0318 1009 1618 2218	4.9E 6.3F 3.4E 2.0F
22 Sa	0021 0714 1441 2116	0357 1053 1711 2306	4.4E 5.8F 2.9E 1.3F
23 Su	0101 0802 1535 2227	0443 1143 1820	3.6E 5.1F 2.5E
24 M	0153 0859 1638 2359	0007 0542 1244 1948	0.8F 2.8E 4.3F 2.3E
25 Tu ◑	0311 1007 1751	0136 0710 1357 2109	0.5F 2.2E 3.8F 2.4E
26 W	0128 0509 1125 1903	0318 0845 1516 2219	0.7F 2.1E 3.6F 2.6E
27 Th	0224 0658 1244 2000	0438 1000 1625 2313	1.3F 2.3E 3.7F 3.0E
28 F	0302 0809 1350 2041	0533 1059 1718 2350	2.1F 2.7E 4.0F 3.5E
29 Sa	0329 0900 1443 2112	0611 1146 1758	2.8F 3.1E 4.3F
30 Su	0351 0940 1526 2134	0015 0639 1222 1829	3.9E 3.6F 3.5E 4.3F
31 M	0407 1016 1605 2152	0035 0703 1253 1858	4.3E 4.4F 3.8E 4.3F

November

Day	Slack h m	Maximum h m	knots
1 Tu	0422 1049 1644 2207	0054 0728 1323 1927	4.7E 5.2F 4.0E 4.0F
2 W ●	0438 1124 1725 2224	0118 0755 1356 1959	5.0E 5.9F 4.2E 3.6F
3 Th	0459 1202 1811 2246	0147 0828 1431 2034	5.3E 6.5F 4.2E 3.2F
4 F	0529 1245 1903 2315	0220 0905 1511 2114	5.4E 6.7F 4.1E 2.7F
5 Sa	0608 1332 1958 2352	0259 0948 1556 2157	5.2E 6.7F 3.8E 2.1F
6 Su	0659 1426 2058	0342 1036 1647 2246	4.9E 6.3F 3.3E 1.6F
7 M	0037 0759 1526 2204	0433 1131 1749 2344	4.4E 5.6F 2.8E 1.1F
8 Tu	0135 0908 1635 2323	0534 1238 1915	3.7E 4.9F 2.4E
9 W ◐	0257 1024 1749	0105 0655 1358 2052	0.8F 3.2E 4.4F 2.4E
10 Th	0047 0452 1147 1859	0250 0833 1519 2202	1.0F 2.9E 4.2F 2.7E
11 F	0151 0644 1310 1954	0415 0956 1628 2253	1.9F 3.1E 4.2F 3.3E
12 Sa	0237 0811 1423 2035	0516 1103 1723 2332	3.1F 3.4E 4.3F 3.9E
13 Su	0314 0917 1526 2107	0604 1157 1809	4.3F 3.8E 4.2F
14 M	0345 1011 1621 2132	0006 0644 1242 1849	4.5E 5.5F 4.0E 3.9F
15 Tu	0411 1057 1710 2153	0038 0721 1323 1926	5.0E 6.3F 4.0E 3.5F
16 W ○	0434 1139 1757 2215	0110 0756 1400 2003	5.2E 6.9F 3.9E 2.9F
17 Th	0458 1218 1842 2240	0142 0832 1438 2040	5.2E 7.0F 3.7E 2.4F
18 F	0526 1257 1928 2311	0217 0908 1517 2119	5.0E 6.8F 3.4E 1.9F
19 Sa	0602 1337 2014 2351	0253 0947 1600 2200	4.6E 6.4F 3.2E 1.5F
20 Su	0646 1419 2103	0334 1029 1648 2247	4.1E 5.9F 2.9E 1.2F
21 M	0040 0737 1505 2157	0421 1116 1749 2342	3.5E 5.3F 2.7E 1.0F
22 Tu	0139 0832 1556 2259	0516 1209 1904	3.0E 4.7F 2.6E
23 W	0250 0929 1651	0054 0627 1308 2013	0.8F 2.5E 4.2F 2.7E
24 Th ◑	0008 0417 1030 1748	0217 0749 1411 2113	0.9F 2.2E 3.8F 2.9E
25 F	0110 0558 1136 1841	0334 0904 1516 2205	1.3F 2.1E 3.5F 3.2E
26 Sa	0156 0727 1247 1926	0438 1010 1615 2248	2.0F 2.2E 3.4F 3.5E
27 Su	0230 0833 1357 2002	0527 1108 1706 2321	2.9F 2.4E 3.3F 3.9E
28 M	0257 0924 1500 2030	0603 1155 1749 2350	3.9F 2.8E 3.1F 4.3E
29 Tu	0319 1005 1556 2054	0633 1234 1827	4.9F 3.2E 3.0F
30 W	0341 1042 1646 2116	0018 0703 1308 1903	4.7E 5.8F 3.5E 2.8F

December

Day	Slack h m	Maximum h m	knots
1 Th	0404 1119 1733 2141	0049 0735 1342 1940	5.0E 6.5F 3.8E 2.6F
2 F ●	0434 1158 1819 2213	0123 0811 1418 2018	5.2E 7.0F 3.9E 2.4F
3 Sa	0512 1240 1906 2252	0202 0851 1459 2100	5.3E 7.1F 3.8E 2.2F
4 Su	0600 1328 1955 2339	0245 0935 1544 2146	5.3E 7.0F 3.6E 2.0F
5 M	0657 1419 2046	0332 1024 1635 2236	5.1E 6.6F 3.3E 1.8F
6 Tu	0035 0758 1512 2141	0425 1118 1734 2334	4.7E 6.0F 3.0E 1.6F
7 W	0145 0900 1608 2242	0525 1218 1847	4.2E 5.4F 2.9E
8 Th ◐	0309 1004 1703 2348	0047 0638 1324 2002	1.5F 3.6E 4.7F 3.0E
9 F	0447 1113 1758	0216 0803 1435 2106	1.8F 3.0E 4.0F 3.3E
10 Sa	0052 0633 1233 1848	0340 0927 1545 2202	2.5F 2.7E 3.5F 3.6E
11 Su	0146 0805 1402 1934	0448 1042 1049 2251	3.6F 2.7E 3.0F 4.1E
12 M	0231 0917 1523 2013	0543 1147 1745 2334	4.8F 2.9E 2.7F 4.4E
13 Tu	0308 1012 1629 2048	0627 1239 1832	5.8F 3.2E 2.5F
14 W	0341 1057 1722 2119	0013 0706 1320 1914	4.7E 6.5F 3.3E 2.2F
15 Th	0410 1136 1807 2150	0049 0742 1356 1953	4.7E 6.9F 3.4E 2.0F
16 F ○	0440 1211 1846 2223	0126 0817 1430 2029	4.7E 6.9F 3.3E 1.8F
17 Sa	0512 1245 1921 2300	0202 0852 1505 2106	4.5E 6.6F 3.2E 1.7F
18 Su	0550 1319 1955 2343	0240 0929 1542 2145	4.3E 6.3F 3.1E 1.6F
19 M	0633 1355 2031	0319 1008 1623 2225	4.1E 5.9F 3.0E 1.5F
20 Tu	0031 0718 1434 2112	0402 1049 1711 2311	3.8E 5.5F 3.0E 1.5F
21 W	0125 0805 1513 2158	0448 1133 1805	3.5E 5.1F 3.0E
22 Th	0224 0852 1553 2247	0005 0541 1218 1902	1.4F 3.0E 4.7F 3.1E
23 F	0333 0940 1634 2340	0108 0644 1307 1956	1.5F 2.5E 4.1F 3.2E
24 Sa ◑	0458 1035 1716	0219 0758 1401 2047	1.7F 2.1E 3.4F 3.3E
25 Su	0031 0636 1146 1759	0330 0915 1505 2137	2.2F 1.8E 2.7F 3.4E
26 M	0117 0803 1318 1842	0432 1029 1613 2224	3.0F 1.9E 2.1F 3.6E
27 Tu	0157 0905 1453 1925	0523 1131 1714 2308	4.0F 2.2E 1.9F 3.9E
28 W	0234 0953 1604 2005	0604 1219 1805 2349	5.1F 2.7E 1.8F 4.3E
29 Th	0309 1034 1655 2043	0641 1257 1847	6.0F 3.1E 1.9F
30 F	0347 1112 1737 2122	0029 0719 1332 1927	4.7E 6.6F 3.4E 2.1F
31 Sa ●	0428 1152 1815 2204	0111 0758 1408 2007	5.1E 7.1F 3.6E 2.3F

Time meridian 135° E. 0000 is midnight. 1200 is noon. Times are not adjusted for Daylight Saving Time.
If three consecutive entries are marked (E) the middle one is not a true maximum but an intermediate value to show the current pattern.

Kurushima Kaikyo, Japan, 2005

F–Flood, Dir. Southward E–Ebb, Dir. Northward

January

Day	Slack h m	Maximum h m	knots	Day	Slack h m	Maximum h m	knots
1 Sa	0203 0827 1501 2124	0519 1143 1810	4.6E 6.0F 4.6E	16 Su	0308 0931 1547 2201	0007 0626 1232 1858	6.1F 6.3E 6.5F 5.7E
2 Su	0249 0909 1540 2208	0005 0603 1223 1853	3.9F 4.3E 5.5F 4.5E	17 M ◐	0406 1020 1626 2252	0056 0719 1317 1945	5.6F 5.1E 5.2F 5.2E
3 M	0344 0959 1626 2259	0052 0656 1311 1944	3.9F 3.9E 5.0F 4.5E	18 Tu	0515 1116 1710 2350	0151 0819 1407 2039	5.0F 4.0E 3.9F 4.6E
4 Tu ◑	0453 1059 1720 2356	0149 0802 1408 2042	3.9F 3.6E 4.3F 4.4E	19 W	0637 1223 1803	0259 0930 1510 2141	4.4F 3.0E 2.7F 4.2E
5 W	0613 1213 1821	0257 0917 1517 2146	4.1F 3.4E 3.8F 4.5E	20 Th	0057 0808 1347 1907	0420 1052 1629 2251	4.2F 2.5E 2.0F 4.0E
6 Th	0059 0737 1335 1925	0413 1035 1634 2250	4.6F 3.6E 3.5F 4.9E	21 F	0209 0928 1515 2018	0542 1213 1751	4.4F 2.6E 2.0F
7 F	0202 0851 1455 2028	0525 1148 1746 2351	5.5F 4.2E 3.7F 5.5E	22 Sa	0316 1029 1623 2123	0000 0648 1320 1855	4.1E 5.0F 3.1E 2.4F
8 Sa	0302 0954 1603 2125	0629 1253 1847	6.7F 5.0E 4.2F	23 Su	0412 1115 1711 2217	0100 0738 1410 1945	4.5E 5.6F 3.7E 3.0F
9 Su	0358 1049 1659 2217	0048 0723 1350 1941	6.3E 7.8F 5.8E 4.8F	24 M	0458 1150 1750 2302	0150 0817 1450 2025	4.9E 6.1F 4.2E 3.5F
10 M ●	0450 1139 1748 2307	0141 0813 1441 2029	7.0E 8.8F 6.4E 5.4F	25 Tu ○	0535 1220 1823 2340	0231 0852 1523 2100	5.2E 6.4F 4.7E 4.0F
11 Tu	0539 1226 1833 2354	0231 0859 1527 2114	7.7E 9.4F 6.9E 5.9F	26 W	0609 1246 1853	0305 0922 1551 2132	5.4E 6.7F 5.1E 4.5F
12 W	0627 1309 1914	0318 0943 1611 2157	8.1E 9.7F 7.1E 6.3F	27 Th	0014 0639 1310 1921	0335 0950 1617 2202	5.6E 6.9F 5.4E 4.8F
13 Th	0041 0713 1350 1954	0405 1026 1652 2240	8.2E 9.4F 7.0E 6.5F	28 F	0047 0708 1334 1948	0404 1019 1641 2233	5.7E 7.0F 5.6E 5.2F
14 F	0128 0759 1429 2035	0451 1108 1733 2323	7.9E 8.8F 6.7E 6.4F	29 Sa	0119 0738 1401 2017	0433 1048 1708 2305	5.7E 7.0F 5.7E 5.4F
15 Sa	0216 0844 1508 2117	0537 1149 1815	7.2E 7.8F 6.3E	30 Su	0154 0811 1430 2049	0505 1120 1738 2340	5.7E 6.7F 5.7E 5.5F
				31 M	0234 0848 1502 2125	0542 1156 1812	5.4E 6.3F 5.6E

February

Day	Slack h m	Maximum h m	knots	Day	Slack h m	Maximum h m	knots
1 Tu	0320 0932 1540 2208	0020 0626 1236 1854	5.4F 4.9E 5.5F 5.3E	16 W ◐	0434 1026 1602 2255	0106 0731 1312 1937	5.5F 3.6E 3.6F 4.5E
2 W ◑	0417 1025 1623 2300	0108 0721 1324 1945	5.2F 4.2E 4.5F 4.8E	17 Th	0549 1123 1644	0202 0837 1401 2038	4.3F 2.3E 2.2F 3.6E
3 Th	0531 1133 1719	0207 0832 1425 2050	4.8F 3.5E 3.4F 4.4E	18 F	0002 0736 1256 1754	0322 1011 1524 2205	3.4F 1.5E 1.2F 3.0E
4 F	0005 0702 1302 1831	0324 0959 1547 2207	4.7F 3.2E 2.7F 4.4E	19 Sa	0129 0916 1504 1946	0513 1158 1730 2340	3.3F 1.8E 1.1F 3.1E
5 Sa	0121 0832 1438 1953	0453 1129 1719 2326	5.1F 3.5E 2.7F 4.8E	20 Su	0254 1017 1617 2117	0635 1311 1850	4.0F 2.7E 2.0F
6 Su	0237 0945 1554 2106	0611 1244 1834	6.2F 4.4E 3.5F	21 M	0357 1056 1700 2216	0052 0725 1356 1938	3.7E 4.8F 3.6E 3.0F
7 M	0345 1042 1650 2208	0035 0713 1343 1931	5.8E 7.4F 5.5E 4.6F	22 Tu	0443 1126 1732 2259	0141 0802 1430 2014	4.5E 5.6F 4.5E 4.0F
8 Tu	0442 1130 1735 2301	0134 0804 1431 2019	6.8E 8.5F 6.4E 5.7F	23 W	0520 1152 1801 2334	0220 0833 1459 2045	5.1E 6.3F 5.3E 4.8F
9 W ●	0533 1211 1815 2349	0225 0849 1513 2102	7.8E 9.2F 7.1E 6.7F	24 Th ○	0551 1216 1827	0251 0901 1524 2114	5.7E 6.9F 5.9E 5.5F
10 Th	0619 1249 1852	0311 0930 1552 2142	8.5E 9.5F 7.5E 7.4F	25 F	0005 0621 1240 1852	0320 0928 1547 2142	6.1E 7.3F 6.3E 6.1F
11 F	0035 0702 1324 1928	0355 1008 1628 2222	8.7E 9.3F 7.6E 7.8F	26 Sa	0035 0650 1304 1917	0347 0956 1611 2211	6.4E 7.5F 6.6E 6.6F
12 Sa	0119 0743 1357 2004	0436 1045 1703 2300	8.4E 8.6F 7.4E 7.7F	27 Su	0105 0721 1329 1944	0415 1024 1636 2241	6.6E 7.4F 6.7E 6.9F
13 Su	0203 0823 1428 2041	0517 1121 1738 2340	7.6E 7.6F 7.0E 7.3F	28 M	0138 0753 1356 2013	0446 1055 1704 2314	6.5E 7.1F 6.6E 6.9F
14 M	0248 0903 1458 2120	0558 1156 1813	6.5E 6.4F 6.3E				
15 Tu	0337 0942 1529 2204	0021 0641 1232 1852	6.5F 5.1E 5.0F 5.5E				

March

Day	Slack h m	Maximum h m	knots	Day	Slack h m	Maximum h m	knots
1 Tu	0216 0829 1426 2047	0521 1128 1735 2351	6.2E 6.4F 6.3E 6.7F	16 W	0307 0905 1437 2120	0605 1152 1802	4.7E 4.6F 5.5E
2 W	0259 0911 1459 2127	0601 1205 1812	5.5E 5.4F 5.8E	17 Th	0355 0942 1506 2205	0026 0647 1225 1839	5.6F 3.3E 3.3F 4.3E
3 Th	0353 1000 1538 2217	0035 0653 1249 1900	6.1F 4.5E 4.2F 5.0E	18 F ◐	0459 1032 1540 2306	0113 0745 1307 1934	4.2F 1.9E 2.0F 3.1E
4 F ◑	0504 1107 1629 2325	0131 0802 1348 2007	5.4F 3.4E 2.9F 4.2E	19 Sa	0647 1211 1644	0221 0928 1423 2118	3.0F 1.2E 0.8F 2.2E
5 Sa	0641 1245 1750	0249 0939 1517 2139	4.7F 2.8E 2.0F 3.9E	20 Su	0039 0842 1449 1930	0429 1131 1714 2318	2.6F 1.6E 0.8F 2.3E
6 Su	0053 0822 1432 1936	0432 1119 1707 2314	4.8F 3.2E 2.2F 4.4E	21 M	0223 0940 1553 2111	0607 1241 1835	3.3F 2.7E 2.0F
7 M	0223 0935 1544 2101	0600 1234 1826	5.8F 4.3E 3.4F	22 Tu	0330 1016 1629 2204	0031 0656 1322 1917	3.2E 4.3F 3.8E 3.3F
8 Tu	0336 1027 1633 2204	0029 0702 1328 1920	5.6E 7.0F 5.5E 5.0F	23 W	0416 1045 1659 2242	0118 0731 1354 1949	4.2E 5.3F 4.9E 4.6F
9 W	0434 1110 1713 2255	0127 0750 1411 2005	6.9E 8.1F 6.6E 6.4F	24 Th	0453 1112 1726 2315	0154 0802 1421 2019	5.1E 6.2F 5.8E 5.6F
10 Th ●	0522 1147 1749 2341	0215 0832 1449 2045	8.0E 8.8F 7.4E 7.6F	25 F	0525 1138 1752 2346	0226 0831 1447 2047	6.0E 6.9F 6.5E 6.6F
11 F	0606 1220 1824	0259 0909 1524 2123	8.6E 8.9F 7.9E 8.4F	26 Sa ○	0557 1203 1817	0255 0859 1512 2116	6.6E 7.3F 7.0E 7.3F
12 Sa	0024 0646 1250 1858	0338 0944 1557 2159	8.7E 8.7F 8.0E 8.7F	27 Su	0016 0629 1229 1843	0324 0929 1537 2145	7.0E 7.5F 7.2E 7.8F
13 Su	0105 0723 1318 1931	0416 1017 1628 2235	8.3E 8.0F 7.8E 8.6F	28 M	0048 0701 1255 1911	0354 0959 1603 2216	7.2E 7.3F 7.3E 8.1F
14 M	0145 0758 1345 2005	0452 1049 1659 2310	7.4E 7.0F 7.3E 7.9F	29 Tu	0122 0736 1323 1941	0427 1030 1632 2250	7.0E 6.8F 7.1E 8.0F
15 Tu	0225 0832 1411 2041	0528 1120 1729 2347	6.1E 5.9F 6.5E 6.9F	30 W	0201 0814 1352 2017	0503 1104 1704 2329	6.4E 6.0F 6.7E 7.5F
				31 Th	0245 0856 1425 2058	0545 1142 1742	5.5E 4.9F 5.9E

Time meridian 135° E. 0000 is midnight. 1200 is noon. Times are not adjusted for Daylight Saving Time.

Kurushima Kaikyo, Japan, 2005

F–Flood, Dir. Southward   E–Ebb, Dir. Northward

April

Day	Slack h m	Maximum h m	knots
1 F		0014	6.7F
	0340	0638	4.4E
	0948	1227	3.7F
	1505	1832	5.0E
	2151		
2 Sa ◑		0111	5.7F
	0452	0752	3.3E
	1059	1329	2.5F
	1601	1945	4.0E
	2303		
3 Su		0231	4.8F
	0630	0931	2.8E
	1240	1505	1.8F
	1737	2129	3.7E
4 M	0038	0417	4.7F
	0805	1105	3.4E
	1417	1657	2.5F
	1934	2305	4.3E
5 Tu	0211	0543	5.5F
	0911	1212	4.5E
	1519	1811	4.0F
	2056		
6 W		0017	5.6E
	0323	0642	6.5F
	0959	1302	5.7E
	1605	1902	5.7F
	2156		
7 Th		0112	6.8E
	0419	0728	7.3F
	1039	1343	6.7E
	1643	1945	7.2F
	2245		
8 F		0159	7.7E
	0506	0808	7.8F
	1113	1419	7.4E
	1719	2024	8.3F
	2329		
9 Sa ●		0241	8.1E
	0548	0844	7.8F
	1144	1453	7.9E
	1754	2100	8.9F
10 Su	0011	0319	8.0E
	0625	0917	7.5F
	1213	1524	7.9E
	1828	2135	9.0F
11 M	0049	0354	7.5E
	0700	0948	6.9F
	1239	1554	7.7E
	1901	2210	0.7F
12 Tu	0127	0429	6.5E
	0733	1018	6.1F
	1305	1623	7.1E
	1934	2243	7.9F
13 W	0204	0501	5.4E
	0805	1048	5.1F
	1330	1652	6.3E
	2008	2317	6.8F
14 Th	0242	0535	4.2E
	0836	1118	4.0F
	1357	1721	5.3E
	2044	2353	5.5F
15 F	0323	0613	3.1E
	0913	1153	3.0F
	1426	1755	4.1E
	2124		

Day	Slack h m	Maximum h m	knots
16 Sa ◐		0035	4.3F
	0416	0707	2.0E
	1005	1236	1.9F
	1503	1845	2.8E
	2217		
17 Su		0133	3.1F
	0540	0843	1.5E
	1142	1352	0.9F
	1613	2030	1.9E
	2340		
18 M		0314	2.5F
	0726	1034	1.8E
	1353	1626	1.0F
	1904	2234	1.9E
19 Tu	0124	0505	2.9F
	0833	1145	2.8E
	1502	1753	2.2F
	2041	2349	2.8E
20 W	0241	0605	3.8F
	0917	1229	4.0E
	1541	1837	3.6F
	2133		
21 Th		0038	3.9E
	0334	0647	4.9F
	0952	1305	5.1E
	1613	1912	5.0F
	2212		
22 F		0118	5.0E
	0416	0722	5.8F
	1024	1336	6.0E
	1643	1944	6.2F
	2247		
23 Sa		0153	5.9E
	0454	0756	6.5F
	1055	1406	6.8E
	1712	2016	7.3F
	2321		
24 Su ○		0227	6.7E
	0531	0829	6.9F
	1124	1435	7.3E
	1741	2048	8.1F
	2356		
25 M		0300	7.1E
	0608	0901	7.0F
	1153	1504	7.5E
	1811	2121	8.7F
26 Tu	0032	0335	7.2E
	0645	0935	6.7F
	1223	1535	7.5E
	1843	2156	8.8F
27 W	0110	0412	6.9E
	0723	1010	6.2F
	1254	1608	7.3E
	1918	2233	8.6F
28 Th	0152	0453	6.3E
	0805	1047	5.4F
	1327	1644	6.8E
	1958	2315	8.0F
29 F	0240	0540	5.4E
	0851	1129	4.5F
	1405	1728	6.0E
	2044		
30 Sa		0003	7.1F
	0336	0636	4.4E
	0946	1218	3.5F
	1452	1824	5.0F
	2141		

May

Day	Slack h m	Maximum h m	knots
1 Su ◑		0102	6.0F
	0446	0750	3.6E
	1057	1324	2.6F
	1559	1943	4.1E
	2254		
2 M		0218	5.0F
	0611	0915	3.4E
	1222	1456	2.4F
	1741	2119	3.9E
3 Tu	0023	0352	4.7F
	0730	1034	3.9E
	1341	1634	3.2F
	1925	2247	4.4E
4 W	0150	0512	5.1F
	0832	1136	4.7E
	1441	1744	4.6F
	2043	2356	5.4E
5 Th	0302	0612	5.7F
	0920	1226	5.7E
	1529	1837	6.1F
	2142		
6 F		0052	6.2E
	0359	0659	6.1F
	0959	1309	6.5E
	1611	1922	7.4F
	2232		
7 Sa		0139	6.8E
	0446	0740	6.4F
	1035	1347	7.1E
	1650	2002	8.2F
	2317		
8 Su ●		0222	7.0E
	0528	0817	6.3F
	1106	1422	7.4E
	1727	2039	8.7F
	2359		
9 M		0301	6.8E
	0606	0851	6.0F
	1136	1455	7.4E
	1802	2115	8.6F
10 Tu	0037	0337	6.2E
	0641	0923	5.6F
	1205	1526	7.1E
	1836	2149	8.1F
11 W	0114	0411	5.5E
	0713	0954	5.0F
	1232	1556	6.5E
	1910	2222	7.4F
12 Th	0148	0444	4.7E
	0746	1025	4.3F
	1300	1625	5.8E
	1943	2255	6.5F
13 F	0223	0517	3.9E
	0820	1058	3.6F
	1331	1656	4.9E
	2017	2330	5.5F
14 Sa	0300	0556	3.2E
	0901	1135	2.8F
	1406	1732	3.9E
	2055		
15 Su		0011	4.6F
	0344	0647	2.6E
	0955	1223	2.1F
	1452	1822	2.9E
	2141		

Day	Slack h m	Maximum h m	knots
16 M ◐		0100	3.7F
	0441	0758	2.3E
	1110	1331	1.6F
	1606	1946	2.1E
	2246		
17 Tu		0210	3.1F
	0555	0920	2.5E
	1236	1510	1.6F
	1807	2126	2.0E
18 W	0010	0338	3.1F
	0707	1029	3.2E
	1347	1642	2.5F
	1945	2246	2.6E
19 Th	0133	0454	3.6F
	0805	1124	4.1E
	1439	1742	3.7F
	2047	2346	3.6E
20 F	0240	0551	4.3F
	0852	1208	5.0E
	1520	1827	5.1F
	2135		
21 Sa		0035	4.6E
	0335	0638	5.1F
	0933	1248	5.9E
	1557	1907	6.4F
	2217		
22 Su		0119	5.6E
	0423	0719	5.7F
	1011	1325	6.6E
	1633	1945	7.5F
	2258		
23 M		0200	6.3E
	0507	0759	6.0F
	1047	1401	7.2E
	1709	2023	8.5F
	2339		
24 Tu ○		0241	6.8E
	0550	0838	6.1F
	1122	1437	7.5E
	1745	2102	9.0F
25 W	0020	0323	6.9E
	0633	0917	6.0F
	1158	1515	7.6E
	1824	2142	9.2F
26 Th	0104	0405	6.7E
	0716	0956	5.7F
	1235	1554	7.4E
	1905	2223	9.0F
27 F	0149	0450	6.3E
	0800	1038	5.2F
	1315	1637	6.9E
	1949	2308	8.4F
28 Sa	0238	0539	5.6E
	0848	1123	4.6F
	1400	1727	6.3E
	2039	2357	7.5F
29 Su	0331	0634	5.0E
	0941	1215	4.0F
	1454	1826	5.5E
	2136		
30 M ◑		0053	6.5F
	0430	0736	4.4E
	1040	1317	3.6F
	1604	1937	4.7E
	2243		
31 Tu		0158	5.5F
	0535	0843	4.2E
	1146	1432	3.5F
	1732	2057	4.3E
	2359		

June

Day	Slack h m	Maximum h m	knots
1 W		0312	4.8F
	0640	0949	4.4E
	1253	1555	4.0F
	1904	2216	4.4E
2 Th	0118	0428	4.5F
	0739	1050	4.8E
	1355	1708	4.9F
	2022	2326	4.7E
3 F	0231	0532	4.5F
	0830	1144	5.4E
	1450	1808	6.0F
	2126		
4 Sa		0027	5.1E
	0334	0626	4.6F
	0915	1233	6.0E
	1539	1858	6.9F
	2220		
5 Su		0120	5.4E
	0426	0712	4.7F
	0955	1317	6.4E
	1624	1942	7.5F
	2308		
6 M		0207	5.5E
	0511	0753	4.7F
	1032	1357	6.6E
	1705	2023	7.8F
	2352		
7 Tu ●		0249	5.4E
	0552	0830	4.6F
	1107	1434	6.6E
	1744	2100	7.7F
8 W	0030	0328	5.1E
	0629	0905	4.4F
	1141	1509	6.3E
	1821	2135	7.4F
9 Th	0106	0403	4.8E
	0704	0939	4.2F
	1214	1542	5.9E
	1855	2209	6.9F
10 F	0138	0436	4.4E
	0738	1013	3.9F
	1247	1613	5.3E
	1928	2241	6.3F
11 Sa	0209	0509	4.1E
	0813	1048	3.6F
	1322	1040	4.7E
	2000	2314	5.8F
12 Su	0240	0544	3.8E
	0852	1126	3.2F
	1401	1723	4.1E
	2035	2351	5.2F
13 M	0315	0624	3.6E
	0936	1210	2.9F
	1448	1808	3.5E
	2115		
14 Tu		0032	4.7F
	0356	0713	3.5E
	1028	1303	2.7F
	1549	1906	3.0E
	2205		
15 W ◐		0123	4.1F
	0447	0810	3.6E
	1127	1407	2.7F
	1707	2019	2.7E
	2309		

Day	Slack h m	Maximum h m	knots
16 Th		0225	3.7F
	0545	0912	3.8E
	1229	1522	3.0F
	1834	2136	2.8E
17 F	0024	0337	3.6F
	0647	1012	4.2E
	1328	1635	3.8F
	1951	2248	3.3E
18 Sa	0142	0449	3.7F
	0747	1109	4.8E
	1422	1738	5.0F
	2055	2352	4.1E
19 Su	0253	0552	4.1F
	0841	1200	5.5E
	1512	1831	6.2F
	2150		
20 M		0049	4.9E
	0355	0646	4.6F
	0929	1249	6.2E
	1559	1919	7.4F
	2240		
21 Tu		0141	5.7E
	0449	0735	5.0F
	1015	1335	6.8E
	1644	2005	8.4F
	2328		
22 W ○		0229	6.3E
	0539	0821	5.4F
	1059	1420	7.3E
	1729	2049	9.1F
23 Th	0014	0316	6.7E
	0624	0905	5.6F
	1142	1505	7.6E
	1813	2133	9.4F
24 F	0059	0401	6.8E
	0709	0947	5.7F
	1226	1549	7.7E
	1859	2217	9.4F
25 Sa	0144	0446	6.6E
	0752	1031	5.7F
	1311	1636	7.5E
	1946	2301	8.9F
26 Su	0228	0531	6.3E
	0835	1116	5.5F
	1400	1726	7.0E
	2035	2347	8.1F
27 M	0313	0617	5.9E
	0921	1204	5.3F
	1454	1819	6.3E
	2126		
28 Tu		0035	7.0F
	0359	0707	5.4E
	1009	1257	5.0F
	1557	1918	5.5E
	2222		
29 W ◑		0127	5.8F
	0447	0800	5.0E
	1103	1358	4.7F
	1709	2024	4.6E
	2325		
30 Th		0225	4.7F
	0539	0857	4.8E
	1202	1508	4.6F
	1832	2136	4.0E

Time meridian 135° E. 0000 is midnight. 1200 is noon. Times are not adjusted for Daylight Saving Time.

Kurushima Kaikyo, Japan, 2005

F–Flood, Dir. Southward E–Ebb, Dir. Northward

July

Day	Slack h m	Maximum h m	knots
1 F	0035	0332	3.7F
	0635	0958	4.7E
	1306	1625	4.8F
	1956	2251	3.7E
2 Sa	0153	0444	3.2F
	0732	1059	4.8E
	1411	1737	5.3F
	2111		
3 Su		0003	3.7E
	0308	0552	3.1F
	0829	1159	5.1E
	1511	1839	6.0F
	2213		
4 M		0107	4.0E
	0411	0649	3.2F
	0922	1254	5.4E
	1605	1930	6.5F
	2305		
5 Tu		0200	4.3E
	0503	0738	3.5F
	1010	1343	5.6E
	1653	2014	6.8F
	2349		
6 W ●		0246	4.5E
	0547	0821	3.7F
	1054	1426	5.8E
	1735	2053	6.9F
7 Th	0026	0324	4.7E
	0624	0859	3.9F
	1134	1505	5.7E
	1812	2127	6.9F
8 F	0057	0358	4.7E
	0658	0933	4.1F
	1211	1538	5.6E
	1845	2158	6.7F
9 Sa	0124	0428	4.8E
	0729	1006	4.2F
	1246	1609	5.3E
	1916	2227	6.5F
10 Su	0149	0455	4.8E
	0800	1039	4.2F
	1320	1639	5.0E
	1945	2257	6.3F
11 M	0215	0523	4.8E
	0831	1112	4.3F
	1356	1711	4.8E
	2016	2328	6.0F
12 Tu	0243	0553	4.8E
	0905	1148	4.2F
	1436	1747	4.4E
	2051		
13 W		0003	5.6F
	0316	0629	4.7E
	0943	1229	4.1F
	1523	1831	4.0E
	2133		
14 Th		0043	5.1F
	0354	0711	4.6E
	1028	1318	4.0F
	1620	1926	3.6E
	2224		
15 F ◐		0132	4.4F
	0440	0802	4.4E
	1119	1417	4.0F
	1732	2034	3.2E
	2329		
16 Sa		0232	3.7F
	0535	0903	4.3E
	1219	1529	4.1F
	1857	2154	3.1E
17 Su	0050	0347	3.1F
	0641	1011	4.4E
	1326	1649	4.7F
	2020	2315	3.5E
18 M	0219	0510	3.1F
	0751	1119	4.8E
	1432	1801	5.8F
	2130		
19 Tu		0028	4.2E
	0337	0621	3.6F
	0856	1223	5.6E
	1533	1901	7.0F
	2229		
20 W		0129	5.2E
	0439	0719	4.3F
	0954	1319	6.4E
	1628	1953	8.2F
	2320		
21 Th ○		0221	6.1E
	0529	0810	5.1F
	1047	1411	7.3E
	1719	2040	9.1F
22 F	0005	0307	6.7E
	0613	0855	5.8F
	1135	1459	7.9E
	1807	2124	9.6F
23 Sa	0048	0350	7.2E
	0654	0938	6.4F
	1222	1545	8.3E
	1853	2206	9.6F
24 Su	0128	0430	7.3E
	0733	1019	6.8F
	1308	1630	8.3E
	1938	2247	9.2F
25 M	0206	0510	7.2E
	0811	1101	7.0F
	1355	1715	7.8E
	2022	2327	8.4F
26 Tu	0243	0549	6.8E
	0850	1144	6.8F
	1444	1801	7.0E
	2107		
27 W		0007	7.2F
	0319	0629	6.3E
	0932	1229	6.4F
	1537	1850	5.9E
	2154		
28 Th ◑		0049	5.9F
	0356	0713	5.7E
	1019	1319	5.7F
	1639	1945	4.6E
	2245		
29 F		0135	4.4F
	0436	0802	5.0E
	1112	1419	4.9F
	1755	2052	3.4E
	2347		
30 Sa		0231	3.1F
	0524	0902	4.4E
	1217	1536	4.4F
	1928	2214	2.6E
31 Su	0109	0347	2.1F
	0627	1015	4.0E
	1333	1708	4.3F
	2100	2344	2.5E

August

Day	Slack h m	Maximum h m	knots
1 M	0248	0521	1.8F
	0746	1133	4.0E
	1450	1827	4.8F
	2211		
2 Tu		0102	3.0E
	0407	0638	2.2F
	0903	1243	4.4E
	1554	1924	5.5F
	2302		
3 W		0158	3.7E
	0500	0734	2.9F
	1005	1339	4.9E
	1646	2008	6.0F
	2341		
4 Th		0239	4.3E
	0540	0817	3.6F
	1055	1423	5.3E
	1727	2043	6.5F
5 F ●	0011	0313	4.8E
	0613	0852	4.2F
	1135	1500	5.6E
	1802	2114	6.7F
6 Sa	0036	0341	5.2E
	0642	0923	4.7F
	1210	1531	5.7E
	1832	2141	6.8F
7 Su	0059	0406	5.5E
	0709	0953	5.1F
	1241	1558	5.7E
	1900	2207	6.9F
8 M	0121	0429	5.7E
	0734	1021	5.4F
	1311	1624	5.7E
	1927	2234	6.9F
9 Tu	0143	0452	5.8E
	0800	1050	5.6F
	1342	1651	5.6E
	1956	2302	6.7F
10 W	0208	0517	5.9E
	0828	1121	5.7F
	1416	1722	5.4E
	2028	2333	6.3F
11 Th	0237	0547	5.7E
	0900	1156	5.6F
	1456	1759	5.0E
	2105		
12 F		0009	5.6F
	0309	0622	5.4E
	0937	1238	5.3F
	1545	1846	4.3E
	2151		
13 Sa ◐		0050	4.6F
	0347	0705	4.9E
	1023	1329	4.9F
	1649	1948	3.5E
	2250		
14 Su		0142	3.5F
	0435	0804	4.3E
	1123	1438	4.5F
	1816	2112	2.9E
15 M	0013	0257	2.5F
	0541	0922	4.0E
	1239	1610	4.5F
	1956	2251	3.0E
16 Tu	0159	0438	2.2F
	0711	1050	4.2E
	1403	1740	5.4F
	2118		
17 W		0017	3.8E
	0328	0607	2.9F
	0837	1208	5.1E
	1517	1849	6.7F
	2218		
18 Th		0119	5.0E
	0428	0709	4.1F
	0945	1311	6.3E
	1618	1942	8.0F
	2307		
19 F		0209	6.1E
	0514	0759	5.5F
	1041	1404	7.5E
	1711	2028	9.0F
	2349		
20 Sa ○		0251	7.1E
	0553	0842	6.7F
	1129	1451	8.4E
	1758	2109	9.5F
21 Su	0027	0329	7.7E
	0630	0922	7.6F
	1215	1534	8.9E
	1841	2147	9.5F
22 M	0101	0405	7.9E
	0705	1001	8.2F
	1259	1615	8.8E
	1923	2224	9.1F
23 Tu	0134	0440	7.9E
	0741	1039	8.4F
	1342	1655	8.2E
	2003	2300	8.2F
24 W	0205	0514	7.5E
	0817	1118	8.1F
	1426	1736	7.2E
	2042	2335	7.0F
25 Th	0235	0549	6.9E
	0855	1158	7.3F
	1514	1818	5.8E
	2122		
26 F		0011	5.6F
	0305	0626	6.0E
	0937	1241	6.3F
	1607	1906	4.3E
	2205		
27 Sa ◑		0049	4.1F
	0338	0709	5.0E
	1026	1333	5.0F
	1716	2007	2.8E
	2258		
28 Su		0135	2.6F
	0417	0806	3.9E
	1129	1445	3.8F
	1857	2137	1.8E
29 M	0023	0247	1.4F
	0518	0932	3.1E
	1254	1638	3.3F
	2049	2330	1.8E
30 Tu	0239	0502	1.0F
	0717	1116	3.0E
	1430	1816	3.9F
	2159		
31 W		0053	2.6E
	0402	0636	1.9F
	0902	1236	3.6E
	1542	1912	4.7F
	2242		

September

Day	Slack h m	Maximum h m	knots
1 Th		0142	3.6E
	0446	0727	3.0F
	1007	1331	4.4E
	1633	1951	5.5F
	2314		
2 F		0217	4.5E
	0519	0805	4.0F
	1052	1411	5.1E
	1711	2022	6.1F
	2340		
3 Sa		0246	5.3E
	0547	0835	4.9F
	1128	1443	5.7E
	1743	2050	6.6F
4 Su ●	0003	0311	5.9E
	0613	0903	5.6F
	1158	1511	6.0E
	1811	2115	7.0F
5 M	0024	0333	6.3E
	0637	0929	6.2F
	1226	1536	6.3E
	1838	2141	7.1F
6 Tu	0046	0355	6.5E
	0701	0956	6.6F
	1253	1601	6.4E
	1905	2207	7.1F
7 W	0109	0417	6.6E
	0725	1023	6.9F
	1323	1628	6.3E
	1935	2235	6.9F
8 Th	0133	0441	6.6E
	0752	1053	6.9F
	1356	1659	6.1E
	2007	2305	6.3F
9 F	0200	0509	6.4E
	0822	1127	6.8F
	1435	1735	5.5E
	2044	2339	5.5F
10 Sa	0230	0542	5.9E
	0858	1207	6.3F
	1522	1819	4.6E
	2129		
11 Su ◐		0019	4.4F
	0305	0623	5.2E
	0943	1257	5.5F
	1624	1920	3.5E
	2228		
12 M		0110	3.1F
	0350	0721	4.3E
	1043	1405	4.7F
	1753	2050	2.7E
	2356		
13 Tu		0227	2.0F
	0459	0850	3.6E
	1207	1544	4.4F
	1940	2239	2.8E
14 W	0152	0423	1.9F
	0652	1035	3.9E
	1344	1725	5.2F
	2102		
15 Th		0003	3.9E
	0315	0556	3.0F
	0831	1159	5.0E
	1505	1834	6.5F
	2159		
16 F		0101	5.2E
	0407	0656	4.6F
	0939	1301	6.4E
	1607	1925	7.7F
	2244		
17 Sa		0146	6.4E
	0448	0742	6.2F
	1033	1352	7.7E
	1659	2008	8.5F
	2322		
18 Su ○		0225	7.3E
	0525	0823	7.6F
	1120	1436	8.6E
	1743	2047	8.9F
	2356		
19 M		0300	8.0E
	0600	0901	8.7F
	1203	1517	8.9E
	1825	2123	8.8F
20 Tu	0027	0334	8.3E
	0635	0938	9.2F
	1245	1555	8.7E
	1903	2157	8.3F
21 W	0057	0406	8.2E
	0709	1015	9.2F
	1326	1633	7.9E
	1940	2230	7.4F
22 Th	0125	0438	7.8E
	0744	1051	8.6F
	1407	1711	6.8E
	2016	2303	6.3F
23 F	0153	0510	7.0E
	0821	1129	7.7F
	1450	1749	5.4E
	2052	2335	5.0F
24 Sa	0220	0544	6.0E
	0900	1208	6.4F
	1538	1832	3.8E
	2130		
25 Su ◑		0011	3.6F
	0250	0622	4.8E
	0945	1254	4.9F
	1639	1929	2.4E
	2220		
26 M		0053	2.3F
	0325	0715	3.5E
	1044	1358	3.5F
	1816	2102	1.5E
	2349		
27 Tu		0203	1.1F
	0426	0853	2.4E
	1211	1553	2.7F
	2012	2302	1.6E
28 W	0218	0445	0.8F
	0707	1055	2.3E
	1359	1746	3.2F
	2119		
29 Th		0020	2.6E
	0333	0619	2.0F
	0857	1215	3.1E
	1515	1841	4.1F
	2159		
30 F		0105	3.7E
	0412	0704	3.3F
	0954	1306	4.0E
	1604	1917	5.0F
	2229		

Time meridian 135° E. 0000 is midnight. 1200 is noon. Times are not adjusted for Daylight Saving Time.

Kurushima Kaikyo, Japan, 2005

F–Flood, Dir. Southward E–Ebb, Dir. Northward

October

Day	Slack h m	Maximum h m	knots
1 Sa		0138	4.8E
	0443	0737	4.5F
	1034	1343	4.9E
	1641	1948	5.7F
	2255		
2 Su		0206	5.6E
	0510	0806	5.5F
	1106	1414	5.6E
	1713	2016	6.3F
	2320		
3 M ●		0231	6.3E
	0535	0834	6.4F
	1135	1443	6.2E
	1744	2044	6.8F
	2344		
4 Tu		0255	6.7E
	0600	0901	7.1F
	1204	1510	6.5E
	1813	2111	7.0F
5 W	0008	0318	7.0E
	0625	0929	7.6F
	1233	1538	6.7E
	1844	2140	6.9F
6 Th	0033	0343	7.1E
	0651	0958	7.9F
	1305	1608	6.6E
	1916	2210	6.5F
7 F	0059	0409	7.0E
	0720	1030	7.9F
	1341	1641	6.3E
	1952	2242	5.9F
8 Sa	0128	0439	6.6E
	0752	1106	7.6F
	1422	1720	5.5E
	2031	2317	5.0F
9 Su	0159	0514	6.0E
	0831	1147	6.9F
	1511	1807	4.6E
	2119		
10 M		0000	3.9F
	0236	0559	5.2E
	0919	1239	6.0F
	1614	1913	3.5E
	2223		
11 Tu ◐		0055	2.8F
	0326	0703	4.2E
	1023	1349	5.0F
	1741	2045	2.9E
	2353		
12 W		0218	1.9F
	0449	0841	3.6E
	1152	1526	4.5F
	1919	2222	3.2E
13 Th	0133	0411	2.2F
	0649	1024	3.9E
	1329	1702	5.1F
	2034	2336	4.2E
14 F	0245	0536	3.6F
	0823	1143	5.1E
	1448	1810	6.1F
	2127		
15 Sa		0030	5.4E
	0335	0634	5.3F
	0928	1243	6.4E
	1550	1900	7.0F
	2210		
16 Su		0114	6.5E
	0416	0719	6.9F
	1020	1333	7.4E
	1641	1943	7.6F
	2247		
17 M ○		0153	7.4E
	0454	0800	8.2F
	1107	1417	8.1E
	1725	2021	7.7F
	2319		
18 Tu		0229	7.9E
	0530	0839	9.1F
	1150	1458	8.2E
	1805	2056	7.6F
	2350		
19 W		0302	8.2E
	0605	0916	9.4F
	1231	1536	7.8E
	1843	2130	7.1F
20 Th	0019	0335	8.0E
	0641	0952	9.2F
	1311	1613	7.0E
	1918	2202	6.3F
21 F	0047	0406	7.5E
	0717	1028	8.5F
	1351	1649	5.9E
	1953	2235	5.4F
22 Sa	0116	0438	6.7E
	0753	1104	7.5F
	1431	1726	4.7E
	2028	2308	4.3F
23 Su	0145	0511	5.7E
	0831	1142	6.2F
	1514	1808	3.5E
	2108	2344	3.2F
24 M	0217	0549	4.5E
	0913	1225	4.8F
	1607	1902	2.5E
	2201		
25 Tu ◑		0029	2.1F
	0259	0641	3.2E
	1006	1321	3.5F
	1721	2026	1.8E
	2324		
26 W		0140	1.2F
	0410	0815	2.1E
	1121	1447	2.7F
	1854	2204	2.0E
27 Th	0117	0351	1.1F
	0641	1009	2.0E
	1259	1635	2.8F
	2005	2317	2.8E
28 F	0234	0531	2.1F
	0824	1129	2.6E
	1420	1743	3.5F
	2053		
29 Sa		0007	3.8E
	0320	0620	3.4F
	0921	1222	3.5E
	1516	1828	4.3F
	2130		
30 Su		0045	4.8E
	0354	0657	4.7F
	1001	1303	4.4E
	1600	1904	5.1F
	2203		
31 M		0117	5.6E
	0425	0729	5.8F
	1036	1338	5.3E
	1638	1938	5.8F
	2233		

November

Day	Slack h m	Maximum h m	knots
1 Tu		0147	6.3E
	0454	0800	6.8F
	1109	1411	6.0E
	1714	2010	6.2F
	2303		
2 W ●		0216	6.8E
	0523	0831	7.6F
	1141	1444	6.5E
	1750	2042	6.4F
	2332		
3 Th		0245	7.1E
	0553	0904	8.2F
	1216	1518	6.7E
	1826	2116	6.3F
4 F	0001	0314	7.2E
	0623	0937	8.5F
	1253	1554	6.6E
	1904	2150	5.9F
5 Sa	0032	0346	7.1E
	0657	1013	8.4F
	1333	1633	6.2E
	1944	2226	5.3F
6 Su	0105	0422	6.7E
	0735	1053	8.1F
	1417	1717	5.5E
	2029	2307	4.6F
7 M	0142	0503	6.1E
	0819	1139	7.3F
	1509	1809	4.7E
	2120	2354	3.7F
8 Tu	0227	0555	5.3E
	0911	1232	6.4F
	1611	1914	4.0E
	2223		
9 W ◐		0053	3.0F
	0327	0705	4.4E
	1017	1339	5.4F
	1725	2032	3.6E
	2338		
10 Th		0211	2.6F
	0455	0834	4.0E
	1138	1502	4.9F
	1843	2149	3.8F
11 F	0056	0345	3.1F
	0640	1003	4.2E
	1305	1627	4.9F
	1950	2256	4.5E
12 Sa	0201	0505	4.3F
	0805	1118	5.0E
	1423	1735	5.3F
	2043	2351	5.4E
13 Su	0255	0605	5.7F
	0911	1219	5.9E
	1527	1829	5.8F
	2128		
14 M		0037	6.3E
	0341	0654	7.1F
	1006	1312	6.6E
	1620	1914	6.1F
	2206		
15 Tu		0119	7.0E
	0423	0738	8.2F
	1054	1358	6.9E
	1706	1954	6.2F
	2242		
16 W ○		0158	7.4E
	0503	0818	8.8F
	1139	1441	6.9E
	1747	2031	6.0F
	2315		
17 Th		0235	7.6E
	0542	0857	8.9F
	1221	1521	6.5E
	1826	2107	5.6F
	2347		
18 F		0310	7.4E
	0620	0935	8.6F
	1301	1559	5.9E
	1902	2141	5.1F
19 Sa	0019	0344	6.9E
	0657	1011	7.9F
	1340	1636	5.1E
	1938	2216	4.5F
20 Su	0051	0418	6.2E
	0734	1047	7.0F
	1417	1714	4.3E
	2015	2251	3.8F
21 M	0125	0453	5.2E
	0811	1123	6.0F
	1455	1754	3.6E
	2057	2330	3.1F
22 Tu	0204	0533	4.2E
	0850	1203	5.0F
	1536	1842	3.0E
	2147		
23 W		0017	2.5F
	0252	0623	3.2E
	0935	1249	4.1F
	1626	1942	2.7E
	2250		
24 Th ◑		0117	2.0F
	0400	0733	2.4E
	1031	1348	3.3F
	1726	2051	2.7E
25 F	0005	0239	1.8F
	0542	0859	2.1E
	1143	1503	3.0F
	1831	2158	3.1E
26 Sa	0115	0409	2.3F
	0720	1018	2.3E
	1302	1620	3.2F
	1931	2255	3.8E
27 Su	0211	0517	3.4F
	0828	1122	3.0E
	1412	1723	3.7F
	2022	2342	4.6E
28 M	0257	0606	4.5F
	0919	1215	3.9E
	1511	1813	4.3F
	2106		
29 Tu		0025	5.3E
	0336	0648	5.7F
	1002	1300	4.7E
	1602	1858	4.9F
	2146		
30 W		0104	6.0E
	0413	0728	6.8F
	1043	1343	5.5E
	1648	1939	5.3F
	2224		

December

Day	Slack h m	Maximum h m	knots
1 Th		0141	6.6E
	0450	0806	7.7F
	1123	1424	6.1E
	1732	2018	5.5F
	2301		
2 F ●		0218	7.0E
	0527	0844	8.4F
	1204	1505	6.4E
	1815	2058	5.6F
	2337		
3 Sa		0255	7.2E
	0605	0924	8.8F
	1246	1547	6.4E
	1858	2138	5.4F
4 Su	0015	0335	7.2E
	0645	1005	8.9F
	1330	1631	6.2E
	1941	2219	5.1F
5 M	0055	0417	6.9E
	0729	1048	8.5F
	1416	1717	5.8E
	2026	2303	4.8F
6 Tu	0139	0504	6.5E
	0816	1134	7.9F
	1505	1808	5.3E
	2115	2351	4.4F
7 W	0231	0559	5.8E
	0909	1225	7.0F
	1557	1903	4.9E
	2208		
8 Th ◐		0047	4.0F
	0333	0702	5.2E
	1009	1322	6.0F
	1654	2003	4.6E
	2306		
9 F		0152	3.9F
	0450	0815	4.6E
	1117	1428	5.1F
	1754	2105	4.5E
10 Sa	0010	0307	4.1F
	0617	0931	4.4E
	1233	1540	4.6F
	1854	2207	4.8E
11 Su	0113	0424	4.8F
	0741	1046	4.5E
	1350	1651	4.3F
	1950	2306	5.2E
12 M	0213	0532	5.7F
	0853	1153	4.9E
	1459	1752	4.4F
	2041		
13 Tu		0000	5.7E
	0307	0629	6.7F
	0953	1253	5.2E
	1559	1845	4.5F
	2127		
14 W		0049	6.2E
	0358	0719	7.4F
	1047	1345	5.4E
	1651	1932	4.5F
	2210		
15 Th		0135	6.6E
	0444	0805	7.9F
	1135	1432	5.5E
	1736	2014	4.6F
	2250		
16 F ○		0217	6.7E
	0528	0846	8.0F
	1217	1515	5.3E
	1817	2053	4.5F
	2328		
17 Sa		0258	6.6E
	0608	0924	7.8F
	1256	1554	5.1E
	1855	2130	4.4F
18 Su	0006	0335	6.2E
	0646	1000	7.3F
	1331	1630	4.8E
	1930	2206	4.2F
19 M	0043	0411	5.7E
	0722	1034	6.7F
	1403	1705	4.5E
	2006	2242	3.9F
20 Tu	0120	0446	5.1E
	0756	1107	6.1F
	1433	1738	4.2E
	2043	2319	3.7F
21 W	0200	0522	4.4E
	0830	1141	5.5F
	1504	1814	4.0E
	2123		
22 Th		0000	3.4F
	0244	0602	3.8E
	0906	1218	4.9F
	1538	1854	3.8E
	2207		
23 F		0046	3.1F
	0336	0650	3.2E
	0948	1301	4.3F
	1619	1941	3.7E
	2258		
24 Sa ◑		0140	2.9F
	0441	0750	2.7E
	1040	1352	3.8F
	1708	2036	3.7E
	2354		
25 Su		0245	3.0F
	0600	0900	2.5E
	1146	1455	3.3F
	1805	2135	3.9E
26 M	0053	0359	3.4F
	0721	1014	2.7E
	1303	1608	3.2F
	1906	2235	4.3E
27 Tu	0152	0508	4.3F
	0831	1124	3.3E
	1421	1720	3.4F
	2006	2332	4.8E
28 W	0246	0608	5.4F
	0931	1226	4.1E
	1531	1821	3.8F
	2102		
29 Th		0025	5.5E
	0336	0659	6.5F
	1023	1322	4.9E
	1630	1914	4.3F
	2152		
30 F		0114	6.1E
	0424	0747	7.6F
	1111	1412	5.7E
	1721	2002	4.8F
	2239		
31 Sa ●		0201	6.7E
	0510	0832	8.5F
	1157	1459	6.3E
	1807	2047	5.2F
	2324		

Time meridian 135° E. 0000 is midnight. 1200 is noon. Times are not adjusted for Daylight Saving Time.

Kanmon Kaikyo (Hayatomo Seto), Japan, 2005

F–Flood, Dir. Westward E–Ebb, Dir. Eastward

January

Day	Slack (h m)	Maximum (h m)	knots
1 Sa	0203	0622	7.5E
	0943	1257	6.5F
	1633	1922	5.2E
	2214		
2 Su		0044	4.1F
	0247	0702	6.9E
	1020	1335	6.2F
	1706	2008	5.3E
	2318		
3 M		0141	3.9F
	0344	0747	6.2E
	1102	1417	5.7F
	1739	2100	5.5E
4 Tu ◑	0030	0250	3.9F
	0503	0845	5.4E
	1153	1504	5.1F
	1813	2201	5.8E
5 W	0140	0412	4.3F
	0653	1006	4.5E
	1257	1600	4.6F
	1850	2306	6.4E
6 Th	0241	0535	5.2F
	0853	1142	4.2E
	1413	1705	4.1F
	1931		
7 F		0006	7.0E
	0336	0643	6.1F
	1031	1300	4.4E
	1531	1809	3.9F
	2018		
8 Sa		0101	7.7E
	0428	0741	7.0F
	1140	1403	4.8E
	1636	1907	4.1F
	2109		
9 Su		0153	8.3E
	0518	0834	7.7F
	1235	1458	5.1E
	1729	2000	4.5F
	2203		
10 M ●		0245	8.8E
	0607	0923	8.1F
	1324	1547	5.5E
	1814	2051	4.9F
	2257		
11 Tu		0335	9.1E
	0655	1011	8.3F
	1410	1634	5.7E
	1858	2141	5.2F
	2350		
12 W		0424	9.2E
	0743	1057	8.3F
	1452	1718	5.9E
	1942	2232	5.5F
13 Th	0042	0511	9.1E
	0829	1141	8.0F
	1530	1801	6.0E
	2029	2322	5.5F
14 F	0135	0557	8.8E
	0913	1223	7.7F
	1605	1844	6.2E
	2123		
15 Sa		0013	5.4F
	0229	0641	8.1E
	0956	1305	7.1F
	1635	1927	6.2E
	2223		
16 Su		0105	5.1F
	0326	0725	7.1E
	1037	1344	6.4F
	1704	2012	6.2E
	2330		
17 M ◐		0203	4.7F
	0434	0812	5.9E
	1117	1424	5.6F
	1731	2103	6.2E
18 Tu	0045	0313	4.3F
	0607	0912	4.4E
	1202	1507	4.6F
	1758	2204	6.2E
19 W	0201	0446	4.3F
	0826	1045	2.9E
	1305	1602	3.5F
	1829	2311	6.3E
20 Th	0311	0622	4.9F
	1041	1236	2.5E
	1456	1721	2.5F
	1911		
21 F		0015	6.6E
	0408	0729	5.6F
	1147	1357	3.1E
	1642	1836	2.3F
	2011		
22 Sa		0111	7.0E
	0454	0817	6.2F
	1223	1445	3.8E
	1734	1934	2.8F
	2114		
23 Su		0159	7.3E
	0533	0853	6.6F
	1251	1518	4.4E
	1803	2018	3.5F
	2209		
24 M		0241	7.7E
	0609	0925	6.9F
	1314	1547	4.8E
	1825	2054	4.2F
	2255		
25 Tu ○		0319	8.0E
	0643	0956	7.1F
	1338	1614	5.2E
	1848	2128	4.7F
	2336		
26 W		0355	8.2E
	0715	1025	7.2F
	1401	1641	5.5E
	1914	2200	5.1F
27 Th	0014	0429	8.3E
	0745	1055	7.2F
	1425	1711	5.7E
	1943	2234	5.3F
28 F	0051	0502	8.2E
	0815	1124	7.1F
	1449	1742	6.0E
	2016	2310	5.4F
29 Sa	0128	0535	8.0E
	0843	1154	7.0F
	1512	1814	6.1E
	2053	2347	5.5F
30 Su	0208	0609	7.6E
	0912	1224	6.7F
	1535	1848	6.3E
	2136		
31 M		0029	5.4F
	0254	0645	7.0E
	0943	1254	6.4F
	1558	1924	6.4E
	2226		

February

Day	Slack (h m)	Maximum (h m)	knots
1 Tu		0117	5.3F
	0348	0724	6.2E
	1017	1325	5.8F
	1621	2003	6.4E
	2324		
2 W ◑		0213	5.1F
	0501	0813	5.1E
	1057	1359	5.0F
	1647	2052	6.4E
3 Th	0032	0326	5.1F
	0648	0925	3.8E
	1150	1442	4.1F
	1718	2158	6.5E
4 F	0149	0500	5.4F
	0912	1111	2.9E
	1317	1551	3.0F
	1802	2318	6.8E
5 Sa	0306	0629	6.1F
	1056	1257	3.2E
	1524	1730	2.5F
	1915		
6 Su		0035	7.3E
	0413	0739	6.9F
	1155	1410	4.2E
	1646	1857	3.2F
	2046		
7 M		0144	7.9E
	0510	0833	7.6F
	1239	1501	5.0E
	1733	2004	4.3F
	2202		
8 Tu		0243	8.5E
	0601	0919	8.0F
	1317	1543	5.7E
	1811	2058	5.2F
	2305		
9 W ●		0335	8.9E
	0648	1001	8.2F
	1349	1622	6.2E
	1849	2145	5.9F
10 Th	0000	0420	9.1E
	0731	1040	8.2F
	1418	1659	6.6E
	1928	2230	6.4F
11 F	0053	0502	8.9E
	0811	1117	7.9F
	1444	1735	6.9E
	2011	2313	6.5F
12 Sa	0143	0542	8.5E
	0847	1152	7.5F
	1507	1811	7.1E
	2056	2357	6.5F
13 Su	0235	0620	7.7E
	0921	1225	6.9F
	1529	1847	7.1E
	2146		
14 M		0041	6.1F
	0330	0658	6.6E
	0952	1255	6.2F
	1549	1923	7.0E
	2240		
15 Tu		0129	5.6F
	0435	0737	5.3E
	1020	1322	5.2F
	1607	2001	6.7E
	2340		
16 W		0226	5.0F
	0604	0824	3.5E
	1045	1347	4.1F
◐	1622	2047	6.3E
17 Th	0052	0352	4.5F
	0848	0950	1.0E
	1059	1411	2.7F
	1628	2155	5.9E
18 F	0214	0601	4.7F
	1151	1322	1.0E
		1517	0.9E
		2325	5.9E
19 Sa	0331	0722	5.4F
	1202	1424	2.9E
		1830	1.3E
20 Su		0048	6.2E
	0429	0807	6.0F
	1218	1447	3.9E
	1758	1939	2.3F
	2100		
21 M		0148	6.7E
	0513	0841	6.5F
	1235	1509	4.5E
	1759	2019	3.6F
	2208		
22 Tu		0234	7.2E
	0551	0909	6.8F
	1253	1530	5.1E
	1811	2051	4.6F
	2257		
23 W		0311	7.7E
	0623	0936	7.0F
	1310	1553	5.6E
	1829	2121	5.3F
	2338		
24 Th ○		0344	7.9E
	0653	1002	7.2F
	1328	1618	6.1E
	1852	2151	5.8F
25 F	0015	0416	8.1E
	0720	1028	7.2F
	1345	1646	6.5E
	1920	2223	6.2F
26 Sa	0053	0448	8.0E
	0747	1055	7.2F
	1402	1714	6.8E
	1951	2257	6.5F
27 Su	0131	0520	7.8E
	0812	1121	7.0F
	1419	1744	7.1E
	2025	2333	6.6F
28 M	0214	0553	7.3E
	0840	1148	6.7F
	1438	1815	7.2E
	2104		

March

Day	Slack (h m)	Maximum (h m)	knots
1 Tu		0011	6.6F
	0302	0628	6.7E
	0908	1214	6.2F
	1458	1846	7.2E
	2148		
2 W		0055	6.4F
	0400	0707	5.7E
	0940	1240	5.5F
	1519	1921	7.1E
	2240		
3 Th		0147	6.1F
	0519	0755	4.4E
	1014	1308	4.7F
	1541	2003	6.8E
	2344		
4 F ◑		0258	5.7F
	0718	0907	2.8E
	1058	1344	3.5F
	1605	2107	6.5E
5 Sa	0108	0442	5.7F
	0955	1113	1.8E
	1253	1459	1.8F
	1639	2243	6.4E
6 Su	0241	0627	6.3F
	1110	1319	3.2E
	1613	1733	1.5F
	1844		
7 M		0027	6.8E
	0359	0735	7.0F
	1149	1411	4.5E
	1656	1913	3.5F
	2100		
8 Tu		0144	7.5E
	0459	0823	7.5F
	1220	1450	5.4E
	1727	2010	5.0F
	2218		
9 W		0240	8.2E
	0548	0902	7.8F
	1246	1524	6.2E
	1759	2055	6.0F
	2318		
10 Th ●		0326	8.5E
	0630	0938	7.9F
	1308	1558	6.9E
	1832	2137	6.8F
11 F	0011	0407	8.6E
	0708	1012	7.8F
	1328	1631	7.3E
	1909	2217	7.2F
12 Sa	0101	0445	8.3E
	0743	1045	7.5F
	1347	1704	7.6E
	1947	2256	7.3F
13 Su	0152	0521	7.7E
	0814	1115	7.0F
	1405	1736	7.7E
	2028	2335	7.2F
14 M	0244	0556	6.9E
	0842	1143	6.4F
	1424	1807	7.7E
	2111		
15 Tu		0015	6.8F
	0339	0631	5.9E
	0908	1208	5.7F
	1443	1838	7.4E
	2156		
16 W		0058	6.3F
	0442	0707	4.6E
	0931	1230	4.8F
	1458	1909	7.0E
	2245		
17 Th		0147	5.7F
	0605	0750	2.9E
	0947	1248	3.7F
	1507	1945	6.4E
	2344		
18 F		0258	5.0F
		0908	1.0F
		1259	2.4F
◐	1458	2038	5.7E
19 Sa	0102	0512	4.8F
	1140	2215	5.2E
20 Su	0233	0645	5.3F
	1129	1405	3.1E
		1824	1.8E
21 M		0016	5.4E
	0346	0734	5.8F
	1141	1418	4.1E
	1737	1925	2.5F
	2052		
22 Tu		0127	6.1E
	0438	0809	6.3F
	1154	1437	4.9E
	1731	2001	4.1F
	2203		
23 W		0212	6.7E
	0518	0837	6.6F
	1209	1457	5.6E
	1743	2031	5.1F
	2253		
24 Th		0249	7.2E
	0551	0902	6.8F
	1223	1521	6.2E
	1802	2101	5.9F
	2335		
25 F		0322	7.5E
	0620	0928	7.0F
	1237	1546	6.8E
	1826	2131	6.6F
26 Sa	0014	0354	7.6E
	0647	0953	7.0F
	1251	1613	7.2E
○	1854	2204	7.0F
27 Su	0055	0427	7.5E
	0713	1019	6.9F
	1306	1642	7.6E
	1925	2238	7.3F
28 M	0138	0500	7.3E
	0740	1045	6.6F
	1323	1712	7.8E
	2000	2315	7.4F
29 Tu	0225	0535	6.8E
	0808	1111	6.3F
	1344	1743	7.9E
	2039	2355	7.4F
30 W	0320	0613	6.0E
	0839	1138	5.7F
	1406	1815	7.8E
	2123		
31 Th		0040	7.1F
	0425	0656	5.0E
	0912	1205	5.0F
	1429	1851	7.5E
	2215		

Time meridian 135° E. 0000 is midnight. 1200 is noon. Times are not adjusted for Daylight Saving Time.
If three consecutive entries are marked (F) or (E) the middle one is not a true maximum but an intermediate value to show the current pattern.

Kanmon Kaikyo (Hayatomo Seto), Japan, 2005

F–Flood, Dir. Westward E–Ebb, Dir. Eastward

April

Day	Slack (h m)	Maximum (h m)	knots
1 F		0134	6.7F
	0551	0750	3.7E
	0951	1236	4.0F
	1454	1936	7.0E
	2320		
2 Sa		0249	6.2F
	0750	0914	2.3E
	1054	1322	2.6F
◑	1520	2045	6.5E
3 Su	0045	0437	6.1F
	0945	1138	2.5E
	1447	1523	0.5F
	1557	2232	6.1E
4 M	0220	0611	6.4F
	1035	1302	4.0E
	1608	1757	2.5F
	1928		
5 Tu		0021	6.5E
	0338	0711	6.9F
	1107	1345	5.1E
	1637	1910	4.4F
	2118		
6 W		0133	7.1E
	0437	0755	7.2F
	1133	1421	6.1E
	1707	1959	5.7F
	2230		
7 Th		0224	7.5E
	0524	0833	7.4F
	1153	1454	6.9E
	1739	2041	6.7F
	2329		
8 F		0308	7.7E
	0604	0906	7.3F
	1210	1527	7.5E
	1812	2120	7.3F
9 Sa	0022	0347	7.6E
	0639	0938	7.1F
	1228	1559	7.9E
●	1848	2158	7.6F
10 Su	0112	0424	7.3E
	0710	1008	6.8F
	1246	1630	8.1E
	1924	2235	7.6F
11 M	0202	0459	6.7E
	0739	1036	6.3F
	1307	1700	8.1E
	2002	2313	7.5F
12 Tu	0252	0533	6.0E
	0807	1103	5.7F
	1328	1730	8.0E
	2041	2351	7.2F
13 W	0344	0608	5.1E
	0835	1128	5.0F
	1348	1800	7.7E
	2122		
14 Th		0032	6.7F
	0441	0645	4.1E
	0904	1152	4.3F
	1406	1830	7.2E
	2205		
15 F		0117	6.2F
	0549	0729	2.9E
	0935	1217	3.3F
	1420	1906	6.6E
	2256		
16 Sa		0217	5.6F
	0730	0844	1.5E
	1033	1247	2.0F
◐	1420	1953	5.9E
17 Su	0000	0353	5.1F
	0926	1153	2.0E
		1404	1.6E
		2111	5.2E
18 M	0123	0535	5.2F
	1005	1253	3.3E
		1726	1.2E
		2311	5.0E
19 Tu	0244	0637	5.6F
	1028	1322	4.3E
	1638	1841	2.8F
	2025		
20 W		0042	5.4E
	0345	0717	5.9F
	1047	1347	5.2E
	1647	1924	4.4F
	2144		
21 Th		0135	6.0E
	0430	0749	6.2F
	1104	1413	6.0E
	1705	1959	5.5F
	2240		
22 F		0216	6.5E
	0507	0817	6.4F
	1118	1439	6.7E
	1729	2032	6.4F
	2327		
23 Sa		0253	6.8E
	0540	0844	6.5F
	1133	1508	7.4E
	1757	2106	7.1F
24 Su	0013	0328	6.9E
	0610	0912	6.4F
	1149	1538	7.9E
○	1828	2141	7.6F
25 M	0059	0404	6.9E
	0640	0940	6.3F
	1208	1609	8.2E
	1902	2218	7.9F
26 Tu	0147	0441	6.6E
	0711	1009	6.0F
	1231	1642	8.4E
	1941	2259	8.0F
27 W	0241	0520	6.1E
	0745	1039	5.6F
	1257	1717	8.4E
	2023	2342	7.8F
28 Th	0340	0603	5.4E
	0822	1111	5.1F
	1326	1754	8.2E
	2111		
29 F		0032	7.5F
	0447	0653	4.5E
	0906	1147	4.3F
	1359	1838	7.8E
	2206		
30 Sa		0132	7.0F
	0605	0757	3.6F
	1008	1235	3.4F
	1436	1931	7.2E
	2312		

May

Day	Slack (h m)	Maximum (h m)	knots
1 Su		0249	6.6F
	0732	0926	3.2E
	1202	1355	2.3F
◑	1531	2046	6.5E
2 M	0031	0417	6.4F
	0843	1106	3.8E
	1420	1601	2.2F
	1730	2224	6.1E
3 Tu	0154	0534	6.4F
	0929	1215	4.9E
	1524	1743	3.7F
	1949	2359	6.2E
4 W	0307	0631	6.5F
	1002	1303	5.8E
	1605	1849	5.1F
	2125		
5 Th		0109	6.4E
	0405	0717	6.6F
	1027	1342	6.7E
	1642	1939	6.1F
	2239		
6 F		0202	6.6E
	0453	0755	6.5F
	1047	1418	7.3E
	1717	2022	6.9F
	2340		
7 Sa		0246	6.6E
	0533	0829	6.3F
	1107	1452	7.8E
	1753	2101	7.4F
8 Su	0034	0326	6.4E
	0608	0901	6.1F
	1128	1525	8.1E
●	1829	2139	7.6F
9 M	0122	0403	6.1E
	0641	0931	5.8F
	1151	1557	8.3E
	1905	2216	7.6F
10 Tu	0207	0439	5.7E
	0712	1000	5.4F
	1216	1629	8.3E
	1941	2253	7.5F
11 W	0252	0514	5.3E
	0745	1029	4.9F
	1242	1659	8.1E
	2019	2332	7.3F
12 Th	0336	0550	4.7E
	0819	1058	4.4F
	1308	1731	7.8E
	2059		
13 F		0011	6.9F
	0423	0629	4.2E
	0859	1129	3.9F
	1335	1805	7.4E
	2140		
14 Sa		0056	6.5F
	0515	0716	3.6E
	0950	1206	3.2F
	1403	1844	6.8E
	2225		
15 Su		0147	6.0F
	0614	0817	3.2E
	1108	1257	2.3F
	1433	1930	6.2E
	2317		
16 M		0250	5.6F
	0717	0944	3.2E
	1313	1417	1.3F
◐	1519	2032	5.6E
17 Tu	0020	0402	5.3F
	0811	1109	3.8E
	1445	1606	1.7F
	1726	2155	5.1E
18 W	0130	0511	5.3F
	0850	1204	4.7E
	1523	1736	3.2F
	1940	2332	5.0E
19 Th	0236	0605	5.4F
	0918	1244	5.6E
	1554	1835	4.6F
	2112		
20 F		0044	5.3E
	0331	0646	5.5F
	0941	1319	6.4E
	1623	1920	5.7F
	2221		
21 Sa		0136	5.7E
	0417	0722	5.6F
	1002	1353	7.1E
	1655	2000	6.6F
	2319		
22 Su		0220	6.0E
	0458	0756	5.7F
	1024	1427	7.8E
	1729	2040	7.3F
23 M	0011	0302	6.1E
	0536	0829	5.6F
	1048	1503	8.3E
	1805	2120	7.8F
24 Tu	0103	0343	6.1E
	0613	0902	5.5F
	1117	1540	8.6E
○	1846	2202	8.1F
25 W	0155	0426	5.9E
	0652	0937	5.3F
	1149	1619	8.7E
	1929	2248	8.1F
26 Th	0250	0511	5.5E
	0733	1016	5.0F
	1225	1701	8.7E
	2017	2337	8.0F
27 F	0347	0600	5.1E
	0821	1059	4.6F
	1307	1746	8.5E
	2109		
28 Sa		0031	7.7F
	0447	0655	4.7E
	0919	1152	4.1F
	1355	1837	8.0E
	2205		
29 Su		0130	7.3F
	0547	0757	4.5E
	1034	1258	3.6F
	1454	1935	7.4E
	2305		
30 M		0234	6.9F
	0644	0905	4.6E
	1207	1420	3.4F
◑	1612	2041	6.7E
31 Tu	0011	0339	6.5F
	0732	1016	5.0E
	1334	1548	3.6F
	1755	2159	6.0E

June

Day	Slack (h m)	Maximum (h m)	knots
1 W	0118	0442	6.1F
	0812	1120	5.7E
	1440	1713	4.4F
	1948	2324	5.6E
2 Th	0224	0539	5.8F
	0844	1214	6.4E
	1533	1823	5.3F
	2128		
3 F		0038	5.4E
	0324	0629	5.5F
	0912	1300	7.0E
	1617	1918	6.1F
	2249		
4 Sa		0137	5.3E
	0418	0712	5.3F
	0939	1341	7.5E
	1658	2004	6.7F
	2353		
5 Su		0226	5.3E
	0505	0750	5.1F
	1006	1419	7.9E
	1736	2046	7.1F
6 M	0043	0309	5.2E
	0546	0826	4.9F
	1035	1455	8.1E
	1814	2125	7.3F
7 Tu	0125	0348	5.1E
	0623	0900	4.7F
	1106	1530	8.2E
●	1850	2203	7.4F
8 W	0203	0425	5.0E
	0659	0933	4.6F
	1140	1605	8.2E
	1927	2240	7.3F
9 Th	0239	0501	4.9E
	0734	1006	4.4F
	1214	1639	8.1E
	2005	2317	7.2F
10 F	0316	0538	4.8E
	0812	1041	4.2F
	1249	1715	7.9E
	2042	2356	6.9F
11 Sa	0353	0616	4.6E
	0853	1120	4.0F
	1326	1751	7.6E
	2121		
12 Su		0036	6.7F
	0432	0658	4.5E
	0942	1204	3.6F
	1405	1830	7.1E
	2200		
13 M		0117	6.4F
	0513	0744	4.5E
	1042	1256	3.3F
	1450	1913	6.6E
	2241		
14 Tu		0202	6.0F
	0553	0836	4.5E
	1155	1358	3.0F
	1548	2002	6.0E
	2327		
15 W		0249	5.6F
	0632	0935	4.8E
	1308	1511	3.2F
◐	1711	2102	5.3E
16 Th	0021	0341	5.3F
	0708	1036	5.3E
	1408	1630	3.8F
	1856	2221	4.8E
17 F	0122	0436	5.0F
	0740	1132	6.0E
	1457	1743	4.8F
	2039	2347	4.6E
18 Sa	0226	0530	4.8F
	0811	1221	6.7E
	1541	1841	5.8F
	2206		
19 Su		0055	4.8E
	0327	0620	4.7F
	0843	1305	7.3E
	1623	1932	6.7F
	2315		
20 M		0151	5.0E
	0422	0706	4.7F
	0918	1349	7.9E
	1705	2019	7.4F
21 Tu	0013	0242	5.3E
	0511	0749	4.7F
	0957	1433	8.4E
	1750	2106	7.8F
22 W	0106	0330	5.4E
	0558	0833	4.8F
	1040	1519	8.7E
○	1836	2154	8.1F
23 Th	0158	0419	5.4E
	0643	0919	4.8F
	1126	1606	8.9E
	1925	2243	8.2F
24 F	0248	0507	5.4E
	0730	1009	4.9F
	1217	1656	8.9E
	2015	2333	8.0F
25 Sa	0338	0557	5.4E
	0821	1104	4.8F
	1311	1746	8.7E
	2105		
26 Su		0023	7.8F
	0424	0646	5.5E
	0918	1202	4.8F
	1408	1836	8.3E
	2155		
27 M		0112	7.4F
	0506	0736	5.6E
	1024	1303	4.7F
	1512	1927	7.6E
	2246		
28 Tu		0201	6.9F
	0544	0828	5.8E
	1136	1407	4.5F
	1625	2022	6.7E
	2337		
29 W		0251	6.3F
	0618	0924	6.0E
	1249	1519	4.5F
◑	1756	2125	5.6E
30 Th	0030	0341	5.6F
	0650	1023	6.3E
	1359	1639	4.8F
	1945	2244	4.6E

Time meridian 135° E. 0000 is midnight. 1200 is noon. Times are not adjusted for Daylight Saving Time.
If three consecutive entries are marked (F) or (E) the middle one is not a true maximum but an intermediate value to show the current pattern.

Kanmon Kaikyo (Hayatomo Seto), Japan, 2005

F–Flood, Dir. Westward E–Ebb, Dir. Eastward

July

Day	Slack h m	Maximum h m	knots
1 F	0130	0436	4.9F
	0722	1122	6.6E
	1501	1758	5.3F
	2140		
2 Sa		0009	4.0E
	0239	0534	4.2F
	0755	1217	7.0E
	1555	1903	5.9F
	2308		
3 Su		0121	3.9E
	0352	0629	3.8F
	0833	1306	7.3E
	1642	1956	6.4F
4 M	0007	0219	4.1E
	0455	0719	3.7F
	0914	1352	7.6E
	1724	2040	6.8F
5 Tu	0049	0305	4.4E
	0543	0804	3.7F
	0958	1434	7.8E
	1803	2119	7.0F
6 W ●	0122	0344	4.6E
	0619	0844	4.0F
	1042	1514	8.0E
	1841	2155	7.1F
7 Th	0151	0418	4.8E
	0651	0922	4.2F
	1125	1553	8.1E
	1916	2230	7.2F
8 F	0219	0451	5.0E
	0723	0958	4.4F
	1206	1629	8.1E
	1951	2304	7.1F
9 Sa	0248	0523	5.1E
	0756	1035	4.6F
	1246	1705	8.0E
	2025	2337	7.0F
10 Su	0317	0556	5.3E
	0833	1114	4.6F
	1325	1740	7.7E
	2058		
11 M		0011	6.8F
	0346	0631	5.4E
	0914	1155	4.6F
	1406	1816	7.4E
	2130		
12 Tu		0044	6.6F
	0415	0708	5.5E
	1002	1240	4.4F
	1451	1853	6.9E
	2203		
13 W		0117	6.3F
	0441	0747	5.6E
	1056	1329	4.3F
	1544	1933	6.2E
	2239		
14 Th		0153	5.9F
	0508	0830	5.8E
	1156	1427	4.3F
	1654	2022	5.4E
	2320		
15 F ◐		0230	5.3F
	0534	0920	6.0E
	1258	1536	4.5F
	1828	2128	4.5E
16 Sa	0011	0314	4.7F
	0604	1019	6.3E
	1400	1655	5.1F
	2023	2257	3.8E
17 Su	0119	0411	4.1F
	0641	1122	6.7E
	1500	1810	5.8F
	2210		
18 M		0024	3.7E
	0243	0519	3.6F
	0728	1223	7.3E
	1557	1914	6.6F
	2323		
19 Tu		0136	4.1E
	0403	0626	3.6F
	0825	1321	7.8E
	1650	2010	7.3F
20 W	0019	0236	4.6E
	0504	0728	4.0F
	0927	1417	8.3E
	1741	2102	7.8F
21 Th ○	0107	0328	5.0E
	0552	0825	4.5F
	1029	1512	8.8E
	1831	2150	8.1F
22 F	0151	0414	5.5E
	0636	0920	5.0F
	1127	1604	9.0E
	1919	2235	8.2F
23 Sa	0231	0458	5.8E
	0720	1012	5.5F
	1224	1652	9.0E
	2005	2319	8.1F
24 Su	0308	0540	6.1E
	0807	1104	5.7F
	1319	1738	8.8E
	2050		
25 M		0001	7.8F
	0340	0622	6.4E
	0859	1155	5.8F
	1417	1823	8.3E
	2132		
26 Tu		0041	7.4F
	0409	0703	6.5E
	0955	1246	5.7F
	1517	1907	7.4E
	2212		
27 W		0120	6.7F
	0436	0746	6.6E
	1057	1341	5.4F
	1627	1953	6.3E
	2252		
28 Th ◑		0158	5.9F
	0501	0832	6.6E
	1204	1444	5.1F
	1755	2047	4.8E
	2332		
29 F		0236	4.9F
	0526	0925	6.6E
	1314	1605	4.9F
	1955	2205	3.2E
30 Sa	0021	0321	3.8F
	0553	1028	6.5E
	1426	1741	5.1F
	2217	2356	2.3E
31 Su	0151	0428	2.7F
	0628	1136	6.6E
	1532	1902	5.7F
	2339		

August

Day	Slack h m	Maximum h m	knots
1 M		0135	2.8E
	0409	0557	2.1F
	0725	1241	6.8E
	1627	1958	6.2F
2 Tu	0019	0233	3.6E
	0520	0711	2.5F
	0842	1338	7.1E
	1713	2039	6.6F
3 W	0047	0309	4.2E
	0551	0805	3.3F
	0948	1427	7.4E
	1753	2113	6.9F
4 Th	0109	0338	4.7E
	0614	0845	4.0F
	1041	1509	7.7E
	1829	2144	7.0F
5 F ●	0130	0405	5.1E
	0636	0920	4.6F
	1125	1546	7.9E
	1901	2214	7.1F
6 Sa	0150	0432	5.4E
	0702	0953	5.1F
	1205	1620	8.0E
	1932	2242	7.1F
7 Su	0211	0459	5.7E
	0731	1026	5.4F
	1244	1652	8.0E
	2000	2311	7.1F
8 M	0233	0529	6.0E
	0803	1100	5.5F
	1322	1724	7.8E
	2028	2338	6.9F
9 Tu	0254	0559	6.2E
	0840	1137	5.6F
	1402	1757	7.4E
	2055		
10 W		0006	6.7F
	0314	0630	6.4E
	0920	1216	5.6F
	1446	1830	6.8E
	2123		
11 Th		0034	6.3F
	0333	0703	6.5E
	1005	1259	5.5F
	1539	1907	6.1E
	2153		
12 F		0101	5.8F
	0353	0739	6.5E
	1056	1350	5.3F
	1647	1951	5.1E
	2227		
13 Sa ◐		0131	5.2F
	0416	0821	6.5E
	1157	1454	5.2F
	1824	2053	3.8E
	2310		
14 Su		0206	4.3F
	0442	0916	6.5E
	1309	1620	5.3F
	2040	2226	2.7E
15 M	0018	0301	3.3F
	0518	1032	6.6E
	1427	1754	5.9F
	2233		
16 Tu		0018	2.7E
	0229	0437	2.5F
	0622	1154	7.0E
	1539	1911	6.6F
	2334		
17 W		0143	3.6E
	0413	0617	2.9F
	0803	1310	7.6E
	1640	2008	7.3F
18 Th	0017	0237	4.6E
	0505	0733	4.0F
	0931	1415	8.2E
	1733	2055	7.8F
19 F	0053	0320	5.4E
	0545	0831	5.1F
	1039	1510	8.7E
	1821	2137	8.1F
20 Sa ○	0125	0359	6.0E
	0623	0921	5.9F
	1137	1557	8.9E
	1905	2216	8.1F
21 Su	0153	0436	6.5E
	0703	1007	6.4F
	1232	1641	8.9E
	1945	2253	8.0F
22 M	0218	0513	7.0E
	0746	1052	6.7F
	1325	1722	8.5E
	2023	2329	7.6F
23 Tu	0241	0549	7.2E
	0832	1136	6.7F
	1421	1801	7.8E
	2058		
24 W		0002	7.0F
	0303	0625	7.3E
	0922	1222	6.5F
	1521	1841	6.8E
	2130		
25 Th		0034	6.3F
	0324	0702	7.2E
	1016	1311	6.1F
	1631	1922	5.5E
	2200		
26 F		0103	5.4F
	0345	0741	7.0E
	1115	1408	5.5F
	1801	2011	3.8E
	2227		
27 Sa ◑		0130	4.3F
	0403	0826	6.6E
	1223	1529	5.0F
	2027	2131	1.5E
	2246		
28 Su		0157	2.9F
	0414	0928	6.1E
	1343	1731	5.0F
	2339		
29 M		0041	1.0E
	0209	0255	0.5F
	0335	1057	5.9E
	1503	1857	5.6F
	2348		
30 Tu		0203	2.8E
		0604	1.0E
		1225	6.1E
	1607	1947	6.1F
31 W	0006	0230	3.8E
	0537	0721	2.4F
	0842	1332	6.6E
	1655	2023	6.5F

September

Day	Slack h m	Maximum h m	knots
1 Th	0022	0253	4.6E
	0541	0804	3.7F
	0954	1419	7.1E
	1734	2053	6.8F
2 F	0038	0315	5.1E
	0554	0837	4.6F
	1044	1457	7.5E
	1807	2120	7.0F
3 Sa	0053	0338	5.6E
	0613	0907	5.4F
	1126	1530	7.7E
	1837	2146	7.1F
4 Su ●	0108	0402	6.1E
	0636	0937	5.9F
	1204	1601	7.8E
	1903	2211	7.1F
5 M	0123	0428	6.5E
	0703	1008	6.2F
	1241	1632	7.8E
	1928	2236	7.0F
6 Tu	0139	0455	6.8E
	0734	1040	6.5F
	1319	1703	7.5E
	1953	2301	6.8F
7 W	0155	0524	7.1E
	0806	1115	6.6F
	1400	1735	7.1E
	2017	2326	6.5F
8 Th	0211	0553	7.2E
	0843	1152	6.5F
	1447	1808	6.5E
	2044	2350	6.1F
9 F	0230	0624	7.2E
	0924	1233	6.4F
	1542	1846	5.6E
	2112		
10 Sa		0015	5.5F
	0250	0656	7.1E
	1012	1322	6.1F
	1656	1930	4.5E
	2144		
11 Su ◐		0040	4.7F
	0312	0735	6.9E
	1112	1426	5.8F
	1846	2034	3.0E
	2223		
12 M		0113	3.7F
	0336	0831	6.5E
	1230	1601	5.6F
	2118	2225	1.6E
	2347		
13 Tu		0216	2.3F
	0408	0959	6.3E
	1403	1752	6.0F
	2241		
14 W		0043	2.7E
	0327	0443	1.5F
	0552	1145	6.6E
	1524	1905	6.8F
	2320		
15 Th		0142	4.1E
	0423	0636	3.2F
	0824	1310	7.3E
	1627	1954	7.4F
	2350		
16 F		0222	5.2E
	0457	0740	4.8F
	0948	1411	8.0E
	1718	2035	7.7F
17 Sa	0016	0258	6.1E
	0530	0828	6.0F
	1051	1459	8.4E
	1802	2112	7.9F
18 Su ○	0038	0332	6.8E
	0605	0911	6.8F
	1146	1542	8.5E
	1841	2147	7.8F
19 M	0057	0406	7.4E
	0642	0952	7.3F
	1239	1622	8.3E
	1917	2219	7.5F
20 Tu	0117	0440	7.7E
	0723	1033	7.5F
	1332	1700	7.8E
	1949	2251	7.1F
21 W	0137	0513	7.9E
	0805	1114	7.4F
	1428	1737	7.0E
	2019	2320	6.5F
22 Th	0157	0547	7.9E
	0849	1156	7.1F
	1527	1814	5.9E
	2047	2347	5.7F
23 F	0218	0619	7.6E
	0937	1241	6.6F
	1635	1853	4.6E
	2113		
24 Sa		0012	4.8F
	0237	0653	7.2E
	1029	1333	5.9F
	1801	1940	3.0E
	2134		
25 Su ◑		0033	3.7F
	0250	0732	6.6E
	1130	1447	5.2F
		2108	0.5F
26 M		0051	2.2F
	0245	0827	5.9E
	1247	1657	5.0F
	2259		
27 Tu		1005	5.4E
	1417	1824	5.4F
	2305		
28 W		0137	3.3E
		0604	1.2E
		1201	5.5E
	1531	1915	5.9F
	2319		
29 Th		0156	4.2E
	0511	0706	2.8F
	0841	1311	6.1E
	1623	1950	6.3F
	2333		
30 F		0217	5.0E
	0512	0744	4.2F
	0951	1357	6.6E
	1702	2019	6.6F
	2346		

Time meridian 135° E. 0000 is midnight. 1200 is noon. Times are not adjusted for Daylight Saving Time.
If three consecutive entries are marked (F) or (E) the middle one is not a true maximum but an intermediate value to show the current pattern.

Kanmon Kaikyo (Hayatomo Seto), Japan, 2005

F–Flood, Dir. Westward   E–Ebb, Dir. Eastward

October

Day	Slack (h m)	Maximum (h m)	knots
1 Sa		0239	5.7E
	0526	0816	5.2F
	1041	1433	7.0E
	1735	2045	6.7F
	2359		
2 Su		0302	6.3E
	0546	0845	6.0F
	1123	1505	7.3E
	1803	2109	6.8F
3 M ●	0012	0327	6.8E
	0610	0915	6.6F
	1202	1537	7.4E
	1829	2133	6.8F
4 Tu	0025	0353	7.3E
	0637	0946	7.0F
	1241	1608	7.3E
	1853	2157	6.7F
5 W	0040	0421	7.6E
	0706	1018	7.2F
	1321	1640	7.1E
	1918	2221	6.5F
6 Th	0056	0449	7.8E
	0739	1053	7.3F
	1405	1714	6.6E
	1944	2246	6.1F
7 F	0115	0519	7.9E
	0816	1131	7.3F
	1456	1750	6.0E
	2013	2311	5.6F
8 Sa	0136	0551	7.8E
	0857	1214	7.1F
	1558	1830	5.1E
	2045	2336	5.0F
9 Su	0200	0625	7.5E
	0947	1305	6.7F
	1717	1921	3.9E
	2122		
10 M		0005	4.1F
	0225	0708	7.1E
	1048	1413	6.2F
	1906	2035	2.5E
	2219		
11 Tu ◐		0040	3.0F
	0252	0810	6.6E
	1208	1554	5.9F
	2102	2245	2.2E
12 W	0130	0235	1.1F
	0332	0948	6.2E
	1343	1735	6.2F
	2159		
13 Th		0026	3.7E
	0333	0512	2.1F
	0639	1140	6.4E
	1504	1839	6.7F
	2233		
14 F		0113	5.0E
	0406	0636	4.1F
	0843	1259	7.0E
	1605	1926	7.1F
	2259		
15 Sa		0151	6.0E
	0439	0729	5.6F
	1000	1355	7.5E
	1655	2004	7.3F
	2320		
16 Su		0226	6.9E
	0512	0814	6.6F
	1101	1441	7.7E
	1736	2039	7.3F
	2338		
17 M ○		0300	7.5E
	0547	0855	7.3F
	1157	1522	7.7E
	1813	2112	7.1F
	2357		
18 Tu		0334	8.0E
	0624	0934	7.7F
	1250	1601	7.4E
	1846	2143	6.8F
19 W	0017	0406	8.3E
	0702	1013	7.8F
	1342	1638	6.8E
	1917	2212	6.3F
20 Th	0039	0439	8.4E
	0742	1053	7.7F
	1435	1715	6.1E
	1946	2240	5.7F
21 F	0102	0511	8.2E
	0823	1133	7.4F
	1529	1751	5.2E
	2016	2306	5.0F
22 Sa	0124	0542	7.9E
	0906	1216	6.9F
	1627	1830	4.2E
	2047	2332	4.2F
23 Su	0145	0615	7.4E
	0953	1304	6.2F
	1737	1918	3.0E
	2123	2359	3.2F
24 M	0201	0652	6.7E
	1046	1406	5.6F
	1916	2041	1.8E
	2239		
25 Tu ◑		0034	1.7F
	0202	0741	6.0E
	1151	1543	5.2F
	2059	2325	2.4E
26 W		0201	1.9E
		0900	5.3E
	1312	1717	5.2F
	2139		
27 Th		0030	3.6E
		0512	1.1E
		1100	5.0E
	1431	1818	5.5F
	2204		
28 F		0102	4.6E
	0426	0626	2.9F
	0812	1226	5.4E
	1532	1900	5.8F
	2223		
29 Sa		0129	5.4E
	0436	0711	4.4F
	0933	1319	5.9E
	1617	1932	6.0F
	2239		
30 Su		0154	6.2E
	0455	0745	5.5F
	1030	1400	6.3E
	1653	1959	6.2F
	2254		
31 M		0220	6.9E
	0518	0817	6.3F
	1117	1435	6.6E
	1724	2025	6.3F
	2308		

November

Day	Slack (h m)	Maximum (h m)	knots
1 Tu		0248	7.4E
	0544	0850	7.0F
	1200	1510	6.7E
	1753	2051	6.2F
	2324		
2 W ●		0317	7.9E
	0613	0923	7.4F
	1243	1544	6.7E
	1821	2117	6.1F
	2342		
3 Th		0347	8.3E
	0645	0958	7.7F
	1328	1619	6.5E
	1850	2144	5.9F
4 F	0004	0419	8.4E
	0721	1036	7.8F
	1415	1656	6.1E
	1921	2211	5.6F
5 Sa	0029	0453	8.5E
	0801	1117	7.7F
	1509	1737	5.5E
	1956	2241	5.1F
6 Su	0058	0529	8.3E
	0846	1204	7.4F
	1611	1824	4.7E
	2037	2316	4.5F
7 M	0129	0610	7.9E
	0938	1300	7.0F
	1724	1923	3.9E
	2133		
8 Tu		0001	3.7F
	0206	0700	7.4E
	1040	1410	6.6F
	1846	2041	3.3E
	2309		
9 W ◐		0112	2.6F
	0255	0808	6.8E
	1154	1534	6.3F
	2000	2218	3.7E
10 Th	0135	0310	2.0F
	0433	0938	6.3E
	1316	1654	6.2F
	2050	2337	4.7E
11 F	0253	0501	3.2F
	0658	1116	6.1E
	1431	1757	6.4F
	2126		
12 Sa		0030	5.8E
	0338	0617	4.8F
	0847	1234	6.4E
	1534	1846	6.4F
	2153		
13 Su		0113	6.7E
	0418	0711	6.0F
	1009	1333	6.6E
	1625	1926	6.4F
	2216		
14 M		0151	7.4E
	0455	0758	6.8F
	1115	1421	6.6E
	1708	2003	6.3F
	2237		
15 Tu		0227	8.0E
	0533	0839	7.4F
	1211	1503	6.5E
	1746	2036	6.1F
	2301		
16 W ○		0302	8.4E
	0610	0919	7.7F
	1302	1543	6.3E
	1820	2108	5.8F
	2326		
17 Th		0336	8.6E
	0648	0958	7.8F
	1349	1621	5.9E
	1853	2139	5.4F
	2353		
18 F		0410	8.6E
	0727	1037	7.7F
	1434	1658	5.4E
	1926	2209	5.0F
19 Sa	0021	0443	8.4E
	0806	1116	7.4F
	1518	1735	4.9E
	2001	2239	4.5F
20 Su	0050	0517	8.1E
	0847	1157	7.0F
	1604	1815	4.3E
	2041	2311	4.0F
21 M	0119	0551	7.6E
	0928	1241	6.5F
	1653	1901	3.8E
	2131	2348	3.2F
22 Tu	0147	0629	7.1E
	1013	1331	6.0F
	1749	1959	3.4E
	2246		
23 W		0037	2.3F
	0217	0713	6.4E
	1103	1429	5.6F
	1848	2118	3.4E
24 Th ◑	0054	0155	1.1F
	0252	0811	5.7E
	1202	1537	5.2F
	1941	2246	4.0E
25 F	0245	0344	1.1F
	0441	0930	5.0E
	1311	1647	5.1F
	2022	2346	4.8E
26 Sa	0322	0521	2.8F
	0715	1111	4.8E
	1419	1744	5.1F
	2053		
27 Su		0028	5.6E
	0351	0624	4.2F
	0900	1227	5.0E
	1517	1828	5.2F
	2117		
28 M		0103	6.4E
	0419	0710	5.4F
	1015	1320	5.4E
	1605	1904	5.3F
	2139		
29 Tu		0135	7.2E
	0448	0749	6.3F
	1112	1404	5.7E
	1645	1937	5.3F
	2201		
30 W		0209	7.8E
	0520	0826	7.1F
	1201	1444	5.9E
	1722	2009	5.4F
	2225		

December

Day	Slack (h m)	Maximum (h m)	knots
1 Th		0243	8.3E
	0554	0904	7.6F
	1247	1523	5.9E
	1757	2041	5.4F
	2253		
2 F ●		0319	8.6E
	0631	0944	7.9F
	1333	1604	5.8E
	1832	2114	5.3F
	2325		
3 Sa		0357	8.8E
	0712	1026	8.0F
	1422	1647	5.6E
	1911	2150	5.1F
4 Su	0002	0437	8.8E
	0756	1112	7.9F
	1513	1733	5.3E
	1953	2231	4.8F
5 M	0042	0521	8.7E
	0844	1202	7.7F
	1608	1824	4.9E
	2044	2319	4.4F
6 Tu	0128	0609	8.3E
	0936	1256	7.3F
	1705	1920	4.7E
	2149		
7 W		0020	3.9F
	0221	0702	7.8E
	1031	1354	6.9F
	1759	2023	4.6E
	2314		
8 Th ◐		0134	3.5F
	0328	0802	7.1E
	1132	1455	6.4F
	1848	2130	5.0E
9 F	0048	0259	3.4F
	0458	0913	6.3E
	1237	1558	6.0F
	1930	2239	5.6E
10 Sa	0206	0429	4.0F
	0655	1038	5.6E
	1340	1700	5.6F
	2006	2340	6.3E
11 Su	0307	0550	5.0F
	0850	1203	5.3E
	1452	1756	5.4F
	2039		
12 M		0031	7.0E
	0357	0654	5.9F
	1025	1311	5.2E
	1553	1845	5.1F
	2110		
13 Tu		0117	7.6E
	0442	0746	6.6F
	1135	1406	5.3E
	1647	1928	4.9F
	2141		
14 W		0159	8.0E
	0523	0831	7.1F
	1228	1453	5.3E
	1731	2007	4.8F
	2214		
15 Th		0238	8.3E
	0602	0912	7.4F
	1311	1534	5.3E
	1810	2044	4.7F
	2249		
16 F ○		0316	8.5E
	0640	0951	7.5F
	1349	1613	5.2E
	1845	2119	4.7F
	2325		
17 Sa		0352	8.5E
	0718	1029	7.5F
	1424	1649	5.1E
	1919	2153	4.6F
18 Su	0002	0428	8.4E
	0755	1106	7.3F
	1457	1725	5.0E
	1955	2228	4.5F
19 M	0038	0504	8.2E
	0832	1143	7.1F
	1531	1801	4.9E
	2033	2305	4.3F
20 Tu	0115	0539	7.9E
	0908	1220	6.8F
	1606	1839	4.8E
	2118	2346	4.0F
21 W	0153	0615	7.4E
	0945	1258	6.4F
	1642	1920	4.7E
	2212		
22 Th		0033	3.5F
	0232	0653	6.9E
	1022	1336	6.0F
	1718	2006	4.7E
	2321		
23 F		0129	3.1F
	0320	0735	6.2E
	1102	1417	5.6F
	1753	2100	4.9E
24 Sa ◑	0041	0237	2.9F
	0428	0829	5.3E
	1149	1504	5.1F
	1828	2204	5.2E
25 Su	0154	0400	3.2F
	0615	0943	4.5E
	1248	1558	4.6F
	1901	2307	5.8E
26 M	0250	0525	4.1F
	0821	1120	4.0E
	1358	1659	4.2F
	1936		
27 Tu		0001	6.5E
	0335	0630	5.2F
	1003	1238	4.2E
	1511	1756	4.1F
	2013		
28 W		0048	7.1E
	0417	0723	6.2F
	1112	1337	4.6E
	1612	1847	4.2F
	2053		
29 Th		0133	7.8E
	0458	0809	7.0F
	1204	1427	4.9E
	1703	1933	4.4F
	2135		
30 F		0217	8.3E
	0540	0853	7.5F
	1251	1514	5.2E
	1746	2017	4.6F
	2221		
31 Sa ●		0301	8.7E
	0623	0938	7.9F
	1335	1559	5.4E
	1826	2101	4.9F
	2309		

Time meridian 135° E. 0000 is midnight. 1200 is noon. Times are not adjusted for Daylight Saving Time.
If three consecutive entries are marked (F) or (E) the middle one is not a true maximum but an intermediate value to show the current pattern.

Changjiang Entrance, China, 2005

F–Flood, Dir. 305° True E–Ebb, Dir. 125° True

January

Day	Slack h m	Maximum h m	knots
1 Sa		0050	1.7F
	0338	0659	2.1E
	0948	1302	2.9F
	1634	1949	2.4E
	2259		
2 Su		0131	1.7F
	0424	0741	1.9E
	1030	1343	2.5F
	1710	2029	2.2E
	2340		
3 M		0219	1.6F
	0519	0831	1.6E
	1120	1431	2.2F
	1750	2113	2.0E
4 Tu ◑	0027	0316	1.7F
	0627	0934	1.4E
	1225	1529	1.8F
	1838	2207	1.9E
5 W	0120	0422	1.8F
	0747	1053	1.4E
	1349	1640	1.4F
	1935	2310	1.8E
6 Th	0218	0533	2.1F
	0908	1219	1.6E
	1521	1757	1.3F
	2041		
7 F		0018	1.9E
	0318	0640	2.6F
	1019	1336	2.0E
	1642	1908	1.4F
	2148		
8 Sa		0124	2.1E
	0416	0739	3.1F
	1120	1439	2.5E
	1747	2010	1.6F
	2249		
9 Su		0224	2.4E
	0511	0833	3.7F
	1214	1533	3.1E
	1842	2104	1.9F
	2345		
10 M ●		0318	2.7E
	0603	0923	4.2F
	1303	1623	3.5E
	1929	2152	2.2F
11 Tu	0036	0408	3.0E
	0653	1010	4.5F
	1350	1709	3.7E
	2014	2237	2.4F
12 W	0125	0456	3.2E
	0741	1055	4.5F
	1435	1753	3.8E
	2056	2322	2.5F
13 Th	0213	0543	3.2E
	0828	1138	4.4F
	1518	1835	3.6E
	2137		
14 F		0006	2.5F
	0302	0629	3.0E
	0914	1222	4.0F
	1558	1916	3.4E
	2218		
15 Sa		0050	2.4F
	0353	0714	2.7E
	1000	1305	3.4F
	1638	1956	3.0E
	2259		
16 Su		0137	2.3F
	0447	0802	2.3E
	1049	1351	2.7F
	1716	2037	2.5E
	2342		
17 M ◐		0228	2.0F
	0547	0855	1.8E
	1143	1440	2.0F
	1755	2121	2.1E
18 Tu	0029	0326	1.8F
	0657	0957	1.4E
	1249	1538	1.4F
	1839	2212	1.7E
19 W	0122	0434	1.7F
	0818	1117	1.2E
	1418	1653	0.9F
	1936	2316	1.4E
20 Th	0222	0550	1.8F
	0940	1245	1.2E
	1558	1820	0.7F
	2050		
21 F		0028	1.3E
	0326	0701	2.0F
	1045	1357	1.5E
	1715	1935	0.9F
	2204		
22 Sa		0134	1.5E
	0425	0757	2.3F
	1136	1450	1.9E
	1807	2029	1.1F
	2303		
23 Su		0228	1.7E
	0515	0843	2.7F
	1218	1533	2.3E
	1846	2111	1.4F
	2349		
24 M		0312	2.0E
	0559	0922	3.1F
	1255	1609	2.6E
	1919	2146	1.7F
25 Tu ○	0029	0351	2.3E
	0638	0957	3.4F
	1329	1643	2.9E
	1950	2219	2.0F
26 W	0105	0428	2.6E
	0715	1029	3.6F
	1401	1715	3.0E
	2019	2249	2.2F
27 Th	0139	0503	2.8E
	0750	1101	3.7F
	1432	1746	3.1E
	2047	2320	2.3F
28 F	0213	0536	2.8E
	0824	1131	3.7F
	1501	1816	3.1E
	2116	2350	2.5F
29 Sa	0248	0610	2.8E
	0858	1202	3.5F
	1529	1846	3.0E
	2145		
30 Su		0022	2.5F
	0324	0644	2.7E
	0933	1234	3.2F
	1557	1915	2.9E
	2215		
31 M		0058	2.5F
	0404	0721	2.4E
	1011	1309	2.8F
	1626	1946	2.6E
	2248		

February

Day	Slack h m	Maximum h m	knots
1 Tu		0137	2.4F
	0450	0802	2.1E
	1056	1350	2.3F
	1658	2020	2.3E
	2326		
2 W ◑		0224	2.3F
	0547	0854	1.8E
	1152	1440	1.7F
	1736	2103	1.9E
3 Th	0013	0324	2.1F
	0700	1006	1.5E
	1312	1547	1.1F
	1828	2204	1.6E
4 F	0114	0442	2.1F
	0832	1145	1.4E
	1503	1720	0.8F
	1945	2330	1.5E
5 Sa	0232	0610	2.3F
	1001	1322	1.8E
	1644	1857	0.9F
	2122		
6 Su		0102	1.6E
	0351	0727	2.8F
	1111	1433	2.4E
	1750	2008	1.3F
	2242		
7 M		0215	2.1E
	0500	0827	3.5F
	1206	1528	3.0E
	1838	2101	1.9F
	2344		
8 Tu		0314	2.6E
	0558	0916	4.0F
	1254	1614	3.5E
	1919	2146	2.4F
9 W ●	0036	0405	3.1E
	0649	1000	4.4F
	1336	1656	3.8E
	1956	2227	2.8F
10 Th	0123	0451	3.4E
	0735	1041	4.5F
	1415	1735	3.9E
	2032	2306	3.1F
11 F	0208	0534	3.5E
	0819	1120	4.3F
	1451	1811	3.8E
	2106	2344	3.2F
12 Sa	0253	0615	3.4E
	0902	1158	3.9F
	1525	1844	3.6E
	2140		
13 Su		0022	3.1F
	0337	0655	3.1E
	0943	1236	3.3F
	1556	1917	3.2E
	2213		
14 M		0101	2.9F
	0422	0735	2.6E
	1025	1314	2.6F
	1626	1948	2.7E
	2246		
15 Tu		0142	2.6F
	0510	0816	2.1E
	1110	1353	1.9F
	1655	2020	2.1E
	2322		
16 W ◐		0228	2.2F
	0605	0905	1.5E
	1206	1439	1.2F
	1727	2057	1.6E
17 Th	0004	0325	1.8F
	0717	1015	1.1E
	1331	1546	0.6F
	1810	2154	1.1E
18 F	0103	0448	1.5F
	0852	1202	0.9E
	1551	1744	0.3F
	1941	2330	0.9E
19 Sa	0229	0628	1.6F
	1019	1341	1.3E
	1721	1929	0.5F
	2143		
20 Su		0110	1.0E
	0356	0741	2.0F
	1118	1438	1.7E
	1802	2024	0.9F
	2256		
21 M		0215	1.4E
	0459	0829	2.4F
	1200	1518	2.2E
	1832	2100	1.4F
	2342		
22 Tu		0301	1.9E
	0546	0906	2.9F
	1235	1552	2.6E
	1859	2131	1.9F
23 W	0020	0340	2.4E
	0625	0939	3.3F
	1306	1623	3.0E
	1924	2159	2.3F
24 Th ○	0054	0415	2.8E
	0701	1008	3.6F
	1335	1652	3.2E
	1949	2227	2.7F
25 F	0126	0449	3.0E
	0735	1037	3.7F
	1402	1720	3.4E
	2014	2254	2.9F
26 Sa	0159	0521	3.2E
	0808	1105	3.7F
	1428	1747	3.4E
	2040	2322	3.1F
27 Su	0232	0554	3.2E
	0842	1135	3.5F
	1454	1814	3.3E
	2106	2353	3.2F
28 M	0307	0627	3.1E
	0917	1206	3.2F
	1519	1841	3.1E
	2134		

March

Day	Slack h m	Maximum h m	knots
1 Tu		0026	3.2F
	0345	0702	2.8E
	0955	1240	2.7F
	1546	1909	2.8E
	2205		
2 W		0103	3.0F
	0428	0741	2.5E
	1039	1318	2.1F
	1615	1940	2.4E
	2240		
3 Th		0147	2.8F
	0521	0830	2.0E
	1135	1406	1.5F
	1650	2019	1.9E
	2324		
4 F ◑		0245	2.4F
	0632	0940	1.5E
	1259	1514	0.8F
	1739	2119	1.4E
5 Sa	0027	0408	2.1F
	0810	1129	1.4E
	1511	1708	0.5F
	1913	2303	1.1E
6 Su	0200	0554	2.2F
	0948	1316	1.8E
	1650	1900	0.8F
	2123		
7 M		0056	1.4E
	0339	0718	2.7F
	1058	1425	2.4E
	1742	2005	1.4F
	2247		
8 Tu		0213	2.0E
	0454	0816	3.3F
	1150	1514	3.0E
	1821	2051	2.1F
	2344		
9 W		0309	2.6E
	0551	0902	3.8F
	1233	1556	3.5E
	1856	2131	2.8F
10 Th ●	0032	0357	3.2E
	0640	0943	4.1F
	1311	1633	3.8E
	1928	2207	3.3F
11 F	0115	0439	3.5E
	0724	1020	4.2F
	1345	1707	3.9E
	1959	2242	3.6F
12 Sa	0156	0519	3.7E
	0805	1056	4.0F
	1417	1739	3.8E
	2030	2317	3.7F
13 Su	0236	0556	3.6E
	0845	1131	3.5F
	1446	1809	3.5E
	2059	2351	3.7F
14 M	0316	0632	3.3E
	0923	1205	3.0F
	1513	1836	3.1E
	2128		
15 Tu		0026	3.4F
	0355	0707	2.8E
	1002	1239	2.3F
	1540	1903	2.6E
	2157		
16 W		0102	3.0F
	0437	0743	2.3E
	1044	1315	1.7F
	1605	1930	2.1E
	2227		
17 Th		0141	2.5F
	0524	0825	1.7E
	1135	1357	1.0F
	1632	2001	1.5E
	2303		
18 F ◐		0231	1.9F
	0625	0925	1.2E
	1257	1457	0.4F
	1706	2048	1.0E
	2354		
19 Sa		0348	1.5F
	0757	1113	0.9E
	1542	1709	0.1F
	1838	2230	0.6E
20 Su	0127	0546	1.4F
	0937	1310	1.2E
	1702	1909	0.4F
	2124		
21 M		0041	0.8E
	0320	0710	1.8F
	1042	1409	1.7E
	1734	2000	1.0F
	2240		
22 Tu		0152	1.3E
	0433	0800	2.3F
	1125	1447	2.2E
	1759	2034	1.5F
	2324		
23 W		0239	1.9E
	0522	0837	2.7F
	1159	1519	2.6E
	1823	2102	2.1F
24 Th	0001	0318	2.4E
	0603	0908	3.1F
	1229	1549	3.0E
	1847	2129	2.6F
25 F	0034	0354	2.8E
	0640	0938	3.4F
	1257	1618	3.3E
	1911	2156	3.1F
26 Sa ○	0107	0428	3.2E
	0715	1007	3.5F
	1324	1646	3.4E
	1936	2224	3.5F
27 Su	0140	0502	3.4E
	0750	1036	3.5F
	1350	1714	3.5E
	2002	2253	3.7F
28 M	0215	0535	3.5E
	0826	1107	3.3F
	1417	1742	3.4E
	2030	2325	3.8F
29 Tu	0251	0610	3.3E
	0904	1141	2.9F
	1444	1810	3.1E
	2059	2359	3.7F
30 W	0331	0647	3.0E
	0945	1217	2.4F
	1513	1840	2.8E
	2132		
31 Th		0039	3.4F
	0416	0729	2.6E
	1032	1259	1.8F
	1545	1914	2.3E
	2209		

Time meridian 120° E. 0000 is midnight. 1200 is noon. Times are not adjusted for Daylight Saving Time.

Changjiang Entrance, China, 2005

F–Flood, Dir. 305° True   E–Ebb, Dir. 125° True

April

Day	Slack h m	Maximum h m	knots
1 F		0126	3.0F
	0511	0821	2.1E
	1134	1351	1.2F
	1623	1957	1.8E
	2256		
2 Sa ◑		0227	2.5F
	0623	0936	1.6E
	1310	1511	0.6F
	1723	2105	1.3E
3 Su	0006	0356	2.1F
	0758	1124	1.5E
	1515	1715	0.5F
	1927	2303	1.0E
4 M	0149	0542	2.1F
	0928	1259	1.9E
	1630	1852	1.0F
	2132		
5 Tu		0052	1.4E
	0330	0701	2.5F
	1033	1401	2.5E
	1714	1949	1.7F
	2245		
6 W		0203	2.0E
	0443	0756	3.0F
	1122	1447	3.0E
	1750	2031	2.4F
	2337		
7 Th		0256	2.6E
	0539	0840	3.4F
	1202	1527	3.3E
	1823	2107	3.1F
8 F	0021	0341	3.1E
	0626	0919	3.5F
	1236	1602	3.6E
	1853	2142	3.6F
9 Sa ●	0101	0421	3.4E
	0709	0955	3.5F
	1308	1634	3.6E
	1923	2215	3.9F
10 Su	0139	0459	3.6E
	0749	1030	3.3F
	1338	1704	3.5E
	1952	2248	4.0F
11 M	0217	0534	3.5E
	0027	1103	3.0F
	1407	1732	3.3E
	2020	2321	3.9F
12 Tu	0254	0608	3.2E
	0905	1137	2.5F
	1435	1800	2.9E
	2049	2355	3.6F
13 W	0331	0641	2.8E
	0943	1212	2.0F
	1502	1826	2.5E
	2118		
14 Th		0031	3.1F
	0410	0717	2.3E
	1025	1249	1.4F
	1530	1855	2.0E
	2149		
15 F		0110	2.6F
	0455	0758	1.8E
	1117	1333	0.9F
	1601	1928	1.5E
	2225		
16 Sa ◐		0159	2.1F
	0551	0855	1.4E
	1235	1436	0.5F
	1645	2017	1.0E
	2315		
17 Su		0309	1.6F
	0708	1025	1.1E
	1438	1627	0.3F
	1821	2149	0.7E
18 M	0040	0450	1.5F
	0836	1208	1.3E
	1559	1816	0.5F
	2043	2352	0.8E
19 Tu	0230	0616	1.6F
	0944	1313	1.6E
	1638	1914	1.1F
	2203		
20 W		0110	1.2E
	0350	0713	2.0F
	1032	1357	2.1E
	1707	1951	1.7F
	2252		
21 Th		0203	1.8E
	0447	0754	2.4F
	1109	1433	2.5E
	1734	2022	2.3F
	2331		
22 F		0246	2.3E
	0533	0830	2.7F
	1142	1506	2.8E
	1801	2052	2.9F
23 Sa	0007	0325	2.8E
	0614	0903	3.0F
	1213	1538	3.1E
	1828	2122	3.4F
24 Su ○	0043	0403	3.2E
	0653	0935	3.1F
	1243	1609	3.3E
	1857	2153	3.8F
25 M	0120	0440	3.5E
	0733	1009	3.0F
	1313	1641	3.3E
	1927	2227	4.1F
26 Tu	0158	0518	3.5E
	0813	1045	2.8F
	1344	1713	3.2E
	1959	2303	4.1F
27 W	0239	0557	3.4E
	0855	1122	2.5F
	1416	1747	3.0E
	2033	2342	4.0F
28 Th	0323	0639	3.1E
	0942	1204	2.0F
	1452	1823	2.7E
	2112		
29 F		0027	3.6F
	0412	0725	2.7E
	1035	1253	1.5F
	1532	1904	2.2E
	2156		
30 Sa		0119	3.1F
	0510	0822	2.2E
	1144	1355	1.0F
	1624	1957	1.7E
	2250		

May

Day	Slack h m	Maximum h m	knots
1 Su ◑		0225	2.6F
	0620	0937	1.9E
	1313	1521	0.8F
	1745	2116	1.3E
2 M	0005	0349	2.2F
	0740	1106	1.8E
	1443	1704	0.9F
	1943	2301	1.2E
3 Tu	0142	0519	2.2F
	0855	1224	2.1E
	1546	1824	1.4F
	2123		
4 W		0034	1.5E
	0314	0632	2.3F
	0955	1323	2.4E
	1632	1919	2.0F
	2230		
5 Th		0141	2.0E
	0425	0727	2.5F
	1042	1410	2.8E
	1710	2002	2.6F
	2320		
6 F		0235	2.5E
	0522	0813	2.7F
	1122	1450	3.0E
	1744	2040	3.2F
7 Sa	0004	0320	2.9E
	0610	0852	2.8F
	1158	1526	3.1E
	1816	2115	3.6F
8 Su ●	0043	0400	3.1E
	0653	0930	2.7F
	1231	1559	3.1E
	1847	2149	3.8F
9 M	0121	0438	3.3E
	0733	1005	2.6F
	1303	1631	3.0E
	1918	2223	3.9F
10 Tu	0157	0513	3.2E
	0812	1041	2.4F
	1334	1701	2.9E
	1948	2257	3.8F
11 W	0234	0547	3.0E
	0860	1117	2.1F
	1405	1731	2.6E
	2019	2333	3.5F
12 Th	0312	0622	2.7E
	0930	1154	1.7F
	1437	1801	2.3E
	2052		
13 F		0011	3.2F
	0352	0659	2.4E
	1013	1234	1.4F
	1511	1835	2.0E
	2127		
14 Sa		0052	2.7F
	0436	0742	2.0E
	1103	1321	1.0F
	1551	1914	1.6E
	2207		
15 Su		0141	2.3F
	0526	0834	1.7E
	1206	1420	0.7F
	1644	2005	1.2E
	2258		
16 M ◐		0240	1.9F
	0625	0940	1.5E
	1321	1537	0.6F
	1804	2119	1.0E
17 Tu	0007	0352	1.7F
	0730	1054	1.5E
	1430	1700	0.8F
	1945	2251	0.9E
18 W	0133	0507	1.7F
	0832	1159	1.7E
	1522	1806	1.2F
	2107		
19 Th		0012	1.2E
	0256	0611	1.8F
	0924	1252	2.0E
	1601	1855	1.8F
	2207		
20 F		0116	1.7E
	0403	0703	2.0F
	1009	1336	2.3E
	1637	1935	2.4F
	2254		
21 Sa		0208	2.2E
	0459	0747	2.2F
	1049	1417	2.5E
	1711	2012	3.0F
	2337		
22 Su		0255	2.7E
	0549	0828	2.4F
	1127	1456	2.8E
	1745	2049	3.5F
23 M	0019	0339	3.1E
	0635	0908	2.5F
	1204	1534	3.0E
	1821	2127	3.9F
24 Tu ○	0102	0422	3.3E
	0720	0948	2.5F
	1241	1613	3.1E
	1858	2207	4.2F
25 W	0146	0505	3.5E
	0805	1030	2.4F
	1320	1653	3.0E
	1937	2249	4.2F
26 Th	0232	0549	3.4E
	0852	1114	2.2F
	1401	1734	2.9E
	2019	2334	4.1F
27 F	0320	0635	3.2E
	0942	1202	1.9F
	1445	1818	2.6E
	2105		
28 Sa		0023	3.8F
	0411	0725	2.9E
	1037	1255	1.6F
	1536	1907	2.3E
	2155		
29 Su		0118	3.3F
	0506	0821	2.5E
	1139	1358	1.3F
	1639	2006	1.9E
	2253		
30 M ◑		0219	2.8F
	0605	0923	2.3E
	1246	1511	1.2F
	1759	2118	1.6E
31 Tu	0003	0329	2.4F
	0708	1031	2.1E
	1352	1630	1.3F
	1932	2241	1.5E

June

Day	Slack h m	Maximum h m	knots
1 W	0124	0443	2.1F
	0809	1136	2.2E
	1451	1742	1.7F
	2057		
2 Th		0003	1.6E
	0248	0553	2.0F
	0906	1235	2.2E
	1541	1841	2.1F
	2205		
3 F		0112	1.9E
	0402	0653	1.9F
	0956	1326	2.3E
	1625	1930	2.6F
	2259		
4 Sa		0210	2.2E
	0504	0744	2.0F
	1041	1411	2.5E
	1704	2012	3.0F
	2345		
5 Su		0258	2.5E
	0555	0829	2.0F
	1121	1451	2.5E
	1741	2051	3.3F
6 M	0026	0341	2.7E
	0641	0910	2.0F
	1159	1529	2.6E
	1816	2128	3.5F
7 Tu ●	0104	0420	2.9E
	0722	0949	2.0F
	1236	1604	2.6E
	1851	2205	3.6F
8 W	0142	0457	2.9E
	0802	1026	1.9F
	1311	1638	2.5E
	1926	2242	3.5F
9 Th	0220	0533	2.8E
	0840	1104	1.8F
	1347	1713	2.5E
	2001	2319	3.4F
10 F	0258	0609	2.7E
	0918	1143	1.7F
	1424	1748	2.3E
	2038	2358	3.2F
11 Sa	0336	0646	2.5E
	0950	1222	1.5F
	1502	1825	2.1E
	2116		
12 Su		0038	2.9F
	0416	0726	2.3E
	1040	1305	1.4F
	1545	1905	1.9E
	2156		
13 M		0120	2.6F
	0457	0808	2.1E
	1126	1352	1.2F
	1635	1952	1.6E
	2242		
14 Tu		0206	2.3F
	0540	0855	1.9E
	1215	1447	1.2F
	1736	2047	1.4E
	2336		
15 W ◐		0258	2.0F
	0626	0946	1.8E
	1306	1548	1.3F
	1848	2154	1.3E
16 Th ◐	0042	0359	1.7F
	0716	1042	1.8E
	1358	1652	1.5F
	2005	2310	1.3E
17 F	0200	0504	1.6F
	0809	1139	1.8E
	1448	1752	1.9F
	2115		
18 Sa		0024	1.6E
	0319	0608	1.6F
	0903	1234	2.0E
	1535	1846	2.4F
	2216		
19 Su		0130	2.0E
	0429	0706	1.7F
	0955	1328	2.2E
	1622	1936	2.9F
	2310		
20 M		0227	2.4E
	0530	0759	1.8F
	1046	1418	2.4E
	1708	2023	3.4F
21 Tu	0000	0320	2.9E
	0623	0849	2.0F
	1134	1507	2.7E
	1753	2109	3.9F
22 W ○	0049	0409	3.2E
	0713	0936	2.1F
	1221	1555	2.8E
	1840	2155	4.2F
23 Th	0137	0456	3.4E
	0801	1023	2.2F
	1308	1642	3.0E
	1926	2242	4.3F
24 F	0225	0543	3.5E
	0848	1110	2.2F
	1356	1729	3.0E
	2014	2329	4.2F
25 Sa	0313	0629	3.4E
	0935	1158	2.1F
	1447	1817	2.8E
	2102		
26 Su		0017	4.0F
	0400	0716	3.2E
	1023	1248	2.0F
	1541	1907	2.6E
	2153		
27 M		0106	3.5F
	0447	0803	3.0E
	1111	1342	1.9F
	1641	2001	2.3E
	2247		
28 Tu		0159	3.0F
	0534	0853	2.6E
	1202	1441	1.8F
	1748	2101	1.9E
	2347		
29 W ◑		0255	2.4F
	0623	0945	2.3E
	1255	1545	1.8F
	1904	2209	1.6E
30 Th	0057	0359	1.9F
	0714	1041	2.1E
	1350	1653	1.9F
	2023	2325	1.5E

Time meridian 120° E. 0000 is midnight. 1200 is noon. Times are not adjusted for Daylight Saving Time.

Changjiang Entrance, China, 2005

F–Flood, Dir. 305° True E–Ebb, Dir. 125° True

July

Day	Slack h m	Maximum h m	knots
1 F	0217	0508	1.5F
	0807	1140	1.9E
	1445	1758	2.1F
	2136		
2 Sa		0040	1.6E
	0340	0617	1.3F
	0905	1239	1.8E
	1538	1857	2.3F
	2237		
3 Su		0146	1.8E
	0452	0720	1.3F
	1001	1334	1.9E
	1627	1949	2.6F
	2328		
4 M		0241	2.1E
	0549	0814	1.4F
	1053	1423	2.0E
	1713	2034	2.9F
5 Tu	0013	0327	2.3E
	0636	0900	1.5F
	1140	1508	2.1E
	1756	2116	3.1F
6 W ●	0053	0408	2.6E
	0716	0940	1.7F
	1221	1548	2.3E
	1836	2154	3.3F
7 Th	0130	0445	2.7E
	0753	1018	1.8F
	1300	1626	2.4E
	1914	2231	3.5F
8 F	0206	0520	2.8E
	0827	1053	1.9F
	1337	1702	2.5E
	1951	2307	3.5F
9 Sa	0241	0555	2.8E
	0901	1128	1.9F
	1414	1738	2.5E
	2027	2341	3.4F
10 Su	0315	0628	2.8E
	0933	1203	1.9F
	1452	1814	2.4E
	2103		
11 M		0016	3.2F
	0348	0701	2.7E
	1006	1239	1.9F
	1531	1851	2.3E
	2140		
12 Tu		0051	2.9F
	0420	0735	2.5E
	1040	1316	1.9F
	1614	1931	2.1E
	2220		
13 W		0128	2.6F
	0453	0809	2.3E
	1116	1358	1.8F
	1703	2015	1.8E
	2305		
14 Th ◐		0210	2.2F
	0527	0847	2.1E
	1156	1447	1.8F
	1801	2108	1.6E
15 F	0001	0300	1.8F
	0607	0932	1.9E
	1242	1544	1.9F
	1910	2215	1.4E
16 Sa	0113	0402	1.4F
	0656	1027	1.7E
	1336	1651	2.0F
	2027	2337	1.5E
17 Su	0242	0517	1.1F
	0758	1134	1.7E
	1437	1802	2.3F
	2144		
18 M		0100	1.8E
	0411	0635	1.1F
	0909	1246	1.8E
	1540	1909	2.8F
	2251		
19 Tu		0210	2.2E
	0523	0743	1.4F
	1018	1352	2.1E
	1641	2007	3.3F
	2348		
20 W		0309	2.7E
	0620	0841	1.7F
	1119	1452	2.4E
	1738	2100	3.8F
21 Th ○	0039	0400	3.2E
	0708	0930	2.0F
	1213	1545	2.8E
	1830	2148	4.3F
22 F	0127	0447	3.5E
	0752	1016	2.4F
	1304	1636	3.1E
	1920	2233	4.4F
23 Sa	0212	0531	3.7E
	0833	1100	2.6F
	1353	1723	3.3E
	2008	2317	4.4F
24 Su	0254	0613	3.7E
	0913	1143	2.7F
	1443	1810	3.2E
	2055	2359	4.1F
25 M	0335	0653	3.6E
	0953	1227	2.7F
	1533	1856	3.0E
	2141		
26 Tu		0043	3.6F
	0414	0733	3.3E
	1032	1313	2.6F
	1627	1943	2.7E
	2230		
27 W		0127	3.0F
	0452	0812	2.8E
	1113	1402	2.4F
	1724	2033	2.2E
	2322		
28 Th ◑		0215	2.2F
	0529	0853	2.4E
	1157	1455	2.2F
	1828	2130	1.7E
29 F	0024	0309	1.5F
	0610	0940	1.9E
	1246	1558	2.0F
	1942	2242	1.4E
30 Sa	0145	0418	1.0F
	0701	1039	1.5E
	1344	1712	1.9F
	2104		
31 Su		0009	1.3E
	0326	0545	0.7F
	0812	1152	1.3E
	1451	1830	1.9F
	2218		

August

Day	Slack h m	Maximum h m	knots
1 M		0130	1.5E
	0454	0709	0.7F
	0933	1306	1.4E
	1558	1935	2.2F
	2316		
2 Tu		0232	1.8E
	0552	0811	1.0F
	1041	1408	1.6E
	1655	2025	2.6F
3 W	0001	0318	2.2E
	0632	0856	1.3F
	1132	1456	1.9E
	1743	2107	3.0F
4 Th	0040	0356	2.5E
	0706	0932	1.7F
	1214	1538	2.2E
	1825	2143	3.3F
5 F ●	0114	0430	2.8E
	0736	1005	2.0F
	1251	1615	2.5E
	1903	2216	3.5F
6 Sa	0146	0502	3.0E
	0804	1035	2.2F
	1326	1650	2.7E
	1938	2247	3.6F
7 Su	0216	0532	3.1E
	0832	1105	2.4F
	1401	1724	2.8E
	2012	2317	3.6F
8 M	0245	0601	3.1E
	0859	1135	2.5F
	1436	1758	2.8E
	2045	2346	3.4F
9 Tu	0312	0629	3.0E
	0926	1206	2.6F
	1511	1831	2.7E
	2120		
10 W		0017	3.1F
	0338	0657	2.8E
	0953	1238	2.6F
	1549	1906	2.5E
	2156		
11 Th		0050	2.7F
	0405	0725	2.6E
	1023	1314	2.5F
	1632	1944	2.2E
	2237		
12 F		0127	2.2F
	0433	0756	2.3E
	1058	1357	2.4F
	1722	2030	1.9E
	2328		
13 Sa ◐		0212	1.7F
	0507	0834	2.0E
	1139	1450	2.2F
	1827	2133	1.5E
14 Su	0040	0312	1.1F
	0552	0927	1.6E
	1234	1602	2.1F
	1952	2303	1.4E
15 M	0224	0439	0.7F
	0702	1046	1.4E
	1349	1731	2.2F
	2124		
16 Tu		0044	1.6E
	0413	0623	0.8F
	0842	1222	1.5E
	1513	1855	2.6F
	2239		
17 W		0202	2.2E
	0524	0741	1.2F
	1011	1342	1.9E
	1628	1959	3.2F
	2337		
18 Th		0300	2.8E
	0613	0836	1.7F
	1116	1445	2.4E
	1730	2050	3.8F
19 F	0025	0348	3.3E
	0654	0921	2.3F
	1210	1539	2.9E
	1823	2135	4.3F
20 Sa ○	0108	0430	3.7E
	0731	1002	2.8F
	1258	1627	3.3E
	1911	2216	4.5F
21 Su	0148	0510	3.9E
	0806	1041	3.2F
	1345	1712	3.6E
	1956	2256	4.4F
22 M	0225	0546	4.0E
	0841	1120	3.4F
	1431	1754	3.5E
	2040	2335	4.0F
23 Tu	0300	0622	3.7E
	0915	1159	3.4F
	1517	1836	3.3E
	2123		
24 W		0013	3.4F
	0332	0655	3.4E
	0948	1238	3.2F
	1603	1917	2.9E
	2207		
25 Th		0052	2.7F
	0403	0728	2.9E
	1023	1320	2.9F
	1653	2000	2.3E
	2254		
26 F ◑		0133	2.0F
	0434	0801	2.3E
	1050	1406	2.5F
	1748	2050	1.7E
	2351		
27 Sa		0220	1.2F
	0506	0839	1.7E
	1142	1503	2.0F
	1858	2157	1.2E
28 Su	0115	0327	0.6F
	0549	0933	1.2E
	1239	1624	1.6F
	2030	2339	1.0E
29 M	0331	0522	0.3F
	0717	1107	0.9E
	1405	1805	1.6F
	2158		
30 Tu		0120	1.3E
	0503	0709	0.5F
	0923	1250	1.0E
	1535	1922	2.0F
	2259		
31 W		0220	1.7E
	0545	0806	0.9F
	1038	1357	1.4E
	1641	2011	2.4F
	2342		

September

Day	Slack h m	Maximum h m	knots
1 Th		0301	2.1E
	0615	0843	1.4F
	1125	1445	1.9E
	1730	2049	2.9F
2 F	0016	0335	2.5E
	0641	0914	1.9F
	1203	1523	2.3E
	1810	2121	3.2F
3 Sa	0047	0405	2.9E
	0706	0942	2.3F
	1237	1558	2.7E
	1845	2151	3.5F
4 Su ●	0115	0433	3.1E
	0730	1009	2.7F
	1310	1632	3.0E
	1919	2219	3.6F
5 M	0142	0500	3.3E
	0754	1036	3.0F
	1342	1704	3.1E
	1952	2247	3.5F
6 Tu	0207	0527	3.3E
	0819	1103	3.2F
	1415	1736	3.1E
	2025	2315	3.3F
7 W	0231	0553	3.2E
	0844	1132	3.2F
	1449	1809	3.0E
	2059	2344	3.0F
8 Th	0255	0619	3.0E
	0910	1203	3.2F
	1525	1842	2.8E
	2135		
9 F		0016	2.6F
	0320	0645	2.7E
	0939	1238	3.0F
	1606	1919	2.4E
	2217		
10 Sa		0053	2.1F
	0348	0715	2.3E
	1012	1319	2.8F
	1655	2005	2.0E
	2309		
11 Su ◐		0137	1.4F
	0420	0751	1.9E
	1053	1413	2.4F
	1801	2109	1.6E
12 M	0026	0240	0.8F
	0505	0845	1.4E
	1151	1530	2.1F
	1933	2249	1.4E
13 Tu	0232	0427	0.4F
	0631	1021	1.1E
	1319	1716	2.1F
	2112		
14 W		0039	1.6E
	0418	0626	0.7F
	0845	1217	1.3E
	1501	1845	2.5F
	2225		
15 Th		0151	2.3E
	0513	0736	1.3F
	1015	1338	1.9E
	1621	1947	3.2F
	2318		
16 F		0243	2.9E
	0552	0823	2.0F
	1115	1438	2.5E
	1721	2034	3.7F
17 Sa	0002	0326	3.4E
	0627	0903	2.8F
	1204	1528	3.1E
	1812	2116	4.1F
18 Su ○	0041	0404	3.8E
	0700	0941	3.3F
	1248	1613	3.5E
	1858	2154	4.2F
19 M	0116	0440	4.0E
	0732	1017	3.8F
	1331	1655	3.7E
	1942	2231	4.0F
20 Tu	0149	0514	3.9E
	0804	1052	4.0F
	1413	1734	3.7E
	2023	2307	3.6F
21 W	0220	0545	3.7E
	0835	1128	3.9F
	1455	1812	3.4E
	2104	2343	3.0F
22 Th	0249	0615	3.2E
	0906	1204	3.6F
	1537	1850	3.0E
	2145		
23 F		0019	2.3F
	0317	0644	2.7E
	0937	1242	3.2F
	1621	1929	2.4E
	2230		
24 Sa		0057	1.6F
	0345	0714	2.2E
	1009	1324	2.6F
	1711	2013	1.8E
	2324		
25 Su ◑		0141	1.0F
	0414	0747	1.6E
	1047	1416	2.0F
	1815	2116	1.2E
26 M	0051	0247	0.4F
	0451	0837	1.0E
	1141	1538	1.5F
	1947	2305	1.0E
27 Tu	0330	0501	0.2F
	0635	1024	0.7E
	1318	1732	1.5F
	2122		
28 W		0054	1.2E
	0442	0652	0.5F
	0912	1229	0.8E
	1507	1853	1.8F
	2225		
29 Th		0150	1.7E
	0514	0743	1.0F
	1024	1336	1.3E
	1617	1943	2.2F
	2306		
30 F		0228	2.2E
	0539	0816	1.6F
	1108	1422	1.9E
	1706	2020	2.7F
	2340		

Time meridian 120° E. 0000 is midnight. 1200 is noon. Times are not adjusted for Daylight Saving Time.

Changjiang Entrance, China, 2005

F–Flood, Dir. 305° True   E–Ebb, Dir. 125° True

October

Day	Slack (h m)	Maximum (h m)	knots
1 Sa		0259	2.6E
	0603	0844	2.2F
	1143	1500	2.4E
	1746	2051	3.0F
2 Su	0009	0328	2.9E
	0626	0911	2.7F
	1216	1535	2.8E
	1823	2120	3.2F
3 M ●	0036	0356	3.2E
	0650	0937	3.1F
	1249	1608	3.1E
	1857	2148	3.3F
4 Tu	0102	0423	3.3E
	0714	1004	3.5F
	1321	1641	3.3E
	1932	2216	3.3F
5 W	0127	0451	3.3E
	0739	1032	3.7F
	1354	1715	3.3E
	2006	2246	3.1F
6 Th	0152	0518	3.2E
	0805	1102	3.8F
	1429	1748	3.2E
	2043	2317	2.8F
7 F	0218	0546	3.0E
	0834	1135	3.7F
	1507	1824	3.0E
	2122	2352	2.3F
8 Sa	0245	0615	2.7E
	0905	1212	3.4F
	1550	1904	2.6E
	2207		
9 Su		0031	1.8F
	0316	0647	2.3E
	0941	1257	3.0F
	1642	1953	2.1E
	2305		
10 M		0121	1.2F
	0353	0728	1.8E
	1026	1355	2.5F
	1751	2103	1.7E
11 Tu ◐	0033	0234	0.6F
	0449	0831	1.3E
	1131	1518	2.1F
	1921	2244	1.5E
12 W	0234	0433	0.5F
	0644	1021	1.0E
	1309	1703	2.1F
	2051		
13 Th		0021	1.9E
	0355	0617	0.9F
	0854	1213	1.3E
	1452	1827	2.5F
	2159		
14 F		0126	2.4E
	0442	0718	1.6F
	1012	1328	1.9E
	1610	1925	2.9F
	2249		
15 Sa		0214	2.9E
	0519	0802	2.4F
	1107	1425	2.6E
	1709	2012	3.3F
	2330		
16 Su		0255	3.3E
	0553	0840	3.1F
	1153	1512	3.1E
	1759	2052	3.5F
17 M ○	0007	0333	3.6E
	0625	0916	3.7F
	1235	1555	3.5E
	1844	2130	3.5F
18 Tu	0040	0407	3.7E
	0657	0951	4.1F
	1315	1635	3.6E
	1927	2206	3.3F
19 W	0112	0439	3.6E
	0727	1025	4.2F
	1355	1713	3.6E
	2007	2241	3.0F
20 Th	0142	0510	3.3E
	0758	1100	4.1F
	1434	1749	3.3E
	2047	2317	2.5F
21 F	0212	0540	3.0E
	0829	1136	3.7F
	1513	1826	2.9E
	2128	2353	2.0F
22 Sa	0241	0609	2.5E
	0900	1213	3.3F
	1555	1903	2.4E
	2211		
23 Su		0032	1.4F
	0311	0639	2.0E
	0933	1254	2.7F
	1642	1946	1.9E
	2305		
24 M		0118	0.9F
	0345	0714	1.5E
	1011	1345	2.1F
	1739	2045	1.4E
25 Tu ◑	0025	0224	0.5F
	0432	0805	1.0E
	1104	1457	1.6F
	1856	2215	1.1E
26 W	0222	0414	0.3F
	0614	0941	0.7E
	1230	1635	1.5F
	2021	2353	1.3E
27 Th	0339	0559	0.6F
	0833	1141	0.8E
	1418	1800	1.6F
	2127		
28 F		0055	1.6E
	0418	0657	1.1F
	0950	1256	1.2E
	1536	1857	1.9F
	2214		
29 Sa		0137	2.0E
	0447	0735	1.7F
	1038	1346	1.8E
	1632	1938	2.3F
	2250		
30 Su		0212	2.4E
	0514	0806	2.3F
	1116	1428	2.3E
	1718	2014	2.6F
	2322		
31 M		0244	2.7E
	0540	0835	2.8F
	1151	1507	2.7E
	1758	2046	2.8F
	2352		

November

Day	Slack (h m)	Maximum (h m)	knots
1 Tu		0315	3.0E
	0606	0904	3.3F
	1226	1544	3.1E
	1837	2118	2.8F
2 W ●	0021	0346	3.1E
	0634	0934	3.7F
	1301	1620	3.3E
	1915	2150	2.8F
3 Th	0050	0418	3.1E
	0703	1006	4.0F
	1337	1657	3.4E
	1953	2223	2.7F
4 F	0120	0449	3.1E
	0735	1040	4.0F
	1416	1734	3.3E
	2034	2300	2.4F
5 Sa	0151	0522	2.9E
	0808	1118	3.9F
	1458	1814	3.1E
	2118	2340	2.0F
6 Su	0225	0557	2.6E
	0846	1200	3.6F
	1545	1859	2.7E
	2208		
7 M		0026	1.6F
	0304	0637	2.2E
	0928	1250	3.2F
	1639	1953	2.3E
	2311		
8 Tu		0123	1.1F
	0354	0727	1.8E
	1020	1351	2.7F
	1744	2101	2.0E
9 W ◐	0031	0240	0.8F
	0507	0839	1.4E
	1128	1508	2.3F
	1859	2224	1.9E
10 Th	0159	0418	0.9F
	0658	1018	1.2E
	1300	1637	2.2F
	2015	2344	2.0E
11 F	0308	0545	1.3F
	0845	1154	1.4E
	1435	1755	2.3F
	2118		
12 Sa		0046	2.4E
	0358	0647	2.0F
	0958	1308	1.9E
	1552	1856	2.5F
	2209		
13 Su		0137	2.7E
	0439	0734	2.6F
	1053	1406	2.4E
	1654	1945	2.6F
	2252		
14 M		0220	3.0E
	0515	0814	3.2F
	1139	1454	2.9E
	1746	2028	2.7F
	2331		
15 Tu		0259	3.1E
	0550	0852	3.7F
	1221	1538	3.2E
	1833	2108	2.7F
16 W ○	0006	0335	3.2E
	0623	0928	4.0F
	1301	1618	3.3E
	1915	2146	2.6F
17 Th	0040	0409	3.1E
	0657	1004	4.1F
	1339	1656	3.3E
	1956	2223	2.4F
18 F	0113	0442	2.9E
	0729	1040	3.9F
	1418	1732	3.1E
	2036	2300	2.1F
19 Sa	0146	0514	2.7E
	0803	1117	3.7F
	1456	1808	2.8E
	2116	2338	1.8F
20 Su	0220	0546	2.4E
	0837	1156	3.3F
	1536	1846	2.5E
	2159		
21 M		0019	1.4F
	0255	0620	2.0E
	0913	1237	2.9F
	1620	1928	2.1E
	2248		
22 Tu		0105	1.1F
	0336	0700	1.7E
	0953	1324	2.4F
	1708	2018	1.7E
	2346		
23 W		0202	0.8F
	0429	0749	1.3E
	1042	1420	2.0F
	1803	2119	1.5E
24 Th ◑	0056	0314	0.7F
	0546	0900	1.0E
	1148	1528	1.7F
	1905	2230	1.5E
25 F	0204	0437	0.8F
	0726	1030	0.9E
	1312	1643	1.5F
	2007	2335	1.6E
26 Sa	0258	0546	1.2F
	0853	1154	1.1E
	1438	1751	1.6F
	2102		
27 Su		0030	1.8E
	0340	0638	1.7F
	0954	1259	1.5E
	1548	1846	1.7F
	2148		
28 M		0115	2.1E
	0416	0720	2.3F
	1042	1352	2.0E
	1646	1931	1.9F
	2230		
29 Tu		0156	2.3E
	0450	0757	2.8F
	1124	1438	2.5E
	1735	2013	2.1F
	2308		
30 W		0235	2.6E
	0525	0833	3.3F
	1204	1521	2.9E
	1820	2052	2.2F
	2345		

December

Day	Slack (h m)	Maximum (h m)	knots
1 Th ●		0313	2.8E
	0600	0910	3.8F
	1245	1603	3.2E
	1904	2131	2.3F
2 F	0022	0352	2.9E
	0637	0948	4.1F
	1326	1645	3.4E
	1947	2211	2.3F
3 Sa	0059	0431	2.9E
	0715	1028	4.2F
	1409	1727	3.4E
	2031	2253	2.2F
4 Su	0139	0511	2.9E
	0756	1111	4.1F
	1455	1811	3.3E
	2118	2338	2.0F
5 M	0221	0553	2.7E
	0840	1157	3.9F
	1542	1858	3.0E
	2208		
6 Tu		0027	1.8F
	0309	0640	2.4E
	0928	1247	3.5F
	1632	1949	2.7E
	2303		
7 W		0123	1.5F
	0406	0734	2.1E
	1021	1343	3.0F
	1726	2045	2.5E
8 Th ◐	0003	0228	1.4F
	0518	0838	1.7E
	1124	1446	2.6F
	1824	2147	2.3E
9 F	0106	0342	1.4F
	0646	0957	1.5E
	1240	1557	2.1F
	1924	2253	2.2E
10 Sa	0208	0459	1.7F
	0817	1122	1.5E
	1406	1712	1.9F
	2024	2356	2.2E
11 Su	0304	0607	2.1F
	0934	1239	1.7E
	1530	1820	1.8F
	2121		
12 M		0053	2.3E
	0353	0703	2.5F
	1036	1344	2.1E
	1640	1919	1.8F
	2212		
13 Tu		0144	2.4E
	0438	0750	3.0F
	1126	1438	2.4E
	1738	2009	1.8F
	2258		
14 W		0229	2.5E
	0519	0833	3.3F
	1210	1525	2.7E
	1827	2054	1.9F
	2340		
15 Th		0310	2.6E
	0558	0913	3.6F
	1251	1606	2.9E
	1910	2135	2.0F
16 F ○	0020	0349	2.6E
	0636	0951	3.7F
	1330	1644	3.0E
	1950	2214	2.0F
17 Sa	0057	0425	2.6E
	0713	1029	3.7F
	1407	1721	3.0E
	2028	2252	1.9F
18 Su	0134	0500	2.6E
	0749	1107	3.6F
	1444	1757	2.8E
	2106	2329	1.8F
19 M	0211	0536	2.5E
	0826	1144	3.4F
	1521	1833	2.7E
	2143		
20 Tu		0008	1.7F
	0249	0612	2.3E
	0903	1222	3.2F
	1558	1909	2.5E
	2221		
21 W		0047	1.5F
	0330	0650	2.0E
	0941	1301	2.8F
	1635	1948	2.2E
	2302		
22 Th		0131	1.4F
	0416	0732	1.8E
	1023	1342	2.4F
	1714	2029	2.0E
	2345		
23 F		0219	1.3F
	0512	0822	1.5E
	1111	1429	2.0F
	1755	2115	1.8E
24 Sa ◑	0033	0315	1.3F
	0620	0924	1.2E
	1212	1524	1.6F
	1841	2207	1.7E
25 Su	0123	0419	1.4F
	0738	1040	1.1E
	1329	1630	1.4F
	1934	2306	1.6E
26 M	0216	0525	1.7F
	0855	1201	1.3E
	1456	1741	1.2F
	2032		
27 Tu		0007	1.7E
	0309	0626	2.1F
	1002	1313	1.7E
	1614	1847	1.3F
	2131		
28 W		0105	1.9E
	0359	0719	2.6F
	1057	1413	2.1E
	1718	1945	1.5F
	2227		
29 Th		0159	2.1E
	0448	0808	3.1F
	1147	1505	2.6E
	1811	2035	1.7F
	2317		
30 F		0249	2.4E
	0535	0854	3.6F
	1234	1552	3.0E
	1859	2122	1.9F
31 Sa ●	0005	0336	2.7E
	0621	0939	4.1F
	1319	1638	3.4E
	1944	2206	2.1F

Time meridian 120° E. 0000 is midnight. 1200 is noon. Times are not adjusted for Daylight Saving Time.

Wusong Kou, China, 2005

F–Flood, Dir. 290° True E–Ebb, Dir. 110° True

January

Day	Slack h m	Maximum h m	knots
1 Sa	0043	0301	2.0F
	0555	0943	1.5E
	1210	1506	2.8F
	1817	2224	2.3E
2 Su	0121	0344	1.9F
	0642	1025	1.3E
	1255	1551	2.4F
	1856	2259	2.0E
3 M	0201	0432	1.9F
	0738	1113	1.2E
	1350	1643	2.0F
	1941	2337	1.8E
4 Tu	0246	0529	1.9F
	0841	1214	1.2E
	1501	1747	1.6F
◑	2035		
5 W		0021	1.5E
	0337	0633	2.0F
	0950	1334	1.3E
	1629	1902	1.4F
	2139		
6 Th		0119	1.4E
	0434	0740	2.2F
	1056	1501	1.6E
	1756	2017	1.5F
	2250		
7 F		0238	1.3E
	0535	0843	2.6F
	1158	1616	2.1E
	1909	2123	1.7F
	2359		
8 Sa		0403	1.3E
	0635	0941	3.1F
	1253	1719	2.6E
	2009	2221	2.1F
9 Su	0101	0514	1.5E
	0731	1033	3.7F
	1344	1815	3.1E
	2101	2312	2.5F
10 M	0154	0614	1.7E
	0825	1122	4.2F
	1433	1906	3.5E
●	2149	2359	2.8F
11 Tu	0243	0709	2.0E
	0915	1209	4.5F
	1519	1954	3.7E
	2234		
12 W		0044	3.0F
	0330	0758	2.1E
	1004	1253	4.6F
	1604	2038	3.8E
	2316		
13 Th		0127	3.1F
	0416	0844	2.2E
	1051	1337	4.5F
	1649	2121	3.7E
	2358		
14 F		0210	3.1F
	0504	0929	2.2E
	1137	1420	4.1F
	1733	2201	3.3E
15 Sa	0038	0253	2.9F
	0555	1012	2.1E
	1225	1505	3.5F
	1817	2240	2.9E
16 Su	0118	0339	2.6F
	0651	1056	1.9E
	1315	1552	2.8F
	1901	2318	2.3E
17 M	0159	0428	2.3F
	0752	1144	1.6E
	1411	1645	2.0F
◐	1948	2357	1.8E
18 Tu	0243	0525	2.0F
	0858	1241	1.4E
	1521	1749	1.4F
	2043		
19 W		0042	1.4E
	0331	0630	1.8F
	1010	1352	1.3E
	1649	1908	1.0F
	2152		
20 Th		0141	1.0E
	0429	0743	1.8F
	1121	1511	1.4E
	1820	2033	0.9F
	2315		
21 F		0255	0.9E
	0532	0853	1.9F
	1224	1619	1.7E
	1930	2145	1.1F
22 Sa	0031	0403	1.0E
	0634	0953	2.2F
	1314	1714	2.0E
	2022	2238	1.4F
23 Su	0126	0500	1.1E
	0729	1041	2.6F
	1357	1800	2.3E
	2103	2320	1.7F
24 M	0208	0548	1.4E
	0816	1122	3.0F
	1434	1842	2.6E
	2138	2355	2.1F
25 Tu	0243	0633	1.6E
	0857	1157	3.3F
	1509	1920	2.9E
○	2210		
26 W		0027	2.4F
	0316	0714	1.8E
	0936	1230	3.6F
	1542	1955	3.0E
	2241		
27 Th		0057	2.6F
	0347	0753	2.0E
	1012	1302	3.8F
	1613	2029	3.1E
	2310		
28 F		0127	2.8F
	0418	0829	2.1E
	1048	1333	3.7F
	1643	2101	3.0E
	2340		
29 Sa		0158	2.8F
	0451	0904	2.1E
	1123	1406	3.6F
	1712	2131	2.8E
30 Su	0009	0231	2.9F
	0526	0938	2.1E
	1200	1441	3.3F
	1742	2159	2.6E
31 M	0039	0306	2.8F
	0605	1010	1.9E
	1240	1519	2.8F
	1816	2224	2.2E

February

Day	Slack h m	Maximum h m	knots
1 Tu	0111	0346	2.6F
	0650	1045	1.8E
	1326	1604	2.3F
	1854	2245	1.8E
2 W	0146	0434	2.4F
	0743	1128	1.6E
	1425	1659	1.7F
◑	1941	2310	1.5E
3 Th	0228	0533	2.2F
	0849	1235	1.4E
	1548	1812	1.2F
	2044	2357	1.1E
4 F	0324	0647	2.1F
	1009	1416	1.5E
	1732	1944	1.0F
	2212		
5 Sa		0132	0.9E
	0442	0811	2.2F
	1133	1553	1.9E
	1902	2113	1.2F
	2351		
6 Su		0339	0.9E
	0607	0927	2.7F
	1242	1708	2.4E
	2006	2219	1.8F
7 M	0104	0508	1.3E
	0721	1027	3.3F
	1337	1808	3.0E
	2056	2309	2.4F
8 Tu	0156	0614	1.7E
	0822	1117	4.0F
	1425	1859	3.5E
	2139	2352	3.0F
9 W	0240	0708	2.2E
	0914	1201	4.5F
	1507	1944	3.8E
●	2218		
10 Th		0031	3.5F
	0321	0755	2.5E
	1001	1242	4.7F
	1548	2025	3.8E
	2254		
11 F		0109	3.7F
	0402	0836	2.8E
	1045	1321	4.6F
	1626	2102	3.7E
	2329		
12 Sa		0146	3.8F
	0443	0914	2.8E
	1127	1359	4.2F
	1704	2135	3.3E
13 Su	0002	0223	3.6F
	0527	0950	2.7E
	1209	1438	3.6F
	1742	2206	2.8E
14 M	0034	0302	3.3F
	0612	1025	2.4E
	1251	1518	2.8F
	1819	2232	2.2E
15 Tu	0105	0342	2.8F
	0700	1100	2.0E
	1338	1602	2.0F
	1857	2256	1.7E
16 W	0136	0426	2.3F
	0753	1141	1.6E
	1435	1654	1.3F
◐	1941	2322	1.1E
17 Th	0212	0521	1.8F
	0859	1240	1.3E
	1559	1809	0.7F
	2045		
18 F		0011	0.7E
	0300	0639	1.4F
	1025	1413	1.2E
	1801	2010	0.5F
	2246		
19 Sa		0150	0.5E
	0425	0824	1.4F
	1153	1549	1.4E
	1927	2146	0.8F
20 Su	0033	0335	0.6E
	0606	0942	1.8F
	1256	1656	1.8E
	2014	2236	1.3F
21 M	0126	0448	1.0E
	0717	1032	2.3F
	1341	1746	2.2E
	2048	2310	1.8F
22 Tu	0202	0543	1.4E
	0808	1109	2.8F
	1417	1827	2.6E
	2118	2338	2.3F
23 W	0231	0628	1.8E
	0849	1141	3.3F
	1449	1903	2.9E
	2145		
24 Th		0005	2.7F
	0258	0708	2.1E
	0926	1211	3.7F
○	1517	1938	3.1E
	2212		
25 F		0032	3.1F
	0325	0745	2.4E
	1001	1240	3.9F
	1545	2009	3.2E
	2238		
26 Sa		0059	3.4F
	0353	0819	2.6E
	1035	1309	4.0F
	1611	2040	3.1E
	2305		
27 Su		0128	3.6F
	0423	0852	2.7E
	1110	1340	3.8F
	1638	2108	2.9E
	2332		
28 M		0158	3.7F
	0455	0923	2.7E
	1145	1414	3.5F
	1708	2133	2.6E
	2359		

March

Day	Slack h m	Maximum h m	knots
1 Tu		0232	3.5F
	0531	0953	2.5E
	1224	1450	3.0F
	1740	2154	2.1E
2 W	0027	0309	3.2F
	0614	1024	2.3E
	1308	1532	2.3F
	1818	2208	1.7E
3 Th	0058	0353	2.8F
	0704	1102	1.9E
	1404	1625	1.6F
	1903	2224	1.3E
4 F	0134	0448	2.3F
	0809	1204	1.6E
	1527	1739	1.0F
◑	2009	2311	0.8E
5 Sa	0227	0608	1.9F
	0938	1349	1.5E
	1726	1932	0.7F
	2207		
6 Su		0115	0.6E
	0405	0754	1.9F
	1116	1540	1.8E
	1859	2113	1.1F
7 M	0007	0346	0.8E
	0600	0920	2.4F
	1231	1658	2.4E
	1955	2213	1.9F
8 Tu	0109	0512	1.3E
	0721	1019	3.2F
	1324	1756	2.9E
	2039	2256	2.7F
9 W	0152	0612	1.9E
	0820	1105	3.9F
	1407	1844	3.3E
	2116	2334	3.4F
10 Th	0229	0700	2.5E
	0908	1145	4.3F
	1446	1925	3.6E
●	2151		
11 F		0010	3.9F
	0304	0741	3.0E
	0952	1223	4.6F
	1522	2002	3.6E
	2223		
12 Sa		0044	4.2F
	0341	0818	3.2E
	1032	1259	4.4F
	1557	2035	3.4E
	2253		
13 Su		0118	4.3F
	0418	0852	3.2E
	1111	1334	4.0F
	1632	2104	3.0E
	2322		
14 M		0151	4.1F
	0456	0923	3.1E
	1149	1410	3.4F
	1706	2128	2.5E
	2349		
15 Tu		0225	3.7F
	0536	0952	2.8E
	1227	1446	2.7F
	1741	2148	2.0E
16 W	0015	0300	3.1F
	0618	1020	2.3E
	1308	1525	1.9F
	1817	2204	1.5E
17 Th	0039	0338	2.5F
	0705	1053	1.9E
	1359	1611	1.1F
	1858	2224	1.0E
18 F	0106	0423	1.8F
	0803	1142	1.4E
	1517	1722	0.5F
◐	2000	2308	0.6E
19 Sa	0142	0535	1.3F
	0928	1311	1.1E
	1738	1955	0.3F
	2231		
20 Su		0054	0.3E
	0311	0749	1.1F
	1110	1510	1.3E
	1903	2130	0.8F
21 M	0026	0314	0.5E
	0540	0918	1.5F
	1222	1626	1.7E
	1943	2212	1.3F
22 Tu	0111	0434	0.9E
	0700	1007	2.1F
	1309	1715	2.1E
	2014	2242	1.9F
23 W	0141	0526	1.5E
	0751	1043	2.7F
	1345	1756	2.5E
	2042	2308	2.5F
24 Th	0207	0610	2.0E
	0831	1113	3.2F
	1415	1834	2.8E
	2108	2334	3.1F
25 F	0232	0649	2.4E
	0908	1142	3.6F
	1442	1909	3.0E
	2134	2359	3.6F
26 Sa	0257	0725	2.8E
	0943	1212	3.9F
	1509	1941	3.0E
○	2201		
27 Su		0027	4.0F
	0324	0759	3.1E
	1018	1242	3.9F
	1536	2012	2.9E
	2228		
28 M		0057	4.2F
	0354	0832	3.2E
	1054	1314	3.8F
	1605	2041	2.7E
	2255		
29 Tu		0128	4.2F
	0428	0904	3.1E
	1131	1349	3.4F
	1637	2108	2.3E
	2323		
30 W		0203	4.0F
	0506	0937	2.9E
	1211	1427	2.9F
	1712	2131	1.9E
	2353		
31 Th		0241	3.5F
	0550	1010	2.6E
	1257	1511	2.2F
	1754	2149	1.5E

Time meridian 120° E. 0000 is midnight. 1200 is noon. Times are not adjusted for Daylight Saving Time.

Wusong Kou, China, 2005

F–Flood, Dir. 290° True E–Ebb, Dir. 110° True

April

Day	Slack h m	Maximum h m	knots
1 F	0024	0325	3.0F
	0643	1052	2.2E
	1358	1608	1.4F
	1846	2211	1.0E
2 Sa ◑	0102	0424	2.3F
	0752	1157	1.8E
	1527	1733	0.8F
	2012	2315	0.6E
3 Su	0202	0552	1.8F
	0923	1340	1.6E
	1720	1935	0.8F
	2238		
4 M		0149	0.5E
	0404	0744	1.8F
	1059	1523	1.9E
	1837	2101	1.4F
5 Tu	0014	0350	0.9E
	0600	0905	2.3F
	1210	1635	2.4E
	1927	2153	2.2F
6 W	0101	0502	1.6E
	0714	1000	3.0F
	1300	1731	2.8E
	2008	2233	3.0F
7 Th	0137	0555	2.2E
	0809	1044	3.5F
	1341	1817	3.0E
	2044	2309	3.7F
8 F	0210	0640	2.8E
	0855	1123	3.9F
	1418	1857	3.1E
	2116	2343	4.2F
9 Sa ●	0243	0719	3.2E
	0937	1200	4.0F
	1452	1932	3.1E
	2147		
10 Su		0016	4.4F
	0316	0755	3.4E
	1015	1235	3.9F
	1526	2003	2.8E
	2216		
11 M		0049	4.4F
	0352	0826	3.4E
	1052	1309	3.5F
	1600	2029	2.5E
	2243		
12 Tu		0121	4.2F
	0428	0855	3.2E
	1129	1344	3.0F
	1635	2051	2.1E
	2308		
13 W		0153	3.7F
	0507	0922	2.9E
	1206	1419	2.4F
	1711	2110	1.7E
	2333		
14 Th		0227	3.2F
	0548	0949	2.5E
	1246	1459	1.7F
	1750	2128	1.3E
	2357		
15 F		0304	2.5F
	0634	1021	2.0E
	1336	1546	1.1F
	1836	2154	0.9E
16 Sa ◐	0025	0349	1.9F
	0729	1107	1.6E
	1450	1701	0.6F
	1949	2240	0.6E
17 Su	0105	0500	1.4F
	0844	1224	1.3E
	1642	1910	0.5F
	2200		
18 M		0018	0.3E
	0234	0658	1.1F
	1014	1411	1.3E
	1802	2038	0.9F
	2346		
19 Tu		0233	0.5E
	0459	0829	1.4F
	1129	1529	1.6E
	1848	2125	1.4F
20 W	0034	0355	1.0E
	0624	0924	1.9F
	1220	1624	1.9E
	1922	2158	2.0F
21 Th	0105	0451	1.5E
	0719	1003	2.4F
	1259	1710	2.2E
	1953	2227	2.7F
22 F	0131	0538	2.1E
	0804	1038	2.9F
	1331	1753	2.4E
	2022	2256	3.3F
23 Sa	0158	0620	2.6E
	0844	1110	3.3F
	1401	1832	2.6E
	2052	2325	3.9F
24 Su ○	0225	0659	3.0E
	0922	1143	3.5F
	1431	1909	2.6E
	2122	2356	4.3F
25 M	0256	0736	3.3E
	1000	1217	3.6F
	1503	1943	2.5E
	2152		
26 Tu		0029	4.5F
	0330	0812	3.4E
	1038	1252	3.5F
	1537	2016	2.3E
	2223		
27 W		0103	4.4F
	0408	0847	3.4E
	1119	1330	3.1F
	1614	2047	2.0E
	2255		
28 Th		0141	4.1F
	0451	0924	3.2E
	1204	1413	2.5F
	1657	2117	1.7E
	2330		
29 F		0224	3.6F
	0541	1004	2.8E
	1255	1503	1.9F
	1748	2149	1.3E
30 Sa	0008	0314	3.0F
	0639	1053	2.4E
	1401	1607	1.4F
	1856	2233	0.9E

May

Day	Slack h m	Maximum h m	knots
1 Su ◑	0058	0419	2.4F
	0748	1200	2.0E
	1524	1736	1.0F
	2040	2359	0.6E
2 M	0216	0547	1.9F
	0910	1328	1.9E
	1649	1914	1.2F
	2239		
3 Tu		0158	0.8E
	0409	0722	1.9F
	1032	1450	2.0E
	1755	2028	1.7F
	2352		
4 W		0328	1.2E
	0546	0837	2.2F
	1138	1557	2.2E
	1846	2120	2.4F
5 Th	0037	0434	1.8E
	0657	0933	2.6F
	1228	1653	2.4E
	1927	2202	3.0F
6 F	0113	0528	2.3E
	0752	1019	3.0F
	1310	1741	2.5E
	2004	2240	3.6F
7 Sa	0146	0613	2.8E
	0838	1059	3.2F
	1347	1822	2.5E
	2038	2315	4.0F
8 Su ●	0219	0653	3.1E
	0919	1136	3.3F
	1423	1858	2.4E
	2110	2349	4.2F
9 M	0254	0728	3.2E
	0958	1212	3.2F
	1458	1929	2.2E
	2139		
10 Tu		0022	4.2F
	0329	0759	3.2E
	1035	1248	2.9F
	1534	1955	2.0E
	2208		
11 W		0056	3.9F
	0407	0829	3.1E
	1112	1324	2.5F
	1612	2018	1.8E
	2236		
12 Th		0130	3.5F
	0447	0858	2.8E
	1150	1402	2.1F
	1652	2041	1.5E
	2304		
13 F		0206	3.1F
	0529	0928	2.5E
	1232	1444	1.6F
	1736	2107	1.3E
	2335		
14 Sa		0246	2.6F
	0615	1003	2.2E
	1321	1535	1.2F
	1829	2141	1.0E
15 Su	0011	0335	2.1F
	0707	1049	1.8E
	1423	1642	0.9F
	1938	2232	0.7E
16 M ◐	0101	0439	1.6F
	0806	1151	1.6E
	1535	1804	0.9F
	2108	2353	0.6E
17 Tu	0221	0600	1.4F
	0914	1306	1.5E
	1641	1921	1.1F
	2235		
18 W		0134	0.7E
	0406	0721	1.4F
	1021	1416	1.6E
	1733	2019	1.5F
	2334		
19 Th		0257	1.0E
	0534	0826	1.7F
	1118	1516	1.7E
	1817	2103	2.1F
20 F	0014	0402	1.5E
	0639	0916	2.0F
	1204	1611	1.9E
	1856	2141	2.7F
21 Sa	0048	0457	2.0E
	0732	0959	2.4F
	1244	1703	2.0E
	1933	2217	3.3F
22 Su	0121	0546	2.5E
	0818	1038	2.8F
	1322	1752	2.1E
	2009	2252	3.8F
23 M	0156	0631	3.0E
	0902	1117	3.0F
	1359	1835	2.1E
	2045	2329	4.2F
24 Tu ○	0233	0713	3.3E
	0944	1156	3.1F
	1438	1916	2.1E
	2122		
25 W		0007	4.4F
	0313	0754	3.5E
	1028	1237	3.0F
	1519	1955	2.0E
	2200		
26 Th		0047	4.4F
	0358	0835	3.5E
	1113	1320	2.8F
	1605	2034	1.9E
	2239		
27 F		0131	4.1F
	0446	0917	3.3E
	1201	1408	2.4F
	1655	2114	1.6E
	2322		
28 Sa		0218	3.7F
	0539	1003	3.0E
	1255	1502	2.0F
	1755	2200	1.4E
29 Su	0011	0312	3.2F
	0636	1054	2.7E
	1354	1605	1.7F
	1908	2258	1.1E
30 M ◑	0110	0415	2.6F
	0738	1154	2.4E
	1459	1717	1.5F
	2037		
31 Tu		0011	1.0E
	0225	0527	2.2F
	0846	1259	2.1E
	1603	1833	1.6F
	2205		

June

Day	Slack h m	Maximum h m	knots
1 W		0133	1.1E
	0354	0646	1.9F
	0955	1404	2.0E
	1702	1943	2.0F
	2313		
2 Th		0250	1.5E
	0520	0800	1.9F
	1058	1507	1.9E
	1754	2040	2.4F
3 F	0004	0357	1.8E
	0633	0901	2.0F
	1152	1605	1.9E
	1841	2128	2.8F
4 Sa	0045	0455	2.2E
	0731	0952	2.2F
	1239	1657	1.9E
	1922	2210	3.2F
5 Su	0123	0543	2.5E
	0821	1037	2.4F
	1321	1742	1.8E
	2000	2249	3.5F
6 M	0200	0625	2.8E
	0904	1117	2.5F
	1401	1822	1.8E
	2036	2327	3.7F
7 Tu ●	0237	0703	2.9E
	0944	1156	2.5F
	1440	1856	1.7E
	2110		
8 W		0004	3.7F
	0315	0737	3.0E
	1023	1234	2.4F
	1519	1926	1.7E
	2144		
9 Th		0040	3.6F
	0354	0810	2.9E
	1101	1313	2.2F
	1559	1955	1.6E
	2217		
10 F		0117	3.4F
	0434	0843	2.8E
	1139	1352	2.0F
	1641	2026	1.5E
	2252		
11 Sa		0156	3.1F
	0516	0917	2.6E
	1219	1433	1.8F
	1724	2100	1.4E
	2329		
12 Su		0236	2.8F
	0558	0954	2.4E
	1301	1518	1.6F
	1812	2140	1.2E
13 M	0010	0320	2.4F
	0641	1034	2.1E
	1345	1607	1.4F
	1908	2228	1.1E
14 Tu	0059	0409	2.1F
	0726	1119	1.9E
	1433	1702	1.4F
	2011	2327	1.0E
15 W ◐	0200	0507	1.8F
	0816	1208	1.7E
	1523	1802	1.5F
	2118		
16 Th		0036	1.0E
	0316	0613	1.6F
	0911	1301	1.6E
	1613	1904	1.7F
	2221		
17 F		0151	1.2E
	0439	0722	1.5F
	1009	1358	1.5E
	1704	2000	2.1F
	2315		
18 Sa		0306	1.5E
	0555	0826	1.6F
	1106	1459	1.5E
	1754	2052	2.5F
19 Su	0003	0414	1.9E
	0700	0921	1.9F
	1200	1606	1.5E
	1842	2140	3.1F
20 M	0048	0513	2.4E
	0756	1012	2.2F
	1250	1709	1.6E
	1929	2225	3.5F
21 Tu	0133	0605	2.9E
	0847	1059	2.5F
	1338	1805	1.7E
	2016	2310	4.0F
22 W ○	0218	0655	3.2E
	0935	1145	2.7F
	1426	1855	1.8E
	2102	2355	4.3F
23 Th	0305	0741	3.5E
	1022	1230	2.8F
	1513	1943	1.9E
	2148		
24 F		0040	4.4F
	0352	0827	3.6E
	1108	1316	2.8F
	1602	2030	1.9E
	2235		
25 Sa		0126	4.3F
	0441	0913	3.5E
	1154	1403	2.7F
	1654	2117	1.9E
	2323		
26 Su		0213	4.0F
	0530	0958	3.3E
	1241	1452	2.5F
	1751	2206	1.8E
27 M	0014	0303	3.5F
	0621	1044	3.0E
	1329	1544	2.3F
	1855	2257	1.7E
28 Tu	0110	0356	2.9F
	0714	1131	2.6E
	1418	1641	2.2F
	2006	2353	1.6E
29 W ◑	0212	0455	2.3F
	0809	1220	2.2E
	1509	1743	2.1F
	2118		
30 Th		0055	1.5E
	0326	0603	1.8F
	0908	1312	1.8E
	1602	1850	2.1F
	2226		

Time meridian 120° E. 0000 is midnight. 1200 is noon. Times are not adjusted for Daylight Saving Time.

Wusong Kou, China, 2005

F–Flood, Dir. 290° True E–Ebb, Dir. 110° True

July

Day	Slack h m	Maximum h m	knots
1 F		0204	1.5E
	0447	0717	1.5F
	1010	1409	1.5E
	1655	1954	2.2F
	2326		
2 Sa		0315	1.7E
	0607	0828	1.4F
	1114	1510	1.4E
	1749	2053	2.4F
3 Su	0018	0420	2.0E
	0714	0930	1.5F
	1214	1610	1.3E
	1840	2145	2.7F
4 M	0105	0515	2.2E
	0809	1023	1.7F
	1306	1703	1.3E
	1928	2232	2.9F
5 Tu	0147	0602	2.4E
	0855	1108	1.9F
	1352	1750	1.4E
	2012	2315	3.2F
6 W	0227	0644	2.6E
	0936	1150	2.0F
	1433	1831	1.5E
●	2053	2355	3.3F
7 Th	0306	0723	2.8E
	1014	1227	2.2F
	1511	1910	1.6E
	2132		
8 F		0032	3.5F
	0343	0759	2.8E
	1049	1303	2.3F
	1548	1947	1.7E
	2210		
9 Sa		0108	3.5F
	0420	0833	2.8E
	1123	1338	2.3F
	1625	2025	1.7E
	2247		
10 Su		0143	3.4F
	0456	0907	2.8E
	1156	1412	2.3F
	1703	2102	1.7E
	2324		
11 M		0217	3.2F
	0531	0940	2.6E
	1229	1448	2.2F
	1744	2139	1.6E
12 Tu	0003	0254	2.9F
	0605	1012	2.4E
	1302	1525	2.1F
	1827	2217	1.6E
13 W	0045	0334	2.5F
	0641	1044	2.1E
	1337	1608	2.1F
	1916	2259	1.5E
14 Th	0134	0420	2.1F
	0721	1117	1.9E
	1415	1657	2.0F
◐	2010	2347	1.4E
15 F	0233	0515	1.7F
	0807	1152	1.6E
	1459	1754	2.0F
	2112		
16 Sa		0051	1.4E
	0349	0622	1.4F
	0904	1238	1.3E
	1551	1859	2.1F
	2218		
17 Su		0212	1.5E
	0516	0738	1.3F
	1013	1344	1.1E
	1651	2006	2.3F
	2324		
18 M		0335	1.8E
	0637	0852	1.4F
	1127	1510	1.1E
	1756	2111	2.7F
19 Tu	0025	0446	2.3E
	0744	0957	1.7F
	1235	1634	1.2E
	1900	2209	3.2F
20 W	0120	0547	2.8E
	0839	1051	2.2F
	1332	1745	1.5E
	1959	2301	3.8F
21 Th	0209	0643	3.2E
	0928	1139	2.6F
	1422	1846	1.8E
○	2053	2348	4.3F
22 F	0256	0733	3.5E
	1013	1223	3.0F
	1508	1939	2.1E
	2144		
23 Sa		0033	4.5F
	0341	0819	3.7E
	1055	1305	3.2F
	1554	2027	2.3E
	2232		
24 Su		0116	4.6F
	0425	0902	3.7E
	1135	1347	3.3F
	1641	2112	2.4E
	2319		
25 M		0159	4.3F
	0509	0942	3.5E
	1214	1429	3.3F
	1730	2155	2.4E
26 Tu	0006	0242	3.8F
	0552	1020	3.1E
	1252	1512	3.0F
	1824	2238	2.3E
27 W	0055	0328	3.1F
	0637	1057	2.6E
	1331	1559	2.7F
	1922	2322	2.0E
28 Th	0148	0418	2.3F
	0723	1134	2.0E
	1412	1651	2.4F
◑	2025		
29 F		0012	1.7E
	0252	0516	1.6F
	0814	1214	1.5E
	1457	1752	2.0F
	2132		
30 Sa		0115	1.5E
	0412	0630	1.1F
	0918	1306	1.1E
	1550	1904	1.9F
	2245		
31 Su		0233	1.5E
	0546	0800	0.9F
	1042	1417	0.9E
	1656	2022	1.9F
	2355		

August

Day	Slack h m	Maximum h m	knots
1 M		0350	1.6E
	0708	0922	0.9F
	1206	1532	0.9E
	1806	2131	2.1F
2 Tu	0052	0454	1.9E
	0806	1022	1.3F
	1307	1638	1.0E
	1909	2225	2.5F
3 W	0138	0546	2.2E
	0850	1106	1.7F
	1351	1733	1.2E
	2002	2309	2.9F
4 Th	0217	0631	2.5E
	0926	1142	2.0F
	1427	1822	1.5E
	2046	2346	3.3F
5 F	0253	0710	2.7E
	0958	1215	2.4F
	1459	1904	1.7E
●	2126		
6 Sa		0019	3.5F
	0326	0745	2.9E
	1027	1244	2.6F
	1531	1943	2.0E
	2202		
7 Su		0050	3.7F
	0357	0817	3.0E
	1055	1313	2.8F
	1603	2019	2.1E
	2237		
8 M		0120	3.7F
	0426	0848	2.9E
	1123	1342	2.9F
	1635	2053	2.2E
	2312		
9 Tu		0151	3.6F
	0455	0917	2.8E
	1150	1412	3.0F
	1709	2126	2.2E
	2347		
10 W		0223	3.3F
	0524	0944	2.5E
	1218	1445	2.9F
	1745	2158	2.1E
11 Th	0025	0258	2.8F
	0555	1009	2.2E
	1247	1521	2.7F
	1827	2231	1.9E
12 F	0107	0339	2.3F
	0630	1031	1.8E
	1318	1604	2.5F
	1915	2310	1.7E
13 Sa	0200	0428	1.8F
	0713	1052	1.4E
	1355	1657	2.2F
◐	2015		
14 Su		0005	1.5E
	0313	0534	1.2F
	0809	1129	1.1E
	1444	1805	2.0F
	2131		
15 M		0131	1.5E
	0452	0703	0.9F
	0934	1251	0.8E
	1555	1931	2.0F
	2257		
16 Tu		0309	1.7E
	0630	0842	1.0F
	1120	1446	0.8E
	1726	2056	2.4F
17 W	0012	0430	2.2E
	0739	0954	1.6F
	1239	1625	1.1E
	1848	2202	3.0F
18 Th	0110	0537	2.7E
	0831	1046	2.2F
	1332	1742	1.5E
	1955	2254	3.8F
19 F	0158	0632	3.2E
	0914	1129	2.9F
	1416	1842	2.0E
	2050	2338	4.3F
20 Sa	0240	0720	3.6E
	0953	1208	3.5F
	1456	1932	2.5E
○	2139		
21 Su		0019	4.7F
	0320	0802	3.8E
	1030	1246	3.9F
	1537	2016	2.8E
	2224		
22 M		0058	4.7F
	0359	0840	3.7E
	1104	1323	4.0F
	1618	2056	3.0E
	2307		
23 Tu		0137	4.4F
	0438	0915	3.4E
	1138	1400	4.0F
	1701	2133	2.9E
	2350		
24 W		0216	3.8F
	0516	0948	2.9E
	1210	1437	3.6F
	1748	2209	2.7E
25 Th	0033	0256	3.0F
	0555	1017	2.4E
	1242	1517	3.2F
	1837	2246	2.3E
26 F	0120	0339	2.2F
	0635	1044	1.8E
	1314	1601	2.6F
◑	1932	2327	1.9E
27 Sa	0217	0431	1.4F
	0720	1112	1.2E
	1350	1655	2.0F
	2037		
28 Su		0023	1.5E
	0338	0544	0.7F
	0825	1157	0.7E
	1437	1811	1.5F
	2202		
29 M		0150	1.3E
	0538	0747	0.5F
	1026	1329	0.5E
	1600	1959	1.4F
	2330		
30 Tu		0328	1.4E
	0707	0927	0.8F
	1216	1517	0.6E
	1746	2123	1.8F
31 W	0036	0438	1.8E
	0756	1018	1.3F
	1309	1632	0.9E
	1901	2216	2.3F

September

Day	Slack h m	Maximum h m	knots
1 Th	0122	0529	2.1E
	0832	1053	1.8F
	1344	1726	1.3E
	1953	2254	2.8F
2 F	0158	0610	2.5E
	0901	1123	2.3F
	1413	1811	1.7E
	2035	2326	3.3F
3 Sa	0230	0646	2.7E
	0928	1150	2.8F
	1440	1851	2.1E
	2112	2356	3.6F
4 Su	0258	0719	2.9E
	0954	1216	3.1F
	1507	1927	2.4E
●	2146		
5 M		0024	3.8F
	0325	0751	3.0E
	1019	1242	3.5F
	1534	2002	2.6E
	2220		
6 Tu		0052	3.9F
	0351	0820	2.9E
	1044	1309	3.6F
	1603	2034	2.7E
	2253		
7 W		0122	3.7F
	0417	0848	2.7E
	1110	1337	3.7F
	1634	2105	2.7E
	2328		
8 Th		0153	3.4F
	0444	0914	2.4E
	1136	1408	3.5F
	1708	2136	2.5E
9 F	0004	0227	2.9F
	0515	0937	2.0E
	1202	1443	3.3F
	1747	2207	2.3E
10 Sa	0046	0306	2.3F
	0550	0953	1.6E
	1231	1523	2.8F
	1835	2243	2.0E
11 Su	0138	0354	1.6F
	0632	1005	1.2E
	1305	1614	2.3F
◐	1936	2338	1.6E
12 M	0253	0502	1.0F
	0734	1046	0.8E
	1352	1727	1.9F
	2101		
13 Tu		0110	1.5E
	0447	0652	0.7F
	0929	1239	0.5E
	1520	1914	1.8F
	2241		
14 W		0258	1.7E
	0626	0843	1.0F
	1138	1459	0.7E
	1720	2049	2.2F
	2359		
15 Th		0419	2.3E
	0726	0947	1.8F
	1243	1632	1.2E
	1848	2151	3.0F
16 F	0054	0521	2.8E
	0810	1031	2.6F
	1326	1738	1.8E
	1951	2239	3.7F
17 Sa	0138	0612	3.2E
	0848	1110	3.4F
	1402	1830	2.4E
	2042	2321	4.3F
18 Su	0217	0657	3.5E
	0924	1146	4.0F
	1438	1915	2.9E
○	2128	2359	4.6F
19 M	0254	0737	3.5E
	0957	1220	4.4F
	1515	1956	3.3E
	2210		
20 Tu		0036	4.5F
	0329	0812	3.4E
	1028	1254	4.5F
	1552	2032	3.4E
	2250		
21 W		0112	4.2F
	0404	0843	3.0E
	1059	1329	4.4F
	1631	2106	3.2E
	2330		
22 Th		0148	3.5F
	0440	0912	2.5E
	1127	1403	4.0F
	1713	2138	2.9E
23 F	0010	0226	2.8F
	0516	0935	2.0E
	1155	1439	3.3F
	1757	2210	2.5E
24 Sa	0054	0306	2.0F
	0554	0955	1.5E
	1222	1518	2.6F
	1847	2245	2.0E
25 Su	0147	0354	1.2F
	0638	1015	1.0E
	1250	1606	1.9F
◑	1949	2335	1.5E
26 M	0309	0510	0.5F
	0748	1058	0.5E
	1329	1724	1.3F
	2115		
27 Tu		0105	1.2E
	0524	0740	0.4F
	1025	1250	0.3E
	1507	1934	1.2F
	2254		
28 W		0300	1.3E
	0643	0910	0.8F
	1213	1508	0.5E
	1729	2100	1.6F
29 Th	0006	0408	1.7E
	0723	0953	1.4F
	1255	1617	1.0E
	1844	2150	2.1F
30 F	0052	0454	2.1E
	0754	1024	1.9F
	1324	1705	1.5E
	1934	2226	2.6F

Time meridian 120° E. 0000 is midnight. 1200 is noon. Times are not adjusted for Daylight Saving Time.

Wusong Kou, China, 2005

F–Flood, Dir. 290° True   E–Ebb, Dir. 110° True

October

Day	Slack (h m)	Maximum (h m)	knots
1 Sa	0127	0534	2.4E
	0822	1051	2.5F
	1349	1747	2.0E
	2015	2257	3.1F
2 Su	0156	0610	2.6E
	0847	1117	3.1F
	1414	1826	2.4E
	2051	2326	3.5F
3 M	0223	0645	2.8E
	0913	1142	3.5F
	1439	1903	2.7E
●	2126	2354	3.7F
4 Tu	0249	0718	2.8E
	0939	1208	3.9F
	1505	1938	3.0E
	2200		
5 W		0023	3.7F
	0315	0749	2.7E
	1005	1236	4.1F
	1533	2011	3.1E
	2234		
6 Th		0054	3.6F
	0342	0819	2.5E
	1031	1306	4.1F
	1604	2043	3.1E
	2310		
7 F		0127	3.3F
	0411	0846	2.2E
	1059	1339	3.9F
	1640	2115	2.9E
	2349		
8 Sa		0203	2.8F
	0445	0910	1.8E
	1127	1415	3.5F
	1722	2149	2.5E
9 Su	0033	0244	2.2F
	0524	0929	1.4E
	1157	1457	3.0F
	1813	2230	2.2E
10 M	0129	0337	1.5F
	0613	0948	1.0E
	1234	1551	2.4F
	1918	2329	1.8E
11 Tu	0251	0455	0.9F
	0733	1047	0.6E
	1329	1713	1.0F
◐	2046		
12 W		0102	1.6E
	0440	0654	0.8F
	0959	1308	0.5E
	1521	1903	1.8F
	2223		
13 Th		0242	1.9E
	0601	0828	1.3F
	1143	1508	0.9E
	1721	2031	2.2F
	2337		
14 F		0356	2.3E
	0654	0923	2.1F
	1233	1623	1.5E
	1841	2131	2.8F
15 Sa	0030	0453	2.7E
	0737	1006	2.9F
	1310	1721	2.2E
	1940	2218	3.4F
16 Su	0113	0543	2.9E
	0814	1044	3.7F
	1344	1810	2.7E
	2029	2259	3.9F
17 M	0150	0626	3.0E
	0848	1119	4.2F
	1418	1853	3.2E
○	2113	2336	4.0F
18 Tu	0226	0705	3.0E
	0921	1153	4.6F
	1452	1931	3.4E
	2154		
19 W		0013	4.0F
	0300	0739	2.8E
	0952	1227	4.6F
	1528	2006	3.5E
	2233		
20 Th		0049	3.6F
	0335	0809	2.5E
	1021	1300	4.4F
	1606	2038	3.3E
	2311		
21 F		0125	3.1F
	0411	0835	2.1E
	1049	1335	3.9F
	1645	2108	3.0E
	2351		
22 Sa		0202	2.4F
	0448	0855	1.7E
	1116	1410	3.3F
	1728	2138	2.5E
23 Su	0033	0243	1.7F
	0528	0913	1.3E
	1142	1449	2.6F
	1816	2212	2.0E
24 M	0125	0333	1.1F
	0618	0938	0.9E
	1211	1537	2.0F
	1914	2259	1.6E
25 Tu	0241	0450	0.6F
	0735	1026	0.5E
	1254	1651	1.4F
◑	2029		
26 W		0018	1.3E
	0427	0651	0.6F
	0040	1213	0.3E
	1432	1842	1.2F
	2158		
27 Th		0202	1.3E
	0542	0818	0.9F
	1132	1428	0.6E
	1649	2010	1.4F
	2313		
28 F		0312	1.6E
	0627	0907	1.4F
	1219	1538	1.0E
	1809	2107	1.8F
29 Sa	0005	0401	1.9E
	0701	0941	2.0F
	1250	1629	1.5E
	1903	2148	2.3F
30 Su	0044	0445	2.1E
	0731	1011	2.6F
	1317	1714	2.1E
	1947	2222	2.7F
31 M	0116	0526	2.3E
	0800	1039	3.2F
	1343	1756	2.5E
	2027	2254	3.0F

November

Day	Slack (h m)	Maximum (h m)	knots
1 Tu	0145	0605	2.4E
	0829	1107	3.7F
	1409	1835	2.9E
	2104	2325	3.3F
2 W	0214	0642	2.4E
	0858	1137	4.1F
	1438	1912	3.2E
●	2141	2358	3.4F
3 Th	0244	0718	2.3E
	0928	1208	4.3F
	1510	1949	3.3E
	2218		
4 F		0032	3.3F
	0315	0752	2.2E
	0959	1242	4.3F
	1545	2025	3.3E
	2258		
5 Sa		0109	3.0F
	0350	0824	1.9E
	1031	1318	4.1F
	1626	2101	3.1E
	2340		
6 Su		0149	2.5F
	0430	0854	1.6E
	1105	1359	3.7F
	1712	2141	2.8E
7 M	0029	0236	2.0F
	0518	0925	1.3E
	1143	1447	3.1F
	1806	2227	2.4E
8 Tu	0130	0335	1.5F
	0620	1007	0.9E
	1230	1547	2.5F
	1911	2329	2.1E
9 W	0245	0454	1.1F
	0754	1125	0.7E
	1341	1706	2.0F
◐	2029		
10 Th		0049	1.9E
	0407	0629	1.2F
	0955	1316	0.7E
	1525	1838	1.9F
	2152		
11 F		0209	2.0E
	0516	0750	1.6F
	1110	1448	1.2E
	1707	2000	2.1F
	2303		
12 Sa		0317	2.2E
	0610	0848	2.3F
	1209	1558	1.7E
	1824	2103	2.4F
	2358		
13 Su		0415	2.3E
	0655	0934	3.0F
	1248	1655	2.3E
	1924	2153	2.8F
14 M	0043	0506	2.4E
	0735	1015	3.6F
	1323	1745	2.8E
	2014	2236	3.1F
15 Tu	0123	0552	2.4E
	0811	1052	4.0F
	1358	1828	3.1E
	2058	2315	3.2F
16 W	0201	0632	2.3E
	0846	1128	4.3F
	1434	1907	3.3E
○	2140	2353	3.2F
17 Th	0237	0707	2.2E
	0918	1204	4.3F
	1510	1942	3.3E
	2219		
18 F		0031	3.0F
	0314	0737	2.0E
	0949	1239	4.1F
	1548	2014	3.2E
	2258		
19 Sa		0109	2.6F
	0353	0802	1.8E
	1020	1315	3.7F
	1629	2045	2.9E
	2337		
20 Su		0148	2.2F
	0433	0825	1.5E
	1050	1353	3.3F
	1711	2117	2.6E
21 M	0020	0231	1.7F
	0517	0849	1.3E
	1122	1435	2.7F
	1757	2152	2.2E
22 Tu	0108	0321	1.3F
	0609	0923	1.0E
	1159	1522	2.2F
	1847	2236	1.9E
23 W	0205	0423	1.0F
	0715	1014	0.7E
	1249	1622	1.7F
	1944	2334	1.6E
24 Th	0311	0538	0.9F
	0843	1136	0.6E
	1405	1736	1.4F
◑	2049		
25 F		0045	1.5E
	0415	0654	1.1F
	1015	1317	0.7E
	1546	1856	1.3F
	2158		
26 Sa		0154	1.5E
	0508	0756	1.4F
	1118	1438	1.0E
	1715	2005	1.5F
	2258		
27 Su		0253	1.6E
	0552	0843	1.9F
	1202	1541	1.5E
	1823	2058	1.8F
	2348		
28 M		0346	1.7E
	0632	0922	2.5F
	1237	1635	2.0E
	1916	2142	2.1F
29 Tu	0030	0436	1.8E
	0709	0958	3.0F
	1309	1723	2.5E
	2002	2222	2.4F
30 W	0108	0524	1.9E
	0746	1034	3.5F
	1343	1808	2.9E
	2045	2300	2.7F

December

Day	Slack (h m)	Maximum (h m)	knots
1 Th	0145	0609	2.0E
	0822	1110	3.9F
	1418	1850	3.2E
●	2127	2339	2.8F
2 F	0223	0651	2.0E
	0859	1148	4.2F
	1456	1931	3.4E
	2209		
3 Sa		0019	2.9F
	0302	0731	1.9E
	0937	1228	4.3F
	1537	2012	3.4E
	2252		
4 Su		0100	2.8F
	0344	0811	1.8E
	1017	1310	4.2F
	1622	2054	3.3E
	2338		
5 M		0146	2.5F
	0430	0851	1.7E
	1059	1355	3.9F
	1710	2139	3.1E
6 Tu	0027	0235	2.2F
	0523	0936	1.5E
	1146	1445	3.4F
	1802	2227	2.8E
7 W	0121	0331	2.0F
	0627	1028	1.3E
	1240	1541	2.9F
	1859	2320	2.5E
8 Th	0219	0435	1.8F
	0747	1135	1.1E
	1347	1646	2.4F
◐	2001		
9 F		0019	2.2E
	0319	0546	1.7F
	0917	1251	1.2E
	1510	1800	2.0F
	2108		
10 Sa		0123	2.0E
	0420	0659	1.9F
	1035	1410	1.4E
	1640	1919	1.8F
	2217		
11 Su		0226	1.9E
	0516	0804	2.3F
	1135	1523	1.8E
	1801	2029	1.9F
	2319		
12 M		0328	1.8E
	0607	0859	2.7F
	1223	1626	2.2E
	1907	2127	2.0F
13 Tu	0014	0426	1.8E
	0654	0947	3.1F
	1305	1720	2.5E
	2002	2217	2.2F
14 W	0101	0517	1.8E
	0737	1030	3.5F
	1344	1806	2.8E
	2049	2301	2.4F
15 Th	0145	0601	1.7E
	0817	1111	3.7F
	1423	1847	3.0E
	2131	2343	2.4F
16 F	0226	0640	1.7E
	0854	1150	3.8F
	1501	1925	3.0E
○	2211		
17 Sa		0022	2.4F
	0305	0713	1.7E
	0931	1229	3.8F
	1540	2000	3.0E
	2250		
18 Su		0102	2.3F
	0345	0743	1.6E
	1006	1307	3.6F
	1619	2033	2.9E
	2328		
19 M		0141	2.2F
	0425	0814	1.6E
	1042	1345	3.4F
	1659	2106	2.7E
20 Tu	0005	0220	2.0F
	0506	0846	1.5E
	1119	1424	3.0F
	1738	2140	2.5E
21 W	0044	0301	1.8F
	0551	0924	1.3E
	1158	1504	2.7F
	1818	2216	2.2E
22 Th	0123	0344	1.6F
	0641	1009	1.2E
	1243	1548	2.2F
	1900	2255	2.0E
23 F	0205	0433	1.5F
	0740	1102	1.1E
	1337	1638	1.8F
	1945	2339	1.7E
24 Sa	0250	0528	1.5F
	0846	1207	1.0E
	1446	1738	1.5F
◑	2036		
25 Su		0028	1.5E
	0338	0629	1.6F
	0954	1323	1.1E
	1609	1849	1.3F
	2136		
26 M		0127	1.3E
	0430	0731	1.8F
	1050	1442	1.3E
	1733	1959	1.3F
	2240		
27 Tu		0233	1.3E
	0523	0828	2.2F
	1149	1552	1.7E
	1844	2101	1.5F
	2342		
28 W		0342	1.3E
	0616	0919	2.6F
	1237	1652	2.2E
	1942	2155	1.8F
29 Th	0038	0445	1.4E
	0706	1007	3.2F
	1322	1744	2.7E
	2032	2244	2.1F
30 F	0128	0541	1.6E
	0755	1053	3.6F
	1406	1834	3.1E
	2119	2329	2.5F
31 Sa	0214	0633	1.7E
	0842	1138	4.1F
	1450	1921	3.4E
●	2204		

Time meridian 120° E. 0000 is midnight. 1200 is noon. Times are not adjusted for Daylight Saving Time.

Basilan Strait (off Zamboanga), Philippines, 2005

F–Flood, Dir. 270° True E–Ebb, Dir. 090° True

January

Day	Slack h m	Maximum h m	knots
1 Sa	0050	0451	4.5E
	0853	1132	2.2F
	1433	1716	1.7E
	2003	2253	1.9F
2 Su	0128	0528	3.9E
	0927	1207	2.0F
	1505	1803	1.9E
	2109	2344	1.4F
3 M	0211	0609	3.1E
	1002	1246	1.8F
	1541	1900	2.1E
	2237		
4 Tu ◑		0053	1.0F
	0308	0701	2.3E
	1040	1331	1.6F
	1622	2010	2.4E
5 W	0027	0229	0.7F
	0439	0811	1.5E
	1123	1426	1.5F
	1712	2129	3.0E
6 Th	0203	0421	0.9F
	0700	0943	0.9E
	1216	1530	1.5F
	1811	2242	3.8E
7 F	0310	0550	1.5F
	0905	1113	0.8E
	1318	1637	1.7F
	1913	2344	4.6E
8 Sa	0403	0653	2.2F
	1022	1221	0.9E
	1419	1738	2.1F
	2013		
9 Su		0039	5.5E
	0451	0743	2.8F
	1113	1313	1.1E
	1514	1832	2.6F
	2111		
10 M ●		0128	6.1E
	0535	0826	3.2F
	1153	1358	1.4E
	1604	1922	3.0F
	2204		
11 Tu		0214	6.6E
	0618	0907	3.5F
	1227	1439	1.7E
	1652	2009	3.4F
	2254		
12 W		0257	6.7E
	0659	0945	3.5F
	1258	1520	2.0E
	1741	2055	3.5F
	2342		
13 Th		0339	6.4E
	0737	1021	3.3F
	1328	1600	2.3E
	1831	2140	3.3F
14 F	0027	0419	5.8E
	0814	1056	3.1F
	1358	1641	2.5E
	1926	2226	3.0F
15 Sa	0112	0458	5.0E
	0847	1130	2.8F
	1427	1724	2.7E
	2027	2315	2.4F
16 Su	0156	0537	3.9E
	0918	1203	2.4F
	1457	1812	2.7E
	2139		
17 M ◐		0010	1.6F
	0241	0616	2.8E
	0947	1237	2.0F
	1530	1907	2.8E
	2310		
18 Tu		0118	0.9F
	0334	0700	1.8E
	1014	1316	1.6F
	1609	2015	2.8E
19 W	0110	0258	0.5F
	0501	0802	0.8E
	1044	1406	1.3F
	1658	2138	2.9E
20 Th	0255	0506	0.6F
	0759	0950	0.3E
	1135	1519	1.1F
	1801	2300	3.3E
21 F	0354	0631	1.1F
	1003	1139	0.3E
	1314	1642	1.1F
	1912		
22 Sa		0003	3.7E
	0435	0719	1.6F
	1045	1242	0.6E
	1439	1750	1.4F
	2017		
23 Su		0051	4.2E
	0508	0753	2.0F
	1111	1323	1.0E
	1535	1840	1.8F
	2111		
24 M		0130	4.6E
	0538	0822	2.3F
	1134	1355	1.4E
	1617	1921	2.2F
	2156		
25 Tu ○		0204	5.0E
	0605	0849	2.6F
	1156	1424	1.7E
	1653	1956	2.5F
	2234		
26 W		0235	5.2E
	0631	0914	2.8F
	1218	1452	2.0E
	1726	2028	2.8F
	2309		
27 Th		0304	5.3E
	0657	0939	2.9F
	1240	1520	2.3E
	1759	2100	2.9F
	2342		
28 F		0332	5.3E
	0721	1003	2.9F
	1301	1547	2.6E
	1834	2132	2.9F
29 Sa	0015	0400	5.0E
	0745	1027	2.9F
	1322	1617	2.8E
	1913	2206	2.8F
30 Su	0049	0428	4.6E
	0808	1051	2.8F
	1343	1648	3.0E
	1957	2243	2.5F
31 M	0125	0458	3.9E
	0831	1116	2.6F
	1406	1724	3.2E
	2050	2327	2.0F

February

Day	Slack h m	Maximum h m	knots
1 Tu	0204	0530	3.1E
	0853	1143	2.3F
	1433	1807	3.3E
	2158		
2 W ◑		0021	1.4F
	0252	0607	2.1E
	0915	1215	2.0F
	1507	1904	3.2E
	2335		
3 Th		0140	0.8F
	0402	0655	1.1E
	0935	1257	1.6F
	1552	2024	3.3E
4 F	0141	0354	0.6F
	0702	0828	0.2E
	0950	1406	1.3F
	1700	2207	3.6E
5 Sa	0312	0602	1.3F
		1104	0.0
		1556	1.2F
	1832	2334	4.4E
6 Su	0407	0703	2.1F
	1056	1229	0.4E
	1402	1729	1.7F
	2000		
7 M		0037	5.2E
	0451	0745	2.7F
	1116	1318	1.1E
	1520	1835	2.4F
	2111		
8 Tu		0127	5.9E
	0530	0821	3.2F
	1139	1357	1.8E
	1616	1927	3.1F
	2208		
9 W ●		0210	6.3E
	0606	0853	3.5F
	1201	1433	2.4E
	1704	2013	3.6F
	2259		
10 Th		0249	6.4E
	0638	0923	3.7F
	1224	1508	3.0E
	1751	2056	3.8F
	2344		
11 F		0326	6.1E
	0708	0951	3.6F
	1247	1542	3.5E
	1837	2137	3.8F
12 Sa	0027	0359	5.4E
	0735	1018	3.4F
	1310	1616	3.8E
	1924	2217	3.4F
13 Su	0107	0431	4.5E
	0758	1043	3.2F
	1333	1651	4.0E
	2014	2258	2.8F
14 M	0145	0501	3.5E
	0818	1108	2.8F
	1356	1727	3.9E
	2109	2341	2.0F
15 Tu	0221	0529	2.5E
	0834	1131	2.3F
	1422	1807	3.6E
	2219		
16 W ◐		0030	1.2F
	0258	0556	1.5E
	0846	1156	1.9F
	1450	1857	3.1E
17 Th	0006	0144	0.4F
	0341	0623	0.6E
	0847	1224	1.3F
	1526	2016	2.7E
18 F	0306	0505	0.2F
		0718	0.1F
		1316	0.7F
	1625	2224	2.7E
19 Sa	0402	0655	0.8F
		1203	0.0
		1611	0.5F
	1822	2355	3.2E
20 Su	0432	0721	1.4F
	1049	1252	0.6E
	1458	1749	1.0F
	2004		
21 M		0045	3.8E
	0458	0744	1.9F
	1056	1320	1.2E
	1547	1841	1.6F
	2107		
22 Tu		0122	4.4E
	0521	0806	2.4F
	1110	1346	1.8E
	1622	1918	2.2F
	2152		
23 W		0152	4.8E
	0543	0827	2.7F
	1126	1410	2.3E
	1653	1951	2.7F
	2230		
24 Th ○		0219	5.1E
	0604	0847	3.0F
	1143	1434	2.8E
	1723	2021	3.1F
	2304		
25 F		0245	5.2E
	0624	0907	3.2F
	1159	1458	3.3E
	1754	2051	3.4F
	2337		
26 Sa		0310	5.1E
	0644	0927	3.3F
	1216	1522	3.7E
	1827	2121	3.4F
27 Su	0010	0336	4.8E
	0703	0947	3.3F
	1234	1548	4.1E
	1903	2154	3.3F
28 M	0044	0401	4.2E
	0721	1007	3.2F
	1253	1617	4.4E
	1944	2229	3.0F

March

Day	Slack h m	Maximum h m	knots
1 Tu	0119	0428	3.5E
	0738	1029	3.0F
	1315	1650	4.4E
	2032	2309	2.4F
2 W	0157	0456	2.6E
	0754	1052	2.7F
	1341	1728	4.3E
	2133	2358	1.7F
3 Th	0241	0526	1.6E
	0806	1119	2.3F
	1413	1819	3.9E
	2306		
4 F ◑		0111	0.9F
	0347	0601	0.6E
	0806	1151	1.7F
	1455	1938	3.4E
5 Sa	0137	0406	0.5F
		0723	0.3F
		1249	1.1F
	1605	2151	3.4E
6 Su	0312	0621	1.3F
		1141	0.1F
		1552	0.8F
	1813	2333	4.1E
7 M	0400	0659	2.1F
	1039	1237	0.7E
	1437	1740	1.5F
	2003		
8 Tu		0033	4.9E
	0437	0730	2.8F
	1046	1312	1.7E
	1540	1841	2.5F
	2116		
9 W		0119	5.5E
	0510	0758	3.3F
	1101	1344	2.6E
	1627	1928	3.3F
	2211		
10 Th ●		0157	5.7E
	0538	0824	3.6F
	1120	1416	3.5E
	1710	2010	3.8F
	2258		
11 F		0231	5.6E
	0604	0849	3.7F
	1139	1446	4.2E
	1751	2048	4.0F
	2341		
12 Sa		0303	5.2E
	0628	0913	3.7F
	1200	1516	4.8E
	1831	2125	4.0F
13 Su	0020	0333	4.6E
	0648	0935	3.6F
	1220	1546	5.0E
	1912	2200	3.6F
14 M	0056	0400	3.8E
	0706	0956	3.3F
	1242	1615	5.0E
	1954	2235	3.0F
15 Tu	0130	0425	2.9E
	0721	1017	3.0F
	1304	1646	4.7E
	2039	2311	2.2F
16 W	0200	0448	2.1E
	0734	1037	2.5F
	1327	1719	4.1E
	2135	2351	1.4F
17 Th	0229	0509	1.3E
	0743	1057	2.0F
	1351	1758	3.4E
	2301		
18 F ◐		0046	0.5F
	0254	0524	0.6E
	0739	1116	1.4F
	1418	1900	2.6E
19 Sa	0332	1121	0.7F
	1451	2138	2.2E
20 Su	0344	0649	0.7F
	1120	1235	0.1E
	1359	1556	0.1F
	1718	2334	2.8E
21 M	0405	0656	1.3F
	1012	1240	0.8E
	1515	1742	0.8F
	1946		
22 Tu		0023	3.4E
	0426	0712	1.9F
	1014	1259	1.6E
	1546	1829	1.6F
	2053		
23 W		0057	4.0E
	0446	0730	2.4F
	1026	1320	2.3E
	1614	1904	2.3F
	2139		
24 Th		0126	4.4E
	0505	0749	2.8F
	1040	1342	3.1E
	1643	1935	2.9F
	2218		
25 F		0152	4.6E
	0524	0807	3.1F
	1054	1405	3.8E
	1712	2006	3.4F
	2254		
26 Sa ○		0218	4.6E
	0542	0826	3.3F
	1110	1428	4.4E
	1743	2036	3.7F
	2329		
27 Su		0243	4.4E
	0559	0845	3.4F
	1127	1453	4.9E
	1816	2107	3.8F
28 M	0004	0309	4.0E
	0617	0905	3.4F
	1147	1520	5.3E
	1852	2141	3.6F
29 Tu	0040	0335	3.4E
	0633	0926	3.3F
	1209	1550	5.5E
	1933	2217	3.2F
30 W	0116	0402	2.7E
	0649	0949	3.1F
	1235	1623	5.3E
	2021	2257	2.6F
31 Th	0156	0430	1.9E
	0704	1014	2.8F
	1304	1703	4.9E
	2123	2348	1.7F

Time meridian 120° E. 0000 is midnight. 1200 is noon. Times are not adjusted for Daylight Saving Time.
If three consecutive entries are marked (F) the middle one is not a true maximum but an intermediate value to show the current pattern.

Basilan Strait (off Zamboanga), Philippines, 2005

F–Flood, Dir. 270° True E–Ebb, Dir. 090° True

April

Day	Slack h m	Maximum h m	knots
1 F	0244	0500	1.1E
	0713	1042	2.3F
	1338	1755	4.2E
	2259		
2 Sa		0109	0.9F
	0420	0538	0.2E
	0652	1114	1.6F
	1421	1920	3.4E
3 Su	0122	0430	0.7F
		0831	0.5F
		1224	0.7F
	1535	2143	3.2E
4 M	0247	0554	1.5F
	0958	1142	0.3E
	1333	1614	0.7F
	1816	2320	3.8E
5 Tu	0331	0627	2.2F
	0944	1220	1.5E
	1500	1744	1.6F
	2006		
6 W		0016	4.3E
	0405	0655	2.7F
	0955	1251	2.6E
	1547	1838	2.5F
	2115		
7 Th		0059	4.6E
	0434	0721	3.1F
	1012	1321	3.6E
	1628	1922	3.3F
	2209		
8 F		0135	4.7E
	0459	0745	3.4F
	1032	1351	4.5E
	1707	2000	3.8F
	2254		
9 Sa		0207	4.4E
	0521	0808	3.5F
	1052	1419	5.2E
	1744	2036	3.9F
	2334		
10 Su		0237	4.0E
	0541	0831	3.6F
	1113	1447	5.6E
	1820	2110	3.8F
11 M	0011	0304	3.5E
	0600	0852	3.4F
	1135	1515	5.7E
	1857	2143	3.5F
12 Tu	0044	0330	2.9E
	0617	0914	3.2F
	1158	1544	5.5E
	1934	2215	2.9F
13 W	0115	0354	2.3E
	0633	0936	2.9F
	1221	1613	5.1E
	2015	2248	2.2F
14 Th	0144	0417	1.7E
	0649	0958	2.5F
	1246	1645	4.4E
	2105	2325	1.5F
15 F	0213	0440	1.1E
	0704	1020	1.9F
	1311	1723	3.6E
	2217		
16 Sa		0018	0.7F
	0251	0504	0.5E
	0711	1043	1.3F
	1337	1819	2.8E
17 Su	0040	0229	0.2F
		0543	0.1F
		1103	0.6F
	1403	2024	2.2E
18 M	0230	0530	0.7F
	0927	1136	0.2E
	1446	1509	0.0
	1534	2238	2.4E
19 Tu	0306	0557	1.2F
	0903	1156	1.1E
	1501	1710	0.6F
	1904	2337	2.9E
20 W	0332	0619	1.8F
	0913	1219	1.9E
	1528	1801	1.5F
	2024		
21 Th		0017	3.4E
	0354	0639	2.2F
	0927	1242	2.8E
	1556	1840	2.2F
	2118		
22 F		0049	3.6E
	0415	0659	2.6F
	0942	1306	3.7E
	1625	1914	2.9F
	2203		
23 Sa		0119	3.7E
	0434	0720	2.9F
	1000	1331	4.6E
	1656	1948	3.4F
	2244		
24 Su		0148	3.6E
	0453	0741	3.2F
	1020	1357	5.3E
	1729	2021	3.7F
	2323		
25 M		0216	3.4E
	0512	0803	3.3F
	1042	1426	5.8E
	1805	2056	3.8F
26 Tu	0002	0245	3.0E
	0531	0827	3.4F
	1108	1457	6.1E
	1844	2132	3.6F
27 W	0041	0314	2.5E
	0550	0853	3.3F
	1137	1531	6.1E
	1929	2212	3.2F
28 Th	0121	0345	1.9E
	0610	0922	3.1F
	1209	1609	5.8E
	2020	2257	2.5F
29 F	0208	0419	1.3E
	0630	0953	2.7F
	1244	1654	5.1E
	2125	2354	1.7F
30 Sa	0310	0500	0.6E
	0649	1029	2.1F
	1324	1751	4.3E
	2252		

May

Day	Slack h m	Maximum h m	knots
1 Su		0121	1.1F
	0526	0609	0.1E
	0653	1120	1.3F
	1413	1917	3.5E
2 M	0036	0331	1.1F
	0813	0908	0.1E
	1006	1329	0.5F
	1540	2116	3.1E
3 Tu	0152	0451	1.6F
	0813	1058	1.1E
	1355	1612	0.7F
	1812	2246	3.2E
4 W	0241	0534	2.1F
	0834	1144	2.2E
	1500	1733	1.6F
	1959	2346	3.3E
5 Th	0316	0607	2.5F
	0856	1220	3.4E
	1545	1828	2.4F
	2110		
6 F		0030	3.4E
	0346	0635	2.8F
	0919	1253	4.4E
	1624	1911	3.0F
	2204		
7 Sa		0108	3.3E
	0411	0702	3.0F
	0943	1324	5.1E
	1701	1949	3.4F
	2250		
8 Su		0141	3.0E
	0434	0727	3.2F
	1007	1353	5.6E
	1736	2024	3.5F
	2329		
9 M		0211	2.7E
	0455	0752	3.2F
	1032	1422	5.9E
	1811	2057	3.4F
10 Tu	0004	0239	2.4E
	0516	0817	3.1F
	1058	1451	5.8E
	1845	2129	3.1F
11 W	0036	0306	2.1E
	0537	0842	3.0F
	1125	1521	5.5E
	1922	2201	2.7F
12 Th	0106	0333	1.8E
	0600	0908	2.7F
	1153	1553	5.1E
	2001	2235	2.2F
13 F	0137	0401	1.4E
	0624	0935	2.3F
	1222	1627	4.5E
	2048	2315	1.6F
14 Sa	0214	0432	1.0E
	0650	1005	1.9F
	1251	1707	3.8E
	2146		
15 Su		0007	1.1F
	0305	0515	0.6E
	0724	1041	1.3F
	1323	1759	3.1E
	2304		
16 M		0126	0.8F
	0440	0635	0.3E
	0834	1137	0.6F
	1400	1918	2.5E
17 Tu	0030	0311	0.8F
	0632	0912	0.5E
	1225	1355	0.2F
	1515	2104	2.3E
18 W	0133	0421	1.1F
	0720	1038	1.3E
	1416	1610	0.5F
	1756	2226	2.3E
19 Th	0215	0502	1.5F
	0749	1121	2.2E
	1457	1721	1.2F
	1942	2322	2.5E
20 F	0247	0534	1.9F
	0814	1154	3.2E
	1532	1811	1.9F
	2054		
21 Sa		0006	2.5E
	0314	0602	2.2F
	0839	1226	4.2E
	1607	1853	2.6F
	2150		
22 Su		0045	2.5E
	0339	0630	2.6F
	0906	1258	5.1E
	1643	1932	3.2F
	2239		
23 M		0120	2.5E
	0403	0658	2.9F
	0935	1331	5.8E
	1720	2011	3.5F
	2324		
24 Tu		0154	2.3E
	0427	0728	3.1F
	1007	1406	6.3E
	1801	2050	3.6F
25 W	0007	0229	2.1E
	0453	0800	3.3F
	1042	1444	6.5E
	1844	2131	3.4F
26 Th	0049	0304	1.8E
	0520	0835	3.3F
	1120	1524	6.4E
	1931	2215	3.1F
27 F	0134	0342	1.5E
	0551	0912	3.1F
	1200	1608	6.0E
	2023	2303	2.6F
28 Sa	0223	0425	1.1E
	0628	0954	2.7F
	1243	1656	5.3E
	2121	2359	2.1F
29 Su	0320	0519	0.8E
	0720	1044	2.0F
	1330	1752	4.4E
	2224		
30 M		0104	1.7F
	0429	0636	0.7E
	0848	1153	1.3F
	1425	1901	3.5E
	2332		
31 Tu		0219	1.5F
	0538	0821	1.0E
	1122	1342	0.7F
	1546	2027	2.8E

June

Day	Slack h m	Maximum h m	knots
1 W	0036	0329	1.6F
	0632	0955	1.8E
	1337	1546	0.8F
	1750	2154	2.4E
2 Th	0129	0424	1.8F
	0714	1058	2.8E
	1449	1714	1.3F
	1944	2304	2.1E
3 F	0212	0508	2.0F
	0750	1144	3.8E
	1539	1816	1.9F
	2105	2359	2.0E
4 Sa	0248	0546	2.2F
	0824	1224	4.6E
	1620	1903	2.4F
	2205		
5 Su		0043	1.9E
	0320	0620	2.4F
	0857	1300	5.1E
	1657	1943	2.8F
	2252		
6 M		0120	1.8E
	0349	0652	2.6F
	0930	1333	5.5E
	1733	2018	2.9F
	2330		
7 Tu		0153	1.7E
	0418	0723	2.7F
	1003	1406	5.6E
	1807	2051	2.9F
8 W	0003	0224	1.7E
	0446	0754	2.7F
	1036	1438	5.6E
	1841	2123	2.7F
9 Th	0033	0254	1.6E
	0516	0826	2.7F
	1108	1511	5.4E
	1916	2156	2.5F
10 F	0103	0325	1.6E
	0547	0857	2.6F
	1141	1544	5.1E
	1953	2230	2.3F
11 Sa	0135	0358	1.4E
	0621	0930	2.3F
	1214	1619	4.7E
	2031	2306	2.0F
12 Su	0210	0435	1.3E
	0700	1006	2.0F
	1247	1657	4.2E
	2113	2347	1.7F
13 M	0251	0519	1.2E
	0750	1048	1.5F
	1322	1739	3.6E
	2158		
14 Tu		0034	1.5F
	0337	0616	1.1E
	0904	1142	1.0F
	1404	1828	3.0E
	2247		
15 W		0126	1.3F
	0426	0730	1.3E
	1054	1302	0.6F
	1501	1932	2.3E
	2338		
16 Th		0223	1.3F
	0516	0853	1.7E
	1257	1450	0.5F
	1643	2051	1.8E
17 F	0029	0320	1.3F
	0604	1004	2.5E
	1416	1631	0.9F
	1853	2212	1.5E
18 Sa	0117	0412	1.5F
	0648	1100	3.4E
	1508	1743	1.5F
	2036	2320	1.4E
19 Su	0200	0459	1.8F
	0732	1147	4.3E
	1553	1839	2.2F
	2150		
20 M		0015	1.4E
	0240	0543	2.1F
	0817	1232	5.2E
	1636	1926	2.8F
	2247		
21 Tu		0101	1.4E
	0317	0625	2.5F
	0902	1315	5.9E
	1719	2010	3.2F
	2334		
22 W		0143	1.5E
	0354	0707	2.9F
	0947	1358	6.4E
	1802	2052	3.4F
23 Th	0016	0224	1.5E
	0433	0750	3.2F
	1033	1441	6.6E
	1846	2134	3.4F
24 F	0055	0304	1.6E
	0514	0833	3.3F
	1119	1524	6.5E
	1930	2215	3.2F
25 Sa	0132	0346	1.7E
	0600	0918	3.2F
	1205	1608	6.1E
	2014	2257	3.0F
26 Su	0210	0431	1.7E
	0654	1006	2.9F
	1251	1653	5.4E
	2058	2340	2.7F
27 M	0248	0521	1.8E
	0758	1058	2.3F
	1339	1739	4.5E
	2140		
28 Tu		0023	2.3F
	0327	0618	1.9E
	0918	1200	1.6F
	1431	1829	3.5E
	2222		
29 W		0110	2.0F
	0410	0726	2.1E
	1103	1319	1.0F
	1534	1927	2.4E
	2305		
30 Th		0200	1.7F
	0455	0843	2.5E
	1303	1503	0.7F
	1711	2041	1.5E
	2350		

Time meridian 120° E. 0000 is midnight. 1200 is noon. Times are not adjusted for Daylight Saving Time.
If three consecutive entries are marked (F) the middle one is not a true maximum but an intermediate value to show the current pattern.

Basilan Strait (off Zamboanga), Philippines, 2005

F–Flood, Dir. 270° True E–Ebb, Dir. 090° True

July

Day	Slack h m	Maximum h m	knots
1 F		0256	1.6F
	0545	1000	3.0E
	1436	1651	0.9F
	1926	2209	0.9E
2 Sa	0041	0356	1.5F
	0638	1106	3.6E
	1536	1810	1.4F
	2117	2329	0.8E
3 Su	0138	0454	1.6F
	0730	1200	4.2E
	1622	1904	1.8F
	2223		
4 M		0028	0.8E
	0233	0546	1.8F
	0821	1245	4.6E
	1700	1945	2.2F
	2305		
5 Tu		0113	1.0E
	0321	0632	2.0F
	0907	1325	5.0E
	1735	2020	2.4F
	2336		
6 W ●		0149	1.2E
	0403	0712	2.3F
	0950	1401	5.2E
	1807	2050	2.5F
7 Th	0002	0221	1.5E
	0440	0749	2.5F
	1029	1435	5.3E
	1838	2119	2.6F
8 F	0027	0252	1.7E
	0516	0824	2.6F
	1105	1506	5.3E
	1907	2148	2.6F
9 Sa	0052	0322	1.8E
	0552	0857	2.7F
	1139	1537	5.2E
	1936	2216	2.6F
10 Su	0118	0353	1.9E
	0628	0930	2.6F
	1212	1608	4.9E
	2005	2244	2.5F
11 M	0144	0425	2.0E
	0708	1005	2.4F
	1245	1639	4.5E
	2033	2313	2.4F
12 Tu	0210	0500	2.1E
	0754	1043	2.1F
	1320	1711	4.0E
	2101	2342	2.2F
13 W	0237	0539	2.2E
	0850	1127	1.7F
	1358	1746	3.2E
	2130		
14 Th ◐		0013	1.9F
	0305	0626	2.4E
	1001	1223	1.2F
	1445	1826	2.4E
	2159		
15 F		0049	1.7F
	0339	0724	2.5E
	1138	1341	0.8F
	1552	1920	1.5E
	2232		
16 Sa		0133	1.4F
	0422	0839	2.8E
	1329	1534	0.7F
	1801	2044	0.7E
	2314		
17 Su		0235	1.3F
	0519	1003	3.4E
	1452	1727	1.1F
	2050	2236	0.4E
18 M	0019	0353	1.3F
	0629	1118	4.1E
	1549	1840	1.8F
	2221		
19 Tu		0001	0.5E
	0142	0508	1.7F
	0741	1218	5.0E
	1637	1930	2.5F
	2307		
20 W		0058	0.8E
	0250	0611	2.2F
	0846	1310	5.8E
	1720	2012	3.0F
	2340		
21 Th ○		0143	1.2E
	0347	0704	2.8F
	0944	1356	6.3E
	1800	2050	3.4F
22 F	0010	0223	1.7E
	0437	0753	3.3F
	1036	1439	6.6E
	1839	2125	3.6F
23 Sa	0038	0302	2.2E
	0526	0838	3.6F
	1125	1520	6.5E
	1915	2159	3.6F
24 Su	0105	0340	2.6E
	0616	0923	3.6F
	1211	1559	6.1E
	1948	2232	3.4F
25 M	0132	0420	2.9E
	0708	1008	3.3F
	1256	1636	5.3E
	2019	2303	3.2F
26 Tu	0200	0500	3.2E
	0805	1055	2.8F
	1339	1713	4.3E
	2048	2334	2.8F
27 W	0228	0543	3.3E
	0909	1145	2.1F
	1423	1749	3.1E
	2113		
28 Th ◑		0005	2.4F
	0258	0632	3.2E
	1027	1245	1.3F
	1511	1827	2.0E
	2134		
29 F		0037	2.0F
	0332	0731	3.1E
	1215	1408	0.6F
	1617	1913	0.9E
	2152		
30 Sa		0117	1.5F
	0415	0850	3.0E
	1428	1627	0.4F
	1918	2041	0.1E
	2158		
31 Su		0220	1.1F
	0515	1027	3.1E
	1546	1827	0.9F
		2317	0.0

August

Day	Slack h m	Maximum h m	knots
1 M		0405	0.9F
	0637	1147	3.6E
	1630	1917	1.4F
	2259		
2 Tu		0037	0.4E
	0216	0533	1.2F
	0758	1242	4.1E
	1704	1950	1.9F
	2311		
3 W		0118	0.9E
	0326	0631	1.7F
	0901	1324	4.5E
	1732	2016	2.2F
	2327		
4 Th		0149	1.3E
	0411	0715	2.1F
	0950	1358	4.9E
	1758	2041	2.5F
	2345		
5 F ●		0217	1.8E
	0448	0751	2.5F
	1030	1428	5.1E
	1822	2104	2.8F
6 Sa	0004	0243	2.2E
	0522	0823	2.8F
	1104	1455	5.2E
	1844	2126	2.9F
7 Su	0023	0309	2.6E
	0554	0853	3.0F
	1137	1522	5.2E
	1906	2148	3.0F
8 M	0042	0335	2.9E
	0627	0924	3.1F
	1208	1547	5.0E
	1927	2209	3.0F
9 Tu	0101	0401	3.2E
	0703	0955	3.0F
	1240	1613	4.5E
	1947	2230	2.9F
10 W	0120	0430	3.4E
	0742	1029	2.7F
	1313	1640	3.9E
	2007	2252	2.7F
11 Th	0140	0501	3.5E
	0828	1107	2.3F
	1349	1708	3.1E
	2026	2315	2.4F
12 F	0202	0537	3.5E
	0925	1152	1.7F
	1429	1738	2.2E
	2043	2341	2.1F
13 Sa ◐	0230	0623	3.4E
	1044	1254	1.0F
	1522	1813	1.2E
	2056		
14 Su		0012	1.7F
	0307	0729	3.3E
	1250	1447	0.5F
	1725	1910	0.2E
	2047		
15 M		0101	1.3F
	0403	0914	3.3E
	1451	1743	0.9F
		2216	0.3F
16 Tu		0252	1.0F
	0537	1103	3.9E
	1551	1851	1.7F
	2307		
17 W		0013	0.2E
	0121	0459	1.3F
	0724	1215	4.7E
	1634	1929	2.5F
	2304		
18 Th		0102	0.9E
	0301	0614	2.1F
	0846	1307	5.6E
	1711	2002	3.1F
	2320		
19 F		0139	1.7E
	0400	0708	2.9F
	0948	1350	6.1E
	1744	2032	3.5F
	2339		
20 Sa ○		0214	2.5E
	0448	0755	3.6F
	1040	1429	6.3E
	1815	2101	3.8F
21 Su	0000	0248	3.3E
	0534	0838	4.0F
	1127	1505	6.1E
	1844	2128	3.8F
22 M	0023	0322	3.9E
	0620	0919	4.0F
	1211	1539	5.6E
	1910	2155	3.7F
23 Tu	0045	0355	4.4E
	0706	1000	3.8F
	1252	1611	4.7E
	1933	2220	3.5F
24 W	0108	0430	4.5E
	0754	1040	3.2F
	1331	1641	3.7E
	1952	2244	3.1F
25 Th	0132	0505	4.4E
	0847	1122	2.4F
	1409	1709	2.6E
	2008	2307	2.7F
26 F ◑	0158	0543	4.1E
	0950	1208	1.5F
	1445	1735	1.6E
	2018	2330	2.1F
27 Sa	0225	0628	3.5E
	1122	1311	0.6F
	1523	1756	0.7E
	2015	2353	1.5F
28 Su	0257	0735	2.9E
	1441	1607	0.1F
		1750	0.1F
29 M		0020	0.9F
	0345	0948	2.8E
	1558	1906	0.7F
30 Tu		0013	0.2F
		0326	0.4F
	0541	1142	3.0E
	1626	1917	1.3F
	2251		
31 W		0048	0.5E
	0250	0538	0.9F
	0749	1236	3.6E
	1650	1935	1.8F
	2245		

September

Day	Slack h m	Maximum h m	knots
1 Th		0113	1.2E
	0342	0633	1.6F
	0858	1313	4.2E
	1711	1954	2.3F
	2255		
2 F		0136	1.8E
	0417	0711	2.2F
	0946	1342	4.6E
	1731	2013	2.6F
	2308		
3 Sa		0159	2.5E
	0448	0743	2.7F
	1024	1409	4.9E
	1750	2032	2.9F
	2324		
4 Su ●		0221	3.0E
	0517	0813	3.1F
	1057	1433	5.0E
	1808	2051	3.1F
	2340		
5 M		0244	3.6E
	0547	0841	3.4F
	1129	1457	4.8E
	1826	2109	3.2F
	2356		
6 Tu		0307	4.0E
	0617	0910	3.5F
	1200	1521	4.5E
	1843	2127	3.2F
7 W	0012	0331	4.4E
	0650	0940	3.4F
	1232	1545	4.1E
	1900	2146	3.2F
8 Th	0030	0357	4.6E
	0726	1012	3.2F
	1305	1610	3.4E
	1915	2206	3.0F
9 F	0050	0426	4.7E
	0808	1048	2.7F
	1340	1635	2.6E
	1930	2227	2.7F
10 Sa	0113	0500	4.5E
	0901	1130	2.0F
	1419	1701	1.8E
	1941	2250	2.4F
11 Su ◐	0142	0543	4.1E
	1017	1229	1.1F
	1510	1729	0.8E
	1943	2317	1.9F
12 M	0218	0647	3.5E
	1233	1434	0.4F
		1807	0.1F
		2353	1.2F
13 Tu	0311	0847	3.2E
	1443	1802	1.0F
		2310	0.4F
14 W		0233	0.6F
	0507	1057	3.7E
	1536	1838	1.8F
	2227		
15 Th		0017	0.5E
	0210	0509	1.2F
	0725	1206	4.5E
	1613	1907	2.6F
	2224		
16 F		0051	1.6E
	0320	0617	2.2F
	0848	1254	5.1E
	1645	1934	3.1F
	2237		
17 Sa		0123	2.6E
	0409	0707	3.1F
	0949	1334	5.5E
	1713	2000	3.5F
	2254		
18 Su ○		0154	3.7E
	0452	0750	3.8F
	1039	1410	5.5E
	1739	2025	3.8F
	2314		
19 M		0225	4.5E
	0534	0830	4.1F
	1124	1443	5.1E
	1803	2049	3.8F
	2336		
20 Tu		0256	5.2E
	0615	0908	4.2F
	1205	1513	4.5E
	1824	2113	3.8F
	2358		
21 W		0327	5.5E
	0656	0945	3.9F
	1244	1542	3.8E
	1843	2136	3.5F
22 Th	0021	0357	5.5E
	0738	1021	3.3F
	1320	1609	2.9E
	1859	2158	3.2F
23 F	0044	0429	5.2E
	0823	1058	2.5F
	1353	1633	2.1E
	1913	2219	2.7F
24 Sa	0109	0502	4.6E
	0917	1138	1.6F
	1423	1654	1.3E
	1922	2240	2.2F
25 Su ◑	0134	0541	3.7E
	1035	1230	0.7F
	1451	1711	0.6E
	1920	2259	1.5F
26 M	0202	0636	2.9E
	1401	1459	0.0
		1641	0.0
		2308	0.8F
27 Tu	0233	0854	2.2E
	1530	1841	0.7F
28 W		0038	0.0
		0309	0.1F
	0424	1115	2.6E
	1552	1841	1.2F
	2154		
29 Th		0027	0.8E
	0307	0529	0.7F
	0728	1209	3.2E
	1611	1856	1.7F
	2154		
30 F		0045	1.6E
	0339	0618	1.5F
	0841	1244	3.7E
	1630	1913	2.2F
	2205		

Time meridian 120° E. 0000 is midnight. 1200 is noon. Times are not adjusted for Daylight Saving Time.
If three consecutive entries are marked (F) the middle one is not a true maximum but an intermediate value to show the current pattern.

Basilan Strait (off Zamboanga), Philippines, 2005

F–Flood, Dir. 270° True E–Ebb, Dir. 090° True

October

Day	Slack h m	Maximum h m	knots
1 Sa		0106	2.4E
	0407	0654	2.2F
	0930	1312	4.1E
	1648	1931	2.6F
	2219		
2 Su		0128	3.2E
	0435	0726	2.8F
	1009	1339	4.2E
	1706	1949	2.9F
	2234		
3 M ●		0150	3.9E
	0504	0755	3.3F
	1045	1404	4.2E
	1723	2007	3.1F
	2249		
4 Tu		0213	4.5E
	0533	0825	3.6F
	1119	1429	4.1E
	1740	2026	3.2F
	2306		
5 W		0237	5.0E
	0603	0854	3.7F
	1153	1453	3.7E
	1757	2045	3.3F
	2325		
6 Th		0302	5.4E
	0637	0926	3.6F
	1227	1518	3.2E
	1813	2105	3.2F
	2346		
7 F		0330	5.5E
	0714	0959	3.3F
	1302	1544	2.6E
	1828	2127	3.1F
8 Sa	0011	0401	5.4E
	0758	1037	2.7F
	1339	1611	1.9E
	1843	2151	2.8F
9 Su	0039	0438	5.1E
	0853	1122	2.0F
	1423	1640	1.2E
	1854	2218	2.4F
10 M	0112	0524	4.4E
	1011	1228	1.1F
	1534	1715	0.4E
	1853	2250	1.8F
11 Tu ◐	0151	0633	3.7E
	1215	1448	0.7F
		1841	0.3F
		2339	1.0F
12 W	0249	0836	3.2E
	1402	1711	1.2F
	2151	2259	0.1E
13 Th	0011	0308	0.5F
	0505	1035	3.4E
	1456	1754	1.9F
	2114	2350	1.2E
14 F	0233	0509	1.2F
	0724	1143	3.9E
	1533	1825	2.5F
	2125		
15 Sa		0025	2.5E
	0326	0612	2.2F
	0845	1230	4.3E
	1604	1853	3.0F
	2144		
16 Su		0057	3.6E
	0409	0659	3.1F
	0945	1310	4.3E
	1631	1919	3.3F
	2205		
17 M ○		0128	4.7E
	0449	0741	3.7F
	1035	1345	4.2E
	1655	1944	3.5F
	2227		
18 Tu		0159	5.4E
	0528	0820	4.0F
	1120	1417	3.8E
	1717	2008	3.6F
	2250		
19 W		0229	5.9E
	0606	0856	3.9F
	1159	1447	3.3E
	1737	2032	3.5F
	2315		
20 Th		0259	6.1E
	0644	0931	3.6F
	1236	1515	2.7E
	1756	2056	3.3F
	2340		
21 F		0330	5.9E
	0723	1005	3.1F
	1309	1541	2.2E
	1814	2120	3.0F
22 Sa	0007	0401	5.4E
	0806	1040	2.4F
	1340	1606	1.6E
	1832	2145	2.6F
23 Su	0034	0434	4.7E
	0855	1118	1.6F
	1412	1632	1.1E
	1849	2210	2.1F
24 M	0102	0513	3.9E
	1002	1210	0.9F
	1451	1700	0.6E
	1904	2237	1.4F
25 Tu ◑	0131	0606	3.0E
	1154	1349	0.3F
	1652	1753	0.1E
	1853	2312	0.7F
26 W	0202	0747	2.3E
	1353	1640	0.6F
	2032	2238	0.2E
27 Th	0118	0220	0.0
	0312	1003	2.3E
	1440	1727	1.1F
	2028	2331	1.0E
28 F	0247	0448	0.5F
	0636	1114	2.6E
	1509	1754	1.6F
	2044	2359	1.9E
29 Sa	0318	0547	1.3F
	0806	1158	3.0E
	1532	1817	2.0F
	2101		
30 Su		0025	2.8E
	0348	0628	2.0F
	0905	1233	3.2E
	1554	1839	2.4F
	2119		
31 M		0050	3.7E
	0417	0703	2.6F
	0952	1304	3.2E
	1614	1900	2.7F
	2138		

November

Day	Slack h m	Maximum h m	knots
1 Tu		0115	4.5E
	0447	0736	3.1F
	1033	1333	3.2E
	1634	1921	2.9F
	2158		
2 W ●		0141	5.2E
	0518	0809	3.5F
	1112	1401	3.0E
	1653	1944	3.1F
	2221		
3 Th		0209	5.7E
	0552	0842	3.6F
	1150	1429	2.7E
	1711	2008	3.2F
	2247		
4 F		0239	6.0E
	0629	0917	3.5F
	1228	1458	2.3E
	1731	2034	3.2F
	2315		
5 Sa		0312	6.1E
	0711	0955	3.2F
	1307	1529	1.9E
	1751	2102	3.1F
	2348		
6 Su		0349	5.8E
	0758	1038	2.6F
	1351	1602	1.4E
	1813	2134	2.8F
7 M	0023	0432	5.3E
	0856	1129	2.0F
	1446	1642	0.8E
	1838	2211	2.3F
8 Tu	0103	0524	4.6E
	1007	1239	1.4F
	1611	1742	0.3E
	1913	2300	1.6F
9 W ◐	0150	0633	3.8E
	1134	1418	1.2F
	1820	1942	0.2E
	2107		
10 Th		0031	0.8F
	0257	0812	3.2E
	1256	1552	1.4F
	1919	2200	0.9E
11 F	0057	0309	0.6F
	0504	0952	3.0E
	1354	1649	1.9F
	1951	2306	2.0E
12 Sa	0229	0454	1.2F
	0712	1105	3.0E
	1437	1730	2.3F
	2019	2350	3.2E
13 Su	0321	0600	2.0F
	0839	1159	3.0E
	1511	1804	2.6F
	2047		
14 M		0027	4.3E
	0405	0650	2.7F
	0943	1243	2.9E
	1540	1834	2.9F
	2115		
15 Tu		0102	5.2E
	0444	0732	3.2F
	1035	1321	2.7E
	1607	1903	3.1F
	2143		
16 W ○		0135	5.7E
	0522	0811	3.4F
	1119	1355	2.5E
	1632	1932	3.1F
	2212		
17 Th		0207	6.0E
	0600	0847	3.4F
	1157	1426	2.2E
	1656	2001	3.1F
	2242		
18 F		0239	6.0E
	0637	0921	3.1F
	1232	1455	1.9E
	1720	2028	3.0F
	2312		
19 Sa		0312	5.8E
	0715	0955	2.7F
	1303	1524	1.6E
	1745	2058	2.8F
	2344		
20 Su		0345	5.3E
	0756	1030	2.3F
	1335	1554	1.4E
	1813	2128	2.5F
21 M	0015	0421	4.7E
	0840	1109	1.8F
	1410	1628	1.1E
	1845	2201	2.0F
22 Tu	0048	0501	4.0E
	0932	1157	1.3F
	1455	1712	0.8E
	1928	2241	1.4F
23 W	0122	0549	3.4E
	1034	1259	1.0F
	1601	1820	0.6E
	2044	2337	0.8F
24 Th ◑	0201	0653	2.7E
	1145	1419	0.9F
	1725	2015	0.7E
	2329		
25 F		0122	0.3F
	0302	0820	2.2E
	1249	1533	1.0F
	1828	2156	1.3E
26 Sa	0150	0334	0.4F
	0514	0948	2.0E
	1338	1625	1.3F
	1910	2253	2.1E
27 Su	0244	0500	0.9F
	0715	1055	2.0E
	1417	1704	1.6F
	1942	2333	3.0E
28 M	0322	0556	1.6F
	0837	1145	2.1E
	1448	1737	1.9F
	2011		
29 Tu		0007	3.9E
	0357	0641	2.3F
	0938	1227	2.1E
	1516	1808	2.2F
	2041		
30 W		0041	4.8E
	0432	0721	2.8F
	1029	1305	2.0E
	1542	1830	2.5F
	2112		

December

Day	Slack h m	Maximum h m	knots
1 Th ●		0115	5.5E
	0509	0759	3.2F
	1114	1340	1.9E
	1608	1909	2.8F
	2146		
2 F		0150	6.0E
	0548	0838	3.3F
	1156	1414	1.8E
	1634	1942	3.0F
	2223		
3 Sa		0228	6.3E
	0629	0917	3.3F
	1237	1449	1.6E
	1703	2018	3.1F
	2301		
4 Su		0307	6.3E
	0714	0959	3.1F
	1318	1527	1.5E
	1736	2056	3.1F
	2342		
5 M		0349	6.1E
	0801	1043	2.8F
	1402	1609	1.3E
	1816	2138	2.8F
6 Tu	0025	0435	5.6E
	0851	1132	2.4F
	1449	1658	1.1E
	1908	2226	2.3F
7 W	0111	0525	4.8E
	0945	1226	2.1F
	1542	1801	1.1E
	2024	2326	1.7F
8 Th ◐	0202	0622	4.0E
	1040	1326	1.8F
	1639	1922	1.3E
	2220		
9 F		0051	1.0F
	0308	0731	3.1E
	1137	1429	1.7F
	1734	2054	1.8E
10 Sa	0038	0244	0.7F
	0447	0853	2.3E
	1232	1530	1.8F
	1823	2211	2.7E
11 Su	0214	0431	1.0F
	0653	1016	1.8E
	1322	1623	1.9F
	1908	2310	3.6E
12 M	0315	0548	1.6F
	0836	1124	1.6E
	1406	1710	2.1F
	1950	2358	4.4E
13 Tu	0403	0645	2.2F
	0950	1219	1.5E
	1446	1752	2.3F
	2030		
14 W		0040	5.0E
	0445	0730	2.6F
	1045	1303	1.4E
	1523	1831	2.4F
	2109		
15 Th		0119	5.5E
	0523	0809	2.8F
	1126	1341	1.5E
	1557	1907	2.6F
	2147		
16 F ○		0155	5.6E
	0600	0845	2.8F
	1201	1415	1.5E
	1631	1943	2.7F
	2224		
17 Sa		0230	5.6E
	0636	0918	2.7F
	1230	1447	1.5E
	1705	2017	2.7F
	2301		
18 Su		0305	5.5E
	0711	0950	2.6F
	1258	1519	1.6E
	1739	2052	2.7F
	2336		
19 M		0339	5.2E
	0746	1023	2.4F
	1327	1552	1.6E
	1817	2126	2.5F
20 Tu	0011	0413	4.8E
	0821	1056	2.2F
	1358	1628	1.5E
	1859	2203	2.2F
21 W	0045	0448	4.4E
	0856	1132	1.9F
	1431	1709	1.5E
	1949	2244	1.8F
22 Th	0120	0525	3.8E
	0933	1210	1.7F
	1507	1757	1.5E
	2053	2332	1.3F
23 F	0158	0607	3.1E
	1012	1251	1.5F
	1547	1855	1.6E
	2222		
24 Sa ◑		0037	0.8F
	0246	0657	2.4E
	1054	1338	1.3F
	1630	2007	1.9E
25 Su	0015	0208	0.5F
	0401	0804	1.7E
	1140	1431	1.2F
	1716	2122	2.3E
26 M	0153	0356	0.6F
	0608	0928	1.2E
	1231	1529	1.2F
	1805	2229	3.0E
27 Tu	0255	0524	1.1F
	0815	1051	1.0E
	1323	1625	1.4F
	1856	2325	3.9E
28 W	0342	0627	1.8F
	0942	1157	0.9E
	1411	1717	1.7F
	1947		
29 Th		0014	4.7E
	0426	0716	2.4F
	1041	1248	1.0E
	1455	1805	2.1F
	2038		
30 F		0059	5.4E
	0508	0759	2.9F
	1126	1331	1.2E
	1636	1860	2.6F
	2127		
31 Sa ●		0143	6.0E
	0549	0839	3.2F
	1204	1411	1.4E
	1617	1934	2.9F
	2215		

Time meridian 120° E. 0000 is midnight. 1200 is noon. Times are not adjusted for Daylight Saving Time.
If three consecutive entries are marked (F) the middle one is not a true maximum but an intermediate value to show the current pattern.

Pages 146 through 163 intentionally omitted by NOAA

See Important Notices page xii

Pages 146 through 163 intentionally omitted by NOAA

See Important Notices page xii

Pages 146 through 163 intentionally omitted by NOAA

See Important Notices page xii

Pages 146 through 163 intentionally omitted by NOAA

See Important Notices page xii

Pages 146 through 163 intentionally omitted by NOAA

See Important Notices page xii

Pages 146 through 163 intentionally omitted by NOAA

See Important Notices page xii

Pages 146 through 163 intentionally omitted by NOAA

See Important Notices page xii

Pages 146 through 163 intentionally omitted by NOAA

See Important Notices page xii

Pages 146 through 163 intentionally omitted by NOAA

See Important Notices page xii

Pages 146 through 163 intentionally omitted by NOAA

See Important Notices page xii

Pages 146 through 163 intentionally omitted by NOAA

See Important Notices page xii

Pages 146 through 163 intentionally omitted by NOAA

See Important Notices page xii

Pages 146 through 163 intentionally omitted by NOAA

See Important Notices page xii

Pages 146 through 163 intentionally omitted by NOAA

See Important Notices page xii

Pages 146 through 163 intentionally omitted by NOAA

See Important Notices page xii

Pages 146 through 163 intentionally omitted by NOAA

See Important Notices page xii

Pages 146 through 163 intentionally omitted by NOAA

See Important Notices page xii

Pages 146 through 163 intentionally omitted by NOAA

See Important Notices page xii

EXTRA CURRENTS, 2005

Snow Passage, Alaska

January

	Slack h m	Maximum h m	knots
19	2040	2230	1.7E

February

	Slack h m	Maximum h m	knots
18	2118	2308	1.6E

March

	Slack h m	Maximum h m	knots
19	2052	2243	1.4E
20	2137	2338	1.7E

April

	Slack h m	Maximum h m	knots
17	2016	2209	1.4E
		2309	1.3E

May

	Slack h m	Maximum h m	knots
15		2251	1.1E
16		2348	1.3E
17	2009	2218	1.5E

June

	Slack h m	Maximum h m	knots
15	1743	2135	1.5E
		2323	1.0E
30		2141	1.6E

July

	Slack h m	Maximum h m	knots
13		2300	0.7E
14		2354	0.7E

August

	Slack h m	Maximum h m	knots
29	1835	2241	2.4E

September

	Slack h m	Maximum h m	knots
27	1821	2215	2.1E
28	2045	2312	2.2E

October

	Slack h m	Maximum h m	knots
26		1810	*
		1853	*
		2148	1.6E
27	2040	2245	1.7E

November

	Slack h m	Maximum h m	knots
25		1210	1.4E
	1427	1618	0.7F
		1820	0.4F
		1919	0.4F
	024	2213	1.3E

December

	Slack h m	Maximum h m	knots
23	1851	2042	0.9E
24		1802	*
		1902	*
		2137	1.1E

Wrangell Narrows, Alaska

March

	Slack h m	Maximum h m	knots
18		2323	*
19	2310		

April

	Slack h m	Maximum h m	knots
16		2246	0.3E
17	2056		

May

	Slack h m	Maximum h m	knots
15		2203	0.6E

October

	Slack h m	Maximum h m	knots
11	1932	2308	1.3E
25	1901	2250	0.8E

November

	Slack h m	Maximum h m	knots
23	1822	2004	1.1E

TABLE 2. — CURRENT DIFFERENCES AND OTHER CONSTANTS AND ROTARY TIDAL CURRENTS

EXPLANATION OF TABLE

In this publication, reference stations are those for which daily predictions are listed in Table 1. Those stations appearing in Table 2 are called subordinate stations. The principal purpose of Table 2 is to present data that will enable one to determine the approximate times of minimum currents (slack waters) and the times and speeds of maximum currents at numerous subordinate stations on the Pacific Coast of North America and Asia. By applying the specific corrections given in Table 2 to the predicted times and speeds of the current at the appropriate reference station, reasonable approximations of the current at the subordinate station may be compiled.

Locations and Depths.—Because the latitude and longitude are listed according to the exactness recorded in the original survey records, the locations of the subordinate stations are presented in varying degrees of accuracy. Since a minute of latitude in nearly equivalent to a mile, a location given to the nearest minute may not indicate the exact position of the station. This should be noted, especially in the case of a narrow stream, where the nearest minute of latitude or longitude may locate a station inland. In such cases, unless the description locates the station elsewhere, reference is made to the current in the center of the channel. In some instances, the charts may not present a convenient name for locating a station. In those cases, the position may be described by a bearing from some prominent place on the chart.

Although current measurements may have been recorded at various depths in the past, the data listed here for most of the subordinate stations are mean values determined to have been representative of the current at each location. For that reason, no specific current meter depths for those stations are given in Table 2. In recent years, however, new data from individual meter depths at a given location have been published and subsequent new data also may be presented in a similar manner.

Since most of the current data in Table 2 came from meters suspended from survey vessels or anchored buoys, the listed depths are those measured downward from the surface. Some later data have come from meters anchored at fixed depths from the bottom. Those meter positions were defined as depths below chart datum. Such defined depths in this and subsequent editions will be accompanied by the small letter "d."

Minimum Currents.—Between the maximum flood and maximum ebb phases, the current may or may not diminish to a true slack water or zero speed stage. For that reason, the all-inclusive terms, "minimum before flood" and "minimum before ebb" are used in the Table 2 heading rather than "slack water." Average speeds and directions of the minimums are given where they are known. Dashes are used where the values are unknown or unreliable and should not be interpreted as zero speed values.

Maximum Currents.—Near the coast and in inland tidal waters, the current increases from minimum current (slack water) for a period of about 3 hours until the maximum speed or the strength of the current is reached. The speed then decreases for another period of about 3 hours when minimum current is again reached and the current begins a similar cycle in the opposite direction. The current that flows toward the coast or up a stream is known as the flood current; the opposite flow is known as the ebb current. Table 2 lists the average speeds and directions of the maximum floods and maximum ebbs. The directions are given in degrees, true, reading clockwise from 000° at north to 359° and are the directions toward which the current flow.

Differences and speed Ratios.—Table 2 contains mean time differences by which the reader can compile approximate times for the minimum and maximum current phases at the subordinate stations. Time differences for those phases should be applied to the corresponding phases at the reference station. It will be seen upon inspection that some subordinate stations exhibit either a double flood or a double ebb stage, or both. Explanations of these stages can be found in the glossary located elsewhere in this publication. In those cases, a separate time difference is listed for each of the three flood (or ebb) phases and these should be applied only to the daily maximum flood (or ebb) phase at the reference station. The results obtained by the application of the time differences will be based upon the time meridian shown above the name of the subordinate station. Differences of time meridians between a subordinate station and its reference station have been accounted for and no further adjustment by the reader is needed. Summer or daylight-saving time is not used in this publication.

The speed ratios are used to compile approximations of the daily current speeds at the subordinate stations and refer only to the maximum floods and ebbs. No attempt is made to predict the speeds of the minimum currents. Normally, these ratios should be applied to the corresponding maximum current phases at the reference station. As mentioned above, however, some subordinate stations may exhibit either a double flood or a double ebb or both. As with the time differences, separate ratios are listed for each of the three flood (or ebb phases) and should be applied only to the daily maximum flood (or ebb) speed at the reference station. It should be noted that although the speed of a given current phase at a subordinate station is obtained by reference to the corresponding phase at the reference station, the directions of the current at the two places may differ considerably. Table 2 lists the average directions of the various current phases at the subordinate stations.

Rotary Tidal Currents.—Table 5 contains listings of data for those stations which exhibited rotary current patterns. Briefly, a rotary current can be described as one which flows continually with the direction of flow changing through all points of the compass during the tidal period. A more complete description can be found in the glossary located elsewhere in this publication. The average speeds and directions are listed in hour increments as referred to the predicted times of a particular current phase at a reference station in Table 1. The Moon, at times of new, full, or perigee may increase speeds 15 to 20 percent above average; or 30 to 40 percent if perigee occurs at or near the time of new or full Moon. Conversely, the Moon at times of quadrature or apogee may decrease the speeds 15 to 20 percent or 30 to 40 percent if they occur together. Near average speeds may be expected when apogee occurs near or at new or full Moon, or when perigee occurs at or near quadrature. The directions of the currents are given in degrees true, reading clockwise from 000° at north to 359° and are the directions toward which the current flows.

Example of the use of Table 2.—Suppose we wish to calculate the approximate times of the minimum currents and the times and speeds of the maximum currents on a particular morning at the location listed as Cordova, Orca Inlet. From Table 2 we learn that the reference station is Wrangell Narrows whose predicted currents for the morning are listed below. Currents for Cordova can be approximated by using the Table 2 corrections as shown below.

	Minimum before flood	*Maximum flood*		*Minimum before ebb*	*Maximum ebb*	
	h.m.	*h.m.*	*kn.*	*h.m.*	*h.m.*	*kn.*
Wrangell Narrows...............................	0011	0243	2.8	0613	0912	2.8
Table 2 corrections.............................	–023	+019	x0.5 ratio	+023	+016	x0.3 ratio
Cordova...	2348*	0302	1.4	0636	0928	0.8

* this minimum current phase is seen to occur just before midnight of the previous day.

Table 2 list the mean values of the minimum current phases as 0.0 knots; therefore, no directions are given. The average directions of the maximum flood and maximum ebb are 212° true and 026° true, respectively.

NOTE.—subordinate locations referencing Iloilo, San Bernardino Strait, San Juanico, and Cebu Harbor were included only for future consideration. See IMPORTANT NOTICE on page VI.

TABLE 2 – CURRENT DIFFERENCES AND OTHER CONSTANTS

No.	PLACE	Meter Depth	POSITION Latitude	POSITION Longitude	TIME DIFFERENCES Min. before Flood	TIME DIFFERENCES Flood	TIME DIFFERENCES Min. before Ebb	TIME DIFFERENCES Ebb	SPEED RATIOS Flood	SPEED RATIOS Ebb	Minimum before Flood knots	Minimum before Flood Dir.	Maximum Flood knots	Maximum Flood Dir.	Minimum before Ebb knots	Minimum before Ebb Dir.	Maximum Ebb knots	Maximum Ebb Dir.
		ft	North	West	h m	h m	h m	h m			knots	Dir.	knots	Dir.	knots	Dir.	knots	Dir.
	BAY of PANAMA Time meridian, 75° W				on San Francisco Bay Ent., p.8													
1	Bayoneta I., 1.5 miles W of, Perlas Is		8° 30'	79° 05'	+1 07	+2 02	+2 21	+1 22	0.6	0.4	--	--	1.6	005°	--	--	1.5	200°
6	Chame Bay Entrance, near Chame Point		8° 39'	79° 43'	+2 10	+2 31	+2 10	+2 45	0.6	0.4	--	--	1.8	210°	--	--	1.4	065°
	COSTA RICA Time meridian, 90° W																	
11	Puntarenas, Gulf of Nicoya		9° 58'	84° 49'	+0 01	---	+0 44	---	--	--	--	--	--	300	--	--	--	--
	LOWER CALIFORNIA Time meridian, 105° W																	
16	Magdalena Bay entrance		24° 32'	112° 02'	−4 43	−3 52	−3 46	−3 45	0.4	0.3	--	--	1.3	035	--	--	1.0	--
	SAN CLEMENTE ISLAND Time meridian, 120° W				on San Diego Bay Ent., p.4													
21	China Point Light, 20 miles SSW of		32° 29'	118° 32'	---	---	---	---	0.3	0.1	--	--	0.4	315	--	--	0.2	115°
	SAN DIEGO BAY																	
26	Point Loma Light, 0.8 nmi. east of	15d	32° 39.95'	117° 13.57'	−0 18	−0 43	−0 05	+0 45	0.5	0.4	0.1	265°	0.6	328°	0.1	241°	0.6	174°
	. . . do.	33d	32° 39.95'	117° 13.57'	−1 08	−0 46	−0 08	−0 23	0.4	0.2	0.1	241°	0.5	333°	0.1	086°	0.3	166°
31	SAN DIEGO BAY ENTRANCE		32° 40.90'	117° 13.80'		*Daily predictions*					--	--	1.2	355°	--	--	1.5	173°
36	Ballast Point, south of	5d	32° 41.07'	117° 13.93'	−1 04	−1 02	−1 01	−2 03	0.4	0.2	0.1	335°	0.5	044°	--	--	0.4	238°
41	Ballast Point, 100 yards north of		32° 41'	117° 14'	−0 27	−0 24	−0 23	−0 02	1.0	0.9	--	--	1.2	325°	--	--	1.4	133°
46	Ballast Point, 0.55 nmi. north of	14d	32° 41.75'	117° 13.95'	−0 05	−0 39	−0 34	+0 24	0.5	0.4	--	--	0.6	354°	--	--	0.6	193°
	. . . do.	34d	32° 41.75'	117° 13.95'	−0 44	+0 10	+0 03	−0 03	0.5	0.2	--	--	0.6	344°	--	--	0.3	147°
51	Quarantine Station, La Playa		32° 42'	117° 14'	−0 03	+0 15	+0 10	+0 20	0.8	0.8	--	--	1.0	021°	--	--	1.2	200°
56	North Island	14d	32° 42.78'	117° 12.77'	−0 26	−0 56	−0 54	−0 03	0.5	0.4	--	--	0.6	062°	--	--	0.6	245°
	. . . do.	34d	32° 42.78'	117° 12.77'	−0 43	−1 05	−0 44	−0 33	0.3	0.2	--	--	0.4	077°	--	--	0.3	241°
61	Harbor Island (east end), SSW of	15d	32° 43.15'	117° 11.50'	+0 29	+0 09	−0 24	+0 23	0.3	0.3	0.1	031°	0.4	097°	--	--	0.4	293°
66	San Diego, 0.5 mile west of		32° 43'	117° 11'	−0 16	−0 08	−0 12	−0 12	0.6	0.5	--	--	0.7	121°	--	--	0.7	304°
71	Airport CGS, 0.3 nmi. SE of	14d	32° 43.32'	117° 10.67'		Current weak and variable												
76	B St. Pier (San Diego) <1>	34d	32° 43.02'	117° 10.58'		Current weak and variable												
81	G St. Pier (San Diego), 0.22 nmi. SW of	14d	32° 42.50'	117° 10.65'	+0 10	+0 20	−0 03	+0 41	0.3	0.3	--	--	0.4	139°	0.1	237°	0.5	304°
	. . . do.	37d	32° 42.50'	117° 10.65'	−0 23	+0 21	−0 18	+0 58	0.3	0.2	--	--	0.3	125°	--	--	0.3	301°
86	Fifth Avenue Marina Entrance	11d	32° 42.33'	117° 09.92'	−0 12	+0 09	+1 46	+1 07	0.1	--	--	--	0.2	031°	--	--	--	--
91	Coronado, off northeast end	14d	32° 41.88'	117° 09.83'	−0 24	−0 58	−0 51	+0 09	0.7	0.5	--	--	0.8	128°	0.1	037°	0.7	317°
	. . . do.	38d	32° 41.88'	117° 09.83'	−0 41	−0 59	−1 01	+0 03	0.5	0.4	--	--	0.6	130°	--	--	0.5	319°
96	28th St. Pier (San Diego), 0.92 nmi. SW	7d	32° 40.48'	117° 08.97'	−0 44	−1 17	−1 10	−0 05	0.2	0.1	--	--	0.2	182°	--	--	0.2	351°
101	28th St. Pier (San Diego), 0.35 nmi. SW	14d	32° 40.97'	117° 08.57'	−0 14	+0 15	+0 15	+0 13	0.3	0.2	--	--	0.4	133°	--	--	0.3	317°
	. . . do.	28d	32° 40.97'	117° 08.57'	+0 01	+0 24	+0 15	+0 35	0.2	0.2	--	--	0.3	148°	--	--	0.3	328°
106	National City		32° 39'	117° 07'	+0 23	0 00	−0 02	+0 50	0.4	0.4	--	--	0.5	166°	--	--	0.6	002°
111	National City, WSW of Pier 12	32d	32° 39.73'	117° 07.53'	+0 22	+0 34	+0 34	+0 58	0.2	0.2	--	--	0.2	178°	--	--	0.2	351°
116	Sweetwater Channel, southwest of	14d	32° 38.70'	117° 07.37'	+0 29	−0 33	−0 05	+0 46	0.1	0.2	0.1	259°	0.2	203°	--	--	0.3	348°
	CALIFORNIA COAST				on San Francisco Bay Ent., p.8													
121	San Pedro Channel <2>		33° 36'	118° 16'	---	---	---	---	--	--	--	--	--	--	--	--	--	--
126	Los Angeles and Long Beach Harbors <3>		---	---	---	---	---	---	--	--	--	--	--	--	--	--	--	--
131	El Segundo, Santa Monica Bay <4>		33° 54'	118° 26'	---	---	---	---	--	--	--	--	0.6	330	--	--	--	--
136	Point Arguello		34° 34'	120° 40'	−2 21	−2 21	−2 21	−2 21	0.2	0.2	--	--	0.5	005	--	--	0.5	185°
141	Point San Luis		35° 09'	120° 46'	−2 01	−2 01	−2 01	−2 01	0.2	0.2	--	--	0.5	305	--	--	0.5	125°
146	Point Piedras Blancas		35° 40'	121° 18'	−1 29	−1 29	−1 29	−1 29	0.2	0.2	--	--	0.5	315	--	--	0.5	135°
151	Point Sur		36° 18'	121° 55'	−1 11	−1 11	−1 11	−1 11	0.2	0.2	--	--	0.5	325	--	--	0.5	145°

Endnotes can be found at the end of table 2.

TABLE 2 – CURRENT DIFFERENCES AND OTHER CONSTANTS

No.	PLACE	Meter Depth	POSITION Latitude	Longitude	TIME DIFFERENCES Min. before Flood	Flood	Min. before Ebb	Ebb	SPEED RATIOS Flood	Ebb	AVERAGE SPEEDS AND DIRECTIONS Minimum before Flood knots	Dir.	Maximum Flood knots	Dir.	Minimum before Ebb knots	Dir.	Maximum Ebb knots	Dir.
		ft	North	West	h m	h m	h m	h m										
	MONTEREY BAY Time meridian, 120° W				on San Francisco Bay Ent., p.8													
156	Point Pinos		36° 38'	121° 57'	−1 01	−1 01	−1 01	−1 01	0.2	0.2	– –	– –	0.5	035°	– –	– –	0.5	215°
161	Point Santa Cruz, 2 miles south of		36° 55'	122° 01'	Current weak and variable													
	CALIFORNIA COAST–cont.																	
166	Ano Nuevo Island, 2 miles SW of		37° 05'	122° 22'	Current weak and variable													
171	Point Montara, 2 miles west of		37° 32'	122° 34'	Current weak and variable													
	GOLDEN GATE and APPROACHES <5>																	
176	Point Lobos, 8.7 miles WSW of <6>		37° 45'	122° 42'	−0 30	−0 30	−0 30	−0 30	0.1	0.1	– –	– –	0.4	135°	– –	– –	0.4	315°
181	Point Bonita Lt., 5.27 nmi. WSW of	39d	37° 48.27'	122° 38.33'	−1 52	−1 41	−1 41	−1 19	0.2	0.2	– –	– –	0.4	095°	0.1	196°	0.5	266°
186	Point Lobos, 5 miles west of <7>		37° 47'	122° 37'	– – –	−0 10	– – –	−0 32	0.2	0.2	– –	– –	0.6	020°	– –	– –	0.8	236°
191	Point Lobos, 3.73 nmi. W of	46d	37° 47.25'	122° 35.32'	−2 14	−1 09	−1 10	−0 49	0.3	0.2	– –	– –	0.8	092°	– –	– –	0.5	241°
196	Point Lobos, 2.5 miles west of <7>		37° 46.37'	122° 34.90'	– – –	+0 26	– – –	−1 26	0.3	0.5	– –	– –	0.9	080°	– –	– –	1.7	269°
201	Point Lobos, 1.3 nmi. SW of	46d	37° 46.30'	122° 32.13'	−1 24	−0 57	−0 23	−0 24	0.4	0.3	– –	– –	1.0	008°	– –	– –	1.1	182°
206	South Channel		37° 45'	122° 32'	−1 29	−1 27	−1 04	−1 10	0.4	0.4	– –	– –	1.2	023°	– –	– –	1.4	180°
211	Point Lobos, 5.47 nmi. SW of	39d	37° 43.23'	122° 35.87'	−1 59	−2 08	−1 41	−1 28	0.2	0.2	0.2	110°	0.6	048°	– –	– –	0.8	211°
216	Bonita Channel approach <7>		37° 50'	122° 37'	– – –	−0 16	– – –	−1 14	0.2	0.2	– –	– –	0.5	061°	– –	– –	0.6	290°
221	Bonita Channel, off Tennessee Cove <7>		37° 50.05'	122° 33.78'	– – –	−0 40	– – –	−1 14	0.4	0.2	– –	– –	1.1	112°	– –	– –	0.8	325°
226	Bonita Channel, off Point Bonita		37° 48.95'	122° 32.13'	– – –	−0 28	– – –	−2 14	0.4	0.1	– –	– –	1.3	115°	– –	– –	0.3	355°
231	Point Bonita, 0.8 nmi. NE of	22d	37° 49.25'	122° 30.97'	−4 49	−3 54	−5 11	−4 19	0.3	0.3	0.2	323°	0.8	064°	0.1	326°	0.9	267°
	. . . do.	41d	37° 49.25'	122° 30.97'	−5 10	−4 29	−5 35	−4 19	0.2	0.2	0.1	171°	0.6	111°	0.2	174°	0.7	247°
236	Point Bonita Lt., 0.4 nmi. SSE of	43d	37° 48.72'	122° 31.27'	−0 46	−1 06	−3 10	−1 13	0.1	0.4	– –	– –	0.3	104°	0.1	162°	1.4	229°
241	Point Bonita, 0.95 nmi. SSE of	22d	37° 48.07'	122° 31.13'	−0 28	−0 38	−0 28	−0 51	0.7	0.8	0.1	146°	2.1	072°	0.5	142°	2.6	239°
	. . . do.	42d	37° 48.07'	122° 31.13'	−0 32	−0 38	−0 17	−0 46	0.7	0.6	0.2	144°	2.1	072°	0.3	159°	2.2	243°
246	Mile Rock Lt., 0.2 nmi. NW of	15d	37° 47.72'	122° 30.68'	−0 33	−0 31	−0 32	−0 51	0.8	0.8	0.4	143°	2.5	063°	– –	– –	2.6	230°
	. . . do.	35d	37° 47.72'	122° 30.68'	−0 46	−0 34	−0 27	−0 52	0.9	0.7	0.5	135°	2.5	054°	0.1	340°	2.3	231°
251	Point Diablo, 0.2 mile SE of		37° 49.07'	122° 29.80'	−0 07	−0 59	−1 37	−0 10	0.6	0.9	– –	– –	1.8	082°	– –	– –	3.0	263°
256	Baker Beach (South Bay), 0.3 nmi. NW of	31d	37° 47.87'	122° 29.31'	−5 04	−4 21	−2 49	−3 10	0.4	0.3	0.2	113°	1.2	038°	0.2	161°	0.9	208°
	. . . do.	50d	37° 47.87'	122° 29.31'	−5 09	−4 20	−2 23	−3 12	0.3	0.2	0.1	120°	1.0	043°	0.2	162°	0.7	215°
261	Fort Point, 0.3 nmi. west of	75d	37° 48.55'	122° 28.97'	−2 08	−0 47	+0 12	+0 02	0.5	0.2	0.6	342°	1.4	044°	0.2	101°	0.7	335°
266	SAN FRANCISCO BAY ENT. (Outside)		37° 48.63'	122° 30.13'		*Daily predictions*					– –	– –	2.9	065°	– –	– –	3.4	245°
					on Golden Gate Bridge, p.12													
271	GOLDEN GATE BRIDGE	22d	37° 49.75'	122° 27.73'		*Daily predictions*					– –	– –	2.8	055°	0.4	144°	1.7	237°
	. . . do.	35d	37° 49.75'	122° 27.73'	−0 08	−0 02	−0 02	−0 11	1.0	1.0	– –	– –	2.8	055°	0.4	142°	1.7	232°
	. . . do.	48d	37° 49.75'	122° 27.73'	−0 13	−0 04	−0 03	−0 15	1.0	1.0	– –	– –	2.8	053°	0.3	140°	1.7	228°
276	Golden Gate Bridge 0.46 nmi. east of	20d	37° 49.20'	122° 28.37'	+0 29	+0 10	+0 02	−0 14	1.0	1.3	– –	– –	2.7	060°	– –	– –	2.2	250°
	. . . do.	39d	37° 49.20'	122° 28.37'	−0 03	−0 19	−0 04	−0 23	0.9	1.3	– –	– –	2.5	074°	– –	– –	2.2	251°
	. . . do.	69d	37° 49.20'	122° 28.37'	−0 03	−0 24	−0 02	−0 16	0.9	1.2	– –	– –	2.5	075°	– –	– –	2.0	257°
281	Golden Gate Bridge, 0.8 mile east of		37° 49.3'	122° 27.7'	+0 37	+0 14	+0 25	+0 34	0.9	1.3	– –	– –	2.5	070°	– –	– –	2.3	256°
286	Fort Point, 0.5 nmi. east of	55d	37° 48.7'	122° 27.98'	−0 48	−2 48	−0 32	−1 17	0.2	0.7	0.2	039°	0.6	099°	0.5	251°	1.3	287°
	SAN FRANCISCO BAY, South <8>																	
291	Alcatraz Island, 0.2 mile west of		37° 49.67'	122° 25.82'	+0 09	−0 14	+0 22	+0 34	0.8	1.2	– –	– –	2.3	070°	– –	– –	2.1	266°
296	Alcatraz Island, southwest of	20d	37° 48.87'	122° 25.92'	−0 12	−0 37	−0 45	−0 32	0.5	0.9	– –	– –	1.4	080°	– –	– –	1.6	260°
	. . . do.	42d	37° 48.87'	122° 25.92'	−0 20	−0 34	−0 24	−0 31	0.4	0.8	– –	– –	1.2	082°	0.1	167°	1.4	251°
	. . . do.	68d	37° 48.87'	122° 25.92'	−0 42	−0 27	−0 25	−0 29	0.3	0.6	0.1	338°	0.9	080°	0.1	163°	1.0	252°
301	Alcatraz Island, south of		37° 49.10'	122° 25.32'	+0 10	−0 04	+0 05	−0 07	0.5	1.1	– –	– –	1.5	093°	– –	– –	1.9	286°
306	Alcatraz Island, 0.5 mile north of <9>		37° 50.1'	122° 25.2'	+1 08	+0 38	−0 11	+0 24	0.7	1.4	– –	– –	1.8	066°	– –	– –	2.4	284°
311	Alcatraz Island, 0.8 mile east of		37° 49.10'	122° 24.03'	−0 18	−1 10	−0 43	−0 03	0.4	1.2	– –	– –	1.1	134°	– –	– –	2.1	314°

Endnotes can be found at the end of table 2.

TABLE 2 – CURRENT DIFFERENCES AND OTHER CONSTANTS

No.	PLACE	Meter Depth	POSITION Latitude	POSITION Longitude	TIME DIFFERENCES Min. before Flood	TIME DIFFERENCES Flood	TIME DIFFERENCES Min. before Ebb	TIME DIFFERENCES Ebb	SPEED RATIOS Flood	SPEED RATIOS Ebb	Minimum before Flood knots	Minimum before Flood Dir.	Maximum Flood knots	Maximum Flood Dir.	Minimum before Ebb knots	Minimum before Ebb Dir.	Maximum Ebb knots	Maximum Ebb Dir.
		ft	North	West	h m	h m	h m	h m			knots	Dir.	knots	Dir.	knots	Dir.	knots	Dir.
	SAN FRANCISCO BAY, South <8>–cont. Time meridian, 120° W																	
					on Golden Gate Bridge, p.12													
316	Treasure Island, 0.78 NW of	10d	37° 50.25'	122° 23.38'	−1 00	−1 00	−0 30	−1 33	0.4	0.6	0.1	185°	1.2	122°	0.2	016°	1.0	292°
	... do	20d	37° 50.25'	122° 23.38'	−1 44	−0 37	−0 31	−1 26	0.3	0.4	– –	– –	0.9	122°	0.1	023°	0.7	299°
321	Treasure Island, 0.8 mile west of	8	37° 49.3	122° 23.5'	−0 06	−0 38	−0 41	−0 10	0.4	1.4	– –	– –	1.2	148°	– –	– –	2.5	334°
326	Treasure Island, 0.2 mile west of	8	37° 49.3	122° 22.7'	−0 59	−0 51	−0 34	−0 39	0.5	1.1	– –	– –	1.3	172°	– –	– –	1.9	343°
					on Oakland, p.16													
331	Yerba Buena Island, W of (midchannel)		37° 48.6'	122° 23'	+1 11	−0 03	+0 01	+1 07	1.0	2.0	– –	– –	1.4	168°	– –	– –	2.5	316°
336	OAKLAND, YERBA BUENA ISLAND	13d	37° 48.59'	122° 21.04'		*Daily predictions*					– –	– –	1.4	167°	0.1	254°	1.3	338°
	... do.	29d	37° 48.59'	122° 21.04'	−0 38	−0 04	−0 01	−0 07	0.7	0.6	– –	– –	1.0	163°	0.1	080°	0.8	355°
341	San Francisco–Oakland Bay Bridge <5>		– – –	– – –	– – –	– – –	– – –	– – –	– –	– –	– –	– –	– –	– –	– –	– –	– –	– –
346	Treasure Island, 0.5 mile north of	8d	37° 50.43'	122° 22.10'	+0 31	+0 02	+0 22	+0 42	0.8	0.9	– –	– –	1.1	118°	– –	– –	1.2	304°
351	Treasure Island, 0.85 nmi. east of	11d	37° 49.50'	122° 20.78'	−0 13	+0 16	+0 24	−0 26	0.6	0.5	– –	– –	0.8	161°	– –	– –	0.7	340°
356	Treasure Island, 0.3 mile east of	8d	37° 49.6'	122° 21.3'	+1 12	+0 13	+0 00	+1 02	1.0	1.2	– –	– –	1.4	140°	– –	– –	1.5	327°
361	Yerba Buena Island, 0.3 nmi. SE of	23d	37° 48.25'	122° 21.43'	+0 31	+0 05	+0 07	+0 21	0.5	0.8	0.1	047°	0.6	159°	0.1	259°	1.0	316°
366	Oakland Outer Harbor entrance	8d	37° 48.4'	122° 20.7'	−0 04	−0 07	+0 22	+0 09	1.1	1.3	– –	– –	1.6	172°	– –	– –	1.7	336°
371	Oakland Inner Harbor entrance	8d	37° 48.1'	122° 20.4'	−0 11	−0 12	−0 01	−0 09	0.7	1.2	– –	– –	1.0	178°	– –	– –	1.5	338°
376	Oakland Inner Harbor channel	8d	37° 47.7'	122° 19.1'	+0 25	−0 21	+0 05	+0 31	0.4	0.6	– –	– –	0.5	122°	– –	– –	0.7	318°
381	Oakland Inner Harbor Reach	33d	37° 47.37'	122° 17.15'	−0 55	+0 01	−0 05	−0 12	0.2	0.1	– –	– –	0.3	082°	– –	– –	0.2	255°
386	Oakland Harbor, Webster Street		37° 47.30'	122° 16.47'	+0 08	−0 32	−0 18	−0 03	0.6	0.8	– –	– –	0.9	120°	– –	– –	1.0	300°
391	Oakland Harbor, High Street Bridge	6	37° 45.38'	122° 13.48'	+0 22	−0 49	−0 18	−0 14	1.0	1.1	– –	– –	1.4	150°	– –	– –	1.4	330°
396	Oakland 7th St. Marine, 0.6 nmi. SSW of	21d	37° 47.67'	122° 20.65'	−0 01	+0 27	−0 02	+0 03	0.8	0.6	– –	– –	1.1	154°	– –	– –	0.8	342°
					on Golden Gate Bridge, p.12													
401	Rincon Point, 0.57 nmi. east of	20d	37° 47.42'	122° 22.40'	−0 11	−0 33	−0 22	−0 21	0.7	1.2	– –	– –	2.0	141°	– –	– –	2.1	332°
	... do.	40d	37° 47.42'	122° 22.40'	−0 23	−0 34	−0 17	−0 36	0.6	0.9	– –	– –	1.7	130°	– –	– –	1.6	314°
	... do.	58d	37° 47.42'	122° 22.40'	−0 30	−0 28	−0 22	−0 32	0.5	0.6	– –	– –	1.2	151°	– –	– –	1.0	333°
406	Rincon Point, midbay	11	37° 47.00'	122° 21.23'	−0 27	−0 34	−0 40	−0 33	0.8	1.2	– –	– –	2.1	166°	– –	– –	2.1	322°
411	Mission Rock, 0.6 mile east of		37° 46.4'	122° 22.1'	−0 14	−0 22	−0 10	−0 03	0.9	1.3	– –	– –	2.5	160°	– –	– –	2.3	320°
416	Mission Rock, 1.3 mile east of	8	37° 46.5'	122° 21.2'	−0 04	−0 32	−0 04	+0 05	0.8	1.2	– –	– –	2.1	182°	– –	– –	2.0	344°
421	Mission Rock, 2.0 miles east of		37° 46.6'	122° 20.3'	−0 54	−0 52	−0 20	−0 29	0.6	1.3	– –	– –	1.7	142°	– –	– –	2.3	330°
426	Potrero Point, 1.08 nmi. east of	20d	37° 45.45'	122° 21.47'	−0 13	−0 26	−0 08	−0 13	0.6	1.0	– –	– –	1.7	169°	– –	– –	1.8	342°
	... do.	49d	37° 45.45'	122° 21.47'	−0 20	−0 33	−0 07	+0 04	0.4	0.6	– –	– –	1.2	139°	– –	– –	1.1	327°
431	Potrero Point, 2 miles east of		37° 45'	122° 20'	−0 13	−0 45	+0 01	−0 11	0.6	0.9	– –	– –	1.6	159°	– –	– –	1.6	328°
436	Alameda Radar Tower, 0.9 nmi. SSW of	4d	37° 44.73'	122° 16.98'	−1 00	−1 15	−1 20	−1 21	0.2	0.4	– –	– –	0.5	132°	– –	– –	0.7	309°
441	Point Avisadero, 0.3 mile east of		37° 43.8'	122° 20.9'	−0 06	−0 28	−0 05	+0 11	0.6	1.1	– –	– –	1.6	156°	– –	– –	1.9	337°
446	Point Avisadero, 1 mile east of		37° 43.8'	122° 20.2'	−0 08	−0 05	+0 33	+0 24	0.6	0.9	– –	– –	1.6	154°	– –	– –	1.5	352°
451	Point Avisadero, 2 miles east of		37° 43.9'	122° 18.8'	−0 08	−0 13	−0 02	+0 07	0.5	0.9	– –	– –	1.5	148°	– –	– –	1.6	335°
456	Point Avisadero, 0.6 nm. ESE of	20d	37° 43.38'	122° 19.43'	−0 08	−0 32	−0 06	−0 08	0.5	0.8	– –	– –	1.4	140°	0.1	064°	1.4	335°
	... do.	37d	37° 43.38'	122° 19.43'	−0 13	−0 30	−0 02	−0 12	0.4	0.5	– –	– –	1.1	132°	0.1	066°	0.9	327°
461	Point Avisadero, 1.25 nmi. SSE of	20d	37° 42.47'	122° 20.97'	−0 21	−0 22	−0 21	−0 21	0.4	0.6	– –	– –	1.2	175°	0.1	274°	1.1	004°
	.. do.	32d	37° 42.47'	122° 20.97'	−0 19	−0 13	−0 25	−0 17	0.3	0.5	– –	– –	0.9	186°	– –	– –	0.8	008°
466	Oakland Airport, southwest of	3d	37° 40.93'	122° 13.85'	−0 41	−0 53	−0 31	−0 47	0.3	0.4	0.1	212°	0.7	125°	– –	– –	0.7	304°
471	Sierra Point, 1.3 miles ENE of		37° 41.08'	122° 21.40'	−0 23	−0 41	−0 17	−0 09	0.4	0.6	– –	– –	1.0	180°	– –	– –	1.0	003°
476	Sierra Point, 1.2 nmi. east of	18d	37° 40.68'	122° 19.05'	−0 18	−0 37	−0 18	+0 09	0.3	0.4	0.1	072°	0.9	210°	0.1	279°	0.8	358°
481	Oyster Point, 2.8 miles east of		37° 39.9'	122° 19.4'	−0 05	−0 27	−0 05	+0 15	0.4	0.7	– –	– –	1.0	172°	– –	– –	1.2	345°
486	Sierra Point, 4.4 miles east of		37° 40.4'	122° 17.7'	−0 24	−0 41	−0 01	−0 02	0.4	0.5	– –	– –	1.0	152°	– –	– –	0.8	329°
491	Point San Bruno, 0.51 nmi. east of	10d	37° 39.25'	122° 21.83'	−0 52	−0 34	−0 13	−0 36	0.2	0.3	– –	– –	0.6	174°	– –	– –	0.5	359°
496	Mulford Gardens Chan. "Buoy 2", SSW of	5d	37° 39.03'	122° 13.63'	−0 51	−0 59	−0 42	−0 42	0.3	0.4	0.1	221°	0.8	125°	– –	– –	0.7	315°
501	Little Coyote Pt., 3.4 nmi. NNE of	4d	37° 37.43'	122° 13.88'	−0 42	−1 01	−0 42	−0 30	0.2	0.3	0.1	220°	0.6	128°	– –	– –	0.6	310°
506	Little Coyote Pt., 3.1 nmi. ENE of	4d	37° 35.88'	122° 12.33'	−1 28	−1 41	−0 50	−1 18	0.2	0.3	0.2	230°	0.5	135°	– –	– –	0.5	317°
511	Little Coyote Pt., 1.2 nmi. NE of	10d	37° 35.42'	122° 14.92'	+0 15	−0 33	−0 03	+0 20	0.5	0.9	– –	– –	1.5	121°	– –	– –	1.5	303°
	... do.	20d	37° 35.42'	122° 14.92'	+0 07	−0 34	+0 01	+0 18	0.5	0.8	0.1	220°	1.3	122°	– –	– –	1.4	307°
	... do.	39d	37° 35.42'	122° 14.92'	0 00	−0 28	−0 04	+0 22	0.3	0.6	0.1	230°	0.9	138°	– –	– –	1.1	310°
516	San Mateo Bridge		37° 35.2'	122° 15.2'	+0 34	−0 04	+0 33	+0 46	0.5	0.9	– –	– –	1.5	142°	– –	– –	1.5	312°

Endnotes can be found at the end of table 2.

TABLE 2 – CURRENT DIFFERENCES AND OTHER CONSTANTS

No.	PLACE	Meter Depth	POSITION		TIME DIFFERENCES				SPEED RATIOS		AVERAGE SPEEDS AND DIRECTIONS							
			Latitude	Longitude	Min. before Flood	Flood	Min. before Ebb	Ebb	Flood	Ebb	Minimum before Flood		Maximum Flood		Minimum before Ebb		Maximum Ebb	
		ft	North	West	h m	h m	h m	h m			knots	Dir.	knots	Dir.	knots	Dir.	knots	Dir.
	SAN FRANCISCO BAY, South <8>–cont. Time meridian, 120° W				on Golden Gate Bridge, p.12													
521	Redwood Pt., Blair I., 1.15 nmi. NNE of	19d	37° 33.48'	122° 11.93'	−0 22	−0 28	+0 06	+0 25	0.5	0.9	– –	– –	1.5	120°	– –	– –	1.5	310°
	. . . do.	42d	37° 33.48'	122° 11.93'	+0 07	−0 34	−0 02	+0 25	0.4	0.6	0.1	020°	1.0	107°	0.1	021°	1.1	300°
526	Dumbarton Highway Bridge		37° 30.6'	122° 07.2'	+0 27	+0 10	+0 20	+0 39	0.5	0.6	– –	– –	1.5	151°	– –	– –	1.1	338°
531	Dumbarton Hwy. Bridge, 0.28 nmi. SE of	25d	37° 30.08'	122° 06.93'	+0 03	−0 01	+0 10	−0 06	0.5	0.7	0.1	227°	1.4	141°	– –	– –	1.3	319°
	. . . do.	42d	37° 30.08'	122° 06.93'	−0 12	+0 03	+0 09	−0 14	0.4	0.6	0.1	056°	1.2	137°	– –	– –	1.0	322°
536	Dumbarton Point, 1.15 nmi. SE of	17d	37° 29.25'	122° 04.88'	+0 12	−0 27	+0 20	+1 25	0.4	0.5	– –	– –	1.0	131°	– –	– –	0.8	320°
541	Dumbarton Point, 2.25 miles SE of		37° 28.5'	122° 04.2'	+0 32	−0 13	+0 08	+1 27	0.4	0.6	– –	– –	1.2	127°	– –	– –	1.1	302°
546	Yellow Bluff, 0.8 nmi. NE of	19d	37° 50.73'	122° 27.43'	−0 03	−0 35	−0 29	−0 28	0.6	0.9	0.1	302°	1.6	013°	0.1	300°	1.5	211°
	. . . do.	51d	37° 50.73'	122° 27.43'	−0 18	−0 38	+0 02	−0 02	0.4	0.5	– –	– –	1.0	012°	0.3	297°	0.9	226°
	SAN FRANCISCO BAY, North																	
551	Yellow Bluff, 0.8 mile east of	8	37° 50.1'	122° 27.3'	+0 19	+0 15	+0 20	+0 35	1.0	1.4	– –	– –	2.9	022°	– –	– –	2.3	257°
556	Point Cavallo, 1.3 miles east of	8	37° 49.9'	122° 26.6'	+0 26	+0 25	+0 40	+0 35	0.9	1.5	– –	– –	2.4	087°	– –	– –	2.6	256°
561	Point Blunt, Angel I., 0.5 nmi. SW of	21d	37° 50.73'	122° 25.38'	+0 51	+0 08	+0 01	+0 24	0.4	1.0	– –	– –	1.2	077°	0.2	326°	1.7	258°
	. . . do.	40d	37° 50.73'	122° 25.38'	−0 03	−0 13	+0 25	+0 24	0.4	0.9	0.1	349°	1.2	082°	– –	– –	1.5	258°
566	Point Blunt, Angel I., 0.25 mile S of	8	37° 50.8'	122° 25.0'	+0 07	+0 12	+0 22	+0 59	0.6	1.3	– –	– –	1.7	103°	– –	– –	2.2	258°
571	Pt. Blunt, Angel I., 0.8 mi. SE of <10>		37° 50.7'	122° 24.3'	−0 11	+0 01	+0 19	−0 21	0.4	0.8	– –	– –	1.0	086°	– –	– –	1.3	297°
576	Pt. Blunt, Angel I., 0.25 nmi. east of	21d	37° 51.17'	122° 24.73'	+1 40	−0 07	+0 15	+2 14	0.4	1.0	0.2	050°	1.0	049°	0.1	124°	1.7	185°
	. . . do.	41d	37° 51.17'	122° 24.73'	+0 50	−0 33	+0 32	+1 53	0.2	0.8	0.1	102°	0.7	037°	0.1	309°	1.4	194°
581	Angel Island, off Quarry Point		37° 51.8'	122° 24.8'	+2 06	+0 21	+1 22	+1 38	0.2	1.1	– –	– –	0.5	027°	– –	– –	1.9	130°
586	Angel Island, 0.75 mile east of		37° 51.8'	122° 24.1'	+1 25	+1 09	+1 33	+1 50	0.4	0.5	– –	– –	1.1	351°	– –	– –	0.9	192°
591	Point Simpson, Angel I., 1.05 nmi. E of	21d	37° 52.43'	122° 24.17'	+0 48	+1 07	+1 01	+0 24	0.5	0.5	– –	– –	1.4	332°	0.3	246°	0.8	177°
	. . . do.	42d	37° 52.43'	122° 24.17'	+0 24	+1 18	+0 59	+0 24	0.3	0.3	0.1	262°	0.9	336°	0.1	244°	0.5	169°
596	Fleming Point, 1.72 nmi. SW of	3d	37° 52.62'	122° 21.53'	−1 09	−1 03	−0 26	−0 56	0.2	0.3	0.1	178°	0.5	082°	0.1	351°	0.4	259°
601	Richardson Bay entrance	6	37° 51.4'	122° 28.2'	−3 06	−4 11	−3 29	−2 34	0.3	0.3	– –	– –	0.9	074°	– –	– –	0.5	225°
606	Raccoon Strait, off Point Stuart	15	37° 51.67'	122° 27.12'	+0 06	+0 07	−0 04	−0 05	0.5	1.1	– –	– –	1.3	014°	– –	– –	1.9	226°
611	Raccoon Strait	28d	37° 52.3'	122° 26.52'	−0 06	−0 40	−0 20	−0 19	0.7	1.0	0.2	332°	1.8	047°	0.1	312°	1.8	237°
	. . . do.	48d	37° 52.3'	122° 26.52'	−0 12	−0 54	−0 10	−0 21	0.7	1.0	– –	– –	1.8	046°	– –	– –	1.6	238°
616	Raccoon Strait, off Hospital Cove	8	37° 52.3'	122° 26.3'	– – –	−0 39	−0 08	−0 07	0.5	0.9	– –	– –	1.5	037°	– –	– –	1.6	223°
621	Bluff Point, 0.1 mile east of	8	37° 53.0'	122° 26.1'	+0 59	−0 27	+0 16	+0 41	0.6	1.1	– –	– –	1.7	009°	– –	– –	1.9	147°
626	Bluff Point, 1.15 nmi. east of	21d	37° 53.23'	122° 24.78'	+0 50	+0 45	+0 56	+0 24	0.4	0.6	0.2	085°	1.0	336°	0.3	236°	1.0	167°
631	Southampton Shoal Light, 0.2 mile E of	10	37° 52.95'	122° 23.75'	+0 59	+0 54	+0 51	+0 16	0.3	0.6	– –	– –	0.9	019°	– –	– –	1.1	188°
					on Richmond, p.20													
636	Point Chauncey, 1.3 miles east of	8	37° 53.5'	122° 25.1'	+0 06	+0 22	+0 42	+0 00	0.8	0.8	– –	– –	1.3	340°	– –	– –	1.4	162°
641	Point Chauncey, 0.75 nmi. NW of	19d	37° 54.18'	122° 27.53'	+0 08	−0 46	+0 10	−0 09	0.7	0.6	– –	– –	1.1	317°	– –	– –	1.0	131°
646	Point Chauncey, 1.25 nmi. north of	21d	37° 54.90'	122° 26.87'	+0 25	−0 11	+0 17	+0 03	1.0	1.1	0.1	074°	1.6	336°	– –	– –	1.8	159°
	. . . do.	33d	37° 54.90'	122° 26.87'	−0 10	−0 47	+0 22	+0 18	0.7	0.7	0.2	068°	1.1	355°	0.1	053°	1.1	144°
651	Point Potrero Reach (buoy "10")		37° 54.18'	122° 22.35'	Current weak and variable													
656	Point Richmond, 0.5 mile west of	6d	37° 54.3'	122° 24.0'	−0 16	+0 01	+0 10	−0 12	0.5	0.5	– –	– –	0.8	332°	– –	– –	0.9	142°
661	Point Richmond, 0.8 nmi. NNW of	4d	37° 55.25'	122° 23.80'	−0 48	−1 55	−1 14	−2 02	0.2	0.2	– –	– –	0.4	325°	– –	– –	0.4	147°
666	RICHMOND	14d	37° 55.76'	122° 25.50'	*Daily Predictions*						0.1	058°	1.6	328°	– –	– –	1.7	147°
	. . . do.	7d	37° 55.76'	122° 25.50'	+0 16	+0 10	−0 03	+0 04	1.0	1.1	0.1	058°	1.6	324°	0.1	235°	1.9	150°
	. . . do.	31d	37° 55.76'	122° 25.20'	−0 33	−0 13	+0 07	−0 06	1.0	0.8	– –	– –	1.6	333°	0.1	063°	1.4	145°
	. . . do.	43d	37° 55.76'	122° 25.20'	−0 51	−0 30	+0 10	−0 07	0.8	0.7	0.1	235°	1.3	330°	0.1	067°	1.1	149°
671	Red Rock, east of	11	37° 55.77'	122° 25.70'	+0 06	+0 03	+0 28	+0 07	0.8	0.8	– –	– –	1.3	318°	– –	– –	1.4	175°
676	Red Rock, 0.60 nmi. NNE of	17d	37° 56.40'	122° 25.60'	−0 18	−0 37	+0 10	−0 33	1.1	0.9	0.1	060°	1.8	337°	– –	– –	1.5	155°
	. . . do.	23d	37° 56.40'	122° 25.60'	−0 33	−0 22	+0 04	−0 38	1.0	0.8	– –	– –	1.5	334°	0.1	067°	1.3	145°
	. . . do.	38d	37° 56.40'	122° 25.60'	−0 46	−0 21	+0 16	−0 34	0.9	0.6	– –	– –	1.4	335°	0.1	080°	1.0	145°
681	Point San Quentin, 0.82 nmi. east of	15d	37° 56.47'	122° 27.70'	−0 09	+0 02	+0 02	−0 46	0.4	0.5	– –	– –	0.7	013°	0.1	282°	0.8	182°
686	Point San Quentin, 1.3 nmi. east of	23d	37° 56.53'	122° 27.16'	+0 20	+0 19	+0 35	+0 23	0.8	0.9	– –	– –	1.2	005°	– –	– –	1.5	181°
	. . . do.	38d	37° 56.53'	122° 27.16'	−0 05	+0 19	+0 26	+0 11	0.6	0.5	– –	– –	1.0	010°	– –	– –	0.9	186°
691	Point San Quentin, 1.9 miles east of	8	37° 57.0'	122° 26.4'	+0 52	+0 26	+0 28	+0 37	0.9	1.3	– –	– –	1.4	014°	– –	– –	2.1	168°

Endnotes can be found at the end of table 2.

TABLE 2 – CURRENT DIFFERENCES AND OTHER CONSTANTS

No.	PLACE	Meter Depth	POSITION		TIME DIFFERENCES				SPEED RATIOS		AVERAGE SPEEDS AND DIRECTIONS							
			Latitude	Longitude	Min. before Flood	Flood	Min. before Ebb	Ebb	Flood	Ebb	Minimum before Flood		Maximum Flood		Minimum before Ebb		Maximum Ebb	
		ft	North	West	h m	h m	h m	h m			knots	Dir.	knots	Dir.	knots	Dir.	knots	Dir.
	SAN PABLO BAY Time meridian, 120° W				on Richmond, p.20													
696	Point San Pablo, midchannel	8	37° 58.12'	122° 26.37'	+0 55	+0 45	+0 45	+0 40	1.1	1.3	--	--	1.8	014°	--	--	2.2	180°
701	Point San Pedro, 0.55 nmi. SE of	20d	37° 58.78'	122° 26.20'	+0 28	+0 23	+0 39	+0 30	1.3	1.5	0.1	095°	2.1	016°	0.1	282°	2.4	192°
	. . . co.	40d	37° 58.78'	122° 26.20'	+0 12	−0 11	+0 37	+2 25	1.1	1.2	0.1	294°	1.8	014°	0.1	110°	2.0	205°
	. . . co.	63d	37° 58.78'	122° 26.20'	−0 14	+0 19	+0 35	+0 20	0.7	0.8	0.3	315°	1.1	014°	0.1	111°	1.4	236°
706	Pinole Point, 1.18 nmi. west of	19d	38° 00.48'	122° 23.38'	+0 07	−0 25	+0 07	+0 06	0.5	0.6	--	--	0.9	043°	0.1	130°	0.9	218°
711	Pinole Point, 3.0 nmi. WNW of	8d	38° 01.60'	122° 25.48'	+0 01	−0 19	+0 05	+0 03	0.5	0.5	--	--	0.8	007°	--	--	0.9	185°
716	Pinole Point, 1.27 nmi. NNW of	33d	38° 01 85'	122° 22.63'	+0 47	+0 41	+0 50	+1 04	0.6	0.6	--	--	1.0	054°	--	--	1.0	226°
721	Pinole Point, 1.42 nmi. NNW of	21d	38° 02 03'	122° 22.75'	+1 09	+0 51	+0 52	+1 11	0.7	0.8	--	--	1.1	052°	--	--	1.3	233°
726	Petaluma River approach (Buoys 3 & 4)	5d	38° 03.68'	122° 25.47'	+0 02	−0 35	+0 11	+0 16	0.4	0.4	--	--	0.6	357°	--	--	0.6	179°
731	Petaluma River approach	4d	38° 04.2'	122° 25.2'	−0 01	−0 24	−0 15	−0 06	0.4	0.4	--	--	0.6	018°	--	--	0.7	186°
736	Petaluma River entrance	7d	38° 06.63'	122° 29.58'	+0 10	+0 05	−0 19	−1 00	0.5	0.6	--	--	0.8	277°	--	--	1.0	095°
741	Wilson Point, 1.55 nmi. north of	10d	38° 02.25'	122° 09.03'	+0 54	+0 57	+1 10	+1 08	0.5	0.5	--	--	0.7	071°	--	--	0.8	253°
746	Wilson Point, 3.90 nmi. NNW of	4d	38° 04.47'	122° 20.55'	−0 08	−0 44	+0 08	+0 22	0.3	0.3	0.1	314°	0.5	046°	0.1	145°	0.5	237°
					on Carquinez Strait, p.24													
751	Davis Point, 1.0 nmi. NW of	20d	38° 03.72'	122° 16.60'	−0 13	−1 19	−0 20	−0 23	0.7	0.8	--	--	1.4	072°	--	--	1.8	260°
756	Davis Point, (midchannel)		38° 03.80'	122° 15.5'	+0 05	−0 18	−0 13	−0 32	0.8	0.9	--	--	1.6	091°	--	--	2.1	249°
	CARQUINEZ STRAIT																	
761	Mare Island Strait ent., between dikes		38° 04.23'	122° 14.72'	−2 07	−1 28	−1 22	−2 02	0.6	0.8	--	--	1.3	040°	--	--	1.7	210°
766	Mare Island Strait (Buoy "4")	20d	38° 04.45'	122° 14.57'	−2 07	−0 55	−1 19	−3 01	0.4	0.2	--	--	0.8	342°	0.1	261°	0.5	177°
771	Mare Island Strait, off South Vallejo		38° 05.00'	122° 15.00'	−1 57	−1 27	−1 34	−2 32	0.7	0.8	--	--	1.4	325°	--	--	1.8	166°
776	CARQUINEZ STRAIT (west end, bridge)	20d	38° 03.68'	122° 13.10'		*Daily predictions*					--	--	2.1	103°	--	--	2.2	283°
781	Martinez Marina, 0.65 nmi. NW of	20d	38° 01.98'	122° 08.98'	+0 06	+0 07	+0 14	+0 04	0.8	0.6	0.1	013°	1.7	091°	0.1	176°	1.4	272°
786	Martinez Marina, 0.50 nmi. west of	30d	38° 01.72'	122° 08.92'	−0 19	−0 33	+0 31	+0 17	0.8	0.8	--	--	1.6	089°	--	--	1.7	271°
791	Martinez Marina, 0.61 nmi. NNW of	23d	38° 02.18'	122° 08.68'	−0 38	−0 25	−0 09	−0 54	0.6	0.6	--	--	1.2	085°	--	--	1.3	266°
796	Army Pt. Pier Lt. 0.2 nm . SE of	21d	38° 02.33'	122° 08.02'	+0 11	−0 03	−0 12	+0 14	0.8	1.1	--	--	1.7	063°	--	--	2.4	238°
	. . . do.	41d	38° 02.33'	122° 08.02'	−0 19	+0 12	−0 02	+0 09	0.6	0.5	--	--	1.3	038°	--	--	1.2	245°
	SUISUN BAY				on Benicia Bridge, p.28													
	Montezuma Slough																	
801	1 mile inside west entrance	22d	38° 08.83'	122° 03.38'	−0 24	−0 39	−0 17	−1 42	0.5	0.5	--	--	0.8	016°	--	--	0.7	191°
806	West entrance	6d	38° 07.92'	122° 03.48'	−0 10	−0 55	+0 06	−1 02	0.9	0.9	--	--	1.4	135°	--	--	1.4	315°
	Suisun Slough																	
811	1 mile inside entrance	15d	38° 08.27'	122° 04.88'	−0 54	−0 52	−0 31	−1 42	0.3	0.4	--	--	0.5	034°	--	--	0.5	202°
816	Entrance	3d	38° 07.27'	122° 04.05'	−0 29	−0 59	−0 24	−0 58	0.8	0.9	--	--	1.2	290°	--	--	1.3	110°
821	0.5 nmi. east of entrance	4d	38° 07.13'	122° 03.20'	−0 21	−1 07	−0 06	+0 07	0.5	0.3	--	--	0.7	358°	--	--	0.4	156°
826	BENICIA BRIDGE	11d	38° 02.49'	122° 07.53'		*Daily predictions*					--	--	1.5	047°	--	--	1.5	230°
831	Avon Pier, 0.15 nmi. north of	30d	38° 03.10'	122° 05.42'	+0 06	+0 32	+0 37	+0 10	0.9	0.7	--	--	1.3	059°	--	--	1.0	237°
836	Pt. Edith, 1.7 nmi. NNW of	24d	38° 04.72'	122° 05.03'	+0 05	+0 14	+0 38	+0 06	0.8	0.5	--	--	1.1	035°	0.1	296°	0.8	219°
841	Seal Island, south of	24d	38° 03.20'	122° 02.97'	+0 15	+0 08	+1 09	+0 37	0.6	0.6	--	--	0.9	069°	--	--	0.9	271°
846	Roe Island, south of	6d	38° 03.95'	122° 02.10'	+0 42	+0 29	+0 29	+0 18	0.9	1.1	--	--	1.3	090°	--	--	1.7	270°
851	Roe Island, Gilbert Pt., 0.15 nmi. NW of	16d	38° 04.42'	122° 01.30'	+1 08	+0 54	+1 05	+1 01	0.6	0.8	--	--	0.9	105°	--	--	1.2	283°
856	Suisun Cutoff	24d	38° 05.33'	122° 00.43'	+1 41	+1 53	+1 42	+1 04	0.7	0.4	--	--	1.0	126°	--	--	0.7	298°
861	Middle Point Lt., 0.18 nmi. NNW of	20d	38° 03.45'	121° 59.57'	+0 40	+0 22	+1 12	+0 55	1.0	0.8	--	--	1.5	097°	--	--	1.2	271°
	. . . do.	38d	38° 03.45'	121° 59.57'	+0 30	+0 32	+1 12	+0 59	0.8	0.6	--	--	1.3	130°	--	--	0.8	307°
866	Stake Point, 0.9 nmi. NNW of	4d	38° 03.88'	121° 57.33'	+0 14	+0 56	+0 48	+0 28	0.5	0.4	0.2	002°	0.8	077°	--	--	0.6	283°
871	Simmons Pt., Chipps Is., 0.6 nmi. ESE of	12d	38° 03.87'	121° 55.30'	+1 17	+1 16	+1 11	+1 01	1.1	1.2	--	--	1.7	101°	--	--	1.7	279°
	. . . do.	34d	38° 03.87'	121° 55.30'	+1 13	+1 16	+1 09	+1 12	0.9	1.0	0.1	007°	1.4	084°	0.1	006°	1.4	284°
876	Spoonbill Creek, near bridge	3d	38° 03.53'	121° 54.28'	+0 12	+0 27	+1 17	+0 36	1.0	0.9	--	--	1.5	105°	--	--	1.3	285°
881	Montezuma Slough, east end, near bridge	6d	38° 04.67'	121° 53.03'	+2 30	+2 25	+2 23	+2 22	0.7	0.8	--	--	1.0	135°	--	--	1.2	315°
886	New York Slough, 0.6miles E of Pt. Emmet	7d	38° 01.95'	121° 52.17'	+1 39	+1 26	+1 51	+1 43	0.8	0.9	--	--	1.2	110°	--	--	1.3	295°
891	New York Slough, Winter Island	15d	38° 01.70'	121° 50.78'	+1 02	+0 45	+1 00	+1 10	0.8	0.8	--	--	1.1	122°	--	--	1.2	302°

Endnotes can be found at the end of table 2.

TABLE 2 – CURRENT DIFFERENCES AND OTHER CONSTANTS

No.	PLACE	Meter Depth	POSITION		TIME DIFFERENCES				SPEED RATIOS		AVERAGE SPEEDS AND DIRECTIONS							
			Latitude	Longitude	Min. before Flood	Flood	Min. before Ebb	Ebb	Flood	Ebb	Minimum before Flood		Maximum Flood		Minimum before Ebb		Maximum Ebb	
		ft	North	West	h m	h m	h m	h m			knots	Dir.	knots	Dir.	knots	Dir.	knots	Dir.
	SACRAMENTO RIVER <11> Time meridian, 120° W				on Benicia Bridge, p.28													
896	Entrance, 0.7 mile SW of Chain Island	7d	38° 03.50'	121° 52.23'	+1 27	+1 33	+1 48	+1 30	0.8	0.9	--	--	1.2	055°	--	--	1.3	212°
901	Point Sacramento, 0.3 mile NE OF	7d	38° 04'	121° 50'	+1 15	+1 05	+1 37	+1 06	0.7	0.5	--	--	1.1	098°	--	--	0.8	286°
906	Sherman Island East, 0.2 mile north of	14d	38° 03.52'	121° 48.25'	+1 22	+1 12	+1 23	+0 29	0.7	0.8	--	--	1.0	094°	--	--	1.1	270°
	. . . do.	25d	38° 03.52'	121° 48.25'	+2 11	+1 12	+1 23	+0 55	0.6	0.6	--	--	0.8	079°	--	--	0.9	274°
	SAN JOAQUIN RIVER <12>																	
911	Pt. San Joaquin, 0.45 nmi. ENE of	18d	38° 03.70'	121° 51.00'	+1 24	+1 37	+1 44	+1 13	1.0	0.6	0.1	006°	1.5	108°	0.1	194°	1.0	282°
	. . . do.	33d	38° 03.70'	121° 51.00'	+1 23	+1 47	+1 39	+0 54	0.7	0.5	--	--	1.1	112°	--	--	0.8	275°
916	Point Beenar, 0.7 nmi. north of	5d	38° 02.53'	121° 50.28'	+2 35	+2 09	+1 54	+1 21	0.4	0.4	--	--	0.6	163°	--	--	0.6	349°
921	Point Beenar, 100 yds., NE of	14d	38° 01.95'	121° 50.13'	+2 35	+2 09	+2 42	+2 24	0.7	0.9	--	--	1.0	137°	0.1	231°	1.3	314°
	. . . do.	22d	38° 01.95'	121° 50.13'	+2 29	+2 09	+2 45	+2 25	0.5	0.7	--	--	0.8	142°	--	--	1.1	324°
926	Antioch Point, 0.3 mile east of		38° 02'	121° 49'	+2 22	+2 33	+2 13	+2 00	1.0	0.9	--	--	1.5	128°	--	--	1.4	304°
931	West Island Lt., 0.5 mile SE of	5d	38° 01.27'	121° 45.72'	+2 19	+2 24	+2 11	+2 08	0.3	0.5	--	--	0.4	090°	--	--	0.7	270°
936	Vulcan Island, 0.5 mile east of	17d	37° 59.12'	121° 23.45'	+3 38	+3 34	+3 39	+3 29	0.5	0.3	--	--	0.7	135°	--	--	0.4	315°
941	Brandt Bridge		37° 51.85'	121° 19.38'	+5 40	+5 28	+5 02	+5 20	0.5	0.7	--	--	0.7	135°	--	--	1.0	315°
	CALIFORNIA COAST–cont.				on San Francisco Bay Ent., p.8													
946	Point Reyes <13>		38° 00'	123° 02'	–1 12	–1 12	–1 12	–1 12	0.3	0.3	--	--	1.1	320°	--	--	1.1	140°
951	Salt Point <13>		38° 34'	123° 21'	–1 20	–1 20	–1 20	–1 20	0.3	0.3	--	--	0.9	325°	--	--	0.9	145°
956	Point Arena <13>		38° 57'	123° 45'	–1 29	–1 29	–1 29	–1 29	0.3	0.3	--	--	1.1	330°	--	--	1.1	150°
961	Point Cabrillo <13>		39° 21'	123° 50'	–1 38	–1 38	–1 38	–1 38	0.3	0.3	--	--	1.0	345°	--	--	1.0	165°
966	Cape Vizcaino <13>		39° 44'	123° 50'	–1 48	–1 48	–1 48	–1 48	0.3	0.3	--	--	0.9	325°	--	--	0.9	145°
971	Point Delgada <13>		40° 00'	124° 04'	–1 37	–1 37	–1 37	–1 37	0.3	0.3	--	--	1.0	325°	--	--	1.0	145°
976	Punta Gorda <13>		40° 15'	124° 22'	–1 36	–1 36	–1 36	–1 36	0.3	0.3	--	--	1.1	335°	--	--	1.1	155°
981	Cape Mendocino Light. 4.6 mi. W of <14>		40° 26'	124° 30'	---	---	---	---	--	--	--	--	--	--	--	--	--	--
986	Table Bluff Light <13>		40° 42'	124° 17'	–1 11	–1 11	–1 11	–1 11	0.3	0.3	--	--	0.8	010°	--	--	0.8	190°
	HUMBOLDT BAY																	
991	Humboldt Bay entrance		40° 45.48'	124° 13.80'	–0 52	–0 51	–1 09	–0 44	0.6	0.6	--	--	1.6	124°	--	--	2.0	312°
996	North Spit, 0.15 n.mi. SW of	15d	40° 45.47'	124° 13.37'	–1 44	–2 40	–3 18	–1 18	0.2	0.5	0.1	155°	0.6	073°	--	--	1.6	243°
	. . .do.	32d	40° 45.47'	124° 13.37'	–1 58	–2 52	–3 28	–1 25	0.2	0.4	--	--	0.5	070°	--	--	1.3	257°
1001	South Spit, 0.1 n.mi. NE of	15d	40° 44.87'	124° 13.45'	–1 53	–1 35	–2 04	–1 14	0.3	0.4	--	--	0.8	178°	0.1	086°	1.2	002°
1006	North Bay Channel, west of Eureka	15d	40° 47.23'	124° 11.56'	–1 23	–0 42	–1 17	–1 13	0.5	0.5	--	--	1.6	021°	--	--	1.8	197°
	. . .do.	28d	40° 47.23'	124° 11.56'	–1 47	–0 53	–1 19	–1 24	0.5	0.4	--	--	1.4	013°	0.1	286°	1.4	200°
1011	North Bay Channel at Fairhaven	13d	40° 47.24'	124° 11.66'	–1 45	–0 38	–1 28	–1 27	0.5	0.5	--	--	1.3	030°	--	--	1.8	216°
1016	North Bay Channel at Samoa Channel	15d	40° 47.77'	124° 11.24'	–1 49	–0 35	–1 17	–1 29	0.4	0.4	--	--	1.2	015°	--	--	1.3	196°
	CALIFORNIA COAST–cont.																	
1021	Trinidad Head <13>		41° 03'	124° 10'	–0 57	–0 57	–0 57	–0 57	0.3	0.3	--	--	1.0	005°	--	--	1.0	185°
1026	Redding Rock Light <13>		41° 21'	124° 11'	–0 52	–0 52	–0 52	–0 52	0.3	0.3	--	--	0.9	010°	--	--	0.9	190°
1031	St. George Reef <13>		41° 49'	124° 20'	–0 41	–0 41	–0 41	–0 41	0.3	0.3	--	--	1.0	005°	--	--	1.0	185°

Endnotes can be found at the end of table 2.

TABLE 2 – CURRENT DIFFERENCES AND OTHER CONSTANTS

No.	PLACE	Meter Depth	POSITION Latitude	POSITION Longitude	TIME DIFFERENCES Min. before Flood	TIME DIFFERENCES Flood	TIME DIFFERENCES Min. before Ebb	TIME DIFFERENCES Ebb	SPEED RATIOS Flood	SPEED RATIOS Ebb	Minimum before Flood knots	Minimum before Flood Dir.	Maximum Flood knots	Maximum Flood Dir.	Minimum before Ebb knots	Minimum before Ebb Dir.	Maximum Ebb knots	Maximum Ebb Dir.
		ft	North	West	h m	h m	h m	h m			knots	Dir.	knots	Dir.	knots	Dir.	knots	Dir.
	OREGON COAST Time meridian, 120° W				on San Francisco Bay Ent., p.8													
1036	Cape Sebastian <13>		42° 20'	124° 26'	−0 26	−0 26	−0 26	−0 26	0.3	0.3	--	--	1.1	355°	--	--	1.1	175°
1041	Cape Blanco <13>		42° 50'	124° 35'	−0 10	−0 10	−0 10	−0 10	0.3	0.3	--	--	1.1	010°	--	--	1.1	190°
					on Wrangell Narrows, p.80													
1046	Coquille River entrance		43° 07.30'	124° 25.18'	0 00	−0 25	−0 27	−0 17	0.4	0.6	--	--	1.4	091°	--	--	1.2	290°
1051	Coos Bay entrance		43° 21.30'	124° 20.47'	+0 36	+0 18	+0 18	+0 39	0.6	1.1	--	--	1.8	100°	--	--	2.2	280°
1056	Umpqua River entrance		43° 40.70'	124° 11.60'	+0 46	+0 28	+0 28	+0 49	0.3	0.5	--	--	0.8	010°	--	--	1.0	190°
1061	Heceta Head <13>		44° 08'	124° 08'	+1 06	+0 48	+0 48	+1 09	0.4	0.7	--	--	1.2	005°	--	--	1.4	185°
	YAQUINA BAY																	
1066	Yaquina Bay entrance		44° 37'	124° 04'	−0 20	−0 04	−0 17	−0 55	0.8	1.1	--	--	2.4	050°	--	--	2.3	235°
1071	Highway Bridge		44° 37 40'	124° 03.42'	−0 31	+0 10	−0 40	+0 07	0.6	1.0	--	--	1.9	044°	--	--	2.1	222°
1076	Yaquina, Yaquina River		44° 36 12'	124° 00.68'	−0 09	+0 10	−0 33	+0 27	0.3	0.5	--	--	1.0	184°	--	--	1.1	002°
1081	Yaquina River, 1 mile below Toledo		44° 36.03'	123° 56.50'	+0 33	+0 47	+0 14	+0 58	0.4	0.7	--	--	1.4	332°	--	--	1.4	132°
	TILLAMOOK BAY																	
1086	Tillamook Bay entrance		45° 33.73'	123° 56.30'	+0 40	+0 13	−0 05	−0 16	0.9	1.3	--	--	3.0	141°	--	--	2.6	305°
	COLUMBIA RIVER and APPROACHES <15>				on Grays Harbor Ent., p.32													
1091	Lighted Horn Buoy R"C" <14>		46° 11'	124° 11'	---	---	---	---	--	--	--	--	--	--	--	--	--	--
1096	Sand Island Tower, southeast of	15	46° 15.17'	123° 59.45'	+1 12	+1 03	+0 38	+0 07	1.6	1.6	0.1	016°	3.0	107°	0.3	191°	4.4	275°
1101	Clatsop Spit, NNE of	15	46° 14.77'	123° 59.65'	+0 45	+0 56	+0 56	+0 20	1.4	1.1	0.1	032°	2.6	114°	0.2	205°	3.2	289°
1106	Sand Island Tower, southeast of	15	46° 15.47'	123° 59.67'	+0 39	+0 33	+0 27	−0 44	1.1	0.7	0.2	015°	2.1	092°	0.1	184°	1.9	262°
1111	Sand Island Tower, southeast of	23	46° 15.72'	123° 59.88'	+0 28	+0 08	+0 20	+0 14	0.6	0.3	--	--	1.2	008°	--	--	0.8	202°
1116	Sand Island, SSE of	12	46° 15.33'	123° 58.08'	+0 18	−0 25	−0 16	−0 32	0.3	0.4	--	--	0.6	097°	--	--	1.1	265°
1121	Clatsop Spit, northeast of	12	46° 14.35'	123° 59.88'	---	−0 01	−1 16	+0 36	0.4	1.0	--	--	0.8	131°	0.1	045°	2.8	313°
1126	Astoria Range	12	46° 11.92'	123° 49.42'	---	+1 19	+0 38	+0 52	0.6	1.0	--	--	1.2	088°	0.2	175°	2.8	243°
1131	Youngs Bay Entrance	17	46° 11.18'	123° 53.27'	+1 46	+1 45	+1 28	−0 38	0.9	0.4	0.2	172°	1.7	093°	0.1	006°	1.1	260°
								+0 40		0.3							0.8	266°
								+2 45		0.5							1.4	261°
1136	Youngs Bay Bridge	9	46° 10.67'	123° 52.10'	+0 10	−0 14	+0 50	+0 28	0.4	0.3	--	--	0.8	135°	0.1	222°	0.8	320°
1141	Hammond, northeast of ship channel	15	46° 12.67'	123° 56.07'	+0 57	+0 24	+0 26	+0 01	0.4	0.6	0.1	219°	0.8	134°	0.1	230°	1.7	307°
1146	McGowan, SSW of	14	46° 14.37'	123° 54.92'	+1 28	+1 14	+0 21	+0 23	0.9	1.6	--	--	1.7	107°	--	--	4.3	282°
1151	Point Ellice, east of	17	46° 14.50'	123° 50.90'	+0 55	+0 25	+0 51	+0 11	0.8	0.9	--	--	1.6	065°	0.1	336°	2.4	254°
1156	Point Adams, NNE OF	14	46° 13.67'	123° 58.05'	+1 16	+0 13	+0 26	+1 11	0.8	0.9	0.3	202°	1.6	139°	0.4	210°	2.5	297°
1161	Chinook Point, WSW of	14	46° 14.53'	123° 57.85'	+0 29	+0 25	+0 45	−0 21	1.2	0.9	0.2	200°	2.2	117°	0.4	199°	2.4	287°
1166	Tongue Point, northwest of	15	46° 13.15'	123° 46.00'	+2 31	+1 09	+0 24	+0 46	0.4	0.9	--	--	0.8	077°	--	--	2.4	249°
1171	Altoona, SSW of	22	46° 15.55'	123° 39.40'	---	+1 23	+1 19	+2 22	0.3	0.7	--	--	0.5	061°	--	--	1.9	243°
1176	Woody Island Channel	15	46° 14.37'	123° 40.40'	+1 12	+1 09	+1 36	+1 15	0.5	0.3	--	--	1.0	118°	--	--	0.9	298°
1181	Woody Island Channel (off Seal Island)	12	46° 13.05'	123° 37.75'	+2 38	+1 53	+1 25	+2 53	0.3	0.4	0.1	156°	0.5	081°	0.1	358°	1.2	247°
1186	Three Tree Point, 5 miles WSW of	14	46° 15.90'	123° 32.10'	+4 29	+2 40	---	+2 36	0.1	0.9	--	--	0.2	080°	--	--	2.4	258°
1191	Quinn Island, Prairie Channel	8	46° 14.23'	123° 30.20'	+3 19	+1 52	+0 55	+1 53	0.3	0.5	--	--	0.5	097°	--	--	1.3	281°
1196	Clifton Channel	10	46° 13.07'	123° 27.92'	+3 41	+2 27	+1 21	+2 17	0.3	0.4	--	--	0.5	118°	--	--	1.1	301°
1201	Tenasillahe Island, northwest of <16>	22	46° 14.60'	123° 26.10'	---	---	---	+2 47	--	0.8	--	--	--	--	--	--	2.2	329°
1206	Hunting Island, south of	20	46° 12.43'	123° 24.25'	+4 20	+2 56	+1 38	+3 20	0.2	0.4	0.1	206°	0.3	125°	--	--	1.1	295°
1211	Puget Island, Wauna Range <17>	23	46° 10.45'	123° 25.38'	---	---	---	−3 22	--	0.1	--	--	--	--	--	--	0.1	315°
								+2 57		0.7							2.0	313°
1216	Puget Island, Westport Turn & Range <17>	22	46° 08.67'	123° 20.38'	---	---	---	−2 38	--	--	--	--	--	--	--	--	--	--
								+4 01		0.7							1.9	269°
1221	Cathlamet Channel, SE of Nassa Point	19	46° 09.37'	123° 18.90'	+5 16	+3 23	+1 39	+5 38	0.1	0.6	0.1	221°	0.2	103°	--	--	1.6	278°
1226	Oak Point Channel <17>	15	46° 11.08'	123° 11.03'	---	---	---	−1 59	--	0.4	--	--	--	--	--	--	1.1	235°
								+5 35		0.9							2.5	230°

Endnotes can be found at the end of table 2.

TABLE 2 – CURRENT DIFFERENCES AND OTHER CONSTANTS

No.	PLACE	Meter Depth	POSITION		TIME DIFFERENCES				SPEED RATIOS		AVERAGE SPEEDS AND DIRECTIONS							
			Latitude	Longitude	Min. before Flood	Flood	Min. before Ebb	Ebb	Flood	Ebb	Minimum before Flood		Maximum Flood		Minimum before Ebb		Maximum Ebb	
		ft	North	West	h m	h m	h m	h m			knots	Dir.	knots	Dir.	knots	Dir.	knots	Dir.
	COLUMBIA RIVER and APPROACHES <15>—cont. Time meridian, 120° W				on Grays Harbor Ent., p.32													
1231	Stella, southwest of <17>	18	46° 11.13'	123° 07.45'	– – –	– – –	– – –	−1 54 +5 43	– –	– – 0.7	– –	– –	– –	– –	– –	– –	0.1 1.9	295 291°
1236	Walker Island Channel <17>	20	46° 09.17'	123° 02.57'	– – –	– – –	– – –	−1 29 +5 47	– –	0.1 0.7	– –	– –	– –	– –	– –	– –	0.2 2.0	319 313°
1241	Walker Island, south of	12	46° 08.47'	123° 02.75'	+5 46	+4 15	+2 41	+5 09	0.2	0.4	– –	– –	0.4	148°	– –	– –	1.2	330°
1246	Slaughters Channel <17>	18	46° 07.42'	122° 59.22'	– – –	– – –	– – –	−1 15 +5 51	– –	0.2 0.7	– –	– –	– –	– –	– –	– –	0.6 2.0	306 306°
1251	Cottonwood Island, west of <17>	34	46° 04.28'	122° 53.45'	– – –	– – –	– – –	−1 02 +5 58	– –	0.1 0.6	– –	– –	– –	– –	– –	– –	0.4 1.5	325 328°
1256	Kalma Upper Range <17>	13	46° 00.17'	122° 51.08'	– – –	– – –	– – –	−0 35 +5 58	– –	0.3 0.7	– –	– –	– –	– –	– –	– –	0.8 2.0	343 342°
	WASHINGTON COAST																	
1261	South Bend, Willapa River		46° 39.88'	123° 48.08'	+0 19	+0 20	+0 24	−0 06	0.6	0.5	– –	– –	1.2	090°	– –	– –	1.4	270°
	GRAYS HARBOR																	
1266	Entrance, 0.2 mile south of north jetty		46° 55.58'	124° 09.68'	−0 28	+0 16	+0 10	−0 47	1.3	0.6	– –	– –	2.5	070°	– –	– –	1.7	243°
1271	Entrance, Point Chehalis Range		46° 54.50'	124° 09.33'	+0 08	−0 11	−0 22	−0 21	0.9	1.1	– –	– –	1.7	092°	– –	– –	3.1	268°
1276	Entrance, 0.6 mile WNW of Westport		46° 54.88'	124° 07.50'	0 00	0 00	+0 06	0 00	1.2	0.7	– –	– –	2.2	044°	– –	– –	1.9	238°
1281	GRAYS HARBOR ENTRANCE		46° 55.35'	124° 07.98'	*Daily predictions*						– –	– –	1.9	061°	– –	– –	2.8	242°
1286	Entrance, 1.1 miles NW of Westport		46° 55'	124° 08'	+0 04	−0 06	−0 05	−0 23	0.9	0.9	– –	– –	1.8	078°	– –	– –	2.5	233°
1291	Channel, 1.5 miles north of Westport		46° 56'	124° 06'	−0 22	−0 04	−0 04	−0 25	0.7	0.6	– –	– –	1.4	036°	– –	– –	1.8	226°
1296	Channel, 2.1 miles NNE of Westport		46° 56'	124° 05'	−0 02	0 00	−0 02	+0 02	0.6	0.5	– –	– –	1.2	021°	– –	– –	1.3	249°
1301	Aberdeen, Chehalis River <18>		46° 58'	123° 49'	– – –	+0 35	– – –	+0 28	– –	– –	– –	– –	– –	042	– –	– –	– –	230°
1306	Westport, channel 0.4 mile NE of		46° 54.85'	124° 06.50'	−0 41	−0 04	−0 19	−0 36	1.0	0.7	– –	– –	1.9	113°	– –	– –	1.9	310°
	WASHINGTON–BRITISH COLUMBIA COAST				on San Francisco Bay Ent., p.8													
1311	Quillayute River entrance		47° 55'	124° 38'	−0 10	−0 14	−0 26	−0 18	0.1	0.4	– –	– –	0.3	015°	– –	– –	1.3	345°
1316	Cape Alava, 4.4 miles west of <14>		48° 10'	124° 50'	– – –	– – –	– – –	– – –	– –	– –	– –	– –	– –	– –	– –	– –	– –	– –
1321	Swiftsure Bank <14>		48° 33'	125° 00'	– – –	– – –	– – –	– – –	– –	– –	– –	– –	– –	– –	– –	– –	– –	– –
1326	Vancouver Island, west coast <19>		– – –	– – –	– – –	– – –	– – –	– – –	– –	– –	– –	– –	– –	– –	– –	– –	– –	– –
	STRAIT of JUAN DE FUCA				on Strait of Juan de Fuca Ent.,p.36<20>													
1331	STRAIT OF JUAN DE FUCA ENTRANCE		48° 27'	124° 35'	*Daily predictions*						– –	– –	0.6	115°	– –	– –	1.5	290°
1336	Pillar Point		48° 16'	124° 04'	−0 35	+0 06	+1 27	+0 52	1.2	1.2	– –	– –	1.4	100°	– –	– –	0.9	280°
					on Race Rocks, p.40 <20>													
1341	Angeles Point, 2.3 miles north of		48° 12'	123° 33'	+1 22	+0 21	−0 32	+0 13	0.8	1.1	– –	– –	1.0	078°	– –	– –	2.2	260°
1346	RACE ROCKS, 4 miles south of		48° 14'	123° 32'	*Daily predictions*						– –	– –	1.2	091°	– –	– –	2.0	271°
1351	Race Rocks, 0.5 mile southeast of		48° 17'	123° 31'	−1 10	– – –	−1 23	– – –	– –	– –	– –	– –	– –	– –	– –	– –	– –	– –
1356	Race Passage		48° 18'	123° 32'	−1 21	– – –	−1 53	– – –	– –	– –	– –	– –	– –	– –	– –	– –	– –	– –
1361	Port Angeles		48° 08'	123° 25'	Current weak and variable													
					on Admiralty Inlet, p.44 <20>													
1366	Ediz Hook Light, 1.4 miles southeast of		48° 07'	123° 24'	Current weak and variable													
1371	Ediz Hook Light, 1.2 miles north of		48° 10'	123° 25'	−0 32	−0 19	−0 05	−0 05	0.5	0.4	– –	– –	0.8	080°	– –	– –	1.1	295°
1376	Ediz Hook Light, 5.3 miles ENE of		48° 11'	123° 17'	+0 39	+0 12	−0 07	−0 14	0.6	0.8	– –	– –	1.0	055°	– –	– –	2.0	215°
1381	Trial Island, 5.2 miles SSW of		48° 19'	123° 22'	−0 08	+0 39	+1 22	+0 55	0.7	0.5	– –	– –	1.1	045°	– –	– –	1.4	235°
1386	New Dungeness Light, 2.8 miles NNW of		48° 14'	123° 08'	+0 57	+0 27	−0 13	+0 16	0.3	0.5	– –	– –	0.5	075°	– –	– –	1.2	255°
1391	New Dungeness Light, 6 miles NNE of		48° 16'	123° 03'	+0 51	+0 55	+0 27	+0 36	0.3	0.4	– –	– –	0.5	050°	– –	– –	1.1	255°

Endnotes can be found at the end of table 2.

TABLE 2 – CURRENT DIFFERENCES AND OTHER CONSTANTS

No.	PLACE	Meter Depth	POSITION Latitude	POSITION Longitude	TIME DIFFERENCES Min. before Flood	TIME DIFFERENCES Flood	TIME DIFFERENCES Min. before Ebb	TIME DIFFERENCES Ebb	SPEED RATIOS Flood	SPEED RATIOS Ebb	Minimum before Flood knots	Minimum before Flood Dir.	Maximum Flood knots	Maximum Flood Dir.	Minimum before Ebb knots	Minimum before Ebb Dir.	Maximum Ebb knots	Maximum Ebb Dir.
		ft	North	West	h m	h m	h m	h m			knots	Dir.	knots	Dir.	knots	Dir.	knots	Dir.
	STRAIT of JUAN DE FUCA–cont. Time meridian, 120° W				**on Admiralty Inlet, p.44 <20>**													
1396	Discovery Island, 7.6 miles SSE of		48° 18'	123° 10'	+0 27	+0 15	+0 51	+0 30	0.4	0.4	--	--	0.7	035°	--	--	1.0	260°
1401	Discovery Island, 3 miles SSE of		48° 23'	123° 12'	+0 23	+0 15	+0 02	+0 40	0.6	0.9	--	--	0.9	025°	--	--	2.3	250°
1406	Cattle Point, 2.8 miles SSW of <21>		48° 24'	123° 00'	−1 17	+0 52	+1 04	−0 42	0.4	0.2	--	--	0.6	046°	0.2	126°	0.4	187°
1411	Cattle Point, 5 miles SSW of		48° 23'	123° 01'	+1 14	+1 11	+1 20	+0 44	0.6	0.3	--	--	0.9	120°	--	--	0.9	210°
1416	Violet Point, 3.7 miles north of <22>		48° 11'	122° 55'	---	+0 22	−0 36	+0 09	0.2	0.5	--	--	0.4	100°	--	--	1.2	270°
1421	Violet Point, 3.2 miles northwest of		48° 10'	122° 58'	−0 05	−0 32	−0 08	−0 17	0.3	0.4	--	--	0.6	120°	--	--	1.0	325°
1426	Kamen Point, 1.3 miles southwest of		48° 06'	122° 58'	−1 10	−0 52	−1 06	−0 34	0.2	0.4	--	--	0.3	125°	--	--	1.0	265°
1431	Discovery Bay entrance <23>		48° 06'	122° 54'	---	---	---	---	--	--	--	--	--	--	--	--	--	--
1436	Smith Island, 2 miles east of <24>		48° 19'	122° 48'	+1 22	+0 51	---	+0 19	0.2	0.2	--	--	0.4	--	--	--	0.5	220
1441	Smith Island, 1.4 miles SSW of		48° 18'	122° 51'	−0 06	+0 03	+0 29	+0 25	0.4	0.4	--	--	0.7	090°	--	--	1.0	280°
1446	Smith Island, 3.7 miles ESE of <25>		48° 18'	122° 45'	---	+1 11	---	+1 35	--	0.3	--	--	--	--	--	--	0.9	225
1451	Point Partridge, 1.6 miles NW of <26>		48° 15'	122° 48'	---	---	---	---	--	0.4	--	--	--	--	--	--	1.1	175
1456	Point Partridge, 3.7 miles west of		48° 14'	122° 52'	+1 27	+0 17	−0 24	+0 42	0.2	0.8	--	--	0.4	140°	--	--	2.1	250°
	ADMIRALTY INLET																	
1461	Point Wilson, 1.1 miles NW of		48° 10'	122° 46'	−1 53	−0 58	−0 08	−0 54	1.5	1.1	--	--	2.4	085°	--	--	2.8	285°
1466	Point Wilson, 0.5 mi., northeast of <27>		48° 09'	122° 45'	−1 08	−1 27	−0 47	−0 48	1.4	1.0	--	--	2.2	114°	--	--	2.6	298°
1471	Point Wilson, 1.4 miles northeast of <27>		48° 10'	122° 44'	−0 28	−0 17	+0 41	+0 07	2.0	1.3	--	--	3.4	112°	--	--	3.5	297°
1476	Point Wilson, 2.3 miles NE of <27>		48° 10'	122° 42'	−1 29	−0 07	+0 33	−0 07	1.0	1.0	--	--	2.0	143°	--	--	2.3	323°
1481	Admiralty Head, 0.5 mile west of		48° 09'	122° 42'	−0 31	−0 03	+0 01	−0 07	1.3	1.2	--	--	2.1	145°	--	--	3.1	025°
1486	Point Wilson, 0.8 mile east of		48° 09'	122° 44'	−0 53	−0 20	+0 27	−0 54	1.5	1.0	--	--	2.5	165°	--	--	2.6	280°
1491	Port Townsend, 0.5 mile S of Pt. Hudson		48° 07'	122° 45'	Current weak and variable						--	--	0.3	238°	--	--	0.5	048°
1496	Point Hudson, 0.5 mile east of		48° 07'	122° 44'	−3 21	−2 36	−2 42	−2 26	0.8	0.2	--	--	1.2	115°	--	--	0.6	010°
	Marrowstone Point																	
1501	1.1 miles northwest of		48° 07'	122° 42'	−3 31	−2 20	−1 02	−1 42	0.8	0.5	--	--	1.3	100°	--	--	1.3	275°
1506	0.4 mile northeast of <27>		48° 05'	122° 41'	−1 20	−1 03	−0 04	−1 03	1.1	1.1	--	--	2.4	122°	--	--	3.1	338°
1511	0.3 mile northeast of		48° 05'	122° 41'	−0 53	−1 36	−1 13	−0 13	1.2	1.1	--	--	2.0	170°	--	--	2.8	015°
1516	1.6 miles northeast of <27>		48° 07'	122° 40'	−0 16	+0 07	−0 03	−0 17	1.2	1.2	--	--	2.3	152°	--	--	2.6	344°
1521	2.5 miles northeast of <28>		48° 03'	122° 38'	---	---	---	---	--	--	--	--	--	--	--	--	--	--
1526	Nodule Point, 0.5 mile southeast of		48° 02'	122° 40'	−1 27	−0 47	−0 59	−0 24	1.2	1.0	--	--	2.0	160°	--	--	2.5	339°
1531	ADMIRALTY INLET (off Bush Point)		48° 02'	122° 38'	*Daily predictions*						--	--	1.6	179°	--	--	2.6	003°
1536	Bush Point Light, 0.5 mile NW of		48° 02'	122° 37'	−0 32	−0 21	−0 09	−0 35	1.1	1.1	--	--	1.7	141°	--	--	2.9	309°
1541	Mutiny Bay, 3.3 miles SE of Bush Point		47° 59.25'	122° 33.50'	---	−2 11	---	−2 25	0.6	0.4	--	--	1.0	133	--	--	1.1	354°
1546	Olele Point, 1.8 miles ENE of <9>		47° 59'	122° 38'	−0 34	−0 31	−0 34	−0 37	0.5	0.5	--	--	0.8	167°	--	--	1.3	352°
					on Deception Pass, p.52													
1551	Port Townsend Canal		48° 02'	122° 44'	+0 06	−0 40	−0 46	−0 31	0.5	0.4	--	--	2.6	150°	--	--	2.9	330°
	HOOD CANAL				**on Admiralty Inlet, p.44 <20>**													
1556	Foulweather Bluff		47° 55.90'	122° 38.33'	0 00	−0 24	−0 15	−0 25	0.4	0.4	--	--	0.7	140°	--	--	0.9	325°
1561	Port Gamble Bay, 0.5 mile N of entrance		47° 51.87'	122° 34.63'	Current weak and variable													
1566	Port Gamble Bay entrance		47° 51.27'	122° 34.63'	−1 03	−0 39	+0 04	−0 14	0.6	0.3	--	--	0.9	185°	--	--	0.7	000°
1571	Port Gamble Bay		47° 50.00'	122° 34.53'	Current weak and variable													
1576	South Point		47° 49'	122° 41'	---	−0 44	---	−0 29	0.4	0.4	--	--	0.6	218	--	--	1.0	040°
1581	Hazel Point		47° 41.62'	122° 45.52'	---	−0 54	---	−0 52	0.3	0.3	--	--	0.4	183	--	--	0.8	005°
1586	Chinom Point		47° 32'	123° 02'	Current weak and variable													
1591	The Great Bend		47° 21.30'	123° 01.80'	---	−1 06	---	−0 50	0.3	0.2	--	--	0.4	049	--	--	0.5	251°

Endnotes can be found at the end of table 2.

TABLE 2 – CURRENT DIFFERENCES AND OTHER CONSTANTS

No.	PLACE	Meter Depth	POSITION		TIME DIFFERENCES				SPEED RATIOS		AVERAGE SPEEDS AND DIRECTIONS							
			Latitude	Longitude	Min. before Flood	Flood	Min. before Ebb	Ebb	Flood	Ebb	Minimum before Flood		Maximum Flood		Minimum before Ebb		Maximum Ebb	
		ft	North	West	h m	h m	h m	h m			knots	Dir.	knots	Dir.	knots	Dir.	knots	Dir.
	PUGET SOUND Time meridian, 120° W				on Admiralty Inlet, p.44 <20>													
1596	Useless Bay		47° 58.70'	122° 29.72'	Current weak and variable													
1601	Foulweather Bluff		47° 57.25'	122° 34.75'	+0 09	+0 01	+0 14	−0 18	0.9	0.7	– –	– –	1.5	115°	– –	– –	1.8	335°
1606	Edmonds, 2.7 miles WSW of		47° 48.38'	122° 26.67'	+0 44	+0 06	+0 13	+0 19	0.1	0.2	– –	– –	0.2	170°	– –	– –	0.5	000°
1611	Apple Cove Point, 0.5 mile east of		47° 49'	122° 28'	– – –	−0 11	– – –	−0 29	0.3	0.3	– –	– –	0.5	168	– –	– –	0.8	008°
1616	President Point, 1.5 miles east of		47° 45.72'	122° 26.10'	– – –	−0 38	– – –	−0 41	0.2	0.2	– –	– –	0.3	203	– –	– –	0.5	024°
1621	Port Madison entrance		47° 44.13'	122° 29.50'	Current weak and variable													
1626	Agate Passage, north end		47° 43.32'	122° 33.30'	−1 28	−1 00	−0 18	−0 59	0.8	0.7	– –	– –	1.2	230°	– –	– –	1.8	032°
1631	Agate Passage, south end <27>		47° 42.77'	122° 33.93'	−1 25	−0 53	0 00	−0 47	2.0	1.4	– –	– –	3.3	216°	– –	– –	3.6	037°
1636	Port Orchard		47° 38.25'	122° 35.08'	Current weak and variable													
1641	Port Orchard, off Keyport		47° 42.02'	122° 36.50'	Current weak and variable													
1646	Liberty Bay entrance, Port Orchard		47° 42.40'	122° 37.65'	– – –	−0 06	– – –	+0 22	0.4	0.3	– –	– –	0.7	280	– –	– –	0.8	113°
1651	Shilshole Bay		47° 41'	122° 25'	Current weak and variable													
1656	West Point, 0.3 mile west of		47° 39.67'	122° 26.33'	−0 44	−0 43	−0 16	−0 47	0.4	0.3	– –	– –	0.7	225°	– –	– –	0.7	015°
1661	West Point, 1.2 miles west of		47° 39.57'	122° 27.92'	Current weak and variable													
1666	Elliott Bay entrance		47° 36.72'	122° 24.43'	Current weak and variable													
1671	Alki Point, 0.3 mile west of		47° 34.53'	122° 25.68'	−0 36	−0 44	−0 13	−0 39	0.3	0.2	– –	– –	0.5	160°	– –	– –	0.5	330°
1676	Restoration Point, 0.6 mile ESE of		47° 34.85'	122° 27.97'	– – –	−0 38	– – –	−0 06	0.3	0.3	– –	– –	0.5	135	– –	– –	0.7	034°
	Rich Passage																	
1681	Approach, north of Blake Island		47° 33.67'	122° 30.06'	– – –	−0 17	– – –	+0 44	0.1	0.3	– –	– –	0.2	301°	– –	– –	0.9	071°
1686	East end		47° 34'	122° 32'	– – –	+0 37	– – –	+0 51	0.5	0.4	– –	– –	0.8	321°	– –	– –	1.1	143°
1691	Off Pleasant Beach		47° 35'	122° 32'	−0 14	−0 15	+0 36	+0 43	0.8	1.1	– –	– –	1.3	330°	– –	– –	2.8	132°
1696	West end		47° 35.40'	122° 33.73'	−0 44	−0 08	+0 46	+0 22	1.5	1.2	– –	– –	2.4	238°	– –	– –	3.1	055°
1701	Port Orchard, southwest of Waterman		47° 34'	122° 36'	Current weak and variable													
1706	Sinclair Inlet		47° 33.2'	122° 38.02'	Current weak and variable													
1711	Port Washington Narrows, south ent		47° 34'	122° 37'	−0 09	+0 13	+0 55	+1 08	0.6	0.7	– –	– –	1.0	326°	– –	– –	1.7	154°
1716	Port Washington Narrows, north ent <27>		47° 36.08'	122° 39.70'	−0 25	+0 19	+1 08	+1 15	1.4	0.8	– –	– –	2.2	330°	– –	– –	2.2	156°
1721	Blake Island, southwest of		47° 31.50'	122° 29.97'	−2 58	−2 37	−1 06	−0 25	0.2	0.2	– –	– –	0.3	131°	– –	– –	0.5	326°
					on The Narrows, p.48													
1726	Colvos Passage <29>		– – –	– – –	– – –	– – –	– – –	+0 49	– –	0.4	– –	– –	– –	– –	– –	– –	1.1	016
1731	East Passage		– – –	– – –	Current weak and variable													
1736	Quartermaster Harbor entrance		47° 21.18'	122° 28.85'	– – –	−0 31	– – –	−0 09	0.1	0.1	– –	– –	0.4	349	– –	– –	0.4	167°
1741	Commencement Bay		– – –	– – –	Current weak and variable													
1746	Dalco Passage <30>		47° 19.50'	122° 31.45'	– – –	−0 11	– – –	– – –	0.4	– –	– –	– –	1.3	290	– –	– –	– –	– –
1751	Gig Harbor entrance		47° 19.55'	122° 34.48'	−0 39	−0 57	−0 46	+0 11	0.3	0.4	– –	– –	0.9	016°	– –	– –	1.2	176°
	The Narrows																	
1756	THE NARROWS, north end (midstream)		47° 18.37'	122° 32.98'	*Daily predictions*						– –	– –	3.2	136°	– –	– –	2.8	334°
1761	North End (east side) <31>		47° 18.50'	122° 32.50'	– – –	– – –	– – –	+0 23	– –	1.4	– –	– –	– –	– –	– –	– –	3.8	340
1766	North End (west side) <32>		47° 18.27'	122° 33.42'	– – –	+0 08	+0 05	−1 28	0.9	0.4	– –	– –	3.0	143°	– –	– –	1.1	324°
1771	0.1 mile east of Point Evans <33>		47° 17.15'	122° 32.67'	−0 06	−0 06	−0 28	+0 02	1.3	1.2	– –	– –	4.3	201°	– –	– –	3.4	023°
1776	South end (midstream) <34>		47° 15.65'	122° 33.50'	−0 05	+0 04	−0 16	−0 05	1.2	1.1	– –	– –	3.8	218°	– –	– –	3.1	022°
1781	Hale Passage, east end <31>		47° 14.78'	122° 35.85'	– – –	– – –	– – –	−2 44	– –	0.3	– –	– –	– –	– –	– –	– –	0.7	093
1786	Hale Passage, west end		47° 16.67'	122° 39.73'	0 00	−1 12	−2 20	−1 00	0.4	0.6	– –	– –	1.4	299°	– –	– –	1.8	119°
1791	Carr Inlet		– – –	– – –	Current weak and variable													
1796	Gibson Point, 0.8 mile east of		47° 13.08'	122° 35.37'	+0 28	+0 43	+0 08	+0 15	0.7	0.6	– –	– –	2.1	203°	– –	– –	1.8	029°
1801	Cormorant Passage <35>		47° 09.22'	122° 37.78'	– – –	– – –	– – –	+2 06	– –	0.3	– –	– –	– –	– –	– –	– –	0.7	026
1806	Nisqually Reach, E of Sandy Point <36>		47° 08.93'	122° 39.50'	– – –	– – –	– – –	+1 17	– –	0.3	– –	– –	– –	– –	– –	– –	0.8	036
1811	Nisqually Reach		47° 07'	122° 42'	+0 26	+0 24	+0 20	+0 31	0.3	0.4	– –	– –	1.1	259°	– –	– –	1.1	044°
1816	Balch Passage		47° 11.25'	122° 41.83'	−0 09	−1 07	−0 40	−0 40	0.4	0.8	– –	– –	1.1	296°	– –	– –	2.2	107°
1821	Pitt Passage, east of Pitt Island		47° 13.42'	122° 42.95'	−0 55	−1 16	−1 43	−1 24	0.3	0.5	– –	– –	0.9	204°	– –	– –	1.4	028°
1826	Drayton Passage <31>		47° 10.35'	122° 44.50'	– – –	– – –	– – –	−1 44	– –	0.1	– –	– –	– –	– –	– –	– –	0.4	030
1831	Devils Head, west of <38>		47° 09.63'	122° 47.38'	– – –	– – –	– – –	+0 22	– –	0.2	– –	– –	– –	– –	– –	– –	0.6	158
1836	Dana Passage		47° 09.80'	122° 52.07'	+0 17	−0 09	−0 06	−0 12	0.5	0.8	– –	– –	1.5	249°	– –	– –	2.2	076°
1841	*Budd Inlet entrance*		47° 08.37'	122° 55.03'	– – –	−0 42	– – –	−0 19	0.2	0.1	– –	– –	0.7	236	– –	– –	0.4	031°
1846	Olympia, Budd Inlet		47° 05'	122° 55'	Current weak and variable													

Endnotes can be found at the end of table 2.

TABLE 2 – CURRENT DIFFERENCES AND OTHER CONSTANTS

No.	PLACE	Meter Depth	POSITION Latitude	POSITION Longitude	TIME DIFFERENCES Min. before Flood	TIME DIFFERENCES Flood	TIME DIFFERENCES Min. before Ebb	TIME DIFFERENCES Ebb	SPEED RATIOS Flood	SPEED RATIOS Ebb	Minimum before Flood knots	Minimum before Flood Dir.	Maximum Flood knots	Maximum Flood Dir.	Minimum before Ebb knots	Minimum before Ebb Dir.	Maximum Ebb knots	Maximum Ebb Dir.
		ft	North	West	h m	h m	h m	h m			knots	Dir.	knots	Dir.	knots	Dir.	knots	Dir.
	PUGET SOUND–cont. Time meridian, 120° W				on The Narrows, p.48													
1851	Eld Inlet entrance		47° 08.78'	122° 56.00'	−0 04	+0 09	−0 36	−0 10	0.3	0.2	– –	– –	0.9	224°	– –	– –	0.6	028°
1856	Unsal Point, 0.3 mile south of <31>		47° 09.57'	122° 53.68'	– – –	– – –	– – –	−1 04	– –	0.4	– –	– –	– –	– –	– –	– –	1.2	137
1861	Peale Passage, south end		47° 10.50'	122° 53.23'	– – –	−0 09	– – –	−0 28	0.1	0.1	– –	– –	0.4	341	– –	– –	0.4	173°
1866	Peale Passage, north end		47° 13.35'	122° 55.22'	−0 14	−0 23	−0 56	−0 30	0.2	0.4	– –	– –	0.5	319	– –	– –	1.2	145°
1871	Squaxin Passage, north of Hunter Point		47° 10.62'	122° 55.15'	−0 42	−0 32	−0 33	−0 45	0.4	0.4	– –	– –	1.4	286	– –	– –	1.2	121°
1876	Totten Inlet entrance		47° 11.33'	122° 56.70'	−0 19	+0 02	−0 25	−0 09	0.5	0.4	– –	– –	1.7	243	– –	– –	1.0	054°
1881	Hammersley Inlet, 0.8 mile east of Libby Point		47° 12.15'	122° 58.47'	+1 00	+0 37	+0 05	+1 06	0.7	0.9	– –	– –	2.3	283°	– –	– –	2.4	102°
1886	Hammersley Inlet, west of Skookum Point		47° 12.42'	123° 02.37'	+1 14	+1 29	+0 57	+1 04	0.5	0.6	– –	– –	1.7	287°	– –	– –	1.6	101°
1891	Pickering Passage, south end		47° 13.17'	122° 56.08'	+1 05	+1 06	+0 21	+0 53	0.4	0.5	– –	– –	1.3	190°	– –	– –	1.3	013°
1896	Pickering Passage, off Graham Point		47° 14.90'	122° 55.53'	+1 17	+1 03	+0 24	+1 20	0.3	0.5	– –	– –	1.1	199°	– –	– –	1.3	034°
1901	Pickering Passage, north end		47° 18.35'	122° 51.05'	+1 24	+1 03	+1 11	+1 33	0.2	0.2	– –	– –	0.6	250°	– –	– –	0.6	067°
1906	Case Inlet, 1 mile SE of McMicken Island		47° 14.30'	122° 50.62'	Current weak and variable													
	POSSESSION SOUND–SKAGIT BAY				on Admiralty Inlet, p.44 <40>													
1911	Possession Sound		– – –	– – –	Current weak and variable													
1916	Port Susan		48° 06'	122° 22'	Current weak and variable													
1921	Saratoga Passage		– – –	– – –	Current weak and variable													
1926	Skagit Bay, 1 mile north of Rocky Point		48° 16'	122° 32'	– – –	+0 56	– – –	+1 47	0.4	0.4	– –	– –	0.6	060	– –	– –	1.0	236°
					on Deception Pass, p.52													
1931	Skagit Bay, 1 mi. S of Goat Island <41>		48° 20.67'	122° 32.62'	−1 07	−1 26	−0 53	−0 53	0.3	0.2	– –	– –	1.8	139	– –	– –	1.4	312°
1936	Skagit Bay, channel SW of Hope Island		48° 23.58'	122° 34.80'	−0 20	−0 40	−0 08	−0 15	0.4	0.3	– –	– –	2.3	165	– –	– –	2.0	344°
	ROSARIO STRAIT <40>																	
1941	Yokeko Point, Deception Pass		48° 24.77'	122° 36.82'	+0 04	−0 24	−0 25	+0 17	0.4	0.4	– –	– –	2.1	064°	– –	– –	2.7	222°
1946	DECEPTION PASS, (Narrows)		48° 24.37'	122° 38.58'	*Daily predictions*						– –	– –	5.2	090°	– –	– –	6.6	270°
					on Rosario Strait, p.56 <40>													
1951	Deception Island, 1.0 mile west of		48° 24.22'	122° 41.85'	– – –	+1 14	– – –	−1 23	0.6	0.5	– –	– –	0.6	035	– –	– –	1.0	210°
1956	Iceberg Point, 2.1 miles SSW of		48° 23'	122° 55'	−0 34	−0 05	+1 21	−0 10	1.0	0.4	– –	– –	1.1	010	– –	– –	0.8	260°
1961	Colville Island, 1 mile SSE of		48° 24'	122° 49'	+0 19	+0 31	+0 46	+0 07	1.0	1.2	– –	– –	1.1	055°	– –	– –	2.3	235°
1966	Colville Island, 1.4 miles east of		48° 25'	122° 47'	+0 03	+0 14	+0 17	−0 14	1.4	1.0	– –	– –	1.6	055°	– –	– –	1.9	215°
1971	Deception Island, 2.7 miles west of		48° 24.75'	122° 44.37'	+1 08	+0 50	+0 38	−0 34	0.8	0.7	– –	– –	0.9	015°	– –	– –	1.3	190°
1976	W Point, Whidbey I., 1.8 mi. SW of <9>		48° 23'	122° 42'	– – –	+0 46	– – –	−2 23	0.6	0.5	– –	– –	0.6	005	– –	– –	1.0	207°
1981	Deception Island, 1.3 miles NW of		48° 25.18'	122° 41.9'	– – –	+0 04	– – –	−2 29	1.2	0.6	– –	– –	1.3	017	– –	– –	1.1	161°
1986	ROSARIO STRAIT		48° 27.50'	122° 47.00'	*Daily predictions*						– –	– –	1.1	335°	– –	– –	1.9	175°
1991	Lopez Pass		48° 28.78'	122° 49.12'	−1 13	−1 33	−0 39	−1 16	1.4	1.0	– –	– –	1.6	275	– –	– –	1.9	085°
1996	Burrows Bay, 0.5 mile east of Allan I		48° 27.77'	122° 40.97'	−3 01	−0 48	+0 22	−0 43	0.9	0.2	– –	– –	1.0	022	– –	– –	0.4	209°
2001	Burrows I.–Allan I., Passage between		48° 28.30'	122° 41.98'	−2 05	−0 34	+0 09	−0 57	2.0	0.7	– –	– –	2.2	304	– –	– –	1.4	096°
2006	Burrows Island Light, 0.3 miles WNW of		48° 29'	122° 44'	+0 29	+0 30	+0 13	+0 16	1.0	1.1	– –	– –	1.1	015°	– –	– –	2.1	200°
2011	Ficalgo Head–Burrows Island, between		48° 29.33'	122° 42.20'	– – –	−1 03	−0 32	−1 57	1.6	0.5	– –	– –	1.8	270	– –	– –	0.9	090°
2016	Green Point, 0.8 mile northwest of		48° 30.28'	122° 42.37'	−0 55	−0 30	−0 08	−1 02	1.8	1.0	– –	– –	2.0	020	– –	– –	1.9	190°
2021	Shannon Point, 2.0 miles west of		48° 30.63'	122° 43.83'	−0 19	+0 49	+0 50	−0 16	1.2	1.0	– –	– –	1.4	005°	– –	– –	1.8	190°
2026	Fauntleroy Point Light, 0.8 mile ESE of		48° 31.20'	122° 46.18'	+0 08	+0 16	−1 10	−0 33	1.1	0.7	– –	– –	1.2	310°	– –	– –	1.4	125°
2031	Thatcher Pass		48° 31.65'	122° 48.18'	+0 43	+1 01	+1 26	+0 51	0.4	0.5	– –	– –	0.4	300°	– –	– –	0.9	075°
2036	Frost–Willow Island, between		48° 32.35'	122° 49.85'	+0 46	+1 14	+0 19	+0 29	0.6	0.4	– –	– –	0.6	010°	– –	– –	0.8	126°
2041	Strawberry Island, 0.8 mile west of		48° 33.67'	122° 45.25'	+0 43	+0 46	+0 37	+0 30	1.4	1.4	– –	– –	1.5	020°	– –	– –	2.6	190°
2046	Peavine Pass, west entrance		48° 35.22'	122° 49.20'	−0 42	−0 41	−0 26	−0 52	1.6	1.2	– –	– –	1.7	055	– –	– –	2.2	285°
2051	Obstruction Pass Light, 0.4 mile NW of		48° 36.22'	122° 48.80'	−0 59	−0 44	−0 38	−0 49	1.2	0.5	– –	– –	1.3	100	– –	– –	1.0	270°
2056	Peapod Rocks Light, 1.2 miles south of		48° 37.33'	122° 44.83'	+0 08	+1 12	+0 49	−0 15	1.2	1.0	– –	– –	1.3	030°	– –	– –	1.9	215°
2061	Barnes Island, 0.8 mile southwest of		48° 41.15'	122° 47.33'	+0 37	+1 20	−0 07	+0 08	0.6	0.5	– –	– –	0.6	315°	– –	– –	0.9	140°
2066	Raccoon Point, 0.6 mile NNE of		48° 42.38'	122° 49.75'	−0 36	−0 45	−1 41	−0 20	0.6	0.4	– –	– –	0.6	286	– –	– –	0.8	101°

Endnotes can be found at the end of table 2.

TABLE 2 – CURRENT DIFFERENCES AND OTHER CONSTANTS

No.	PLACE	Meter Depth	POSITION Latitude	POSITION Longitude	TIME DIFFERENCES Min. before Flood	TIME DIFFERENCES Flood	TIME DIFFERENCES Min. before Ebb	TIME DIFFERENCES Ebb	SPEED RATIOS Flood	SPEED RATIOS Ebb	Minimum before Flood knots	Minimum before Flood Dir.	Maximum Flood knots	Maximum Flood Dir.	Minimum before Ebb knots	Minimum before Ebb Dir.	Maximum Ebb knots	Maximum Ebb Dir.
		ft	North	West	h m	h m	h m	h m										
	ROSARIO STRAIT <40>–cont. Time meridian, 120° W				on Rosario Strait, p.56 <40>													
2071	Parker Reef Light, 0.5 mile north of		48° 43.98'	122° 53.40'	+1 31	+1 02	+1 14	+1 02	1.0	0.8	--	--	1.1	067°	--	--	1.5	278°
2076	Matia Island, 0.8 mile west of		48° 44.93'	122° 51.95'	+0 22	+0 26	−0 26	+0 14	1.1	0.8	--	--	1.2	350°	--	--	1.5	206°
2081	Guemes Channel, west entrance		48° 31.27'	122° 39.13'	−0 21	−0 33	−1 24	−0 36	0.8	1.1	--	--	0.9	095°	--	--	2.1	255°
2086	Padilla Bay, 0.8 mi. NW of March Pt <42>		48° 31'	122° 35'	---	---	---	---	--	--	--	--	--	--	--	--	--	--
2091	Guemes Island, 0.5 mile east of		48° 32'	122° 34'	Current weak and variable													
2096	Huckleberry Island, 0.5 mile north of		48° 32.75'	122° 33.98'	−3 17	+0 18	+0 36	−1 10	0.7	0.3	--	--	0.8	006°	--	--	0.6	253°
2101	William Point Light, 0.8 mile W of		48° 35.05'	122° 34.77'	---	+0 36	---	+0 23	0.1	0.2	--	--	0.1	--	--	--	0.3	230
2106	Eliza Island, 0.5 mile southeast of		48° 39'	122° 34'	Current weak and variable													
2111	Bellingham Channel, off Cypress I. Light		48° 33.62'	122° 39.82'	−0 48	+0 08	−0 49	−0 51	1.1	1.2	--	--	1.2	045°	--	--	2.2	185°
2116	Towhead Island, 0.4 mile east of		48° 36.73'	122° 42.13'	−1 55	+1 13	+2 43	+0 01	0.7	0.2	--	--	0.8	315°	--	--	0.4	125°
2121	Sinclair Island Light, 0.6 mile SE of		48° 36.17'	122° 38.92'	+0 16	+0 48	+0 22	+0 18	1.3	0.8	--	--	1.4	045°	0 0	--	1.6	210°
2126	Sinclair Island, 1 mile northeast of <43>		48° 38.67'	122° 39.50'	---	---	---	+1 23	--	0.4	--	--	--	--	--	--	0.8	110
2131	Lawrence Point, Orcas I., 1.3 mi. NE of		48° 40.70'	122° 42.87'	+0 56	+0 58	+0 59	+0 55	1.2	0.8	--	--	1.4	345°	--	--	1.4	145°
2136	Hale Passage, 0.5 mile SE of Lummi Point		48° 43.88'	122° 40.67'	−0 43	+0 05	+0 20	−0 11	1.0	0.5	--	--	1.1	350°	--	--	1.0	145°
2141	Clark Island, 1.6 miles north of		48° 43.88'	122° 46.40'	+0 47	+1 14	+0 44	−0 02	0.6	0.6	--	--	0.7	335°	--	--	1.1	150°
2146	Matia Island, 1.4 miles north of		48° 46.33'	122° 50.97'	+0 02	+0 57	+0 59	+0 08	0.6	0.3	--	--	0.6	330°	--	--	0.6	190°
2151	Parker Reef Light, 1 mile north of		48° 44.52'	122° 53.67'	---	+1 38	---	−0 52	0.9	0.4	--	--	1.0	065	--	--	0.7	265°
2156	Puffin Island Light, 4.8 miles north of		48° 49.33'	122° 48.50'	+0 18	+1 30	+0 55	+0 11	0.4	0.4	--	--	0.5	325°	--	--	0.8	210°
2161	Neptune Beach, 1.2 miles NW of <44>		48° 49.47'	122° 43.58'	---	+1 43	---	---	0.4	--	--	--	0.4	359	--	--	--	--
2166	Cherry Point, 1.5 miles SE of		48° 50.70'	122° 43.50'	---	+1 04	---	+0 13	0.3	0.2	--	--	0.3	326	--	--	0.3	116°
	SAN JUAN CHANNEL				on San Juan Channel, p.60													
2171	Cattle Point, 1.2 miles southeast of		48° 26.03'	122° 56.82'	+0 11	−0 20	+0 34	−0 01	0.3	0.9	--	--	0.8	340°	--	--	2.4	195°
2176	SAN JUAN CHANNEL (south entrance)		48° 27.68'	122° 57.05'	*Daily predictions*						--	--	2.6	010°	--	--	2.6	180°
2181	Kings Point, Lopez Island, 1 mile NNW of		48° 29.00'	122° 57.35'	+0 51	−0 07	+0 27	+0 36	0.6	0.5	--	--	1.6	020°	--	--	1.3	185°
2186	Pear Point, 1.1 miles east of		48° 30.68'	122° 57.17'	+0 40	+1 09	−0 10	+1 01	0.4	0.5	--	--	1.0	359°	--	--	1.4	203°
2191	Turn Rock Light, 1.9 miles northwest of		48° 33.40'	122° 59.90'	+1 19	+1 22	+0 20	−0 01	0.4	0.5	--	--	0.9	330°	--	--	1.3	135°
2196	Crane Island, south of, Wasp Passage		48° 35.37'	122° 59.92'	−0 10	+0 35	+0 29	+0 07	0.2	0.1	--	--	0.4	288°	--	--	0.3	075°
2201	Wasp Passage Light, 0.5 mile WSW of		48° 35.53'	122° 59.37'	+0 19	+0 28	+0 15	−0 15	0.5	0.4	--	--	1.2	300°	--	--	1.1	110°
2206	Spring Passage, south entrance		48° 36.68'	123° 02.05'	+0 04	−1 09	−0 43	−0 13	0.4	0.4	--	--	0.9	010°	--	--	1.0	150°
2211	Limestone Point, Spieden Channel		48° 37.58'	123° 06.55'	+0 23	−0 12	−1 00	+0 26	0.7	1.2	--	--	1.8	085°	--	--	3.2	283°
2216	Point Disney, 1.6 miles east of		48° 40.37'	123° 00.37'	+2 36	+1 31	+0 48	+1 51	0.2	0.8	--	--	0.5	020°	--	--	2.2	230°
2221	Harney Channel		48° 35.45'	122° 55.23'	+0 15	+0 21	+0 11	+0 05	0.2	0.3	--	--	0.6	250°	--	--	0.7	120°
2226	East Sound entrance		48° 35.22'	122° 51.42'	Current weak and variable													
2231	East Sound, 0.2 mile SW of Rosario Point		48° 38.65'	122° 52.88'	Current weak and variable													
	HARO STRAIT and BOUNDARY PASS				on Admiralty Inlet, p.44 <40>													
2236	Discovery Island, 3.3 miles northeast of		48° 27'	123° 09'	+1 16	+1 03	+0 59	+0 59	0.8	0.6	--	--	1.3	345°	--	--	1.6	170°
2241	Kellett Bluff, west of		48° 35.35'	123° 13.50'	+0 01	+0 20	+0 36	+0 20	1.0	0.8	--	--	1.6	000°	--	--	2.1	170°
2246	Turn Point, Boundary Pass		48° 41.72'	123° 14.13'	+0 26	+0 51	+1 18	+0 26	0.4	0.6	--	--	0.7	080°	--	--	1.6	260°
2251	Skipjack Island, 2 miles NNE of		48° 46'	123° 01'	+0 26	+0 34	+0 49	−0 02	1.6	1.4	0.5	295°	1.7	041°	0.2	120°	2.7	203°
					on San Juan Channel, p.60													
2256	Johns Island, 0.8 mile north of		48° 41'	123° 09'	−0 15	−0 31	+0 06	+0 49	0.2	0.2	--	--	0.6	090°	--	--	0.6	350°
2261	Waldron Island, 1.7 miles west of		48° 42.25'	123° 06.52'	+0 37	+0 36	+1 10	+1 05	0.5	0.4	--	--	1.4	040°	--	--	1.2	260°
2266	Skipjack Island, 1.5 miles northwest of		48° 44.97'	123° 03.65'	+1 17	+1 19	+1 08	+1 12	0.3	0.5	--	--	0.8	035°	--	--	1.4	290°
2271	Point Hammond, 1.1 miles northwest of		48° 43.92'	123° 01.52'	+1 11	+1 05	+0 27	+1 13	0.2	0.9	--	--	0.6	055°	--	--	2.4	255°
					on Rosario Strait, p.56 <40>													
2276	Alden Point, Patos Island, 2 miles S of		48° 45.47'	122° 58.82'	+0 09	−0 26	+0 15	−0 53	0.9	1.1	--	--	1.0	025°	--	--	2.1	185°
2281	Patos Island Light, 1.4 miles west of		48° 47.33'	123° 00.20'	+0 19	+0 30	+0 40	−0 02	0.8	1.0	--	--	0.8	065°	--	--	2.0	180°
2286	Toe Point, Patos Island, 0.5 mile S of		48° 46.70'	122° 56.45'	−2 31	+0 49	+0 51	−0 47	1.4	0.8	--	--	1.6	045°	--	--	1.6	270°

Endnotes can be found at the end of table 2.

TABLE 2 – CURRENT DIFFERENCES AND OTHER CONSTANTS

No.	PLACE	Meter Depth	POSITION		TIME DIFFERENCES				SPEED RATIOS		AVERAGE SPEEDS AND DIRECTIONS							
			Latitude	Longitude	Min. before Flood	Flood	Min. before Ebb	Ebb	Flood	Ebb	Minimum before Flood		Maximum Flood		Minimum before Ebb		Maximum Ebb	
		ft	North	West	h m	h m	h m	h m			knots	Dir.	knots	Dir.	knots	Dir.	knots	Dir.
	GEORGIA STRAIT Time meridian, 120° W																	
					on Admiralty Inlet, p.44 <40>													
2291	Drayton Harbor Entrance		48° 59.45'	122° 46.07'	−0 27	−0 11	+0 48	+0 13	0.6	0.4	--	--	1.0	133°	0.0	--	1.0	313°
					on Active Pass, p.64													
2296	Sucia I., 0.5 mile WNW of Lewson Bluff		48° 46'	122° 56'	+0 54	+0 26	+0 53	+1 15	0.2	0.4	--	--	0.8	025°	--	--	1.4	205°
2301	Sansum Narrows		48° 47.00	123° 33.37'	+0 24	+0 24	−0 37	−0 37	0.5	0.5	--	--	1.7	000°	--	--	1.7	180°
2306	ACTIVE PASS		48° 52.78'	123° 17.75'		Daily predictions					--	--	3.3	045°	--	--	3.7	225°
2311	Porlier Pass		49° 00.65'	123° 35.50'	−0 20	−0 24	−0 28	−0 24	1.3	1.3	--	--	4.3	045°	--	--	4.7	225°
2316	Gabriola Pass		49° 07.70'	123° 42.15'	−0 35	−0 33	−0 32	−0 33	1.1	1.1	--	--	4.0	090°	--	--	4.0	270°
2321	Dodd Narrows		49° 08'	123° 49'	−0 45	−0 45	−0 45	−0 45	1.4	1.4	--	--	4.8	315°	--	--	5.2	135°
					on Burrard Inlet, p.68													
2326	BURRARD INLET, First Narrows		49° 19'	123° 08'		Daily predictions					--	--	3.7	135°	--	--	3.7	315°
2331	Second Narrows, Burrard Inlet		49° 18'	123° 01'	−0 18	−0 11	−0 04	−0 11	0.9	0.9	--	--	3.3	090°	--	--	3.3	270°
2336	Seechelt Rapids		49° 45'	123° 55'	+1 35	+1 24	+1 12	+1 24	1.8	1.8	--	--	6.5	150°	--	--	6.5	330°
					on Seymour Narrows, p.72													
2341	Stevens Pass		49° 31'	124° 31'	+0 15	+0 15	+0 15	+0 15	0.2	0.2	--	--	2.2	310°	--	--	2.2	130°
2346	Cape Lazo		49° 43'	124° 48'	+0 15	+0 15	+0 15	+0 15	0.2	0.2	--	--	2.0	355°	--	--	2.0	175°
2351	Kuhushan Point		49° 53'	125° 04'	+0 10	+0 10	+0 10	+0 10	0.2	0.2	--	--	2.0	325°	--	--	2.0	145°
2356	Shelter Point		49° 57'	125° 10'	+0 10	+0 10	+0 10	+0 10	0.2	0.2	--	--	2.0	145°	--	--	2.0	325°
	DISCOVERY PASSAGE																	
2361	Off Cape Mudge		50° 00'	125° 14'	+0 15	+0 15	+0 15	+0 15	0.5	0.5	--	--	5.0	165°	--	--	5.0	345°
2366	Orange Point		50° 04'	125° 17'	+0 10	+0 10	+0 10	+0 10	0.5	0.5	--	--	5.0	145°	--	--	5.0	325°
2371	Race Point		50° 07'	125° 20'	+0 05	+0 05	+0 05	+0 05	0.7	0.7	--	--	6.5	125°	--	--	6.5	305°
2376	SEYMOUR NARROWS		50° 08'	125° 21'		Daily predictions					--	--	9.2	180°	--	--	9.8	000°
2381	Separation Head		50° 11'	125° 22'	−0 05	−0 05	−0 05	−0 05	0.4	0.4	--	--	3.4	170°	--	--	3.6	350°
2386	Moriarty Point		50° 16'	125° 25'	−0 10	−0 10	−0 10	−0 10	0.3	0.3	--	--	2.5	170°	--	--	2.5	350°
2391	Chatham Point		50° 20'	125° 27'	−0 20	−0 20	−0 20	−0 20	0.3	0.3	--	--	2.5	165°	--	--	2.5	345°
	JOHNSTONE STRAIT																	
2396	Ripple Point		50° 22'	125° 35'	−0 40	−0 40	−0 40	−0 40	0.4	0.4	--	--	3.4	105°	--	--	3.6	285°
2401	Camp Point		50° 24'	125° 51'	−1 00	−1 00	−1 00	−1 00	0.4	0.4	--	--	3.4	090°	--	--	3.6	270°
2406	Race Passage <45>		50° 23'	125° 53'	−0 58	−0 58	−0 58	−0 58	0.5	0.5	--	--	4.8	110°	--	--	5.2	290°
2411	Current Passage		50° 25'	125° 54'	−1 00	−1 00	−1 00	−1 00	0.5	0.5	--	--	4.8	120°	--	--	5.2	300°
2416	Ransom Point		50° 28'	126° 06'	−1 00	−1 00	−1 00	−1 00	0.3	0.3	--	--	2.5	110°	--	--	2.5	290°
2421	Off Broken Island		50° 30'	126° 17'	−1 00	−1 00	−1 00	−1 00	0.3	0.3	--	--	2.5	100°	--	--	2.5	280°
2426	Robson Bight (off)		50° 30'	126° 35'	−1 15	−1 15	−1 15	−1 15	0.3	0.3	--	--	2.5	100°	--	--	2.5	280°
2431	Ella Point, Weynton Passage		50° 33'	126° 48'	−1 25	−1 25	−1 25	−1 25	0.4	0.4	--	--	3.9	105°	--	--	4.1	285°
	BROUGHTON STRAIT																	
2436	Pearse Passage, Cormorant Island		50° 35'	126° 54'	−1 30	−1 30	−1 30	−1 30	0.4	0.4	--	--	3.9	165°	--	--	4.1	345°
2441	Leonard Point, Cormorant Island		50° 35'	126° 58'	−1 35	−1 35	−1 35	−1 35	0.3	0.3	--	--	2.5	090°	--	--	2.5	270°
2446	Ledge Point		50° 36'	127° 04'	−1 40	−1 40	−1 40	−1 40	0.3	0.3	--	--	2.5	110°	--	--	2.5	290°
2451	Pulteney Point		50° 37'	127° 10'	−1 45	−1 45	−1 45	−1 45	0.3	0.3	--	--	2.5	120°	--	--	2.5	300°

Endnotes can be found at the end of table 2.

TABLE 2 – CURRENT DIFFERENCES AND OTHER CONSTANTS

No.	PLACE	Meter Depth	POSITION Latitude	POSITION Longitude	TIME DIFFERENCES Min. before Flood	TIME DIFFERENCES Flood	TIME DIFFERENCES Min. before Ebb	TIME DIFFERENCES Ebb	SPEED RATIOS Flood	SPEED RATIOS Ebb	Minimum before Flood knots	Minimum before Flood Dir.	Maximum Flood knots	Maximum Flood Dir.	Minimum before Ebb knots	Minimum before Ebb Dir.	Maximum Ebb knots	Maximum Ebb Dir.
		ft	North	West	h m	h m	h m	h m										
	QUEEN CHARLOTTE STRAIT Time meridian, 120° W				on Seymour Narrows, p.72													
2456	False Head, 2 miles north from		50° 41'	127° 17'	−2 20	−2 20	−2 20	−2 20	0.3	0.3	--	--	2.5	130°	--	--	2.5	310°
2461	Dillon Point, 1 mile north		50° 46'	127° 25'	−2 30	−2 30	−2 30	−2 30	0.3	0.3	--	--	2.5	110°	--	--	2.5	290°
2466	Gordon Channel		50° 55'	127° 40'	−2 40	−2 40	−2 40	−2 40	0.3	0.3	--	--	2.5	125°	--	--	2.5	305°
	GOLETAS CHANNEL																	
2471	Duval Point		50° 48'	127° 30'	−3 05	−3 05	−3 05	−3 05	0.3	0.3	--	--	2.5	110°	--	--	2.5	290°
2476	Boxer Point		50° 49'	127° 39'	−3 15	−3 15	−3 15	−3 15	0.3	0.3	--	--	2.5	110°	--	--	2.5	290°
2481	Lemon Point		50° 51'	127° 46'	−3 20	−3 20	−3 20	−3 20	0.3	0.3	--	--	2.5	110°	--	--	2.5	290°
2486	Heath Point		50° 53'	127° 53'	−3 25	−3 25	−3 25	−3 25	0.3	0.3	--	--	3.0	110°	--	--	3.0	290°
2491	Nawhitti Bar		50° 54'	128° 00'	−4 30	−4 38	−4 47	−4 38	0.4	0.4	--	--	4.0	100°	--	--	4.0	280°
	PASSAGES NORTH OF VANCOUVER ISLAND																	
2496	Surge Narrows, Okisollo Channel		50° 14'	125° 10'	−0 45	−0 45	−0 45	−0 45	0.7	0.7	--	--	7.0	140°	--	--	7.0	320°
2501	Hole In The Wall, Okisollo Channel		50° 18'	125° 13'	−0 55	−0 55	−0 50	−0 55	0.8	0.8	--	--	7.5	050°	--	--	7.5	230°
2506	Rapids, near Barnes Bay, Okisollo Chan		50° 19'	125° 16'	−0 50	−0 55	−0 55	−0 55	0.7	0.7	--	--	6.5	072°	--	--	6.5	252°
2511	Arran Rapids, north of Stuart Island		50° 25'	125° 08'	−0 45	−0 45	−0 45	−0 45	0.7	0.7	--	--	7.0	065°	--	--	7.0	245°
2516	Yuculta Rapids, SW of Stuart Island		50° 21'	125° 09'	−0 40	−0 40	−0 40	−0 40	0.5	0.5	--	--	5.0	145°	--	--	5.0	325°
2521	Godwin Point, Cordero Island		50° 28'	125° 25'	−0 55	−0 55	−0 55	−0 55	0.2	0.2	--	--	2.2	050°	--	--	2.2	230°
2526	Shell Point, Blind Channel		50° 26'	125° 31'	−1 10	−1 10	−1 10	−1 10	0.5	0.5	--	--	5.0	355°	--	--	5.0	175°
2531	Green Point Rapids, Cordero Channel		50° 27'	125° 31'	−1 25	−1 30	−1 35	−1 30	0.5	0.5	--	--	5.0	130°	--	--	5.0	310°
2536	Whirlpool Rapids, Wellbore Channel		50° 27'	125° 47'	−1 50	−1 50	−1 50	−1 50	0.6	0.6	--	--	6.0	185°	--	--	6.0	005°
2541	Shaw Point, Sunderland Channel		50° 28'	125° 56'	−1 05	−1 05	−1 05	−1 05	0.2	0.2	--	--	1.5	060°	--	--	1.5	240°
2546	Root Point, Chatham Channel		50° 35'	126° 12'	−1 05	−1 05	−1 05	−1 05	0.6	0.6	--	--	5.5	110°	--	--	5.5	290°
2551	Littleton Point, Chatham Channel		50° 37'	126° 17'	−1 05	−1 05	−1 05	−1 05	0.4	0.4	--	--	3.5	130°	--	--	3.5	310°
2556	Ripple Bluff, Knight Inlet		50° 38'	126° 31'	−1 15	−1 15	−1 15	−1 15	0.3	0.3	--	--	2.5	105°	--	--	2.5	285°
2561	Owl Island, main ent. to Knight Inlet		50° 38'	126° 41'	−1 20	−1 20	−1 20	−1 20	0.3	0.3	--	--	2.5	120°	--	--	2.5	300°
	HECATE STRAIT and CHATHAM SOUND				on Wrangell Narrows, p.80													
2566	Meyers Narrows, Meyers Passage		52° 37'	128° 39'	−1 00	−0 56	−0 54	−0 35	0.9	1.1	--	--	2.2	090°	--	--	2.2	270°
2571	Otter Passage, Nepean Sound		53° 08'	129° 45'	−0 19	−0 22	−0 26	−0 01	--	2.1	--	--	--	050	--	--	4.4	230°
2576	Grenville Channel (narrow portion) <46>		53° 36'	129° 41'	+1 23	+1 17	+1 09	+1 38	0.4	1.6	--	--	1.4	320°	--	--	3.4	140°
2581	Skeena River, Middle Passage		54° 06'	130° 13'	+1 39	+1 31	+1 21	+1 52	0.7	1.1	--	--	2.2	120°	--	--	2.2	310°
2586	Casey Point, Prince Rupert Harbor		54° 16'	130° 22'	+1 37	+1 29	+1 19	+1 50	--	1.0	--	--	--	340°	--	--	2.1	160°
2591	Tuck Narrows, Prince Rupert Harbor		54° 24'	130° 15'	+0 24	+0 28	+0 30	+0 49	1.6	2.3	--	--	5.2	325°	--	--	4.8	145°
2596	Between Rose Spit and Overfall Shoal		54° 14'	131° 35'	+0 14	+0 18	+0 20	+0 39	0.9	1.3	--	--	2.8	145°	--	--	2.8	325°
	DIXON ENTRANCE																	
2601	Naden Harbor, Alexandria Narrows		54° 02.19'	132° 34.44'	+0 33	+0 30	+0 26	+0 51	0.4	0.9	--	--	1.3	205°	--	--	1.8	025°
2606	Masset Harbor, 5 miles Inside		54° 01'	132° 10'	+3 09	+2 59	+2 47	+3 20	1.3	2.1	--	--	4.0	145°	--	--	4.5	335°
	Time meridian, 135° W																	
2611	Cape Muzon, 4 miles south from		54° 36'	132° 41'	−0 39	−0 47	−0 57	−0 26	0.8	1.2	--	--	2.4	045°	--	--	2.4	225°
2616	Point Marsh, 5 miles south from		54° 38'	132° 18'	−0 39	−0 47	−0 57	−0 26	0.7	1.1	--	--	2.3	035°	--	--	2.3	215°
2621	Cape Chacon, 4 miles south from		54° 37'	132° 01'	−0 34	−0 42	−0 52	−0 21	0.9	1.3	--	--	2.8	040°	--	--	2.8	220°
2626	West Devil Rock, 2 miles north of		54° 42'	131° 36'	−0 34	−0 42	−0 52	−0 21	0.8	1.2	--	--	2.4	035°	--	--	2.4	215°
2631	Barren Island Light, 2 miles south from		54° 42'	131° 21'	−0 29	−0 37	−0 47	−0 16	0.7	1.1	--	--	2.3	040°	--	--	2.3	220°

Endnotes can be found at the end of table 2.

TABLE 2 – CURRENT DIFFERENCES AND OTHER CONSTANTS

No.	PLACE	Meter Depth	POSITION Latitude	POSITION Longitude	TIME DIFFERENCES Min. before Flood	TIME DIFFERENCES Flood	TIME DIFFERENCES Min. before Ebb	TIME DIFFERENCES Ebb	SPEED RATIOS Flood	SPEED RATIOS Ebb	Minimum before Flood knots	Minimum before Flood Dir.	Maximum Flood knots	Maximum Flood Dir.	Minimum before Ebb knots	Minimum before Ebb Dir.	Maximum Ebb knots	Maximum Ebb Dir.
		ft	North	West	h m	h m	h m	h m			knots	Dir.	knots	Dir.	knots	Dir.	knots	Dir.
	DIXON ENTRANCE–cont. Time meridian, 120° W				on Wrangell Narrows, p.80													
2636	East Devil Rock, 1 mile north of		54° 42'	131° 05'	+0 31	+0 23	+0 13	+0 44	0.7	1.1	– –	– –	2.2	085°	– –	– –	2.2	265°
2641	Between Dundas Island and Cape Fox		54° 42'	130° 50'	+0 14	+0 18	+0 20	+0 39	0.6	1.4	– –	– –	2.0	090°	– –	– –	3.0	270°
2646	Tongass Islands, east of, Nakat Bay		54° 46.94	130° 44.23'	–1 31	–2 40	–1 38	–1 31	0.3	0.6	– –	– –	0.8	328°	– –	– –	1.2	157°
2651	Boston Islands, 1 mile south from		54° 41'	130° 34'	–0 29	–0 37	–0 47	–0 16	0.7	1.1	– –	– –	2.0	085°	– –	– –	2.0	265°
2656	Portland Inlet, 2 miles S. of Wales Pt		54° 40'	130° 29'	+0 36	+0 28	+0 18	+0 49	0.8	1.2	– –	– –	2.4	040°	– –	– –	2.4	220°
	PEARSE CANAL																	
2661	Haystack Island, Tongass Passage		54° 43'	130° 37'	+0 31	+0 23	+0 13	+0 44	0.8	1.2	– –	– –	2.5	340°	– –	– –	2.5	160°
2666	Point Phipp		54° 47'	130° 38'	+0 31	+0 23	+0 13	+0 44	0.9	1.3	– –	– –	2.8	065°	– –	– –	2.8	245°
2671	Narrows		54° 50.08'	130° 29.13'	+0 34	+0 26	+0 31	+0 47	0.9	1.3	– –	– –	2.8	045°	– –	– –	2.8	225°
	Time meridian, 135° W																	
2676	Blaine Point		55° 02'	130° 13'	–0 19	–0 27	–0 37	–0 06	0.6	1.0	– –	– –	2.0	025°	– –	– –	2.0	205°
	PORTLAND CANAL Time meridian, 120° W																	
2681	Wales Point, 2 miles south from		54° 40'	130° 29'	+0 36	+0 28	+0 18	+0 49	0.8	1.2	– –	– –	2.4	040°	– –	– –	2.4	220°
2686	Cliff Point, 1 mile east from		54° 48'	130° 19'	+0 36	+0 28	+0 18	+0 49	0.7	1.1	– –	– –	2.2	045°	– –	– –	2.2	225°
2691	Flat Point , 1 mile east of		54° 55'	130° 10'	+0 41	+0 33	+0 23	+0 54	0.6	1.0	– –	– –	2.0	030°	– –	– –	2.0	210°
2696	Tree Point, 1 mile east of		55° 02'	130° 10'	+0 41	+0 33	+0 23	+0 54	0.6	0.9	– –	– –	1.8	345°	– –	– –	1.8	165°
2701	Dickens Point		55° 09'	130° 09'	+0 41	+0 33	+0 23	+0 54	0.5	0.7	– –	– –	1.5	020°	– –	– –	1.5	200°
2706	White Point		55° 34'	130° 07'	+0 46	+0 38	+0 28	+0 59	0.3	0.5	– –	– –	1.0	350°	– –	– –	1.0	170°
	Time meridian, 135° W																	
2711	Camp Point		55° 17'	129° 59'	–0 14	–0 22	–0 32	–0 01	0.4	0.6	– –	– –	1.3	000°	– –	– –	1.3	180°
2716	Miners Point		55° 43'	130° 09'	–0 09	–0 17	–0 27	+0 04	0.3	0.4	– –	– –	0.8	350°	– –	– –	0.8	170°
2721	Lion Point		55° 53'	130° 02'	–0 09	–0 17	–0 27	+0 04	0.2	0.2	– –	– –	0.5	025°	– –	– –	0.5	205°
	REVILLAGIGEDO CHANNEL																	
2726	Duke Point, 3.5 miles northeast of		54° 57'	131° 06'	Current weak and variable						0.0	– –	0.5	353°	– –	– –	0.4	168°
2731	Midcy Point, 2.9 miles ENE of		55° 11'	131° 15'	Current weak and variable						– –	– –	0.3	265°	– –	– –	0.2	115°
2736	Walker Island, 1.1 miles north of		55° 12	131° 20'	Current weak and variable						– –	– –	0.3	320°	– –	– –	0.2	105°
2741	Angle Point, 0.5 mile southwest of <47>		55° 14	131° 26'	– – –	+1 27	– – –	– – –	0.1	– –	– –	– –	0.3	290	– –	– –	– –	– –
2746	Reef Point, 0.7 mile northeast of		55° 15	131° 28'	Current weak and variable													
2751	Race Point, 0.7 mile ENE of		55° 17.15'	131° 32.76'	+0 49	+0 22	+0 10	+0 18	0.2	0.2	– –	– –	0.5	145°	– –	– –	0.4	300°
	CARROLL INLET																	
2756	Carroll Point, 0.7 mile northwest of		55° 18'	131° 30'	Current weak and variable													
	TONGASS NARROWS <48>																	
2761	Saxman Spire, 0.2 mile south of		55° 18.93'	131° 36.16'	–0 32	–2 00	–2 02	–0 12	0.2	0.3	– –	– –	0.6	320°	– –	– –	0.7	110°
2766	Ketchikan		55° 20.17'	131° 38.65'	–0 50	+0 08	+0 04	–0 41	0.3	0.1	– –	– –	0.8	310°	– –	– –	0.2	120°
2771	Rosa Reef, 0.5 mile north of <50>		55° 25'	131° 48'	Current weak and variable						– –	– –	0.1	150°	– –	– –	0.1	325°
2776	Point Higgins, 1 mile west of <51>		55° 27'	131° 52'	– – –	+2 02	– – –	+0 11	0.1	0.1	– –	– –	0.4	010°	– –	– –	0.2	200°

Endnotes can be found at the end of table 2.

TABLE 2 – CURRENT DIFFERENCES AND OTHER CONSTANTS

No.	PLACE	Meter Depth	POSITION Latitude	POSITION Longitude	TIME DIFFERENCES Min. before Flood	Flood	Min. before Ebb	Ebb	SPEED RATIOS Flood	Ebb	AVERAGE SPEEDS AND DIRECTIONS Minimum before Flood knots	Dir.	Maximum Flood knots	Dir.	Minimum before Ebb knots	Dir.	Maximum Ebb knots	Dir.
		ft	North	West	h m	h m	h m	h m										
	FELICE STRAIT Time meridian, 135° W				on Wrangell Narrows, p.80													
2781	Hotspur Island, 0.5 mile southeast of		54° 58'	131° 29'	−0 53	−1 29	−1 06	−0 56	0.3	0.4	--	--	1.0	040°	--	--	0.9	220°
2786	Point Davidson, 1 mile south of		54° 59'	131° 36'	−0 29	−0 37	−0 47	−0 16	0.6	1.0	--	--	2.0	060°	--	--	2.0	240°
2791	Harris Island		55° 00'	131° 32'	−0 29	−0 37	−0 47	−0 16	1.3	1.8	--	--	4.2	055°	--	--	3.8	235°
2796	Ajax Reef		55° 00'	131° 28'	−0 24	−0 32	−0 42	−0 11	0.9	1.4	--	--	3.0	095°	--	--	3.0	275°
2801	Snipe Island Light, 1.2 miles SW of <52>		55° 00'	131° 25'	---	---	---	−0 22	--	0.5	--	--	--	--	--	--	1.1	250
2806	Snipe Island		55° 00'	131° 23'	−0 19	−0 27	−0 37	−0 06	1.3	1.8	--	--	4.2	070°	--	--	3.8	250°
2811	Grass Rock, Tamgas Harbor entrance		55° 01.30'	131° 31.34'	−0 24	−0 32	−0 42	−0 11	0.8	1.2	--	--	2.5	015°	--	--	2.5	195°
2816	Indian Reef		55° 02'	131° 21'	−0 19	−0 27	−0 37	−0 06	1.1	1.6	--	--	3.6	030°	--	--	3.4	210°
2821	Indian Rock Buoy, 1.3 miles east of <53>		55° 02'	131° 18'	---	---	---	−0 53	--	0.3	--	--	--	--	--	--	0.5	195
2826	Indian Rock Buoy, 0.3 miles NW of <53>		55° 02'	131° 21'	---	---	---	−1 37	--	0.5	--	--	--	--	--	--	0.9	195
2831	Kwain Bay,2.0 miles east of		55° 05'	131° 19'	Current weak and variable													
2836	Beaver Creek, Mary Island, 0.6 mile W of		55° 05'	131° 15'	Current weak and variable													
2841	Customhouse Cove, 1 mile west of		55° 06'	131° 16'	−0 19	−0 27	−0 37	−0 06	0.6	1.0	--	--	2.0	020°	--	--	2.0	200°
	NICHOLS PASSAGE																	
2846	Hid Reef, 2.7 miles south of		55° 02'	131° 40'	−0 18	−0 54	−0 40	+0 12	0.2	0.2	--	--	0.7	000°	--	--	0.4	190°
2851	Point McCartey Light		55° 07'	131° 42'	−0 24	−0 32	−0 42	−0 11	0.6	1.0	--	--	2.0	040°	--	--	2.0	220°
2856	Point McCartey Light, 1.5 miles east of		55° 07'	131° 40'	Current weak and variable													
2861	Village Point, Metlakatla, 0.2 mile N of		55° 08'	131° 34'	Current weak and variable													
2863	Wharburton Island		55° 08.01'	131° 37.72'	−0 24	−0 32	−0 42	−0 11	0.7	1.1	--	--	2.2	025°	--	--	2.2	205°
2871	Driest Point		55° 11'	131° 36'	−0 19	−0 27	−0 37	−0 06	0.6	0.9	--	--	1.9	355°	--	--	1.9	175°
2873	Bostwick Point, 0.6 mile southeast of		55° 13'	131° 41'	Current weak and variable													
2881	Bostwick Point, 2.1 miles east of		55° 13'	131° 38'	Current weak and variable													
2886	Blank Point		55° 14.92'	131° 40.16'	−0 14	−0 22	−0 32	−0 01	0.7	1.1	--	--	2.3	010°	--	--	2.3	190°
289-	Walden Rocks, 0.6 mile north of		55° 17'	131° 36'	+0 14	−1 09	−2 22	−0 35	0.3	0.8	--	--	1.0	025°	--	--	1.6	175°
2896	Walden Rocks		55° 17'	131° 37'	−0 09	−0 17	−0 27	+0 04	0.9	1.3	--	--	2.8	025°	--	--	2.8	205°
	BEHM CANAL																	
2901	Point Sykes		55° 12'	131° 07'	−0 19	−0 27	−0 37	−0 06	0.3	0.5	--	--	1.0	040°	--	--	1.0	220°
2906	Point Nelson		55° 18'	130° 57'	Current weak and variable													
2911	Short Pass		55° 22.88'	130° 58.55'	Current weak and variable						--	--	0.3	047°	--	--	0.5	222°
2916	Behm Narrows		55° 54.82'	131° 31.92'	−1 41	−3 07	−3 41	−2 58	0.1	0.6	--	--	0.2	062°	--	--	1.2	264°
2921	Helm Point		55° 36'	131° 50'	−0 19	−0 27	−0 37	−0 06	0.4	0.6	--	--	1.2	030°	--	--	1.2	210°
2926	Guard Islands, 2 miles northwest of		55° 28'	131° 54'	−0 19	−0 27	−0 37	−0 06	0.4	0.7	--	--	1.4	020°	--	--	1.4	200°
	CLARENCE STRAIT																	
2931	Duke Island, 2.8 miles WSW of <54>		54° 55'	131° 34'	---	−0 58	---	+0 18	0.2	0.6	0.1	000°	0.7	088°	0.3	155°	1.2	238°
2936	Point Davison, 0.8 mile south of <55>		54° 59'	131° 36'	+1 16	−0 31	+0 02	---	0.2	--	--	--	0.7	045°	--	--	--	240°
2941	Grass Rock, 1 mile south of		55° 00'	131° 33'	−0 24	−0 29	−0 02	−0 22	0.2	0.6	--	--	0.8	025°	--	--	1.3	235°
2946	Moira Rock, 2 miles east of		55° 05'	131° 56'	−0 24	−0 32	−0 42	−0 11	0.5	0.7	--	--	1.5	005°	--	--	1.5	185°
2951	Moira Sound entrance		55° 05.64'	132° 00.14'	Current weak and variable													
2956	Halibut Creek, 1 mile east of		55° 15'	131° 58'	Current weak and variable						--	--	0.3	350°	--	--	0.6	160°
2961	Halibut Creek, 4.5 miles east of		55° 14'	131° 52'	---	−0 06	---	+0 17	0.2	0.2	--	--	0.5	010°	--	--	0.5	115°
2966	Cholmondeley Sound entrance		55° 17'	132° 04'	Current weak and variable						--	--	0.1	259°	--	--	0.4	057°
2971	Skin Island, 3 miles east from		55° 18'	131° 59'	−0 19	−0 27	−0 37	−0 06	0.5	0.7	--	--	1.5	350°	--	--	1.5	170°
2976	Grindall Island, 1.3 miles south of <57>		55° 25'	132° 08'	---	---	---	---	--	--	--	--	1.1	270	--	--	--	--
2981	Skowl Arm, Kasaan Bay		55° 26'	132° 19'	Current weak and variable						--	--	0.3	255°	--	--	0.2	045°
2986	Guard Island Lighthouse, 5.1 miles SW of		55° 24'	132° 00'	+0 06	−0 03	−0 12	+0 20	0.2	0.3	--	--	0.7	006°	--	--	0.7	164°
2991	Guard Islands, 2 miles west of		55° 27'	131° 57'	−0 19	−0 27	−0 37	−0 06	0.4	0.7	--	--	1.5	010°	--	--	1.5	190°
2996	Ship Island		55° 36'	132° 15'	−0 14	−0 22	−0 32	−0 01	0.5	0.7	--	--	1.5	335°	--	--	1.5	155°
3001	Narrow Point		55° 47'	132° 25'	−0 09	−0 17	−0 27	+0 04	0.5	0.7	--	--	1.5	330°	--	--	1.5	150°
3006	Mabel Island, 3 miles west from		55° 55'	132° 30'	+0 01	−0 07	−0 17	+0 14	0.5	0.7	--	--	1.5	330°	--	--	1.5	150°
3011	Lincoln Rock Light, 1 mile west from		56° 03'	132° 43'	+0 11	+0 03	−0 07	+0 24	0.5	0.7	--	--	1.5	335°	--	--	1.5	155°

Endnotes can be found at the end of table 2.

TABLE 2 – CURRENT DIFFERENCES AND OTHER CONSTANTS

No.	PLACE	Meter Depth	POSITION		TIME DIFFERENCES				SPEED RATIOS		AVERAGE SPEEDS AND DIRECTIONS							
			Latitude	Longitude	Min. before Flood	Flood	Min. before Ebb	Ebb	Flood	Ebb	Minimum before Flood		Maximum Flood		Minimum before Ebb		Maximum Ebb	
		ft	North	West	h m	h m	h m	h m			knots	Dir.	knots	Dir.	knots	Dir.	knots	Dir.
	ERNEST SOUND Time meridian, 135° W				**on Wrangell Narrows, p.80**													
3016	McHenry Ledge, 1 mile north of		55° 48'	132° 18'	−0 09	−0 17	−0 27	+0 04	0.6	1.0	– –	– –	2.0	045°	– –	– –	2.0	225°
3021	Vixen Point, 3 miles west of		55° 51'	132° 11'	−0 04	−0 12	−0 22	+0 09	0.5	0.8	– –	– –	1.7	065°	0.0	–	– 1.7	245°
3026	Eaton Point		55° 57.17'	132° 05.40'	+0 01	−0 07	−0 17	+0 14	0.7	1.0	– –	– –	2.1	015°	– –	– –	2.1	195°
3031	Niblack Islands		56° 02'	132° 05'	+0 06	−0 02	−0 12	+0 19	0.7	1.0	– –	– –	2.1	005°	– –	– –	2.1	185°
3036	Blanche Rock		56° 05'	132° 05'	+0 11	+0 03	−0 07	+0 24	0.6	1.0	– –	– –	2.0	035°	– –	– –	2.0	215°
3041	Point Warde		56° 11'	131° 58'	+0 16	+0 08	−0 02	+0 29	0.6	1.0	– –	– –	2.0	045°	– –	– –	2.0	225°
	BLAKE CHANNEL and EASTERN PASSAGE																	
3046	Blake Island		56° 13.38'	131° 54.53'	+0 21	+0 13	+0 03	+0 34	0.8	1.2	– –	– –	2.5	330°	– –	– –	2.5	150°
3051	Berg Bay		56° 20.62'	132° 00.46'	+0 26	+0 18	+0 08	+0 39	0.7	1.1	– –	– –	2.2	335°	– –	– –	2.2	155°
3056	The Narrows		56° 22.07'	132° 06.16'	+0 31	+0 23	+0 13	+0 44	0.9	1.4	– –	– –	3.0	225°	– –	– –	3.0	045°
3061	Channel Island, north of		56° 22.51'	132° 10.18'	+0 31	+0 23	+0 13	+0 44	0.6	1.0	– –	– –	2.0	140°	– –	– –	2.0	320°
3066	Mill Creek		56° 27'	132° 13'	+0 31	+0 23	+0 13	+0 44	0.6	0.9	– –	– –	1.8	150°	– –	– –	1.8	330°
3071	Point Highfield		56° 30'	132° 23'	+0 06	−0 02	−0 12	+0 19	0.5	0.7	– –	– –	1.5	095°	– –	– –	1.5	275°
	ZIMOVIA STRAIT																	
3076	Found Island, 1 mile northwest of		56° 07'	132° 06'	+0 11	+0 03	−0 07	+0 24	0.5	0.8	– –	– –	1.7	320°	– –	– –	1.7	140°
3081	No Name Island, near Thoms Place		56° 09'	132° 09'	+0 16	+0 08	−0 02	+0 29	0.5	0.8	– –	– –	1.6	330°	– –	– –	1.6	150°
3086	Village Islands		56° 13'	132° 19'	+0 16	+0 08	−0 02	+0 29	0.5	0.7	– –	– –	1.5	315°	– –	– –	1.5	135°
3091	Young Rock, 2 miles south of		56° 20'	132° 23'	+0 21	+0 13	+0 03	+0 34	0.5	0.8	– –	– –	1.6	165°	– –	– –	1.6	345°
3096	East Point, east of		56° 23'	132° 24'	+0 06	−0 02	−0 12	+0 19	0.5	0.8	– –	– –	1.7	010°	– –	– –	1.7	190°
3101	Wrangell Harbor entrance		56° 28'	132° 24'	Current weak and variable													
3106	Wrangell Harbor, 1.6 miles west of		56° 28'	132° 27'	+2 28	+3 36	+2 16	+0 33	0.3	0.4	0.1	290°	0.8	050°	0.2	125°	0.8	235°
3111	Point Shekesti, 2.3 miles west of		56° 28'	132° 26'	Current weak and variable													
	CLARENCE STRAIT–cont.																	
3116	Key Reef		56° 10'	132° 50'	+0 21	+0 13	+0 03	+0 34	0.5	0.7	– –	– –	1.5	010°	– –	– –	1.5	190°
					on Sergius Narrows, p.84													
3121	Snow Passage, north entrance		56° 18'	133° 02'	+1 01	+0 02	+0 29	+1 07	0.2	0.6	– –	– –	1.4	122°	– –	– –	3.1	282°
3126	Kashevarof Passage, north entrance		56° 15'	133° 03'	+1 03	+0 49	+0 56	+0 50	0.2	0.4	– –	– –	1.5	155°	– –	– –	1.9	335°
3131	SNOW PASSAGE	25d	56° 17.48'	132° 59.54'	*Daily predictions, p.76*						0.1	029°	1.2	120°	– –	– –	2.2	294°
	STIKINE STRAIT				**on Wrangell Narrows, p.80**													
3136	Steamer Point, 1 mile west of		56° 13'	132° 44'	+0 21	+0 13	+0 03	+0 34	0.6	1.0	– –	– –	2.0	040°	– –	– –	2.0	220°
3141	Round Point		56° 17'	132° 37'	+0 21	+0 13	+0 03	+0 34	0.5	0.7	– –	– –	1.5	015°	– –	– –	1.5	195°
3146	South Craig Point		56° 23'	132° 36'	+0 26	+0 18	+0 08	+0 39	0.6	1.0	– –	– –	2.0	010°	– –	– –	2.0	190°
3151	Vank Island, off Neal Point		56° 26.55'	132° 35.51'	+0 26	+0 18	+0 08	+0 39	0.6	1.0	– –	– –	2.0	035°	– –	– –	2.0	215°
	CORDOVA BAY																	
3156	Cape Muzon, 5 miles east of		54° 40'	132° 32'	−0 39	−0 47	−0 57	−0 26	0.4	0.6	– –	– –	1.2	005°	– –	– –	1.2	185°
3161	Dewey Rocks, 2 miles west of		54° 45'	132° 32'	−0 34	−0 42	−0 52	−0 21	0.3	0.5	– –	– –	1.0	005°	– –	– –	1.0	185°
3166	Eureka Channel, off Leading Point		54° 49'	132° 23'	– – –	– – –	– – –	– – –	– –	– –	– –	– –	0.3	028	– –	– –	0.6	235°
3171	Boat Rocks, 2 miles west of		54° 49'	132° 34'	−0 34	−0 42	−0 52	−0 21	0.3	0.5	– –	– –	1.1	005°	– –	– –	1.1	185°
3176	Ship Islands, 2 miles southwest of	7d	54° 53'	132° 33'	– – –	– – –	– – –	– – –	– –	– –	– –	– –	0.4	358	– –	– –	0.2	177°
3181	Webster Point, 1 mile west of		54° 58'	132° 38'	−0 29	−0 37	−0 47	−0 16	0.5	0.8	– –	– –	1.7	005°	– –	– –	1.7	185°
3186	Mellen Rock, 1 mile east of		55° 02'	132° 39'	−0 24	−0 32	−0 42	−0 11	0.6	1.0	– –	– –	2.0	010°	– –	– –	2.0	190°

Endnotes can be found at the end of table 2.

TABLE 2 – CURRENT DIFFERENCES AND OTHER CONSTANTS

No.	PLACE	Meter Depth	POSITION Latitude	POSITION Longitude	TIME DIFFERENCES Min. before Flood	TIME DIFFERENCES Flood	TIME DIFFERENCES Min. before Ebb	TIME DIFFERENCES Ebb	SPEED RATIOS Flood	SPEED RATIOS Ebb	Minimum before Flood knots	Minimum before Flood Dir.	Maximum Flood knots	Maximum Flood Dir.	Minimum before Ebb knots	Minimum before Ebb Dir.	Maximum Ebb knots	Maximum Ebb Dir.
		ft	North	West	h m	h m	h m	h m			knots	Dir.	knots	Dir.	knots	Dir.	knots	Dir.
	TLEVAK STRAIT Time meridian, 135° W				on Wrangell Narrows, p.80													
3191	Shoe Rock, 1 mile north of		54° 58'	132° 44'	−0 29	−0 37	−0 47	−0 16	0.5	0.8	--	--	1.7	285°	--	--	1.7	105°
3196	Grand Island, 2 miles north of		55° 00'	132° 52'	−0 24	−0 32	−0 42	−0 11	0.5	0.7	--	--	1.5	295°	--	--	1.5	115°
3201	High Point, 1 mile east of		55° 01'	132° 56'	−0 24	−0 32	−0 42	−0 11	0.6	0.9	--	--	1.8	340°	--	--	1.8	160°
3206	McFarland Islands		55° 04'	132° 57'	−0 19	−0 27	−0 37	−0 06	0.5	0.8	--	--	1.7	340°	--	--	1.7	160°
3211	Corlies Islands		55° 08'	132° 58'	−0 19	−0 27	−0 37	−0 06	0.5	0.7	--	--	1.5	345°	--	--	1.5	165°
3216	Sukkwan Narrows		55° 11.90'	132° 49.44'	−0 29	−0 45	−0 20	−0 09	0.4	0.6	--	--	1.4	323°	--	--	1.2	130°
3221	The Sentinels, 1 mile west of		55° 11'	133° 01'	−0 19	−0 27	−0 37	−0 06	0.5	0.8	--	--	1.6	335°	--	--	1.6	155°
					on Sergius Narrows, p.84													
3226	Halibut Nose		55° 13'	133° 04'	+0 08	−0 06	+0 01	−0 05	0.2	0.3	--	--	1.5	170°	--	--	1.5	350°
3231	Lively Islands, west of <59>		55° 13.99'	133° 05.54'	−0 07	−0 21	−0 14	−0 20	0.5	0.7	--	--	3.2	175°	--	--	3.2	355°
3236	Tlevak Narrows, Turn Point, east of		55° 15.87'	133° 07.33'	−0 15	−1 06	−0 23	−0 50	0.5	0.9	--	--	3.0	129°	--	--	4.3	288°
	MEARES PASSAGE				on Wrangell Narrows, p.80													
3241	Meares Island, south of		55° 15.47'	133° 10.97'	−0 19	−0 27	−0 37	−0 06	0.7	1.0	--	--	2.1	090°	--	--	2.1	270°
3246	Eagle Point		55° 14'	133° 14'	−0 24	−0 32	−0 42	−0 11	0.6	0.9	--	--	1.8	020°	--	--	1.8	200°
3251	Millar Rocks		55° 12'	133° 15'	−0 29	−0 37	−0 47	−0 16	0.4	0.6	--	--	1.2	030°	--	--	1.2	210°
3256	Diver Islands		55° 11'	133° 17'	−0 29	−0 37	−0 47	−0 16	0.3	0.5	--	--	1.0	025°	--	--	1.0	205°
	ULLOA CHANNEL																	
3261	Cape Flores		55° 21'	133° 19'	−0 19	−0 27	−0 37	−0 06	0.7	1.1	--	--	2.2	150°	--	--	2.2	330°
3266	Point Verde		55° 18.45'	133° 16.69'	−0 19	−0 27	−0 37	−0 06	0.6	0.9	--	--	1.8	160°	--	--	1.8	340°
3271	Waterfall Cannery		55° 18'	133° 15'	−0 19	−0 27	−0 37	−0 06	0.6	1.0	0.0	--	2.0	155°	--	--	2.0	335°
	BUCARELLI BAY to DAVIDSON INLET																	
3276	Cape Bartolome, 2 miles east of		55° 14'	133° 33'	−0 29	−0 37	−0 47	−0 16	0.4	0.6	--	--	1.2	020°	--	--	1.2	200°
3281	Point Rosary, 2 miles west of		55° 16'	133° 31'	−0 29	−0 37	−0 47	−0 16	0.4	0.6	--	--	1.3	015°	--	--	1.3	195°
3286	Point Arboleda, 1 mile west of		55° 19'	133° 29'	−0 24	−0 32	−0 42	−0 11	0.4	0.7	--	--	1.4	020°	--	--	1.4	200°
3291	Cabras Islands, 1 mile west of		55° 21'	133° 25'	−0 24	−0 32	−0 42	−0 11	0.4	0.6	--	--	1.2	045°	--	--	1.2	225°
3296	Cape Flores, 1 mile north of		55° 22'	133° 18'	−0 19	−0 27	−0 37	−0 06	0.5	0.7	--	--	1.5	055°	--	--	1.5	235°
3301	San Juanito Islands, 1 mile south of		55° 23'	133° 15'	−0 19	−0 27	−0 37	−0 06	0.4	0.7	--	--	1.4	060°	--	--	1.4	240°
3306	Toti Island, 1.1 miles west of		55° 25'	133° 09'	−0 19	−0 27	−0 37	−0 06	0.3	0.5	--	--	1.1	070°	--	--	1.1	250°
3311	Coronados Islands, 2 miles west of		55° 26'	133° 10'	−0 14	−0 22	−0 32	−0 01	0.3	0.5	--	--	1.0	025°	--	--	1.0	205°
3316	Craig Cannery		55° 29'	133° 09'	−0 14	−0 22	−0 32	−0 01	0.3	0.5	--	--	1.0	010°	--	--	1.0	190°
					on Sergius Narrows, p.84													
3321	Tonowek Narrows		55° 45.42'	133° 20.35'	+0 58	+0 44	+0 51	+0 45	0.5	0.6	--	--	3.0	030°	--	--	3.0	210°
					on Wrangell Narrows, p.80													
3326	Karheen Passage, 0.5 mile NW of Cob I		55° 48.36'	133° 19.19'	−4 23	−4 54	−5 12	−4 34	0.1	0.2	--	--	0.4	308°	--	--	0.5	140°
3331	Tuxekan Passage, south entrance		55° 46'	133° 15'	---	−0 30	---	−0 40	0.1	0.2	--	--	0.4	060°	--	--	0.4	225°
3336	Tuxekan Passage, 0.2 mile S of Tuxekan		55° 52.96'	133° 14.54'	−3 34	−3 32	−1 42	−3 05	0.3	0.2	--	--	0.8	323°	--	--	0.4	137°
	EL CAPITAN PASSAGE																	
3341	Skookumchuck Pass		55° 54.88'	133° 18.74'	−0 06	−0 19	−0 11	+0 03	0.7	1.0	--	--	2.2	025°	--	--	2.1	207°
3346	Brockman Island, east of		55° 58.01'	133° 18.29'	+0 28	+0 43	−0 13	+0 31	0.1	0.2	--	--	0.3	351°	--	--	0.5	182°
3351	Off Tonga Inlet		55° 58.35'	133° 15.96'	Current weak and variable													
3356	Tenass Island, 0.3 mile SSW of		55° 59.20'	133° 18.29'	+0 06	−0 48	−1 18	+0 05	0.2	0.3	--	--	0.5	039°	--	--	0.6	220°

Endnotes can be found at the end of table 2.

TABLE 2 – CURRENT DIFFERENCES AND OTHER CONSTANTS

No.	PLACE	Meter Depth	POSITION Latitude	POSITION Longitude	TIME DIFFERENCES Min. before Flood	TIME DIFFERENCES Flood	TIME DIFFERENCES Min. before Ebb	TIME DIFFERENCES Ebb	SPEED RATIOS Flood	SPEED RATIOS Ebb	Minimum before Flood knots	Minimum before Flood Dir.	Maximum Flood knots	Maximum Flood Dir.	Minimum before Ebb knots	Minimum before Ebb Dir.	Maximum Ebb knots	Maximum Ebb Dir.
		ft	North	West	h m	h m	h m	h m			knots	Dir.	knots	Dir.	knots	Dir.	knots	Dir.
	EL CAPITAN PASSAGE–cont. Time meridian, 135° W				on Wrangell Narrows, p.80													
	Aneskett Point																	
3361	1.2 miles south of		56° 08'	133° 17'	Current weak and variable													
3366	1 mile WNW of		56° 09.21'	133° 18.04'	Current weak and variable													
3371	The Narrows, west of		56° 09.61'	133° 20.28'	+4 57	+5 37	+5 54	+5 37	0.2	0.1	– –	– –	0.5	065°	– –	– –	0.3	248°
	SUMNER STRAIT																	
3376	Warren Channel		55° 55'	133° 50'	−0 53	−1 05	−0 58	−0 28	0.4	1.2	– –	– –	1.4	345°	– –	– –	2.4	165°
3381	Cora Point, 2 miles east of		55° 55'	134° 03'	−0 34	−0 42	−0 52	−0 21	0.1	1.0	– –	– –	0.2	020°	– –	– –	2.0	200°
3386	Coronation I.–Spanish Is., between		55° 55.57'	134° 07.96'	−0 55	−1 12	−1 53	−1 36	0.4	0.7	– –	– –	1.4	276°	– –	– –	1.4	096°
3391	Decision Passage		55° 59.51'	134° 07.38'	+0 08	−0 24	−0 24	−0 08	0.3	0.6	– –	– –	0.9	055°	– –	– –	1.2	245°
3396	Point St. Albans, 3 miles SSE of		56° 02'	133° 57'	−0 05	−1 14	−1 05	+0 28	0.2	0.4	– –	– –	0.7	078°	– –	– –	0.9	232°
3401	Point St. Albans, 4 miles east of		56° 05'	133° 51'	−0 29	−0 37	−0 46	−0 16	0.1	1.0	– –	– –	0.2	025°	– –	– –	2.0	205°
3406	Ruins Point, 2 miles west of		56° 04'	133° 45'	+0 15	+0 38	−0 48	−0 49	0.2	0.3	– –	– –	0.5	350°	– –	– –	0.6	175°
3411	Shaken Bay Light, 3 miles west of		56° 09'	133° 42'	−0 29	−0 37	−0 47	−0 16	0.1	1.0	– –	– –	0.2	000°	– –	– –	2.0	180°
3416	Shaken Bay entrance		56° 09.83'	133° 37.90'	Current weak and variable													
3421	Shaken Bay Light, 2.8 miles WNW of		56° 10'	133° 42'	−1 21	−1 48	−1 34	−0 36	0.2	0.3	– –	– –	0.6	019°	– –	– –	0.6	182°
3426	Amelius Island, 1 mile east of		56° 10.70'	133° 50.85'	−1 05	−1 54	−2 24	−1 28	0.3	0.6	– –	– –	0.8	044°	– –	– –	1.3	226°
3431	Beauclerc Island Light, 1 mile east of		56° 15'	133° 49'	+0 09	−0 41	−1 30	−1 28	0.1	0.4	– –	– –	0.4	070°	– –	– –	0.9	205°
3436	Labouchere Island, 2.2 miles W of <61>		56° 17'	133° 44'	– – –	– – –	– – –	−0 16	– –	0.4	– –	– –	– –	– –	– –	– –	0.8	160
3441	Sumner Island, 1.8 miles south of		56° 22'	133° 49'	−2 39	−3 33	−0 54	−1 16	0.3	0.3	– –	– –	0.8	010°	– –	– –	0.6	240°
3446	Strait Island, 1 mile southeast of <62>		56° 23'	133° 39'	– – –	−0 43	– – –	−0 51	0.2	1.4	– –	– –	0.5	095	– –	– –	3.0	250°
3451	Vichnefski Rock Light		56° 26'	133° 01'	+0 01	−0 07	−0 17	+0 14	0.2	1.4	– –	– –	0.5	040°	– –	– –	3.0	220°
3456	Station Island, 1 mile south of <62>		56° 28'	132° 46'	– – –	– – –	– – –	+1 22	– –	1.0	– –	– –	0.1	090	– –	– –	2.0	251°
3461	Blaquiere Point, 1 mile SSW of <63>		56° 33'	132° 34'	−0 08	+0 23	−0 20	+0 23	0.3	0.5	– –	– –	1.1	045°	– –	– –	1.0	240°
	KEKU STRAIT																	
3466	Conclusion Island, west of		56° 28'	133° 50'	Current weak and variable													
3471	Conclusion Island, east of		56° 29'	133° 46'	Current weak and variable													
3476	Skiff Island, 2.1 miles northwest of		56° 32'	133° 44'	Current weak and variable													
3481	Southern entrance to Keku Strait		56° 34'	133° 43'	+0 57	+0 13	−0 25	−0 32	0.3	0.6	– –	– –	0.9	005°	– –	– –	1.2	184°
3486	Eagle Island, 0.5 mile southeast of		56° 36.44'	133° 40.93'	−0 04	−0 24	−0 49	+0 43	0.5	1.0	– –	– –	1.6	000°	– –	– –	2.0	130°
3491	Devils Elbow		56° 38.17'	133° 41.37'	−0 18	−0 09	−0 41	+0 13	0.8	0.8	– –	– –	2.4	285°	– –	– –	1.8	090°
3496	Summit Island, west of		56° 40.59'	133° 43.95'	+1 48	+1 43	+2 10	+1 46	0.7	1.2	– –	– –	2.2	185°	– –	– –	2.6	010°
3501	High Island, 1.1 miles south of		56° 42.14'	133° 43.98'	+0 40	+0 40	+0 28	+0 30	0.3	0.5	– –	– –	1.1	150°	– –	– –	1.1	010°
3506	High Island, northwest of		56° 45'	133° 44'	+0 07	+0 49	+0 34	+0 18	0.3	0.4	– –	– –	0.9	210°	– –	– –	0.7	325°
3511	Cucumber Reef, 0.2 mile northwest of		56° 47.40'	133° 46.02'	+0 11	+0 16	+0 04	+0 11	0.4	0.8	– –	– –	1.4	110°	– –	– –	1.9	325°
	WRANGELL NARROWS																	
3516	Point Alexander		56° 30.62'	132° 57.50'	+0 06	−0 02	−0 12	+0 19	0.3	0.5	– –	– –	1.0	005°	– –	– –	1.0	185°
3521	Point Deception		56° 32'	132° 58'	+0 06	−0 02	−0 12	+0 19	0.3	0.5	– –	– –	1.0	000°	– –	– –	1.0	180°
3526	Point Lockwood		56° 33.35'	132° 57.71'	+0 06	−0 02	−0 12	+0 19	0.9	1.4	– –	– –	3.0	000°	– –	– –	3.0	180°
3531	Spike Rock		56° 36.06'	132° 58.56'	+0 06	−0 02	−0 12	+0 19	1.5	2.1	– –	– –	4.7	005°	– –	– –	4.3	185°
3536	South Ledge		56° 37'	132° 58'	+0 06	−0 02	−0 12	+0 19	1.5	2.1	– –	– –	4.7	040°	– –	– –	4.3	220°
3541	Anchor Point		56° 38.37'	132° 55.87'	+0 06	−0 02	−0 12	+0 19	1.1	1.6	– –	– –	3.6	045°	– –	– –	3.4	225°
3546	Vexation Point, Woody Island		56° 39.47'	132° 55.62'	+0 06	−0 02	−0 12	+0 19	0.8	1.2	– –	– –	2.5	005°	– –	– –	2.5	185°
3551	Rock Point		56° 40.53'	132° 56.35'	+0 06	−0 02	−0 12	+0 19	0.3	0.5	– –	– –	1.0	335°	– –	– –	1.0	155°
3556	Green Point		56° 42'	132° 57'	+0 11	+0 03	−0 07	+0 24	0.3	0.5	– –	– –	1.0	185°	– –	– –	1.0	005°
3561	Mountain Point		56° 44'	132° 57'	+0 21	+0 13	+0 03	+0 34	0.6	1.0	– –	– –	2.0	165°	– –	– –	2.0	345°
3566	Blunt Point		56° 46.70'	132° 58.76'	+0 31	+0 23	+0 13	+0 44	1.1	1.6	– –	– –	3.6	160°	– –	– –	3.4	340°
3571	Turn Point		56° 48.47'	132° 59.01'	+0 31	+0 23	+0 13	+0 44	1.4	1.8	– –	– –	4.3	220°	– –	– –	3.8	040°
3576	WRANGELL NARROWS (off Petersburg)		56° 48.98'	132° 57.84'	*Daily predictions*						– –	– –	3.2	246°	0.1	334°	2.1	062°
3581	Prolewy Rocks		56° 49.34'	132° 56.90'	+0 31	+0 23	+0 13	+0 44	1.1	1.6	– –	– –	3.6	240°	– –	– –	3.4	060°

Endnotes can be found at the end of table 2.

TABLE 2 – CURRENT DIFFERENCES AND OTHER CONSTANTS

No.	PLACE	Meter Depth	POSITION Latitude	POSITION Longitude	TIME DIFFERENCES Min. before Flood	TIME DIFFERENCES Flood	TIME DIFFERENCES Min. before Ebb	TIME DIFFERENCES Ebb	SPEED RATIOS Flood	SPEED RATIOS Ebb	Minimum before Flood knots	Minimum before Flood Dir.	Maximum Flood knots	Maximum Flood Dir.	Minimum before Ebb knots	Minimum before Ebb Dir.	Maximum Ebb knots	Maximum Ebb Dir.
		ft	North	West	h m	h m	h m	h m										
	FREDERICK SOUND <64> Time meridian, 135° W				on Wrangell Narrows, p.80													
3586	Cosmos Point, 0.5 mile east of		56° 40'	132° 36'	+1 55	+1 15	+0 25	+0 42	0.1	0.2	--	--	0.4	180°	--	--	0.5	305°
	STEPHENS PASSAGE																	
3591	The Five Fingers, 1 mile east of		57° 16'	133° 36'	+0 11	+0 03	−0 07	+0 24	0.3	0.5	--	--	1.0	020°	--	--	1.0	200°
3596	The Brothers, 2 miles east of		57° 18'	133° 43'	+0 11	+0 03	−0 07	+0 24	0.3	0.5	--	--	1.0	025°	--	--	1.0	205°
3601	Point Gambier, 2 miles east of		57° 26'	133° 46'	+0 16	+0 08	−0 02	+0 29	0.2	0.2	--	--	0.5	005°	--	--	0.5	185°
3606	Point Hugh		57° 37'	133° 46'	+0 21	+0 13	+0 03	+0 34	0.3	0.5	--	--	1.0	355°	--	--	1.0	175°
3611	Tracy Arm	32d	57° 46.78'	133° 46.78'	+0 26	−0 17	−0 12	+0 32	0.3	1.2	0.2	320°	1.0	060°	0.4	--	2.4	226°
3616	Midway Island		57° 50'	133° 50'	+0 21	+0 13	+0 03	+0 34	0.3	0.5	--	--	1.0	335°	--	--	1.0	155°
3621	Grave Point, Taku Harbor		58° 04'	134° 03'	+0 26	+0 18	+0 08	+0 39	0.6	1.0	--	--	2.0	350°	--	--	2.0	170°
3626	Point Arden		58° 09'	134° 08'	+0 26	+0 18	+0 08	+0 39	0.3	0.5	--	--	1.0	355°	--	--	1.0	175°
3631	Taku Inlet, SE of Cooper Point	25d	58° 14.20'	134° 04.58'	Current weak and variable													
	. . . do.	70d	58° 14.20'	134° 04.58'	Current weak and variable													
3636	Taku Inlet, 0.2 mile off Flat Point		58° 20'	134° 03'	+0 48	+0 11	−0 12	+0 07	0.2	0.5	--	--	0.7	039°	--	--	1.0	200°
3641	Taku Inlet, 0.2 mile off Taku Point		58° 24'	134° 01'	+0 53	+0 34	+0 10	−0 08	0.4	0.4	--	--	1.2	357°	--	--	0.9	203°
	Gastineau Channel																	
3646	Point Salisbury, west of	25d	58° 12.45'	134° 14.50'	Current weak and variable													
	. . . do.	70d	58° 12.45'	134° 14.50'	Current weak and variable													
3651	Douglas, north of	15d	58° 16.98'	134° 23.62'	+2 12	+0 44	+0 11	+1 32	0.1	0.2	--	--	0.3	305°	--	--	0.5	144°
	. . . do.	25d	58° 16.98'	134° 23.62'	+1 33	+0 41	+0 27	+1 08	0.1	0.3	--	--	0.4	302°	--	--	0.5	136°
3656	Juneau Hbr. Ferry Pier, SW of	15d	58° 17.75'	134° 24.13'	−2 23	−1 53	−0 19	−0 12	0.1	--	--	--	0.3	302	--	--	0.1	085°
	. . . do.	25d	58° 17.75'	134° 24.13'	−2 39	−3 09	−0 27	−1 23	0.1	--	--	--	0.3	303	--	--	0.1	120°
3661	Juneau Harbor	15d	58° 17.62'	134° 24.40'	Current weak and variable													
	. . . do.	25d	58° 17.62'	134° 24.40'	Current weak and variable													
3666	Union Oil Pier, 0.3 nmi. WNW of	15d	58° 17.47'	134° 24.42'	+3 41	+2 05	+0 30	+0 38	--	0.3	--	--	0.1	306°	--	--	0.6	094°
	. . . do.	25d	58° 17.47'	134° 24.42'	+3 19	+2 03	+0 34	+1 05	0.1	0.3	--	--	0.2	334°	0.1	031°	0.6	102°
3671	West Juneau, NE of	25d	58° 17.78'	134° 25.47'	+0 24	+0 49	+0 32	+1 21	0.1	0.2	--	--	0.3	314°	--	--	0.4	137°
3676	Aurora Basin, SW of	15d	58° 18.30'	134° 26.45'	+1 01	+0 34	+0 10	+0 51	0.2	0.4	0.1	218°	0.5	289°	0.2	233°	0.8	163°
3681	Horse Island, east of	25d	58° 15.38'	134° 04.58'	Current weak and variable													
	. . . do.	70d	58° 15.38'	134° 04.58'	Current weak and variable													
3686	Portland Island, SW of	25d	58° 19.17'	134° 42.63'	Current weak and variable													
	. . . do.	70d	58° 19.17'	134° 42.63'	Current weak and variable													
					on North Inian Pass, p.88													
3691	Saginaw Channel, 2 mi. E of Pt.Retreat	25d	58° 24.30'	134° 53.10'	−0 56	−0 53	+1 12	+0 20	0.3	0.1	0.1	066°	0.8	155°	--	--	0.7	340°
	. . . do.	70d	58° 24.30'	134° 53.10'	−0 56	−0 58	+0 59	+0 31	0.3	0.1	--	--	0.8	149°	--	--	0.6	338°
3696	Sentinel Island, south of	25d	58° 32.17'	134° 56.03'	Current weak and variable													
	. . . do.	70d	58° 32.17'	134° 56.03'	Current weak and variable													
	LYNN CANAL																	
3701	Clear Point, WNW of	25d	58° 14.93'	134° 57.77'	Current weak and variable													
3706	Point Retreat, 1 mile west of		58° 25'	134° 58'	+0 30	+0 30	+0 30	+0 30	0.1	0.1	--	--	0.3	345°	--	--	0.7	165°
3711	Vanderbilt Reef, 2 miles west of <62>		58° 35'	135° 04'	+2 09	+2 01	+1 49	+1 34	0.1	0.2	--	--	0.4	340°	--	--	1.0	151°
3716	Point Bridget, NW of	70d	58° 41.95'	135° 02.00'	Current weak and variable													
3721	Point Sherman, WSW of	70d	58° 50.80'	135° 11.80'	−0 58	−0 22	+1 12	+0 25	0.1	--	--	--	0.2	341°	--	--	0.2	175°
3726	Eldred Rock		58° 58'	135° 14'	+2 00	+2 00	+2 00	+2 00	0.1	0.1	--	--	0.3	350°	--	--	0.7	170°
3731	Glacier Point, Chilkat Inlet	25d	59° 06.28'	135° 22.35'	Current weak and variable													
	. . . do.	70d	59° 06.28'	135° 22.35'	−1 22	−0 24	+0 56	0 00	0.1	0.1	--	--	0.3	325°	--	--	0.4	145°
3736	Seduction Pt., NE of, Chilkoot Inlet	25d	59° 06.40'	135° 14.60'	Current weak and variable													
	. . . do.	70d	59° 06.40'	135° 14.60'	Current weak and variable													
3741	Battery Point, Chilkoot Inlet		59° 13'	135° 21'	+2 05	+2 05	+2 05	+2 05	0.1	0.1	--	--	0.3	345°	--	--	0.7	165°

Endnotes can be found at the end of table 2.

TABLE 2 – CURRENT DIFFERENCES AND OTHER CONSTANTS

No.	PLACE	Meter Depth	POSITION Latitude	POSITION Longitude	TIME DIFFERENCES Min. before Flood	Flood	Min. before Ebb	Ebb	SPEED RATIOS Flood	Ebb	AVERAGE SPEEDS AND DIRECTIONS Minimum before Flood knots	Dir.	Maximum Flood knots	Dir.	Minimum before Ebb knots	Dir.	Maximum Ebb knots	Dir.
		ft	North	West	h m	h m	h m	h m			knots	Dir.	knots	Dir.	knots	Dir.	knots	Dir.
	LYNN CANAL–cont. Time meridian, 135° W				on North Inian Pass, p.88													
3746	Low Point, entrance to Taiya Inlet		59° 16'	135° 22'	+2 05	+2 05	+2 05	+2 05	0.1	0.1	--	--	0.3	005°	--	--	0.7	185°
3751	Tanani Point, Lutak Inlet	70d	56° 16.92'	135° 26.98'	Current weak and variable													
3756	Skagway, Taiya Inlet		59° 27'	135° 20'	+2 10	+2 10	+2 10	+2 10	0.1	0.1	--	--	0.3	015°	--	--	0.7	195°
3761	Skagway River entrance	25d	59° 26.71'	135° 19.72'	Current weak and variable													
	CHATHAM STRAIT				on Wrangell Narrows, p.80													
3766	Hazy Islands		55° 53'	134° 36'	−0 19	−0 27	−0 37	−0 06	0.3	0.5	--	--	1.0	025°	--	--	1.0	205°
3771	Cape Ommaney Light 5 miles east of		56° 10'	134° 31'	−0 14	−0 22	−0 32	−0 01	0.3	0.5	--	--	1.0	005°	--	--	1.0	185°
3776	Fort Walter Light, 3 miles east of		56° 23'	134° 32'	−0 09	−0 17	−0 27	+0 04	0.5	0.7	--	--	1.5	005°	--	--	1.5	185°
3781	Point Ellis, 4 miles west of		56° 34'	134° 27'	−0 04	−0 12	−0 22	+0 09	0.5	0.7	--	--	1.5	350°	--	--	1.5	170°
3786	Kingsmill Point Light, 3 miles west of		56° 50'	134° 31'	+0 01	−0 07	−0 17	+0 14	0.6	1.0	--	--	2.0	355°	--	--	2.0	175°
3791	Point Gardner Light, 2 miles west of		57° 01'	134° 40'	+0 06	−0 02	−0 12	+0 19	0.6	1.0	--	--	2.0	350°	--	--	2.0	170°
3796	Point Caution, 3 miles west of		57° 15'	134° 44'	+0 11	+0 03	−0 07	+0 24	0.6	1.0	--	--	2.0	355°	--	--	2.0	175°
3801	Point Thatcher, 3 miles east of		57° 25'	134° 44'	+0 11	+0 03	−0 07	+0 24	0.5	0.7	--	--	1.5	340°	--	--	1.5	160°
3806	Danger Point Light, 3 miles west of		57° 31'	134° 42'	+0 16	+0 08	−0 02	+0 29	0.5	0.7	--	--	1.5	350°	--	--	1.5	170°
3811	Turn Point, Kootznahoo Inlet		57° 30'	134° 35'	+0 56	+0 48	+0 38	+1 09	2.2	2.9	--	--	6.9	105°	--	--	6.1	285°
3816	South Passage Point, 3 miles east of		57° 46'	134° 50'	+0 16	+0 08	−0 02	+0 29	0.5	0.7	--	--	1.5	175°	--	--	1.5	355°
3821	Point Augusta, ESE of	70d	58° 01.50'	134° 52.40'	−0 02	−0 04	+0 01	+0 54	0.2	0.3	--	--	0.5	344°	--	--	0.6	164°
	SITKA SOUND																	
3826	Biorka Channel		56° 50'	135° 30'	+0 47	+0 20	+0 15	+0 13	0.1	0.2	--	--	0.4	045°	--	--	0.4	232°
3831	Entrance		56° 58'	135° 37'	Current weak and variable						--	--	0.2	057°	--	--	0.2	247°
3836	Vitskari Island, 3 miles northeast of		57° 02'	135° 29'	Current weak and variable						--	--	0.2	107°	--	--	0.1	306°
3841	Eastern Channel		57° 01'	135° 22'	Current weak and variable						--	--	0.1	107°	--	--	0.1	261°
3846	Silver Bay entrance		57° 02'	135° 14'	Current weak and variable						--	--	0.1	055°	--	--	0.1	240°
3851	Sitka Harbor, channel off Harbor Island		57° 03'	135° 20'	−0 58	−1 17	−2 02	−1 16	0.1	0.2	--	--	0.3	333°	--	--	0.4	156°
	KRESTOF SOUND																	
3856	West Channel (narrows)		57° 09'	135° 35'	−0 43	−0 51	−1 01	−0 30	0.4	0.5	--	--	1.3	242°	--	--	1.1	064°
3861	East Channel (narrows)		57° 10'	135° 33'	−0 30	−0 32	−0 48	−0 23	0.4	0.7	--	--	1.3	051°	--	--	1.4	229°
3866	East Channel north entrance <67>		57° 11'	135° 33'	---	−0 50	---	−0 48	0.3	0.4	--	--	0.8	340	--	--	0.8	160°
	NAKWASINA SOUND AND PASSAGE																	
3871	Nakwasina Sound, South entrance		57° 11'	135° 25'	Current weak and variable													
3876	Allan Point, 1.5 miles W of, Nakwasina Passage <66		57° 15'	135° 26'	−1 06	−1 06	−0 59	−0 40	0.6	0.8	--	--	2.0	260°	--	--	1.6	082°
	OLGA STRAIT																	
3881	South end <67>		57° 11'	135° 28'	---	+0 09	---	+0 52	0.3	0.5	--	--	1.0	324°	--	--	1.0	130°
3886	Creek Point, 0.44 n.mi. SE of	15d	57° 12.61'	135° 29.70'	+0 12	−0 24	+0 02	+0 27	0.4	0.6	--	--	1.3	319°	0.1	229°	1.2	141°
	NEVA STRAIT																	
3891	Whitestone Narrows, S of Whitestone Point	16d	57° 14.7'	135° 33.83'	−0 24	−0 30	−0 07	−0 04	0.3	0.4	--	--	1.0	161°	--	--	0.8	338°
	... do.	43d	57° 14.7'	135° 33.83'	−0 25	−0 30	−0 03	+0 06	0.3	0.3	0.1	250°	1.0	163°	--	--	0.7	339°
3896	Wyvill Reef		57° 16'	135° 35'	−0 27	−0 30	−0 04	−0 13	0.5	0.7	--	--	1.6	150°	--	--	1.4	323°
3901	Highwater Island, west of		57° 17'	135° 36'	−0 15	−0 29	−0 11	−0 34	0.3	0.7	--	--	1.0	144°	--	--	1.4	330°
3906	Zeal Point, 0.34 n.mi. SSW of	16d	57° 17.22'	135° 36.47'	+0 09	−0 14	−0 02	+0 02	0.2	0.3	--	--	0.5	163°	--	--	0.6	346°
	... do.	29d	57° 17.22'	135° 36.47'	−0 12	−0 27	−0 07	−0 02	0.2	0.3	--	--	0.5	165°	--	--	0.6	347°
	... do.	48d	57° 17.22'	135° 36.47'	−0 57	−1 02	−0 26	−0 14	0.1	0.2	--	--	0.4	170°	--	--	0.5	345°
3911	Kane Island, 0.29 n.mi. East of		57° 19.33'	135° 39.21'	Current weak and Variable													
3916	North of Kane Islands		57° 20'	135° 40'	Current weak and variable													

Endnotes can be found at the end of table 2.

TABLE 2 – CURRENT DIFFERENCES AND OTHER CONSTANTS

No.	PLACE	Meter Depth	POSITION		TIME DIFFERENCES				SPEED RATIOS		AVERAGE SPEEDS AND DIRECTIONS							
			Latitude	Longitude	Min. before Flood	Flood	Min. before Ebb	Ebb	Flood	Ebb	Minimum before Flood		Maximum Flood		Minimum before Ebb		Maximum Ebb	
		ft	North	West	h m	h m	h m	h m			knots	Dir.	knots	Dir.	knots	Dir.	knots	Dir.
	SALISBURY SOUND Time meridian, 135° W				on Wrangell Narrows, p.80													
3921	Sea Rock, 1 mile north of		57° 21'	135° 53'	−0 19	−0 27	−0 37	−0 06	0.3	0.5	--	--	1.0	065°	--	--	1.0	245°
3926	Kalinin Point, 1 mile north of		57° 21'	135° 48'	−0 19	−0 27	−0 37	−0 06	0.3	0.5	--	--	1.0	080°	--	--	1.0	260°
3931	Sinitsin Island		57° 21'	135° 46'	−0 19	−0 27	−0 37	−0 06	0.5	0.7	--	--	1.5	095°	--	--	1.5	275°
	PERIL STRAIT				on Sergius Narrows, p.84													
3936	Kakul Narrows		57° 22'	135° 41'	+0 33	+0 37	+0 21	+0 24	0.3	0.5	--	--	1.9	046°	--	--	2.4	235°
3941	Suloia Point, 0.32 n.mi. ENE of <122>	26d	57° 23.51'	135° 38.46'	+1 47	+0 43	−0 31	−0 14	0.1	0.2	0.3	035°	0.3	346°	0.1	052°	1.0	113°
	. . . do.	65d	57° 23.51'	135° 38.46'	+1 40	+0 11	−0 15	−0 01	0.1	0.2	0.4	044°	0.5	004°	0.1	057°	1.0	114°
	. . . do.	92d	57° 23.51'	135° 38.46'	+1 33	+0 13	−0 02	+0 58	0.1	0.2	0.4	045°	0.5	004°	--	--	1.0	113°
3946	SERGIUS NARROWS	18d	57° 24.42'	135° 37.87'	*Daily predictions*						--	--	6.3	059°	0.1	154°	4.9	241°
	. . . do.	31d	57° 24.42'	135° 37.87'	0 00	0 00	0 00	0 00	1.0	1.0	0.1	331°	6.3	058°	--	--	4.9	241°
	. . . do.	44d	57° 24.42'	135° 37.87'	0 00	−0 01	0 00	0 00	1.0	1.0	0.1	331°	6.3	058°	--	--	4.9	242°
3951	Point Siroi		57° 25'	135° 35'	+0 31	+0 11	+0 23	+0 15	0.3	0.4	--	--	1.7	059°	--	--	1.9	283°
3956	Middle Point		57° 26'	135° 35'	−0 09	−0 37	−0 35	−0 06	0.2	0.4	--	--	1.4	010°	--	--	2.1	187°
3961	Big Rose Island, 0.2 n.mi. SE of	12d	57° 27.18'	135° 32.24'	+0 02	+0 11	−0 02	−0 24	0.3	0.4	0.1	308°	1.9	042°	0.2	126°	2.2	212°
	. . . do.	32d	57° 27.18'	135° 32.24'	−0 01	−0 11	−0 01	−0 20	0.4	0.4	0.1	307°	2.3	041°	0.2	128°	2.2	212°
	. . . do.	92d	57° 27.18'	135° 32.24'	+0 01	−0 19	−0 01	−0 17	0.3	0.4	0.1	303°	1.9	030°	0.1	129°	1.8	221°
3966	Povorotni Island, 0.23 n.mi. WSW of	10d	58° 30.63'	135° 33.70'	−0 17	+0 15	+0 09	−0 37	0.2	0.2	--	--	0.9	323°	--	--	1.1	159°
	. . . do.	50d	58° 30.63'	135° 33.70'	−0 15	+0 05	+0 00	−0 39	0.2	0.2	--	--	0.9	325°	--	--	1.2	158°
	. . . do.	89d	58° 30.63'	135° 33.70'	+0 06	−0 15	−0 32	−0 30	0.1	0.3	--	--	0.8	328°	--	--	1.3	156°
					on Wrangell Narrows, p.80													
3971	Otstoia Island Light, 1 mile north of		57° 35'	135° 27'	+0 06	−0 02	−0 12	+0 19	0.6	1.0	--	--	2.0	280°	--	--	2.0	100°
3976	Nismeni Point, 1 mile north of		57° 35'	135° 25'	+0 06	−0 02	−0 12	+0 19	0.6	1.0	--	--	2.0	285°	--	--	2.0	105°
3981	Peschani Point, 1 mile east of		57° 32'	135° 18'	+0 06	−0 02	−0 12	+0 19	0.6	1.0	--	--	2.0	325°	--	--	2.0	145°
3986	Point Elizabeth, 1 mile northeast of		57° 31'	135° 16'	+0 06	−0 02	−0 12	+0 19	0.6	1.0	--	--	2.0	315°	--	--	2.0	135°
3991	Point Benham, 1 mile east of		57° 29'	135° 11'	+0 06	−0 02	−0 12	+0 19	0.6	1.0	--	--	2.0	310°	--	--	2.0	130°
3996	False Linderberg Head, 1 mile south of		57° 27'	135° 05'	+0 06	−0 02	−0 12	+0 19	0.6	1.0	--	--	2.0	305°	--	--	2.0	125°
4001	Lindenberg Head		57° 27'	135° 02'	+0 06	−0 02	−0 12	+0 19	0.6	1.0	--	--	2.0	280°	--	--	2.0	100°
4006	Eva Islands		57° 27'	134° 56'	+0 11	+0 03	−0 07	+0 24	0.4	0.6	--	--	1.3	275°	--	--	1.3	095°
4011	Fairway Island		57° 27'	134° 53'	+0 11	+0 03	−0 07	+0 24	0.6	1.0	--	--	2.0	265°	--	--	2.0	085°
4016	Morris Reef, south of		57° 27'	134° 49'	+0 11	+0 03	−0 07	+0 24	0.5	0.7	--	--	1.5	275°	--	--	1.5	095°
	KHAZ BAY																	
4021	Elbow Passage, south of Klag Island <69>		57° 37'	136° 06'	+0 49	---	+0 51	---	--	--	--	--	--	090°	--	--	--	270°
	CROSS SOUND				on North Inian Pass, p.88													
4026	Cape Spencer, 3 miles south of		58° 09'	136° 38'	+0 10	+0 10	+0 10	+0 10	0.4	0.4	--	--	1.2	070°	--	--	2.0	250°
4031	Yakobi Rock, 1 mile west of		58° 05'	136° 36'	+0 10	+0 10	+0 10	+0 10	0.4	0.4	--	--	1.2	045°	--	--	2.0	225°
4036	Port Althrop ent., E of George I <70>		58° 12'	136° 22'	---	---	---	---	--	--	--	--	--	340°	--	--	--	160°
4041	NORTH INIAN PASS		58° 17'	136° 23'	*Daily predictions*						--	--	2.9	075°	--	--	5.1	260°
4046	South Inian Pass		58° 13'	136° 21'	+0 15	+0 15	+0 15	+0 15	1.2	1.2	--	--	3.5	080°	--	--	6.0	260°
	ICY STRAIT and GLACIER BAY																	
4051	North Passage		58° 19'	136° 07'	+0 05	+0 05	+0 05	+0 05	0.7	0.7	--	--	2.0	070°	--	--	3.6	250°
4056	South Passage		58° 14'	136° 06'	+0 05	+0 05	+0 05	+0 05	0.8	0.8	--	--	2.3	085°	--	--	4.1	265°
4061	Beardslee Island, West of, Glacier Bay		58° 28'	136° 02'	−0 20	0 00	+1 30	+1 12	1.8	1.0	--	--	5.3	343°	--	--	5.1	158°
4066	Pleasant Island, 3 miles south of <71>		58° 17'	135° 35'	−1 00	−1 00	−1 00	−1 00	0.1	0.2	--	--	0.2	091°	--	--	1.2	284°

Endnotes can be found at the end of table 2.

TABLE 2 – CURRENT DIFFERENCES AND OTHER CONSTANTS

No.	PLACE	Meter Depth	POSITION		TIME DIFFERENCES				SPEED RATIOS		AVERAGE SPEEDS AND DIRECTIONS							
			Latitude	Longitude	Min. before Flood	Flood	Min. before Ebb	Ebb	Flood	Ebb	Minimum before Flood		Maximum Flood		Minimum before Ebb		Maximum Ebb	
		ft	North	West	h m	h m	h m	h m			knots	Dir.	knots	Dir.	knots	Dir.	knots	Dir.
	LITUYA BAY Time meridian, 135° W				on Wrangell Narrows, p.80													
4071	Lituya Bay entrance <72>		58° 37'	137° 40'	+0 25	+0 53	+0 49	+0 49	1.6	2.0	--	--	5.1	035°	--	--	4.1	210°
	ICY BAY				on Kvichak Bay, p.106													
4076	Point Riou, 2.6 nautical miles SW of	13	59° 51.3'	141° 30.2'	See Table 5.													
4081	Claybluff Point Light, 5.2 nmi. SSW of	14	59° 33.5'	141° 40.2'	See Table 5.													
4086	Claybluff Point Light, 3.5 nmi. south of	75	59° 54.6'	141° 35.7'	−3 02	−2 09	−3 14	−3 20	0.2	0.2	0.2	132°	0.5	104°	0.3	142°	0.5	209°
4091	Claybluff Point Light, 2.3 nmi. SE of	206	59° 56.8'	141° 31.2'	−3 10	−2 44	−3 17	−3 14	0.1	0.1	--	--	0.2	030°	0.1	325°	0.3	215°
4096	Carson Creek Entrance, 1.4 nmi. ESE of	15	59° 59.0'	141° 28.2'	−4 48	−0 36	−0 20	−2 09	0.2	0.0	0.1	154°	0.4	071°	--	--	0.1	164°
4101	Carson Creek Entrance, 3.3 nmi. SE of	78	59° 58.2'	141° 24.8'	−2 49	−1 45	−1 03	−2 03	0.3	0.2	--	--	0.9	067°	0.1	135°	0.4	232°
4106	Carson Creek Entrance, 2.4 nmi. ESE of	50	59° 59.2'	141° 26.2'	−3 00	−1 38	−2 24	−2 06	0.2	0.3	--	--	0.6	054°	0.1	138°	0.7	244°
4111	Kichyatt Point, 1.3 nautical miles NE of	378	60° 02.1'	141° 19.7'	See Table 5.													
	CONTROLLER BAY				on Wrangell Narrows, p.80													
4116	Wingham Island, off northeast corner		60° 03'	144° 23'	+0 42	+1 11	+0 46	+1 06	0.5	0.6	--	--	1.5	068°	--	--	1.2	288°
4121	Kanak Island, southeast of		60° 05'	144° 18'	+0 58	+0 37	+0 38	+0 53	0.5	1.0	--	--	1.7	067°	--	--	2.0	255°
	PRINCE WILLIAM SOUND				on Sergius Narrows, p.84													
4126	Wooded Islands, east of	71d	59° 52.00'	147° 17.20'	+4 06	+3 32	+1 43	+2 46	0.1	0.3	0.2	163°	0.3	082°	0.2	111°	1.4	211°
4131	Box Point, 2.3 miles E of, Montague I		59° 57'	147° 17'	+2 54	+2 16	+1 09	+2 31	--	0.1	--	--	--	--	--	--	0.7	188
4136	Cape Puget, east of, Port Bainbridge	69d	59° 57.22'	148° 19.45'	Current weak and variable						0.1	307°	0.2	333°	--	--	0.3	226°
4141	Montague Strait, NW of Pt. Woodcock	70d	59° 56.87'	147° 51.67'	+3 05	+2 45	+2 24	+3 29	0.1	0.1	--	--	0.4	038°	0.1	326°	0.5	223°
4146	Latouche Passage	70d	59° 58.80'	148° 02.20'	+2 57	+2 15	+0 59	+2 23	0.1	0.2	--	--	0.5	030°	--	--	0.9	220°
4151	Elrington Passage		60° 00'	148° 06'	+2 28	+1 49	+1 47	+2 05	0.3	0.3	--	--	1.6	036°	--	--	1.3	228°
4156	Prince of Wales Passage		60° 02'	148° 08'	+3 03	+2 24	+1 24	+1 49	0.1	0.5	--	--	0.8	001°	--	--	2.5	212°
4161	Bainbridge Passage		60° 07'	148° 13'	+0 52	+1 02	+1 09	+1 09	0.5	0.5	--	--	3.1	051°	--	--	2.4	236°
4166	Bainbridge Passage (mid–passage)	21d	60° 07.60'	148° 12.23'	+1 32	+1 14	+0 54	+1 25	0.3	0.3	0.2	133°	2.0	052°	--	--	1.6	208°
4171	Knight I. Passage, N of Evans Pt. <114>	71d	60° 09.85'	147° 54.13'	---	---	---	+2 19	--	0.1	--	--	--	--	--	--	0.4	145
4176	Cape Hinchinbrook, SW of, Hinchinbrook I	20d	60° 11.20'	146° 44.90'	+2 22	+1 50	+2 06	+3 17	0.1	0.2	0.3	219°	0.5	317°	0.3	243°	0.8	180°
4181	Greene I. and Gilmour Pt., between <115>	70d	60° 16.07'	147° 18.30'	---	+5 54	---	---	0.1	--	--	--	0.5	037	--	--	--	--
4186	Bear Cape and Zaikof Point, between	70d	60° 18.70'	146° 48.80'	+3 42	+3 41	+2 03	+2 14	0.1	0.2	--	--	0.4	352°	0.1	080°	1.1	162°
4191	Montague Point, 4.5 miles northeast of	71d	60° 24.65'	146° 58.10'	See Table 5.													
4196	Knight I. Passage, E of Pt. Newell <116>	75d	60° 26.73'	147° 51.47'	---	---	---	+3 23	--	0.1	--	--	--	--	--	--	0.4	197
4201	Ship Channel, east of Smith Island	69d	60° 32.95'	147° 01.20'	See Table 5.													
4206	Johnston Point, 7 nm. northwest of	71d	60° 33.10'	146° 47.00'	+3 06	+1 33	−0 31	+2 26	--	0.1	0.1	312°	0.1	014°	0.1	044°	0.4	204°
	ORCA BAY																	
4211	Hinchinbrook I. and Hawkins I., between	20d	60° 28.83	146° 20.47'	+3 12	+3 40	+3 30	+3 03	0.2	0.2	--	--	1.1	163°	0.1	069°	1.1	350°
4216	Johnston Pt., 4 mi. N of	20d	60° 33.25	146° 35.80'	See Table 5.													
4221	Gravina Pt. and Makaka Pt., between	20d	60° 34.80	146° 15.20'	See Table 5.													
4226	Salmo Point, N of, Hawkins Island	20d	60° 35.95'	145° 48.40'	Current weak and variable						--	--	0.2	059°	--	--	0.1	236°
4231	Knowles Head, 1.5 miles S of <117>	20d	60° 39.47'	146° 36.35'	---	---	+1 59	+3 32	--	0.1	--	--	--	--	0.1	331	0.3	276°
4236	Port Nellie Juan Entrance	68d	60° 36.53'	148° 07.50'	Current weak and variable						--	--	0.1	190°	--	--	0.1	042°
4241	Eleanor Island–Naked Island, between <73>		60° 36'	147° 31'	---	---	---	---	--	--	--	--	--	--	--	--	--	--
4246	Perry I. and Lone I., between <118>	75d	60° 41.35'	147° 48.00'	Current weak and variable													
4251	Perry Passage		60° 41'	148° 02'	Current weak and variable													
4256	Ship Chan., between Naked I–Goose I	87d	60° 41.60'	147° 02.25'	Current weak and variable						--	--	--	--	--	--	0.2	163
4261	Culross Passage		60° 43'	148° 15'	Current weak and variable													
4266	Wells Passage, north of Point Culross	72d	60° 46.37'	148° 08.40'	+2 29	+2 31	+3 00	+3 19	0.1	0.1	--	--	0.3	266°	--	--	0.3	106°
4271	Storey Island, northeast of	69d	60° 47.35'	147° 17.30'	Current weak and variable						--	--	0.2	282°	--	--	0.1	146°
4276	Bligh I. and Porcupine Pt., between	20d	60° 46.05'	146° 44.80'	Current weak and variable						--	--	0.2	056°	0.1	158°	0.3	270°
4281	Fish Bay, southwest of	20d	60° 47.65'	146° 27.90'	Current weak and variable													
4286	Ship Channel, west of Bligh Island	18d	60° 50.83'	147° 00.70'	Current weak and variable						--	--	0.1	355°	--	--	0.1	124°

Endnotes can be found at the end of table 2.

TABLE 2 – CURRENT DIFFERENCES AND OTHER CONSTANTS

No.	PLACE	Meter Depth	POSITION Latitude	POSITION Longitude	TIME DIFFERENCES Min. before Flood	TIME DIFFERENCES Flood	TIME DIFFERENCES Min. before Ebb	TIME DIFFERENCES Ebb	SPEED RATIOS Flood	SPEED RATIOS Ebb	Minimum before Flood knots	Minimum before Flood Dir.	Maximum Flood knots	Maximum Flood Dir.	Minimum before Ebb knots	Minimum before Ebb Dir.	Maximum Ebb knots	Maximum Ebb Dir.
		ft	North	West	h m	h m	h m	h m										
	ORCA BAY–cont. Time meridian, 135° W				on Sergius Narrows, p.84													
4291	Port Wells, southeast of Battles Bay	70d	60° 53.75'	148° 10.80'	Current weak and variable													
4296	Unakwik Inlet, northeast of Olsen Island	82d	60° 52.90'	147° 31.83'	Current weak and variable													
4301	Glacier Island, west of	69d	60° 53.22'	147° 19.75'	Current weak and variable													
4306	Busby Island, WNW of	15d	60° 53.65'	146° 52.25'	Current weak and variable													
4311	Columbia Bay East Entrance	20d	60° 55.45'	147° 02.75'	+4 06	+3 44	+4 12	+4 12	0.1	0.1	--	--	0.3	297°	--	--	0.4	122°
4316	Valdez Arms, west of Rocky Point	69d	60° 57.65'	146° 49.27'	Current weak and variable						--	--	0.2	048°	--	--	0.1	201°
4321	Valdez Narrows	50	61° 04.57'	146° 39.40'	Current weak and variable													
4326	Jackson Point, northeast of	69	61° 05.70'	146° 21.05'	Current weak and variable													
4331	Shoup Bay, southeast of	19	61° 06.93'	146° 33.30'	Current weak and variable													
	ORCA INLET				on Wrangell Narrows, p.80													
4336	Orca Cannery, 0.8 mile southwest of		60° 34'	145° 45'	+1 05	+1 28	+1 25	+1 24	0.2	0.5	--	--	0.7	170°	--	--	1.1	025°
4341	Salmo Point, Hawkins I., 1.2 miles SE of		60° 36'	145° 45'	+0 34	+0 55	+0 58	+0 30	0.3	0.5	--	--	0.9	160°	--	--	1.1	020°
4346	Salmo Point, Hawkins I., 0.3 mile NW of		60° 37'	145° 47'	Current weak and variable													
4351	Shepard Point, 0.9 mile southwest of		60° 37'	145° 42'	Current weak and variable													
	COOK INLET																	
4356	Cape Douglas, northeast of <120>	22d	58° 53.33'	153° 11.27'	---	+4 15	---	+5 17	--	0.7	--	--	0.7	167	--	--	1.5	167°
4361	Cape Douglas, northeast of	22d	58° 56.45'	152° 53.40'	+3 18	+2 44	+2 46	+3 05	0.1	0.2	--	--	0.3	312°	0.1	241°	0.4	167°
4366	Ushagat Island, northwest of	22d	59° 00.37'	152° 33.80'	+2 59	+2 34	+3 09	+3 04	0.1	0.2	--	--	0.5	283°	--	--	0.5	117°
4371	Kennedy Entrance	22d	59° 04.00'	151° 57.40'	+2 48	+2 48	+2 56	+2 57	0.6	0.8	--	--	2.0	295°	0.1	032°	1.7	129°
4376	Ushagat Island, north of	22d	59° 05.00'	152° 15.30'	+2 24	+3 20	+4 03	+3 10	0.5	0.4	0.3	195°	1.6	292°	0.3	033°	0.9	121°
4381	Cape Elizabeth	22d	59° 07.20'	151° 53.70'	+2 23	+2 45	+3 04	+2 47	0.7	0.9	--	--	2.2	306°	0.1	218°	1.8	136°
4386	Chugach Passage		59° 10'	151° 47'	+0 48	+0 39	+1 47	+0 44	1.0	0.9	--	--	3.1	355°	--	--	1.8	170°
4391	Augustine Island, 14.5 nmi. ESE of	22d	59° 17.80'	152° 54.70'	Current weak and variable													
4396	Augustine Island, northwest of	22d	59° 24.50'	153° 37.22'	+5 55	+5 44	+5 17	+6 26	0.2	0.5	0.2	032°	0.6	060°	0.3	142°	1.0	229°
4401	Chinitna Point, southeast of	22d	59° 34.30'	152° 49.50'	+4 33	+4 31	+4 48	+4 38	0.4	0.6	0.8	292°	1.3	021°	0.5	121°	1.3	209°
4406	Iniskin Bay		59° 40'	153° 27'	+2 10	+2 04	+2 16	+2 21	0.3	0.6	--	--	0.9	358°	--	--	1.2	179°
4411	Kachemak Bay Entrance	22d	59° 33.30'	151° 47.80'	+3 19	+2 25	+2 15	+2 44	0.2	0.4	0.2	100°	0.6	035°	0.1	303°	0.8	210°
4416	Kachemak Bay, west of	22d	59° 34.00'	152° 16.30'	+3 43	+3 22	+3 45	+3 45	0.6	0.8	0.1	277°	1.8	010°	0.1	101°	1.7	188°
4421	Anchor Point, 5 miles northwest of	22d	59° 51.10'	152° 00.50'	+4 06	+3 57	+4 29	+4 33	0.8	1.0	--	--	2.4	022°	--	--	2.1	198°
4426	Anchor Point, WNW of	22d	59° 51.55'	152° 12.30'	+4 36	+4 12	+4 39	+5 04	0.7	1.2	0.4	269°	2.4	017°	0.2	078°	2.5	197°
4431	Chinitna Bay, 12 miles northeast of	22d	59° 54.80'	152° 28.60'	+4 02	+3 35	+4 06	+3 32	0.8	0.9	0.3	109°	2.5	026°	0.1	117°	1.8	196°
4436	Cape Starichkof, northwest of	20d	59° 58.05'	151° 57.05'	+4 28	+3 39	+4 17	+4 32	0.7	1.2	--	--	2.2	019°	0.1	290°	2.5	199°
4441	Cape Ninilchik, west of	20d	60° 01.15'	152° 07.00'	+5 23	+4 34	+4 50	+5 32	0.8	1.7	0.9	315°	2.6	031°	0.4	121°	3.5	214°
4446	Chisik Island, 5 nmi. east of	20d	60° 07.60'	152° 24.10'	+4 25	+4 00	+4 35	+4 57	0.7	1.0	0.1	288°	2.1	017°	--	--	2.0	197°
4451	Tuxedni Channel (South Entrance)	20d	60° 05.80'	152° 34.45'	+3 40	+3 31	+3 44	+3 27	0.6	0.7	0.1	238°	2.0	331°	0.1	211°	1.4	150°
4456	Tuxedni Channel		60° 09'	152° 39'	+3 31	+2 55	+2 22	+2 21	0.3	0.9	--	--	1.1	332°	--	--	1.9	162°
4461	Ninilchik Harbor, northwest of	20d	60° 08.00'	151° 56.05'	+5 10	+4 51	+5 00	+5 26	0.8	1.7	0.8	298°	2.7	029°	0.3	124°	3.5	212°
4466	Redoubt Point, southeast of	20d	60° 13.90'	152° 15.85'	+5 04	+4 55	+5 09	+5 15	0.6	1.2	0.2	284°	2.0	025°	0.3	111°	2.5	198°
4471	Kalgin Island, 10 nmi. southeast of	20d	60° 17.15'	151° 42.70'	+5 21	+4 53	+4 59	+5 46	0.9	1.6	0.2	111°	2.7	033°	--	--	3.3	207°
4476	Cape Kasilof, 3 miles west of		60° 22'	151° 28'	+3 48	+3 14	+5 09	+4 27	0.9	1.1	--	--	3.0	022°	--	--	2.3	206°
4481	Drift River Terminal, SSE of	20d	60° 32.55'	152° 05.95'	+5 50	+5 28	+5 41	+5 55	0.6	1.5	0.2	128°	1.9	051°	0.2	129°	3.1	221°
4486	Kenai, 6 miles southwest of		60° 29'	151° 26'	+4 40	+3 58	+4 56	+4 29	0.8	1.3	--	--	2.4	021°	--	--	2.6	193°
4491	Kenai Packers Cannery Wharf		60° 33'	151° 14'	+4 00	+3 15	+2 11	+2 23	0.2	0.7	--	--	0.7	115°	--	--	1.4	285°
4496	Kenai City Wharf		60° 33'	151° 14'	+3 57	+3 16	+2 55	+4 12	0.2	0.7	--	--	0.5	130°	--	--	1.4	300°
4501	West Foreland, south of	20d	60° 35.90'	151° 44.10'	+5 38	+4 52	+5 32	+5 22	1.0	1.7	0.1	313°	3.1	058°	0.8	145°	3.6	228°
4506	Nikishka	17	60° 41.10'	151° 23.90'	+4 46	+4 28	+4 59	+3 36	1.2	1.3	--	--	3.8	000°	--	--	2.6	180°
4511	Nikishka, 0.8 mile west of	15	60° 41.20'	151° 25.20'	+5 22	+4 37	+5 12	+4 44	1.2	1.7	--	--	3.8	345°	--	--	3.6	175°
4516	West Foreland, 1 nmi. east of	20d	60° 44.35'	151° 38.50'	+6 09	+4 46	+5 26	+6 25	1.1	2.1	0.2	075°	3.6	349°	0.2	077°	4.4	171°
4521	West Foreland, 4 nmi. east of	20d	60° 43.25'	151° 33.60'	+5 42	+5 10	+5 49	+5 33	1.3	2.0	--	--	4.1	003°	0.5	274°	4.1	200°
4526	East Foreland, 1.5 nmi. east of	20d	60° 43.40'	151° 28.00'	+5 39	+5 10	+5 38	+5 28	1.0	1.7	0.1	285°	3.1	006°	0.5	277°	3.5	195°
4531	Middle Ground Shoal, southeast of	20d	60° 50.75'	151° 20.20'	+6 00	+4 53	+5 24	+5 45	1.1	2.1	0.1	141°	3.4	063°	0.4	144°	4.3	231°
4536	North Foreland, southeast of	20d	61° 00.20'	151° 04.70'	+6 18	+5 43	+5 57	+6 30	1.1	1.7	0.5	142°	3.4	062°	0.1	148°	3.4	221°

Endnotes can be found at the end of table 2.

TABLE 2 – CURRENT DIFFERENCES AND OTHER CONSTANTS

No.	PLACE	Meter Depth	POSITION Latitude	POSITION Longitude	TIME DIFFERENCES Min. before Flood	Flood	Min. before Ebb	Ebb	SPEED RATIOS Flood	Ebb	AVERAGE SPEEDS AND DIRECTIONS Minimum before Flood knots	Dir.	Maximum Flood knots	Dir.	Minimum before Ebb knots	Dir.	Maximum Ebb knots	Dir.
		ft	North	West	h m	h m	h m	h m			knots	Dir.	knots	Dir.	knots	Dir.	knots	Dir.
	COOK INLET–cont. Time meridian, 135° W				on Wrangell Narrows, p.80													
4541	Moose Point, NNW of	20d	61° 00.95'	150° 42.00'	+6 21	+4 50	+5 52	+5 46	0.8	1.3	0.1	152°	2.7	061°	– –	– –	2.7	237°
4546	Moose Point, northwest of	20d	61° 04.65'	150° 45.00'	+6 34	+5 01	+6 03	+6 30	0.9	1.3	0.1	349°	2.8	086°	0.1	160°	2.7	255°
4551	Point Possession, northeast of	20d	61° 03.55'	150° 23.00'	+6 52	+5 05	+6 27	+6 05	1.4	2.0	– –	– –	4.5	102°	– –	– –	4.1	275°
4556	Point Possession, WNW of	20d	61° 03.00'	150° 27.70'	+6 33	+5 06	+6 11	+5 57	1.0	1.6	0.1	344°	3.3	074°	– –	– –	3.3	246°
4561	Point Possession, northwest of	20d	61° 05.25'	150° 28.30'	+6 45	+5 36	+6 33	+5 43	1.0	1.4	0.1	358°	3.2	087°	0.1	169°	2.9	255°
4566	Fire Island, west of	20d	61° 09.75'	150° 30.75'	+7 12	+6 19	+6 57	+6 54	1.1	1.8	– –	– –	3.6	093°	0.3	180°	3.8	258°
4571	Fire Island, 1.0nm east of	2d	61° 10.75'	150° 07.53'	+6 28	+5 38	+6 06	+6 40	0.7	0.9	0.2	174°	2.2	095°	0.3	180°	1.8	258°
	...do.	14d	61° 10.75'	150° 07.53'	+6 21	+5 31	+6 00	+6 33	0.6	0.7	– –	– –	2.0	095°	0.3	185°	1.5	266°
	...do.	28d	61° 10.75'	150° 07.53'	+6 15	+5 30	+5 51	+6 25	0.5	0.5	0.2	354°	1.6	095°	0.3	190°	1.1	271°
4576	Fire Island, 1.0nm north of	15d	61° 11.53'	150° 10.55'	+6 36	+5 58	+6 23	+6 32	1.2	1.4	0.1	351°	3.7	077°	0.1	352°	3.0	264°
	...do.	25d	61° 11.53'	150° 10.55'	+6 28	+5 20	+6 20	+6 26	1.1	1.4	0.1	351°	3.3	076°	0.1	352°	2.9	265°
	...do.	41d	61° 11.53'	150° 10.55'	+6 21	+5 22	+6 17	+6 22	0.9	1.3	0.1	353°	3.0	076°	0.2	351°	2.7	265°
4581	Point Campbell, northwest of	20d	61° 11.40'	150° 09.28'	+7 20	+5 32	+7 12	+6 16	0.9	1.1	0.2	173°	2.9	093°	– –	– –	2.3	258°
4586	Point Woronzof, southwest of	20d	61° 11.23'	150° 03.75'	+7 22	+5 44	+7 09	+9 00	0.9	0.9	0.1	145°	2.8	057°	– –	– –	1.8	225°
4591	Point Woronzof, west of	20d	61° 12.42'	150° 03.67'	+7 18	+5 35	+7 04	+6 16	0.9	1.4	0.3	143°	2.9	061°	0.1	321°	3.0	225°
4596	Point Woronzof, 1.2nm NE of	26d	61° 13.01'	149° 59.06'	+7 03	+5 56	+6 06	+6 58	0.4	1.0	0.1	169°	1.3	084°	– –	– –	2.0	262°
	...do.	58d	61° 13.01'	149° 59.06'	+6 56	+5 58	+6 06	+6 51	0.3	0.6	– –	– –	1.0	094°	0.1	181°	1.2	268°
	...do.	85d	61° 13.01'	149° 59.06'	+6 44	+6 23	+6 08	+6 27	0.3	0.5	– –	– –	0.9	106°	0.2	190°	0.9	271°
4601	Anchorage, west of	20d	61° 13.67'	149° 56.90'	+7 32	+5 58	+7 20	+6 20	1.1	1.5	0.5	148°	3.5	081°	0.2	162°	3.1	234°
4606	Anchorage, 0.2 mile offshore <74>	15	61° 13.50'	149° 54.38'	+4 08	+5 12	+5 33	+5 18	0.5	1.2	– –	– –	1.5	028°	– –	– –	2.5	207°
4611	Anchorage Shipdock, northwest of	20d	61° 14.75'	149° 54.50'	+7 49	+6 23	+7 21	+6 12	1.2	1.9	0.1	111°	3.9	028°	– –	– –	4.0	195°
4616	Cairn Point, northwest of (east side)	20d	61° 16.03'	149° 54.05'	+7 45	+6 42	+7 16	+6 47	1.0	2.1	0.2	087°	3.2	359°	0.2	323°	4.4	192°
4621	Cairn Point, northwest of (west side)	20d	61° 16.03'	149° 53.60'	+7 43	+6 31	+7 17	+6 05	1.1	1.8	0.2	105°	3.6	018°	0.2	288°	3.8	198°
4626	Port Mackenzie <121>	13d	61° 16.06'	149° 54.98'	+5 17	+4 59	+5 54	+6 03	0.3	0.9	0.1	279°	0.9	018°	– –	– –	1.8	193°
4631	Knik Arm Mud Flats, south of	10d	61° 19.90'	149° 47.62'	+7 55	+5 49	+6 36	+6 14	0.8	1.5	– –	– –	2.7	074°	0.1	147°	3.1	241°
	SHELIKOF STRAIT and KODIAK ISLAND				on Sergius Narrows, p.84													
4636	Cape Current Narrows, Shuyak Strait		58° 28'	152° 30'	+2 23	+2 09	+2 16	+2 10	0.6	0.8	– –	– –	3.8	275°	– –	– –	3.8	095°
4641	Kukak Bay, Shelikof Strait		58° 20'	154° 07'	Current weak and variable						– –	– –	0.2	298°	– –	– –	0.2	146°
4646	Raspberry Strait <75>		– – –	– – –	– – –	– – –	– – –	– – –	– –	– –	– –	– –	– –	– –	– –	– –	– –	– –
4651	Kupreanof Strait, 0.8 mi. off Chernof Pt		57° 58'	152° 55'	+3 16	+3 32	+3 30	+3 05	0.4	0.3	– –	– –	2.2	298°	– –	– –	1.5	101°
4656	Whale Passage, northwest entrance		57° 56'	152° 52'	+2 24	+2 28	+2 17	+2 08	0.7	0.5	– –	– –	4.6	305°	– –	– –	2.6	115°
4661	Whale Passage, off Bird Point		57° 55'	152° 48'	+2 31	+2 10	+2 15	+2 07	0.7	1.1	– –	– –	4.4	305°	– –	– –	5.2	140°
	KODIAK ISLAND				on Wrangell Narrows, p.80													
4666	Narrow Strait, off Ouzinkie Point		57° 55'	152° 32'	−1 17	−2 20	−1 25	−2 19	0.4	0.7	– –	– –	1.3	326°	– –	– –	1.5	133°
4671	Kodiak Harbor (narrows) Chiniak Bay		57° 47'	152° 24'	+0 49	+0 36	+1 31	+0 58	0.3	0.2	– –	– –	0.9	056°	– –	– –	0.5	228°
					on North Inian Pass, p.88													
4676	Sitkinak Strait		56° 39'	154° 09'	−0 24	−0 00	+1 04	+0 35	0.7	0.5	– –	– –	1.9	300°	– –	– –	2.3	120°
	SHUMAGIN ISLANDS				on Kvichak Bay, p.106													
4681	Popof Strait		55° 20'	160° 31'	−2 32	−2 05	−1 28	−2 11	0.1	0.2	– –	– –	0.2	357°	– –	– –	0.5	199°
4686	Unga Strait (1.4 miles N of Unga Spit)		55° 26'	160° 44'	+5 24	+5 42	+5 24	+5 06	0.5	0.1	– –	– –	1.2	282°	– –	– –	0.2	114°
	ALASKA PENINSULA				on Isanotski Strait, p.92													
4691	Ukolnoi Island, 3.3 miles northeast of		55° 16'	161° 26'	Current weak and variable													
4696	Seal Cape Light, 0.8 mile south of		55° 20'	161° 15'	Current weak and variable													
4701	Amak Island, 5 miles north of		55° 30'	163° 10'	+4 52	+5 31	+4 50	+5 09	0.2	0.2	– –	– –	0.8	337°	– –	– –	0.7	139°
4706	Amak Island, 5 miles southeast of		55° 21'	163° 01'	+4 24	+4 45	+4 32	+4 35	0.2	0.2	– –	– –	0.8	350°	– –	– –	0.7	200°
4711	Bechevin Bay, off the entrance <76>		55° 07'	163° 28'	– – –	+3 46	– – –	+3 52	0.2	0.1	– –	– –	0.8	047°	– –	– –	0.4	265°

Endnotes can be found at the end of table 2.

TABLE 2 – CURRENT DIFFERENCES AND OTHER CONSTANTS

No.	PLACE	Meter Depth	POSITION		TIME DIFFERENCES				SPEED RATIOS		AVERAGE SPEEDS AND DIRECTIONS							
			Latitude	Longitude	Min. before Flood	Flood	Min. before Ebb	Ebb	Flood	Ebb	Minimum before Flood		Maximum Flood		Minimum before Ebb		Maximum Ebb	
		ft	North	West	h m	h m	h m	h m			knots	Dir.	knots	Dir.	knots	Dir.	knots	Dir.
	UNIMAK ISLAND Time meridian, 135° W				on Isanotski Strait, p.92													
4716	Otter Point, off of, north side		55° 04'	163° 47'	+3 36	+3 36	+3 35	+3 53	0.2	0.3	--	--	0.9	089°	--	--	0.8	258°
	ISANOTSKI STRAIT																	
4721	ISANOTSKI STRAIT (False Pass Cnry) <77>		54° 52'	163° 24'		*Daily predictions*					--	--	3.6	358°	--	--	2.8	187°
4726	Bechevin Bay, off Rocky Point		54° 59'	163° 26'	−0 55	−0 26	+0 14	+0 02	0.5	0.5	--	--	1.8	331°	--	--	1.4	139°
	ALEUTIAN ISLANDS <78>				on Unimak Pass, p.98													
4731	Davidson Bank <79>		54° 00'	163° 00'	---	---	---	---	--	--	--	--	--	--	--	--	--	--
4736	UNIMAK PASS (off Scotch Cap)		54° 22'	164° 48'		*Daily predictions*					--	--	3.4	295°	--	--	3.0	105°
4741	Unimak Pass, 11 miles WSW of Sennett Pt		54° 25'	165° 12'	−0 43	+0 41	+1 17	0 00	0.6	0.2	--	--	2.0	301°	--	--	0.7	106°
4746	Unimak Pass, 2.4 miles N of Tanginak I		54° 14'	165° 18'	+0 44	+0 17	−0 48	−0 22	0.4	0.5	--	--	1.3	298°	--	--	1.5	144°
4751	Avatanak Strait		54° 07'	165° 28'	+0 44	+0 52	+0 47	+0 24	1.2	1.1	--	--	4.1	050°	--	--	3.3	258°
					on Akutan Pass, p.102													
4756	Derbin Strait <80>		54° 06'	165° 14'	0 00	−0 06	−0 12	−0 19	1.0	1.0	--	--	5.8	342°	--	--	5.3	148°
4761	Ugamak Strait, off Kaligagan Island <81>		54° 09'	164° 53'	+0 31	+0 13	+0 11	+0 05	0.6	0.8	--	--	3.5	324°	--	--	4.0	116°
4766	Ugamak Strait (North end) <81>		54° 12'	164° 55'	+0 56	+0 39	+1 10	+0 38	0.6	0.3	--	--	3.3	322°	--	--	1.8	122°
4771	AKUTAN PASS		54° 01'	166° 03'		*Daily predictions*					--	--	5.8	294°	--	--	5.3	113°
4776	Baby Pass		53° 59'	166° 04'	−0 03	−0 04	−0 16	−0 02	0.7	0.9	--	--	4.2	303°	--	--	4.8	129°
4781	Unalga Pass		53° 57'	166° 12'	+0 01	−0 07	+0 01	−0 05	1.1	1.0	--	--	6.3	320°	--	--	5.3	128°
					on Unimak Pass, p.98													
4786	Udagak Strait (narrows) <82>		53° 44'	166° 18'	−2 11	−2 21	−2 03	−3 09	0.6	0.3	--	--	2.1	286°	--	--	1.0	142°
4791	Umnak Pass, south approach		53° 15'	167° 55'	−0 38	−1 01	−0 23	−0 14	0.6	1.0	--	--	2.0	341°	--	--	3.0	181°
4796	Konets Head–Emerald Island, between <83>		53° 18'	167° 51'	---	−1 05	−0 55	−2 21	1.0	0.8	--	--	3.4	331°	--	--	2.3	125°
4801	Umnak Pass, southeast of Ship Rock		53° 21'	167° 48'	+0 11	+0 14	+0 20	−0 17	1.1	0.9	--	--	3.7	056°	--	--	2.7	225°
4806	Umnak Pass, northwest of Ship Rock		53° 23'	167° 51'	+0 01	−1 00	−0 28	−1 08	1.1	1.1	--	--	3.7	052°	--	--	3.3	218°
	Time meridian, 150° W				on Isanotski Strait, p.92													
4811	Yunaska Island, 1 mile east of <84>		52° 40'	170° 32'	---	---	---	---	0.5	0.8	--	--	1.9	021°	--	--	2.1	159°
4816	Finch Cove, Seguam Island		52° 23'	172° 23'	---	---	---	---	--	--	--	--	1.4	315°	--	--	--	130°
					on Unimak Pass, p.98													
4821	Fenimore Rock, 1.2 miles southwest of		51° 58'	175° 34'	−1 16	−1 52	−2 17	−1 58	0.9	1.0	--	--	3.2	010°	--	--	3.0	140°
4826	Igitkin Pass, 0.8 mile N of Tanager Pt		51° 57'	175° 52'	−0 41	−1 11	−1 25	−1 49	0.9	0.7	--	--	3.1	040°	--	--	2.1	245°
4831	Chugul Pass, 0.8 mile SW of Tanager Pt		51° 56'	175° 53'	−2 45	−1 54	−0 48	−2 15	0.5	0.5	--	--	1.6	325°	--	--	1.6	160°
4836	Chugul Pass, 2 miles NE of Cape Ruin		51° 56'	175° 56'	+0 30	−0 09	−0 54	−1 09	0.4	0.5	--	--	1.4	335°	--	--	1.5	150°
4841	Chugul Pass, 0.5 mile NE of Cape Ruin		51° 55'	175° 58'	−1 12	−1 12	−0 18	−0 56	0.4	0.6	--	--	1.2	305°	--	--	1.8	120°
					on Isanotski Strait, p.92													
4846	Umak Pass, off Narrows Point		51° 51'	176° 04'	−0 09	−0 18	+0 06	−0 27	0.9	0.8	--	--	3.2	305°	--	--	2.2	130°
					on Akutan Pass, p.102													
4851	Little Tanaga Strait, off Tana Pt <85>		51° 49'	176° 14'	−0 49	−0 54	−0 33	−0 34	0.7	0.6	--	--	2.5	320°	--	--	3.0	130°
4856	Kagalaska Strait, off Galas Point <85>		51° 48'	176° 25'	−1 21	−0 59	−0 07	+0 02	0.7	0.5	--	--	3.9	310°	--	--	2.8	175°

Endnotes can be found at the end of table 2.

TABLE 2 – CURRENT DIFFERENCES AND OTHER CONSTANTS

No.	PLACE	Meter Depth	POSITION		TIME DIFFERENCES				SPEED RATIOS		AVERAGE SPEEDS AND DIRECTIONS							
			Latitude	Longitude	Min. before Flood	Flood	Min. before Ebb	Ebb	Flood	Ebb	Minimum before Flood		Maximum Flood		Minimum before Ebb		Maximum Ebb	
		ft	North	West	h m	h m	h m	h m			knots	Dir.	knots	Dir.	knots	Dir.	knots	Dir.
	ALEUTIAN ISLANDS <78>–cont. Time meridian, 150° W				on Unimak Pass, p.98													
4861	Adak Strait, 1 mile NE of Naga Pt <86>		51° 47'	177° 05'	---	---	-2 05	-2 05	0.6	0.7	--	--	2.0	010°	--	--	2.2	190°
4866	Adak Strait, 4 miles ENE of Naga Point		51° 47'	177° 00'	-0 54	-0 30	-0 44	-0 07	0.6	0.6	--	--	1.9	010°	--	--	1.9	195°
4871	Adak Strait, off Argonne Point <87>		51° 48'	176° 57'	-3 52	-0 44	-1 32	-2 48	1.0	0.6	--	--	2.8	010°	--	--	1.4	200°
4876	Kanaga Pass, 0.3 mile NW of Annoy Rock		51° 43'	177° 48'	-0 03	-0 10	-0 23	-0 57	0.7	0.7	--	--	2.5	000°	--	--	2.2	195°
4881	Kanaga Pass, 2.2 miles NE of Annoy Rock		51° 45'	177° 45'	-0 24	-0 36	-0 54	-1 36	0.8	0.7	--	--	2.6	020°	--	--	2.2	225°
4886	Tanaga Pass, 4 mi. of C. Amagalik <88>		51° 39'	178° 13'	-1 45	-2 29	-1 45	-0 01	0.5	0.5	--	--	--	315°	--	--	--	200°
4891	Ogliuga Island, pass East of, Delarof Is		51° 37'	178° 36'	-2 19	-2 12	-2 23	-2 18	0.2	0.2	--	--	0.7	036°	--	--	0.5	220°
4896	Gareloi I., 0.5 mile SE of, Delarof Is		51° 45'	178° 45'	---	---	---	---	--	--	--	--	--	055°	--	--	--	245°
4901	Ulak Pass, Delarof Islands		51° 19'	179° 02'	-0 20	-0 06	-0 20	-0 38	0.7	0.7	--	--	2.4	326°	--	--	2.2	125°
			North	East														
4906	Fetrel Bank, Semisopochnoi Island <89>		52° 10'	179° 52'	---	---	---	---	--	--	--	--	--	--	--	--	--	--
4911	Amchitka Island, south coast <90>		51° 33'	178° 51'	---	---	---	---	--	--	--	--	0.7	309°	--	--	0.5	124°
4916	Oglala Pass, Rat Islands <91>		51° 42'	178° 31'	---	---	---	---	--	--	--	--	--	--	--	--	2.3	--
4921	Little Sitkin Island, SE coast <90>		51° 54'	178° 32'	---	---	---	---	--	--	--	--	0.6	050°	--	--	0.6	240°
4926	Fat Island Pass, Rat Islands <92>		51° 53'	178° 20'	---	---	---	-3 00	--	0.2	--	--	--	--	--	--	0.6	110°
4931	Krysi Pass, Rat Islands		51° 51'	178° 07'	+0 03	-0 08	-0 29	-0 40	0.7	0.7	--	--	2.4	040°	--	--	2.2	210°
4936	Sea Lion Pass, Rat Islands <93>		51° 54'	177° 54'	+0 06	-0 01	-0 34	-0 23	0.7	0.8	--	--	2.4	012°	--	--	2.4	195°
4941	Tahoma Reef <7>		51° 49'	175° 52'	---	-0 23	---	-1 04	0.3	0.3	--	--	0.7	007°	--	--	0.9	147°
4946	Attu Island, 5 miles NE of Cape Wrangell		52° 59'	172° 32'	---	---	---	---	--	--	--	--	1.4	064°	--	--	0.9	201°
	Time meridian, 135° W		North	West	on Kvichak Bay, p.106													
4951	Cape Lieskof, 3 miles west of		55° 45'	162° 12'	-5 39	-4 39	-4 38	-4 53	0.3	0.3	--	--	0.8	056°	--	--	0.7	248°
	PORT MOLLER																	
4956	Entrance Point, 3 miles west of		56° 00'	160° 39'	-5 04	-4 28	-4 06	-4 34	0.7	0.8	--	--	1.7	174°	--	--	2.0	002°
4961	Entrance Point		55° 59'	160° 35'	-4 57	-4 53	-5 06	-5 27	0.5	0.6	--	--	1.2	180°	--	--	1.6	000°
4966	Harbor Point		55° 55'	160° 36'	-4 28	-4 03	-4 59	-4 26	0.4	0.8	--	--	0.9	158°	--	--	1.9	335°
	HERENDEEN BAY–PORT HEIDEN																	
4971	Hague Channel, east of Doe Point		55° 54'	160° 46'	-5 59	-4 39	-4 23	-5 21	0.9	0.6	--	--	2.3	220°	--	--	1.4	033°
4976	Johnston Channel, off Halftide Rock		55° 50'	160° 47'	-4 27	-4 45	-5 15	-4 24	0.5	0.5	--	--	1.2	179°	--	--	1.3	337°
4981	Port Heiden		56° 59'	158° 53'	-2 23	-1 05	-1 14	-1 15	0.4	0.4	--	--	1.0	067°	--	--	1.0	233°
	KVICHAK BAY																	
4986	KVICHAK BAY (off Naknek River entrance)		58° 42'	157° 15'	*Daily predictions*						--	--	2.5	053°	--	--	2.5	239°
4991	Morakas Point, Naknek River <94>		58° 44'	156° 56'	+2 01	+1 05	+0 04	+1 15	0.5	0.9	--	--	1.1	111°	--	--	2.1	294°
4996	Kvichak, Kvichak River <94>		58° 58'	156° 56'	+2 12	+1 30	+0 39	+1 01	--	--	--	--	1.7	078°	--	--	3.0	259°
	NUSHAGAK BAY and APPROACHES																	
5001	Cape Constantine, 4 miles Southeast of		58° 20'	158° 46'	-2 08	-1 38	-1 05	-1 52	0.6	0.7	--	--	1.6	059°	--	--	1.7	238°
5006	Protection Point, 2.5 miles east of		58° 30'	158° 37'	-0 44	-1 25	-0 40	-1 04	1.0	1.0	--	--	1.9	013°	--	--	3.1	180°
5011	Nushagak Bay entrance		58° 34'	158° 25'	-0 59	+0 03	+0 19	-0 23	1.0	1.0	--	--	2.5	343°	--	--	2.5	180°
5016	Etolin Point, 8.5 miles west of		58° 38'	158° 35'	-0 19	+0 13	+0 08	+0 05	0.9	1.2	--	--	2.3	352°	--	--	2.9	173°
5021	Clarks Point, 1 mile west of		58° 50'	158° 35'	-0 07	+0 34	+0 41	-0 02	1.3	1.4	--	--	3.2	018°	--	--	3.4	213°
5026	Dillingham <95>		59° 02'	158° 28'	+0 55	+1 11	+1 19	+0 26	--	--	--	--	3.4	076°	--	--	3.2	262°

Endnotes can be found at the end of table 2.

TABLE 2 – CURRENT DIFFERENCES AND OTHER CONSTANTS

No.	PLACE	Meter Depth	POSITION Latitude	POSITION Longitude	TIME DIFFERENCES Min. before Flood	TIME DIFFERENCES Flood	TIME DIFFERENCES Min. before Ebb	TIME DIFFERENCES Ebb	SPEED RATIOS Flood	SPEED RATIOS Ebb	Minimum before Flood knots	Minimum before Flood Dir.	Maximum Flood knots	Maximum Flood Dir.	Minimum before Ebb knots	Minimum before Ebb Dir.	Maximum Ebb knots	Maximum Ebb Dir.
		ft	North	West	h m	h m	h m	h m										
	KUSKOKWIM BAY Time meridian, 135° W				**on Kvichak Bay, p.106**													
5031	Goodnews Bay entrance		59° 04'	161° 47'	−6 23	−6 05	−5 26	−6 02	0.9	0.9	--	--	2.3	020°	--	--	2.1	213°
5036	Carter Bay, west of		59° 17'	162° 22'	−5 10	−4 29	−3 44	−4 21	0.6	0.6	--	--	1.5	021°	--	--	1.4	212°
5041	Warehouse Bluff, southwest of		59° 47'	162° 14'	−3 43	−3 21	−3 21	−3 45	0.6	0.8	--	--	1.6	007°	--	--	2.1	188°
5046	Apokak Creek entrance		60° 08'	162° 10'	−5 04	−3 42	−2 04	−2 51	1.1	1.1	--	--	3.4	030°	--	--	2.1	205°
	BERING SEA																	
	Pribilof Islands																	
5051	Walrus Island, 0.5 mile west of <96>		57° 11'	169° 57'	−6 40	−6 47	−6 40	−6 38	0.4	0.5	--	--	0.9	018°	--	--	1.2	210°
5056	St. Paul I.–St. George I., between <97>		56° 52'	169° 56'	---	−9 21	---	---	0.2	--	--	--	0.6	260	--	--	--	--
5061	Otter Island, 7 miles east of <97>		57° 03'	170° 10'	---	−10 44	---	---	0.4	--	--	--	0.9	245	--	--	--	--
5066	SW Pt., St. Paul I., 6 mi. SW <97>		57° 07'	170° 34'	---	−7 02	---	---	0.2	--	--	--	0.5	330	--	--	--	--
5071	SW Point, St. Paul Island, 1 mile off		57° 09'	170° 27'	−8 55	−8 10	−8 55	−8 10	0.8	0.3	--	--	1.9	330°	--	--	0.7	170°
					on San Bernardino Strait, p.158													
5076	Nunivak Island, west coast		60° 02'	167° 15'	+11 38	+10 56	+10 26	+10 20	0.2	0.2	--	--	0.8	311°	--	--	0.8	135°
					on Kvichak Bay, p.106													
5081	Hooper Bay entrance		61° 30'	166° 03'	+10 08	+11 16	+11 14	+11 03	0.7	0.8	--	--	1.7	046°	--	--	2.0	223°
5086	St. Mathew I., southwest coast		60° 21'	172° 43'	+2 24	+3 20	+3 29	+3 07	0.5	0.4	--	--	1.2	292°	--	--	1.0	119°
	St. Lawrence Island																	
5091	4.5 miles SE of Southeast Cape <98>		62° 53'	169° 32'	---	−1 23	---	−2 48	0.3	0.3	--	--	0.8	097	0.2	--	0.7	251°
5096	Apavawook Cape, 1 mile south of		63° 07'	168° 56'	−3 58	−3 14	−3 39	−3 23	0.2	0.4	--	--	0.5	075°	--	--	1.1	272°
5101	Off Northeast Cape		63° 20'	168° 50'	−1 41	−0 03	+0 16	−0 50	0.3	0.3	--	--	0.8	095°	--	--	0.7	258°
5106	Tatik Point, 13 miles off of <99>		63° 23'	172° 18'	---	---	---	---	--	--	--	--	0.2	000	--	--	0.7	190°
5111	Gambell, 13 miles NNW of <100>		65° 00'	172° 01'	---	−1 02	---	−1 19	0.7	0.2	--	--	1.7	050	--	--	0.8	075°
					on Unimak Pass, p.98													
5116	Sledge Island, 2 miles north of <101>		64° 32'	166° 10'	−7 01	---	−6 58	−7 34	0.3	0.2	--	--	1.0	305	--	--	0.5	119°
5121	King Island, 42 miles west of <102>		64° 58'	169° 44'	---	---	---	---	--	--	--	--	0.4	030	--	--	0.2	030°
5126	Fairway Rock, 18.5 miles south of <103>		65° 20'	168° 50'	---	---	---	---	--	--	--	--	0.7	000	--	--	0.5	000°
5131	Fairway Rock, 4.8 miles NNE of <104>		65° 42'	168° 39'	---	---	---	---	--	--	--	--	1.1	020	--	--	0.6	020°
	HAWAIIAN ISLANDS Time meridian, 150° W				**on San Diego Bay Ent., p.4**													
5136	Maui Island <105>		20° 46'	155° 58'	---	---	---	---	--	--	--	--	--	--	--	--	--	--
5141	Alalakeiki Channel, west side <106>		20° 36'	156° 32'	---	---	---	---	--	--	--	--	--	--	--	--	--	--
5146	Alalakeiki Channel, east side <107>		20° 37'	156° 29'	---	---	---	---	--	--	--	--	--	--	--	--	--	--
5151	Maalaea Bay, Maui Island		20° 46'	156° 30'	Current weak and variable						--	--	--	--	--	--	--	--
5156	Auau Channel <108>		20° 53'	156° 43'	---	---	---	−2 58	--	0.7	--	--	--	--	--	--	1.1	073
5161	Kalohi Channel		21° 02'	156° 56'	−3 15	−3 47	−3 38	−3 11	0.4	0.3	--	--	0.5	075°	--	--	0.5	227°
5166	Pailolo Channel <109>		21° 04'	156° 43'	---	---	---	---	--	--	--	--	--	--	--	--	--	--
	TOKYO WAN Time meridian, 135° E		North	East	**on Tokyo Wan Entrance, p.110**													
5171	TOKYO WAN ENTRANCE, (N of Kannon Saki)		35° 17'	139° 44'	*Daily predictions*						--	--	1.1	313°	--	--	1.1	133°

Endnotes can be found at the end of table 2.

TABLE 2 – CURRENT DIFFERENCES AND OTHER CONSTANTS

No.	PLACE	Meter Depth	POSITION Latitude	POSITION Longitude	TIME DIFFERENCES Min. before Flood	TIME DIFFERENCES Flood	TIME DIFFERENCES Min. before Ebb	TIME DIFFERENCES Ebb	SPEED RATIOS Flood	SPEED RATIOS Ebb	Minimum before Flood knots	Minimum before Flood Dir.	Maximum Flood knots	Maximum Flood Dir.	Minimum before Ebb knots	Minimum before Ebb Dir.	Maximum Ebb knots	Maximum Ebb Dir.
		ft	North	East	h m	h m	h m	h m			knots	Dir.	knots	Dir.	knots	Dir.	knots	Dir.
	NAIKAI (INLAND SEA) <110> Time meridian, 135° E					on Naruto, p.118												
5176	NARUTO		34° 14'	134° 39'		*Daily predictions*					– –	– –	6.2	350°	– –	– –	7.6	170°
5181	Muyano Seto		34° 11'	134° 37'	0 00	0 00	0 00	0 00	0.5	0.4	– –	– –	3.0	325°	– –	– –	3.0	140°
5186	Kitadomarino Seto		34° 14'	134° 35'	0 00	0 00	0 00	0 00	0.7	0.6	– –	– –	4.2	020°	– –	– –	4.2	195°
						on Tomogashima Suido, p.114												
5191	TOMOGASHIMA SUIDO (Yura Seto) <111>		34° 16'	135° 00'		*Daily predictions*					– –	– –	2.5	354°	– –	– –	2.5	174°
						on Akashi Kaikyo, p.122												
5196	AKASHI KAIKYO <111>		34° 37'	135° 02'		*Daily predictions*					– –	– –	4.2	302°	– –	– –	4.2	122°
5201	Harima Nada, central part <111>		34° 42'	134° 30'	+0 16	+0 26	+0 30	+0 18	0.1	0.1	– –	– –	0.5	260°	– –	– –	0.5	080°
5206	Eisan Seto, east part <111>		34° 27'	134° 04'	−0 36	−0 13	−0 36	−0 52	0.5	0.5	– –	– –	2.2	244°	– –	– –	2.2	064°
5211	Eisan Seto, west part <111>		34° 20'	133° 39'	−1 11	−0 48	−1 11	−1 27	0.3	0.3	– –	– –	1.4	250°	– –	– –	1.4	070°
						on Kurushima Kaikyo, p.126												
5216	Mihara Seto, north of Kone Shima		34° 20'	133° 04'	+0 20	+0 20	+0 20	+0 20	0.7	0.7	– –	– –	3.8	090°	– –	– –	3.6	270°
5221	KURUSHIMA KAIKYO (middle channel)		34° 07'	133° 00'		*Daily predictions*					– –	– –	5.8	180°	– –	– –	5.2	000°
5226	Kurushima Kaikyo (west channel)		34° 07'	132° 59'	+0 20	+0 20	+0 20	+0 20	1.0	1.0	– –	– –	5.5	180°	– –	– –	5.1	000°
5231	Aki Nada, east part		34° 08'	132° 52'	−0 10	−0 10	−0 10	−0 10	0.4	0.4	– –	– –	2.0	045°	– –	– –	2.0	225°
5236	Tsurushima Suido		33° 56'	132° 40'	+0 10	+0 10	+0 10	+0 10	0.4	0.4	– –	– –	2.0	045°	– –	– –	2.0	225°
5241	Kudako Suido		33° 58'	132° 34'	−0 40	−0 40	−0 40	−0 40	0.7	0.7	– –	– –	3.9	045°	– –	– –	3.7	225°
5246	Nuwa Shima Suido		33° 59'	132° 31'	−1 10	−1 10	−1 10	−1 10	0.7	0.7	– –	– –	3.9	000°	– –	– –	3.7	180°
5251	Moro Shima Suido		33° 57'	132° 29'	−1 30	−1 30	−1 30	−1 30	0.7	0.7	– –	– –	3.8	000°	– –	– –	3.6	180°
5256	Obatake Seto (narrows)		33° 57'	132° 11'	−2 20	−2 20	−2 20	−2 20	0.9	0.9	– –	– –	5.3	090°	– –	– –	4.9	270°
5261	Heigun Suido		33° 50'	132° 12'	−1 20	−1 20	−1 20	−1 20	0.2	0.2	– –	– –	1.1	090°	– –	– –	1.1	270°
5266	Iyo Nada, central part		33° 45'	132° 18'	−0 40	−0 40	−0 40	−0 40	0.2	0.2	– –	– –	1.1	045°	– –	– –	1.1	225°
5271	Suo Nada, west part		33° 52'	131° 11'	−1 40	−1 40	−1 40	−1 40	0.1	0.1	– –	– –	0.8	270°	– –	– –	0.8	090°
5276	Hoyo Kaikyo		33° 18'	131° 59'	−1 40	−1 40	−1 40	−1 40	0.6	0.6	– –	– –	3.5	000°	– –	– –	3.5	180°
5281	Bungo Suido, south end		32° 45'	132° 17'	−2 10	−2 10	−2 10	−2 10	0.2	0.2	– –	– –	1.0	000°	– –	– –	1.0	180°
						on Kanmon Kaikyo, p.130												
5286	KANMON KAIKYO (Hayatomo Seto)		33° 58'	130° 58'		*Daily Predictions*					– –	– –	5.6	270°	– –	– –	5.2	090°
5291	Kanmon Kaikyo (O Seto)		33° 55'	130° 56'	0 00	0 00	0 00	0 00	0.7	0.7	– –	– –	3.7	225°	– –	– –	3.5	045°
	KYUSHU, WEST COAST					on Kurushima Kaikyo, p.126												
5296	Hira Shima, 1.5 miles east of		33° 01'	129° 17'	−0 03	−0 20	−0 13	−0 28	0.4	0.4	– –	– –	2.2	028°	– –	– –	1.9	165°
5301	Yushima Seto, 2.3 miles SE of Dosaki		32° 38'	130° 22'	−1 44	−1 35	−2 01	−2 36	0.4	0.5	– –	– –	2.1	042°	– –	– –	2.8	191°
5306	Hayasaki Kaikyo, 2.7 mi. E of Gotsu Sho		32° 34'	130° 10'	−2 16	−2 10	−2 04	−2 16	1.0	0.9	– –	– –	5.8	121°	– –	– –	4.8	278°
	CHANGJIANG Time meridian, 120° E					on Changjiang Entrance, p.134												
5311	CHANGJIANG ENTRANCE		31° 08.23'	122° 00.47'		*Daily predictions*					– –	– –	2.6	305°	– –	– –	2.5	125°
						on Wusong Kou, p.138												
5316	WUSONG KOU		31° 24.92'	121° 31.98'		*Daily predictions*					– –	– –	2.7	290°	– –	– –	2.4	110°

Endnotes can be found at the end of table 2.

TABLE 2 – CURRENT DIFFERENCES AND OTHER CONSTANTS

No.	PLACE	Meter Depth	POSITION		TIME DIFFERENCES				SPEED RATIOS		AVERAGE SPEEDS AND DIRECTIONS							
			Latitude	Longitude	Min. before Flood	Flood	Min. before Ebb	Ebb	Flood	Ebb	Minimum before Flood		Maximum Flood		Minimum before Ebb		Maximum Ebb	
		ft	North	East	h m	h m	h m	h m			knots	Dir.	knots	Dir.	knots	Dir.	knots	Dir.
	SULU ARCHIPELAGO Time meridian, 120° E				on Basilan Strait, p.142													
5321	BASILAN STRAIT, off Zamboanga <112>		6° 54'	122° 04'	*Daily predictions*						– –	– –	2.2	270°	– –	– –	3.4	090°
5326	Basilan Strait, eastern entrance		6° 42'	122° 20'	+0 12	–0 03	–0 17	0 00	0.4	0.3	– –	– –	0.9	292°	– –	– –	1.4	127°
5331	Tapiantana Channel		6° 23'	122° 00'	–0 17	+0 11	+0 17	–0 43	0.8	0.6	– –	– –	1.5	288°	– –	– –	2.4	117°
5336	Canas Island, 1.5 miles west of		6° 28'	121° 53'	–0 12	–0 18	–0 09	–0 33	0.5	0.3	– –	– –	0.9	328°	– –	– –	1.4	157°
5341	Between Mataja I. and Sicagot I		6° 34'	121° 43'	+1 00	+0 54	+0 32	0 00	0.4	0.4	– –	– –	0.9	341°	– –	– –	1.4	166°
5346	Between Bubuan Island and Linawan Island		6° 20'	121° 57'	+0 01	–0 26	–0 31	–0 12	0.6	0.5	– –	– –	1.3	345°	– –	– –	2.0	150°
5351	Between Linawan I. and Tatalan I		6° 18'	121° 52'	+0 03	–0 04	+0 06	–0 04	0.8	0.6	– –	– –	1.5	323°	– –	– –	2.4	161°
5356	Tatalan Island, 4 miles southeast of		6° 11'	121° 54'	+0 15	+0 12	+0 06	0 00	0.8	0.6	– –	– –	1.5	298°	– –	– –	2.4	136°
5361	Tatalan Island, 7 miles west of		6° 16'	121° 43'	–0 18	+0 05	+0 30	–0 36	0.5	0.3	– –	– –	0.9	326°	– –	– –	1.4	133°
5366	Between Parol I. and Balanguingui I		6° 03'	121° 43'	–0 35	–0 35	–0 35	–1 15	0.8	0.9	– –	– –	1.8	308°	– –	– –	2.7	118°
5371	Between Jolo Island and Sulade Island		5° 54'	120° 49'	0 00	0 00	0 00	–1 15	1.1	0.8	– –	– –	2.0	319°	– –	– –	3.1	117°
5376	Between Kulassein I. and Tubigan I		6° 24'	120° 46'	–0 25	–0 43	–1 00	–0 44	0.6	0.5	– –	– –	1.1	314°	– –	– –	1.7	142°
5381	Between Cap Island and Tubalubac Island		5° 58'	120° 13'	–0 06	+0 12	0 00	–0 27	0.6	0.5	– –	– –	1.3	349°	– –	– –	2.0	169°
	ILOILO STRAIT				on Iloilo Strait, p.146													
5386	Pangasinan Point, 1.5 miles west of		10° 36'	122° 29'	– – –	–0 04	– – –	–0 17	0.4	0.4	– –	– –	0.5	063	– –	– –	1.0	226°
5391	Cabalic Point, 2.1 miles west of		10° 37'	122° 29'	– – –	+0 05	– – –	–0 07	0.4	0.4	– –	– –	0.6	058	– –	– –	1.1	226°
5396	Pituguan, midchannel, 1.5 miles NW of		10° 39'	122° 32'	– – –	+0 18	– – –	–0 14	0.8	0.8	– –	– –	0.7	073	– –	– –	2.2	228°
5401	Panay Club, 0.5 mile south of		10° 41'	122° 33'	–0 43	–0 29	–0 09	–0 43	0.6	0.6	– –	– –	1.2	066°	– –	– –	0.8	255°
5406	ILOILO STRAIT		10° 41'	122° 35'	*Daily predictions*						– –	– –	1.9	055°	– –	– –	1.8	235°
5411	Fort San Pedro, 0.2 mile northeast of		10° 41'	122° 35'	–0 43	–0 37	–1 20	–1 03	0.3	0.3	– –	– –	0.3	041°	– –	– –	0.8	226°
5416	Iloilo River entrance, 0.3 mile NE of		10° 42'	122° 35'	–0 19	–0 25	–0 44	–0 41	0.9	0.9	– –	– –	1.6	039°	– –	– –	1.7	216°
5421	Jaro Point, midchannel off of		10° 43'	122° 36'	+0 01	–0 13	–0 15	–0 37	0.7	0.7	– –	– –	1.4	010°	– –	– –	1.1	200°
5426	Doldol, midchannel north of		10° 46'	122° 39'	–0 27	–0 06	+0 09	–0 11	0.9	0.9	– –	– –	2.1	075°	– –	– –	1.2	240°
5431	Dumangas Point, 1.5 miles south of		10° 45'	122° 44'	–0 19	–0 08	+0 12	–0 14	0.8	0.6	– –	– –	1.5	109°	– –	– –	1.1	279°
5436	Navalas Point, 0.7 mile NNE of		10° 44'	122° 43'	+0 44	+0 12	+0 04	+0 21	0.6	0.7	– –	– –	1.1	129°	– –	– –	1.3	293°
	CEBU HARBOR				on Cebu Harbor, p.150													
5441	CEBU HARBOR, off Cebu City		10° 17'	123° 54'	*Daily predictions*						– –	– –	0.9	078°	– –	– –	1.0	252°
5446	Cebu Harbor, 0.6 mile NE of Opon Light		10° 19'	123° 57'	– – –	–0 11	– – –	+0 19	1.4	1.5	– –	– –	1.3	071°	– –	– –	1.5	244°
5451	Cebu Harbor, east entrance		10° 21'	123° 59'	– – –	–0 19	– – –	–0 14	0.3	0.3	– –	– –	0.3	075	– –	– –	0.3	239°
	HINATUAN PASSAGE				on San Juanico Strait, p.154													
5456	Rasa Island, southwest of		9° 47'	125° 34'	–0 52	+0 04	–0 50	–1 41	3.6	4.0	– –	– –	4.0	315°	– –	– –	4.0	135°
	SAN JUANICO STRAIT																	
5461	SAN JUANICO STRAIT, off Tacloban		11° 16'	125° 00'	*Daily predictions*						– –	– –	1.1	290°	– –	– –	1.0	110°
5466	Cauayan Point, southeast of		11° 20'	124° 58'	–0 29	–0 05	–0 07	–0 20	1.4	1.2	– –	– –	1.6	037°	– –	– –	1.2	225°
5471	San Juanico Strait, off Uban Point		11° 22'	124° 59'	–0 20	–0 40	–0 20	–0 40	1.5	1.6	– –	– –	1.6	000°	– –	– –	1.6	180°
5476	Torre Island, west of		11° 25'	124° 59'	–0 16	–0 17	–0 10	–0 03	1.4	1.0	– –	– –	1.6	339°	– –	– –	1.0	175°
5481	Janabatas Channel		11° 26'	124° 55'	–0 44	–0 01	–0 14	–0 07	1.0	0.5	– –	– –	1.1	276°	– –	– –	0.5	104°
5486	Janabatas Channel		11° 27'	124° 51'	–1 47	–0 12	–0 26	–0 28	1.3	0.3	– –	– –	1.4	266°	– –	– –	0.3	099°
	SAN BERNARDINO STRAIT				on San Bernardino Strait, p.158													
5491	SAN BERNARDINO STRAIT		12° 30'	124° 07'	*Daily predictions*						– –	– –	4.6	225°	– –	– –	3.8	045°

Endnotes can be found at the end of table 2.

ENDNOTES

<1> It is reported that an eddy is usually encountered along the ends of the municipal piers which makes docking difficult.

<2> San Pedro Channel, 7 miles south of Los Angeles Harbor Breakwater. There are two periodic currents here both of which are rotary, turning clockwise, and rather weak. The tidal current has a speed at strength of about 0.2 knot. The other current, due apparently to daily land and sea breezes, has a period of 24 hours and an average speed of about 0.2 knot. The greatest speed during 5 months of observations was 1.5 knots. Currents greater than 1 knot occur infrequently.

<3> In Los Angeles and Long Beach Harbors, the tidal current is weak. Currents can exceed 1 knot in the outer harbor at San Pedro, under strong wind conditions. Also, it is reported that three minute surge waves are responsible for major ship movements and damage.

<4> Observations indicate ebb is very weak.

<5> Large current eddies which cause ships to sheer off course are reported near the foundation piers of Golden Gate Bridge and San Francisco—Oakland Bay Bridge.

<6> See "Coastal Tidal Currents," (Table of Contents).

<7> Current is somewhat rotary, turning clockwise.

<8> SLACK WATER TIME DIFFERENCES FOR PLACES ALONG SAN FRANCISCO PIERS:

STATION or LOCALITY	Latitude	Longitude	Beginning of flood	ebb
	N	W	h. m.	h. m.
Time meridian, 120° W	on SAN FRANCISCO BAY ENTRANCE, p.8			
St. Francis Yacht Club breakwater	37° 48.5'	122° 26.5'	−0 10	−1 50
Aquatic Park, 0.2 mile west of	37° 48.6'	122° 25.7'	−0 35	−2 05
Pier 37	37° 48.6'	122° 24.5'	−1 35	−2 20
Pier 29	37° 48.4'	122° 24.0'	−1 10	−2 20
Pier 7	37° 48.0'	122° 23.6'	−0 55	−2 05
Pier 14	37° 47.7'	122° 23.3'	−0 55	−3 00
Pier 26	37° 47.4'	122° 23.0'	−1 40	−1 50
Pier 38	37° 47.0'	122° 23.0'	−0 25	−2 25
Pier 50	37° 46.4'	122° 22.8'	−1 40	−2 20
Bethlehem Pier No. 8	37° 45.6'	122° 22.7'	−1 20	−1 55
Pier 90, 0.5 mile SE. of	37° 44.5'	122° 22.4'	−1 50	−2 05
Point Avisadero	37° 43.7'	122° 21.3'	−1 25	−0 40
Point Avisadero, 0.8 mile south of	37° 43.0'	122° 21.5'	−1 30	−3 25

<9> Current is somewhat rotary, turning counterclockwise.

<10> Current is somewhat rotary, turning counterclockwise. 4h 25m prior to computed maximum flood the current flows southward with a speed 0.6 of the flood speed at the reference station.

<11> Data do not apply during freshets.

<12> Data do not apply during freshets.

<13> Data approximate.

<14> See "Coastal Tidal Currents," (Table of Contents).

<15> The Columbia River bar can be very dangerous because of sudden and unpredictable current changes accompanied by breakers. It is reported that ebb currents on the north side of the bar attain speeds of 6 to 8 knots and that strong NW winds sometimes cause currents that set north in the area outside the jetties. In the entrance, the currents are variable and may reach a speed of more than 5 knots on the ebb while the flood speed seldom exceeds 4 knots. The tidal current in the river is always modified by the river discharge, sometimes to the extent that the flood current is indiscernible and the current ebbs continuously.

<16> Flood and minimum current data indeterminate.

<17> Observations indicate that the current ebbs continuously at this location. Data are given for the smallest and largest mean ebb values expected. The time differences and speed ratios should be applied to the predicted times of maximum ebb at the reference station.

<18> During period of observations (February) flood was weak, and current was ebbing most of the time with a speed of about 2 knots at times of maximum.

ENDNOTES

<19> Along the west coast of Vancouver Island the current is reported to set always northwestward. It is weakest during westerly winds and strongest with easterly winds, being about a knot in moderate weather.

<20> When predicted flood at Admiralty Inlet, Race Rocks, or Strait of Juan de Fuca Entrance is marked with an (*) the flood speed and the preceding and following slacks at stations referred to them cannot be predicted. The current at most of these stations, however, will be weak at such times. Exceptions are the stations whose speed ratios are footnote reference <27>

<21> Current is rotary, turning clockwise.

<22> Time of minimum before flood is indefinite.

<23> Observations indicate that current is weak with direction variable for the greater part of the tidal cycle. A maximum flood speed of 1 knot in a southerly direction has been observed.

<24> Time of minimum before ebb is indefinite.

<25> Slacks are indefinite. The flood current is weak and variable, possibly ebbing at times.

<26> Current ebbs continuously. Maximum ebb, +5h 15m; minimum ebb, –1h 20m.

<27> Flood speed at strength probably does not become less than a knot.

<28> Current is rotary and erratic. Speeds of 3 knots may be encountered.

<29> Current ebbs most of the time. Time difference is for maximum ebb only. Weak current, flood or ebb, usually occurs about 0.8 hour after maximum flood at The Narrows.

<30> Current floods most of the time. Time difference is for maximum flood only. Weak ebb or slack water usually occurs about 1 hour before maximum ebb at The Narrows.

<31> Current ebbs most of the time. Time difference is for maximum ebb only. Weak flood or slack water usually occurs about 1 1/2 hours before maximum flood at The Narrows.

<32> Current floods most of the time. Time of minimum before flood is indefinite.

<33> Close to the east shore the flood speed is reduced about 1/2 but the ebb speed is only slightly less than at Point Evans.

<34> On the west side the speed of the flood current is 0.6 that of midstream and the ebb begins about 1 hour and 15 minutes earlier. On the east side the current is about the same as in midstream.

<35> Current ebbs most of the time. Time difference is for maximum ebb only. Weak flood or slack water usually occurs about 1 hour after maximum flood at The Narrows.

<36> Current ebbs most of the time. Time difference is for maximum ebb only. Weak flood or slack water usually occurs about the time of maximum flood at the Narrows.

<38> Current ebbs most of the time. Time difference is for maximum ebb only. Weak flood or slack water usually occurs about 1/2 hour after maximum flood at The Narrows.

<40> When predicted flood at Admiralty Inlet or Rosario Strait is marked with an (*) the flood speed and the preceding and following slacks at stations referred to them cannot be predicted. The current at most of these stations, however, will be weak at such times.

<41> Ebb current is irregular at times.

<42> Current is predominantly non–tidal, flowing in a northwesterly direction with a maximum speed of 1 knot.

<43> Current ebbs most of the time. Time difference is for maximum ebb only; slack times are indefinite and flood current is weak and variable.

<44> Time difference is for maximum flood only; slack times are indefinite and ebb current is too variable to be predicted.

<45> Dangerous eddy current and tide rips are reported to occur between Helmcken Island and Ripple Shoal around the time of ebb strength.

ENDNOTES

<46> On the flood, the streams coming from the sea through the north and south entrances meet off Evening Point (Lat. 53° 39' N) and separate on the falling tide about a mile farther northward.

<47> Observations indicate that current usually flows WNW, speed varying from zero to an average strength of 0.3 knot which occurs about 1 hour after time of maximum flood at Wrangell Narrows.

<48> Lewis Point to Guard Island—current too weak to be predicted.

<49> Observations indicate that current usually flows NW, speed varying from zero to an average strength of 1.2 knots which occurs about 45 minutes before time of maximum flood at Wrangell Narrows.

<50> Observations indicate that current usually flows NW, speed varying from zero to an average strength of 0.7 knot which occurs about 2 1/2 hours after time of maximum flood at Wrangell Narrows.

<51> Slacks occurs for a period of several hours before maximum current.

<52> Current usually flows WSW; speed varies from zero to an average of 1.1 knots occurring about 1h 05m earlier than time of maximum ebb at Wrangell Narrows.

<53> Slacks are indefinite. Flood current is too weak or variable to be predicted.

<54> Minimum before flood, 2h 41m before maximum flood; minimum before ebb, 3h 46m before maximum ebb.

<55> Lesser ebb, +0h 50m. The greater ebb may reach a maximum speed then decrease slightly for about 1 1/2 hours before increasing to a second maximum. These time differences are: 1st. maximum, –0h 42m; minimum, +0h 43m; second maximum, +1h 32m; and are referred only to the greater ebb phase at the reference station.

<56> Current too weak and variable to be predicted.

<57> Observations indicate that the current usually flows WNW with a non–tidal current of 0.6 knot.

<58> Currents are materially affected by winds.

<59> Northeast of Lively Island, it is reported that the current sets constantly northwestward, being stronger when the main stream west of the island sets northwestward.

<60> In the section of El Capitan Passage west of Dry Pass the current turns westward about the time of strength of eastward current in Dry Pass, and turns eastward about 1 hour before the time of strength of westward current in Dry Pass.

<61> Time difference is for maximum ebb only. Flood current is very erratic.

<62> Current frequently ebbs throughout the day, especially when moon is in quadrature.

<63> Slacks before flood may be variable.

<64> Observations in Frederick Sound during summer months indicate that the current usually flows northwestward, the speed varying with the tide. It apparently flows southeastward only on large tides

<66> The currents in Nakwasina Passage, except at the location 1 1/2 miles west of Allan Point, are too weak and variable to be predicted.

<67> Slacks are undetermined.

<68> Current is erratic in direction and strength at times.

<69> It is reported that currents are strong and passage is navigable only near time of slack water.

<70> Observations indicate that current usually flows northward, speed varying from zero to an average strength of 2 knots which occurs about 2.3 hours before time of maximum flood at North Inian Pass.

<71> A weak ebb probably occurs at this station when flood speed at North Inian Pass is less than 2 knots.

<72> It is reported that currents are strong and passage is navigable only near time of slack water.

<73> Observations indicate that current usually flows eastward with an average speed of 0.8 knot.

<74> It is reported that close inshore at Anchorage an eddy current flows up Knik Arm during the ebb.

ENDNOTES

<75> The tidal currents in this strait are weak except at the Slough and the Narrows where the speed at strength may amount to 2 or 3 knots on large tides.

<76> Current is rotary, turning clockwise. Minimum current about 0.1 knot, setting 160° true.

<77> Off Whirl Point, the speed of the current is about twice that off the Cannery.

<78> Dangerous tide rips occur in most of the passes in the Aleutian Islands when sea and swell oppose strong currents.

<79> Tidal current is weak and rotary, turning clockwise. Observations indicate a 0.2 knot westerly set.

<80> Ebb speed may not exceed 5.5 knots.

<81> When predicted ebb speed at Akutan Pass is less than 2 knots the current at this station is weak and variable.

<82> When predicted ebb speed at Unimak Pass is less than 1 knot the current at this station is weak and variable.

<83> Flood begins 1 hour before maximum ebb at Unimak Pass.

<84> The current at this station changes directions abruptly and the time of change is unpredictable. Maximum flood occurs about 10 minutes earlier and maximum ebb about 55 minutes earlier than the corresponding currents in Isanotski Strait. Usually there will be a strong current during the period from 1 1/2 hours before to 1 1/2 hours after the predicted times of maximum flood and ebb.

<85> Ratios are for greater flood and greater ebb only. The flood and ebb inequalities are small when the moon is near the equator. At other times there is considerable difference between the two floods and also the two ebbs in a day. The lesser flood may even become a small ebb at extreme declinations.

<86> Time difference for greater ebb and slack before greater ebb. Slack before greater flood and greater flood occur 7 hours and 12 hours respectively after greater ebb. Current floods for about 8 hours after greater flood.

<87> Flood speed ratio is for the 1st flood after greater ebb; the ebb speed ratio is for greater ebb.

<88> For greater flood and greater ebb only. The current is rotary, turning clockwise. At the predicted time of slack before greater flood, the current will run westward with speed of about 1.5 knots. At the predicted times of all other slacks and also lesser flood and lesser ebb (or minimum flood), the current will run northward with speed of about one knot.

<89> Current is rotary, turning clockwise. About 5 hours after time of greater ebb at Unimak Pass, current flows NW, speed ratio 0.4 and about 13 hours after greater ebb at Unimak Pass, current flows SE, speed ratio 0.5.

<90> Current is somewhat rotary, turning clockwise and is too variable to be predicted.

<91> Current is somewhat rotary, turning clockwise and is subject to considerable fluctuation. Approximate predictions are obtained through the following relations to the greater ebb at Unimak Pass: +1 1/2 hours, sets SSW, ratio 0.8; +9 hours, probably weak northerly set; + 18 hours, sets NNE, ratio 0.6.

<92> Current is relatively weak and rotary, turning clockwise. Data is for the greater ebb which is the most consistent phase.

<93> Current is somewhat rotary turning clockwise. At times given for slack, flood begins and slack, ebb begins the current probably flows WNW and ESE respectively, with speed of about 1.5 knot.

<94> The current changes from ebb to flood abruptly and predictions for beginning of flood are approximate only.

<95> Maximum flood 1 knot greater and maximum ebb 0.5 knot greater than corresponding speed at Kvichak Bay.

<96> Current is rotary turning clockwise. At the predicted times of slack before flood or ebb the current will run westward or eastward respectively with speed about 0.2 knot.

<97> Current is rotary turning clockwise. Difference and ratio are for maximum flood current only.

<98> Current is rotary turning clockwise. Midway between flood and ebb current is minimum (about 0.2 knot).

ENDNOTES

<99> Current is rotary turning clockwise. An average maximum speed of about 0.7 knot occurs in a SSW direction.

<100> Current flows in an ENE direction with an average speed of 1.1 knots. All values appearing in the ebb columns are actually those for a minimum flood.

<101> Time differences are for slack before greater flood, slack before greater ebb, and greater ebb. Maximum flood occurs about halfway between the times of the slacks obtained through differences. Speed ratios are for greater flood and greater ebb.

<102> Observations indicate that the current usually flows NNE with an average speed of 0.3 knot. Values in the ebb column are actually those for a minimum flood.

<103> Observations indicate that the current flows in a northerly direction with an average speed of 0.6 knot. Values in the ebb columns are actually those for a minimum flood.

<104> Observations indicate that the current flows in a NNE direction with an average speed of 0.9 knot. Values in the ebb columns are actually those for a minimum flood.

<105> Observations indicate the existence of a permanent current setting north with an average speed of 0.7 knot. Combined with the tidal current, the northward current may have an average speed varying from slack to 1.4 knots. The greatest observed speed off Maui Island was 2.7 knots.

<106> Observations indicate the current usually flows northwest on the west side of the channel near Kahoolawe Island with a maximum speed of 0.7 knot.

<107> Observations indicate that current usually flows SSE on east side of channel near Maui Island with a maximum speed of 0.4 knot.

<108> Current seldom floods. It decreases from maximum ebb to a minimum ebb or slack, then increases to maximum ebb again with no significant flow in the flood direction.

<109> Current sets to northeast with an average speed of about 0.3 knot.

<110> The general pattern of the flow into the Naikai is as follows. From the Kii Suido the flood current flows northward through Tomogashima Suido, Izumi Nada, Naruto and Muyano Seto, and westward through Akashi Kaikyo, Harima Nada and Bisan Seto to Bingo Nada. From the Bungo Suido the flood current flows northward through Hayasui Seto and then divides, one branch flowing westward to Shimonoseki Kaikyo and the other branch northeastward through Iyo Nada, Kudako Suido and environs, and Aki Nada. Continuing, the flood current then flows southward through Kurushima Kaikyo and northeastward through Mihara Seto to Bingo Nada. On the ebb the direction of flow is reversed. Bingo Nada is the area where the currents meet on the flood and separate on the ebb.

<111> The ratios and average speeds and directions are those of spring speeds.

<112> It is reported that the current at the pier at Zamboanga usually sets in a westerly direction.

<113> Current flows continuously in a westerly direction. Differences are for mean maximum speed.

<114> Current ebbs continuously. Differences are for mean maximum ebb only.

<115> Current floods continuously. Differences are for mean maximum flood only.

<116> Slacks are indefinite. Flood current is weak and variable. Differences are for mean maximum ebb only.

<117> Minimum before flood is indefinite. Flood current is weak and variable.

<118> Weak and variable current ebbs continuously in a southeasterly direction.

<119> Slacks are indefinite. Flood current is weak and variable. Differences are for a small ebb current.

<120> Current ebbs continuously with speeds varying from 0.7 knot (shown in the maximum flood column) to 1.5 knots.

<121>Due to distrubances caused by the structure, observed currents within 50 feet of the pier can be significantly different from the predictions.

Table 3.— SPEED OF CURRENT AT ANY TIME

EXPLANATION OF TABLES

Though the predictions in this publication give only the slacks and maximum currents, the speed of the current at any intermediate time can be obtained approximately by the use of this table. Directions for its use are given below the table.

Before using the table for a place listed in table 2, the predictions for the day in question should first be obtained by means of the differences and ratios given in table 2.

The examples below follow the numbered steps in the directions.

Example 1.—Find the speed of the current in San Francisco Bay Entrance (Golden Gate) at 4:00 on a day when the predictions which immediately precede and follow 4:00 are as follows:

(1)	Slack; flood begins	Maximum (Flood)	
	Time	Time	Speed
	2:19	5:25	3.2 knots

Directions under the table indicate Table A is to be used for this station.

(2) Interval between slack and maximum flood is 5:25 – 2:19 = $3^h\ 06^m$. Column heading nearest $3^h\ 06^m$ is $3^h\ 00^m$.

(3) Interval between slack and desired time is 4:00 – 2:19 = $1^h\ 41^m$. Line labeled $1^h\ 40^m$ is nearest $1^h\ 41^m$.

(4) Factor in column $3^h\ 00^m$ and on line $1^h\ 40^m$ is 0.8. The above flood speed of 3.2 knots multiplied by 0.8 gives a flood speed of 2.56 knots (or 2.6 knots, since one decimal is sufficient) for the time desired.

Example 2.—Find the speed of the current in Peril Strait at Kakul Narrows at 15:30 on a day when the predictions (obtained through the difference and ratio in table 2) which immediately precede and follow 15:30 are as follows:

(1)	Maximum (Ebb)		Slack; flood begins
	Time	Speed	Time
	13:59	2.8 knots	16:56

Directions under the table indicate table B is to be used, since this station in table 2 is referred to Sergius Narrows.

(2) Interval between slack and maximum ebb is 16:56 – 13:39 = $3^h\ 17^m$. Hence, use column labeled $3^h\ 20^m$.

(3) Interval between slack and time desired is 16:56 – 15:30 = $1^h\ 26^m$. Hence, use line labeled $1^h\ 20^m$.

(4) Factor in column $3^h\ 20^m$ and on line $1^h\ 20^m$ in 0.7. The above ebb speed of 2.8 knots multiplied by 0.7 gives an ebb speed of 2.0 knots for the desired time.

When the interval between slack and maximum current is greater than $5^h\ 40^m$, enter the table with one–half the interval between slack and maximum current and one–half the interval between slack and the desired time and use the factor thus found.

TABLE 3.—SPEED OF CURRENT AT ANY TIME

TABLE A

Interval between slack and desired time	Interval between slack and maximum current													
h. m.	*h. m.* 1 20	*h. m.* 1 40	*h. m.* 2 00	*h. m.* 2 20	*h. m.* 2 40	*h. m.* 3 00	*h.m.* 3 20	*h.m.* 3 40	*h.m.* 4 00	*h.m.* 4 20	*h.m.* 4 40	*h.m.* 5 00	*h.m.* 5 20	*h.m.* 5 40
	ft.	*ft.*	*ft.*	*ft.*	*ft.*	*ft.*	*ft.*	*ft.*	*ft.*	*ft.*	*ft.*	*ft.*	*ft.*	*ft.*
0 20	0.4	0.3	0.3	0.2	0.2	0.2	0.2	0.1	0.1	0.1	0.1	0.1	0.1	0.1
0 40	0.7	0.6	0.5	0.4	0.4	0.3	0.3	0.3	0.3	0.2	0.2	0.2	0.2	0.2
1 00	0.9	0.8	0.7	0.6	0.6	0.5	0.5	0.4	0.4	0.4	0.3	0.3	0.3	0.3
1 20	1.0	1.0	0.9	0.8	0.7	0.6	0.6	0.5	0.5	0.5	0.4	0.4	0.4	0.4
1 40	----	1.0	1.0	0.9	0.8	0.8	0.7	0.7	0.6	0.6	0.5	0.5	0.5	0.4
2 00	----	----	1.0	1.0	0.9	0.9	0.8	0.8	0.7	0.7	0.6	0.6	0.6	0.5
2 20	----	----	----	1.0	1.0	0.9	0.9	0.8	0.8	0.7	0.7	0.7	0.6	0.6
2 40	----	----	----	----	1.0	1.0	1.0	0.9	0.9	0.8	0.8	0.7	0.7	0.7
3 00	----	----	----	----	----	1.0	1.0	1.0	0.9	0.9	0.8	0.8	0.8	0.7
3 20	----	----	----	----	----	----	1.0	1.0	1.0	0.9	0.9	0.9	0.8	0.8
3 40	----	----	----	----	----	----	----	1.0	1.0	1.0	0.9	0.9	0.9	0.9
4 00	----	----	----	----	----	----	----	----	1.0	1.0	1.0	1.0	0.9	0.9
4 20	----	----	----	----	----	----	----	----	----	1.0	1.0	1.0	1.0	0.9
4 40	----	----	----	----	----	----	----	----	----	----	1.0	1.0	1.0	1.0
5 00	----	----	----	----	----	----	----	----	----	----	----	1.0	1.0	1.0
5 20	----	----	----	----	----	----	----	----	----	----	----	----	1.0	1.0
5 40	----	----	----	----	----	----	----	----	----	----	----	----	----	1.0

TABLE B

Interval between slack and desired time	Interval between slack and maximum current													
h. m.	*h. m.* 1 20	*h. m.* 1 40	*h. m.* 2 00	*h. m.* 2 20	*h. m.* 2 40	*h. m.* 3 00	*h. m.* 3 20	*h. m.* 3 40	*h. m.* 4 00	*h. m.* 4 20	*h. m.* 4 40	*h. m.* 5 00	*h. m.* 5 20	*h. m.* 5 40
	ft.	*ft.*	*ft.*	*ft.*	*ft.*	*ft.*	*ft.*	*ft.*	*ft.*	*ft.*	*ft.*	*ft.*	*ft.*	*ft.*
0 20	0.5	0.4	0.4	0.5	0.3	0.3	0.3	0.3	0.2	0.2	0.2	0.2	0.2	0.2
0 40	0.8	0.7	0.6	0.5	0.5	0.5	0.4	0.4	0.4	0.4	0.3	0.3	0.3	0.3
1 00	0.9	0.8	0.8	0.7	0.7	0.6	0.6	0.5	0.5	0.5	0.4	0.4	0.4	0.4
1 20	1.0	1.0	0.9	0.8	0.8	0.7	0.7	0.6	0.6	0.6	0.5	0.5	0.5	0.5
1 40	----	1.0	1.0	0.9	0.9	0.8	0.8	0.7	0.7	0.7	0.6	0.6	0.6	0.6
2 00	----	----	1.0	1.0	0.9	0.9	0.9	0.8	0.8	0.7	0.7	0.7	0.7	0.6
2 20	----	----	----	1.0	1.0	1.0	0.9	0.9	0.8	0.8	0.8	0.7	0.7	0.7
2 40	----	----	----	----	1.0	1.0	1.0	0.9	0.9	0.9	0.8	0.8	0.8	0.7
3 00	----	----	----	----	----	1.0	1.0	1.0	0.9	0.9	0.9	0.9	0.8	0.8
3 20	----	----	----	----	----	----	1.0	1.0	1.0	1.0	0.9	0.9	0.9	0.9
3 40	----	----	----	----	----	----	----	1.0	1.0	1.0	1.0	0.9	0.9	0.9
4 00	----	----	----	----	----	----	----	----	1.0	1.0	1.0	1.0	0.9	0.9
4 20	----	----	----	----	----	----	----	----	----	1.0	1.0	1.0	1.0	0.9
4 40	----	----	----	----	----	----	----	----	----	----	1.0	1.0	1.0	1.0
5 00	----	----	----	----	----	----	----	----	----	----	----	1.0	1.0	1.0
5 20	----	----	----	----	----	----	----	----	----	----	----	----	1.0	1.0
5 40	----	----	----	----	----	----	----	----	----	----	----	----	----	1.0

Use table A for all places except those listed below for table B.
Use table B for Deception Pass, Seymour Narrows, Sergius Narrows, Isanotski Strait. and all stations in table 2 which are referred to these points.

1. From predictions find the time of slack water and the time and velocity of maximum current (flood or ebb), one of which is immediately before and the other after the time for which the velocity is desired.
2. Find the interval of time between the above slack and maximum current, and enter the top of table A or B with the interval which most nearly agrees with this value.
3. Find the interval of time between the above slack and the time desired, and enter the side of table A or B with the interval which most nearly agrees with this value.
4. Find, in the table, the factor corresponding to the above two intervals, and multiply the maximum velocity by this factor. The result will be the approximate velocity at the time desired.

TABLE 4.—DURATION OF SLACK

The predicted times of slack water given in this publication indicate the instant of zero speed, which is only momentary. There is a period on each side of the slack water, however, during which the current is so weak that for practical purposes it may be considered negligible.

The following tables give, for various maximum currents, the approximate period of time during which weak currents not exceeding 0.1 to 0.5 knot will be encountered. This duration includes the last of the flood or ebb and the beginning of the following ebb or flood, that is, half of the duration will be before and half after the time of slack water.

Table A should be used for all places except those listed below for Table B.

Table B should be used for Cape Cod Canal, Hell Gate, Chesapeake and Delaware Canal, and all stations in Table 2 which are referred to them.

Duration of weak current near time of slack water

TABLE A

Maximum current	*Period with a speed not more than -*				
	0.1 knot	*0.2 knot*	*0.3 knot*	*0.4 knot*	*0.5 knot*
Knots	Minutes	Minutes	Minutes	Minutes	Minutes
1.0	23	46	70	94	120
1.5	15	31	46	62	78
2.0	11	23	35	46	58
3.0	8	15	23	31	38
4.0	6	11	17	23	29
5.0	5	9	14	18	23
6.0	4	8	11	15	19
7.0	3	7	10	13	16
8.0	3	6	9	11	14
9.0	3	5	8	10	13
10.0	2	5	7	9	11

TABLE B

Maximum current	*Period with a speed not more than -*				
	0.1 knot	*0.2 knot*	*0.3 knot*	*0.4 knot*	*0.5 knot*
Knots	*Minutes*	*Minutes*	*Minutes*	*Minutes*	*Minutes*
1.0	13	28	46	66	89
1.5	8	18	28	39	52
2.0	6	13	20	28	36
3.0	4	8	13	18	22
4.0	3	6	9	13	17
5.0	3	5	8	10	13

When there is a difference between the speeds of the maximum flood and ebb preceding and following the slack for which the duration is desired, it will be sufficiently accurate for practical purposes to find a separate duration for each maximum speed and take the average of the two as the duration of the weak current.

TABLE 5.—ROTARY TIDAL CURRENTS

Station Name	Depth	Hourly time increments 0	1	2	3	4	5	6	7	8	9	10	11	
	After Maximum Flood at KVICHAK BAY													
Point Riou, 2.6 nm SW	13	0.58	0.42	0.21	0.12	0.23	0.26	0.37	0.56	0.65	0.74	0.74	0.71	knots
		007	006	005	289	269	262	295	314	337	349	348	352	degrees
Claybluff Point Light, 5.2nm SSW	14	0.05	0.35	0.73	0.90	1.01	1.06	1.04	0.87	0.82	0.70	0.58	0.41	knots
		141	225	237	251	257	261	272	291	306	332	002	037	degrees
Kichyatt Point, 1.3nm NE	378	0.12	0.12	0.08	0.03	0.00	0.00	0.01	0.01	0.03	0.06	0.09	0.11	knots
		289	280	278	261	237	105	068	066	357	333	323	310	degrees
	After Maximum Flood at SERGIUS NARROWS													
Montague Point, 4.5 miles NE	71	0.45	0.55	0.58	0.58	0.57	0.45	0.26	0.12	0.16	0.24	0.24	0.27	knots
		276	288	285	286	287	285	276	245	192	187	209	247	degrees
Ship Channel, east of Smith Island	69	0.17	0.20	0.30	0.32	0.29	0.25	0.25	0.31	0.35	0.37	0.33	0.22	knots
		261	303	326	356	028	059	094	123	142	157	176	203	degrees
Johnston Point, 4 miles N	20	0.27	0.35	0.35	0.36	0.37	0.25	0.17	0.05	0.04	0.05	0.13	0.20	knots
		002	022	035	046	053	054	063	075	222	306	308	328	degrees
Gravina Point and Makaka Point, between	20	0.07	0.12	0.18	0.16	0.13	0.10	0.06	0.06	0.09	0.12	0.11	0.04	knots
		077	090	104	112	120	133	169	231	260	264	273	308	degrees

Tabular values are mean current speed and direction at specific intervals relative to the reference station.

COASTAL TIDAL CURRENTS

EXPLANATION

The term *coastal tidal* current is used here to designate the tidal current found offshore from 5 to 20 miles from the coast. The data were based upon observations made through the cooperation of the U.S. Coast Guard at a number of lightship stations along the Pacific coast from San Francisco to Swiftsure Bank, off the coast of Washington.

Rotary current.—Offshore, away from the immediate influence of the coast, the tidal current is quite different from the current found in inland tidal waters, Instead of setting in one direction for a period of 6 hours and in the opposite direction during the following period of 6 hours, the tidal current offshore changes its direction continually, so that in a period of about 12 ½ hours it will have set in all directions of the compass. The type of current is therefore called a rotary current.

Minimum current.—A characteristic feature of the rotary current is the absence of slack water. Although the current generally varies from hour to hour, this variation from greatest current to least current and back again to greater current does not give rise to a period of slack water. When the speed of the rotary tidal current is least, it is known as the minimum current, and when it is greatest it is known as the maximum current. The minimum and maximum speeds of the rotary current are thus related to each other in the same way as slack and strength of current, a minimum speed of the current following a maximum speed by an interval of about 3 hours and being followed in turn by another maximum after a further interval of 3 hours.

Changes in the tidal current.— The speeds of the tidal current given here are average speeds. Near the times when the Moon is full or new the speeds of the tidal current will be about 20 percent, or one-fifth greater than the average , and near the times of the Moon's first and third quarter the speeds will be smaller than the average by one-fifth.

Effect of wind.— It is to be carefully noted that, when a wind is blowing, the current a vessel will encounter is the resultant of the tidal and wind currents. Only the tidal currents together with the greatest observed speed of the current at each light vessel are given here, and the mariner is cautioned to combine with the tidal current the current brought about by any wind that may be blowing. Wind currents are given under the heading, "Wind-driven Currents".

Direction and Speed of currents.— The direction of the current is true, not magnetic, and is the direction *toward* which the current is setting, while the wind when given is in the direction *from* which it is blowing. The speed of the current is given in knots or nautical miles per hour.

Reference to tides.— The tidal currents on the Pacific coast, like the tides, exhibit the feature known as diurnal inequality; that is, the two floods of a day are unequal and likewise the two ebbs. In the case of the tide the higher of the two high waters of a day is known as higher high water, while the lower of the two is known as lower high water. For the two low waters of a day there are likewise distinctive names, the lower one being known as lower low water while the higher on is known as higher low water. In certain instances it is convenient to refer the currents to the tides, and where this is done the following symbols are used to designate the different tides: HH for higher high water, LH for lower high water, LL for lower low water, and HL for higher low water.

OBSERVATION STATIONS

Point Lobos, 8.7 miles WSW. of (former location of San Francisco Lightship), Calif.—The tidal current here is rotary, turning clockwise, as shown in figure 1, in which the average currents have been referred to each hour of the tides at San Francisco (Golden Gate). The predicted tides for this port will be found in the Tide Tables, West Coast of North and South America, issued annually in advance, by the National Ocean Service. The diurnal inequality here is so great that the current is very largely diurnal; that is during the greater part of the month the current changes direction at the rate of about 15° per hour, giving but one strength of flood and one strength of ebb in a day.

The speed of the tidal current here is generally small, as shown in the following table, which represents the average conditions of figure 1.

Time	Speed	Direction	Time	Speed	Direction
Tide Hrs.	Knot	True	Tide Hrs.	Knot	True
HH–3	0.1	060°	LL–3	0.2	170°
HH–2	0.1	070°	LL–2	0.3	180°
HH–1	0.1	085°	LL–1	0.3	210°
HH	0.1	100°	LL	0.3	240°
HH+1	0.1	120°	LL+1	0.3	275°
HH+2	0.1	145°	LL+2	0.4	300°
HH+3	0.2	160°	LL+3	0.4	325°
LH–2	0.3	000°	HL–2	0.2	110°
LH–1	0.3	015°	HL–1	0.2	125°
LH	0.2	030°	HL	0.2	140°
LH+1	0.2	050°	HL+1	0.2	150°
LH+2	0.2	080°	HL+2	0.1	130°

In the column headed "Time," in the above table, the minus (–) sign before the hours indicates that the time refereed to is before the particular tide, while the plus (+) sign indicates that the time is after the tide. Thus, HH–3 in figure 1 and in the table means 3 hours before higher high water, and LL+1 means 1 hour after lower low water.

COASTAL TIDAL CURRENTS

Tidal Current Curve, former location of San Francisco Lightship

Figure 1

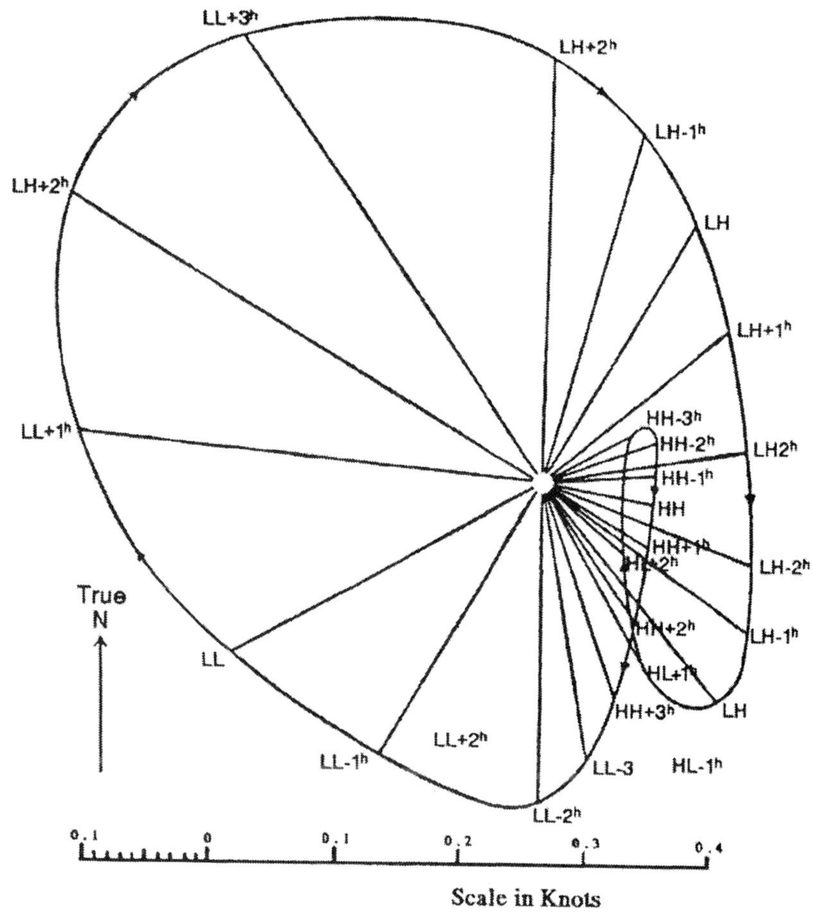

Referred to predicted time of tide at San Francisco (Golden Gate), Calif.

The current observations at this location indicated a permanent current in a northwesterly direction of about 0.1 knot. This was especially noticeable during the winter months. This permanent current, therefore, increases the speed of the tidal currents that set in the northwesterly direction and decreases the speed of the tidal currents setting in the southeasterly direction.

When there is considerable runoff from San Francisco Bay, the combined tidal and nontidal current at the former lightship location generally attains a speed of 1 1/2 knots in a northwesterly direction. The greatest observed speed was 2.9 knots.

Cape Mendocino Light, 4.6 miles west of (former location of Blunts Reef Lightship), Calif.—The tidal current here is rotary, but quite weak, being on the average less than 0.1 knot. At strength of flood the current sets north, and at strength of ebb it sets south. Since the tidal current is weak, it is generally masked by wind currents or other nontidal currents. The observations indicated the existence of a nontidal current setting southwesterly with an average speed of 0.2 knot from March to November and northwesterly with a like average speed from November to March. The greatest observed speed was 3 knots.

Columbia River Approach Lighted Horn Buoy R"C" (former location of the Columbia River Lightship), coast of Oregon.—The tidal current here is rotary, turning clockwise, but rather weak. The speed of the current at strength being about 0.3 knots setting 020° on the flood and 200° on the ebb.

The current from the Columbia River completely masks the flood current; observations showing that there is a notntidal current at the buoy location with an average speed of 0.4 knots setting 235° from February to October; and 295° from October to February. When there is considerable runoff from the river, the combined tidal and nontidal current at the buoy frequently attains a speed of 2 knots or more in a southwesterly direction. The greatest observed speed here is 3.5 knots.

Cape Alava, 4.4 miles west of (former location of Umatilla Reef Lightship), Wash.—The tidal current here is only slightly rotary. Strength of flood comes about one-fourth hour after the strength of flood in the entrance to the Strait of Juan de Fuca, setting 345° with a speed of 0.3 knot. Strength of ebb comes about one-fourth hour after the strength of ebb in the strait and sets 165° with a speed of 0.3 knot.

The tidal current here is generally masked by nontidal currents brought about by winds or other causes. Observations indicated the existence of a nontidal current, setting about 350° with a speed of 0.7 knot from November to April, with the greatest speed during the month of December, when it averaged about 1 knot. From April to November the nontidal current was variable, averaging 0.4 knot, generally in a southeasterly direction. With strong southeasterly winds the combined tidal and nontidal current attains a speed of 2 to 3 knots in a northerly direction. The greatest observed speed was 3.3 knots.

Swiftsure Bank (Latitude 48° 32'N.; Longitude 125° 00'W.).—The tidal current is distinctly rotary, turning clockwise twice each day, as shown in figure 2, in which the average currents have been referred to every hour of the tides at Astoria, Oregon The predicted tides for this port will be found in the Tide Tables, West Coast of North and South America, issued annually in advance, by the National Ocean service. As there is considerable difference between the speeds of the two revolutions which the tidal currents make each day, there are two distinct values for the flood and for the ebb currents, corresponding to the diurnal inequality of the tides.

The speed of the tidal currents here is generally small, being less than 1 knot, as shown in the following table, which represents the average conditions of figure 2.

Time	Speed	Direction	Time	Speed	Direction
Tide Hrs.	*Knot*	*True*	*Tide Hrs.*	*Knot*	*True*
HH–3	0.5	325°	LL–3	0.4	230°
HH–2	0.4	000°	LL–2	0.6	260°
HH–1	0.3	045°	LL–1	0.7	280°
HH	0.4	080°	LL	0.8	295°
HH+1	0.5	110°	LL+1	0.8	310°
HH+2	0.4	135°	LL+2	0.6	335°
HH+3	0.4	170°	LL+3	0.4	020°
LH–2	0.5	060°	HL–2	0.5	175°
LH–1	0.7	085°	HL–1	0.4	225°
LH	0.8	100°	HL	0.5	265°
LH+1	0.9	115°	HL+1	0.6	290°
LH+2	0.8	130°	HL+2	0.6	305°

Figure 2

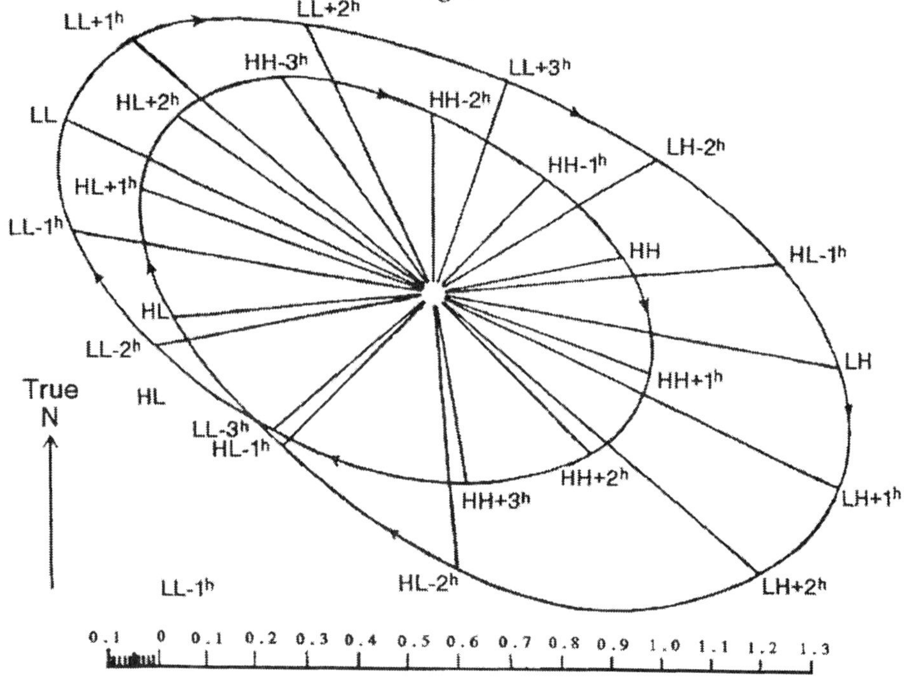

Scale in Knots

Tidal Current Curve, Swiftsure Bank. Referred to predicted time of tide at Astoria, Oregon

In the first column of the above table the letters under "Tide" refer to the different tides of the day, HH standing for higher high water, LH for lower high water, LL for lower low water, and HL for higher low water. The corresponding letters on figure 2 have a similar meaning. The minus (–) sign before the hours indicates that the time referred to is earlier than the particular tide, while the plus (+) sign indicates that the time is after the tide. Thus, HH–3 means 3 hours before higher high water, and LL+1 means 1 hour after lower low water at Astoria, Oregon

It is to be noted that the speeds and directions of the current given in the above table refer only to the tidal current. Observations indicate the existence of a permanent current setting 315° with an average speed of 0.5 knot. This makes the northwesterly currents considerably stronger than the southeasterly. A southeasterly current of as much as 1 *1/2 knots does not occur except with strong westerly or northwesterly winds, while northwesterly currents of 2 knots or more occur frequently. The greatest observed speed at Swiftsure Bank is 3 knots.*

Mau Island, Hawaii (Latitude 20° 46'N.; Longitude 155° 58'W.).—Observations indicate the existence of a permanent current setting north with an average speed of 0.7 knot. Combined with the tidal current, the northward current may have an average speed varying from slack to 1.4 knots. The greatest observed speed off Maui Island was 2.7 knots.

WIND-DRIVEN CURRENTS

A wind continuing for some time will produce a current the speed of which depends on the speed of the wind, and unless the current is deflected by some other cause, the deflective force of the earth's rotation will cause it to set to the right of the direction of the wind in the northern hemisphere and to the left in the southern hemisphere.

The current produced at off-shore locations by local winds of various strengths and directions was investigated from observations made at five lightships from San Francisco to Swiftsure Bank. The observations were made hourly for periods varying from 3 1/2 years to 9 years. The average given below and on the next page may prove helpful in estimating the probable current that may result from various winds at the several locations.

Caution.—There were of course many departures from these averages of speed and direction, for the wind-driven current often depends not only on the length of time the wind blows but also on factors other than the local wind at the time and place of the current. The mariner must not, therefore, assume that the given wind will always produce the indicated current.

It should be remembered, too, that the current which a vessel experiences at any time is the resultant of the combined actions of the tidal current, the wind-driven current, and any other currents such as the California Current or currents due to river dishcharge.

Speed.—The table below shows the average speed of current due to winds of various strengths.

Wind speed (miles per hour)	10	20	30	40	50
*Average current speed (knots) due to wind at following lightship stations:**					
San Francisco	*0.3*	*0.3*	*0.5*	*0.6*	*0.7*
Blunts Reef	*0.2*	*0.3*	*0.4*	*0.7*	*0.8*
Columbia River	*0.4*	*0.5*	*0.6*	*0.8*	*0.8*
Umatilla Reef	*0.2*	*0.6*	*0.9*	*1.0*	*0.9*
Swiftsure Bank	*0.5*	*0.5*	*0.5*	*0.7*	*0.8*

*All of these lightships have since been removed.

***Direction.**—The position of the shoreline with respect to the station influences considerably the direction of the currents due to certain winds. The following table shows for each station the average number of degrees by which the wind-driven current is deflected to the right or left of the wind. Thus at the former location of the San Francisco Lightship the table indicates that with a north wind the wind-driven current flows on the average 061°* west of south, and with an east wind it flows 023° north of west.

WIND-DRIVEN CURRENTS

AVERAGE DEVIATION OF CURRENT TO RIGHT OR LEFT OF WIND DIRECTION

Lightship Station*	SanFrancisco		Blunts Reef		Columbia River		Umatilla Reef		Swiftsure Bank	
	Left	Right	Left	Right	Left	Right	Left	Right	Left	Right
	*	*	*	*	*	*	*	*	*	*
Wind from—										
N	---	061	---	020	---	035	---	044	---	100
NNE	---	027	---	006	---	027	---	018	---	054
NE	---	030	---	010	---	009	---	034	---	048
ENE	---	031	---	032	---	029	---	048	---	033
E	---	023	---	028	---	017	---	052	---	027
ESE	---	029	---	007	---	002	---	038	---	018
SE	---	021	011	---	008	---	---	025	---	009
SSE	---	005	---	013	007	---	---	006	---	001
S	020	---	---	001	019	---	006	---	015	---
SSW	030	---	011	---	044	---	013	---	021	---
SW	049	---	018	---	074	---	032	---	068	---
WSW	040	---	028	---	121	---	052	---	088	---
W	051	---	060	---	---	145	077	---	090	---
WNW	---	033	---	002	---	105	006	---	---	082
NW	---	016	---	031	---	078	---	037	---	130
NNW	---	017	---	043	---	053	---	025	---	111

*All of these lightships have since been removed.

THE COMBINATION OF CURRENTS

In determining from the current tables the speed and direction of the current at any time, it is frequently necessary to combine the tidal current with the wind-driven current. The following methods indicate how the resultant of two or more currents may be easily determined.

Currents in the same direction.—*When two or more currents set in the same direction it is a simple matter to combine them. The resultant current will have a speed which is equal to the sum of all the currents and it will set in the same direction.*

For example, a vessel is near the former location of the San Francisco Lightship at a time when the tidal current is setting 345° with a speed of 0.5 knot, and at the same time a wind of 50 miles per hour is blowing from 150⊃; What current will the vessel be subject to at that time? Since a wind of 50 from 150° will give rise to a current setting 345° with a speed of 0.7 knot, the combined tidal and wind-driven currents will set in the same direction (345°) with a speed of 0.5 + 0.7 = 1.2 knots.

Currents in opposite directions.—*The combination of currents setting in opposite directions is likewise a simple matter. The speed of the smaller current is subtracted from the speed of the greater current, which gives speed of the resultant current; the direction of the resultant current is the same as that of the greater current.*

As an example, let it be required to determine the speed of the current at the former location of the San Francisco Lightship when the tidal current is setting 331° with a speed of 0.5 knot, and when a wind of 45 miles per hour is blowing from the northwest. The current produced by a wind of 45 miles per hour from northwest would set 151° with a speed of 0.6 knot. The tidal and wind currents, therefore, set in opposite directions, the wind current being the stronger. Hence, the resultant current will set in the direction of the wind current (151°) with a speed of 0.6 - 0.5 = 0.1 knot.

Currents in different directions.—*The combination of currents setting at arbitary angles is shown by a graphical method. Taking the combination of two currents as the simplest case, draw a line whose direction and length (to scale) represent the direction and speed of one of the currents to be combined. From the end of this line draw another line (to the same scale) representing the direction and speed of the second current. A line joining the beginning of the first line with the end of the second line represents the direction and speed of the combined current.*

As an example, take the former location of the Umatilla Reef Lightship at a time when the tidal current is 0.4 knot setting 315° and a wind of 50 miles per hour is blowing from 273° . The wind-driven current, according to the preceding section, would be about 0.9 knot setting 025° .

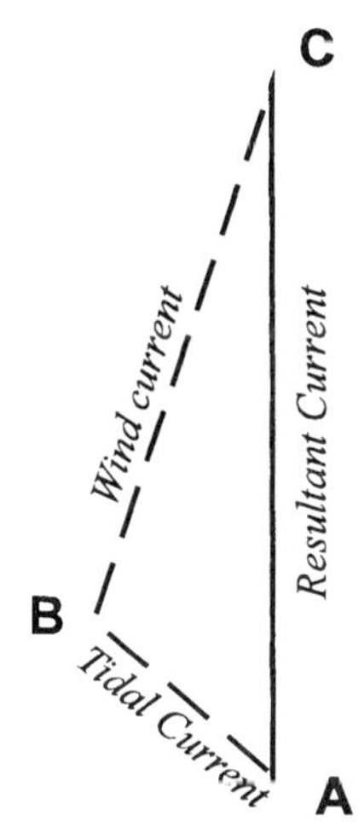

Combination of tidal current and wind-driven current

Using a scale of 2 inches to represent 1 knot draw from point A in the diagram above, the line *AB* 0.8 inches in length directed 315° to represent the tidal current. from *B* draw the line *BC* 1.8 inches in length directed 025° to represent the wind-current. The line AC represents the resultant current, which on being measured, is found to be about 2.2 inches in length directed 005°. Therefore, the combined current sets 005° with a speed of 1.1 knots.

The combination of three or more currents is made in the same way as above, for example, the third current to be combined being drawn from point C. The resultant current is given by joining the origin with the end of the last line. For drawing the lines, a parallel rule and compass rose is convenient. A protractor or polar coordinate paper may also be used.

PUBLICATIONS RELATING TO TIDES AND TIDAL CURRENTS

TIDE TABLES

Advance information relative to the rise and fall of the tide is given in annual tide tables. These tables include the predicted times and heights of high and low waters for every day in the year for a number of reference stations and differences for obtaining similar predictions for numerous other places.

Tide Tables, Central and Western Pacific Ocean and Indian Ocean.
Tide Tables, East Coast of North and South America (Including Greenland).
Tide Tables, Europe and West Coast of Africa (Including the Mediterranean Sea).
Tide Tables, West Coast of North and South America (Including the Hawaiian Islands).

TIDAL CURRENT TABLES

Accompanying the rise and fall of the tide is a periodic horizontal flow of the water known as the tidal current. Advance information relative to these currents is made available in annual tidal current tables which include daily predictions of the times of slack water and the times and velocities of strength of flood and ebb currents for a number of waterways together with differences for obtaining predictions for numerous other places.

Tidal Current Tables, Atlantic Coast of North America.
Tidal Current Tables, Pacific Coast of North America and Asia.

TIDAL CIRCULATION AND WATER LEVEL FORECAST ATLAS

This atlas series provides a composite view of the total tidal phenomenon for major estuaries of the United States. Twelve pairs of hourly charts are used to depict the information. The first of each pair presents contours of mean tidal height throughout the specified estuary. Inserts and a vertical section provide additional tidal information. The second chart presents the speed and direction of the tidal current throughout the area. Corange, cotidal, cospeed, and cophase charts are shown in the following section. Finally, daily tide predictions are provided for important locations, in tabular form, for several years in advance.

Tidal Circulation and Water Level Forecast Atlas, Delaware River and Bay.

OFFICIAL U.S. DATUMS

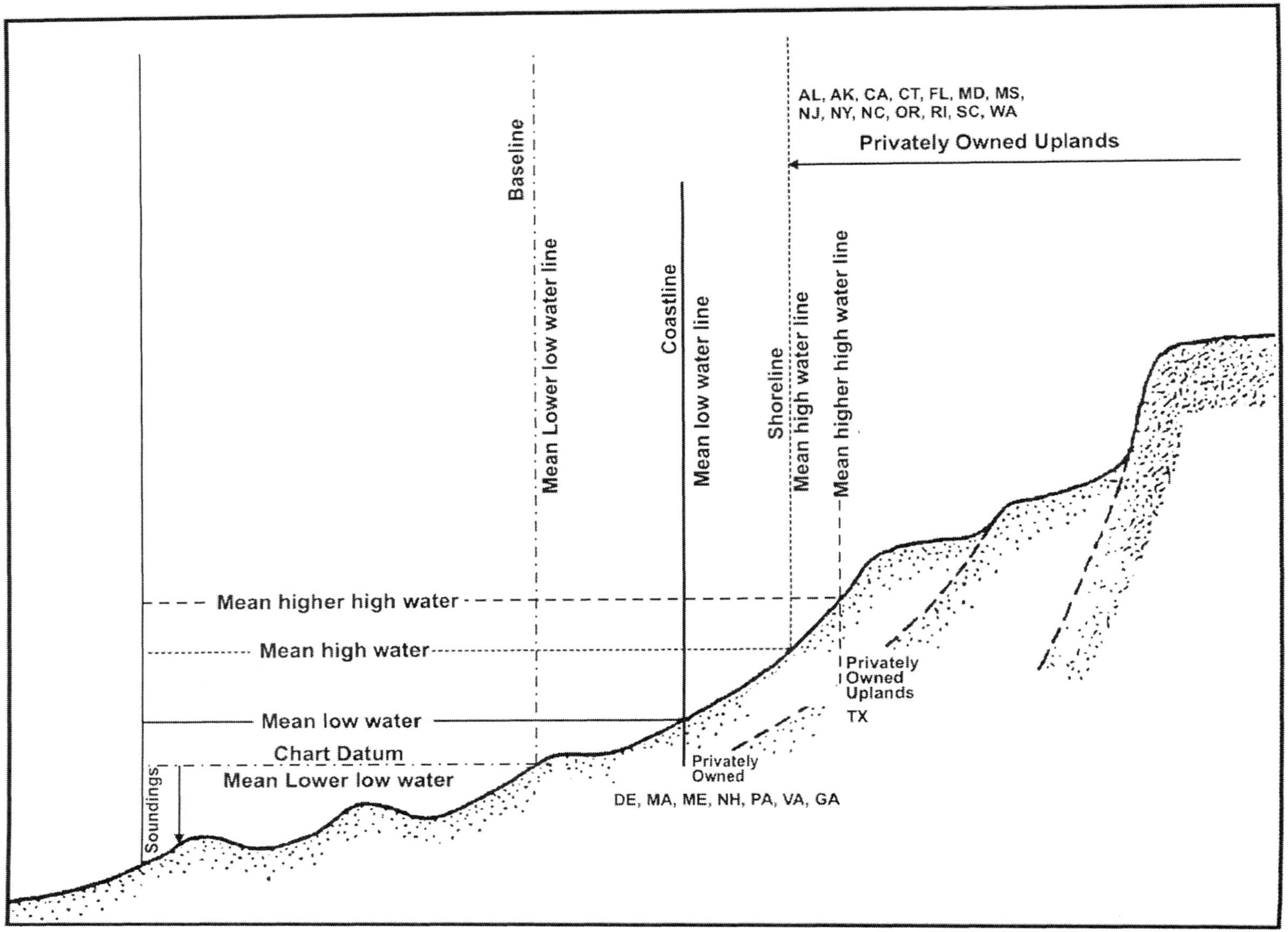

GLOSSARY OF TERMS

ANNUAL INEQUALITY—Seasonal variation in the water level or current, more or less periodic, due chiefly to meteorological causes.

APOGEAN TIDES OR TIDAL CURRENTS—Tides of decreased range or currents of decreased speed occurring monthly as the result of the Moon being in apogee (farthest from the Earth).

AUTOMATIC TIDE GAGE—An instrument that automatically registers the rise and fall of the tide. In some instruments, the registration is accomplished by recording the heights at regular intervals in digital format, in others by a continuous graph in which the height versus corresponding time of the tide is recorded.

BENCH MARK (BM)—A fixed physical object or marks used as reference for a vertical datum. A *tidal bench mark is* one near a tide station to which the tide staff and tidal datums are referred. A *Geodetic bench mark* identifies a surveyed point in the National Geodetic Vertical Network.

CHART DATUM—The tidal datum to which soundings on a chart are referred. It is usually taken to correspond to low water elevation of the tide, and its depression below mean sea level is represented by the symbol Zo.

CURRENT—Generally, a horizontal movement of water. Currents may be classified as *tidal* and *nontidal.* Tidal currents are caused by gravitational interactions between the Sun, Moon, and Earth and are a part of the same general movement of the sea that is manifested in the vertical rise and fall, called *tide*. Nontidal currents include the permanent currents in the general circulatory systems of the sea as well as temporary currents arising from more pronounced meteorological variability.

CURRENT DIFFERENCE—Difference between the time of slack water (or minimum current) or strength of current in any locality and the time of the corresponding phase of the tidal current at a reference station, for which predictions are given in the *Tidal Current Tables*.

CURRENT ELLIPSE—A graphic representation of a rotary current in which the velocity of the current at different hours of the tidal cycle is represented by radius vectors and vectorial angles. A line joining the extremities of the radius vectors will form a curve roughly approximating an ellipse. The cycle is completed in one-half tidal day or in a whole tidal day according to whether the tidal current is of the semidiurnal or the diurnal type. A current of the mixed type will give a curve of two unequal loops each tidal day.

CURRENT METER—An instrument for measuring the speed and direction or just the speed of a current. The measurements are usually Eulerian since the meter is most often fixed or moored at a specific location.

DATUM (vertical)—For marine applications, a base elevation used as a reference from which to reckon heights or depths. It is called a *tidal datum* when defined by a certain phase of the tide. Tidal datums are local datums and should not be extended into areas which have differing topographic features without substantiating measurements. In order that they may be recovered when needed, such datums are referenced to fixed points known as *bench marks*.

DAYLIGHT SAVING TIME—A time used during the summer in some localities in which clocks are advanced 1 hour from the usual standard time.

DIURNAL—Having a period or cycle of approximately 1 tidal day. Thus, the tide is said to be diurnal when only one high water and one low water occur during a tidal day, and the tidal current is said to be diurnal when there is a single flood and single ebb period in the tidal day. A rotary current is diurnal if it changes its direction through all points of the compass once each tidal day.

DIURNAL INEQUALITY—The difference in height of the two high waters or of the two low waters of each day; also the difference in speed between the two flood tidal currents or the two ebb tidal currents of each day. The difference changes with the declination of the Moon and to a lesser extent with the declination of the Sun. In general, the inequality tends to increase with an increasing declination, either north or south, and to diminish as the Moon approaches the Equator. *Mean diurnal high water inequality* (DHQ) is one-half the average difference between the two high waters of each day observed over a specific 19-year Metonic cycle (the National Tidal Datum Epoch). It is obtained by subtracting the mean of all high waters from the mean of the higher high waters. *Mean diurnal low water inequality* (DLQ) is one-half the average difference between the two low waters of each day observed over a specific 19-year Metonic cycle (the National Tidal Datum Epoch). It is obtained by subtracting the mean of the lower low waters from the mean of all low waters. *Tropic high water inequality* (HWQ) is the average difference between the two high waters of the day at the times of the tropic tides. *Tropic low water inequality* (LWQ) is the average difference between the two low waters of the day at the times of the tropic tides. Mean and tropic inequalities as defined above are applicable only when the type of tide is either semidiurnal or mixed. Diurnal inequality is sometimes called *declinational inequality.*

DOUBLE EBB—An ebb tidal current where, after ebb begins, the speed increases to a maximum called *first ebb*; it then decreases, reaching a *minimum ebb* near the middle of the ebb period (and at some places it may actually run in a flood direction for a short period); it then again ebbs to a maximum speed called second ebb after which it decreases to slack water.

DOUBLE FLOOD—A flood tidal current where, after flood begins, the speed increases to a maximum called first flood; it then decreases, reaching a minimum flood near the middle of the flood period (and at some places it may actually run in an ebb direction for a short period); it then again floods to a maximum speed called second flood after which it decreases to slack water.

DOUBLE TIDE—A double-headed tide, that is, a high water consisting of two maxima of nearly the same height separated by a relatively small depression, or a low water consisting of two minima separated by a relatively small elevation. Sometimes, it is called an agger.

DURATION OF FLOOD AND DURATION OF EBB—Duration of flood is the interval of time in which a tidal current is flooding, and the *duration of ebb* is the interval in which it is ebbing. Together they cover, on an average, a period of 12.42 hours for a semidiurnal tidal current or a period of 24.84 hours for a diurnal current. In a normal semidiurnal tidal current, the duration of flood and duration of ebb will each be approximately equal to 6.21 hours, but the times may be modified greatly by the presence of a nontidal flow. In a river the duration of ebb is usually longer than the duration of flood because of the freshwater discharge, especially during the spring when snow and ice melt are the predominant influences.

DURATION OF RISE AND DURATION OF FALL—*Duration of rise* is the interval from low water to high water, and *duration of fall* is the interval from high water to low water. Together they cover, on an average, a period of 12.42 hours for a semidiurnal tide or a period of 24.84 hours for a diurnal tide. In a normal semidiurnal tide, the duration of rise and duration of fall will each be approximately equal to 6.21 hours, but in shallow waters and in rivers there is a tendency for a decrease in the duration of rise and a corresponding increase in the duration of fall.

EBB CURRENT—The movement of a tidal current away from shore or down a tidal river or estuary. In the mixed type of reversing tidal current, the terms *greater ebb* and *lesser ebb* are applied respectively to the ebb tidal currents of greater and lesser speed of each day. The terms *maximum ebb* and *minimum ebb* are applied to the maximum and minimum speeds of a current running continuously ebb, the speed alternately increasing and decreasing without coming to a slack or reversing. The expression maximum ebb is also applicable to any ebb current at the time of greatest speed.

EQUATORIAL TIDAL CURRENTS—Tidal currents occurring semimonthly as a result of the Moon being over the Equator. At these times the tendency of the Moon to produce a diurnal inequality in the tidal current is at a minimum.

EQUATORIAL TIDES—Tides occurring semi monthly as the result of the Moon being over the Equator. At these times the tendency of the Moon to produce a diurnal inequality in the tide is at a minimum.

FLOOD CURRENT—The movement of a tidal current toward the shore or up a tidal river or estuary. In the mixed type of reversing current, the terms *greater flood* and *lesser flood* are applied respectively to the flood currents of greater and lesser speed of each day. The terms *maximum flood* and *minimum flood* are applied to the maximum and minimum speeds of a flood current, the speed of which alternately increases and decreases without coming to a slack or reversing. The expression maximum flood is also applicable to any flood current at the time of greatest speed.

GREAT DIURNAL RANGE (Gt)—The difference in height between mean higher high water and mean lower low water. The expression may also be used in its contracted form, *diurnal range*.

GREENWICH INTERVAL—An interval referred to the transit of the Moon over the meridian of Greenwich as distinguished from the local interval which is referred to the Moon's transit over the local meri-dian. The relation in hours between Greenwich and local intervals may be expressed by the formula:

Greenwich interval = local interval +0.069 L

where L is the west longitude of the local merid-ian in degrees. For east longitude, L is to be con-sidered negative.

GULF COAST LOW WATER DATUM—A chart datum. Specifically, the tidal datum formerly designated for the coastal waters of the Gulf Coast of the United States. It was defined as *mean lower low water* when the type of tide was mixed and *mean low water* when the type of tide was diurnal.

HALF-TIDE LEVEL—See *mean tide level.*

GLOSSARY OF TERMS

HARMONIC ANALYSIS—The mathematical process by which the observed tide or tidal current at any place is separated into basic harmonic constituents.

HARMONIC CONSTANTS—The amplitudes and epochs of the harmonic constituents of the tide or tidal current at any place.

HARMONIC CONSTITUENT—One of the harmonic elements in a mathematical expression for the tide-producing force and in corresponding formulas for the tide or tidal current. Each constituent represents a periodic change or variation in the relative positions of the Earth, Moon, and Sun. A single constituent is usually written in the form $y=A\cos(at+\alpha)$, in which y is a function of time as expressed by the symbol t and is reckoned from a specific origin. The coefficient A is called the amplitude of the constituent and is a measure of its relative importance. The angle $(at+\alpha)$ changes uniformly and its value at any time is called the phase of the constituent. The speed of the constituent is the rate of change in its phase and is represented by the symbol a in the formula. The quantity α is the phase of the constituent at the initial instant from which the time is reckoned. The period of the constituent is the time required for the phase to change through 360° and is the cycle of the astronomical condition represented by the constituent.

HIGH WATER (HW)—The maximum height reached by a rising tide. The height may be due solely to the periodic tidal forces or it may have superimposed upon it the effects of prevailing meteorological conditions. Use of the synonymous term, *high tide*, is discouraged.

HIGHER HIGH WATER (HHW)—The higher of the two high waters of any tidal day.

HIGHER LOW WATER (HLW)—The higher of the two low waters of any tidal day.

HYDRAULIC CURRENT—A current in a channel caused by a difference in the surface level at the two ends. Such a current may be expected in a strait connecting two bodies of water in which the tides differ in time or range. The current in the East River, N.Y., connecting Long Island Sound and New York Harbor, is an example.

KNOT—A unit of speed, one international nautical mile (1,852.0 meters or 6,076.11549 international feet) per hour.

LOW WATER (LW)—The minimum height reached by a falling tide. The height may be due solely to the periodic tidal forces or it may have superimposed upon it the effects of meteorological conditions. Use of the synonymous term, *low tide*, is discouraged.

LOWER HIGH WATER (LHW)—The lower of the two high waters of any tidal day.

LOWER LOW WATER (LLW)—The lower of the two low waters of any tidal day.

LUNAR DAY—The time of the rotation of the Earth with respect to the Moon, or the interval between two successive upper transits of the Moon over the meridian of a place. The mean lunar day is approximately 24.84 solar hours long, or 1.035 times as long as the mean solar day.

LUNAR INTERVAL—The difference in time between the transit of the Moon over the meridian of Greenwich and over a local meridian. The average value of this interval expressed in hours is 0.069 L, in which L is the local longitude in degrees, positive for west longitude and negative for east longitude. The lunar interval equals the difference between the local and Greenwich interval of a tide or current phase.

LUNICURRENT INTERVAL—The interval between the Moon's transit (upper or lower) over the local or Greenwich meridian and a specified phase of the tidal current following the transit. Examples: *strength of flood interval and strength of ebb interval,* which may be abbreviated to *flood interval and ebb interval,* respectively. The interval is described as local or Greenwich according to whether the reference is to the Moon's transit over the local or Greenwich meridian. When not otherwise specified, the reference is assumed to be local.

LUNITIDAL INTERVAL—The interval between the Moon's transit (upper or lower) over the local or Greenwich meridian and the following high or low water. The average of all high water intervals for all phases of the Moon is known as *mean high water lunitidal interval* and is abbreviated to high water interval (HWI). Similarly the *mean low water lunitidal interval* is abbreviated to *low water interval* (LWI). The interval is described as local or Greenwich according to whether the reference is to the transit over the local or Greenwich meridian. When not otherwise specified, the reference is assumed to be local.

MEAN HIGH WATER (MHW)—A tidal datum. The arithmetic mean of the high water heights observed over a specific 19-year Metonic cycle (the National Tidal Datum Epoch). For stations with shorter series, simultaneous observational comparisons are made with a primary control tide station in order to derive the equivalent of a l9-year value.

GLOSSARY OF TERMS

MEAN HIGHER HIGH WATER (MHHW)—A tidal datum. The arithmetic mean of the higher high water heights of a mixed tide observed over a specific l9-year Metonic cycle (the National Tidal Datum Epoch). Only the higher high water of each pair of high waters, or the only high water of a tidal day is included in the mean.

MEAN HIGHER HIGH WATER LINE (MHHWL)—The intersection of the land with the water surface at the elevation of mean higher high water.

MEAN LOW WATER (MLW)—A tidal datum. The arithmetic mean of the low water heights observed over a specific l9-year Metonic cycle (the National Tidal Datum Epoch). For stations with shorter series, simultaneous observational comparisons are made with a primary control tide station in order to derive the equivalent of a l9-year value.

MEAN LOW WATER SPRINGS (MLWS)—A tidal datum. Frequently abbreviated *spring low water.* The arithmetic mean of the low water heights occurring at the time of the spring tides observed over a specific l9-year Metonic cycle (the National Tidal Datum Epoch).

MEAN LOWER LOW WATER (MLLW)—A tidal datum. The arithmetic mean of the lower low water heights of a mixed tide observed over a specific l9-year Metonic cycle (the National Tidal Datum Epoch). Only the lower low water of each pair of low waters, or the only low water of a tidal day is included in the mean.

MEAN RANGE OF TIDE (Mn)—The difference in height between mean high water and mean low water.

MEAN RIVER LEVEL—A tidal datum. The average height of the surface of a tidal river at any point for all stages of the tide observed over a l9-year Metonic cycle (the National Tidal Datum Epoch), usually determined from hourly height readings. In rivers subject to occasional freshets the river level may undergo wide variations, and for practical purposes certain months of the year may be excluded in the determination of tidal datums. For charting purposes, tidal datums for rivers are usually based on observations during selected periods when the river is at or near low water stage.

MEAN SEA LEVEL (MSL)—A tidal datum. The arithmetic mean of hourly water elevations observed over a specific l9-year Metonic cycle (the National Tidal Datum Epoch). Shorter series are specified in the name; e.g., monthly mean sea level and yearly mean sea level.

MEAN TIDE LEVEL (MTL)—Also called half-tide level. A tidal datum midway between mean high water and mean low water.

MIXED TIDE—Type of tide with a large inequality in the high and/or low water heights, with two high waters and two low waters usually occurring each tidal day. In strictness, all tides are mixed but the name is usually applied to the tides intermediate to those predominantly semidiurnal and those predominantly diurnal.

NATIONAL TIDAL DATUM EPOCH—The specific 19-year period adopted by the National Ocean Service as the official time segment over which tide observations are taken and reduced to obtain mean values (e.g., mean lower low water, etc.) for tidal datums. It is necessary for standardization because of periodic and apparent secular trends in sea level. The present National Tidal Datum Epoch is 1960 through 1978. It is reviewed annually for possible revision and must be actively considered for revision every 25 years.

NEAP TIDES OR TIDAL CURRENTS—Tides of decreased range or tidal currents of decreased speed occurring semimonthly as the result of the Moon being in quadrature. The *neap range* (Np) of the tide is the average semidiurnal range occurring at the time of neap tides and is most conveniently computed from the harmonic constants. It is smaller than the mean range where the type of tide is either semidiurnal or mixed and is of no practical significance where the type of tide is diurnal. The average height of the high waters of the neap tides is called *neap high water* or *high water neaps* (MHWN) and the average height of the corresponding low waters is called neap low water or low water neaps (MLWN).

PERIGEAN TIDES OR TIDAL CURRENTS—Tides of increased range or tidal currents of increased speed occurring monthly as the result of the Moon being in perigee or nearest the Earth. The *perigean range* (Pn) of tide is the average semidiurnal range occurring at the time of perigean tides and is most conveniently computed from the harmonic constants. It is larger than the mean range where the type of tide is either semidiurnal or mixed, and is of no practical significance where the type of tide is diurnal.

RANGE OF TIDE—The difference in height between consecutive high and low waters, the *mean range* is the difference in height between mean high water and mean low water. Where the type of tide is diurnal the mean range is the same as the diurnal range. For other ranges, see great diurnal, spring, neap, perigean, apogean, and tropic tides.

REFERENCE STATION—A tide or current station for which independent daily predictions are given in the *Tide Tables and Tidal Current Tables,* and from which corresponding predictions are obtained for subordinate stations by means of differences and ratios.

REVERSING CURRENT—A tidal current which flows alternately in approximately opposite directions with a slack water at each reversal of direction. Currents of this type usually occur in rivers and straits where the direction of flow is more or less restricted to certain channels. When the movement is towards the shore or up a stream, the current is said to be flooding, and when in the opposite direction it is said to be ebbing. The combined flood and ebb movement including the slack water covers, on an average, 12.42 hours for the semidiurnal current. If unaffected by a nontidal flow, the flood and ebb movements will each last about 6 hours, but when combined with such a flow, the durations of flood and ebb may be quite unequal. During the flow in each direction the speed of the current will vary from zero at the time of slack water to a maximum about midway between the slacks.

ROTARY CURRENT—A tidal current that flows continually with the direction of flow changing through all points of the compass during the tidal period. Rotary currents are usually found offshore where the direction of flow is not restricted by any barriers. The tendency for the rotation in direction has its origin in the Coriolis force and, unless modified by local conditions, the change is clockwise in the Northern Hemisphere and counterclockwise in the Southern. The speed of the current usually varies throughout the tidal cycle, passing through the two maxima in approximately opposite directions and the two minima with the direction of the current at approximately 90° from the direction at time of maximum speed.

SEMIDIURNAL—Having a period or cycle of approximately one-half of a tidal day. The predominating type of tide throughout the world is semidiurnal, with two high waters and two low waters each tidal day. The tidal current is said to be semidiurnal when there are two flood and two ebb periods each day.

SET (OF CURRENT)—The direction *towards* which the current flows.

SLACK WATER—The state of a tidal current when its speed is near zero, especially the moment when a reversing current changes direction and its speed is zero. The term is also applied to the entire period of low speed near the time of turning of the current when it is too weak to be of any practical importance in navigation. The relation of the time of slack water to the tidal phases varies in different localities. For standing tidal waves, slack water occurs near the times of high and low water, while for progressive tidal waves, slack water occurs midway between high and low water.

SPRING TIDES OR TIDAL CURRENTS—Tides of increased range or tidal currents of increased speed occurring semimonthly as the result of the Moon being new or full. The *spring range* (Sg) of tide is the average semidiurnal range occurring at the time of spring tides and is most conveniently computed from the harmonic constants. It is larger than the mean range where the type of tide is either semidiurnal or mixed, and is of no practical significance where the type of tide is diurnal. The mean of the high waters of the spring tide is called *spring high water or mean high water springs* (MHWS), and the average height of the corresponding low waters is called *spring low water or mean low water springs* (MLWS).

STAND OF TIDE—Sometimes called a platform tide. An interval at high or low water when there is no sensible change in the height of the tide. The water level is stationary at high and low water for only an instant, but the change in level near these times is so slow that it is not usually perceptible. In general, the duration of the apparent stand will depend upon the range of tide, being longer for a small range than for a large range, but where there is a tendency for a double tide the stand may last for several hours even with a large range of tide.

STANDARD TIME—A kind of time based upon the transit of the Sun over a certain specified meridian, called the *time meridian*, and adopted for use over a considerable area. With a few exceptions, standard time is based upon some meridian which differs by a multiple of 15° from the meridian of Greenwich.

STRENGTH OF CURRENT—Phase of tidal current in which the speed is a maximum; also the speed at this time. Beginning with slack before flood in the period of a reversing tidal current (or minimum before flood in a rotary current), the speed gradually increases to flood strength and then diminishes to slack before ebb (or minimum before ebb in a rotary current), after which the current turns in direction, the speed increases to ebb strength and then diminishes to slack before flood completing the cycle. If it is assumed that the speed throughout the cycle varies as the ordinates of a cosine curve, it can be shown that the average speed for an entire flood or ebb period is equal to $2/\pi$ or 0.6366 of the speed of the corresponding strength of current.

SUBORDINATE CURRENT STATION—(1) A current station from which a relatively short series of observations is reduced by comparison with simultaneous observations from a control current station. (2) A station listed in the *Tidal Current Tables* for which predictions are to be obtained by means of differences and ratios applied to the full predictions at a reference station .

SUBORDINATE TIDE STATION—(1) A tide station from which a relatively short series of observations is reduced by comparison with simultaneous observations from a tide station with a relatively long series of observations. (2) A station listed in the *Tide Tables* for which predictions are to be obtained by means of differences and ratios applied to the full predictions at a reference station.

TIDAL CURRENT TABLES—Tables which give daily predictions of the times and speeds of the tidal currents. These predictions are usually supplemented by current differences and constants through which additional predictions can be obtained for numerous other places.

TIDAL DIFFERENCE—Difference in time or height of a high or low water at a subordinate station and at a reference station for which predictions are given in the *Tide Tables*. The difference, when applied according to sign to the prediction at the reference station, gives the corresponding time or height for the subordinate station .

TIDE—The periodic rise and fall of the water resulting from gravitational interactions between the Sun, Moon, and Earth. The vertical component of the particulate motion of a tidal wave. Although the accompanying horizontal movement of the water is part of the same phenomenon, it is preferable to designate the motion as tidal current.

TIDE TABLES—Tables which give daily predictions of the times and heights of high and low waters. These predictions are usually supplemented by tidal differences and constants through which additional predictions can be obtained for numerous other places.

TIME MERIDIAN—A meridian used as a reference for time.

TROPIC CURRENTS—Tidal currents occurring semimonthly when the effect of the Moon's maximum declination is greatest. At these times the tendency of the Moon to produce a diurnal inequality in the current is at a maximum.

TROPIC RANGES—The *great tropic range* (Gc), or *tropic range*, is the difference in height between tropic higher high water and tropic lower low water. The *small tropic range* (Sc) is the difference in height between tropic lower high water and tropic higher low water. The *mean tropic range* (Mc) is the mean between the great tropic range and the small tropic range. The small tropic range and the mean tropic range are applicable only when the type of tide is semidiurnal or mixed. Tropic ranges are most conveniently computed from the harmonic constants.

TROPIC TIDES—Tides occurring semimonthly when the effect of the Moon's maximum declination is greatest. At these times there is a tendency for an increase in the diurnal range. The tidal datums pertaining to the tropic tides are designated as *tropic higher high water* (TcHHW), *tropic lower high water* (TcLHW), *tropic higher low water* (TcHLW), and *tropic lower low water* (TcLLW).

TYPE OF TIDE—A classification based on characteristic forms of a tide curve. Qualitatively, when the two high waters and two low waters of each tidal day are approximately equal in height, the tide is said to be *semidiurnal*; when there is a relatively large diurnal inequality in the high or low waters or both, it is said to be *mixed*; and when there is only one high water and one low water in each tidal day, it is said to be *diurnal*.

VANISHING TIDE—In a mixed tide with very large diurnal inequality, the lower high water (or higher low water) frequently becomes indistinct (or vanishes) at time of extreme declinations. During these periods the diurnal tide has such overriding dominance that the semidiurnal tide, although still present, cannot be readily seen on the tide curve.

INDEX TO STATIONS
(Numbers refer to table 2)

[Stations marked with an asterisk (*) are reference stations for which daily predictions are given in table 1. Page numbers of reference stations are given in parentheses.]

Q

R